Foundation Mathematics for the Physical Sciences

This tutorial-style textbook develops the basic mathematical tools needed by first- and second-year undergraduates to solve problems in the physical sciences. Students gain hands-on experience through hundreds of worked examples, end-of-section exercises, self-test questions and homework problems.

Each chapter includes a summary of the main results, definitions and formulae. Over 270 worked examples show how to put the tools into practice. Around 170 self-test questions in the footnotes and 300 end-of-section exercises give students an instant check of their understanding. More than 450 end-of-chapter problems allow students to put what they have just learned into practice.

Hints and outline answers to the odd-numbered problems are given at the end of each chapter. Complete solutions to these problems can be found in the accompanying *Student Solution Manual*. Fully worked solutions to all the problems, password-protected for instructors, are available at www.cambridge.org/foundation.

K. F. RILEY read mathematics at the University of Cambridge and proceeded to a Ph.D. there in theoretical and experimental nuclear physics. He became a Research Associate in elementary particle physics at Brookhaven, and then, having taken up a lectureship at the Cavendish Laboratory, Cambridge, continued this research at the Rutherford Laboratory and Stanford; in particular he was involved in the experimental discovery of a number of the early baryonic resonances. As well as having been Senior Tutor at Clare College, where he has taught physics and mathematics for over 40 years, he has served on many committees concerned with the teaching and examining of these subjects at all levels of tertiary and undergraduate education. He is also one of the authors of *200 Puzzling Physics Problems* (Cambridge University Press, 2001).

M. P. HOBSON read natural sciences at the University of Cambridge, specialising in theoretical physics, and remained at the Cavendish Laboratory to complete a Ph.D. in the physics of star formation. As a Research Fellow at Trinity Hall, Cambridge, and subsequently an Advanced Fellow of the Particle Physics and Astronomy Research Council, he developed an interest in cosmology, and in particular in the study of fluctuations in the cosmic microwave background. He was involved in the first detection of these fluctuations using a ground-based interferometer. Currently a University Reader at the Cavendish Laboratory, his research interests include both theoretical and observational aspects of cosmology, and he is the principal author of *General Relativity: An Introduction for Physicists* (Cambridge University Press, 2006). He is also a Director of Studies in Natural Sciences at Trinity Hall and enjoys an active role in the teaching of undergraduate physics and mathematics.

Contents

Preface	*page* xi

1 Arithmetic and geometry   1

1.1	Powers	1
1.2	Exponential and logarithmic functions	7
1.3	Physical dimensions	15
1.4	The binomial expansion	20
1.5	Trigonometric identities	24
1.6	Inequalities	32
	Summary	40
	Problems	42
	Hints and answers	49

2 Preliminary algebra   52

2.1	Polynomials and polynomial equations	53
2.2	Coordinate geometry	64
2.3	Partial fractions	74
2.4	Some particular methods of proof	84
	Summary	91
	Problems	93
	Hints and answers	99

3 Differential calculus   102

3.1	Differentiation	102
3.2	Leibnitz's theorem	112
3.3	Special points of a function	114
3.4	Curvature of a function	116
3.5	Theorems of differentiation	120
3.6	Graphs	124
	Summary	133
	Problems	134
	Hints and answers	138

Contents

4 Integral calculus — 141

- 4.1 Integration — 141
- 4.2 Integration methods — 146
- 4.3 Integration by parts — 152
- 4.4 Reduction formulae — 155
- 4.5 Infinite and improper integrals — 156
- 4.6 Integration in plane polar coordinates — 159
- 4.7 Integral inequalities — 160
- 4.8 Applications of integration — 161
- Summary — 168
- Problems — 170
- Hints and answers — 173

5 Complex numbers and hyperbolic functions — 174

- 5.1 The need for complex numbers — 174
- 5.2 Manipulation of complex numbers — 176
- 5.3 Polar representation of complex numbers — 185
- 5.4 De Moivre's theorem — 189
- 5.5 Complex logarithms and complex powers — 194
- 5.6 Applications to differentiation and integration — 196
- 5.7 Hyperbolic functions — 197
- Summary — 205
- Problems — 206
- Hints and answers — 211

6 Series and limits — 213

- 6.1 Series — 213
- 6.2 Summation of series — 215
- 6.3 Convergence of infinite series — 224
- 6.4 Operations with series — 232
- 6.5 Power series — 233
- 6.6 Taylor series — 238
- 6.7 Evaluation of limits — 244
- Summary — 248
- Problems — 250
- Hints and answers — 257

7 Partial differentiation — 259

- 7.1 Definition of the partial derivative — 259
- 7.2 The total differential and total derivative — 261
- 7.3 Exact and inexact differentials — 264
- 7.4 Useful theorems of partial differentiation — 266

Contents

7.5	The chain rule	267
7.6	Change of variables	268
7.7	Taylor's theorem for many-variable functions	270
7.8	Stationary values of two-variable functions	272
7.9	Stationary values under constraints	276
7.10	Envelopes	282
7.11	Thermodynamic relations	285
7.12	Differentiation of integrals	288
	Summary	290
	Problems	292
	Hints and answers	299

8 Multiple integrals — 301

8.1	Double integrals	301
8.2	Applications of multiple integrals	305
8.3	Change of variables in multiple integrals	315
	Summary	324
	Problems	325
	Hints and answers	329

9 Vector algebra — 331

9.1	Scalars and vectors	331
9.2	Addition, subtraction and multiplication of vectors	332
9.3	Basis vectors, components and magnitudes	336
9.4	Multiplication of two vectors	339
9.5	Triple products	346
9.6	Equations of lines, planes and spheres	348
9.7	Using vectors to find distances	353
9.8	Reciprocal vectors	357
	Summary	359
	Problems	361
	Hints and answers	368

10 Matrices and vector spaces — 369

10.1	Vector spaces	370
10.2	Linear operators	374
10.3	Matrices	376
10.4	Basic matrix algebra	377
10.5	The transpose and conjugates of a matrix	383
10.6	The trace of a matrix	385

Contents

10.7	The determinant of a matrix	386
10.8	The inverse of a matrix	392
10.9	The rank of a matrix	395
10.10	Simultaneous linear equations	397
10.11	Special types of square matrix	408
10.12	Eigenvectors and eigenvalues	412
10.13	Determination of eigenvalues and eigenvectors	418
10.14	Change of basis and similarity transformations	421
10.15	Diagonalisation of matrices	424
10.16	Quadratic and Hermitian forms	427
10.17	The summation convention	432
	Summary	433
	Problems	437
	Hints and answers	445

11 Vector calculus 448

11.1	Differentiation of vectors	448
11.2	Integration of vectors	453
11.3	Vector functions of several arguments	454
11.4	Surfaces	455
11.5	Scalar and vector fields	458
11.6	Vector operators	458
11.7	Vector operator formulae	465
11.8	Cylindrical and spherical polar coordinates	469
11.9	General curvilinear coordinates	476
	Summary	482
	Problems	483
	Hints and answers	490

12 Line, surface and volume integrals 491

12.1	Line integrals	491
12.2	Connectivity of regions	497
12.3	Green's theorem in a plane	498
12.4	Conservative fields and potentials	502
12.5	Surface integrals	504
12.6	Volume integrals	511
12.7	Integral forms for grad, div and curl	513
12.8	Divergence theorem and related theorems	517
12.9	Stokes' theorem and related theorems	523
	Summary	527
	Problems	528
	Hints and answers	534

Contents

13 Laplace transforms — 536

- 13.1 Laplace transforms — 537
- 13.2 The Dirac δ-function and Heaviside step function — 541
- 13.3 Laplace transforms of derivatives and integrals — 544
- 13.4 Other properties of Laplace transforms — 546
- Summary — 549
- Problems — 550
- Hints and answers — 552

14 Ordinary differential equations — 554

- 14.1 General form of solution — 555
- 14.2 First-degree first-order equations — 557
- 14.3 Higher degree first-order equations — 565
- 14.4 Higher order linear ODEs — 569
- 14.5 Linear equations with constant coefficients — 572
- 14.6 Linear recurrence relations — 579
- Summary — 585
- Problems — 587
- Hints and answers — 595

15 Elementary probability — 597

- 15.1 Venn diagrams — 597
- 15.2 Probability — 602
- 15.3 Permutations and combinations — 612
- 15.4 Random variables and distributions — 618
- 15.5 Properties of distributions — 623
- 15.6 Functions of random variables — 628
- 15.7 Important discrete distributions — 632
- 15.8 Important continuous distributions — 643
- 15.9 Joint distributions — 655
- Summary — 661
- Problems — 664
- Hints and answers — 670

A The base for natural logarithms — 673

B Sinusoidal definitions — 676

C Leibnitz's theorem — 679

Contents

D	Summation convention	681
E	Physical constants	684
F	Footnote answers	685

Index   706

Preface

Since *Mathematical Methods for Physics and Engineering* by Riley, Hobson and Bence (Cambridge: Cambridge University Press, 1998), hereafter denoted by *MMPE*, was first published, the range of material it covers has increased with each subsequent edition (2002 and 2006). Most of the additions have been in the form of introductory material covering polynomial equations, partial fractions, binomial expansions, coordinate geometry and a variety of basic methods of proof, though the third edition of *MMPE* also extended the range, but not the general level, of the areas to which the methods developed in the book could be applied. Recent feedback suggests that still further adjustments would be beneficial. In so far as content is concerned, the inclusion of some additional introductory material such as powers, logarithms, the sinusoidal and exponential functions, inequalities and the handling of physical dimensions, would make the starting level of the book better match that of some of its readers.

To incorporate these changes, and others aimed at increasing the user-friendliness of the text, into the current third edition of *MMPE* would inevitably produce a text that would be too ponderous for many students, to say nothing of the problems the physical production and transportation of such a large volume would entail.

For these reasons, we present under the current title, *Foundation Mathematics for the Physical Sciences*, an alternative edition of *MMPE*, one that focuses on the earlier part of a putative extended third edition. It omits those topics that truly are 'methods' and concentrates on the 'mathematical tools' that are used in more advanced texts to build up those methods. The emphasis is very much on developing the basic mathematical concepts that a physical scientist needs, before he or she can narrow their focus onto methods that are particularly appropriate to their chosen field.

One aspect that has remained constant throughout the three editions of *MMPE* is the general style of presentation of a topic – a qualitative introduction, physically based wherever possible, followed by a more formal presentation or proof, and finished with one or two full-worked examples. This format has been well received by reviewers, and there is no reason to depart from its basic structure.

In terms of style, many physical science students appear to be more comfortable with presentations that contain significant amounts of explanation or comment in words, rather than with a series of mathematical equations the last line of which implies 'job done'. We have made changes that move the text in this direction. As is explained below, we also feel that if some of the advantages of small-group face-to-face teaching could be reflected in the written text, many students would find it beneficial.

In keeping with the intention of presenting a more 'gentle' introduction to university-level mathematics for the physical sciences, we have made use of a modest number of appendices. These contain the more formal mathematical developments associated with the material introduced in the early chapters, and, in particular, with that discussed in the introductory chapter on arithmetic and geometry. They can be studied at the points in the main text where references are made to them, or deferred until a greater mathematical fluency has been acquired.

As indicated above, one of the advantages of an oral approach to teaching, apparent to some extent in the lecture situation, and certainly in what are usually known as tutorials,[1] is the opportunity to follow the exposition of any particular point with an immediate short, but probing, question that helps to establish whether or not the student has grasped that point. This facility is not normally available when instruction is through a written medium, without having available at least the equipment necessary to access the contents of a storage disc.

In this book we have tried to go some way towards remedying this by making a non-standard use of footnotes. Some footnotes are used in traditional ways, to add a comment or a pertinent but not essential piece of additional information, to clarify a point by restating it in slightly different terms, or to make reference to another part of the text or an external source. However, about half of the more than 300 footnotes in *this* book contain a question for the reader to answer or an instruction for them to follow; neither will call for a lengthy response, but in both cases an understanding of the associated material in the text will be required. This parallels the sort of follow-up a student might have to supply orally in a small-group tutorial, after a particular aspect of their written work has been discussed.

Naturally, students should attempt to respond to footnote questions using the skills and knowledge they have acquired, re-reading the relevant text if necessary, but if they are unsure of their answer, or wish to feel the satisfaction of having their correct response confirmed, they can consult the specimen answers given in Appendix F. Equally, footnotes in the form of observations will have served their purpose when students are consistently able to say to themselves 'I didn't need that comment – I had already spotted and checked that particular point'.

There are two further features of the present volume that did not appear in *MMPE*. The first of these is that a small set of exercises has been included at the end of each section. The questions posed are straightforward and designed to test whether the student has understood the concepts and procedures described in that section. The questions are not intended as 'drill exercises', with repeated use of the same procedure on marginally different sets of data; each concept is examined only once or twice within the set. There are, nevertheless, a total of more than 300 such exercises. The more demanding questions, and in particular those requiring the synthesis of several ideas from a chapter, are those that appear under the heading of 'Problems' at the end of that chapter; there are more than 450 of these.

The second new feature is the inclusion at the end of each chapter, just before the problems begin, of a summary of the main results of that chapter. For some areas, this takes the form of a tabulation of the various case types that may arise in the context of the chapter; this should help the student to see the parallels between situations which in the main text are presented as a consecutive series of often quite lengthy pieces of mathematical development. It should be said that in such a summary it is not possible to

---

[1] But in Cambridge are called 'supervisions'!

Preface

state every detailed condition attached to each result, and the reader should consider the summaries as reminders and formulae providers, rather than as teaching text; that is the job of the main text and its footnotes. Fortunately, in this volume, occasions on which subtle conditions have to be imposed upon a result are rare.

Finally, we note, for the record, that the format and numbering of the problems associated with the various chapters have not been changed significantly from those in *MMPE*, though naturally only problems related to included topics are retained. This means that abbreviated solutions to all odd-numbered problems can be found in this text. Fully worked solutions to the same problems are available in the companion volume *Student Solution Manual for Foundation Mathematics for the Physical Sciences*; most of them, except for those in the first chapter, can also be found in the *Student Solution Manual for MMPE*.

Fully worked solutions to all problems, both odd- and even-numbered, are available to accredited instructors on the password-protected website www.cambridge.org/foundation. Instructors wishing to have access to the website should contact solutions@cambridge.org for registration details.

1 Arithmetic and geometry

The first two chapters of this book review the basic arithmetic, algebra and geometry of which a working knowledge is presumed in the rest of the text; many students will have at least some familiarity with much, if not all, of it. However, the considerable choice now available in what is to be studied for secondary-education examination purposes means that none of it can be taken for granted. The reader may make a preliminary assessment of which areas need further study or revision by first attempting the problems at the ends of the chapters. Unlike the problems associated with all other chapters, those for the first two are divided into named sections and each problem deals almost exclusively with a single topic.

This opening chapter explains the basic definitions and uses associated with some of the most common mathematical procedures and tools; these are the components from which the mathematical methods developed in more advanced texts are built. So as to keep the explanations as free from detailed mathematical working as possible – and, in some cases, because results from later chapters have to be anticipated – some justifications and proofs have been placed in appendices. The reader who chooses to omit them on a first reading should return to them after the appropriate material has been studied.

The main areas covered in this first chapter are powers and logarithms, inequalities, sinusoidal functions, and trigonometric identities. There is also an important section on the role played by dimensions in the description of physical systems. Topics that are wholly or mainly concerned with algebraic methods have been placed in the second chapter. It contains sections on polynomial equations, the related topic of partial fractions, and some coordinate geometry; the general topic of curve sketching is deferred until methods for locating maxima and minima have been developed in Chapter 3. An introduction to the notions of proof by induction or contradiction is included in Chapter 2, as the examples used to illustrate it are almost entirely algebraic in nature. The same is true of a discussion of the necessary and sufficient conditions for two mathematical statements to be equivalent.

1.1 Powers

If we multiply together n factors each equal to a, we call the result the *n*th *power* of a and write it as a^n. The quantity n, a positive integer in this definition, is called the *index* or *exponent*.

Arithmetic and geometry

The algebraic rules for combining different powers of the same quantity, i.e. combining expressions all of the form a^n, but with different exponents in general, are summarised by the four equations

$$pa^n \pm qa^n = (p \pm q)a^n, \tag{1.1}$$
$$a^m \times a^n = a^{m+n}, \tag{1.2}$$
$$a^m \div a^n = a^{m-n}, \tag{1.3}$$
$$(a^m)^n = a^{mn} = (a^n)^m \quad . \tag{1.4}$$

To these can be added the rules for multiplying and dividing two powers that contain the same exponent:

$$a^n \times b^n = (ab)^n, \tag{1.5}$$
$$a^n \div b^n = \left(\frac{a}{b}\right)^n. \tag{1.6}$$

The multiplication of powers is both commutative and associative. Since these terms are relevant to characterising nearly all mathematical operations, and appear many times in the remainder of this book, we give here a brief discussion of them.

Commutativity

An operation, denoted by $\odot$ say, that acts upon two objects x and y that belong to some particular class of objects, and so produces a result $x \odot y$, is said to be commutative if $x \odot y = y \odot x$ for *all* pairs of objects in the class; loosely speaking, it does not matter in which order the two objects appear. As examples, for real numbers, addition ($x + y = y + x$) and multiplication ($x \times y = y \times x$) are commutative, but subtraction and division are not; the latter two fail to be commutative because $x - y \neq y - x$ and $x/y \neq y/x$.

The same is true with regard to combining powers: when $\odot$ stands for multiplication and x and y are a^m and a^n, then, since $a^m \times a^n = a^n \times a^m$, the operation of multiplication is commutative; but, when $\odot$ stands for division and x and y are as before, the operation is non-commutative because $a^m \div a^n \neq a^n \div a^m$. It might be added that not all forms of multiplication are commutative; for example, if x and y are matrices **A** and **B**, then, in general, **AB** and **BA** are not equal (see Chapter 10).

Associativity

Using the notation of the previous two paragraphs, the operation $\odot$ is said to be associative if $(x \odot y) \odot z = x \odot (y \odot z)$ for *all* triples of objects in the class; here the parentheses indicate that the operations enclosed by them are the first to be carried out within each grouping. Again, as simple examples, for real numbers, addition [$(x + y) + z = x + (y + z)$] and multiplication [$(x \times y) \times z = x \times (y \times z)$] are associative, but subtraction and division are not. Subtraction fails to be associative because $(x - y) - z \neq x - (y - z)$, i.e. $x - y - z \neq x - y + z$; division fails in a similar way.

1.1 Powers

Corresponding results apply to the operations of combining powers. In summary, the multiplication of powers is both commutative and associative; the division of them is neither.[1]

Given the rules set out above for combining powers, and the fact that any non-zero value divided by itself must yield unity, we must have, on setting $n = m$ in (1.3), that

$$1 = a^m \div a^m = a^{m-m} = a^0.$$

Thus, for any $a \neq 0$,

$$a^0 = 1. \quad (1.7)$$

The case in which $a = 0$ is discussed later, when logarithms are considered.

Result (1.7) has already taken us away from our original construction of a power, as the notion of multiplying no factors of a together and obtaining unity is not altogether intuitive; rather we must consider the process of forming a^n as one of multiplying unity n times by a factor of a.

Another consequence of result (1.3), taken together with deduction (1.7), can now be found by setting $m = 0$ in (1.3). Doing this shows that, for $a \neq 0$,

$$\frac{1}{a^n} = a^0 \div a^n = a^{0-n} = a^{-n}. \quad (1.8)$$

In words, the reciprocal of a^n is a^{-n}. The analogy with the construction in the previous paragraph is that a^{-n} is formed by dividing unity n times by a factor of a.

Rule (1.4) allows us to assign a meaning to a^n when n is a general rational number, i.e. n can be written as $n = p/q$ where p and q are integers; n itself is not necessarily an integer. In particular, if we take n to have the form $n = 1/m$, where m is an integer, then the second equality in (1.4) reads

$$a = a^1 = (a^{1/m})^m. \quad (1.9)$$

This shows that the quantity $a^{1/m}$ when raised to the mth power produces the quantity a. This, in turn, implies that $a^{1/m}$ must be interpreted as the mth root of a, otherwise denoted by $\sqrt[m]{a}$. With this identification, the first equality in (1.4) expresses the compatible result that the mth root of a^m is a.

For more general values of p and q, we have that

$$(a^{1/q})^p = a^{p/q} = (a^p)^{1/q}, \quad (1.10)$$

which states that the pth power of the qth root of a is equal to the qth root of the pth power of a.

It should be noted that, so long as only real quantities are allowed, a must be confined to positive values when taking roots in this way; the need for this will be clear from considering the case of a negative and m an even integer. It is possible to find a valid answer for $a^{1/m}$ with a negative if m is an odd integer (or, more generally, if the q in

---

[1] Consider for each of the following operations whether it is commutative and/or associative: (i) $a \odot b = a^2 + b^2$, (ii) $a \odot b = +\sqrt{a^2 + b^2}$, (iii) $a \odot b = a^b$; a and b are real positive numbers.

Arithmetic and geometry

$n = p/q$ is an odd integer), as is shown by the calculation

$$\left(-\tfrac{1}{27}\right)^{-4/3} = (-1)^{-4/3} \left(\tfrac{1}{27}\right)^{-4/3} = (-1)^{-4} \left(\tfrac{1}{3}\right)^{-4}$$
$$= \left(\tfrac{1}{-1}\right)^4 \left(\tfrac{3}{1}\right)^4 = 1 \times 81 = 81.$$

However, both a and p/q could be more general expressions whose signs and values are not fixed, and great care is needed when using anything other than explicit numerical values.

Having established a meaning for a^m when m is either an integer or a rational fraction, we would also wish to attach a mathematical meaning to it when m is not confined to either of these classes, but is *any* real number. Obviously, any general m that is expressed to a finite number of decimal places could be considered formally, but very inconveniently, as a rational fraction; however, there are infinitely many numbers that cannot be expressed in this way, $\sqrt{2}$ and π being just two examples.

This is hardly likely to be a problem for any physically based situation, in which there will always be finite limits on the accuracy with which parameters and measured values can be determined. But, in order to fill the formal gap, a definition of a general power that uses the logarithmic function is adopted for all real values of m. The general properties of logarithms are discussed in Section 1.2, but we state here one that defines a general power of a positive quantity a for any real exponent m:

$$a^m = e^{m \ln a}, \tag{1.11}$$

where $\ln a$ is the logarithm to the base e of a, itself defined by

$$a = e^{\ln a}, \tag{1.12}$$

and e is the value of the exponential function when its argument is unity. As it happens, e itself is irrational (i.e. it cannot be expressed as a rational fraction of the form p/q) and the first seven of the never-ending sequence of figures in its decimal representation are 2.718 281 ...

Such a definition, in terms of functions that have not yet been fully defined or discussed, could be confusing, but most readers will already have had some practical contact with logarithms and should appreciate that the definition can be used to cover all real m. Discussion of the choice of e for the base of the logarithm is deferred until Section 1.2, but with this choice the logarithm is known as a *natural logarithm*.

As a numerical example, consider the value of $7^{0.3}$. This would normally be found directly as 1.792 78 ... by making a few keystrokes on a basic scientific calculator. But what happens inside the calculator essentially follows the procedure given above, and it is instructive to compute the separate steps involved.[2] Set algorithms are used for calculating natural logarithms, $\ln x$, and evaluating exponential functions, e^x, for general values of x. First the value of $\ln 7$ is found as 1.945 91 ... This is then multiplied by 0.3 to yield 0.583 773 ... and then, as the final step, the value of $e^{0.583\,773...}$ is calculated as 1.792 78 ...

---

[2] It is suggested that you do so on your own calculator.

1.1 Powers

Because so many natural relationships between physical quantities express one quantity in terms of the square of another,[3] the most commonly occurring non-integral power that a physical scientist has to deal with is the square root. For practical calculations, with data always of limited accuracy, this causes no difficulty, and even the simplest pocket calculator incorporates a square-root routine. But, for theoretical investigations, procedures that are exact are much to be preferred; so we consider here some methods for dealing with expressions involving square roots.

Written as a power, a square root is of the form $a^{1/2}$, but for the present discussion we will use the notation $\sqrt{a}$. If a is the square of a rational number, then $\sqrt{a}$ is itself a rational number and needs no special attention. However, when a is not such a square, $\sqrt{a}$ is irrational and new considerations arise. It may be that a happens to contain the square of a rational number as a factor; in such a case, the number may be taken out from under the square root sign, but that makes no substantial difference to the situation. For example:

$$\frac{3}{2}\sqrt{\frac{128}{343}} = \frac{3}{2}\frac{8}{7}\sqrt{\frac{2}{7}} = \frac{12}{7}\sqrt{\frac{2}{7}};$$

we started with $\sqrt{128/343}$ which is irrational and, although some simplification has been effected, the resulting expression, $\sqrt{2/7}$, is still irrational. It is almost as if rational and irrational numbers were different species. Square roots that are irrational are particular examples of *surds*. This is a term that covers irrational roots of any order (of the form $a^{1/n}$ for any positive integer n), though we are concerned here only with $n = 2$ and will use the term 'surd' to mean a square root that is irrational.

To emphasise the apparent rational–irrational distinction, consider the simple equation

$$a + b\sqrt{p} = c + d\sqrt{p},$$

where a, b, c and d are rational numbers, whilst $\sqrt{p}$ is irrational and non-zero. We can show that the rational and irrational terms on the two sides can be *separately* equated, i.e. $a = c$ and $b = d$. To do this, suppose, on the contrary, that $b \neq d$. Then the equation can be rearranged as

$$\sqrt{p} = \frac{a - c}{d - b}.$$

But the RHS[4] of this equality is the finite ratio of two rational numbers and so is itself rational; this contradicts the fact that $\sqrt{p}$ is irrational and so shows it was wrong to suppose that $b \neq d$, i.e. b must be equal to d. It then follows immediately, from subtracting $b\sqrt{p}$ from both sides, that, in addition, a is equal to c. To summarise:

$$a + b\sqrt{p} = c + d\sqrt{p} \quad \Rightarrow \quad a = c \quad \text{and} \quad b = d. \tag{1.13}$$

An important tool for handling fractional expressions that involve surds in their denominators is the process of *rationalisation*. This is a procedure that enables an expression of

---

[3] As examples, using standard symbols, $T = \frac{1}{2}mv^2$, $W = RI^2$, $U = \frac{1}{2}CV^2$, $u = \frac{1}{2}\epsilon_0 E^2 + \frac{1}{2}B^2/\mu_0$.
[4] The need to refer to the 'left-hand side' or the 'right-hand side' of an equation occurs so frequently throughout this book, that we almost invariably use the abbreviation LHS or RHS.

Arithmetic and geometry

the general form

$$\frac{a + b\sqrt{p}}{c + d\sqrt{p}},$$

with a, b, c and d rational, to be converted into the (generally) more convenient form $e + f\sqrt{p}$; normally there is no gain to be made unless, though $\sqrt{p}$ is irrational, p itself is rational. The basis of the procedure is the algebraic identity $(x + y)(x - y) = x^2 - y^2$. This identity is used to remove the $\sqrt{p}$ from the denominator, after both numerator and denominator have been multiplied by $c - d\sqrt{p}$ (note the minus sign). Mathematically, the procedure is as follows:

$$\frac{a + b\sqrt{p}}{c + d\sqrt{p}} = \frac{(a + b\sqrt{p})(c - d\sqrt{p})}{(c + d\sqrt{p})(c - d\sqrt{p})} = \frac{ac - bdp + (bc - ad)\sqrt{p}}{c^2 - d^2 p}.$$

This is of the stated form, with the finite[5] rational quantities e and f given by $e = (ac - bdp)/(c^2 - d^2 p)$ and $f = (bc - ad)/(c^2 - d^2 p)$.

As an example to illustrate the procedure, consider the following.

Example Solve the equation

$$a + b\sqrt{28} = \frac{4 + 3\sqrt{7}}{3 - \sqrt{7}}$$

for a and b, (i) by obtaining simultaneous equations for a and b and (ii) by using rationalisation.

(i) Cross-multiplying the given equation and using several of the properties of powers listed at the start of this section, we obtain

$$3a + 3b\sqrt{28} - a\sqrt{7} - b\sqrt{7}\sqrt{28} = 4 + 3\sqrt{7},$$
$$3a + 6b\sqrt{7} - a\sqrt{7} - 7b\sqrt{4} = 4 + 3\sqrt{7},$$
$$3a - 14b + (6b - a)\sqrt{7} = 4 + 3\sqrt{7}.$$

Equating the rational and irrational parts on each side gives the simultaneous equations

$$3a - 14b = 4,$$
$$-a + 6b = 3.$$

These simultaneous equations can now be solved and have the solution $a = 33/2$ and $b = 13/4$.

(ii) Following the rationalisation procedure, the calculation is

$$a + b\sqrt{28} = \frac{4 + 3\sqrt{7}}{3 - \sqrt{7}} = \frac{(4 + 3\sqrt{7})(3 + \sqrt{7})}{(3 - \sqrt{7})(3 + \sqrt{7})}$$
$$= \frac{12 + (3)(7) + (9 + 4)\sqrt{7}}{3^2 - (\sqrt{7})^2} = \frac{33 + 13\sqrt{7}}{2}.$$

Equating the rational and irrational parts on each side gives $a = 33/2$ and $b\sqrt{28} = 13\sqrt{7}/2$, i.e. $b = 13/4$. As they must, the two methods yield the same solution. ◀

---

5 Explain why they cannot be infinite.

1.2 Exponential and logarithmic functions

EXERCISES 1.1

1. Evaluate to three significant figures (s.f.)

   (a) 8^3, (b) 8^{-3}, (c) $8^{1/3}$, (d) $8^{-1/3}$, (e) $(1/8)^{1/3}$, (f) $(1/8)^{-1/3}$.

2. Rationalise the following:

   (a) $\dfrac{2}{\sqrt{5}-1}$, (b) $\dfrac{2-\sqrt{3}}{2+\sqrt{3}}$, (c) $\dfrac{20+\sqrt{12}}{20+\sqrt{48}}$.

3. Rationalisation can be extended to expressions of the form $(a+b\sqrt{q})/(c+d\sqrt{p})$ to produce the form $e+f\sqrt{p}+g\sqrt{q}+h\sqrt{pq}$. Apply the procedure to

   (a) $\dfrac{\sqrt{5}-2}{2-\sqrt{3}}$, (b) $\dfrac{3+\sqrt{15}}{3+\sqrt{3}}$, (c) $\dfrac{(\sqrt{5}-2)(3+\sqrt{15})}{(2-\sqrt{3})(3+\sqrt{3})}$.

   Confirm result (c) by direct multiplication of results (a) and (b).

4. Determine whether each of the operations $\odot$ defined below is commutative and/or associative:
   (a) $a \odot b$ = the highest common factor (h.c.f.) of positive integers a and b.
   (b) For real numbers a, b, c etc., $a \odot b = a + ib$, where $i^2 = -1$. Would your conclusion be different if a, b, c etc. could be complex?
   (c) For all non-negative integers including zero

   $$a \odot b = \begin{cases} 2 & \text{if } ab \text{ is even} \\ 1 & \text{if } ab \text{ is odd or zero.} \end{cases}$$

   (d) For all positive integers (excluding zero)

   $$a \odot b = \begin{cases} 2 & \text{if } ab \text{ is even} \\ 1 & \text{if } ab \text{ is odd or zero.} \end{cases}$$

1.2 Exponential and logarithmic functions

When discussing powers of a real number in the previous section, we made somewhat premature references to logarithms and the exponential function. In this section we introduce these ideas more formally and show how a natural mathematical choice for the 'base' of logarithms arises. This use of the word 'base' is related to the idea of a number base for counting systems, which in everyday life is taken as 10, and in the internal structure of computing systems is binary (base 2), though other bases such as octal (base 8) and hexadecimal (base 16) are frequently used at the interface between such systems and everyday life.[6]

---

[6] The *Ultimate Answer to Life, the Universe and Everything* is 42 when expressed in decimal. Confirm that in other bases it is 101010 (binary), 52 (octal) and 2A (hexadecimal).

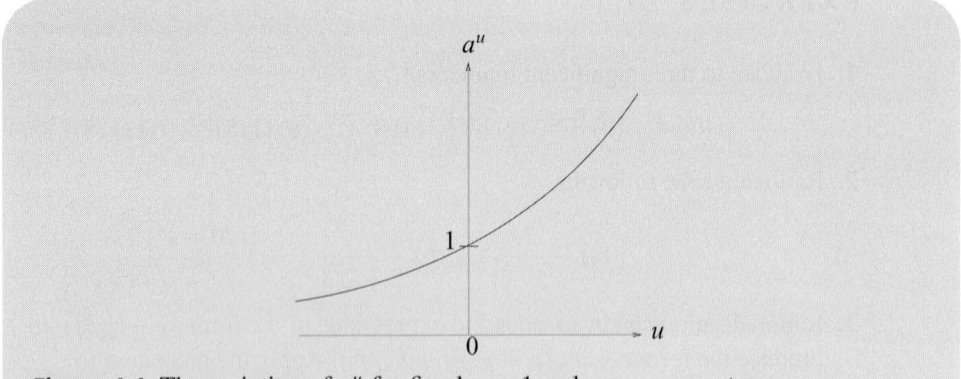

Figure 1.1 The variation of a^u for fixed $a > 1$ and $-\infty < u < +\infty$.

In the context of logarithms, the word *base* will be identified with the quantity we have hitherto denoted by a in expressions of the form a^m. It will become apparent that any positive value of a will do, but we will find that for mathematical purposes the most convenient choice, and therefore the 'natural' one, is for a to have the value denoted by the irrational number e, which is numerically equal to $2.718\,281\ldots$ in ordinary decimal notation.

The usefulness of logarithms for practical calculations depends on the properties expressed in Equations (1.2) and (1.3), namely

$$a^m \times a^n = a^{m+n}, \tag{1.14}$$
$$a^m \div a^n = a^{m-n}. \tag{1.15}$$

These two equations provide a way of reducing multiplication and division calculations to the processes of addition and subtraction (of the corresponding indices) respectively. Before proceeding to this aspect, however, we first define logarithms and then establish some of their general properties.

1.2.1 Logarithms

We start by noting that, for a fixed positive value of a and a variable u, the quantity a^u is a monotonic function of u, which, for $a > 1$, increases from zero for u large and negative, passes through unity at $u = 0$, and becomes arbitrarily large as u becomes large and positive. This is illustrated in Figure 1.1. For $a < 1$, the behaviour of a^u is the reverse of this, but we will restrict our attention to cases in which $a > 1$ and a^u is a monotonically *increasing* function of u.

Since a^u is monotonic and takes all values between 0 and $+\infty$, for any particular positive value of a variable x, we can find a *unique* value, α say, such that

$$a^\alpha = x.$$

This value of α is called the *logarithm* of x to the *base a* and written as

$$\alpha = \log_a x.$$

Thus, the fundamental equality satisfied by a logarithm using any base a is

$$x = a^{\log_a x}. \tag{1.16}$$

1.2 Exponential and logarithmic functions

From setting $x = 1$, and using result (1.7), it follows that

$$\log_a 1 = 0 \tag{1.17}$$

for *any* base a. Further, since $a = a^1$, setting $x = a$ shows that

$$\log_a a = 1. \tag{1.18}$$

It also follows from raising both sides of Equation (1.16) to the nth power[7] that

$$x^n = \left(a^{\log_a x}\right)^n = a^{n \log_a x} \quad \Rightarrow \quad \log_a x^n = n \log_a x. \tag{1.19}$$

It will be clear that, even for a fixed x, the value of a logarithm will depend upon the choice of base. As a concrete example: $\log_{10} 100 = 2$, whilst $\log_2 100 = 6.644$ and $\log_e 100 = 4.605$.

The connection between the logarithms of the same quantity x with respect to two different bases, a and b, is

$$\log_b x = \log_b a \times \log_a x. \tag{1.20}$$

This can be proved by repeated use of (1.16) as follows:

$$b^{\log_b x} = x = a^{\log_a x} = (b^{\log_b a})^{\log_a x} = b^{\log_b a \times \log_a x}.$$

Equating the two indices at the extreme ends of the equality chain yields the stated result. Now setting $x = b$ in (1.20), and recalling that $\log_b b = 1$, shows that

$$\log_b a = \frac{1}{\log_a b}. \tag{1.21}$$

In theoretical work it is not usually necessary to consider bases other than e, but for some practical applications, in engineering in particular, it is useful to note that

$$\log_{10} x = \log_{10} e \times \log_e x \approx 0.4343 \log_e x,$$
$$\log_e x = \log_e 10 \times \log_{10} x \approx 2.3026 \log_{10} x.$$

At this point, a comment on the notation generally used for logarithms employing the various bases is appropriate. Except when dealing with the theory of complex variables, where they have other specialised meanings, the functions $\ln x$ and $\log x$ are normally used to denote $\log_e x$ and $\log_{10} x$, respectively; $\text{Log } x$ is another alternative for $\log_{10} x$. Logarithms employing any base other than e or 10 are normally written in the same way as we have used hitherto.

1.2.2 The exponential function and choice of logarithmic base

Equations (1.14) and (1.15) give clear hints as to how the use of logarithms can be made to turn multiplication and division into addition and subtraction, but they also indicate that it does not matter which base a is used, so long as it is positive and not equal to unity. We have already opted to use a value for a that is greater than 1, but this still leaves an infinity

---

[7] The power n need not be an integer, nor need it be positive, e.g. $\log_{10}(0.3)^{-2.7} = -2.7 \log_{10} 0.3 = (-2.7) \times (-0.5229) = 1.412$. In brief, $0.3^{-2.7} = 10^{1.412} = 25.81$.

of choices. The universal choice of mathematicians is the so-called 'natural' choice of e which, as noted previously, has the value $2.718\,281\ldots$

To see why this is a natural choice requires the use of some elementary calculus, a subject not covered until later in this book (Chapter 3). However, if the reader is already familiar with the notions of derivatives and integrals and the relationships between them, knows (or accepts) that the derivative of x^n is nx^{n-1}, and understands the chain rule for derivatives, then he or she will be able to follow the derivation given in Appendix A. If not, the discussion given below can still be followed, though the three major properties that make e a preferred choice for the logarithmic base will have to be taken on trust until the relevant parts of Chapter 3 have been studied.

We start by defining the *exponential function* $\exp(x)$ of a real variable x. This is simply the sum of an infinite series of terms each of which contains a non-negative integral power n of x, namely x^n, multiplied by a factor that depends upon n in a specific way. A general function of this kind is known as a *power series* in x; such series are discussed in much more detail in Section 6.5. Written both as a formal sum and as an explicit series, the particular function we need is

$$\exp(x) \equiv \sum_{n=0}^{\infty} \frac{x^n}{n!} \equiv 1 + \frac{x}{1!} + \frac{x^2}{2!} + \frac{x^3}{3!} + \cdots \qquad (1.22)$$

For integers m that are ≥ 1, the symbol $m!$ stands for the multiple product $1 \times 2 \times 3 \times \cdots \times m$; it is read either as 'factorial m' or as 'm factorial'. For example, factorial 4 is written as $4!$ and has the value $1 \times 2 \times 3 \times 4 = 24$. The factorial function clearly has the elementary property

$$m \times (m-1)! = m!. \qquad (1.23)$$

The first term in the explicit series for $\exp(x)$, which is given as 1, corresponds to the $n=0$ term in the sum; it is therefore $x^0/0!$. By (1.7), the numerator has the value 1, whatever the value of x. The value of the denominator, $0!$, is also 1, though this will not be obvious. The general definition of $m!$ for m real, but not necessarily a positive integer,[8] involves the gamma function, $\Gamma(n)$, which is defined in Problem 4.13. There it is shown that $0! = \Gamma(1) = 1$. Thus, though it appears to involve x and looks as if it might also involve dividing by zero, $x^0/0!$ has, in fact, the simple value 1 for all x.

It can be shown (see Chapter 6) that, despite the fact that whenever $x > 1$ the quantity x^n grows as n increases, because of the rapidly increasing factors $n!$ in the denominators, the series *always* 'converges'. That is, as more and more terms are added, they become vanishingly small and the total sum becomes arbitrarily close to a definite value (dependent on x); that value is denoted by $\exp(x)$.

It should be noted that definition (1.22) is valid for all real values of x in the range $-\infty < x < +\infty$. At the extremes of the range, $\exp(x) \to 0$ as $x \to -\infty$, and $\exp(x) \to +\infty$ as $x \to +\infty$. In between, it is a monotonically increasing function that has the value 1 when $x = 0$, as is obvious from its definition:

$$\exp(0) = 1. \qquad (1.24)$$

---

[8] It is defined for negative values of m, so long as they are *not* negative integers. A couple of, possibly intriguing, values are $(-\frac{1}{2})! = \sqrt{\pi}$ and $(-\frac{3}{2})! = -2\sqrt{\pi}$.

1.2 Exponential and logarithmic functions

The value of exp(x) that is of particular relevance to the choice of a base[9] for logarithms is exp(1). This is the quantity that is referred to as e and given numerically by the sum of the infinite series obtained by setting $x = 1$ in definition (1.22):

$$e \equiv \exp(1) = \sum_{n=0}^{\infty} \frac{1}{n!} = 1 + \frac{1}{1!} + \frac{1}{2!} + \frac{1}{3!} + \cdots = 2.718\,281\ldots \quad (1.25)$$

If, in our definitions leading to identity (1.16), we set a equal to e (which *is* positive and >1), then we have that the natural logarithm ln x of x satisfies the statement

$$\text{if } e^y = x \text{ then } y = \ln x \text{ and vice versa.} \quad (1.26)$$

In order to study the properties of ln x, we make a slightly different initial definition of a natural logarithm, namely

$$\text{if } \exp(y) = x \text{ then } y = \ln x \text{ and vice versa.} \quad (1.27)$$

Then, by proving that

$$\exp(x) = e^x, \quad (1.28)$$

we show that the validity (by definition) of (1.27) implies that of (1.26) as well. The proof, which is given in Appendix A, uses only the information implied by the definition (1.27) to establish (1.28) and hence (1.26). In the course of the proof, the following important calculus-based properties of the functions exp(x) and ln x are established as by-products:

$$\frac{d \exp(x)}{dx} = \exp(x) \quad \text{or} \quad \frac{d\, e^x}{dx} = e^x, \quad (1.29)$$

$$\frac{d(\ln x)}{dx} = \frac{1}{x}, \quad (1.30)$$

$$\ln x = \int_1^x \frac{1}{u}\, du. \quad (1.31)$$

These three simple, but powerful, properties of logarithms to base e are major reasons for this particular choice of base. They are listed here for future use in later chapters of this main text.

1.2.3 The use of logarithms

Following our extensive discussion of the definition of a logarithm and the connection between the power e^x and the series defining the exponential function exp(x), we return to the practical uses that can be made of logarithms. Nowadays, these are mostly of historical interest, since the invention of small but powerful hand calculators means that nearly all numerical calculations can be carried out at the touch of a few buttons. Nevertheless, it is important that the practical scientist appreciates the mathematical basis of some of these automated procedures.

---

[9] Show that if a general base a is chosen, the graph of $y = a^x$ can be superimposed on that of $y = e^x$ by a simple scaling of the x-axis, $x \to x \ln a$.

Arithmetic and geometry

We first turn our attention to the multiplication and division of powers, as given in Equations (1.14) and (1.15). We have seen that, given any two positive quantities x and y, there are two corresponding unique quantities $\ln x$ and $\ln y$ such that

$$x = e^{\ln x} \quad \text{and} \quad y = e^{\ln y}.$$

It should be remembered that, even though x and y are positive, $\ln x$ and $\ln y$ can be positive or negative, with a negative value for $\ln x$, say, whenever x lies in the range $0 < x < 1$.

With x and y both expressed as powers of e in this way, we can apply property (1.14) to obtain

$$xy = e^{\ln x} e^{\ln y} = e^{\ln x + \ln y}.$$

But, the positive product xy may also be written (uniquely) as a power of e:

$$xy = e^{\ln(xy)}, \tag{1.32}$$

and so, by equating the indices in the two power expressions for xy, we obtain the well-known relationship

$$\ln(xy) = \ln x + \ln y. \tag{1.33}$$

The value of xy can now be recovered using (1.32). In a similar way, property (1.15) leads to the result

$$\ln\left(\frac{x}{y}\right) = \ln x - \ln y, \tag{1.34}$$

with x/y equal to e^u, where u is the RHS of (1.34).

These two results show how the multiplication or division of two positive numbers can be reduced to an addition or subtraction calculation, with no actual multiplication or division required. If either or both of the numbers to be multiplied or divided are negative, then they have to be treated as positive so far as the use of logarithms is concerned, and a separate, but simple, determination of the sign of the answer made.

We next turn to the use of logarithms in connection with the analysis of experimental data. Many experiments in both physics and engineering are aimed at establishing a formula that connects the values of two measured variables, or of verifying a proposed formula and then extracting values for some of its parameters. For the graphical analysis of the experimental data, it is very convenient, whenever possible, to re-cast the expected relationship between the measured quantities into a standard 'straight-line' form. This both helps to give a quick visual impression of whether the plotted data is compatible with the expected relationship and makes the extraction of parameters a routine procedure. If we denote one, or a particular combination, of the physical variables by y say, and another such single or composite variable by x, then a straight-line plot of y against x takes the form

$$y = mx + c. \tag{1.35}$$

The slope m is equal to the ratio $\Delta y/\Delta x$, where Δy is the difference in y-values (positive or negative) corresponding to any arbitrary difference Δx in x-values (again, positive or negative); if x and/or y have physical dimensions (see Section 1.3) associated with

1.2 Exponential and logarithmic functions

them then so does m.[10] The intercept the line makes on the y-axis gives the value of c; its dimensions are the same as those of y.

As a simple example, consider analysing data giving the distance v from a thin lens of an image formed by the lens when the object is placed a distance u from it. The relevant theoretical formula is

$$\frac{1}{u} + \frac{1}{v} = \frac{1}{f},$$

where f is the focal length of the lens. In terms of u and f, v is given by $v = uf/(u-f)$ and a plot of v against u is not very helpful (the reader may find it instructive to sketch it for a fixed positive f).[11] However, if $y = 1/v$ is plotted against $x = 1/u$ then a straight-line plot, as given in (1.35), is obtained. Its slope m should be -1; this can be used either as a check on the accuracy of measurement or as a constraint when drawing the best straight-line fit. The intercept made on the y-axis by the fitted line gives a value for f^{-1}, and hence for f, the focal length of the lens.

There are no logarithms directly involved in this optical example, but if the actual or expected form of the relationship between the two variables is a power law, i.e. one of the form $y = Ax^n$, then it too can be cast into straight-line form by taking the logarithms of both sides. As previously noted, whilst it is normal in mathematical work to use natural logarithms, for practical investigations logarithms to base 10 are often employed. In either case the form is the same, but it needs to be remembered which has been used when recovering the value of A from fitted data. In the mathematical form, the power law relationship becomes

$$\ln y = n \ln x + \ln A. \qquad (1.36)$$

So, a plot of $\ln y$ against $\ln x$ has a slope of n, whilst the intercept on the $\ln y$ axis is $\ln A$, from which A can be found by exponentiation.

Of course, for practical applications, some means of converting x to its logarithm, and of recovering x from its logarithm, has to be available. Historically, these two procedures were carried out using tables of logarithms and anti-logarithms, respectively.

Since numbers are usually presented in decimal form, the logarithms and anti-logarithms given in published tables use base 10, i.e. they are $\log x$ rather than $\ln x$. Only the non-integral part of the logarithm (the mantissa) needs to be provided, as the integral part n can be determined by imagining x written as $x = \xi \times 10^n$ where n is chosen to make ξ lie in the range $1 \leq \xi < 10$. As two examples:

$$\log 365.25 = \log 3.6525 \times 10^2 = 2 + \log 3.6525 = 2 + 0.5626 = 2.5626,$$
$$\log 0.003\,652\,5 = \log 3.6525 \times 10^{-3} = -3 + \log 3.6525 = -3 + 0.5626 = -2.4374.$$

As noted earlier, even the most basic scientific calculator provides these values at the touch of a few buttons, and multiplication and division can be equally easily effected; further, the signs of x, y and the answer are handled automatically.

---

[10] If x and y are dimensionless and equal scales are employed, then m is equal to the tangent of the angle the line makes with the x-axis. See also Section 2.2.1.
[11] And, to make matters worse, the objective of such an experiment is usually that of finding the actual value of f, which is therefore unknown!

Arithmetic and geometry

As illustrated at the end of Section 1.1, logarithms can also be used to evaluate expressions of the form a^m where a is positive and m is a general real number, and not necessarily positive or an integer. As given in Equation (1.11),

$$a^m = e^{m \ln a}. \tag{1.37}$$

Thus, for example, $17^{-0.2}$ is found by first determining that $\ln 17 = 2.833\,21\ldots$, multiplying this by -0.2 to give $-0.566\,64\ldots$, and then evaluating $e^{-0.566\,64\ldots}$ as $0.567\,43$.

When $a = 1$, and consequently $\ln a = 0$, (1.37) reads

$$1^m = e^{m\,0} = e^0 = 1 \tag{1.38}$$

to give the expected result that unity raised to any power is still unity.

An equally expected result is that

$$a^0 = e^{0 \ln a} = e^0 = 1, \tag{1.39}$$

i.e. the zeroth power of any positive quantity is unity; this is also true when a is negative, though we have not proved it here.

Further, we expect that for $a = 0$ and $m \neq 0$, a^m will have the value zero. This is in accord with a natural extension of the prescription, discussed in Section 1.1, that, for integral n, a^n is the result of multiplying unity n times by a factor of a.

Less obvious is the value to be assigned to a^m when both $a = 0$ and $m = 0$. However, the same prescription indicates that the value should be 1, since not multiplying unity by anything must leave it unchanged. The same conclusion can be reached more mathematically by starting from (1.37), taking $a = m = x$ to give $x^x = e^{x \ln x}$ and examining the behaviour of $x \ln x$ as $x \to 0$. By comparing the representation of $\ln x$ as the integral of t^{-1} with the corresponding integral of $t^{-1+\beta}$ for any positive β, it can be shown that $x \ln x$ tends to zero as x tends to zero, and so x^x tends to unity in the same limit. To summarise:

$$0^m = 0 \text{ for } m \neq 0, \text{ but } 0^m = 1 \text{ if } m = 0. \tag{1.40}$$

EXERCISES 1.2

1. Arrange the following expressions into distinct sets, each of which consists of members with a common value:

   (a) $e^{\ln a}$, (b) $a^{\log_a b}$, (c) a^{-2}, (d) $a^{\log_b 1}$, (e) $(1/a^2)^{-1/2}$,

   (f) $\exp(2 \log a)$, (g) $10^{-2 \log a}$, (h) $e^{-2 \ln a}/a^{-1}$, (i) $a^{\log_a 1}$,

   (j) $b^{\log_b a}$, (k) $\ln[\exp(a)]$, (l) $2^{\log_2 2}/10^{\log 2}$, (m) $\log_b b$.

2. Using only the numeric keypad and the $+$, $-$, $=$, $\ln$, $\exp$, x^{-1} and 'answer' keys on a hand calculator, evaluate the following to 4 s.f.:

   (a) $(2.25)^{-2.25}$, (b) $(0.3)^{0.3}$, (c) $(0.3)^{-2.25}$, (d) $(0.3)^{1/2.25}$.

1.3 Physical dimensions

In Section 1.1 we saw how quantities or algebraic expressions that have positive numerical values can be raised to any finite power. So far as arithmetic and algebra are concerned, that is all that is needed. However, when we come to use equations that describe physical situations, and therefore contain symbols that represent physical quantities, we also need to take into account the units in which the quantities are measured. This additional consideration has, in general, two distinct consequences.

The first and obvious one is that all the quantities involved must be expressed in the same system of units. The almost universal choice for scientific purposes is the SI system, though some branches of engineering still use other systems and several areas of physics use derived units that make the values to be manipulated more manageable. Other derived units have less scientific origins.[12] In the SI system the main base units and their abbreviations are the metre (m), the kilogram (kg), the second (s), the ampere (A) and the kelvin (K); they are augmented by the mole for measuring the amount of a substance and the candela for measuring luminous intensity.

In addition, there are many derived units that have specific names of their own, for example the joule (J). However, if need be, the derived units can always be expressed in terms of the base units; in the case of the joule, which is the unit of energy, the equivalence is that 1 J is equal to 1 $kg\,m^2\,s^{-2}$. As another example, 1 V (volt), being the electric potential difference between two points when 1 J of energy is needed to make a current of 1 A flow for 1 s between the points, can be represented as 1 $kg\,m^2\,s^{-3}\,A^{-1}$.

A second, and more fundamental, consequence of the implied presence of appropriate units in equations relating to physical systems is the need to ensure that the 'dimensions' in the various terms in an equation or formula are consistent. This is more fundamental in the sense that it does not depend upon the particular system of units in use – the mean Sun–Earth distance is always a length, whether it is measured in metres, astronomical units or feet and inches. Similarly, a velocity always consists of a length divided by a time, whether it is measured in metres per second or miles per hour. These properties are described by saying that the Earth's distance from the Sun has the dimension of length, whilst its speed has the dimensions of length divided by time.

There is one dimension associated with each of the base units of any system. Purely numerical quantities such as 2, $\frac{1}{3}$, π^2, etc. have no dimensions and affect only the numerical value of an expression; from the point of view of checking the consistency of dimensions, they are to be ignored. For our discussion we will use only the SI system, though references to other systems appear in some examples and problems. We denote the dimensions of a physical quantity X by $[X]$, with those of the five main base units being denoted by the symbols L, M, T, I and Θ as follows:

$$[\text{length}] = L, \quad [\text{mass}] = M, \quad [\text{time}] = T, \quad [\text{current}] = I, \quad [\text{temperature}] = \Theta.$$

---

[12] For interest or amusement, the reader may like to identify, and determine SI values for, the following derived units: (a) barn, (b) denier, (c) hand, (d) hefner, (e) jar, (f) noggin, (g) rod, pole or perch, (h) shake, (i) shed, (j) slug, (k) tog. If help is needed, see G. Woan, *The Cambridge Handbook of Physics Formulas* (Cambridge: Cambridge University Press, 2000).

Arithmetic and geometry

The dimensions of derived quantities are formally obtained by expressing the quantity in terms of its base SI units and then replacing kg by M, etc. Purely numerical quantities, such as those mentioned above, are formally treated as if their dimensions were $L^0 M^0 T^0 I^0 \Theta^0$; this means they can be ignored without appearing to multiply the dimensions of the rest of an expression by zero.

More substantial examples are provided by the two quantities specifically mentioned in an earlier paragraph, energy and voltage. They have dimensions as follows:

$$[E] = M L^2 T^{-2} \quad \text{and} \quad [V] = M L^2 T^{-3} I^{-1}.$$

It should be emphasised that the dimensions of a physical quantity do not depend on the magnitude of that quantity, nor upon the units in which it is measured. Thus, for example, energy has the same dimensions, whether it represents 1 erg or 7.93 MJ.

We now turn to the role of dimensions in the construction of derived quantities and use as a simple example the expression for energy. We have already given the dimensions of energy as $[E] = M L^2 T^{-2}$, and the dimensions of *any* formula that is supposed to be one for energy *must* have this same form. It is almost immediately obvious that the expression for the kinetic energy T of a body of mass m moving with a speed v, namely $T = \frac{1}{2}mv^2$, satisfies this requirement;[13] the formal calculation is as follows:

$$[T] = [\tfrac{1}{2}mv^2] = [\tfrac{1}{2}][m][v^2] = [m][v]^2 = M(L T^{-1})^2 = M L^2 T^{-2}.$$

It will be noticed that the dimensions of a physical quantity obey the same algebraic rules as the symbols that represent that quantity. Thus, in the above illustration, the fact that the velocity appears squared in the expression for the kinetic energy means that the $L T^{-1}$ giving the dimensions of a velocity also appears squared in the dimensions of the energy, i.e. as $L^2 T^{-2}$.

The dimensions of any one physical quantity can contain only integer powers (positive or negative) of the base dimensions, although fractional powers of basic or derived dimensions may appear in formulae; when they do, they again follow the same rules as 'ordinary' powers. For example, the period of oscillation τ of a simple pendulum of length ℓ is given by $\tau = 2\pi(\ell/g)^{1/2}$, where g is the acceleration due to gravity. The dimensional equation reads as follows:

$$[\tau] = [2\pi] \left[\sqrt{\frac{\ell}{g}}\right] = \left(\frac{L}{L T^{-2}}\right)^{1/2} = (T^2)^{1/2} = T,$$

as it should do.

Examination of the dimensions of quantities and combinations of quantities appearing in quoted or derived formulae or equations can be used both positively and negatively.

The constructive use takes the form of dimensional analysis, in which all of the physical variables the investigator thinks might influence a particular phenomenon are formed into combinations that are dimensionless, i.e. for each combination the net index of each of the base units is zero. For the pendulum just considered, such a combination would be $(g\tau^2)/\ell$, as the reader should check. If only one such combination can be formed, then it

---

[13] Check that the formula $\frac{1}{2}kx^2$ for the energy stored in a stretched spring of spring constant k has the correct dimensional form.

1.3 Physical dimensions

must be equal to a constant, but one whose value has to be determined in some other way; for the pendulum it is $4\pi^2$.

If more than one dimensionless combination can be formed, the best that can be said is that some function of these combinations (but not of the individual variables with non-zero dimensions that make up these combinations) is equal to a constant. In some complicated areas of physics and engineering, particularly those involving fluids in motion, this type of analysis is an essential research tool. The following worked example gives some idea of the basic method involved – but hardly produces a previously unknown result!

Example One system of units, proposed by Max Planck and known as *natural units*, is based on five physical constants of nature which are *defined* to have unit value when expressed in those units. The five constants are: c, the speed of light in a vacuum; G, the gravitational constant; k, the Boltzmann constant; $\hbar = h/2\pi$, the Planck constant divided by 2π; $1/4\pi\epsilon_0$, the Coulomb force constant. Use the values and units in Appendix E to find an expression for the natural unit of temperature, T_p, and show that it is approximately equal to 1.4×10^{32} K.

We start by determining the dimensions of each of the five constants; after that we will aim to construct a combination of them that has the dimensions of a temperature, i.e. has just the dimension Θ.

From examining the units of the constants as they are given in the appendix we have

$$[c] = [\text{m s}^{-1}] = LT^{-1},$$
$$[G] = [\text{N kg}^{-2}\,\text{m}^2] = MLT^{-2}\,M^{-2}L^2 = L^3M^{-1}T^{-2},$$
$$[k] = [\text{J K}^{-1}] = ML^2T^{-2}\,\Theta^{-1} = L^2MT^{-2}\,\Theta^{-1},$$
$$[\hbar] = [\text{J s}] = ML^2T^{-2}\,T = L^2MT^{-1}.$$

The value of ϵ_0 is given in farads per metre and to put this into base dimensions would require some additional equation involving either capacitance or ϵ_0 directly. However, it is clear that charge would be involved, and hence so would some power of I, the base dimension of current. But, as it happens, none of the four other 'natural' physical constants includes current in its dimensions, and, as we are trying to construct a pure temperature, no power of ϵ_0 can be a factor in the sought-after combination. We therefore do not need to determine the full dimensions[14] of ϵ_0.

As Θ appears only in $[k]$, and is there as Θ^{-1}, the combination that has just the dimension Θ can only contain k as an overall factor of $1/k$. Thus we assume a combination of the form

$$T_p = \frac{1}{k} c^\alpha G^\beta \hbar^\gamma.$$

Taking dimensions on both sides of the equation gives

$$\Theta = L^{-2}M^{-1}T^2\Theta\; L^\alpha T^{-\alpha}\; L^{3\beta}M^{-\beta}T^{-2\beta}\; L^{2\gamma}M^\gamma T^{-\gamma}.$$

The Θ-dimension has already been arranged to be correct, but equating the powers of L, M and T on the two sides gives the three simultaneous equations

$$0 = -2 + \alpha + 3\beta + 2\gamma,$$
$$0 = -1 - \beta + \gamma,$$
$$0 = 2 - \alpha - 2\beta - \gamma.$$

---

[14] Though the reader may care to, using one of the formulae given in the footnote on p. 5 to show that they are $L^{-3}\,M^{-1}\,T^4\,I^2$.

Arithmetic and geometry

These have solution $\alpha = \frac{5}{2}$, $\beta = -\frac{1}{2}$ and $\gamma = \frac{1}{2}$. Thus

$$T_p = \frac{1}{k}\sqrt{\frac{\hbar c^5}{G}},$$

and its numerical value in SI units is

$$T_p = \frac{1}{1.38 \times 10^{-23}}\sqrt{\frac{6.6 \times 10^{-34} \, (3.0 \times 10^8)^5}{2\pi \, 6.7 \times 10^{-11}}} = 1.4 \times 10^{32} \text{ K},$$

as stated in the question. ◀

We now turn to the more negative and humdrum topic of checking possible equations and formulae for internal dimensional consistency. Suppose that you have derived, or been presented with, an equation purporting to describe a certain physical situation. Before risking examination credit-loss by submitting it for marking, or your academic reputation by publishing, it is well worth checking the equation's dimensional plausibility; finding consistency does not guarantee that the equation is correct, but finding inconsistency *guarantees* that it is *wrong*, and could save a lot of embarrassment. Dimensional aspects that should be checked are

- Both sides of the equation must have exactly the same dimensions.
- Any two items that are added or subtracted must have the same dimensions as each other.
- The arguments of any mathematical functions that can be written as a power series with more than one term must be dimensionless. Examples include the exponential function, the sinusoidal functions, and polynomials.

You should also check that the equation has the expected behaviour for extreme values of the variables and parameters, within the range of validity claimed, as well as for any particularly simple set of values for which the solution can be found by other means. We illustrate some of these checks by means of the following example

---

Example It is claimed that the speed v of waves of wavelength λ travelling on the surface of a liquid, under the influence of both gravity and surface tension, is given by

$$v^2 = ag\lambda + \frac{b\sigma}{\rho\lambda},$$

where ρ and σ are the density and surface tension, respectively, of the liquid. The acceleration due to gravity is $g = 9.81 \text{ m s}^{-2}$ and the coefficient of surface tension is 7.0×10^{-3} in units of joules per square metre; a and b are dimensionless constants. Is the claimed formula plausible?

We first note that, as we are not given any experimental data, and the values of a and b are unknown in any case, the numerical values provided for g and σ are of no help when the possible validity of

1.3 Physical dimensions

the formula is being examined. However, we *can* use the data to establish the dimensions of g and σ, if we do not already know them:

$$[g] = [\text{m s}^{-2}] = LT^{-2}, \quad [\sigma] = [\text{J}][\text{m}^{-2}] = ML^2T^{-2}L^{-2} = MT^{-2}.$$

The dimension of λ, a wavelength, is clearly L and those of the density ρ are ML^{-3}. As the RHS of the formula consists of two combinations of variables that are added, we must next check that they each have the same overall dimensions. Recalling that a and b are dimensionless, we have

$$[ag\lambda] = [a][g][\lambda] = LT^{-2} L = L^2 T^{-2},$$

$$\left[\frac{b\sigma}{\rho\lambda}\right] = \frac{[b][\sigma]}{[\rho][\lambda]} = \frac{MT^{-2}}{ML^{-3} L} = L^2 T^{-2}.$$

As we can see, they do have the same dimensions and therefore can be added together. Furthermore, those dimensions are the same as those on the LHS of the formula, namely $[v^2] = (LT^{-1})^2 = L^2 T^{-2}$. Thus, in summary, no dimensional inconsistencies have been found and the stated formula *could* be a valid one.[15] ◀

The problems at the end of this chapter further illustrate the uses of, and the constraints imposed by, the notion of dimensions, and at the same time introduce equations and formulae from some of the more intriguing areas of quantum and cosmological physics.

EXERCISES 1.3

1. Demonstrate that each of the formulae given below is dimensionally acceptable.
    (a) Bernoulli's equation for the speed v and pressure p at height z in an incompressible ideal fluid of density ρ is

    $$\tfrac{1}{2}\rho v^2 + p + \rho g z = \text{constant}.$$

    (b) The speed v of a wave of wavelength λ travelling through a thin plate of thickness t (in the direction of travel) is

    $$v = \frac{2\pi}{\lambda}\left[\frac{Et^2}{12\rho(1-\sigma^2)}\right]^{1/2}.$$

    Here E is the Young modulus, ρ the density, and σ the (dimensionless) Poisson ratio for the material of the plate. The Young modulus is defined as the ratio of the longitudinal stress (force per unit area) to the longitudinal strain (fractional increase in length) in a thin wire made of the material.

---

[15] In fact, the formula is a valid one for surface waves whose wavelength is much less than the depth of the liquid; a has the value $1/2\pi$ and $b = 2\pi$.

Arithmetic and geometry

(c) The probability density pr(c) of particle speeds c in a classical gas at temperature T is given by

$$\text{pr}(c) = 4\pi c^2 \left(\frac{m}{2\pi kT}\right)^{3/2} \exp\left(-\frac{mc^2}{2kT}\right),$$

where m is the mass of a molecule and k is Boltzmann's constant. The probability density has dimensions TL^{-1}.

2. Below are the names and formulae for three physical constants, together with a set of quoted values. Using the values given in Appendix E, check the formulae and quoted values for numerical and dimensional consistency, and so determine which, if any, have been wrongly quoted (beyond rounding errors).

Fine structure constant $\quad \alpha = \frac{\mu_0 c e^2}{2h} = 7.30 \times 10^{-3}\,\text{s}^{-1}$,

Planck time $\quad t_{\text{Pl}} = \sqrt{\frac{hG}{2\pi c^5}} = 5.39 \times 10^{-42}\,\text{s}$,

Bohr magneton $\quad \mu_{\text{B}} = \frac{eh}{4\pi m_e} = 9.27 \times 10^{-24}\,\text{J T}^{-1}$.

[Note that the force F on a conductor of length ℓ carrying a current i perpendicular to a magnetic field of flux density B is $F = Bi\ell$. The unit of magnetic flux density is the tesla (with symbol T).]

1.4 The binomial expansion

Earlier in this chapter we considered powers of a single quantity or variable, such as a^n, e^n or x^n. We now extend our discussion to functions that are powers of the sum or difference of two terms, e.g. $(x - \alpha)^m$. Later in this book we will find numerous occasions on which we wish to write such a product of repeated factors as a polynomial in x or, more generally, as a sum of terms each of which is a power of x multiplied by a power of α, as opposed to a power of their sum or difference.

To make the discussion general and the result applicable to a wide variety of situations, we will consider the general expansion of $f(x, y) = (x + y)^n$, where x and y may stand for constants, variables or functions but, for the time being, n is a positive integer. It may not be obvious what form the general expansion takes, but some idea can be obtained by carrying out the multiplication explicitly for small values of n. Thus we obtain successively

$(x + y)^1 = x + y,$
$(x + y)^2 = (x + y)(x + y) = x^2 + 2xy + y^2,$
$(x + y)^3 = (x + y)(x^2 + 2xy + y^2) = x^3 + 3x^2y + 3xy^2 + y^3,$
$(x + y)^4 = (x + y)(x^3 + 3x^2y + 3xy^2 + y^3) = x^4 + 4x^3y + 6x^2y^2 + 4xy^3 + y^4.$

This does not *establish* a general formula, but the regularity of the terms in the expansions and the suggestion of a pattern in the coefficients indicate that a general formula for the nth power will have $n + 1$ terms, that the powers of x and y in every term will add up

1.4 The binomial expansion

to n, and that the coefficients of the first and last terms will be unity, whilst those of the second and penultimate terms will be n.[16,17]

In fact, the general expression, the *binomial expansion* for power n, is given by

$$(x+y)^n = \sum_{k=0}^{n} {}^nC_k x^{n-k} y^k, \qquad (1.41)$$

where ${}^nC_k$ is called the *binomial coefficient*. When it is expressed in terms of the factorial functions introduced in Section 1.2 it takes the form $n!/[k!(n-k)!]$ with $0! = 1$. Clearly, simply to make such a statement does not constitute proof of its validity, but, as we will see in Section 1.4.2, Equation (1.41) can be *proved* using a method called induction. Before turning to that proof, we investigate some of the elementary properties of the binomial coefficients.

1.4.1 Binomial coefficients

As stated above, the binomial coefficients are defined by

$$ {}^nC_k \equiv \frac{n!}{k!(n-k)!} \equiv \binom{n}{k} \quad \text{for } 0 \leq k \leq n, \qquad (1.42)$$

where in the second identity we give a common alternative notation for ${}^nC_k$. Obvious properties include

(i) ${}^nC_0 = {}^nC_n = 1$,
(ii) ${}^nC_1 = {}^nC_{n-1} = n$,
(iii) ${}^nC_k = {}^nC_{n-k}$.

We note that, for any given n, the largest coefficient in the binomial expansion is the middle one ($k = n/2$) if n is even; the middle two coefficients $[k = \tfrac{1}{2}(n \pm 1)]$ are equal largest if n is odd. Somewhat less obvious, but a result that will be needed in the next section, is that

$$\begin{aligned}
{}^nC_k + {}^nC_{k-1} &= \frac{n!}{k!(n-k)!} + \frac{n!}{(k-1)!(n-k+1)!} \\
&= \frac{n![(n+1-k)+k]}{k!(n+1-k)!} \\
&= \frac{(n+1)!}{k!(n+1-k)!} = {}^{n+1}C_k.
\end{aligned} \qquad (1.43)$$

An equivalent statement, in which k has been redefined as $k + 1$, is

$$ {}^nC_k + {}^nC_{k+1} = {}^{n+1}C_{k+1}. \qquad (1.44)$$

---

[16] Write down your prediction for the expansion of $(x+y)^5$ and then check it by direct calculation.
[17] One examination paper question read: 'Expand $(x+y)^5$'. The submitted response was, $(x+y)^5 = (x+y)^5 = (x+y)^5 = (x+y)^5 = (x+y)^5 = (x+y)^5 = \ldots$

1.4.2 Proof of the binomial expansion

We are now in a position to *prove* the binomial expansion (1.41). In doing so, we introduce the reader to a procedure applicable to certain types of problems and known as the *method of induction*. The method is discussed much more fully in Section 2.4.1.

We start by *assuming* that (1.41) is true for some positive integer $n = N$, and then proceed to show that, given the assumption, it follows that (1.41) also holds for $n = N + 1$:

$$(x + y)^{N+1} = (x + y) \sum_{k=0}^{N} {}^{N}C_k x^{N-k} y^k$$

$$= \sum_{k=0}^{N} {}^{N}C_k x^{N+1-k} y^k + \sum_{k=0}^{N} {}^{N}C_k x^{N-k} y^{k+1}$$

$$= \sum_{k=0}^{N} {}^{N}C_k x^{N+1-k} y^k + \sum_{j=1}^{N+1} {}^{N}C_{j-1} x^{(N+1)-j} y^j,$$

where in the first line we have used the initial assumption and in the third line have moved the second summation index by unity, by writing $k + 1 = j$. We now separate off the first term of the first sum, ${}^{N}C_0 x^{N+1}$, and write it as ${}^{N+1}C_0 x^{N+1}$; we can do this since, as noted in (i) following (1.42), ${}^{n}C_0 = 1$ for every n. Similarly, the last term of the second summation can be replaced by ${}^{N+1}C_{N+1} y^{N+1}$.

The remaining terms of each of the two summations are now written together, with the summation index denoted by k in both terms.[18] Thus

$$(x + y)^{N+1} = {}^{N+1}C_0 x^{N+1} + \sum_{k=1}^{N} \left({}^{N}C_k + {}^{N}C_{k-1}\right) x^{(N+1)-k} y^k + {}^{N+1}C_{N+1} y^{N+1}$$

$$= {}^{N+1}C_0 x^{N+1} + \sum_{k=1}^{N} {}^{N+1}C_k x^{(N+1)-k} y^k + {}^{N+1}C_{N+1} y^{N+1}$$

$$= \sum_{k=0}^{N+1} {}^{N+1}C_k x^{(N+1)-k} y^k.$$

In going from the first to the second line we have used result (1.43). Now we observe that the final overall equation is just the original assumed result (1.41) but with $n = N + 1$. Thus it has been shown that if the binomial expansion is *assumed* to be true for $n = N$, then it can be *proved* to be true for $n = N + 1$. But it holds trivially for $n = 1$, and therefore for $n = 2$ also. By the same token it is valid for $n = 3, 4, \ldots$, and hence is established for all positive integers n.

1.4.3 Negative and non-integral values of n

Up till now we have restricted n in the binomial expansion to be a positive integer. Negative values can be accommodated, but only at the cost of an infinite series of terms rather than

---

[18] Note that the first summation, having lost its first term, now has an index that runs from 1 to N, and that the second summation, having lost its last term, also has an index that now runs from 1 to N.

1.4 The binomial expansion

the finite one represented by (1.41). For reasons that are intuitively sensible and will be discussed in more detail in Chapter 6, very often we require an expansion in which, at least ultimately, successive terms in the infinite series decrease in magnitude. For this reason, if $|x| > |y|$ and we need to consider $(x + y)^{-m}$, where m itself is a positive integer, then we do so in the form

$$(x + y)^n = (x + y)^{-m} = x^{-m}\left(1 + \frac{y}{x}\right)^{-m}.$$

Since the ratio $|y/x|$ is less than unity, terms containing higher powers of it will be small in magnitude, whilst raising the unit term to any power will not affect its magnitude. If $|y| > |x|$ the roles of the two must be interchanged.

We can now state, but will not explicitly prove, the form of the binomial expansion appropriate to negative values of n (n equal to $-m$):

$$(x + y)^n = (x + y)^{-m} = x^{-m} \sum_{k=0}^{\infty} {}^{-m}C_k \left(\frac{y}{x}\right)^k, \tag{1.45}$$

where the hitherto undefined quantity ${}^{-m}C_k$, which appears to involve factorials of negative numbers, is given by

$${}^{-m}C_k = (-1)^k \frac{m(m+1)\cdots(m+k-1)}{k!} = (-1)^k \frac{(m+k-1)!}{(m-1)!k!} = (-1)^k \, {}^{m+k-1}C_k. \tag{1.46}$$

The binomial coefficient on the extreme right of this equation has its normal meaning and is well defined since $m + k - 1 \geq k$.

Thus we have a definition of binomial coefficients for negative integer values of n in terms of those for positive n. The connection between the two may not be obvious, but they are both formed in the same way in terms of recurrence relations. Whatever the sign of n, or its integral or non-integral nature, the series of coefficients ${}^nC_k$ can be generated by starting with ${}^nC_0 = 1$ and using the recurrence relation

$${}^nC_{k+1} = \frac{n-k}{k+1} \, {}^nC_k. \tag{1.47}$$

The difference between the case of positive integer n and all other cases is that for positive integer n the series terminates when $k = n$, whereas for negative or non-integral n there is no such termination – in line with the infinite series of terms in the corresponding expansions.

Finally, to summarise, Equation (1.47) generates the appropriate coefficients for all values of n, positive or negative, integer or non-integer, with the obvious exception of the case in which $x = -y$ and n is negative.

1.4.4 Relationship with the exponential function

Before we leave the binomial expansion, we use it to establish an alternative representation of the exponential function. The representation takes the form of a limit and is

$$\lim_{n \to \infty} \left(1 + \frac{a}{n}\right)^n = e^a. \tag{1.48}$$

Arithmetic and geometry

The formal definition of a limit is not discussed until Chapter 6, but for our present purposes an intuitive notion of one will suffice. We start by expanding the nth power of $1 + (a/n)$ using the binomial theorem, and remembering that ^nC_k can be written as $[n(n-1)\cdots(n-k+1)]/k!$. This gives

$$\left(1 + \frac{a}{n}\right)^n = 1 + n\frac{a}{n} + \frac{n(n-1)}{2!}\frac{a^2}{n^2} + \cdots + \frac{n(n-1)\cdots(n-k+1)}{k!}\frac{a^k}{n^k} + \cdots$$

This can be rearranged as

$$\left(1 + \frac{a}{n}\right)^n = 1 + a + \frac{(1-n^{-1})}{2!}a^2 + \cdots + \frac{(1-n^{-1})\cdots(1-(k-1)n^{-1})}{k!}a^k + \cdots$$

We now take the limit of both sides as $n \to \infty$; $n^{-1} \to 0$ and all the factors containing it on the RHS tend to unity, leaving

$$\lim_{n \to \infty}\left(1 + \frac{a}{n}\right)^n = 1 + a + \frac{a^2}{2!} + \cdots + \frac{a^k}{k!} + \cdots = \exp(a) = e^a,$$

and thus establishing (1.48).

The most practical example of this result is the way that compound interest on capital A, borrowed or lent, leads to 'exponential growth' of that capital. If the annual rate of interest is a and the interest is paid only at the end of the year, the capital then stands at $A(1 + a)$. However, if it is paid monthly, then the corresponding figure is $A[1 + (a/12)]^{12}$, and if it is paid daily the capital stands at $A[1 + (a/365)]^{365}$ at the end of the year. As the interval between payments becomes shorter the end-of-year capital amount becomes larger. However, it does not increase indefinitely and 'continuous interest payment' results in a capital of Ae^a at the end of the year, and Ae^{na} at the end of n years.[19]

EXERCISES 1.4

1. Evaluate the binomial coefficients (a) $^{-3}C_1$, (b) $^{-5}C_7$, (c) $^{-1}C_k$.
2. Evaluate the binomial coefficients (a) $^{1/2}C_3$, (b) $^{-1/2}C_3$, (c) $^{5/3}C_3$.
3. Demonstrate explicitly the validity of (1.44) for $n = 4$ and $k = 0, 1, 2$. Using only the general simple properties of binomial coefficients, and the simplest of arithmetic, deduce the validity of (1.44) for $k = 3$.

1.5 Trigonometric identities

So many of the applications of mathematics to physics and engineering are concerned with periodic, and in particular sinusoidal, behaviour that a sure and ready handling of

---

[19] Show that capital that attracts an annual interest rate of 5% will exceed twice its initial value more than 400 days earlier if interest is paid continuously rather than yearly (in arrears).

1.5 Trigonometric identities

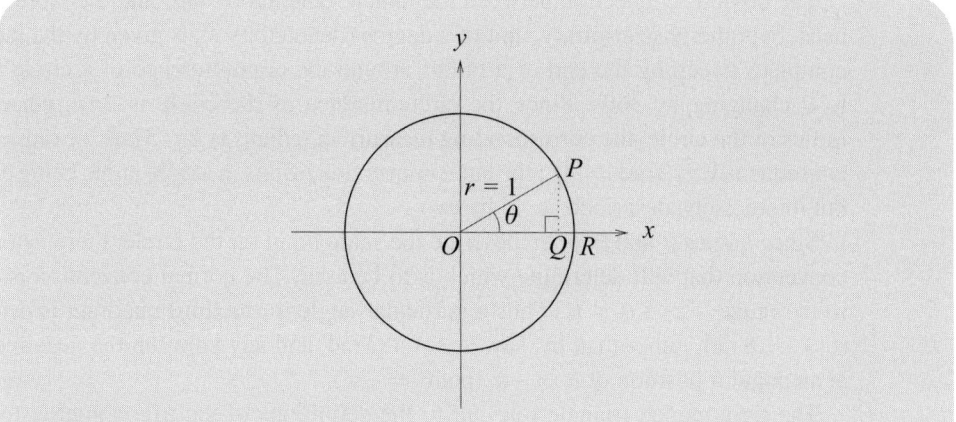

Figure 1.2 The geometric definitions of the basic trigonometric functions.

the corresponding mathematical functions is an essential skill. Even situations with no obvious periodicity are often expressed in terms of periodic functions for the purposes of analysis. Books on mathematical methods devote whole chapters to developing the necessary techniques, and so, as groundwork, we here establish (or remind the reader of) some standard identities with which he or she should be fully familiar, so that the manipulation of expressions containing sinusoids becomes automatic and reliable. So as to emphasise the angular nature of the argument of a sinusoid we will denote it in this section by θ rather than x.

The definitions of the three basic trigonometric functions, the sine, the cosine and the tangent, can be given in either geometric or algebraic forms. In the former, the definitions are in terms of the ratios of the sides of a right-angled triangle, one of whose other angles is θ, as is illustrated in Figure 1.2.

The figure shows a general point P of a circle of unit radius centred on the origin O of a two-dimensional Cartesian coordinate system. The angle θ is that between the direction of the radius of the circle that passes through P and the direction of the positive x-axis.

For mathematical work, angles are measured in radians (rather than degrees), one radian being defined as the magnitude of θ if the position of point P on the circle is such that the arc length RP is equal to the radius of the circle; in this case that arc RP has unit length. It should be noted that it is the distance measured along the circumference of the circle that gives the arc length, *not* the length of the straight-line secant[20] that joins R to P. Thus, for a general circle of radius r, an arc of the circle of length ℓ subtends an angle θ at the centre of the circle given by

$$\ell = r\theta, \tag{1.49}$$

provided that θ is measured in radians.

---

[20] This description of a straight line with a particular property should not be confused with the trigonometric function of the same name defined in Equation (1.55). Show that the length of the secant is $2\sin\theta/2$.

Arithmetic and geometry

The obvious connection between the radian (denoted by rad) and the more commonly used, but otherwise arbitrary, unit of a degree (denoted by °), is given by the fact that one complete sweep by the end of a radius around the circumference of a circle is equated to θ changing by 360°. Since the circumference of the circle is $2\pi r$, where r is the radius of the circle, the corresponding measure in radians is 2π. Thus 2π rad = 360° and consequently $\pi/2$ rad = 90°; the latter conversion results in a right angle being commonly, but imprecisely, described as 'pi by two'.

Since angles θ and $\theta + 2\pi$ describe the same point on the circle, there is a need for a convention that will determine which is to be used. The normal convention is that θ lies in the range $-\pi < \theta \leq \pi$. Thus a particular angle in the third quadrant is described by $\theta = -1.8$ rad, rather than by, say, $\theta = 4.483$ rad, and any point on the negative x-axis is at an angular position of $\theta = +\pi$ (not $\theta = -\pi$).

The right-angled triangle relevant to the definitions of the trigonometric functions is OPQ, where Q is the foot of the perpendicular from P onto the x-axis. With this notation the coordinates of P define the sine and cosine of θ through

$$\sin\theta \equiv \frac{QP}{OP} = \frac{y}{1} = y, \qquad \cos\theta \equiv \frac{OQ}{OP} = \frac{x}{1} = x. \qquad (1.50)$$

It is clear from this geometric definition that both the sine and cosine functions are periodic with period 2π, with, for example, $\sin(\theta + 2\pi n) = \sin\theta$ for any integer n; they are also bounded with $-1 \leq \sin\theta \leq +1$, and similarly for $\cos\theta$. The fact that we have used a unit circle in our definitions, rather than one of general radius r, is irrelevant, since the ratios of the sides of similar triangles are independent of their scales.

The tangent of θ is now defined as the ratio of QP to OQ, or equivalently as the ratio of $\sin\theta$ to $\cos\theta$, i.e.

$$\tan\theta \equiv \frac{QP}{OQ} = \frac{y}{x} = \frac{\sin\theta}{\cos\theta}. \qquad (1.51)$$

It too is periodic, but with a period of π rather than 2π. Unlike the sine and cosine functions from which it is derived, $\tan\theta$ can take any real value in the range $-\infty < \tan\theta < +\infty$.

From these definitions the following symmetry properties are apparent:

$$\sin(-\theta) = -\sin\theta, \qquad \cos(-\theta) = \cos\theta, \qquad \tan(-\theta) = -\tan\theta. \qquad (1.52)$$

The same definitions also imply that $\sin 0 = \cos\pi/2 = \sin\pi = 0$ and that $\cos 0 = \sin\pi/2 = -\cos\pi = 1$. Using these simple values and some of the formulae for multiple angles derived in the next subsection and Chapter 5, the following useful table, giving numerical values for the sinusoids of common angles,[21] can be drawn up.

---

[21] What credit would you give the following (i) for mathematics and (ii) for ingenuity? Solve: $nx = \sin x$ with $n = -0.04657$. Answer: Cancelling $n \neq 0$ from both sides gives $x = \sin x$, i.e. $x = 6$. Check: $\sin 6 = -0.2794 = -0.04657 \times 6$. ✓

1.5 Trigonometric identities

θ (rad)	θ (deg)	$\sin\theta$	$\cos\theta$	$\tan\theta$
0	0	0	1	0
$\pi/6$	30	$1/2$	$\sqrt{3}/2$	$1/\sqrt{3}$
$\pi/4$	45	$1/\sqrt{2}$	$1/\sqrt{2}$	1
$\pi/3$	60	$\sqrt{3}/2$	$1/2$	$\sqrt{3}$
$\pi/2$	90	1	0	∞
$2\pi/3$	120	$\sqrt{3}/2$	$-1/2$	$-\sqrt{3}$
$3\pi/4$	135	$1/\sqrt{2}$	$-1/\sqrt{2}$	-1
$5\pi/6$	150	$1/2$	$-\sqrt{3}/2$	$-1/\sqrt{3}$
π	180	0	-1	0

The algebraic definitions of $\sin\theta$ and $\cos\theta$ are both in the form of power series in θ, with θ measured in radians. The two sums are

$$\sin\theta = \sum_{n=0}^{\infty} \frac{(-1)^n \theta^{2n+1}}{(2n+1)!} = \theta - \frac{\theta^3}{3!} + \frac{\theta^5}{5!} - \cdots \tag{1.53}$$

and

$$\cos\theta = \sum_{n=0}^{\infty} \frac{(-1)^n \theta^{2n}}{(2n)!} = 1 - \frac{\theta^2}{2!} + \frac{\theta^4}{4!} - \cdots \tag{1.54}$$

The general appearance of both definitions is somewhat similar to that of the power series in (1.22) that defines the function $\exp(x)$; this similarity is not coincidental, as will be apparent when complex numbers, and in particular Euler's equation, are studied in Chapter 5. For any particular value of θ, each sum converges to its own specific finite value as more terms are added; these are the values that we denote by $\sin\theta$ and $\cos\theta$. The convergence property holds for all finite values of θ and, in fact, the resulting sums always lie in the range $-1 \leq \sin\theta, \cos\theta \leq 1$ for *all* real θ.

One obvious question that arises is whether the geometric and algebraic definitions of $\sin\theta$ and $\cos\theta$ agree. A demonstration that they are equivalent is given in Appendix B; it employs an approach that is similar to one used in differential calculus and also relies heavily on the compound-angle identities that are derived geometrically in Section 1.5.1. The proof should therefore be returned to after these have been studied.

The *reciprocals* of the basic sinusoidal functions, sine and cosine, have been given the special names of cosecant and secant, respectively. As function names they are abbreviated to cosec and sec; specifically,

$$\operatorname{cosec}\theta = \frac{1}{\sin\theta}, \qquad \sec\theta = \frac{1}{\cos\theta}. \tag{1.55}$$

Care must be taken not to confuse these two functions with the *inverses* of $\sin\theta$ and $\cos\theta$, which are written as $\sin^{-1}$ and $\cos^{-1}$, respectively. Thus

$$y = (\cos x)^{-1} = \frac{1}{\cos x} \Rightarrow y = \sec x, \quad \text{but} \quad y = \cos^{-1} x \Rightarrow \cos y = x.$$

With the aim of avoiding such possible confusion some authors, calculators and computer programs attach the prefix 'a' or 'arc' to the name of a function (rather than add the

Arithmetic and geometry

superscript $^{-1}$ to its end) in order to convert it into its inverse; thus arcsin x is the same function as $\sin^{-1} x$ and atan x is the same as $\tan^{-1} x$.[22]

The well-known basic identity satisfied by the sinusoidal functions $\sin\theta$ and $\cos\theta$ is

$$\cos^2\theta + \sin^2\theta = 1. \qquad (1.56)$$

For $\sin\theta$ and $\cos\theta$ defined geometrically this is an immediate consequence of the theorem due to Pythagoras. If they have been defined algebraically by means of series then the result from Appendix B is needed as a link to the Pythagorean justification; a more direct proof is available using Euler's equation (Chapter 5).

Other standard single-angle formulae derived from (1.56) by dividing through by various powers of $\sin\theta$ and $\cos\theta$ are[23]

$$1 + \tan^2\theta = \sec^2\theta, \qquad (1.57)$$
$$\cot^2\theta + 1 = \operatorname{cosec}^2\theta. \qquad (1.58)$$

1.5.1 Compound-angle identities

The basis for building expressions for the sinusoidal functions of compound angles are those for the sum and difference of just two angles, since all other cases can be built up from these, in principle. Later we will see that a study of complex numbers can provide a more efficient approach in some cases.

To prove the basic formulae for the sine and cosine of a compound angle $A + B$ in terms of the sines and cosines of A and B, we consider the construction shown in Figure 1.3. It shows two sets of axes, Oxy and $Ox'y'$, with a common origin O, but rotated with respect to each other through an angle A. The point P lies on the unit circle centred on the common origin and has coordinates $\cos(A + B), \sin(A + B)$ with respect to the axes Oxy and coordinates $\cos B, \sin B$ with respect to the axes $Ox'y'$.

Parallels to the axes Oxy (dotted lines) and $Ox'y'$ (broken lines) have been drawn through P. Further parallels (MR and RN) to the $Ox'y'$ axes have been drawn through R, the point $(0, \sin(A + B))$ in the Oxy system. That all the angles marked with the symbol • are equal to A follows from the simple geometry of right-angled triangles and crossing lines.

We now determine the coordinates of P in terms of lengths in the figure, expressing those lengths in terms of both sets of coordinates:

(i) $\cos B = x' = TN + NP = MR + NP$
$$= OR \sin A + RP \cos A = \sin(A + B) \sin A + \cos(A + B) \cos A;$$

(ii) $\sin B = y' = OM - TM = OM - NR$
$$= OR \cos A - RP \sin A = \sin(A + B) \cos A - \cos(A + B) \sin A.$$

Now, if equation (i) is multiplied by $\sin A$ and added to equation (ii) multiplied by $\cos A$, the result is

$$\sin A \cos B + \cos A \sin B = \sin(A + B)(\sin^2 A + \cos^2 A) = \sin(A + B).$$

---

[22] If $u = \sec x$ and $v = \sin^{-1} x$, write u in terms of v and v in terms of u.
[23] Derive the less well-known relation $\tan\theta + \cot\theta = \sec\theta \operatorname{cosec}\theta$.

1.5 Trigonometric identities

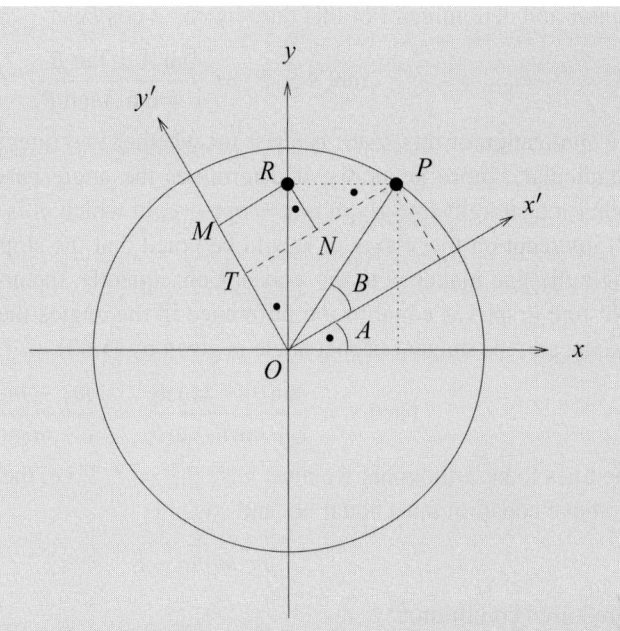

Figure 1.3 Illustration of the compound-angle identities. Refer to the main text for details.

Similarly, if equation (ii) is multiplied by sin A and subtracted from equation (i) multiplied by cos A, the result is

$$\cos A \cos B - \sin A \sin B = \cos(A+B)(\cos^2 A + \sin^2 A) = \cos(A+B).$$

Corresponding graphically based results can be derived for the sines and cosines of the difference of two angles; however, they are more easily obtained by setting B to $-B$ in the previous results and remembering that sin B becomes $-\sin B$ whilst cos B is unchanged. The four results may be summarised by

$$\sin(A \pm B) = \sin A \cos B \pm \cos A \sin B, \tag{1.59}$$
$$\cos(A \pm B) = \cos A \cos B \mp \sin A \sin B. \tag{1.60}$$

The $\mp$ sign (not a $\pm$ sign) on the RHS of the second equation should be noted.[24]

Standard results can be deduced from these by setting one of the two angles equal to π or to $\pi/2$:

$$\sin(\pi - \theta) = \sin\theta, \quad \cos(\pi - \theta) = -\cos\theta, \tag{1.61}$$
$$\sin\left(\tfrac{1}{2}\pi - \theta\right) = \cos\theta, \quad \cos\left(\tfrac{1}{2}\pi - \theta\right) = \sin\theta. \tag{1.62}$$

From these basic results many more can be derived. An immediate deduction, obtained by taking the ratio of the two equations (1.59) and (1.60) and then dividing both the

....

[24] Show formally that, as it must be, (1.56) is satisfied by these expressions for the sine and cosine of $A \pm B$.

numerator and denominator of this ratio by cos A cos B, is

$$\tan(A \pm B) = \frac{\tan A \pm \tan B}{1 \mp \tan A \tan B}. \tag{1.63}$$

One application of this result is a test for whether two lines on a graph are orthogonal (perpendicular); more generally, it determines the angle between them. The standard notation for a straight-line graph is $y = mx + c$, in which m is the slope of the graph and c is its intercept on the y-axis. It should be noted that the slope m is also the tangent of the angle the line makes with the x-axis. Consequently, the angle θ_{12} between two such straight-line graphs is equal to the difference in the angles they individually make with the x-axis, and the tangent of that angle is given by (1.63):

$$\tan \theta_{12} = \frac{\tan \theta_1 - \tan \theta_2}{1 + \tan \theta_1 \tan \theta_2} = \frac{m_1 - m_2}{1 + m_1 m_2}. \tag{1.64}$$

For the lines to be orthogonal we must have $\theta_{12} = \pi/2$, i.e. the final fraction on the RHS of the above equation must equal ∞, and so

$$m_1 m_2 = -1 \tag{1.65}$$

is the required condition.[25]

A kind of inversion of Equations (1.59) and (1.60) enables the sum or difference of two sines or cosines to be expressed as the product of two sinusoids; the procedure is typified by the following. Adding together the expressions given by (1.59) for $\sin(A + B)$ and $\sin(A - B)$ yields

$$\sin(A + B) + \sin(A - B) = 2 \sin A \cos B.$$

If we now write $A + B = C$ and $A - B = D$, this becomes

$$\sin C + \sin D = 2 \sin \left(\frac{C + D}{2} \right) \cos \left(\frac{C - D}{2} \right). \tag{1.66}$$

In a similar way, each of the following equations can be derived:

$$\sin C - \sin D = 2 \cos \left(\frac{C + D}{2} \right) \sin \left(\frac{C - D}{2} \right), \tag{1.67}$$

$$\cos C + \cos D = 2 \cos \left(\frac{C + D}{2} \right) \cos \left(\frac{C - D}{2} \right), \tag{1.68}$$

$$\cos C - \cos D = -2 \sin \left(\frac{C + D}{2} \right) \sin \left(\frac{C - D}{2} \right). \tag{1.69}$$

The minus sign on the right of the last of these equations should be noted; it may help to avoid overlooking this 'oddity' to recall that if $C > D$ then $\cos C < \cos D$.

1.5.2 Double- and half-angle identities

Double-angle and half-angle identities are needed so often in practical calculations that they should be committed to memory by any physical scientist. They can be obtained by

---

[25] Find the equations of the lines through the origin that meet the line $y = 2x + 5$ at angles of $45°$.

1.5 Trigonometric identities

setting B equal to A in results (1.59) and (1.60). When this is done, and use made of Equation (1.56), the following identities are obtained:

$$\sin 2\theta = 2\sin\theta\cos\theta, \tag{1.70}$$

$$\cos 2\theta = \cos^2\theta - \sin^2\theta = 2\cos^2\theta - 1 = 1 - 2\sin^2\theta, \tag{1.71}$$

$$\tan 2\theta = \frac{2\tan\theta}{1-\tan^2\theta}. \tag{1.72}$$

A further set of identities enables sinusoidal functions of θ to be expressed as the ratio of two *polynomial* functions of a variable $t = \tan(\theta/2)$. They are not used in their primary role until Chapter 4, which deals with integration, but we give a derivation of them here for reference.

If $t = \tan(\theta/2)$, then it follows from (1.57) that $1 + t^2 = \sec^2(\theta/2)$ and so $\cos(\theta/2) = (1+t^2)^{-1/2}$, whilst $\sin(\theta/2) = t(1+t^2)^{-1/2}$. Now, using (1.70) and (1.71), we may write[26]

$$\sin\theta = 2\sin\frac{\theta}{2}\cos\frac{\theta}{2} = \frac{2t}{1+t^2}, \tag{1.73}$$

$$\cos\theta = \cos^2\frac{\theta}{2} - \sin^2\frac{\theta}{2} = \frac{1-t^2}{1+t^2}, \tag{1.74}$$

$$\tan\theta = \frac{2t}{1-t^2}. \tag{1.75}$$

It can be shown that the derivative of θ with respect to t takes the algebraic form $2/(1+t^2)$. This completes a package of results that enables expressions involving sinusoids, particularly when they appear as integrands, to be cast in more convenient algebraic forms. The proof of the derivative property and examples of the use of the above results are given in Section 4.2.5.

We conclude this section with a worked example which is of such a commonly occurring form that it might be considered a standard procedure.

Example Solve for θ the equation

$$a\sin\theta + b\cos\theta = k,$$

where a, b and k are given real quantities.

To solve this equation we make use of result (1.59) by setting $a = K\cos\phi$ and $b = K\sin\phi$ for suitable values of K and ϕ. We then have

$$k = K\cos\phi\sin\theta + K\sin\phi\cos\theta = K\sin(\theta+\phi),$$

with

$$K^2 = a^2 + b^2 \quad \text{and} \quad \phi = \tan^{-1}\frac{b}{a}.$$

---

[26] Use result (1.74) and the tabulation on p. 27 to show that $\tan(\pi/12) = 2 - \sqrt{3}$.

Whether ϕ lies in $0 \le \phi \le \pi$ or in $-\pi < \phi < 0$ has to be determined by the individual signs of a and b. The solution is thus

$$\theta = \sin^{-1}\left(\frac{k}{K}\right) - \phi,$$

with K and ϕ as given above. Notice that the inverse sine yields two values in the range $-\pi$ to π and that there is no real solution to the original equation if $|k| > |K| = (a^2 + b^2)^{1/2}$.

EXERCISES 1.5

1. Find, where they exist, the (real) values of

$$\sin\theta, \quad \cos\theta, \quad \sec\theta, \quad \cosec\theta, \quad \sin^{-1}\theta, \quad \cos^{-1}\theta,$$

for (a) $\theta = 1$, (b) $\theta = \pi$, (c) $\theta = \pi^{-1}$.

2. Simplify $(\sec\theta - \tan\theta)(\cosec\theta - \cot\theta)(\sec\theta + \tan\theta)(\cosec\theta + \cot\theta)$.

3. Find the angles at which the graph of $y = 2x + 3$ is met by the graphs of

   (a) $2y = x + 4$, (b) $2y + x + 4 = 0$, (c) $y + 2x + 3 = 0$.

   What is the relationship between the answers to (a) and (c)?

4. If $\theta = \sin^{-1}\left(-\frac{3}{5}\right)$, find, without using a calculator, the values of $\tan 2\theta$ and $\sec 2\theta$.

5. Calculate in surd form (a) $\tan(\pi/8)$ and (b) $\cos(\pi/6)$.

6. Solve, where possible, the equations

   (a) $2\sin\theta + 3\cos\theta = 2$, (b) $2\sin\theta + 3\cos\theta = 4$.

1.6 Inequalities

The behaviour of a physical system is determined by the way the values of the variables chosen to describe it relate to each other. The relationships will normally involve the variables themselves, but might also include terms which describe the rate at which one variable changes as another is altered. The mathematical forms of these relationships are usually referred to as the equations governing the system.

Although the word 'equation' tends to imply that one expression involving some or all of the variables can be equated, i.e. set equal to, a second, but different, such expression, sometimes the most that can be said is that one of the expressions is greater in value than the other. In this section we will study the general properties of this type of relationship, referring to it as an *inequality*.

The notion of ordering amongst the set of real numbers (usually denoted by $\mathcal{R}$) is so natural that statements such as

$$7.5 > 6.3, \quad 3 > -4, \quad 6.3 < 7.5, \quad -4 < 3 \quad \text{and} \quad -13 < -7 < -2$$

1.6 Inequalities

are taken as obvious. Here the symbol $>$ is the mathematical representation of the comparison relation 'greater than' and the inequality $a > b$ is to be interpreted as 'a is greater than b'; correspondingly, $c < d$ represents the statement that c is less than d. Clearly, $a > b$ implies its reverse, $b < a$, and vice versa.

A small extension of this notation are the symbols $\geq$ and $\leq$, which are read as 'greater than or equal to' and 'less than or equal to', respectively. Thus $a \geq b$ means that either a is greater than b or that the two are equal; an equivalent statement is $b \leq a$. Although equality is a possibility, we will continue to refer to a relationship involving $\geq$ and $\leq$ signs as an inequality.

Even though in physically based calculations it is of little practical significance, there is, mathematically, a formal distinction between ranges of a variable defined by $>$ and $<$ signs on the one hand, and by $\geq$ and $\leq$ signs on the other. A range of x given by $a < x < b$ does *not* include the end-points $x = a$ and $x = b$; it is referred to as an *open interval* in x and denoted by (a, b). If the end-points are included, then the range is a *closed interval* and denoted by $[a, b]$. It is also formally possible to have an interval that is open at one end and closed at the other; thus, for example, $(a, b]$ denotes the range $a < x \leq b$.

We are concerned here to develop further general inequality relationships from these basic definitions; these results, expressed algebraically, will be valid for all real numbers, and not just for specific pairs or sets of numerical values. In what follows we take $a, b, c, d, \ldots$ to be arbitrary real numbers or algebraic expressions. Most of the results given below are obvious and are stated with little comment. They will be referred to collectively as *property set* (1).

(a) If $a > b$ and $b > c$, then $a > c$; this is known as the *transitive* property of inequalities.
(b) If $a > b$, then $-a < -b$, i.e. reversing the signs of the expressions on both sides of the inequality gives a valid result provided the $>$ sign is changed into a $<$ sign. Note that this relationship holds even if one or both of a and b are negative.
(c) If $a > b$, then $a + c > b + c$; this relationship holds even if c is negative.
(d) If $a > b$ and $c > d$, then $a + c > b + d$; without further relevant information, nothing can be said about the value of $a + d$ as compared to that of $b + c$. However, it does follow that $a - d > b - c$. This is clear both from logical argument and from formally adding $-c - d$ to both sides of $a + c > b + d$ and appealing to the result in (c) above.
(e) If $a > b$ and $c > 0$, then $ac > bc$; if $c < 0$, then the valid result is that $ac < bc$, with the $>$ sign replaced by a $<$ sign. Result (b) is a particular case of this more general result, one in which $c = -1$. Division of both sides of an inequality by the same quantity is covered by these results, since division by c is equivalent to multiplication by c^{-1}, and c^{-1} has the same sign as c.
(f) If $a > b$ and $c > d$ and *all* four quantities are known to be positive, then $ac > bd$. If any of the quantities could be negative, no general conclusion can be drawn about the relative values of ac and bd. This can illustrated by considering the three valid inequalities $-2 > -6$, $2 > 1$ and $2 > -1$. Multiplying the first two together in the way suggested gives the valid result $-4 > -6$, but doing the same for the first and last produces $-4 > 6$, which is clearly invalid. When any or all of a, b, c and d are even modestly complicated algebraic expressions, rather than all being explicit numerical values, careful investigation is needed before the procedure can be justified.

Arithmetic and geometry

(g) If $a > b$ and a and b have the same sign, then $a^{-1} < b^{-1}$. If they have opposite signs, then clearly $a^{-1} > b^{-1}$ since a is positive and b is negative.

For each result given above there is a corresponding one relating to an inequality involving a 'less than' sign. They are summarised below, as *property set* (2), in the form of a series of mathematical statements in which the symbol $\Rightarrow$ should be read as 'implies that'. The reader is urged to verify each one and so gain some facility in thinking about inequalities.

(a) $a < b$ and $b < c \Rightarrow a < c$.
(b) $a < b \Rightarrow -a > -b$.
(c) $a < b \Rightarrow a + c < b + c$.
(d) $a < b$ and $c < d \Rightarrow a + c < b + d$ and $a - d < b - c$.
(e) $a < b$ and $c > 0 \Rightarrow ac < bc$; $a < b$ and $c < 0 \Rightarrow ac > bc$.
(f) $a < b$ and $c < d$ with $a, b, c, d > 0 \Rightarrow ac < bd$.
(g) $a < b$ and $ab > 0 \Rightarrow a^{-1} > b^{-1}$.

Both property sets, (1) and (2), remain valid if *all* of the $>$ and $<$ signs in any one statement[27] are replaced by $\geq$ and $\leq$ signs, respectively.

It should be noted that partial replacement can produce misleading or even invalid results. For example, if in property (1a) the second $>$ sign were not replaced by $\geq$ then the statement would read

$$\text{if } a \geq b \text{ and } b > c \text{ then } a \geq c.$$

This is not strictly incorrect, in that the correct conclusion, namely $a > c$, is included as a possibility, but it also implies the same for $a = c$, which is wrong. As an even more extreme example, if the third $>$ sign were not replaced by $\geq$, we would obtain

$$\text{if } a \geq b \text{ and } b \geq c \text{ then } a > c.$$

This is clearly wrong in the case $a = b = c$. In summary, if 'greater than' and 'less than' are replaced in a property statement by 'greater than or equal to' and 'less than or equal to', respectively, *every* sign in the statement must be changed.

Having dealt with the basic results governing the addition, subtraction, multiplication and division of inequalities, we should also note the following properties of those inequalities that compare the powers of variables.

Property (3). For *any* real quantity a, $a^2 \geq 0$ with equality if and only if $a = 0$.

Property (4). If $a > b > 0$ then $a^n > b^n$ for $n > 0$, but $a^n < b^n$ if $n < 0$. For n a positive integer, this result follows from repeated application of property (1f) with $c = a$ and $d = b$. The result is valid for any real n, which need not be an integer; in particular, if $a > b > 0$ then $\sqrt{a} > \sqrt{b} > 0$. Of course, if $n = 0$, then $a^n = 1 = b^n$ for any a and b.

Property (5). If $a > 0$ and $m > n > 0$ then $a^m > a^n$ for $a > 1$, but $a^m < a^n$ for $0 < a < 1$. Nothing can be said in general if a is negative. It should be noted that m and n do not have to be integers.

---

[27] This does not apply to $ab > 0$ in (g) of set (2), which merely states in mathematical form that a and b have the same sign.

1.6 Inequalities

We will now illustrate some of the properties of inequalities with a worked example in which we justify each step of the argument, however obvious, by reference to the appropriate property – though one would not normally be so meticulous.

Example In Problem 1.30 it is shown that

$$s = \sin\left(\frac{\pi}{8}\right) = \left(\frac{2-\sqrt{2}}{4}\right)^{1/2}.$$

By noting the values of $(1.4)^2$ and $(1.5)^2$, and without assuming the numerical value of $\sqrt{2}$, show that

$$\frac{1}{2\sqrt{2}} < s < \frac{2}{5}.$$

We first note that $(1.4)^2 = 1.96$ and that $(1.5)^2 = 2.25$. Thus $(1.4)^2 < 2 < (1.5)^2$ and therefore, using property (4), $1.4 < \sqrt{2} < 1.5$.

Now, from squaring both sides of the given expression for s, we have

$$s^2 = \frac{2-\sqrt{2}}{4}.$$

(i) Since $1.4 < \sqrt{2}$, we have from (2b) that $-1.4 > -\sqrt{2}$ and thence from (1c) that $2 - 1.4 > 2 - \sqrt{2}$. Then, reversing this result and using (2e), we may write

$$s^2 = \frac{2-\sqrt{2}}{4} < \frac{2-1.4}{4} = 0.15.$$

We may relate the RHS of this equation to the quoted upper bound for s by noting that $(2/5)^2 = 0.16$ and that $0.15 < 0.16$. Thus, $s^2 < 0.15 < 0.16$ and so, using (4), $s < 2/5$.

(ii) Starting again, this time from the second inequality, $\sqrt{2} < 1.5$, we have by a similar chain of argument that

$$s^2 = \frac{2-\sqrt{2}}{4} > \frac{2-1.5}{4} = \frac{1}{8}.$$

Applying result (4), as before, yields the inequality $s > 1/2\sqrt{2}$.

The results from (i) and (ii) together establish the stated double inequality. ◀

As a second simple example, one that uses the almost trivial property (3) to deduce a significant result, we will now show that the arithmetic mean of two positive quantities, a and b, is always greater than or equal to their geometric mean. The arithmetic mean is defined in the natural way as $\frac{1}{2}(a+b)$, whilst the geometric mean is $\sqrt{ab}$.

We consider the quantity $a - b$ and, starting with property (3), proceed as follows:

$$(a-b)^2 \geq 0, \quad \text{using (3)}$$
$$a^2 - 2ab + b^2 \geq 0,$$
$$a^2 + 2ab + b^2 \geq 4ab, \quad \text{using (1c)}$$
$$(a+b)^2 \geq 4ab,$$
$$a+b \geq 2\sqrt{ab} \quad \text{using (4),}$$
$$\tfrac{1}{2}(a+b) \geq \sqrt{ab} \quad \text{using (1e).}$$

Arithmetic and geometry

This last line gives the stated result – that the arithmetic mean is greater than or equal to the geometric mean. According to property (3), equality will only be obtained when $a - b = 0$, i.e. $a = b$; clearly, then, both means have value a. A simple extension of this result[28] is obtained by taking $a = x$ and $b = \alpha/x$ with both α and x positive; the result then shows that the minimum value of $x + \alpha x^{-1}$ is $2\sqrt{\alpha}$ and that that is achieved when $x = \sqrt{\alpha}$. As a particular case, the sum of a positive number and its reciprocal can never be less than 2.

As we have already seen in connection with the manipulation of inequalities, it is important to be able to establish whether a given quantity or expression can ever take negative values and, if so, over what range or ranges of the variables on which it depends. The general question of whether some particular function can take a zero value has applications throughout science and determines such things as the stability and optimisation of physical systems.

If a function $f(x)$ is greater than zero for all values of x then $f(x)$ is said to be *positive definite*. Correspondingly, if $f(x)$ is always less than zero it is described as *negative definite*. If $f(x)$ can be zero as well as being positive, i.e. $f(x) \geq 0$, rather than simply $f(x) > 0$, but can never be negative, then f is described mathematically as *positive semi-definite*;[29] in a similar way, if $f(x) \leq 0$ for all x it is a negative semi-definite function.

The analysis to determine whether or not a general function can have zero value is usually very complicated, and normally involves at least the use of calculus for analytically expressed functions, and of numerical investigation for tabulated ones. However, in the case of a quadratic function, $f(x)$, the third inequality property can be used in a simple way to determine whether or not $f(x)$ can ever have zero value.

Let $f(x)$ have the form $f(x) = ax^2 + bx + c$, with $a \neq 0$. For the moment, let a be restricted to $a > 0$, and consider the following algebraic rearrangement:

$$f(x) = ax^2 + bx + c$$
$$= a\left(x^2 + \frac{2bx}{2a}\right) + a\left(\frac{b}{2a}\right)^2 - a\left(\frac{b}{2a}\right)^2 + c$$
$$= a\left(x + \frac{b}{2a}\right)^2 - \frac{b^2}{4a} + c. \quad (1.80)$$

Now, by inequality property (3), the first term in the final line can never be less than zero and is only equal to zero when $x = -b/2a$. The minimum value of $f(x)$ is therefore that of the constant $c - (b^2/4a)$. This is positive or negative according as $c > b^2/4a$ or $c < b^2/4a$, respectively. Since $a > 0$, these conditions can be written more neatly as $b^2 < 4ac$ and $b^2 > 4ac$ (again respectively).

Thus, $f(x)$ is positive definite if $b^2 < 4ac$, but has some range of x (clearly including $x = -b/2a$) in which it is negative if $b^2 > 4ac$. If $b^2 = 4ac$ then $f(x)$ consists of the single squared term $a(x + b/2a)^2$ and takes the value zero at $x = -b/2a$; $f(x)$ is then positive semi-definite.

---

[28] Obtain another by showing that $ax^n + bx^{-n}$, with a, b and x all positive, is always $\geq 2\sqrt{ab}$ with the minimum value realised when $x = (b/a)^{1/2n}$.

[29] A physical analogy might be the kinetic energy function of a classical system. This can be positive or zero, but never negative.

1.6 Inequalities

A similar analysis applies if $a < 0$, but $f(x)$ is then negative definite if $b^2/4a > c$, i.e. $b^2 < 4ac$ (recall that a is negative). A negative semi-definite quadratic function has $a < 0$ and $b^2 = 4ac$.

We will now apply these results to six quadratic functions and determine the nature of each in this respect. Our tests will be in algebraic form, but you may find it helpful to sketch each curve and verify that it behaves in the way indicated.

Example For each of the following quadratic functions determine whether it is positive or negative definite, or positive or negative semi-definite, or none of these.

(a) $2x^2 + 6x + 5$, (b) $-x^2 + 6x - 9$, (c) $-2x^2 - 6x - 7$,
(d) $x^2 + x - 6$, (e) $3x^2 - 12x + 12$ (f) $4x^2 - 20x + 16$.

We set out in tabular form the equation, the test on the sign of a, the comparison of 'b^2' with '$4ac$' and the conclusion reached using the criteria just derived.

(a) $2x^2 + 6x + 5$	$2 > 0$	$6^2 < 4(2)(5)$	positive definite
(b) $-x^2 + 6x - 9$	$-1 < 0$	$6^2 = 4(-1)(-9)$	negative semi-definite
(c) $-2x^2 - 6x - 7$	$-2 < 0$	$(-6)^2 < 4(-2)(-7)$	negative definite
(d) $x^2 + x - 6$	$1 > 0$	$1^2 > 4(1)(-6)$	none
(e) $3x^2 - 12x + 12$	$3 > 0$	$(-12)^2 = 4(3)(12)$	positive semi-definite
(f) $4x^2 - 20x + 16$	$4 > 0$	$(-20)^2 > 4(4)(16)$	none.

As already indicated, it may help to sketch the graphs of some of the curves. ◀

It is sometimes possible to prove equality between two quantities, even if the given relationships between them involve only inequalities – though the latter must be inequalities of the 'greater than or equal' type. The essence of the method is contained in the following statement: if it can be shown that $a \geq b$ and also that $b \geq a$, then it follows that $a = b$.

Substantial relevant examples do not arise naturally in the current text, but do so when more advanced topics are studied, in particular for topics which involve establishing the number of members of a set of objects that have a particular property. However, the following contrived example will illustrate the method.

Example Suppose that the quantity E has been shown to satisfy the two inequalities

$$E + b^2 \geq a^2,$$
$$\frac{E}{a+b} + b \leq a,$$

where a and b are both positive. Show that, taken together, the two inequalities are equivalent to an equality that determines the value of E.

The two inequalities can be rewritten as

$$E \geq a^2 - b^2, \qquad (*)$$

$$\frac{E}{a+b} \leq a - b.$$

Now, since a and b are both positive, $a + b > 0$ and so the second inequality above can be multiplied by $a + b$ on both sides without invalidating the relationship:

$$E \leq (a-b)(a+b) = a^2 - b^2 \qquad (**)$$

Thus, combining (*) and (**) yields

$$a^2 - b^2 \leq E \leq a^2 - b^2.$$

Finally, since this double inequality states that E is both 'greater than or equal to $a^2 - b^2$' and 'less than or equal to $a^2 - b^2$', it can only be equal to $a^2 - b^2$. Thus a unique value for E has been determined by the two inequalities. ◂

We conclude this section on inequalities by working through two examples that are in the form of equations to be solved for an unknown x, but in which the comparator between the two sides of the equation is a less than or greater than sign, rather than the more usual equals sign. Correspondingly, we must expect that the solution will be one or more ranges for x, rather than one or more specific values of x.

Example Solve the equation

$$\frac{2}{x-3} > \frac{1}{2-x}.$$

In order to obtain a solution in the form $x > a$ or $x < b$, we need to multiply the given inequality through by $f(x) = (x - 3)(2 - x)$ and it is essential that we take account of whether $f(x)$ is positive or negative, since, in accordance with property (1e), the inequality must be reversed if f is negative.

It is clear that $f(x)$ will change its sign at both $x = 2$ and $x = 3$, and that for $2 < x < 3$ both factors are negative, meaning that $f(x)$ is positive. Consequently, when both sides of the original inequality are multiplied by $(x - 3)(2 - x)$ the $>$ sign must be replaced by a $<$ sign, except when $2 < x < 3$. We must therefore analyse these two regimes separately.

(i) The range $2 < x < 3$. Here, no change of inequality sign occurs and the resulting equation is

$$2(2-x) > (x-3) \quad \Rightarrow \quad 4 - 2x > x - 3 \quad \Rightarrow \quad 7 > 3x \quad \Rightarrow \quad x < \tfrac{7}{3}.$$

The allowed range of x for this case is therefore determined by both $2 < x < 3$ and $x < \tfrac{7}{3}$. Thus the original equation is valid for $2 < x < \tfrac{7}{3}$.

(ii) The ranges $x < 2$ and $x > 3$. With the inequality sign change incorporated, but the rest of the algebra unaltered, we conclude that we must have $x > \tfrac{7}{3}$ for the original inequality to be valid. When this is combined with the ranges under consideration, we see that only the $x > 3$ range is acceptable.

In summary, the solution to the original equation is that either (i) $2 < x < \tfrac{7}{3}$ or (ii) $x > 3$. ◂

1.6 Inequalities

Our final example can be tackled in two ways and, as their workings are short, it is instructive to include both.

Example Solve the equation

$$x(1-x) < \tfrac{1}{4}.$$

(i) We first arrange the equation so that x appears only in a term that is squared, and then examine the implications of the fact that the squared term cannot be negative:

$$x(1-x) < \tfrac{1}{4},$$
$$-x^2 + x - \tfrac{1}{4} < 0,$$
$$-(x-\tfrac{1}{2})^2 + \tfrac{1}{4} - \tfrac{1}{4} < 0,$$
$$-(x-\tfrac{1}{2})^2 < 0.$$

This final inequality is true for *any* x except $x = \tfrac{1}{2}$; this same statement is thus the required solution.

(ii) It is clear that the critical points to consider are $x = 0$ and $x = 1$ and we divide the whole x-range into three parts.

(a) For $x < 0$, the first factor on the LHS is negative, whilst the second is positive. The product is therefore negative and clearly less than the positive quantity $\tfrac{1}{4}$.

(b) For $0 < x < 1$, both factors are positive. The product is zero at $x = 0$ and at $x = 1$, but positive in between. By symmetry, the product is maximal when $x = \tfrac{1}{2}$, and then has the value $\tfrac{1}{2} \times \tfrac{1}{2} = \tfrac{1}{4}$. For x-values other than this, the product is clearly less than $\tfrac{1}{4}$.

(c) For $x > 1$ we have that the first factor is positive, whilst the second is negative. The product is therefore negative and the conclusion is the same as that for $x < 0$.

Combining the three results from (a), (b) and (c), we see that the solution to the equation is 'any x except $x = \tfrac{1}{2}$' – in agreement with the conclusion reached in (i). ◀

EXERCISES 1.6

1. Determine the interval common to the three intervals $[-3, 3]$, $[-1, 5]$ and $(2, 4)$.

2. Find the range of x for which $x^2 \leq 2(x + 12)$.

3. (a) If a, b and c are non-zero positive numbers with $a > b > c$, prove that

$$cb(a + 1) < ac(b + 1) < ab(c + 1).$$

(b) Show further that the result becomes

$$cb(a + 1) > ac(b + 1) > ab(c + 1)$$

if all three numbers are non-zero negative numbers.

(c) Demonstrate by means of a counter-example (e.g. $a = 2$, $b = -1$ and $c = -2$) that neither result holds if a, b and c are of mixed signs.

Arithmetic and geometry

4. Identify the error in the following 'proof' that $2 > 3$. Let a and b be positive numbers with $a > b$. Then

$$\ln a > \ln b \implies (b-a)\ln a > (b-a)\ln b$$
$$\implies \ln a^{(b-a)} > \ln b^{(b-a)} \implies a^{(b-a)} > b^{(b-a)}$$
$$\implies \left(\frac{a}{b}\right)^{(b-a)} > 1 \implies \left(\frac{b}{a}\right)^{(a-b)} > 1.$$

Now set $a = 3$ and $b = 2$, giving $\left(\frac{2}{3}\right)^1 > 1$, i.e. $2 > 3$.

5. Determine by algebraic means the range(s) of x for which $\dfrac{7}{x-4} > x + 2$.

SUMMARY

1. *Logarithms and the exponential function*
   - For a logarithm to any base a (> 0), $x = a^{\log_a x}$ and
   $$\log_a x^n = n \log_a x, \text{ where } n \text{ is any real number.}$$
   - For $x > 0$, its natural logarithm, $\log_e x \equiv \ln x$ is defined by $x = e^{\ln x}$, where $e = \exp(1)$ and
   $$e^x = \exp(x) = \sum_{n=0}^{\infty} \frac{x^n}{n!}.$$
   - The exponential function and natural logarithm have the properties
   $$\frac{d}{dx}(e^x) = e^x, \quad \frac{d}{dx}(\ln x) = \frac{1}{x}, \quad \ln x = \int_1^x \frac{1}{u}\,du,$$
   $$\ln(xy) = \ln x + \ln y, \quad \ln\left(\frac{x}{y}\right) = \ln x - \ln y.$$

2. *Rational and irrational numbers*
   - If $\sqrt{p}$ is irrational, $a + b\sqrt{p} = c + d\sqrt{p}$ implies that $a = c$ and $b = d$.
   - To rationalise $(a + b\sqrt{p})^{-1}$, write it as
   $$\frac{(a - b\sqrt{p})}{(a + b\sqrt{p})(a - b\sqrt{p})} = \frac{a - b\sqrt{p}}{a^2 - b^2 p}.$$

3. *Physical dimensions*
   The base units are [length] $= L$, [mass] $= M$, [time] $= T$, [current] $= I$, [temperature] $= \Theta$.
   - The dimensions of any one physical quantity can contain only integer powers of the base units.
   - The dimension of a constant is zero for each base unit.

Summary

- The dimensions of a product ab are the sums of the dimensions of a and b, separately for each base unit.
- The dimensions of $1/a$ are the negatives of the dimensions of a for each base unit.
- All terms in any physically acceptable equation (including the individual terms in any sum or implied series) must have the same set of base-unit dimensions.

4. *Binomial expansion*
    - For any integer $n > 0$,
      (i)
      $$(x+y)^n = \sum_{k=0}^{n} {}^nC_k x^{n-k} y^k, \text{ with } {}^nC_k = \frac{n!}{k!(n-k)!};$$
      (ii)
      $$(x+y)^{-n} = x^{-n} \sum_{k=0}^{\infty} {}^{-n}C_k \left(\frac{y}{x}\right)^k,$$
      with ${}^{-n}C_k = (-1)^k \times {}^{n+k-1}C_k$ and $|x| > |y|$.
    - For any n, positive or negative, integer or non-integer, and $|x| > |y|$,
      $$(x+y)^n = x^n \sum_{k=0}^{\infty} {}^nC_k \left(\frac{y}{x}\right)^k, \text{ with } {}^nC_0 = 1 \text{ and } {}^nC_{k+1} = \frac{n-k}{k+1} {}^nC_k.$$

5. *Trigonometry*
    - With θ measured in radians, in Figure 1.2,
      $$\sin\theta = \frac{QP}{OP} = \sum_{n=0}^{\infty} \frac{(-1)^n \theta^{2n+1}}{(2n+1)!}, \quad \cos\theta = \frac{OQ}{OP} = \sum_{n=0}^{\infty} \frac{(-1)^n \theta^{2n}}{(2n)!}.$$
    - $\cos^2\theta + \sin^2\theta = 1$, $1 + \tan^2\theta = \sec^2\theta$, $1 + \cot^2\theta = \operatorname{cosec}^2\theta$.
    - $\sin(A \pm B) = \sin A \cos B \pm \cos A \sin B,$
      $\cos(A \pm B) = \cos A \cos B \mp \sin A \sin B.$
    - $\sin 2\theta = 2\sin\theta\cos\theta,$
      $\cos 2\theta = \cos^2\theta - \sin^2\theta = 2\cos^2\theta - 1 = 1 - 2\sin^2\theta.$
    - If $t = \tan(\theta/2)$, then
      $$\sin\theta = \frac{2t}{1+t^2}, \quad \cos\theta = \frac{1-t^2}{1+t^2}, \quad \tan\theta = \frac{2t}{1-t^2}.$$

6. *Inequalities*
    - *Warning*: Multiplying an inequality on both sides by a negative quantity (explicit or implicit) reverses the sign of the inequality.
    - If $a \geq b$ and $b \geq a$, then $a = b$.

Arithmetic and geometry

PROBLEMS

For this particular chapter nearly all of the numerical problems that follow can be solved simply using a calculator. However, you are likely to obtain a better grasp of the mathematical principles involved, and gain valuable experience in order-of-magnitude estimates, if, where it is so indicated, you do not use one.

Symbols not explicitly defined in the problems have the meanings indicated in the list of constants given in Appendix E. The dimensions of the constants should not be deduced directly from the units quoted there, unless the problem indicates that they should be.

Powers and logarithms

1.1. Evaluate the following[30] to 3 s.f.:

(a) e^π,   (b) π^e,   (c) $\log_{10}(\log_2 32)$,   (d) $\log_2(\log_{10} 32)$.

1.2. Simplify the following without using a calculator:

(a) $\dfrac{\sqrt[3]{27}\, 8^{1/2}}{\sqrt{10}\, 3^{-1}}$,   (b) $\dfrac{\ln 10000 - \ln 100}{\ln 10 - \ln 1000}$.

1.3. Find the number for which the cube of its square root is equal to twice the square of its cube root.

1.4. Rationalise the following fractions so that no surd appears in a denominator:

(a) $\dfrac{6}{\sqrt{27}}$,   (b) $\dfrac{4}{3 - \sqrt{11}}$,   (c) $\dfrac{8 - \sqrt{7}}{8 + \sqrt{11}}$.

1.5. By applying the rationalisation procedure twice, show that

$$\frac{131}{3 - \sqrt{5} + \sqrt{7}} = 9 - 11\sqrt{5} + 7\sqrt{7} + 6\sqrt{35}.$$

1.6. Prove that if p and q are distinct primes, with neither equal to 1, then it is not possible to find a rational number a such that $\sqrt{p} = a\sqrt{q}$.

1.7. Solve the following for x:

(a) $x = 1 + \ln x$,   (b) $\ln x = 2 + 4\ln 3$,   (c) $\ln(\ln x) = 1$.

1.8. Evaluate the following:

(a) $\dfrac{8!}{(4!)^2}$,   (b) $\dfrac{5!}{0!\,1!\,2!\,3!\,4!}$,   (c) $\dfrac{(2n)!}{2^n\,(n!)}$.

---

[30] From parts (a) and (b), the question arises as to whether, if $a < b$, there is a definite inequality relationship between a^b and b^a. It can be shown (as a by-product of Problem 3.14) that: if $a < b < e$, then $a^b < b^a$; if $e < a < b$, then $a^b > b^a$; if $a < e < b$ then no general conclusion can be drawn, e.g. $2^5 > 5^2$ but $2^3 < 3^2$.

Problems

1.9. Express $(2n + 1)(2n + 3)(2n + 5)\ldots(4n - 3)(4n - 1)$ in terms of factorials.

1.10. Show, by direct calculation from its series definition, that $\exp(-1)$ lies in the range
$$0.366\,66 < \tfrac{11}{30} < \exp(-1) < \tfrac{53}{144} < 0.368\,06,$$
and verify that it does so using a calculator.

1.11. Measured quantities x and y are known to be connected by the formula
$$y = \frac{ax}{x^2 + b},$$
where a and b are constants. Pairs of values obtained experimentally are

x:	2.0	3.0	4.0	5.0	6.0
y:	0.32	0.29	0.25	0.21	0.18

Use these data to make best estimates of the values of y that would be obtained for (a) $x = 7.0$, and (b) $x = -3.5$. As measured by fractional error, which estimate is likely to be the more accurate?

1.12. Two physical quantities x and y are connected by the equation
$$y^{1/2} = \frac{x}{ax^{1/2} + b},$$
and measured pairs of values for x and y are as follows:

x:	10	12	16	20
y:	409	196	114	94

Determine the best values for a and b by graphical means, and (if you have one available) using a built-in calculator routine that makes a least-squares fit to an appropriate straight line.

1.13. The variation with the absolute temperature T of the thermionic emission current i from a heated surface (in the absence of space charge effects) is said to be given by
$$i = AT^2 e^{-BT},$$
where A and B are both independent of T. How would you plot experimental measurements of i as a function of T so as to check this relationship and then extract values for A and B?

1.14. It is shown in the text that the exponential function $\exp(x)$ is identical in value to the power e^x for all x. It therefore follows from the relation $e^{2x} = e^x \times e^x$ that
$$\exp(2x) = [\exp(x)]^2. \qquad (*)$$
Write both sides of this equation in terms of the relevant series and, by considering $\exp(x)$ as $1 + p(x)$, verify $(*)$ term-by-term up to and including the cubic term in x.

Arithmetic and geometry

Dimensions

1.15. Three very different lengths that appear in quantum physics and cosmology are the Planck length ℓ_p, the Compton wavelength λ_m, and the Schwarzschild radius r_s. Given that

$$\ell_p = \sqrt{\frac{hG}{2\pi c^n}}, \qquad \lambda_m = \frac{h}{mc}, \qquad r_s = \frac{2GM}{c^2},$$

where m and M are masses, calculate the dimensions of the gravitational constant G and those of the Planck constant h. Deduce the value of n in the formula for the Planck length.

1.16. Use the fact that the electrostatic force acting between two electric charges, q_1 and q_2, is given by $q_1 q_2 / 4\pi \epsilon_0 r^2$, where r is the distance between the charges, to determine the dimensions of ϵ_0, the permittivity of a vacuum. In the table of constants (Appendix E), ϵ_0 is expressed in units of farads per metre. Express one farad in terms of the SI base units.

1.17. According to Bohr's theory of the hydrogen atom, the ionisation energy of hydrogen is $m_e e^4 / 8 \epsilon_0^2 h^2$. Using Appendix E, show that this expression does have the dimensions of an energy and that its value when expressed in electron-volts is 13.8 eV.

1.18. It is stated in physics textbooks that:

In a diatomic gas in equilibrium at (absolute) temperature T, the fraction of molecules that have angular momentum $\ell \hbar$, where ℓ is a non-negative integer, is proportional to

$$(2\ell + 1) \exp\left[-\frac{\ell(\ell+1)\hbar^2}{2I\,kT}\right].$$

Here, $\hbar$ is the Planck constant divided by 2π; I is the moment of inertia of the molecule and is given by $\frac{1}{2}mr^2$, where m is the mass of each of the two atoms in a molecule and r is their distance apart.

Check the dimensions of all the combinations of physical variables appearing in the statement and show that they are 'fit for purpose'.

1.19. The electrical conductivity σ of a metal is measured in siemens per metre (S m^{-1}), where 1 S is the unit of conductance of an electrical component and is equivalent to 1 A V^{-1}. The Wiedemann–Franz law states that at absolute temperature T, and under certain conditions, σ is related to the thermal conductivity λ of the metal by the equation

$$\frac{\lambda}{\sigma T} = \frac{\pi^2}{3}\left(\frac{k}{e}\right)^2.$$

Verify that this equation is dimensionally acceptable and, using Appendix E, estimate the thermal conductivity of copper at room temperature, given that its electrical conductivity is $5.6 \times 10^7 \, \text{S m}^{-1}$.

1.20. The radiation energy emitted per unit time by unit area of a 'black body' at temperature T is σT^4, where σ is the Stefan–Boltzmann constant. The same constant can also be expressed in terms of k, h and c as $\mu k^\alpha h^\beta c^\gamma$, where μ, α, β and γ are dimensionless constants. Using Appendix E, determine the numerical value of μ.

1.21. The following is a student's proposed formula for the energy flux S (the magnitude of the so-called Poynting vector) associated with an electromagnetic wave in a vacuum, the electric field strength of the wave being E and the associated magnetic flux density being B:

$$S = \frac{1}{2}\left[\left(\frac{\epsilon_0}{\mu_0}\right)^{1/2} E^2 + \left(\frac{\mu_0}{\epsilon_0}\right)^{1/2} B^2\right].$$

The dimensions of ϵ_0, the permittivity of free space, are $M^{-1}L^{-3}T^4I^2$, and those of its permeability μ_0 are $MLT^{-2}I^{-2}$. Given, further, that the force acting on a rod of length ℓ that carries a current I at right angles to a field of magnetic flux density B is $BI\ell$, determine whether the student's formula could be correct and, if not, locate the error as closely as possible.

Binomial expansion

1.22. Use a binomial expansion to evaluate $1/\sqrt{4.2}$ to five places of decimals, and compare it with a more accurate answer obtained using a calculator.

1.23. Evaluate those of the following that are defined: (a) $^5C_3$, (b) $^3C_5$, (c) $^{-5}C_3$, (d) $^{-3}C_5$.

1.24. By choosing appropriate values for x and y in Equation (1.41), prove the following identities, and verify them directly for some (modest but non-trivial) value of n.

(a) $\binom{n}{0} + \binom{n}{1} + \binom{n}{2} + \cdots + \binom{n}{n} = 2^n,$

(b) $\binom{n}{0} - \binom{n}{1} + \binom{n}{2} - \cdots + (-1)^n \binom{n}{n} = 0.$

1.25. By applying the binomial expansion directly to the identity

$$(x + y)^p (x + y)^q \equiv (x + y)^{p+q},$$

prove the result

$$\sum_{t=0}^{r} {}^pC_{r-t}\,{}^qC_t = {}^{p+q}C_r = \sum_{t=0}^{r} {}^pC_t\,{}^qC_{r-t}$$

which gives a formula for combining terms from two sets of binomial coefficients in a particular way (a kind of 'convolution', for readers who are already familiar with this term).

Trigonometric identities

1.26. Using the table of functions of common angles given in the main text, verify the formulae for $\sin(A+B)$ and $\tan(A+B)$ as follows. Compute $\sin 120°$ and $\tan 120°$ in each of the cases (i) $A = 60°$, $B = 60°$, (ii) $A = 90°$, $B = 30°$, (iii) $A = 180°$, $B = -60°$.

1.27. Prove that

$$\cos\frac{\pi}{12} = \frac{\sqrt{3}+1}{2\sqrt{2}}$$

by considering
(a) the sum of the sines of $\pi/3$ and $\pi/6$,
(b) the sine of the sum of $\pi/3$ and $\pi/4$.

1.28. The following problems are based on the half-angle formulae.
(a) Use the fact that $\sin(\pi/6) = 1/2$ to prove that $\tan(\pi/12) = 2 - \sqrt{3}$.
(b) Use the result of (a) to show further that $\tan(\pi/24) = q(2-q)$ where $q^2 = 2 + \sqrt{3}$.

1.29. Find the real solutions of
(a) $3\sin\theta - 4\cos\theta = 2$,
(b) $4\sin\theta + 3\cos\theta = 6$,
(c) $12\sin\theta - 5\cos\theta = -6$.

1.30. If $s = \sin(\pi/8)$, prove that

$$8s^4 - 8s^2 + 1 = 0,$$

and hence show that $s = [(2-\sqrt{2})/4]^{1/2}$.

1.31. Find all the solutions of

$$\sin\theta + \sin 4\theta = \sin 2\theta + \sin 3\theta$$

that lie in the range $-\pi < \theta \le \pi$. What is the multiplicity of the solution $\theta = 0$?

Inequalities

1.32. By considering the squares of the expressions on their LHSs, prove inequalities (i) and (ii) and then answer parts (iii) and (iv).

Problems

(i) $\sqrt{5} + \sqrt{3} < 4$.
(ii) $\sqrt{7} + \sqrt{5} < 2\sqrt{6}$.
(iii) Prove the general algebraic result involving $\sqrt{n+1} + \sqrt{n-1}$ suggested by inequalities (i) and (ii).
(iv) Without using a calculator, show that $\sqrt{40004} - \sqrt{10002} > 100$.

1.33. Starting from the double inequality $n - 1 < n < n + 1$, show that, for $n \geq 1$,

$$\sqrt{n} - \sqrt{n-1} > \frac{1}{2\sqrt{n}} > \sqrt{n+1} - \sqrt{n}.$$

Deduce that $\sum_{n=1}^{99} n^{-1/2}$ lies in the interval $(18, 6\sqrt{11})$.

1.34. For each of the following quadratic functions determine whether it is positive or negative definite, or positive or negative semi-definite, or none of these.

(a) $2x^2 + 6x + 3$, (b) $-x^2 + 7x - 13$, (c) $-2x^2 - 6x$,
(d) $x^2 - 6$, (e) $3x^2 + 12x + 12$, (f) $4x^2 - 15x + 16$.

1.35. By finding suitable values for A and B in the function $A(x - 3)^2 + B(x - 7)^2$ show that $f(x) = 6x^2 - 68x + 214$ cannot be zero for any value of x. Further, by rearranging the expression for $f(x)$, show that its actual minimum value is $64/3$.

1.36. Given that $a > b > 0$, prove algebraically that

$$\frac{a}{b} > \frac{a+c}{b+c}$$

whenever $c > 0$ and for $c < 0$ when $|c| > b$, but that for $c < 0$ with $|c| < b$ the $>$ sign should be replaced by a $<$ sign.

[*Note*: It may help to illustrate these results graphically on an annotated sketch of $(a + c)/(b + c)$ as a function of c, for fixed a and b. For definiteness the values $a = 4$, $b = 3$ could be used.]

1.37. For the pair of inequalities

$$ax + by > e,$$
$$cx + dy < f,$$

in which $a, b, \ldots, f$ are all positive, consider the following calculation:

$$d(ax + by) > de, \quad b(cx + dy) < bf, \quad \text{using (1e) and (2e)}$$
$$\Rightarrow \quad d(ax + by) - b(cx + dy) > de - bf, \quad \text{using (1d)}$$
$$\Rightarrow \quad x(ad - bc) > de - bf,$$
$$\Rightarrow \quad x > \frac{de - bf}{ad - bc}. \quad (*)$$

For the two particular cases

(i) $\begin{cases} 2x + 3y > 12 \\ 3x + 4y < 25 \end{cases}$ and (ii) $\begin{cases} 5x + 4y > 29 \\ 3x + 4y < 25 \end{cases}$

verify that for $x = 5$ and $y = 2$ all four inequalities are valid. Now show that deduction $(*)$ is not a valid statement in case (i), although in case (ii) it is. Explain why this is so and how the calculation should be corrected in the former case.

1.38. Solve, separately, the following equations for x.

(a) $\dfrac{3}{2-x} > \dfrac{4}{1-x}$, (b) $\left|\dfrac{5x+3}{x-2}\right| < 1$.

1.39. Determine the range(s) of x that simultaneously satisfy the three inequalities

(i) $x^2 - 6 \leq x$, (ii) $|x - 1| \geq 1$, (iii) $x^2 + 2 > 3$.

1.40. Determine the range(s) of x for which real values of x and y satisfy

$$x^2 + y^2 \leq 1 \text{ with } y \text{ in } (0.6, 2.0) \text{ and } |x| \text{ in } [0.5, 2.0],$$

expressing the interval(s) in bracket notation.

Commutativity and associativity

1.41. The 'group' of symmetry operations on an equilateral triangle has six elements. They are (clockwise) rotations (about an axis perpendicular to its plane and passing through its centre) by 0, $2\pi/3$ and $-2\pi/3$, and denoted respectively by A, B and C, together with the reflections of the same triangle in the bisectors of each of the three sides, denoted by K, L and M (see Figure 1.4).

The product $X \odot Y$ is defined as the single element from amongst A, B, ..., M that is equivalent to first applying operation Y to the triangle, and then applying operation X to the result. Thus, as examples, $A \odot X = X = X \odot A$ for any A, $B \odot C = A$, $B \odot K = M$ and $L \odot C = K$. These results have been

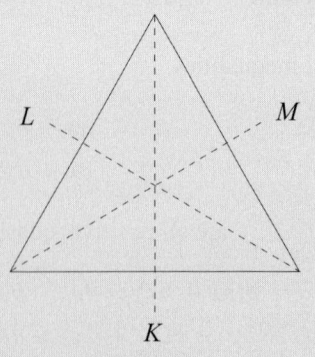

Figure 1.4 Reflections in the three perpendicular bisectors of the sides of an equilateral triangle take the triangle into itself.

entered into the 6 × 6 'multiplication table' for the group, which has row-headings X and column-headings Y.

$y=$ \\ $x=$	A	B	C	K	L	M
A	A	B	C	K	L	M
B	B		A	M		
C	C					
K	K					
L	L		K			
M	M					

Complete the table and then use it to decide whether $\odot$ is (a) commutative and (b) associative.

HINTS AND ANSWERS

1.1. (a) 23.1, (b) 22.5, (c) $\log_{10}(\log_2 32) = \log_{10} 5 = 0.699$, (d) $x = [\ln(\log_{10} 32)]/\ln 2 = 0.590$.

1.3. $a^{3/2} = 2a^{2/3} \Rightarrow a^{3/2 - 2/3} = 2$, i.e. $a^{5/6} = 2 \Rightarrow a = e^{(6 \ln 2)/5} = 2.297\ldots$

1.5. Multiply numerator and denominator by $3 + \sqrt{5} - \sqrt{7}$ to obtain a denominator of $-3 + 2\sqrt{35}$, then rationalise again using $(3 + 2\sqrt{35})$ to obtain the stated result.

1.7. (a) $x = 1$ by direct inspection, or $e^x = e^1 e^{\ln x} = ex$ then $x = 1$ by inspection. (b) $x = e^{2+4\ln 3} = e^2 e^{4\ln 3} = 3^4 e^2 = 81 e^2$. (c) $\ln(\ln x) = 1 \Rightarrow \ln x = e \Rightarrow x = e^e = 15.15$.

1.9. Write the product as $[(4n)!/(2n)!] \div [(2n+2)(2n+4)\cdots(4n-2)(4n)]$; take a factor of 2^n out of the second square bracket, and express what is left in terms of factorials. $[(4n)!\, n!]/[(2n)!\,(2n)!\, 2^n]$.

1.11. Plot or calculate a least-squares fit of either x^2 versus x/y or xy versus y/x to obtain $a \approx 1.19$ and $b \approx 3.4$. (a) 0.16; (b) -0.27. Estimate (b) is the more accurate because, using the fact that $y(-x) = -y(x)$, it is effectively obtained by *inter*polation rather than *extra*polation.

1.13. The equation can be rearranged to read $\ln(i/T^2) = \ln A - BT$, and so $y = \ln(i/T^2)$ should be plotted against T, obtaining a straight-line graph if the relationship is valid. If so, the (negative) slope of the graph gives B and the intercept y_0 on the y-axis gives A as $A = e^{y_0}$.

1.15. From the Schwarzschild radius formula, $[G] = [c^2 r_s/M] = L^3 T^{-2} M^{-1}$, and from the Compton wavelength, $[h] = [\lambda m c] = L^2 T^{-1} M$. Then, from the Planck length

Arithmetic and geometry

formula

$$[c^n] = \left[\frac{hG}{l_p^2}\right] = \frac{L^2T^{-1}M\,L^3T^{-2}M^{-1}}{L^2} = L^3T^{-3},$$

from which, since $[c] = LT^{-1}$, it follows that $n = 3$.

1.17. Recalling that $[\text{joule}] = ML^2T^{-2}$,

$$[E] = \frac{M\,I^4T^4}{I^4T^8M^{-2}L^{-6}\,J^2T^2} = \frac{ML^2}{T^2} = [\text{energy}].$$

The numerical value is 2.2×10^{-18} J, which is $2.2 \times 10^{-18} \div 1.60 \times 10^{-19} = 13.8$ when expressed in electron-volts.

1.19. $[\lambda] = MLT^{-3}\theta^{-1}$; $[\sigma] = M^{-1}L^{-3}T^3I^2$; thus, $[\lambda/\sigma T] = M^2L^4T^{-6}I^{-2}\theta^{-2}$. $[k/e] = (ML^2T^{-2}\theta^{-1})/(IT)$ leading to $[(k/e)^2] = M^2L^4T^{-6}I^{-2}\theta^{-2}$ and making the stated formula dimensionally acceptable. Taking room temperature as 293 K gives an estimate for λ of 402 W m^{-1} K^{-1}.

1.21. From the force on a current-carrying rod, $[B] = MT^{-2}I^{-1}$. Energy flux has dimensions $(ML^2T^{-2})L^{-2}T^{-1} = MT^{-3}$. The dimensions of ϵ_0 are $L^{-3}M^{-1}T^4I^2$ and those of μ_0 are $MLT^{-2}I^{-2}$. The 'E^2' term has dimensions MT^{-3}, whilst those of the 'B^2' term are $L^2M^3T^{-7}I^{-4}$. Thus the electric term is compatible with an energy flux, but the magnetic one is not.[31]

1.23. Write $1/\sqrt{4.2}$ as $\frac{1}{2}(1+0.05)^{-1/2} = \cdots = 0.487949218$, using terms up to $(0.05)^3$.

1.25. Write each term $(x+y)^m$ in the form $\sum_{s=0}^m {}^mC_s x^s y^{m-s}$, then consider all the terms in the product of sums on the LHS that lead to terms containing x^r (they will all contain y^{p+q-r} as well); these are of the form ${}^pC_{r-t}x^{r-t}y^{p-r+t} \times {}^qC_t x^t y^{q-t}$. The sum of all these terms must also give the coefficient of x^r in the expansion of $(x+y)^{p+q}$, i.e. ${}^{p+q}C_r$. The right-hand equality follows, either by symmetry or by interchanging the roles of p and q.

1.27. (a) If $t = \tan(\pi/12)$, then $\frac{1}{2} = 2t/(1+t^2)$, leading to $t^2 - 4t + 1 = 0$ and $t = 2 \pm \sqrt{3}$. The minus sign is indicated, since $t < \tan(\pi/4) = 1$.
(b) If $u = \tan(\pi/24)$, then $2 - \sqrt{3} = t = 2u/(1-u^2)$, which rationalises to give $1 - u^2 = 2q^2 u$. This quadratic equation has the (positive) solution $u = -q^2 + \sqrt{q^4+1} = -q^2 + 2\sqrt{2+\sqrt{3}} = q(2-q)$.

1.29. Use the square of the equation $\sin(\pi/4) = 2s(1-s^2)^{1/2}$. Determine the relevant root sign by noting that $\pi/8 < \pi/4$.

1.31. Consider the squares of the three quantities and deduce that $s_1 < s_2 < s_3$. The largest value is expected when a is as close as possible to $17/2$, i.e. 8.5 (8 or 9 if a has to be an integer).

---

[31] The correct 'B^2' term is $[\mu_0^{-1}B^2]/[2(\epsilon_0\mu_0)^{1/2}]$.

Hints and answers

1.33. See the hints (in order) in the question. Sum all terms from $n = 1$ to $n = 99$; on the left and right, most terms cancel in pairs, leaving

$$\sqrt{99} - \sqrt{0} > \sum_{n=1}^{99} \frac{1}{2\sqrt{n}} > \sqrt{100} - \sqrt{1}.$$

Write $\sqrt{99}$ as $3\sqrt{11}$.

1.35. Equating the coefficients of x^2 and x gives $A + B = 6$ and $-6A - 14B = -68$, yielding $A = 2$ and $B = 4$. Thus, $f(x) = 2(x-3)^2 + 4(x-7)^2$; this is necessarily > 0, since each term is ≥ 0 and they cannot be zero together. Write

$$f(x) = 6\left[x^2 - \frac{68}{6}x + \left(\frac{34}{6}\right)^2\right] - 6\left(\frac{34}{6}\right)^2 + 214.$$

The first term is a perfect square, which is therefore ≥ 0, and the second term gives the minimum value as $214 - 6(34/6)^2 = \frac{64}{3}$.

1.37. In case (i), the expression '$ad - bc$' has the value $(2)(4) - (3)(3) = -1$ and is therefore negative; this means that the inequality should have been reversed when the line (∗) was derived. In case (ii), the corresponding factor is $(5)(4) - (4)(3) = 8$, i.e. positive, and the division by $ad - bc$ was carried out correctly.

1.39. $-2 \leq x < -1$ and $2 \leq x \leq 3$.

1.41. (a) Not commutative. For example $K \odot L = B$ but $L \odot K = C$. (b) Associative. Consider a sample of each of the following cases: (i) one or more elements is A; (ii) $B \odot C \odot P$ and $B \odot P \odot C$, where P is one of K, L, and M; (iii) $K \odot L \odot Q$ and $K \odot Q \odot L$, where Q is either B or C.

2 Preliminary algebra

It is normal practice when starting the mathematical investigation of a physical problem to assign algebraic symbols to the quantity or quantities whose values are sought, either numerically or as explicit algebraic expressions. For the sake of definiteness, in this chapter, our discussion will be in terms of a single quantity, which we will denote by x most of the time. The extension to two or more quantities is straightforward in principle, but usually entails much longer calculations, or a significant increase in complexity when graphical methods are used.

Once the sought-for quantity x has been identified and named, subsequent steps in the analysis involve applying a combination of known laws, consistency conditions and (possibly) given constraints to derive one or more equations satisfied by x. These equations may take many forms, ranging from a simple polynomial equation to, say, a partial differential equation with several boundary conditions. Some of the more complicated possibilities are treated in the later chapters of this book, but for the present we will be concerned with techniques for the solution of relatively straightforward algebraic equations.

When algebraic equations are to be solved, it is nearly always useful to be able to make plots showing how the functions, $f_i(x)$, involved in the problem change as their argument x is varied; here i is simply a label that identifies which particular function is being considered. These plots (or *graphs*) often give a good enough approximation to the required solution, particularly when what is needed is the value of x that satisfies a particular condition such as $f_1(x) = 0$ or $f_1(x) = f_2(x)$; the former is determined by the points at which a plot of $f_1(x)$ crosses the x-axis and the latter by the values of x at which the plots of $f_1(x)$ and $f_2(x)$ intersect each other.

In order to make accurate enough sketches for these purposes, it is important to be able to recognise the main features that will be possessed by the plots of given or deduced functions $f_i(x)$, without having to make detailed calculations for many values of x. Even if a precise (rather than approximate) value of x is required, and it is to be found using, say, numerical methods, preliminary sketches are always advisable, so that an appropriate numerical method may be selected, or a good starting point can be chosen for methods that depend upon successive approximation.

Much of what is needed for drawing adequate sketches could be discussed at this point – and we will be sketching some graphs later in this chapter. However, as indicated in the introduction to Chapter 1, we will defer a general discussion of graph-sketching

2.1 Polynomials and polynomial equations

until the end of Chapter 3, so that the benefits that differential calculus has to offer can be included.

2.1 Polynomials and polynomial equations

Firstly we consider the simplest type of equation, a *polynomial equation*, in which a *polynomial* expression in x, denoted by $f(x)$, is set equal to zero and thereby forms an equation which is satisfied by particular values of x, called the *roots* of the equation:

$$f(x) = a_n x^n + a_{n-1} x^{n-1} + \cdots + a_1 x + a_0 = 0. \tag{2.1}$$

Here n is an integer > 0, called the *degree* of both the polynomial and the equation, and the known coefficients $a_0, a_1, \ldots, a_n$ are real quantities with $a_n \neq 0$.

Equations such as (2.1) arise frequently in physical problems, the coefficients a_i being determined by the physical properties of the system under study. What is needed is to find some or all of the roots of (2.1), i.e. the x-values, α_k, that satisfy $f(\alpha_k) = 0$; here k is an index that, as we shall see later, can take up to n different values, i.e. $k = 1, 2, \ldots, n$. The roots of the polynomial equation can equally well be described as the zeros of the polynomial. When they are *real*, they correspond to the points at which a graph of $f(x)$ crosses the x-axis. Roots that are complex (see Chapter 5) do not have such a graphical interpretation.

For polynomial equations containing powers of x greater than x^4, general methods giving explicit expressions for the roots α_k do not exist. Even for $n = 3$ and $n = 4$ the prescriptions for obtaining the roots are sufficiently complicated that it is usually preferable to obtain exact or approximate values by other methods. Only for $n = 1$ and $n = 2$ can closed-form solutions be given. These results will be well known to the reader, but they are given here for the sake of completeness. For $n = 1$, (2.1) reduces to the *linear* equation

$$a_1 x + a_0 = 0; \tag{2.2}$$

the one and only solution (root) is $\alpha_1 = -a_0/a_1$.

For $n = 2$, (2.1) reduces to the *quadratic* equation

$$a_2 x^2 + a_1 x + a_0 = 0; \tag{2.3}$$

the two roots α_1 and α_2 are given by

$$\alpha_{1,2} = \frac{-a_1 \pm \sqrt{a_1^2 - 4 a_2 a_0}}{2 a_2}. \tag{2.4}$$

When discussing specifically quadratic equations, as opposed to more general polynomial equations, it is usual to write the equation in one of the two notations

$$ax^2 + bx + c = 0, \qquad ax^2 + 2bx + c = 0, \tag{2.5}$$

Preliminary algebra

with respective explicit pairs of solutions

$$\alpha_{1,2} = \frac{-b \pm \sqrt{b^2 - 4ac}}{2a}, \qquad \alpha_{1,2} = \frac{-b \pm \sqrt{b^2 - ac}}{a}. \tag{2.6}$$

This result can be proved, using the first of these notations, by setting the RHS of the rearrangement expressed by Equation (1.80) equal to zero, moving the term $(-b^2/4a) + c$ to the LHS, and dividing through by a. The equation then reads

$$\frac{b^2}{4a^2} - \frac{c}{a} = \left(x + \frac{b}{2a}\right)^2.$$

After the LHS has been written as $(b^2 - 4ac)/4a^2$, taking the square root of both sides yields

$$\pm\sqrt{\frac{b^2 - 4ac}{4a^2}} = x + \frac{b}{2a},$$

from which the first result in (2.6) follows directly.

Of course, the two notations given above are entirely equivalent[1] and the only important point is to associate each form of answer with the corresponding form of equation; most people keep to one form, to avoid any possible confusion.

If the value of the quantity appearing under the square root sign is positive then both roots are real; if it is negative then the roots form a complex conjugate pair, i.e. they are of the form $p \pm iq$ with p and q real (see Chapter 5); if it has zero value then the two roots are equal and special considerations usually arise.

The proof of the general form of the solution to a quadratic equation given above involves a process known as *completing the square*, a procedure that is of sufficient importance in some more advanced work, particularly in connection with the integration of functions of the general form e^{-x^2}, that it is worth explaining its basis and demonstrating it explicitly. To do so we will use the second form of the general quadratic equation given in (2.5) and obtain its solution.

The unknown x appears explicitly in two terms of the equation, once as x^2 and once linearly. The purpose of completing the square is to reduce this to a single explicit appearance, and to do this we use the identity

$$(x + k)^2 = (x + k)(x + k) = x^2 + 2kx + k^2. \tag{2.7}$$

Like the LHS of (2.5), the RHS of this identity contains terms in x^2 and x; they appear in the ratio 1 to $2k$. In the quadratic equation the corresponding ratio is a to $2b$, and so if k is taken as $k = b/a$ then the x-dependent terms in (2.5) and (2.7) will be proportional to one another. If, in addition, we write c as $c + ak^2 - ak^2$, all of the terms on the RHS of (2.7) will be present in the right proportions and we can replace them by the 'completed square' from the LHS of (2.7). We then have an equation that contains x explicitly only once, and this can be rearranged to give an explicit formula for x. As a series of algebraic

---

[1] Set b in the first form equal to 2β in both the equation and its solution, and verify that this is so.

2.1 Polynomials and polynomial equations

steps, the calculation is

$$ax^2 + 2bx + c = 0,$$

$$a\left(x^2 + \frac{2b}{a}x\right) + c = 0,$$

$$a\left(x^2 + \frac{2b}{a}x + \frac{b^2}{a^2}\right) - \frac{b^2}{a} + c = 0,$$

$$\left(x + \frac{b}{a}\right)^2 = \frac{b^2}{a^2} - \frac{c}{a}.$$

Taking the square root of both sides (and hence generating two solutions) gives

$$x + \frac{b}{a} = \pm\sqrt{\frac{b^2}{a^2} - \frac{c}{a}} = \pm\frac{\sqrt{b^2 - ac}}{a},$$

leading to the general solution

$$x = \frac{-b \pm \sqrt{b^2 - ac}}{a}$$

quoted in the second equation in (2.6).

Thus, since linear and quadratic equations can be completely dealt with in a cut-and-dried way, we turn to methods for obtaining partial information about the roots of higher-degree polynomial equations. In some circumstances the knowledge that an equation has a root lying in a certain range, or that it has no real roots at all, is all that is actually required. For example, in the design of electronic circuits it is necessary to know whether the current in a proposed circuit will break into spontaneous oscillation. To test this, it is sufficient to establish whether a certain polynomial equation, whose coefficients are determined by the physical parameters of the circuit, has a root with a positive real part (see Chapter 5); complete determination of all the roots is not needed for this purpose. If the complete set of roots of a polynomial equation is required, it can usually be obtained to any desired accuracy using numerical methods.

There is no explicit step-by-step approach to finding the roots of a general polynomial equation such as (2.1). In most cases analytic methods yield only information *about* the roots, rather than their exact values. To explain the relevant techniques we will consider a particular example, 'thinking aloud' on paper and expanding on special points about methods and lines of reasoning. In more routine situations such comment would be absent and the whole process briefer and more tightly focused.

2.1.1 Example: the cubic case

Let us investigate the roots of the equation

$$g(x) = 4x^3 + 3x^2 - 6x - 1 = 0 \tag{2.8}$$

or, in an alternative phrasing, investigate the zeros of $g(x)$.

We note first of all that this is a *cubic* equation. It can be seen that, for x large and positive, $g(x)$ will be large and positive and, equally, that, for x large and negative, $g(x)$ will be large and negative. Therefore, intuitively (or, more formally, by continuity) $g(x)$

must cross the x-axis at least once and so $g(x) = 0$ must have at least one real root. Furthermore, it can be shown that if $f(x)$ is an nth-degree polynomial then the graph of $f(x)$ must cross the x-axis an even or odd number of times as x varies between $-\infty$ and $+\infty$, according to whether n itself is even or odd.[2] Thus a polynomial of odd degree always has at least one real root, but one of even degree may have no real root. A small complication, discussed later in this section, occurs when repeated roots arise.

Having established that $g(x) = 0$ has at least one real root, we may ask how many real roots it *could* have. To answer this we need one of the fundamental theorems of algebra, mentioned above:

an nth-degree polynomial equation has exactly n roots.

It should be noted that this does not imply that there are n *real* roots (only that there are not more than n); some of the roots may be of the form $p + iq$.

To make the above theorem plausible and to see what is meant by repeated roots, let us suppose that the nth-degree polynomial equation $f(x) = 0$, Equation (2.1), has r roots $\alpha_1, \alpha_2, \ldots, \alpha_r$, considered distinct for the moment. This implies that $f(\alpha_k) = 0$, for $k = 1, 2, \ldots, r$, and that $f(x)$ vanishes only when x is equal to one of these r values α_k. But the same can be said for the function

$$F(x) = A(x - \alpha_1)(x - \alpha_2) \cdots (x - \alpha_r), \tag{2.9}$$

in which A is a non-zero constant; $F(x)$ can clearly be multiplied out to form an rth-degree polynomial expression.

We now call upon a second fundamental result in algebra: that if two polynomial functions $f(x)$ and $F(x)$ have equal values for *all* values of x in some finite range, then their coefficients are equal on a term-by-term basis. In other words, we can equate the coefficients of each and every power of x in the explicit expressions for $f(x)$ and $F(x)$. This result is essentially the same as that used for certain functions of two variables in Appendix B. Applying it to (2.9) and (2.1), and, in particular, to the highest power of x, we have $Ax^r \equiv a_n x^n$ and thus that $r = n$ and $A = a_n$. As r is both equal to n and to the number of roots of $f(x) = 0$, we conclude that the nth-degree polynomial $f(x) = 0$ has n roots. (Although this line of reasoning may make the theorem plausible, it does not constitute a proof since we have not shown that it is permissible to write $f(x)$ in the form of Equation (2.9).)

We next note that the condition $f(\alpha_k) = 0$ for $k = 1, 2, \ldots, r$ could also be met if (2.9) were replaced by

$$F(x) = A(x - \alpha_1)^{m_1}(x - \alpha_2)^{m_2} \cdots (x - \alpha_r)^{m_r}, \tag{2.10}$$

with $A = a_n$. In (2.10) the m_k are integers ≥ 1 and are known as the multiplicities of the roots, m_k being the multiplicity of α_k. Expanding the RHS leads to a polynomial of degree $m_1 + m_2 + \cdots + m_r$. This sum must be equal to n. Thus, if any of the m_k are greater than unity then the number of *distinct* roots, r, is less than n; the total number of roots remains

---

[2] Note that for even n the sign of $g(x)$ is the same for $x \to -\infty$ as it is for $x \to +\infty$. Thus if the sign changes at all, it must change an even number of times. The converse argument applies if n is odd.

2.1 Polynomials and polynomial equations

at n, but one or more of the α_k counts more than once. For example, the equation

$$F(x) = A(x - \alpha_1)^2(x - \alpha_2)^3(x - \alpha_3)(x - \alpha_4) = 0$$

has exactly seven roots, α_1 being a double root and α_2 a triple root, whilst α_3 and α_4 are unrepeated (*simple*) roots.

We can now say that our particular equation (2.8) has either one or three real roots, but in the latter case it may be that not all the roots are distinct. To decide how many real roots the equation has, we need to anticipate two ideas from Chapter 3. The first of these is the notion of the derivative of a function and the second is a result known as Rolle's theorem.

The *derivative* $f'(x)$ of a function $f(x)$ measures the slope of the tangent to the graph of $f(x)$ at that value of x (see Figure 3.1 in the next chapter). For the moment, the reader with no prior knowledge of calculus is asked to accept that the derivative of ax^n has the value nax^{n-1}; a full formal derivation of this result is given as a worked example on p. 105. The derivative of a constant, formally an $n = 0$ polynomial, is zero.[3] It is also the case that the derivative of a sum of individual terms is the sum of their individual derivatives. With these results or assumptions, as the case may be, the reader will see that the derivative $g'(x)$ of the curve $g(x) = 4x^3 + 3x^2 - 6x - 1$ is given by $g'(x) = 12x^2 + 6x - 6$. Similar expressions for the derivatives of other polynomials are used later in this chapter.

Rolle's theorem states that if $f(x)$ has equal values at two different values of x then at some point between these two x-values its derivative is equal to zero; i.e. the tangent to its graph is parallel to the x-axis at that point. Although included for a somewhat different purpose, the graph in Figure 3.2 conveniently illustrates the situation. For our present discussion, we may suppose that the points A and C on the graph correspond to equal values of $f(x)$; let them occur at x-values x_A and x_C. At A, $f(x)$ is a decreasing function of x and the slope of the tangent to the curve (i.e. the derivative of $f(x)$) is negative; at C, where $f(x)$ is an increasing function, the slope is positive. Since it varies smoothly as x goes from x_A to x_C, at some point it must pass through zero, i.e. the tangent must be parallel to the x-axis. In Figure 3.2 this occurs at B and demonstrates the validity of Rolle's theorem – in mathematical form, since $f(x_A) = f(x_C)$ there is an x_B with $x_A < x_B < x_C$ such that $f'(x_B) = 0$.

As already noted, as x increases from x_A to x_C the slope of the tangent at x, and hence the derivative $f'(x)$, increases, from a negative value at A, through zero at B, to a positive value at C. Thus $f'(x)$ is monotonically increasing in the region of B and so *its* derivative, known as the *second derivative* of $f(x)$ and denoted by $f''(x)$, must be positive. This, i.e. $f'(x_B) = 0$ and $f''(x_B) > 0$, is a general characteristic of a function whose graph has a (local) minimum at $x = x_B$. For a (local) maximum, the corresponding criteria are $f'(x) = 0$ and $f''(x) < 0$.

Having briefly mentioned the derivative of a function and Rolle's theorem, we now use them to establish whether $g(x)$ has one or three real zeros. If $g(x) = 0$ does have three real roots α_k, i.e. $g(\alpha_k) = 0$ for $k = 1, 2, 3$, then it follows from Rolle's theorem that between

---

[3] Explain why the derivative of the general polynomial (2.1) does not depend upon the value of a_0, relating your explanation to a typical sketch graph.

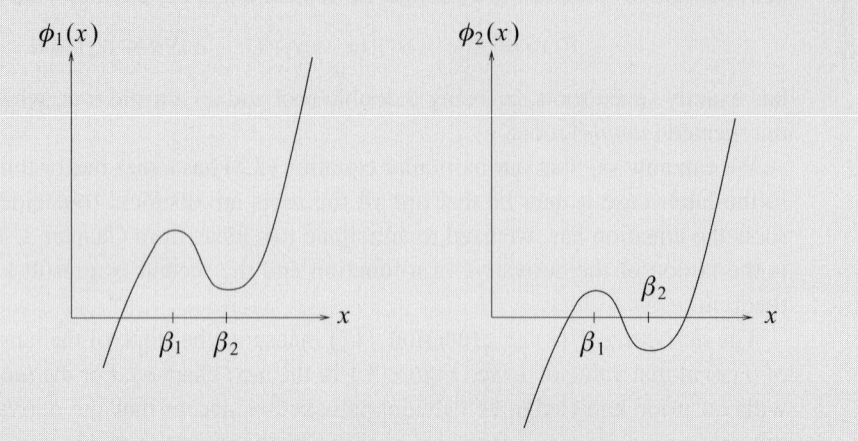

Figure 2.1 Two curves $\phi_1(x)$ and $\phi_2(x)$, both with zero derivatives at the same values of x, but with different numbers of real solutions to $\phi_i(x) = 0$.

any consecutive pair of them (say α_1 and α_2) there must be some real value of x at which $g'(x) = 0$. Similarly, there must be a further zero of $g'(x)$ lying between α_2 and α_3. Thus a *necessary* condition for three real roots of $g(x) = 0$ is that $g'(x) = 0$ itself has two real roots.

However, this condition on the number of roots of $g'(x) = 0$, whilst necessary, is not *sufficient* to guarantee three real roots of $g(x) = 0$. This can be seen by inspecting the cubic curves in Figure 2.1. For each of the two functions $\phi_1(x)$ and $\phi_2(x)$, the derivative is equal to zero at both $x = \beta_1$ and $x = \beta_2$. Clearly, though, $\phi_2(x) = 0$ has three real roots whilst $\phi_1(x) = 0$ has only one. It is easy to see that the crucial difference is that $\phi_1(\beta_1)$ and $\phi_1(\beta_2)$ have the same sign, whilst $\phi_2(\beta_1)$ and $\phi_2(\beta_2)$ have opposite signs.

It will be apparent that for some equations, $\phi(x) = 0$ say, $\phi'(x)$ equals zero at a value of x for which $\phi(x)$ is also zero. Then the graph of $\phi(x)$ just touches the x-axis. When this happens the value of x so found is, in fact, a double real root of the polynomial equation (corresponding to one of the m_k in (2.10) having the value 2) and must be counted twice when determining the number of real roots.

Finally, then, we are in a position to decide the number of real roots of the equation

$$g(x) = 4x^3 + 3x^2 - 6x - 1 = 0.$$

The equation $g'(x) = 0$, with $g'(x) = 12x^2 + 6x - 6$, is a quadratic equation with explicit solutions[4]

$$\beta_{1,2} = \frac{-3 \pm \sqrt{9 + 72}}{12},$$

so that $\beta_1 = \frac{1}{2}$ and $\beta_2 = -1$. The corresponding values of $g(x)$ are $g(\beta_1) = -\frac{11}{4}$ and $g(\beta_2) = 4$, which are of opposite sign. This indicates that $4x^3 + 3x^2 - 6x - 1 = 0$ has

---

[4] The two roots β_1, β_2 are written as $\beta_{1,2}$. By convention β_1 refers to the upper symbol in $\pm$, β_2 to the lower symbol.

three real roots, one lying in the range $-1 < x < \frac{1}{2}$ and the others one on each side of that range.

The techniques we have developed above have been used to tackle a cubic equation, but they can be applied to polynomial equations $f(x) = 0$ of degree greater than 3. However, much of the analysis centres around the equation $f'(x) = 0$, and this itself, being then a polynomial equation of degree 3 or more, either has no closed-form general solution or one that is complicated to evaluate. Thus the amount of information that can be obtained about the roots of $f(x) = 0$ is correspondingly reduced.

2.1.2 A more general case

To illustrate what can (and cannot) be done in the more general case we now investigate as far as possible the real roots of

$$f(x) = x^7 + 5x^6 + x^4 - x^3 + x^2 - 2 = 0.$$

The following points can be made.

(i) This is a seventh-degree polynomial equation; therefore, the number of real roots is 1, 3, 5 or 7.
(ii) $f(0)$ is negative whilst $f(\infty) = +\infty$, so there must be at least one positive root. It also follows that the total number of real roots to the right of $x = 0$ must be odd. Further, since the overall total number of such roots is odd, the number to the left must be even (0, 2, 4 or 6).[5]
(iii) Recalling that the derivative of Ax^n is nAx^{n-1}, we can write the equation $f'(x) = 0$ as

$$f'(x) = 7x^6 + 30x^5 + 4x^3 - 3x^2 + 2x - 0 = x(7x^5 + 30x^4 + 4x^2 - 3x + 2) = 0.$$

Since all terms contain a common factor of x, $x = 0$ is a root of $f'(x) = 0$. The derivative of $f'(x)$, denoted by $f''(x)$ and calculated in the same way, is equal to $42x^5 + 150x^4 + 12x^2 - 6x + 2$; at $x = 0$, $f''(x) = 2$. As noted earlier (and shown more formally in Section 3.3), the joint conditions $f'(0) = 0$ and $f''(0) > 0$ indicate that $f(x)$ has a minimum at $x = 0$.

This is about all that can be deduced by *simple* analytic methods in this case, although some further progress can be made in the ways indicated in Problem 2.3.

There are, in fact, more sophisticated tests that examine the relative signs of successive terms in an equation such as (2.1), and in quantities derived from them, to place limits on the numbers and positions of roots. But they are not prerequisites for the remainder of this book and will not be pursued further here.

We conclude this section with a worked example which demonstrates that the practical application of the ideas developed so far can be both short and decisive.

---

[5] Deduce that there must be at least one, but not more than three, maxima at negative values of x.

Example For what values of k, if any, does

$$f(x) = x^3 - 3x^2 + 6x + k = 0$$

have three real roots?

Firstly we study the equation $f'(x) = 0$, i.e. $3x^2 - 6x + 6 = 0$. This is a quadratic equation but, using (2.6), because $6^2 < 4 \times 3 \times 6$, it can have no real roots. Therefore, it follows immediately that $f(x)$ has no maximum or minimum; consequently, $f(x) = 0$ cannot have more than one real root, whatever the value of k (though it is bound to have one). ◀

2.1.3 Factorising polynomials

In Section 2.1 we saw how a polynomial with r given distinct zeros α_k could be constructed as the product of factors containing those zeros:

$$\begin{aligned} f(x) &= a_n(x - \alpha_1)^{m_1}(x - \alpha_2)^{m_2} \cdots (x - \alpha_r)^{m_r} \\ &= a_n x^n + a_{n-1} x^{n-1} + \cdots + a_1 x + a_0, \end{aligned} \quad (2.11)$$

with $m_1 + m_2 + \cdots + m_r = n$, the degree of the polynomial. It will cause no loss of generality in what follows to suppose that all the zeros are simple, i.e. all $m_k = 1$ and $r = n$, and this we will do.

Sometimes it is desirable to be able to reverse this process, in particular when one exact zero has been found by some method and the remaining zeros are to be investigated. Suppose that we have located one zero, α; it is then possible to write (2.11) as

$$f(x) = (x - \alpha) f_1(x), \quad (2.12)$$

where $f_1(x)$ is a polynomial of degree $n - 1$. How can we find $f_1(x)$? The procedure is much more complicated to describe in a general form than to carry out for an equation with given numerical coefficients a_i. If such manipulations are too complicated to be carried out mentally, they could be laid out along the lines of an algebraic 'long division' sum. However, a more compact form of calculation is as follows. Write $f_1(x)$ as

$$f_1(x) = b_{n-1} x^{n-1} + b_{n-2} x^{n-2} + b_{n-3} x^{n-3} + \cdots + b_1 x + b_0.$$

Substitution of this form into (2.12) and subsequent comparison of the coefficients of x^p for $p = n, n - 1, \ldots, 1, 0$ with those in the second line of (2.11) generates the series of equations

$$\begin{aligned} b_{n-1} &= a_n, \\ b_{n-2} - \alpha b_{n-1} &= a_{n-1}, \\ b_{n-3} - \alpha b_{n-2} &= a_{n-2}, \\ &\vdots \\ b_0 - \alpha b_1 &= a_1, \\ -\alpha b_0 &= a_0. \end{aligned}$$

2.1 Polynomials and polynomial equations

These can be solved successively for the b_j, starting either from the top or from the bottom of the series. In either case the final equation used serves as a check; if it is not satisfied, at least one mistake has been made in the computation – or α is not a zero of $f(x) = 0$. We now illustrate this procedure with a worked example.

Example Determine by inspection the simple roots of the equation

$$f(x) = 3x^4 - x^3 - 10x^2 - 2x + 4 = 0$$

and hence, by factorisation, find the rest of its roots.

From the pattern of coefficients ($\sum_{r=0}^{n} a_r(-1)^r = 0$) it can be seen that $x = -1$ is a solution to the equation. We therefore write

$$f(x) = (x+1)(b_3 x^3 + b_2 x^2 + b_1 x + b_0),$$

and, from equating, in order, the coefficients of x^4, x^3, x^2, x and finally the constant term, we have

$$b_3 = 3,$$
$$b_2 + b_3 = -1,$$
$$b_1 + b_2 = -10,$$
$$b_0 + b_1 = -2,$$
$$b_0 = 4.$$

These equations give $b_3 = 3$, $b_2 = -4$, $b_1 = -6$, $b_0 = 4$ (check) and so

$$f(x) = (x+1)f_1(x) = (x+1)(3x^3 - 4x^2 - 6x + 4).$$

We now note that $f_1(x) = 0$ if x is set equal to 2. Thus $x - 2$ is a factor of $f_1(x)$, which therefore can be written as

$$f_1(x) = (x-2)f_2(x) = (x-2)(c_2 x^2 + c_1 x + c_0),$$

where, from a similar calculation, we conclude that

$$c_2 = 3,$$
$$c_1 - 2c_2 = -4,$$
$$c_0 - 2c_1 = -6,$$
$$-2c_0 = 4.$$

These equations determine $f_2(x)$ as $3x^2 + 2x - 2$. Since $f_2(x) = 0$ is a quadratic equation, its solutions can be written explicitly as

$$x = \frac{-1 \pm \sqrt{1+6}}{3}.$$

Thus the four roots of $f(x) = 0$ are $-1, 2, \frac{1}{3}(-1+\sqrt{7})$ and $\frac{1}{3}(-1-\sqrt{7})$. ◀

2.1.4 Properties of roots

From the fact that a polynomial equation can be written in any of the alternative forms

$$f(x) = a_n x^n + a_{n-1} x^{n-1} + \cdots + a_1 x + a_0 = 0,$$
$$f(x) = a_n (x - \alpha_1)^{m_1} (x - \alpha_2)^{m_2} \cdots (x - \alpha_r)^{m_r} = 0,$$
$$f(x) = a_n (x - \alpha_1)(x - \alpha_2) \cdots (x - \alpha_n) = 0,$$

it follows that it must be possible to express the coefficients a_i in terms of the roots α_k. To take the most obvious example, comparison of the constant terms (formally the coefficient of x^0) in the first and third expressions shows that

$$a_n (-\alpha_1)(-\alpha_2) \cdots (-\alpha_n) = a_0,$$

or, using the product notation[6]

$$\prod_{k=1}^{n} \alpha_k = (-1)^n \frac{a_0}{a_n}. \tag{2.13}$$

Only slightly less obvious is a result obtained by comparing the coefficients of x^{n-1} in the same two expressions for the polynomial:

$$\sum_{k=1}^{n} \alpha_k = -\frac{a_{n-1}}{a_n}. \tag{2.14}$$

Comparing the coefficients of other powers of x yields further results, though they are of less general use than the two just given. One such, which the reader may wish to derive, is

$$\sum_{j=1}^{n} \sum_{k>j}^{n} \alpha_j \alpha_k = \frac{a_{n-2}}{a_n}. \tag{2.15}$$

In the case of a quadratic equation these root properties are used sufficiently often that they are worth stating explicitly, as follows. If the roots of the quadratic equation $ax^2 + bx + c = 0$ are α_1 and α_2 then[7]

$$\alpha_1 + \alpha_2 = -\frac{b}{a},$$
$$\alpha_1 \alpha_2 = \frac{c}{a}.$$

If the alternative standard form for the quadratic is used, b is replaced by $2b$ in both the equation and the first of these results.

---

[6] The symbol $\prod_{n=0}^{N} a_n$ denotes the multiple direct product $a_0 \times a_1 \times a_2 \times \cdots \times a_N$.
[7] Express the standard identity $x^2 - y^2 = (x + y)(x - y)$ in these terms.

2.1 Polynomials and polynomial equations

Example Find a cubic equation whose roots are -4, 3 and 5.

From results (2.13)–(2.15) we can compute that, arbitrarily setting $a_3 = 1$,

$$-a_2 = \sum_{k=1}^{3} \alpha_k = 4, \qquad a_1 = \sum_{j=1}^{3}\sum_{k>j}^{3} \alpha_j \alpha_k = -17, \qquad a_0 = (-1)^3 \prod_{k=1}^{3} \alpha_k = 60.$$

Thus a possible cubic equation is $x^3 + (-4)x^2 + (-17)x + (60) = 0$. Of course, any multiple of $x^3 - 4x^2 - 17x + 60 = 0$ will do just as well.

A direct calculation starting from a factored product would read:

$$(x - (-4))(x - 3)(x - 5) = 0,$$
$$(x + 4)(x^2 - 3x - 5x + 15) = 0,$$
$$(x + 4)(x^2 - 8x + 15) = 0,$$
$$x^3 - 8x^2 + 15x + 4x^2 - 32x + 60 = 0,$$
$$x^3 - 4x^2 - 17x + 60 = 0,$$

and would be at least as quick in this case, since the second and fourth lines would probably not be written down. ◀

EXERCISES 2.1

1. Solve the following quadratic equations using the general formula:

   (a) $x^2 + x - 6 = 0$,   (b) $x^2 + x + 6 = 0$,   (c) $x^2 - 6x + 9 = 0$,

   (d) $2x^2 + 7x - 9 = 0$,   (e) $2x^2 + 7x + 9 = 0$,   (f) $-2x^2 + 7x - 9 = 0$.

2. Solve equation (a) of the previous exercise, $x^2 + x - 6 = 0$, by 'completing the square'.

3. For what value of k will $3x^2 + 12x + 2 = k$ have a repeated root?

4. Determine how many distinct real roots the cubic equation

   $$2x^3 - 3x^2 - 72x + k = 0$$

   has (a) if $k = 208$ and (b) if $k = -135$. Deduce how many real roots it will have (c) if $k = 200$ and (d) if $k = -200$.

5. Given that one of the zeros of $f(x) = 2x^3 + x^2 - 15x - 18$ is a low negative integer, express $f(x)$ as a product of linear factors.

6. If α and β are the roots of quadratic equation $a_2 x^2 + a_1 x + a_0 = 0$, find an expression for $\alpha^2 + \beta^2$. Verify your answer explicitly for the equation $2x^3 + 7x + 3 = 0$.

64 Preliminary algebra

2.2 Coordinate geometry

In this section we are not so much concerned with solving given equations for an unknown quantity x, as with establishing the functions, $f_i(x)$, whose graphs have particular geometrical shapes. The motivation for this is that the shapes, though historically derived from geometrical figures, have many practical applications in both physics and engineering. The straight line, showing the linear dependence of one variable, y, on another, x, and represented as $y = y(x)$, pervades the whole of science, whilst the conic sections have many applications in astronomy, as well as in the fields of satellite and communications technology.

2.2.1 Linear graphs

We have already mentioned in Section 1.2.3 the standard form for a straight-line graph, namely

$$y = mx + c, \qquad (2.16)$$

which represents a linear relationship between the independent variable x and the dependent variable y. The slope m of the graph measures the rate at which y changes as x is changed[8] and is determined graphically by selecting two points on the line, (x_1, y_1) and (x_2, y_2) and calculating

$$m = \frac{y_2 - y_1}{x_2 - x_1}. \qquad (2.17)$$

If m is negative, then y decreases as x increases. Although the plotting of measured pairs of variables x and y is usually helpful for getting an overall impression of measured data, it is normal to calculate the slope, as well as other parameters, using linear regression analysis. This not only yields the best values for the quantities to be determined, it also gives a measure of the uncertainty of each and an objective assessment of how well the data fits the expected x–y relationship. Facilities for doing this are built into most scientific calculators.

If one or both of x and y are measured in physical units, u_x and u_y respectively, then m will have units of u_y/u_x (or $u_y u_x^{-1}$) attached to it; clearly, if the two units are the same, then m is just a number. It is often loosely said that 'the slope of the straight line is equal to the tangent of the angle the line makes with the x-axis'. This must not be taken too literally, as the physical angle on the graph paper depends upon the scales chosen for the x- and y-axes and, in any case, tangents are dimensionless numbers, whereas in general m will not be.

The other parameter in (2.16) is c. This gives the intercept the line makes on the y-axis (when $x = 0$); its units are the same as those of y.

An alternative form for the equation of a straight line is

$$ax + by + k = 0, \qquad (2.18)$$

---

[8] Reconcile this with what is said about derivatives on p. 57.

2.2 Coordinate geometry

to which (2.16) is clearly connected by

$$m = -\frac{a}{b} \quad \text{and} \quad c = -\frac{k}{b}.$$

This form treats x and y on a more symmetrical basis, the intercepts on the two axes being $-k/a$ and $-k/b$ respectively.

For completeness, we repeat here a procedure, already described in Section 1.2.3, that allows a power relationship between two variables, i.e. one of the form $y = Ax^n$, to be represented in straight-line form. This is done by taking the logarithms of both sides of the relationship. Whilst it is normal in mathematical work to use natural logarithms (ln), for practical investigations logarithms to base 10 (log) are often employed. In either case the form is the same, but it needs to be remembered which has been used when recovering the value of A from fitted data. In the mathematical (base e) form, the power relationship becomes

$$\ln y = n \ln x + \ln A. \tag{2.19}$$

Thus, if $\ln y$ is plotted as a function of $\ln x$, the slope of the resulting straight-line graph gives the power n. The intercept on the $\ln y$ axis is $\ln A$, which yields A, either by exponentiation or by taking antilogarithms.

To ease the calculational effort involved when relationships of the form $y = Ax^n$ or $y = Ae^{kx}$ are to be investigated, special graph papers can be employed. For these papers the scales of one or both of the axes are logarithmic (as opposed to linear) so that, for example, each increase in value of a measured quantity by a factor of 10 corresponds to the same physical length on the paper, whether that increase is from 0.001 to 0.01 or from 100 to 1000. With scales graduated in this way, raw numbers can be plotted directly on the paper without first finding their logarithms, though some care is needed for the accurate plotting of measurements that lie between values specifically marked on the scale.[9] For a $y = Ax^n$ plot, log–log paper, for which both scales are logarithmic, would be used, whilst log–linear paper would be appropriate for $y = Ae^{kx}$.

2.2.2 Conic sections

As well as the straight-line graph, standard coordinate forms of two-dimensional curves that students should know and recognise are the ones concerned with the *conic sections* – so called because they can all be obtained by taking suitable plane sections across a (double) cone. As examples, a section perpendicular to the cone's axis produces a circle, a section parallel to, but not containing, one of the lines lying on the cone's surface produces a parabola, and a section parallel or nearly parallel to the axis of the cone produces a (two-branched) hyperbola, or a pair of intersecting lines if it happens to pass through the cone's vertex.

Because the conic sections can take many different orientations and scalings their general equation, a quadratic function of x and y, is complex:

$$Ax^2 + By^2 + Cxy + Dx + Ey + F = 0. \tag{2.20}$$

---

[9] For example, the value that corresponds to the physical mid-point between the 10 and 100 markers is $\sqrt{10} \times 10 = 31.6$, not 50 or 55.

Preliminary algebra

However, each can be represented by one of four generic forms: an ellipse ($C^2 < 4AB$), a parabola ($C^2 = 4AB$), a hyperbola ($C^2 > 4AB$) or, the degenerate form, a pair of straight lines. If they are reduced to their standard representations, in which their axes of symmetry are made to coincide with the coordinate axes, the first three take the forms

$$\frac{(x-\alpha)^2}{a^2} + \frac{(y-\beta)^2}{b^2} = 1 \quad \text{(ellipse)}, \tag{2.21}$$

$$(y-\beta)^2 = 4a(x-\alpha) \quad \text{(parabola)}, \tag{2.22}$$

$$\frac{(x-\alpha)^2}{a^2} - \frac{(y-\beta)^2}{b^2} = 1 \quad \text{(hyperbola)}. \tag{2.23}$$

Here, (α, β) gives the position of the 'centre' of the curve; this is usually taken as the origin $(0, 0)$ when this does not conflict with any given coordinate system or imposed conditions.

The parabolic equation given here is that for a curve symmetric about a line parallel to the x-axis. The 'nose' of the parabola (see the curve drawn in Figure 2.2) is at the point (α, β) and for $x \geq \alpha$, i.e. to the right of $x = \alpha$, there are two values of y that lie on the curve, one a distance $\sqrt{4a(x-\alpha)}$ above the line $y = \beta$ and the other the same distance below it. The parameter a is conventionally taken as positive; and if the parabola lies to the left of $x = \alpha$, the RHS of the equation is then written as $4a(\alpha - x)$. For a parabola that has its symmetry axis parallel to the y-axis, the roles of x and y are interchanged and the equation reads $(x - \alpha)^2 = 4a(y - \beta)$ for a parabola that is U-shaped, or $(x - \alpha)^2 = 4a(\beta - y)$ for one that has the general form of an inverted U.

In Equation (2.21) with $\alpha = \beta = 0$, which represents a standard ellipse centred on the origin, its symmetry about each of the two coordinate axes is reflected in the fact that x and y only appear in the forms x^2 and y^2. Since both terms on the LHS of the equation are necessarily non-negative, the maximum value of $|y|$ occurs when $x = 0$, and vice versa. Thus the four points $(0, \pm b)$ and $(\pm a, 0)$ mark the extreme values for y and x respectively; the corresponding distances from the origin, b and a, are called the semi-axes of the ellipse.

Of course, the circle is the special case of an ellipse in which $b = a$ and the equation for a circle centred on (α, β) takes the form

$$(x - \alpha)^2 + (y - \beta)^2 = a^2. \tag{2.24}$$

The distinguishing characteristic of this equation is that when it is expressed in the form (2.20) the coefficients of x^2 and y^2 are equal and that of xy is zero; this property is not changed by any reorientation or scaling and so acts to identify a general conic as a circle.

Since (2.24) contains three parameters, α, β and a, any three points, P, Q and R, specify a unique circle that passes through them, each contributing one of the three simultaneous linear equations needed to solve for the three parameters. This is also obvious geometrically, as the centre C of the required circle must lie at the intersection of the perpendicular bisectors of any two of the chords PQ, QR, and PR; the radius of the circle is equal to the common value of CP, CQ and CR.

Definitions of the conic sections in terms of geometrical properties are also available; for example, a parabola can be defined as the locus of a point that is always at the same distance from a given straight line (the *directrix*) as it is from a given point (the *focus*).

2.2 Coordinate geometry

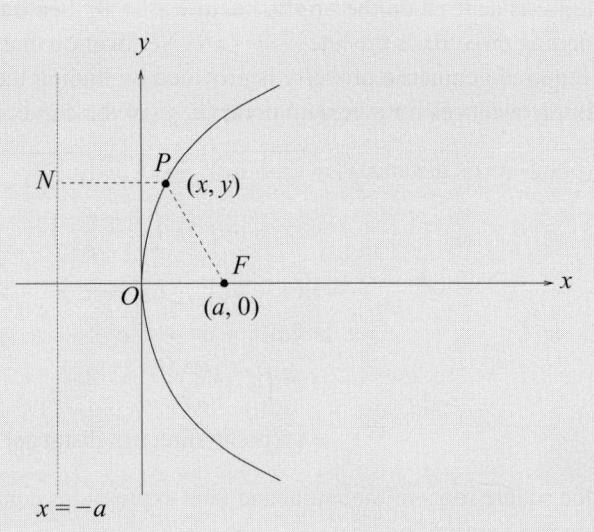

Figure 2.2 Construction of a parabola using the point $(a, 0)$ as the focus and the line $x = -a$ as the directrix.

When these properties are expressed in Cartesian coordinates the above equations are obtained. For a circle, the defining property is that all points on the curve are a distance a from (α, β); (2.24) expresses this requirement very directly. In the following worked example we derive the equation for a parabola.

Example Find the equation of a parabola that has the line $x = -a$ as its directrix and the point $(a, 0)$ as its focus.

Figure 2.2 shows the situation in Cartesian coordinates.
Expressing the defining requirement that PN and PF are equal in length gives
$$(x + a) = [(x - a)^2 + y^2]^{1/2} \quad \Rightarrow \quad (x + a)^2 = (x - a)^2 + y^2$$
which, on expansion of the squared terms, immediately gives $y^2 = 4ax$. This is (2.22) with α and β both set equal to zero. ◀

Although the algebra is more complicated, the same method can be used to derive the equations for the ellipse and the hyperbola. In these cases the distance from the fixed point, the focus, is a definite fraction, e, known as the *eccentricity*, of the distance from the fixed line, the directrix. Associated with any particular ellipse or hyperbola, there are two symmetrically placed pairs of focus plus directrix.

In the case of an ellipse, for which $0 < e < 1$, the eccentricity e provides a quantitative measure of how 'out of round' the ellipse is. The lengths of the semi-axes of the ellipse, a and b, (with $a^2 \geq b^2$) are related to e through

$$e^2 = \frac{a^2 - b^2}{a^2} \quad \text{or} \quad b^2 = a^2(1 - e^2). \tag{2.25}$$

Preliminary algebra

If the ellipse is centred on the origin, i.e. $\alpha = \beta = 0$, then one focus is $(-ae, 0)$ and the corresponding directrix is the line $x = -a/e$. Verification that the curve defined by (2.21) has the required geometric property is provided by finding the ratio of the squares of the two defining distances for a general point (x, y) of the curve, as follows:

$$\begin{aligned}
(\text{point–focus distance})^2 &= (x + ae)^2 + y^2 \\
&= (x + ae)^2 + \left(1 - \frac{x^2}{a^2}\right) b^2, \qquad \text{using (2.21),} \\
&= (x + ae)^2 + (a^2 - x^2)(1 - e^2), \qquad \text{using (2.25),} \\
&= 2aex + a^2 + x^2 e^2 \\
&= e^2 \left(x + \frac{a}{e}\right)^2 \\
&= e^2 (\text{point–directrix distance})^2.
\end{aligned}$$

Taking the square roots of the initial and final expressions confirms that the algebraic and geometric prescriptions agree.

The circle is a special case of an ellipse for which $a = b$ and $e = 0$. Formally for a circle, the two foci coincide at $(a \times 0, 0)$, i.e. at the origin, and the directrices become lines parallel to the y-axis at $x = \pm\infty$.

For a hyperbola, the positive parameter b^2 that appears in the equation for an ellipse is replaced by the negative parameter $-b^2$ as in (2.23). The same algebraic definition (2.25) of e^2 still applies, but now gives e^2 as $(a^2 + b^2)/a^2$, which is > 1; this is qualitatively in accord with the geometric definition.[10] A calculation similar to the one above shows that the two also agree quantitatively.

One aspect of the hyperbola that has no obvious counterpart in the ellipse is its behaviour for large values of x and y. For such values, (2.23) amounts to stating that the difference between two very large values is 1. If the values are large enough, the 1 can be ignored and the resulting equation can, after taking square roots, be written as

$$y - \beta = \pm \frac{b}{a}(x - \alpha). \qquad (2.26)$$

This is the equation of a pair of straight lines passing through the point (α, β) and having slopes $\pm b/a$. These lines are called the *asymptotes* of the hyperbola, and although the hyperbola never intersects them, it gets arbitrarily close to both for large enough values of $|x|$ and hence also of $|y|$. A typical hyperbola and its asymptotes are shown in Figure 2.3.

As we have already seen, the parabola is something of a special case, having a precise value, rather than a range, for its eccentricity. We have already given a derivation of its equation, based on its geometric definition. Since, with $e = 1$, the parabola lies at the boundary between an ellipse, $e < 1$, and a hyperbola, $e > 1$, the equation for a parabola should also be derivable as the limiting case of an ellipse as $e \to 1$. This is done in Problem 2.16 at the end of this chapter.

---

[10] The reader may find it instructive to sketch a standard hyperbolic curve and identify the positions of the foci and directrices, showing that they are given in terms of a and e by the same expressions as for an ellipse.

2.2 Coordinate geometry

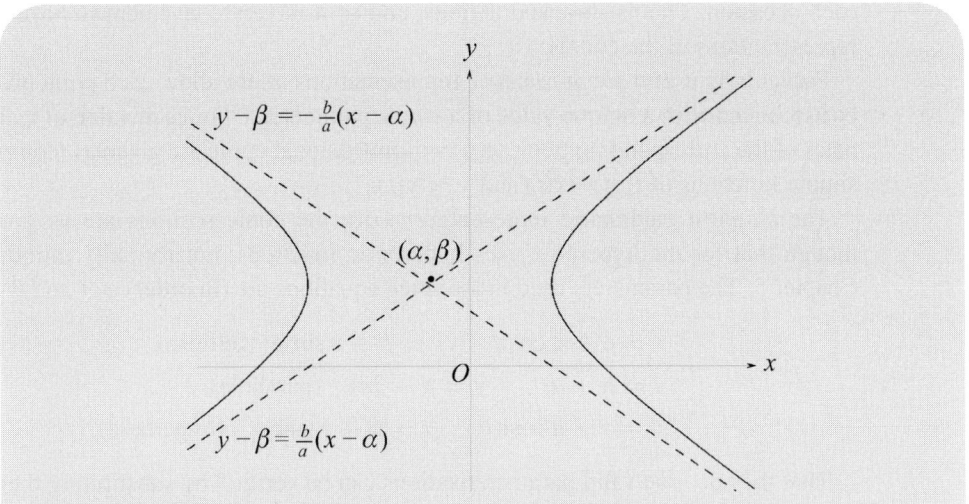

Figure 2.3 A hyperbola centred on (α, β), together with its asymptotes.

As a final example, illustrating several topics from this subsection, we now prove the well-known result that the angle subtended by a diameter at any point on a circle is a right angle.

Example Taking the diameter to be the line joining $Q = (-a, 0)$ and $R = (a, 0)$ and the point P to be any point on the circle $x^2 + y^2 = a^2$, prove that angle QPR is a right angle.

If P is the point (x, y), the slope of the line QP is

$$m_1 = \frac{y - 0}{x - (-a)} = \frac{y}{x + a}.$$

That of RP is

$$m_2 = \frac{y - 0}{x - (a)} = \frac{y}{x - a}.$$

Thus

$$m_1 m_2 = \frac{y^2}{x^2 - a^2}.$$

But, since P is on the circle, $y^2 = a^2 - x^2$ and consequently $m_1 m_2 = -1$. From result (1.65) this implies that QP and RP are orthogonal and that QPR is therefore a right angle. Note that this is true for *any* point P on the circle. ◀

2.2.3 Parametric equations

The equation for each of the conic section curves discussed in Section 2.2.2 involves two variables, x and y. However, they are not independent, since if one of them is given then all associated values of the other can be determined. But finding one or more values of y when x, say, is given, involves, even in this simple case, solving a quadratic equation on

Preliminary algebra

each occasion. This is slow and tedious, and so it is very convenient to have alternative representations of the curves.

Particularly useful are *parametric* representations; these allow each point on a curve to be associated with a unique value of a *single* parameter t. For each value of t, the coordinates of the corresponding point on a two-dimensional curve are given as (comparatively) simple functions of t, $x = x(t)$ and $y = y(t)$.

The simplest parametric representations for the conic sections are as given below, though that for the hyperbola uses hyperbolic functions, not formally introduced until Chapter 5. The parameters used in the three equations are (in order) ϕ, t and θ.

$$x = \alpha + a\cos\phi, \quad y = \beta + b\sin\phi \quad \text{(ellipse)}, \tag{2.27}$$
$$x = \alpha + at^2, \quad y = \beta + 2at \quad \text{(parabola)}, \tag{2.28}$$
$$x = \alpha + a\cosh\theta, \quad y = \beta + b\sinh\theta \quad \text{(hyperbola)}. \tag{2.29}$$

That they do give valid parameterisations can be verified by substituting them into the standard forms (2.21)–(2.23); in each case the standard form is equivalent to an algebraic or trigonometric identity satisfied by the parameter or by functions of it. The parameter identity corresponding to the hyperbola is $\cosh^2\theta - \sinh^2\theta = 1$.

As an example, consider the parametric form of the parabola. The parameterisation given above in Equation (2.28) can be rearranged as

$$t^2 = \frac{x - \alpha}{a} \quad \text{and} \quad t = \frac{y - \beta}{2a}.$$

When t is eliminated, by equating the first equality to the square of the second one, we obtain

$$\frac{x - \alpha}{a} = \frac{(y - \beta)^2}{4a^2},$$

which is easily rearranged to give Equation (2.22).

It will be noticed that for each value of a parameter there is a unique point on the associated curve; the converse is not necessarily true, since a curve that crosses itself will have at least two values of t corresponding to each point of crossing.[11]

Parameterisation need not be restricted to curves lying in a plane, and, indeed, parameterisation generally shows to best advantage when applied to curves or surfaces in three dimensions. Where a surface is involved, two parameters are needed; for example, the quadric surface

$$\frac{x^2}{a^2} + \frac{y^2}{b^2} - \frac{z^2}{c^2} = 1,$$

which has elliptical intersections with $z = $ constant planes, and hyperbolic ones with planes of constant x or y, can be parameterised by

$$x = a\cosh\psi\cos\phi, \quad y = b\cosh\psi\sin\phi, \quad z = c\sinh\psi,$$

---

[11] Sketch the closed curve given by $a^4 y^2 = 4b^2 x^2 (a^2 - x^2)$ and parameterised (check this) by $x = a\sin t$ and $y = b\sin 2t$. Note that the origin corresponds to $t = \pi$ as well as to $t = 0$.

2.2 Coordinate geometry

as the reader may wish to verify. Parameterisations are not necessarily unique, and, for example, the same quadric surface could have been parameterised by

$$x = a \sec\psi \cos\phi, \qquad y = b \sec\psi \sin\phi, \qquad z = c \tan\psi.$$

Sometimes the appropriate parameter arises naturally during the construction of the curve. This is the case when determining the flight path, under constant gravitational acceleration g, of a projectile launched from the ground, as in the following example.

Example At time $t = 0$ a projectile is launched from the origin of the x–y plane, with an initial speed v_0 and at an angle θ to the ground (the x-axis). Its coordinates at a subsequent time t can be shown to be given by

$$x = v_0 \cos\theta\, t,$$
$$y = v_0 \sin\theta\, t - \tfrac{1}{2}gt^2.$$

Show that these equations can be arranged in the form of a parametric description of a parabola and deduce salient features of the flight path.

The parametric form for a parabola, Equation (2.28), has one coordinate linear in the parameter, which we here denote by s, as t is already in use; the other is quadratic in s but with no linear term. Clearly, x must be the first of these, and y must be rearranged to contain only s^2 and a constant. We therefore rewrite the expression for y as

$$-\frac{g}{2}\left(t - \frac{v_0 \sin\theta}{g}\right)^2 + \frac{v_0^2 \sin^2\theta}{2g}$$

by both subtracting and adding $v_0^2 \sin^2\theta / 2g$, the quantity needed to complete the square. To obtain the required form, we must take s as $s = t - (v_0 \sin\theta / g)$, giving

$$y = -\frac{g}{2} s^2 + \frac{v_0^2 \sin^2\theta}{2g}. \qquad (*)$$

But x also has to be rewritten in terms of s:

$$x = v_0 \cos\theta \left(t - \frac{v_0 \sin\theta}{g}\right) + \frac{v_0^2 \sin\theta \cos\theta}{g}$$
$$= v_0 \cos\theta\, s + \frac{v_0^2 \sin\theta \cos\theta}{g}.$$

This is still not in quite the right form, namely Equation (2.28), as the coefficients multiplying s^2 and s need to be in the ratio a to $2a$. Consequently, we multiply the x equation through by $-g/v_0 \cos\theta$ so that it reads

$$-\frac{gx}{v_0 \cos\theta} = -gs - v_0 \sin\theta. \qquad (**)$$

Taken together, $(*)$ and $(**)$ imply that the trajectory is a parabola of the form $(X - \alpha)^2 = 4A(Y - \beta)$, with $A = -g/2$; note the (reversed) roles of X and Y. Specifically,

$$\left(-\frac{gx}{v_0 \cos\theta} + v_0 \sin\theta\right)^2 = 4\left(\frac{-g}{2}\right)\left(y - \frac{v_0^2 \sin^2\theta}{2g}\right),$$

i.e. $\qquad \left(\dfrac{v_0^2 \sin\theta \cos\theta}{g} - x\right)^2 = \dfrac{2v_0^2 \cos^2\theta}{g}\left(\dfrac{v_0^2 \sin^2\theta}{2g} - y\right).$

From this final form we can see that the nose of the inverted parabola (y appears with a minus sign), which occurs at $s = 0$ (i.e. at a time $t = v_0 \sin\theta/g$), is positioned at $x = v_0^2 \sin\theta \cos\theta/g = v_0^2 \sin 2\theta/2g$ (one half of the total range) and $y = v_0^2 \sin^2\theta/2g$ (the maximum height reached). The parabola constant 'a' is given by $v_0^2 \cos^2\theta/2g$, i.e. it varies as the square of the horizontal component of the projectile's launch velocity. We can also see that the maximum horizontal range for a given v_0 is obtained by maximising $v_0^2 \sin 2\theta/g$, i.e. by choosing $\theta = \pi/4$, leading to a maximum range of $v_0^2/2g$.

◀

2.2.4 Plane polar coordinates

Although Cartesian coordinates, x and y, are usually the natural choice for the graphical representation of the mathematical connection between two variables, notionally represented by $y = y(x)$, it is sometimes more convenient to use a different coordinate system, particularly when directions as observed from one particular fixed point, or cyclic behaviour of the variables, plays an important part.

The most common non-Cartesian two-dimensional system is that of *plane polar coordinates*. In it, the position of a point P is specified by its distance ρ from a fixed point O together with the angle ϕ that the line OP makes with a fixed direction; in this system, P has coordinates (ρ, ϕ).

If a plane polar coordinate system and a Cartesian system are made to have a common origin, and the fixed direction in the plane polar system is made to coincide with the x-axis of the Cartesian system, as shown in Figure 2.4, then an immediate one-to-one connection can be made between the two sets of coordinates:

$$x = \rho \cos\phi \text{ and } y = \rho \sin\phi, \qquad (2.30)$$

$$\rho = \sqrt{x^2 + y^2} \text{ and } \phi = \tan^{-1}\frac{y}{x}, \qquad (2.31)$$

where, in the final equation, account needs to be taken of the individual signs of the numerator and denominator in the fraction, so that ϕ is placed in the correct quadrant of the two-dimensional ρ–ϕ plane. This can be done formally by replacing the final equation by a pair of 'simultaneous equations',

$$\cos\phi = \frac{x}{\sqrt{x^2 + y^2}} \quad \text{and} \quad \sin\phi = \frac{y}{\sqrt{x^2 + y^2}}, \qquad (2.32)$$

to be solved for ϕ.

It will be seen that prescription (2.31) for ρ indicates that it is never negative, and this is the appropriate restriction to apply when plane polar coordinates are incorporated as the first two components of the three-dimensional coordinate system, (ρ, ϕ, z), known as cylindrical polar coordinates. However, for the two-dimensional ρ–ϕ coordinate system, negative values of ρ are sometimes used, particularly when they allow a single formula to specify the connection between ρ and ϕ for the full range of values of ϕ. When this convention is in use, the point (ρ, ϕ) is the same point as $(-\rho, \phi + \pi)$; when it is not in use, care must be taken that any formula for ρ in terms of ϕ does not generate

2.2 Coordinate geometry

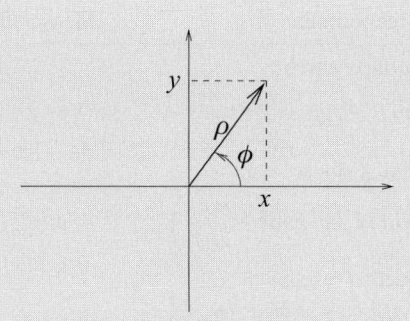

Figure 2.4 The relationship between Cartesian and plane polar coordinates.

negative values – this usually means giving different prescriptions for different parts of the ϕ-range.[12]

Some practice at recognising curves expressed in polar coordinates, and at converting equations between polar and Cartesian coordinates, is provided by the problems at the end of this chapter, but we will conclude this section with a simple example.

Example By converting it to Cartesian coordinates, identify the curve represented by the polar equation

$$\rho = 2a \cos \phi.$$

From (2.31) and (2.32) we can substitute for ρ and $\cos \phi$ and obtain

$$\sqrt{x^2 + y^2} = 2a \frac{x}{\sqrt{x^2 + y^2}}.$$

After cross-multiplication, this can be manipulated by 'completing the square' as follows:

$$x^2 + y^2 - 2ax = 0,$$
$$(x - a)^2 - a^2 + y^2 = 0,$$
$$(x - a)^2 + y^2 = a^2.$$

The final line shows that the given equation is the polar description of a circle of radius a centred on the Cartesian point $(a, 0)$. ◀

EXERCISES 2.2

1. In the table below are given some formulae taken from various areas of physics. Given experimental data for the stated measured quantities and values for the indicated known constants, determine how the data should be plotted in order to obtain a 'straight-line graph'. State explicitly how values for the unknown quantities given in the final column may be deduced from the graph.

---

[12] Sketch the curve $\rho = a \sin \phi$ for $0 \leq \phi < 2\pi$, both with and without the convention, showing that it has two loops in the latter case, but only one in the former.

Preliminary algebra

Area/formula	Measured	Known	Unknown
Surface waves $v(\lambda) = ag\lambda + \dfrac{b\sigma}{\rho\lambda}$	λ, v	g, σ, ρ	a, b
Classical gas $n(v) = Av^2 \exp\left(-\dfrac{mv^2}{2kT}\right)$	v, n	m, k	A, T
Nuclear resonance $\sigma(E) = \dfrac{(2\ell + 1)\lambda^2 \Gamma^2}{\pi[4(E - E_0)^2 + \Gamma^2]}$	E, σ	λ, E_0	ℓ, Γ

2. Without attempting to plot them (except as a check if you have a suitable plotting calculator), state the general shapes of the curves represented by the following equations:

   (a) $2x^2 + 4y^2 + 6xy + 3x + 4y + 2 = 0$,
   (b) $2x^2 + 4y^2 - 6xy - 3x - 4y - 2 = 0$,
   (c) $3x^2 + 3y^2 + 3x + 4y + 2 = 0$,
   (d) $3x^2 + 3y^2 - 6xy - 3x - 4y - 2 = 0$.

3. If $f(x) = 2x^2 - 6y^2 + xy + 2x - 17y - 12 = 0$ is to represent a pair of straight lines, one of which has equation $x + 2y + 3 = 0$, what must be the equation of the other line? Verify that $f(x) = 0$ does, indeed, represent a pair of straight lines.

4. Find the equation of the circle that passes through the three points $(0, 0)$, $(7, -1)$ and $(-1, 3)$.

5. Determine the curve parameterised in Cartesian coordinates by

$$x = a\frac{2}{1+t^2}, \quad y = b\frac{2t}{1+t^2}, \quad -\infty < t < \infty,$$

and make a rough annotated sketch of it.

6. Express the following Cartesian curves in plane polar coordinates and hence sketch them:

   (a) $x^2 - y^2 = 1$, (b) $ay^2 = (x^2 + y^2)^{3/2}$, (c) $4(x^2 + y^2)^3 = a^2(2x^2 - y^2)^2$.

   Note whether or not you are using the 'negative-ρ' convention.

2.3 Partial fractions

In subsequent chapters, and in particular when we come to study integration in Chapter 4, we will need to express a function $f(x)$ that is the ratio of two polynomials in a more manageable form. To remove some potential complexity from our discussion we will

2.3 Partial fractions

assume that all the coefficients in the polynomials are real, although this is not an essential simplification.

Our main concern in this section will be that indicated, namely expressing the ratio of two polynomials in a more convenient form, which will generally mean expressing it as the sum of terms each of which is the ratio of two very simple polynomials – the simpler the better, with a constant (a zero-degree polynomial) in the numerator, if at all possible. However, before we proceed to do so, it is instructive to consider the (generally) more straightforward reverse process, in which a number of terms, each of which is the ratio of two simple polynomials, are combined to make a single, but more complicated, ratio.

If we wish to express

$$f(x) = \frac{p(x)}{q(x)} + \frac{r(x)}{s(x)}$$

as a single fraction, it is clear that we can multiply the first fraction by unity in the form $s(x)/s(x)$, and the second by unity in the form $q(x)/q(x)$. Nothing can have changed in the values $f(x)$ takes as a function of x, but now both denominators are the same and so the two numerators can be added together to give

$$f(x) = \frac{p(x)s(x) + r(x)q(x)}{q(x)s(x)},$$

i.e. a single fraction as the ratio of two polynomials. To take a concrete example, we express

$$\frac{6}{x+2} - \frac{2}{x+1}$$

as a single fraction as follows:

$$\frac{6}{x+2} - \frac{2}{x+1} = \frac{6(x+1) - 2(x+2)}{(x+2)(x+1)} = \frac{4x+2}{x^2+3x+2}. \tag{2.33}$$

The approach can be extended in an obvious way to more than two terms and to cases where some of the polynomials in the various numerators are equal or have factors in common. The general rule is that each term is both multiplied and divided by whatever is needed to make the original denominator equal to the lowest common multiple of all the denominators. The latter, abbreviated to l.c.m., is the 'smallest' algebraic expression that contains each of the denominators as a factor.

A common multiple can always be found by multiplying together all the numerators, but this may be more complicated than it need be, since it may contain repeated polynomial factors that are raised to higher powers than necessary. When all superfluous factors have been removed, what is left is the appropriate *lowest* common multiple. Thus the l.c.m. of $x + 3$, $(x - 4)^2$, $x^2 - 3x - 4$ and $x^2 + 6x + 9$ can be found from their direct product by noting that the fourth term is the square of the first one, which therefore need not be included explicitly, and that the third term can be factorised into $(x + 1)(x - 4)$. Since $(x - 4)$ is already present in $(x - 4)^2$, it need not be explicitly included either. Thus the l.c.m. of the four factors is $(x - 4)^2(x + 1)(x + 3)^2$; this fifth-degree polynomial could be multiplied out to give $x^5 - x^4 - 25x^3 + x^2 + 168x + 44$, but it is usually much more convenient to have such polynomials in factored form.[13]

Before returning to the more difficult task of separating a single complicated fraction into a number of simpler ones, we show a worked example of the reverse combination process.

Example Express
$$f(x) = 3x + \frac{2}{x} - \frac{4}{x+1} + \frac{1}{(x+1)^2}$$
as a single ratio of polynomials.

We first note that because of the presence of a positive power of x with no explicit denominator (formally the denominator is unity) we must expect a final answer in which the degree of the numerator will be greater than that of the denominator.

Noting that $x + 1$ is contained in $(x + 1)^2$ and therefore need not be included explicitly, we see that the l.c.m. of the denominators is $1 \times x \times (x + 1)^2$. To make all the denominators the same, we have to multiply both the numerator and denominator of the first term by $x(x + 1)^2$, those of the second term by $(x + 1)^2$, those of the third term by $x(x + 1)$ and those of the final term by x. Thus the calculation is

$$f(x) = 3x + \frac{2}{x} - \frac{4}{x+1} + \frac{1}{(x+1)^2}$$
$$= \frac{3x \cdot x \cdot (x+1)^2 + 2 \cdot (x+1)^2 - 4 \cdot x \cdot (x+1) + 1 \cdot x}{x(x+1)^2}$$
$$= \frac{3x^4 + 6x^3 + 3x^2 + 2x^2 + 4x + 2 - 4x^2 - 4x + x}{x(x+1)^2}$$
$$= \frac{3x^4 + 6x^3 + x^2 + x + 2}{x(x+1)^2} \quad \text{or} \quad \frac{3x^4 + 6x^3 + x^2 + x + 2}{x^3 + 2x^2 + x}.$$

As expected, the degree of the numerator (4) is greater than that of the denominator (3). ◀

2.3.1 The general method

We now return to the main purpose of this section, having seen the kind of terms we might expect to appear when trying to turn a single ratio of complicated polynomials into a more tractable form. As will become all too apparent, the behaviour of $f(x)$ is crucially determined by the location of the zeros of its denominator, i.e. if $f(x)$ is written as $f(x) = g(x)/h(x)$ where both $g(x)$ and $h(x)$ are polynomials,[14] then $f(x)$ changes extremely rapidly when x is close to those values α_i that are the roots of $h(x) = 0$. To make such behaviour explicit, we write $f(x)$ as a sum of terms such as $A/(x - \alpha)^n$, in

---

[13] Find, in factored form, the l.c.m. of the four quadratic functions $x^2 + 2x - 3$, $x^2 + 4x + 3$, $x^2 + 6x + 9$ and $x^2 - 1$, and show that in unfactored form it is $x^4 + 6x^3 + 8x^2 - 6x - 9$.

[14] It is assumed that the ratio has been reduced so that $g(x)$ and $h(x)$ do not contain any common factors, i.e. there is no value of x that makes both vanish at the same time. We may also assume without any loss of generality that the coefficient of the highest power of x in $h(x)$ has been made equal to unity, by, if necessary, dividing both numerator and denominator by the coefficient of this highest power.

2.3 Partial fractions

which A is a constant, α is one of the α_i that satisfy $h(\alpha_i) = 0$ and n is a positive integer; each possible α_i gives rise to one or more such terms in the sum. Writing a function in this way is known as expressing it in *partial fractions*.

Suppose, for the sake of definiteness, that we wish to express the function

$$f(x) = \frac{4x + 2}{x^2 + 3x + 2}$$

in partial fractions, i.e. to write it as

$$f(x) = \frac{g(x)}{h(x)} = \frac{4x + 2}{x^2 + 3x + 2} = \frac{A_1}{(x - \alpha_1)^{n_1}} + \frac{A_2}{(x - \alpha_2)^{n_2}} + \cdots. \qquad (2.34)$$

The first question that arises is that of how many terms there should be on the RHS. Although some complications occur when $h(x)$ has repeated roots (these are considered below) it is clear that, since $h(x)$ is a quadratic polynomial, $f(x)$ will only become infinite at the *two* values of x, α_1 and α_2 say, that make $h(x) = 0$. Consequently, the RHS can only become infinite at the same two values of x and therefore contains only two partial fractions – these are the ones that happen to be shown explicitly in (2.34). This argument can be trivially extended [again temporarily ignoring the possibility of repeated roots of $h(x)$] to show that if $h(x)$ is a polynomial of degree n then there should be n terms on the RHS, each containing a different root α_i of the equation $h(\alpha_i) = 0$.

A second general question concerns the appropriate values of the n_i. This is answered by putting the RHS over a common denominator in the way discussed at the start of this section. As shown there, the common denominator will have to be the product $(x - \alpha_1)^{n_1}(x - \alpha_2)^{n_2} \cdots$. Comparison of the highest power of x in this new RHS with the same power in $h(x)$ shows that $n_1 + n_2 + \cdots = n$. This result holds whether or not $h(x) = 0$ has repeated roots and, although we do not give a rigorous proof, strongly suggests the following correct conclusions.

- The number of terms on the RHS is equal to the number of *distinct* roots of $h(x) = 0$, each term having a different root α_i in its denominator $(x - \alpha_i)^{n_i}$.
- If α_i is a multiple root of $h(x) = 0$ then the value to be assigned to n_i in (2.34) is that of m_i when $h(x)$ is written in the product form (2.10). Further, as discussed on p. 81, A_i has to be replaced by a polynomial of degree $m_i - 1$. This is also formally true for non-repeated roots, since then both m_i and n_i are equal to unity and $m_i - 1$ is equal to zero; a polynomial of degree zero is simply a constant.

Returning to our specific example we note that the denominator $h(x)$ has zeros at $x = \alpha_1 = -1$ and $x = \alpha_2 = -2$; these x-values are the simple (non-repeated) roots of $h(x) = 0$. Thus the partial fraction expansion will be of the form

$$\frac{4x + 2}{x^2 + 3x + 2} = \frac{A_1}{x + 1} + \frac{A_2}{x + 2}. \qquad (2.35)$$

The reader will probably have noticed that the LHS of this expansion is the final result given in (2.33); so we may expect (actually require, if the expansion is to be unique) that the partial fraction expansion will consist of the two terms that were combined to give (2.33).

Preliminary algebra

We now list several methods available for determining the coefficients A_1 and A_2. We also remind the reader that, as with all the explicit examples and techniques described, these methods are to be considered as models for the handling of any ratio of polynomials, with or without characteristics that make it a special case.

(i) In the way described at the start of this section, the RHS can be put over a common denominator, in this case $(x + 1)(x + 2)$, and then the coefficients of the various powers of x can be equated in the numerators on both sides of the equation. This leads to

$$4x + 2 = A_1(x + 2) + A_2(x + 1),$$
$$\Rightarrow \quad 4 = A_1 + A_2$$
$$\text{and} \quad 2 = 2A_1 + A_2.$$

Solving the simultaneous equations for A_1 and A_2 gives $A_1 = -2$ and $A_2 = 6$.

(ii) A second method is to substitute two (or more generally n) different values of x into each side of (2.35) and so obtain two (or n) simultaneous equations for the two (or n) constants A_i. To justify this practical way of proceeding it is necessary, strictly speaking, to appeal to method (i) above, which establishes that there are *unique* values for A_1 and A_2. If the values for A_1 and A_2 were not unique, but varied according to the particular values of x used, the expansion would be meaningless, as it would not be valid for *all* x.

It is normally very convenient to take zero as one of the values of x, but of course any set will do. Suppose in the present case that we use the values $x = 0$ and $x = 1$ and substitute in (2.35). The resulting equations are

$$\frac{2}{2} = \frac{A_1}{1} + \frac{A_2}{2},$$
$$\frac{6}{6} = \frac{A_1}{2} + \frac{A_2}{3},$$

which on solution give $A_1 = -2$ and $A_2 = 6$, as before. The reader can easily verify that any other pair of values for x (except for a pair that includes α_1 or α_2) gives the same values for A_1 and A_2.

(iii) The very reason why method (ii) fails if x is chosen as one of the roots α_i of $h(x) = 0$ can be made a basis for determining the values of the A_i corresponding to non-multiple roots (but see p. 82), without having to solve simultaneous equations. The method is conceptually more difficult than the other methods presented here and, strictly, needs results from the theory of complex variables to justify it. However, we give a practical 'cookbook' recipe for determining the coefficients, an illustrative example, and a qualitative justification for the procedure.

(a) To determine the coefficient A_k, imagine the denominator $h(x)$ written as the product $(x - \alpha_1)(x - \alpha_2) \cdots (x - \alpha_n)$, with any m-fold repeated root (which cannot include α_k) giving rise to m factors in parentheses.

2.3 Partial fractions

(b) Now set x equal to α_k and evaluate the expression obtained after omitting from $h(x)$ the factor that reads $\alpha_k - \alpha_k$; as the root is non-multiple, it will appear only once.

(c) Divide the value so obtained into $g(\alpha_k)$; the result is the required coefficient A_k.

For our specific example we find that in step (a) that $h(x) = (x+1)(x+2)$ and that, when evaluating A_1, step (b) yields $-1 + 2$, i.e. 1. Since $g(-1) = 4(-1) + 2 = -2$, step (c) gives A_1 as $(-2)/(1)$, i.e in agreement with our other evaluations. In a similar way A_2 is evaluated as $(-6)/(-1) = 6$.

The qualitative justification for the procedure is as follows. In the region of x close to $x = \alpha_k$, the behaviour of $f(x)$ is totally dominated by the behaviour of the individual factor $(x - \alpha_k)^{-1}$, the other factors and the numerator hardly changing as x changes. The same must be true of the partial fractions representation of $f(x)$, with the term $A_k/(x - \alpha_k)$ tending to infinity whilst all other terms become negligible by comparison. Moreover, the factor F multiplying $(x - \alpha_k)^{-1}$ must be the same in both cases. But, in the first case F is the complete expression for $f(x)$, apart from the $(x - \alpha_k)^{-1}$ factor, evaluated at $x = \alpha_k$, whilst in the second F is simply A_k. Now, knowing from method (i) that the coefficients A_k are uniquely determined, we can conclude that A_k is given by removing the factor $(x - \alpha_k)$ from the denominator of $f(x)$ and then evaluating what is left at $x = \alpha_k$.

Thus, in summary, any one of the three methods listed above shows that

$$\frac{4x+2}{x^2+3x+2} = \frac{-2}{x+1} + \frac{6}{x+2},$$

in accord with our expectations. The best method to use in any particular circumstance will depend on the complexity, in terms of the degrees of the polynomials and the multiplicities of the roots of the denominator, of the function being considered and, to some extent, on the individual inclinations of the student; some prefer lengthy but straightforward solution of simultaneous equations, whilst others feel more at home carrying through shorter but more abstract calculations in their heads.

2.3.2 Complications and special cases

Having established the basic method for partial fractions, we now show, through further worked examples, how some complications are dealt with by extensions to the procedure. These extensions are introduced one at a time, but of course in any practical application more than one may be involved.

The degree of the numerator is greater than or equal to that of the denominator

Although we have not specifically mentioned the fact, it will be apparent from trying to apply method (i) of the previous subsection to such a case that if the degree of the numerator (m) is not less than that of the denominator (n) then the ratio of two polynomials cannot be expressed in partial fractions.[15]

---

[15] To demonstrate this, try applying method (i) to the polynomial fraction $(x^2 + 2x + 1)/(x^2 + 3x + 2)$. Note that it is not possible to equate coefficients of the x^2 term.

Preliminary algebra

To get round this difficulty it is necessary to start by dividing the denominator $h(x)$ into the numerator $g(x)$. Doing so yields a further polynomial, which we denote by $s(x)$, together with a polynomial ratio in which the degree of the numerator *is* less than that of the denominator. This ratio, $t(x)$, *will* be expandable in partial fractions. As a formula,

$$f(x) = \frac{g(x)}{h(x)} = s(x) + t(x) \equiv s(x) + \frac{r(x)}{h(x)}. \tag{2.36}$$

It is apparent that the polynomial $r(x)$ is the *remainder* obtained when $g(x)$ is divided by $h(x)$, and, in general, will be a polynomial of degree $n - 1$. It is also clear that the polynomial $s(x)$ will be of degree $m - n$. The actual division process can be set out as an algebraic long-division sum, but is probably more easily handled by writing (2.36) in the form

$$g(x) = s(x)h(x) + r(x) \tag{2.37}$$

or, more explicitly, as

$$\begin{aligned} g(x) = (s_{m-n}x^{m-n} + s_{m-n-1}x^{m-n-1} + \cdots + s_0)h(x) \\ + (r_{n-1}x^{n-1} + r_{n-2}x^{n-2} + \cdots + r_0) \end{aligned} \tag{2.38}$$

and then equating coefficients.

We illustrate this procedure with the following worked example.

Example Find the partial fraction decomposition of the function

$$f(x) = \frac{x^3 + 3x^2 + 2x + 1}{x^2 - x - 6}.$$

Since the degree of the numerator is 3 and that of the denominator is 2, a preliminary long division is necessary. The polynomial $s(x)$ resulting from the division will have degree $3 - 2 = 1$ and the remainder $r(x)$ will be of degree $2 - 1 = 1$ (or less). Thus we write

$$x^3 + 3x^2 + 2x + 1 = (s_1 x + s_0)(x^2 - x - 6) + (r_1 x + r_0).$$

From equating the coefficients of the various powers of x on the two sides of the equation, starting with the highest, we now obtain the simultaneous equations

$$\begin{aligned} 1 &= s_1, \\ 3 &= s_0 - s_1, \\ 2 &= -s_0 - 6s_1 + r_1, \\ 1 &= -6s_0 + r_0. \end{aligned}$$

These are readily solved, in the given order, to yield $s_1 = 1$, $s_0 = 4$, $r_1 = 12$ and $r_0 = 25$. Thus $f(x)$ can be written as

$$f(x) = x + 4 + \frac{12x + 25}{x^2 - x - 6}.$$

2.3 Partial fractions

The last term can now be decomposed into partial fractions as previously. The zeros of the denominator are at $x = 3$ and $x = -2$ and the application of any method from the previous subsection yields the respective constants as $A_1 = 12\frac{1}{5}$ and $A_2 = -\frac{1}{5}$. Thus we have

$$x + 4 + \frac{61}{5(x-3)} - \frac{1}{5(x+2)}.$$

as the final partial fraction decomposition of $f(x)$.

Factors of the form $a^2 + x^2$ in the denominator

We have so far assumed that the roots of $h(x) = 0$, needed for the factorisation of the denominator of $f(x)$, can always be found. In principle they always can be, but in some cases they are not real. Consider, for example, attempting to express in partial fractions a polynomial ratio whose denominator is $h(x) = x^3 - x^2 + 2x - 2$. Clearly $x = 1$ is a zero of $h(x)$, and so a first factorisation is $(x-1)(x^2 + 2)$. However, we cannot make any further progress because the factor $x^2 + 2$ cannot be expressed as $(x - \alpha)(x - \beta)$ for any real α and β.

Complex numbers are introduced later in this book (Chapter 5) and, when the reader has studied them, he or she may wish to justify the procedure set out below. It can be shown to be equivalent to that already given, but the zeros of $h(x)$ are now allowed to be complex and terms that are complex conjugates of each other are combined to leave only real terms.

Since quadratic factors of the form $a^2 + x^2$ that appear in $h(x)$ cannot be reduced to the product of two linear factors, partial fraction expansions including them need to have numerators in the corresponding terms that are not simply constants A_i but linear functions of x, i.e. of the form $B_i x + C_i$. Thus, in the expansion, linear terms (first-degree polynomials) in the denominator have constants (zero-degree polynomials) in their numerators, whilst quadratic terms (second-degree polynomials) in the denominator have linear terms (first-degree polynomials) in their numerators. As a symbolic formula, the partial fraction expansion of

$$\frac{g(x)}{(x - \alpha_1)(x - \alpha_2) \cdots (x - \alpha_p)(x^2 + a_1^2)(x^2 + a_2^2) \cdots (x^2 + a_q^2)}$$

should take the form

$$\frac{A_1}{x - \alpha_1} + \frac{A_2}{x - \alpha_2} + \cdots + \frac{A_p}{x - \alpha_p} + \frac{B_1 x + C_1}{x^2 + a_1^2} + \frac{B_2 x + C_2}{x^2 + a_2^2} + \cdots + \frac{B_q x + C_q}{x^2 + a_q^2}.$$

Of course, the degree of $g(x)$ must be less than $p + 2q$; if it is not, an initial division must be carried out as demonstrated earlier.

Repeated factors in the denominator

Consider trying (incorrectly) to expand

$$f(x) = \frac{x - 4}{(x+1)(x-2)^2}$$

Preliminary algebra

in partial fraction form as follows:

$$\frac{x-4}{(x+1)(x-2)^2} = \frac{A_1}{x+1} + \frac{A_2}{(x-2)^2}.$$

Multiplying both sides of this supposed equality by $(x + 1)(x − 2)^2$ produces an equation whose LHS is linear in x, whilst its RHS is quadratic. This is clearly wrong and so an expansion in the above form cannot be valid. The correction we must make is very similar to that needed in the previous subsection, namely that, since $(x − 2)^2$ is a quadratic polynomial, the numerator of the term containing it must be a first-degree polynomial, and not simply a constant.

The correct form for the part of the expansion containing the repeated root is therefore $(Bx + C)/(x − 2)^2$. Using this form and either of methods (i) or (ii) for determining the constants gives the full partial fraction expansion as

$$\frac{x-4}{(x+1)(x-2)^2} = -\frac{5}{9(x+1)} + \frac{5x-16}{9(x-2)^2},$$

as the reader may verify.

Since any term of the form $(Bx + C)/(x − \alpha)^2$ can be written as

$$\frac{B(x-\alpha) + C + B\alpha}{(x-\alpha)^2} = \frac{B}{x-\alpha} + \frac{C+B\alpha}{(x-\alpha)^2},$$

and similarly for multiply repeated roots, an alternative form for the part of the partial fraction expansion containing a repeated root α is

$$\frac{D_1}{x-\alpha} + \frac{D_2}{(x-\alpha)^2} + \cdots + \frac{D_p}{(x-\alpha)^p}. \qquad (2.39)$$

In this form, all x-dependence has disappeared from the numerators, but at the expense of $p − 1$ additional terms; the total number of constants to be determined remains unchanged, as it must.

When describing possible methods of determining the constants in a partial fraction expansion, we implied that method (iii), p. 78, which avoids the need to solve simultaneous equations, is restricted to terms involving non-repeated roots. In fact, it can be applied in repeated-root situations, when the expansion is put in the form (2.39), but only to find the constant in the term involving the largest inverse power of $x − \alpha$, i.e. D_p in (2.39).

We conclude this section with a more protracted worked example that contains all three of the complications discussed.

Example Resolve the following expression $F(x)$ into partial fractions:

$$F(x) = \frac{x^5 - 2x^4 - x^3 + 5x^2 - 46x + 100}{(x^2 + 6)(x - 2)^2}.$$

2.3 Partial fractions

We note that the degree of the denominator (4) is not greater than that of the numerator (5), and so we must start by dividing the latter by the former. It follows, from the difference in degrees and the coefficients of the highest powers in each, that the result will be a linear expression $s_1 x + s_0$ with the coefficient s_1 equal to 1. Thus the numerator of $F(x)$ must be expressible as

$$(x + s_0)(x^4 - 4x^3 + 10x^2 - 24x + 24) + (r_3 x^3 + r_2 x^2 + r_1 x + r_0),$$

where the second factor in parentheses is the denominator of $F(x)$ multiplied out and written as a polynomial. Equating the coefficients of x^4 gives $-2 = -4 + s_0$ and fixes s_0 as 2. Equating the coefficients of powers of x less than the fourth gives equations involving the coefficients r_i. Putting those from the original numerator on the LHS and those from the above reformulation with s_0 set equal to 2 on the RHS, we obtain

$$-1 = -8 + 10 + r_3,$$
$$5 = -24 + 20 + r_2,$$
$$-46 = 24 - 48 + r_1,$$
$$100 = 48 + r_0.$$

These give $r_3 = -3$, $r_2 = 9$, $r_1 = -22$ and $r_0 = 52$, and so the remainder polynomial $r(x)$ can be constructed and $F(x)$ written as

$$F(x) = x + 2 + \frac{-3x^3 + 9x^2 - 22x + 52}{(x^2 + 6)(x - 2)^2} \equiv x + 2 + f(x).$$

The polynomial ratio $f(x)$ can now be expressed in partial fraction form, noting that its denominator contains both a term of the form $x^2 + a^2$ and a repeated root. Thus

$$f(x) = \frac{Bx + C}{x^2 + 6} + \frac{D_1}{x - 2} + \frac{D_2}{(x - 2)^2}.$$

We could now put the RHS of this equation over the common denominator $(x^2 + 6)(x - 2)^2$ and find B, C, D_1 and D_2 by equating coefficients of powers of x. It is quicker, however, to use a mixture of methods (iii) and (ii). Method (iii) gives D_2 as $(-24 + 36 - 44 + 52)/(4 + 6) = 2$. We choose to evaluate the other coefficients by method (ii), and setting $x = 0$, $x = 1$ and $x = -1$ gives respectively

$$\frac{52}{24} = \frac{C}{6} - \frac{D_1}{2} + \frac{2}{4},$$
$$\frac{36}{7} = \frac{B + C}{7} - D_1 + 2,$$
$$\frac{86}{63} = \frac{C - B}{7} - \frac{D_1}{3} + \frac{2}{9}.$$

These equations reduce to

$$4C - 12 D_1 = 40,$$
$$B + C - 7 D_1 = 22,$$
$$-9B + 9C - 21 D_1 = 72,$$

with solution $B = 0, C = 1, D_1 = -3$.

Thus, finally, we may write

$$F(x) = x + 2 + \frac{1}{x^2 + 6} - \frac{3}{x - 2} + \frac{2}{(x - 2)^2}.$$

as the partial fraction expansion of the original function.

Preliminary algebra

EXERCISES 2.3

1. Write the following as the ratio of two polynomials (expressed in its lowest terms):

    (a) $\dfrac{3}{2x+1} - \dfrac{4x}{2-3x}$,   (b) $2 + \dfrac{1}{x+1} + \dfrac{3}{(x+1)^2}$,   (c) $\dfrac{1}{x+1} - \dfrac{1}{x-1} + \dfrac{2x^2}{x^2-1}$.

2. Without doing any calculations, write out the expected forms for the partial fraction expansions of the following:

    (a) $\dfrac{x^2 - 2x + 3}{(x-2)(x+3)(x+6)}$,   (b) $\dfrac{5x^4 + 3x^3 + x^2 - 2x + 3}{(x-2)(x^2+6x+9)(x+6)}$,   (c) $\dfrac{x^3}{(x+2)(x^2+4)}$.

3. Explain why evaluating the numerators in case (a) of the previous exercise is more efficiently done using method (iii) of the text, rather than either of methods (i) or (ii). Use method (iii) to determine the full partial fraction expansion.

2.4 Some particular methods of proof

Much of the mathematics used by physicists and engineers is concerned with obtaining a particular value, formula or function from a given set of data and stated conditions. However, just as it is essential in physics to formulate the basic laws and so be able to set boundaries on what can or cannot happen,[16] so it is important in mathematics to be able to state general propositions about the outcomes that are or are not possible. To this end one attempts to establish theorems that state in as general a way as possible mathematical results that apply to particular types of situation. In this section we describe two methods that can sometimes be used to prove particular classes of theorems.

The two general methods of proof are known as proof by induction (which has already been met in Section 1.4.2, in connection with the proof of the binomial expansion) and proof by contradiction. They share the common characteristic that at an early stage in the proof an assumption is made that a particular (unproven) statement is true; the consequences of that assumption are then explored. In an inductive proof the conclusion is reached that the assumption is self-consistent and has other equally consistent but broader implications, which are then applied to establish the general validity of the assumption. A proof by contradiction, however, establishes an internal inconsistency and thus shows that the assumption is unsustainable; the natural consequence of this is that the negative of the assumption is established as true.

Later in this book use will be made of these methods of proof to explore new territory, e.g. to examine the properties of vector spaces and matrices. However, at this stage we will draw our illustrative and test examples from the material covered in Chapter 1, from the earlier sections of this chapter and from other topics in elementary algebra and number theory.

---

[16] Two obvious examples taken from classical physics are *conservation of energy* and the *second law of thermodynamics*, the latter stating that the entropy of a closed system never decreases.

2.4 Some particular methods of proof

2.4.1 Proof by induction

The proof of the binomial expansion given in Section 1.4.2 has already shown the way in which an inductive proof is carried through. It also indicates the main limitation of the method, namely that only an initially supposed result can be proved. Thus the method of induction is of no use for *deducing* a previously unknown result; a putative equation or result has to be arrived at by some other means, usually by noticing patterns or by trial and error using simple values of the variables involved. It will also be clear that propositions that can be proved by induction are limited to those containing a parameter that takes a range (usually infinite) of integer values.

For a proposition involving a parameter n, the five steps in a proof using induction are as follows.

(i) Formulate the supposed result for general n.
(ii) Suppose (i) to be true for $n = N$ (or more generally for all values of $n \leq N$; see below), where N is restricted to lie in the stated range.
(iii) Show, using only proven results and supposition (ii), that proposition (i) is true for $n = N + 1$.
(iv) Demonstrate directly, and without any assumptions, that proposition (i) is true when n takes the lowest value in its range.
(v) It then follows from (iii) and (iv) that the proposition is valid for all values of n in the stated range.

It should be noted that, although many proofs at stage (iii) require the validity of the proposition only for $n = N$, some require it for all n less than or equal to N – hence the form of inequality given in parentheses in the stage (ii) assumption.

To illustrate further the method of induction, we now apply it to two worked examples; the first concerns the sum of the squares of the first n natural numbers.

Example Prove that the sum of the squares of the first n natural numbers is given by

$$\sum_{r=1}^{n} r^2 = \tfrac{1}{6} n(n+1)(2n+1). \qquad (2.40)$$

As previously, we start by assuming the result is true for $n = N$ and then consider the sum when n has been increased to $N + 1$, writing it as the sum of the first N terms plus one additional term, the square of $N + 1$. This gives us

$$\sum_{r=1}^{N+1} r^2 = \sum_{r=1}^{N} r^2 + (N+1)^2$$
$$= \tfrac{1}{6} N(N+1)(2N+1) + (N+1)^2$$
$$= \tfrac{1}{6}(N+1)[N(2N+1) + 6N + 6]$$
$$= \tfrac{1}{6}(N+1)[(2N+3)(N+2)]$$
$$= \tfrac{1}{6}(N+1)[(N+1)+1][2(N+1)+1].$$

Preliminary algebra

The original assumption (with n equal to N) was used in the second line; the remaining lines are purely algebraic manipulation, aimed at factorising the RHS of the equality. We now note that the equality represented by the final line is precisely the original assumption, but with N replaced by $N + 1$.

To complete the proof we only have to verify (2.40) directly for $n = 1$. This is trivially done and establishes the result not only for $n = 1$, but, by virtue of what has just been proved, for all positive n. The same and related results are obtained by a different method in Section 6.2.5. ◂

Our second example is somewhat more complex and involves two nested proofs by induction: whilst trying to establish the main result by induction, we find that we are faced with a second proposition which itself requires an inductive proof.

Example Show that $Q(n) = n^4 + 2n^3 + 2n^2 + n$ is divisible by 6 (without remainder) for all positive integer values of n.

Again we start by assuming the result is true for some particular value N of n, whilst noting that it is trivially true for $n = 0$. We next examine $Q(N + 1)$, writing each of its terms as a binomial expansion:

$$Q(N + 1) = (N + 1)^4 + 2(N + 1)^3 + 2(N + 1)^2 + (N + 1)$$
$$= (N^4 + 4N^3 + 6N^2 + 4N + 1) + 2(N^3 + 3N^2 + 3N + 1)$$
$$+ 2(N^2 + 2N + 1) + (N + 1)$$
$$= (N^4 + 2N^3 + 2N^2 + N) + (4N^3 + 12N^2 + 14N + 6).$$

Now, by our assumption, the group of terms within the first parentheses in the last line is divisible by 6 and, clearly, so are the terms $12N^2$ and 6 within the second parentheses. Thus it comes down to deciding whether $4N^3 + 14N$ is divisible by 6 – or equivalently, whether $R(N) = 2N^3 + 7N$ is divisible by 3.

To settle this latter question we try using a second inductive proof and assume that $R(N)$ is divisible by 3 for $N = M$, whilst again noting that the proposition is trivially true for $N = M = 0$. This time we examine $R(M + 1)$:

$$R(M + 1) = 2(M + 1)^3 + 7(M + 1)$$
$$= 2(M^3 + 3M^2 + 3M + 1) + 7(M + 1)$$
$$= (2M^3 + 7M) + 3(2M^2 + 2M + 3).$$

By assumption, the first group of terms in the last line is divisible by 3 and the second group is patently so. We thus conclude that $R(N)$ is divisible by 3 for all $N \geq M$, and taking $M = 0$ shows that it is divisible by 3 for all N.[17]

We can now return to the main proposition and conclude that since $R(N) = 2N^3 + 7N$ is divisible by 3, $4N^3 + 12N^2 + 14N + 6$ is divisible by 6. This in turn establishes that the divisibility of $Q(N + 1)$ by 6 follows from the assumption that $Q(N)$ divides by 6. Since $Q(0)$ clearly divides by 6, the proposition in the question is established for all values of n. ◂

---

[17] Come to the same conclusion in a more *ad hoc* manner by considering $2N^3 + 7N$ in the form $2(N - 1)N(N + 1) + 9N$.

2.4 Some particular methods of proof

2.4.2 Proof by contradiction

The second general line of proof, but again one that is normally only useful when the result is already suspected, is proof by contradiction. We met an elementary example of this type of proof when establishing result (1.13) about equations containing surds. The questions the method can attempt to answer are only those that can be expressed in a proposition that is either true or false. Clearly, it could be argued that any mathematical result can be so expressed but, if the proposition is no more than a guess, the chances of success are negligible. Valid propositions containing even modest formulae are either the result of true inspiration or, much more normally, yet another reworking of an old chestnut!

The essence of the method is to exploit the fact that mathematics is required to be self-consistent, so that, for example, two calculations of the same quantity, starting from the same given data but proceeding by different methods, must give the same answer. Equally, it must not be possible to follow a line of reasoning and draw a conclusion that contradicts either the input data or any other conclusion based upon the same data.

It is this requirement on which the method of proof by contradiction is based. The crux of the method is to assume that the proposition to be proved is *not* true, and then use this incorrect assumption and 'watertight' reasoning to draw a conclusion that contradicts the assumption. The only way out of the self-contradiction is then to conclude that the assumption was indeed false and therefore that the proposition is true.

It must be emphasised that once a (false) contrary assumption has been made, every subsequent conclusion in the argument *must* follow of necessity. Proof by contradiction fails if at any stage we have to admit 'this may or may not be the case'. That is, each step in the argument must be a *necessary* consequence of results that precede it (taken together with the assumption), rather than simply a *possible* consequence.

It should also be added that if no contradiction can be found using sound reasoning based on the assumption, then no conclusion can be drawn about either the proposition or its negative and some other approach must be tried.

We illustrate the general method with an example in which the mathematical reasoning is straightforward, so that attention can be focused on the structure of the proof.

Example A rational number r is a fraction $r = p/q$ in which p and q are integers with q positive. Further, r is expressed in its lowest terms, any integer common factor of p and q having been divided out.
Prove that the square root of an integer m cannot be a rational number, unless the square root itself is an integer.

We begin by supposing that the stated result is *not* true and that we *can* write an equation

$$\sqrt{m} = r = \frac{p}{q} \quad \text{for integers } m, p, q \text{ with } q \neq 1.$$

The requirement that $q \neq 1$ reflects the fact that r is not an integer. It then follows that $p^2 = mq^2$. But, since r is expressed in its lowest terms, p and q, and hence p^2 and q^2, have no factors in common. However, m is an integer; this is only possible if $q = 1$ and $p^2 = m$. This conclusion contradicts the requirement that $q \neq 1$ and so leads to the conclusion that it was wrong to suppose that $\sqrt{m}$ can be expressed as a non-integer rational number. This completes the proof of the statement in the question. ◀

Preliminary algebra

Our second worked example, also taken from elementary number theory, involves slightly more complicated mathematical reasoning but again exhibits the structure associated with this type of proof.

Example The prime integers p_i are labelled in ascending order, thus $p_1 = 1$, $p_2 = 2$, $p_5 = 7$, etc. Show that there is no largest prime number.

Assume, on the contrary, that there is a largest prime and let it be p_N. Consider now the number q formed by multiplying together all the primes from p_1 to p_N and then adding one to the product, i.e.

$$q = p_1 p_2 \cdots p_N + 1.$$

By our assumption p_N is the largest prime, and so no number can have a prime factor greater than this. However, for every prime p_i, $i = 1, 2, \ldots, N$, the quotient q/p_i has the form $M_i + (1/p_i)$ with M_i an integer and $1/p_i$ a non-integer whose magnitude is less than unity for all $i > 1$. This means that q/p_i cannot be an integer and so p_i cannot be a divisor of q.

Since q is not divisible by any of the (assumed) finite set of primes, it must be itself a prime. As q is also clearly greater than p_N, we have a contradiction. This shows that our assumption that there is a largest prime integer must be false, and so it follows that there is no largest prime integer.

It should be noted that the given construction for q does not generate all the primes that actually exist (e.g. for $N = 3$, the construction gives q as 7, rather than the next actual prime value of 5), but this does not matter for the purposes of our proof by contradiction. ◂

2.4.3 Necessary and sufficient conditions

As the final topic in this second preparatory chapter, we consider briefly the notion of, and distinction between, necessary and sufficient conditions in the context of proving a mathematical proposition. In ordinary English the distinction is well defined, and that distinction is maintained in mathematics. However, in the authors' experience, students tend to overlook it and assume (wrongly) that, having proved that the validity of proposition A implies the truth of proposition B, it follows by 'reversing the argument' that the validity of B automatically implies that of A.

As an example, let proposition A be that an integer N is divisible without remainder by 6, and proposition B be that N is divisible without remainder by 2. Clearly, if A is true then it follows that B is true, i.e. A is a sufficient condition for B. It is not however a necessary condition, as is trivially shown by taking N as 8. Conversely, the same value of N shows that whilst the validity of B is a necessary condition for A to hold, it is not sufficient.

An alternative terminology to 'necessary' and 'sufficient' often employed by mathematicians is that of 'if' and 'only if', particularly in the combination 'if and only if' which is usually written as IFF or denoted by a double-headed arrow $\Longleftrightarrow$. The equivalent statements can be summarised by

A if B A is true if B is true *or* $B \Longrightarrow A$,
            B is a sufficient condition for A $B \Longrightarrow A$,

2.4 Some particular methods of proof

A only if B	A is true only if B is true *or* B is a necessary consequence of A	$A \implies B$, $A \implies B$,
A IFF B	A is true if and only if B is true *or* A and B necessarily imply each other	$B \iff A$, $B \iff A$.

The notions of necessary and sufficient conditions extend naturally to cover chains of propositions. We do not need to develop this aspect formally, but we note that both 'is a sufficient condition for' and 'is a necessary condition for' are *transitive* relationships. A relationship $\odot$ between two entities is said to be transitive if $A \odot B$ and $B \odot C$ together imply that $A \odot C$ for all $A, B, C, \ldots$ belonging to some particular set of entities. The reader should satisfy themselves that necessary and sufficient conditions, taken separately as well as together, have this property.[18]

Although at this stage in the book we are able to employ for illustrative purposes only simple and fairly obvious results, the following example is given as a model of how necessary and sufficient conditions should be proved. The essential point is that for the second part of the proof (whether it be the 'necessary' part or the 'sufficient' part) one needs to start again from scratch; more often than not, the lines of the second part of the proof will *not* be simply those of the first written in reverse order.

Example Prove that (A) a function $f(x)$ is a quadratic polynomial with zeros at $x = 2$ and $x = 3$ if and only if (B) the function $f(x)$ has the form $\lambda(x^2 - 5x + 6)$ with λ a non-zero constant.

(1) Assume A, i.e. that $f(x)$ is a quadratic polynomial with zeros at $x = 2$ and $x = 3$. Let its form be $ax^2 + bx + c$ with $a \neq 0$. Then, on substituting the two values of x known to make $f(x)$ have value zero, we have

$$4a + 2b + c = 0,$$
$$9a + 3b + c = 0,$$

and subtraction shows that $5a + b = 0$ and $b = -5a$. Substitution of this into the first of the above equations gives $c = -4a - 2b = -4a + 10a = 6a$. Thus, it follows that

$$f(x) = a(x^2 - 5x + 6) \quad \text{with} \quad a \neq 0,$$

and establishes the 'A only if B' part of the stated result, i.e. if A is true, then so is B.

(2) Now assume that $f(x)$ has the form $\lambda(x^2 - 5x + 6)$ with λ a non-zero constant. Firstly we note that $f(x)$ is a quadratic polynomial, and so it only remains to show that it has zeros at $x = 2$ and $x = 3$. This can be done by straight substitution

$$f(2) = 2^2 - 5(2) + 6 = 0 \quad \text{and} \quad f(3) = 3^2 - 5(3) + 6 = 0.$$

Thus $f(x)$ is a quadratic polynomial and it does have zeroes at $x = 2$ and $x = 3$. This establishes the second ('A if B') part of the result, i.e if B is true, then so is A. Thus we have shown that the assumption of either condition implies the validity of the other and the proof is complete. ◀

---

[18] Which of the following relationships are transitive with respect to the set indicated: (i) 'divides into exactly without remainder' (integers), (ii) 'has no factor in common with' (integers) and (iii) 'whose magnitude differs by less than 1 from that of' (real numbers)?

Preliminary algebra

It should be noted that the propositions have to be carefully and precisely formulated. If, for example, the word 'quadratic' were omitted from A, statement B would still be a sufficient condition for A but not a necessary one, since $f(x)$ could then be $x^3 - 4x^2 + x + 6$ and A would not require B. Omitting the constant λ from the stated form of $f(x)$ in B has the same effect. Conversely, if A were to state that $f(x) = 3(x - 2)(x - 3)$ then B would be a necessary condition for A but not a sufficient one.

EXERCISES 2.4

1. Write, in the form of a summation over a dummy index r, an expression for the sum $S_{odd}(n)$ of the first n odd integers. Construct an inductive proof that shows that $S_{odd}(n) = n^2$.

2. Prove, using induction, that $8^n - 2^n$ is divisible by 6 for all n.

3. The Hungarian mathematician George Pólya put forward the following 'proof' that *all horses are the same colour* and asked students to find the error in the argument.
    (i) Assume that all horses in any set of n horses have the same colour.
    (ii) The statement is clearly true if there is only one horse.
    (iii) Now take any set of $n + 1$ horses and number them $1, 2, \ldots, n, n + 1$.
    (iv) Consider the two sets of horses consisting of numbers 1 to n and 2 to $n + 1$. Each set contains only n horses and so, by assumption (i), each set contains horses of only one colour.
    (v) But the two sets overlap and so the colour must be the same in each set.
    (vi) Thus all $n + 1$ horses have the same colour.
    In view of observation (ii) and the deduction in (vi), it follows by induction that the assumed statement (i) is true for all n, and hence all horses have the same colour.
    Where *is* the error in the argument?

4. Prove that there is no convex plane polygon with more than three acute angles. [Consider the sum of the external angles.]

5. Prove, using a formal method of contradiction argument, the obvious result that the difference between the squares of two positive integers, i.e. excluding zero, can never be unity.

6. Show that there is no rational solution to the equation $x^3 + x + 3 = 0$. Note that if $r = p/q$ with p and q having no common factor, p and q cannot both be even.

7. Between each of the pairs of statements in the two columns below, place the appropriate symbol $\Rightarrow$, $\Leftarrow$, $\Leftrightarrow$ or $\times$ to show the implications of one of the pair for the other.

$x \geq y$	$y < x$
x is an odd integer ≥ 3	x is a prime integer
P is a son of Q	Q is a parent of P
Having a married sister with children	Being an uncle
Some academics write books	Some authors are academics
$y > x$	$x \leq y$
P being a necessary condition for Q	Q being a sufficient condition for P

8. Prove that a necessary and sufficient condition for x to be equal to the difference between the squares of two natural (positive) integers is that x is equal to the product of two integers whose difference is an even integer.

SUMMARY

1. *Quadratic equations*
$$ax^2 + bx + c = 0, \qquad ax^2 + 2bx + c = 0,$$
have respective solutions
$$\alpha_{1,2} = \frac{-b \pm \sqrt{b^2 - 4ac}}{2a}, \qquad \alpha_{1,2} = \frac{-b \pm \sqrt{b^2 - ac}}{a}.$$

2. *Polynomial equations with real coefficients*
   - An nth-degree polynomial has exactly n zeros, but they are not necessarily real, nor necessarily distinct.
   - An nth-degree polynomial has an odd or even number of real zeros according to whether n is odd or even, respectively.
   - For the nth-degree polynomial equation
$$a_n x^n + a_{n-1} x^{n-1} + \cdots + a_1 x + a_0 = 0, \qquad a_n \neq 0$$
with roots $\alpha_1, \alpha_2, \ldots, \alpha_n$,
$$\prod_{k=1}^{n} \alpha_k = (-1)^n \frac{a_0}{a_n}, \qquad \sum_{k=1}^{n} \alpha_k = -\frac{a_{n-1}}{a_n}.$$

3. *Coordinate geometry*
   - Straight line: $y = mx + c$ or $ax + by + k = 0$.
   - The condition for two straight lines to be orthogonal is $m_1 m_2 = -1$.
   - Conic sections 'centred' on (α, β) and their parameterisations:

Conic	Equation	x	y
circle	$(x - \alpha)^2 + (y - \beta)^2 = a^2$	$\alpha + a \cos\phi$	$\beta + a \sin\phi$
ellipse	$\dfrac{(x - \alpha)^2}{a^2} + \dfrac{(y - \beta)^2}{b^2} = 1$	$\alpha + a \cos\phi$	$\beta + b \sin\phi$
parabola	$(y - \beta)^2 = 4a(x - \alpha)$	$\alpha + at^2$	$\beta + 2at$
hyperbola	$\dfrac{(x - \alpha)^2}{a^2} - \dfrac{(y - \beta)^2}{b^2} = 1$	$\alpha + a \cosh\phi$	$\beta + b \sinh\phi$

4. *Plane polar coordinates*

$$x = \rho \cos\phi, \qquad y = \rho \sin\phi,$$

$$\rho = \sqrt{x^2 + y^2}, \qquad \cos\phi = \frac{x}{\sqrt{x^2 + y^2}}, \qquad \sin\phi = \frac{y}{\sqrt{x^2 + y^2}}.$$

5. *Partial fractions expansion*
   For the representation of $f(x) = g(x)/h(x)$, with $g(x)$ a polynomial and $h(x) = (x - \alpha_1)(x - \alpha_2) \cdots (x - \alpha_n)$:
   - With the α_i all different,

   $$f(x) = \sum_{k=1}^{n} \frac{A_k}{x - \alpha_k}, \quad \text{where } A_k = \frac{g(\alpha_k)}{\prod_{j \neq k}^{n}(\alpha_k - \alpha_j)}.$$

   - If the degree of $g(x)$ is $\geq m$, the degree of $h(x)$, then $f(x)$ must first be written as

   $$f(x) = s(x) + \frac{r(x)}{h(x)}, \quad \text{where } \begin{cases} s(x) \text{ is a polynomial,} \\ \text{the degree of } r(x) \text{ is } < m. \end{cases}$$

   - If $h(x)$ contains a factor $a^2 + x^2$, then the corresponding term in the expansion takes the form $(Ax + b)/(a^2 + x^2)$.
   - If $h(x) = 0$ has a repeated root α, i.e. $h(x)$ contains a factor $(x - \alpha)^p$, then the expansion must contain

   $$\text{either} \quad \frac{A_0 + A_1 x + \cdots + A_{p-1} x^{p-1}}{(x - \alpha)^p}$$

   $$\text{or} \quad \frac{B_1}{x - \alpha} + \frac{B_2}{(x - \alpha)^2} + \cdots + \frac{B_p}{(x - \alpha)^p}.$$

6. *Proof by induction* (on n)
   (i) Assume the proposition is true for $n = N$ (or for all $n \leq N$).
   (ii) Use (i) to prove the proposition is then true for $n = N + 1$ (or for all $n \leq N + 1$).
   (iii) Show by observation, or by direct calculation without assumptions, that the proposition is true for the lowest n in its range.
   (iv) Conclude that the proposition is true for all n in its range.

7. *Proof by contradiction*
   (i) Assume the proposition is *not* true.
   (ii) Show, using *only* conclusions that *necessarily* follow from their predecessors and the assumption, that this leads to a contradiction.
   (iii) Conclude that the proposition is true.
   Warning: Failure to find a contradiction gives no information as to whether or not the proposition is true.

Problems

8. *Necessary and sufficient conditions*
- A if B   B is a sufficient condition for A $B \Rightarrow A$
- A only if B   B is a necessary consequence of A $A \Rightarrow B$
- A IFF B   A and B necessarily imply each other $A \Leftrightarrow B$
- *Warning*: Necessary and sufficient condition proofs nearly always require two separate chains of argument. The second part of the proof is usually *not* the lines of the first part written in reverse order.

PROBLEMS

Polynomial equations

2.1. Continue the investigation of Equation (2.8), namely

$$g(x) = 4x^3 + 3x^2 - 6x - 1 = 0,$$

as follows.
(a) Make a table of values of $g(x)$ for integer values of x between -2 and 2. Use it and the information derived in the text to draw a graph and so determine the roots of $g(x) = 0$ as accurately as possible.
(b) Find one accurate root of $g(x) = 0$ by inspection and hence determine precise values for the other two roots.
(c) Show that $f(x) = 4x^3 + 3x^2 - 6x + k = 0$ has only one real root unless $-5 \leq k \leq \frac{7}{4}$.

2.2. Determine how the number of real roots of the equation

$$g(x) = 4x^3 - 17x^2 + 10x + k = 0$$

depends upon k. Are there any cases for which the equation has exactly two distinct real roots?

2.3. Continue the analysis of the polynomial equation

$$f(x) = x^7 + 5x^6 + x^4 - x^3 + x^2 - 2 = 0,$$

investigated in Section 2.1.2, as follows.
(a) By writing the fifth-degree polynomial appearing in the expression for $f'(x)$ in the form $7x^5 + 30x^4 + a(x-b)^2 + c$, show that there is in fact only one positive root of $f(x) = 0$.
(b) By evaluating $f(1)$, $f(0)$ and $f(-1)$, and by inspecting the form of $f(x)$ for negative values of x, determine what you can about the positions of the real roots of $f(x) = 0$.

Preliminary algebra

2.4. Given that $x = 2$ is one root of
$$g(x) = 2x^4 + 4x^3 - 9x^2 - 11x - 6 = 0,$$
use factorisation to determine how many real roots it has.

2.5. Construct the quadratic equations that have the following pairs of roots:
(a) $-6, -3$; (b) $0, 4$; (c) $2, 2$; (d) $3 + 2i, 3 - 2i$, where $i^2 = -1$.

2.6. If α and β are the roots of the equation $x^2 - 6x + 2 = 0$, evaluate $g(\alpha)$ and $g(\beta)$, where
$$g(x) = \frac{x^2 + 2x - 8}{x - 3},$$
giving your answer in its simplest form in terms of integers and surds.

2.7. Use the results of (i) Equation (2.14), (ii) Equation (2.13) and (iii) Equation (2.15) to prove that if the roots of $3x^3 - x^2 - 10x + 8 = 0$ are α_1, α_2 and α_3 then
(a) $\alpha_1^{-1} + \alpha_2^{-1} + \alpha_3^{-1} = 5/4$,
(b) $\alpha_1^2 + \alpha_2^2 + \alpha_3^2 = 61/9$,
(c) $\alpha_1^3 + \alpha_2^3 + \alpha_3^3 = -125/27$.
(d) Convince yourself that eliminating (say) α_2 and α_3 from (i), (ii) and (iii) does *not* give a simple explicit way of finding α_1.

2.8. Determine the shapes, i.e. the height-to-width ratios, of A4 and foolscap folio writing papers, given the following information. (i) When a sheet of A4 paper in portrait orientation is folded in two it becomes an A5 sheet in landscape orientation; the A series of writing papers all have the same shape. (ii) If a foolscap folio sheet is cut once across its width so as to produce a square, what is left has the same shape as the original.

2.9. The product of two numbers, α and β, is equal to λ times their sum, and their ratio is equal to μ times their sum. Find explicit expressions for α and β in terms of λ and μ.

Coordinate geometry

2.10. Obtain in the form (2.20) the equations that describe the following:
(a) a circle of radius 5 with its centre at $(1, -1)$;
(b) the line $2x + 3y + 4 = 0$ and the line orthogonal to it that passes through $(1, 1)$;
(c) an ellipse of eccentricity 0.6 with centre $(1, 1)$ and its major axis of length 10 parallel to the y-axis.

2.11. Determine the forms of the conic sections described by the following equations:
(a) $x^2 + y^2 + 6x + 8y = 0$;
(b) $9x^2 - 4y^2 - 54x - 16y + 29 = 0$;

Problems

(c) $2x^2 + 2y^2 + 5xy - 4x + y - 6 = 0$;
(d) $x^2 + y^2 + 2xy - 8x + 8y = 0$.

2.12. Find the equation of the circle that passes through the three points $(5, -8)$, $(6, -1)$ and $(2, 1)$.

2.13. A paraboloid of revolution whose focus is a distance a from its 'nose' rests symmetrically on the inside of a vertical cone $\rho = bz$, with their axes coincident. Find the distance between the nose of the paraboloid and the vertex of the cone.

2.14. For the general conic section, as given in Equation (2.20), namely
$$Ax^2 + By^2 + Cxy + Dx + Ey + F = 0,$$
investigate the possibility of straight-line asymptotes as follows. Try a solution of the form $y = mx + k$, with m and k both real and finite, as an approximate solution when $|x|$ and $|y|$ both tend to ∞. Show that requiring the coefficient of the terms in x^2 to vanish implies that $C^2 \geq 4AB$ and that m must take one of two particular values. Now, from consideration of the linear term in x, find an expression for k and conclude that a strict inequality must hold, i.e. $C^2 > 4AB$; deduce that, of the three non-degenerate conic sections, only the hyperbola has real asymptotes.

Use your results to find the asymptotes of the conic section whose equation is
$$4x^2 + y^2 - 5xy + 2x - 3y - 4 = 0.$$
Deduce the coordinates of the 'centre' of the conic and, using a rough sketch, determine in which two of the four sectors defined by the asymptotes the conic lies.

2.15. The foci of the ellipse
$$\frac{x^2}{a^2} + \frac{y^2}{b^2} = 1$$
with eccentricity e are the two points $(-ae, 0)$ and $(ae, 0)$. Show that the sum of the distances from *any* point on the ellipse to the foci is $2a$. [The constancy of the sum of the distances from two fixed points can be used as an alternative defining property of an ellipse.]

2.16. The process of obtaining the standard form for a parabola from that for an ellipse consists of two major elements: (i) allowing the major axis to become infinitely long, i.e. $a \to \infty$; (ii) moving the origin of coordinates so that it is coincident with 'the left-hand end' P of the ellipse, rather than with its centre.
   (a) Write down a new equation for the standard ellipse in terms of a, e, y and X only; here X is a new x-coordinate defined so that the point P is at $X = 0$.
   (b) Note that the distance between the point P and the focus F nearest to it is $a(1 - e)$; denote this distance by A.

(c) Arrange the equation, expressed in terms of X, y, a, A and e, so that it contains a term X^2/a.
(d) Finally, let $a \to \infty$ and $e \to 1$, to obtain $y^2 = 4AX$ for the equation of a parabola that passes through the origin and has the point $(A, 0)$ as its focus.

2.17. Describe and sketch the following parametrically defined curves.
(a) $x = t$, $y = t^{-1}$,   (b) $x = \cos t$, $y = \sin t$, $z = t$,
(c) $x = t^3 - 3t$, $y = t^2 - 1$,   (d) $x = a(t - \sin t)$, $y = a(1 - \cos t)$,
(e) $x = a\cos^3 t$, $y = a\sin^3 t$,   (f) $x = 4\cos 3t$, $y = 3\cos 2t$.

2.18. Show that the three-dimensional cubic curve that is parameterised as

$$x = a + b\lambda, \qquad y = a\lambda + b\lambda^2, \qquad z = -\lambda^3,$$

where λ is real, lies in the surface $y^3 + azx^2 + bxyz = 0$. Would parameterisations

(i) $x = a - b\lambda$, $y = -a\lambda + b\lambda^2$, $z = \lambda^3$,
(ii) $x = -a - b\lambda$, $y = -a\lambda - b\lambda^2$, $z = \lambda^3$

do just as well?

2.19. Show that the locus of points in three-dimensional Cartesian space given by the parameterisation

$$x = au(3 - u^2), \qquad y = 3au^2, \qquad z = au(3 + u^2),$$

lies on the intersection of the surfaces $y^3 + 27axz - 81a^2y = 0$ and $y = \lambda(z - x)/(z + x)$, where λ is a constant you should determine.

2.20. A particular curve in the x–y plane, which has origin O, is known as the *cissoid of Diocles* and is generated as follows.
(a) Draw a circle of unit diameter centred on $(\frac{1}{2}, 0)$ and a chord OP through a point P of the circle. Let the extended chord cut the line $x = 1$ at Q.
(b) On the chord OP mark off OR equal in length to PQ.
(c) As the point P traverses the circle the point R traces out the cissoid.
Find a parameterisation of the cissoid (i) in geometric terms using the angle ϕ that the chord makes with the x-axis and (ii) in algebraic terms using t, the y-coordinate of Q. Show that the equation of the cissoid is

$$x(x^2 + y^2) = y^2.$$

2.21. Identify the following curves, each given in plane polar coordinates.

(a) $\rho = 2a \sin \phi$,   (b) $\rho = a + b\phi$,   (c) $\rho \sin(\phi - \alpha) = p$,

where all symbols other than ρ and ϕ signify constants.

2.22. Sketch the following curves, each given in plane polar coordinates. Where it is relevant, use the convention that allows negative values for ρ.

Problems

(a) Lemniscate of Bernoulli: $\rho^2 = a^2 \cos 2\phi$, where $\cos 2\phi \geq 0$ and $\rho = 0$ otherwise,
(b) 'flower': $\rho = a \sin 3\phi$,
(c) 'flower': $\rho = a|\sin 3\phi|$,
(d) cardioid: $\rho = a(1 - \sin \phi)$,
(e) limaçon: $\rho = a(\frac{1}{2} - \sin \phi)$.

2.23. Show that the equation of a standard ellipse with major axis $2a$ and eccentricity e can be expressed in the form

$$\rho = a \left(\frac{1 - e^2}{1 - e^2 \cos^2 \phi} \right)^{1/2},$$

using plane polar coordinates with their origin at the centre of the ellipse. [*Note*: The usual plane polar description of an ellipse is $\rho = \ell(1 + e \cos \phi)^{-1}$, but this is referred to a coordinate system centred on a focus of the ellipse.]

Partial fractions

2.24. Express the following as the ratio of two polynomials, with the denominator in factored form:

(a) $\dfrac{2}{x+3} - \dfrac{x+4}{x-2}$,   (b) $\dfrac{1}{x+2} + \dfrac{2}{(x+2)^2} + \dfrac{3}{(x+2)^3}$,

(c) $x + 3 + \dfrac{2}{(x-2)^2} - \dfrac{4}{x+1}$,   (d) $\dfrac{A}{x^2 - x - 2} + \dfrac{B}{x^2 + x - 6} + \dfrac{C}{x^2 + 2x + 1}$.

2.25. Resolve the following into partial fractions using the three methods given in Section 2.3, verifying that the same decomposition is obtained by each method:

(a) $\dfrac{2x+1}{x^2 + 3x - 10}$,   (b) $\dfrac{4}{x^2 - 3x}$.

2.26. Express the following in partial fraction form:

(a) $\dfrac{2x^3 - 5x + 1}{x^2 - 2x - 8}$,   (b) $\dfrac{x^2 + x - 1}{x^2 + x - 2}$.

2.27. Rearrange the following functions in partial fraction form:

(a) $\dfrac{x - 6}{x^3 - x^2 + 4x - 4}$,   (b) $\dfrac{x^3 + 3x^2 + x + 19}{x^4 + 10x^2 + 9}$.

2.28. Resolve the following into partial fractions in such a way that x does not appear in any numerator:

(a) $\dfrac{2x^2 + x + 1}{(x-1)^2(x+3)}$,   (b) $\dfrac{x^2 - 2}{x^3 + 8x^2 + 16x}$,   (c) $\dfrac{x^3 - x - 1}{(x+3)^3(x+1)}$.

Preliminary algebra

Proof by induction and contradiction

2.29. Prove by induction that

$$\sum_{r=1}^{n} r = \frac{1}{2}n(n+1) \quad \text{and} \quad \sum_{r=1}^{n} r^3 = \frac{1}{4}n^2(n+1)^2.$$

2.30. Prove by induction that

$$1 + r + r^2 + \cdots + r^k + \cdots + r^n = \frac{1 - r^{n+1}}{1 - r}.$$

2.31. Prove that $3^{2n} + 7$, where n is a non-negative integer, is divisible by 8.

2.32. If a sequence of terms, u_n, satisfies the recurrence relation $u_{n+1} = (1-x)u_n + nx$, with $u_1 = 0$, show, by induction, that, for $n \geq 1$,

$$u_n = \frac{1}{x}[nx - 1 + (1-x)^n].$$

2.33. Establish the values of k for which the binomial coefficient ^pC_k is divisible by p when p is a prime number. Use your result and the method of induction to prove that $n^p - n$ is divisible by p for all integers n and all prime numbers p. Deduce that $n^5 - n$ is divisible by 30 for any integer n.

2.34. An arithmetic progression of integers a_n is one in which $a_n = a_0 + nd$, where a_0 and d are integers and n takes successive values $0, 1, 2, \ldots$.
   (a) Show that if any one term of the progression is the cube of an integer then so are infinitely many others.
   (b) Show that no cube of an integer can be expressed as $7n + 5$ for some positive integer n.

2.35. Prove, by the method of contradiction, that the equation

$$x^n + a_{n-1}x^{n-1} + \cdots + a_1 x + a_0 = 0,$$

in which all the coefficients a_i are integers, cannot have a rational root, unless that root is an integer. Deduce that any integral root must be a divisor of a_0 and hence find all rational roots of
   (a) $x^4 + 6x^3 + 4x^2 + 5x + 4 = 0$,
   (b) $x^4 + 5x^3 + 2x^2 - 10x + 6 = 0$.

Necessary and sufficient conditions

2.36. Prove that the equation $ax^2 + bx + c = 0$, in which a, b and c are real and $a > 0$, has two real distinct solutions IFF $b^2 > 4ac$.

2.37. For the real variable x, show that a sufficient, but not necessary, condition for $f(x) = x(x+1)(2x+1)$ to be divisible by 6 is that x is an integer.

2.38. Given that at least one of a and b and that at least one of c and d are non-zero, show that $ad = bc$ is both a necessary and sufficient condition for the equations

$$ax + by = 0,$$
$$cx + dy = 0$$

to have a solution in which at least one of x and y is non-zero.

2.39. The coefficients a_i in the polynomial $Q(x) = a_4x^4 + a_3x^3 + a_2x^2 + a_1x$ are all integers. Show that $Q(n)$ is divisible by 24 for all integers $n \geq 0$ if and only if all of the following conditions are satisfied:
  (i) $2a_4 + a_3$ is divisible by 4;
  (ii) $a_4 + a_2$ is divisible by 12;
  (iii) $a_4 + a_3 + a_2 + a_1$ is divisible by 24.

HINTS AND ANSWERS

2.1. (b) The roots are 1, $\frac{1}{8}(-7 + \sqrt{33}) = -0.1569$, $\frac{1}{8}(-7 - \sqrt{33}) = -1.593$. (c) -5 and $\frac{7}{4}$ are the values of k that make $f(-1)$ and $f(\frac{1}{2})$ equal to zero.

2.3. (a) $a = 4$, $b = \frac{3}{8}$ and $c = \frac{23}{16}$ are all positive. Therefore $f'(x) > 0$ for all $x > 0$. (b) $f(1) = 5$, $f(0) = -2$ and $f(-1) = 5$, and so there is at least one root in each of the ranges $0 < x < 1$ and $-1 < x < 0$. $(x^7 + 5x^6) + (x^4 - x^3) + (x^2 - 2)$ is positive definite for $-5 < x < -\sqrt{2}$. There are therefore no roots in this range, but there must be one to the left of $x = -5$.

2.5. (a) $x^2 + 9x + 18 = 0$; (b) $x^2 - 4x = 0$; (c) $x^2 - 4x + 4 = 0$; (d) $x^2 - 6x + 13 = 0$.

2.7. (a) Write as $\dfrac{\alpha_2\alpha_3 + \alpha_1\alpha_3 + \alpha_2\alpha_1}{\alpha_1\alpha_2\alpha_3}$.
  (b) Write as $(\alpha_1 + \alpha_2 + \alpha_3)^2 - 2(\alpha_1\alpha_2 + \alpha_2\alpha_3 + \alpha_3\alpha_1)$.
  (c) Write as $(\alpha_1 + \alpha_2 + \alpha_3)^3 - 3(\alpha_1 + \alpha_2 + \alpha_3)(\alpha_1\alpha_2 + \alpha_2\alpha_3 + \alpha_3\alpha_1) + 3\alpha_1\alpha_2\alpha_3$.
  (d) No answer is available as it cannot be done. All manipulation is complicated and, at best, leads back to the original equation. Unfortunately, this is a 'proof by frustration', rather than one by contradiction.

2.9. $\alpha = \lambda/[1 - (\lambda\mu)^{1/2}]$, $\beta = \sqrt{\lambda/\mu}$.

2.11. (a) A circle of radius 5 centred on $(-3, -4)$.
  (b) A hyperbola with 'centre' $(3, -2)$ and 'semi-axes' 2 and 3.
  (c) The expression factorises into two lines, $x + 2y - 3 = 0$ and $2x + y + 2 = 0$.
  (d) Write the expression as $(x + y)^2 = 8(x - y)$ to see that it represents a parabola passing through the origin, with the line $x + y = 0$ as its axis of symmetry.

2.13. If the 'nose' is at $z = z_0$, then, on the ring of contact, $b^2z^2 = \rho^2 = 4a(z - z_0)$ must have a double root, leading to $z_0 = a/b^2$.

Preliminary algebra

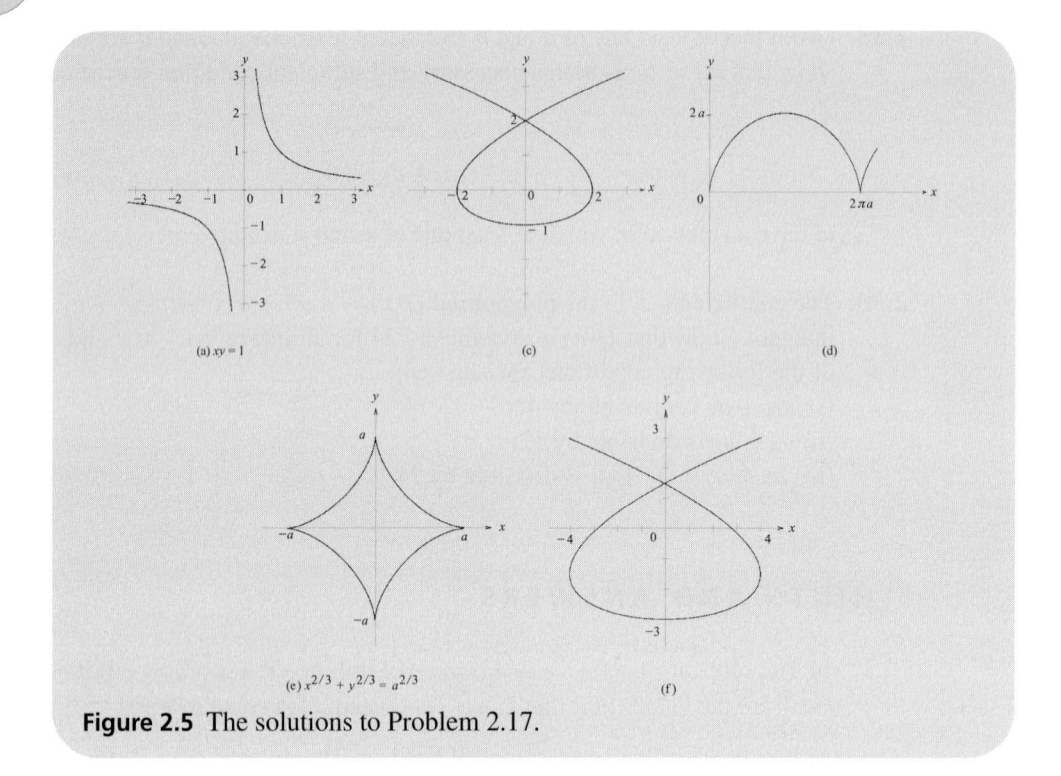

Figure 2.5 The solutions to Problem 2.17.

(a) $xy = 1$
(c)
(d)
(e) $x^{2/3} + y^{2/3} = a^{2/3}$
(f)

2.15. Show that the sum is given by $s = [(x+ae)^2 + y^2]^{1/2} + [(x-ae)^2 + y^2]^{1/2}$ with $y^2 = (1-e^2)(a^2 - x^2)$. This leads to $s = 2a$, whatever the value of x.

2.17. For the two-dimensional curves, see Figure 2.5.
(a) Rectangular hyperbola, $xy = 1$.
(b) Spiral on a cylindrical surface of unit radius with its axis along the z-axis. The spiral has pitch 2π.
(c) Crosses the x-axis at $x = \pm 2$; crosses the y-axis at $y = -1$ and $y = 2$ (twice). Asymptotically $y = x^{2/3}$.
(d) Cycloid of 'amplitude' $2a$ and 'period' $2\pi a$. At a cusp the tangent to the curve is vertical.
(e) $(x/a)^{2/3} + (y/a)^{2/3} = 1$, giving the astroid $x^{2/3} + y^{2/3} = a^{2/3}$.
(f) Closed curve limited by $x = \pm 4$, $y = \pm 3$, which reverses and then retraces its initial path after $t = \pi$.

2.19. Substitution for x, y and z in the equation for the first surface produces an identity, as it does in the equation of the second surface, for all u, provided $\lambda = 9a$. The parameterised curve therefore lies on the intersection.

2.21. (a) A circle of radius a centred on $(0, a)$. (b) An equiangular spiral that starts at $(a, 0)$ and whose radius increases uniformly by $2\pi b$ for each turn of the spiral.
(c) A straight line parallel to the direction $\phi = \alpha$ and whose perpendicular distance from the origin is p.

Hints and answers

2.23. Recall that $b^2 = a^2(1 - e^2)$ and write $y^2 = \rho^2(1 - \cos^2\phi)$.

2.25. (a) $\dfrac{5}{7(x-2)} + \dfrac{9}{7(x+5)}$, (b) $-\dfrac{4}{3x} + \dfrac{4}{3(x-3)}$.

2.27. (a) $\dfrac{x+2}{x^2+4} - \dfrac{1}{x-1}$, (b) $\dfrac{x+1}{x^2+9} + \dfrac{2}{x^2+1}$.

2.29. Look for factors common to the $n = N$ sum and the additional $n = N+1$ term, so as to reduce the sum for $n = N+1$ to a single term.

2.31. Write 3^{2n} as $8m - 7$.

2.33. Divisible for $k = 1, 2, \ldots, p-1$. Expand $(n+1)^p$ as $n^p + \sum_1^{p-1} {}^pC_k n^k + 1$. Apply the stated result for $p = 5$. Note that $n^5 - n = n(n-1)(n+1)(n^2+1)$; the product of any three consecutive integers must divide by both 2 and 3.

2.35. By assuming $x = p/q$ with $q \neq 1$, show that a fraction $-p^n/q$ is equal to an integer $a_{n-1}p^{n-1} + \cdots + a_1 pq^{n-2} + a_0 q^{n-1}$. This is a contradiction, and is only resolved if $q = 1$ and the root is an integer.
(a) The only possible candidates are $\pm 1, \pm 2, \pm 4$. None of these is a root.
(b) The only possible candidates are $\pm 1, \pm 2, \pm 3, \pm 6$. Only -3 is a root.

2.37. $f(x)$ can be written as $x(x+1)(x+2) + x(x+1)(x-1)$. Each term consists of the product of three consecutive integers, of which one must therefore divide by 2 and (a different) one by 3. Thus each term separately divides by 6, and so therefore does $f(x)$. Note that if x is the root of $2x^3 + 3x^2 + x - 24 = 0$ that lies near the non-integer value $x = 1.826$, then $x(x+1)(2x+1) = 24$ and therefore divides by 6.

2.39. Note that, for example, the condition for $6a_4 + a_3$ to be divisible by 4 is the same as the condition for $2a_4 + a_3$ to be divisible by 4.
For the necessary (only if) part of the proof set $n = 1, 2, 3$ and take integer combinations of the resulting equations.
For the sufficient (if) part of the proof use the stated conditions to prove the proposition by induction. Note that $n^3 - n$ is divisible by 6 and that $n^2 + n$ is even.

3 Differential calculus

This and the next chapter are concerned with the formalism of probably the most widely used mathematical technique in the physical sciences, namely the calculus. The current chapter deals with the process of differentiation whilst Chapter 4 is concerned with its inverse process, integration. The topics covered are essential for the remainder of the book; once studied, the contents of the two chapters serve as reference material, should that be needed. Readers who have had previous experience of differentiation and integration should ensure full familiarity by looking at the worked examples in the main text and by attempting the problems at the ends of the two chapters.

Also included in this chapter is a section on curve sketching. Most of the mathematics needed as background to this important skill for applied physical scientists was covered in the first two chapters, but delaying our main discussion of it until the end of this chapter allows the location and characterisation of turning points to be included amongst the techniques available.

3.1 Differentiation

Differentiation is the process of determining how quickly or slowly a function varies, as the quantity on which it depends, its *argument*, is changed. More specifically, it is the procedure for obtaining an expression (numerical or algebraic) for the rate of change of the function with respect to its argument. Familiar examples of rates of change include acceleration (the rate of change of velocity) and chemical reaction rate (the rate of change of chemical composition). Both acceleration and reaction rate give a measure of the change of a quantity with respect to time. However, differentiation may also be applied to changes with respect to other quantities, for example the change in pressure with respect to a change in temperature.

Although it will not be apparent from what we have said so far, differentiation is in fact a limiting process; that is, it deals only with the infinitesimal change in one quantity resulting from an infinitesimal change in another.

3.1.1 Differentiation from first principles

Let us consider a function $f(x)$ that depends on only one variable, x, together with numerical constants, e.g. $f(x) = 3x^2$ or $f(x) = \sin x$ or $f(x) = 2 + 3/x$. Figure 3.1 shows the graph of such a function. Near any particular point, P, the value of the function changes by an amount Δf, say, as x changes by a small amount Δx. The slope of the

3.1 Differentiation

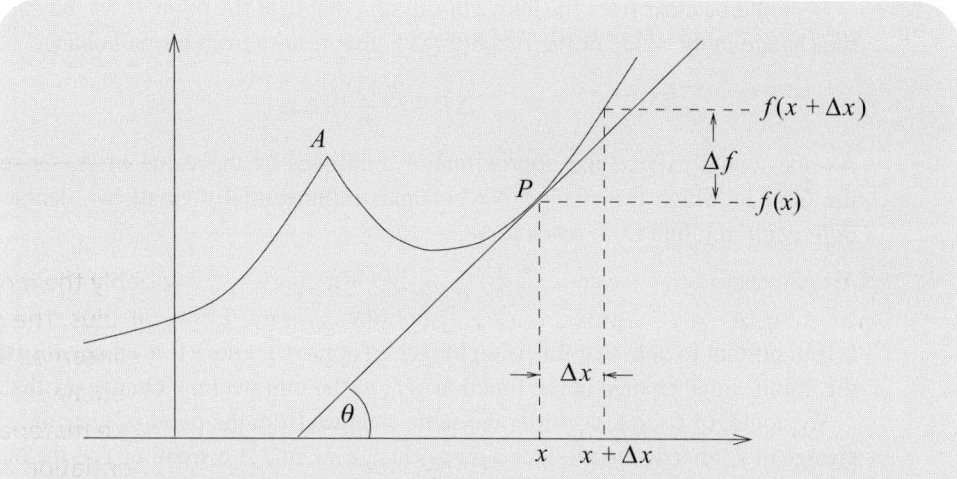

Figure 3.1 The graph of a function $f(x)$ showing that the gradient or slope of the function at P, given by $\tan\theta$, is approximately equal to $\Delta f/\Delta x$.

tangent to the graph[1] of $f(x)$ at P is then approximately $\Delta f/\Delta x$, and the change in the value of the function is $\Delta f = f(x + \Delta x) - f(x)$. In order to calculate the true value of the gradient, or *first derivative*, of the function at P, we must let Δx become infinitesimally small. We therefore define the first derivative of $f(x)$ as

$$f'(x) \equiv \frac{df(x)}{dx} \equiv \lim_{\Delta x \to 0} \frac{f(x + \Delta x) - f(x)}{\Delta x}, \tag{3.1}$$

provided that the limit exists.

The value of the limit, and hence of $f'(x)$, will depend in almost all cases on the value of x. However, because it sometimes causes initial confusion, it should be emphasised that, even though the symbol Δx appears three times on the RHS of the definition, $f'(x)$ does not depend on the value of any 'Δx'; as a result of the limiting process Δx has disappeared from the RHS – or can be thought of as having been reduced to the standard value of zero. If the limit in (3.1) does exist at the point $x = a$, then the function is said to be differentiable at a; otherwise it is said to be non-differentiable at a. The formal concept of a limit and its existence or non-existence is discussed in Chapter 6; for our present purposes we will adopt an intuitive approach.

In definition (3.1), we require that the same limit is obtained, whether Δx tends to zero through positive values or through negative values. A function that is differentiable at a is necessarily continuous at a (there must be no jump in the value of the function at a), though the converse is not necessarily true. This latter assertion is illustrated in Figure 3.1: the function is continuous at the 'kink' A, but the two limits of the gradients as Δx tends to zero through positive and negative values are different. Consequently, the function is not differentiable at A.

---

[1] The distinction between the tangent to a graph and the tangent of an angle should be noted. See also the remark on p. 64 about the relationship between the slope of a straight line on a graph and the tangent of the angle it makes with the x-axis.

Differential calculus

It should be clear from the above discussion that near the point P we may approximate the change in the value of the function, Δf, that results from a small change Δx in x by

$$\Delta f \approx \frac{df(x)}{dx} \Delta x. \tag{3.2}$$

As one would expect, the approximation improves as the value of Δx is reduced. In the limit in which the change Δx becomes infinitesimally small, we denote it by the *differential* dx, and (3.2) reads

$$df = \frac{df(x)}{dx} dx. \tag{3.3}$$

It is important to note that this is no longer an approximation. It is an *equality* that relates the infinitesimal change in the function, df, to the infinitesimal change dx that causes it.

We could, of course, consider the same changes from the point of view of asking what change in x is needed to produce a given change df in f, i.e. treating f as the independent variable and x as the dependent one. We would then write the equation corresponding to (3.3) as

$$dx = \frac{dx}{df} df.$$

But (3.3) itself can be rearranged to read

$$dx = df \left(\frac{df}{dx}\right)^{-1}.$$

Comparing these last two equations shows the general result

$$\frac{dx}{df} = \left(\frac{df}{dx}\right)^{-1}, \tag{3.4}$$

i.e. that the derivative of x with respect to f and that of f with respect to x are reciprocals.[2]

So far we have discussed only the first derivative of a function. However, we can also define the *second derivative* as the gradient of the gradient of a function. Again we use the definition (3.1), but now with $f(x)$ replaced by $f'(x)$. Hence the second derivative is defined by

$$f''(x) \equiv \lim_{\Delta x \to 0} \frac{f'(x + \Delta x) - f'(x)}{\Delta x}, \tag{3.5}$$

provided that the limit exists. A physical example of a second derivative is the second derivative of the distance travelled by a particle with respect to time. Since the first derivative of distance travelled gives the particle's speed, the second derivative gives its acceleration.

We can continue in this manner, the nth derivative of the function $f(x)$ being defined by

$$f^{(n)}(x) \equiv \lim_{\Delta x \to 0} \frac{f^{(n-1)}(x + \Delta x) - f^{(n-1)}(x)}{\Delta x}. \tag{3.6}$$

---

[2] Accepting the statement on p. 57 that the derivative of $f = Ax^n$ with respect to x is $df/dx = nAx^{n-1}$, verify (3.4) when $f^2 = 4x^3$.

3.1 Differentiation

It should be noted that with this notation $f'(x) \equiv f^{(1)}(x)$, $f''(x) \equiv f^{(2)}(x)$, etc., and that formally $f^{(0)}(x) \equiv f(x)$.

All of this should be familiar to the reader, though perhaps not with such formal definitions. The following example shows the differentiation of $f(x) = Ax^n$ from first principles; the expression for the derivative has already been used twice, but the following proof does not rely on any result derived from that usage. In practical applications, however, first-principle derivations are cumbersome and it is desirable simply to remember the derivatives of standard simple functions; the techniques given in the remainder of this section can then be applied to find more complicated derivatives.

Example Find from first principles the derivative with respect to x of $f(x) = Ax^n$.

In order to use definition (3.1) we will need to examine the behaviour of $A(x + \Delta x)^n - Ax^n$ for small values of Δx. In particular, we must determine the leading non-vanishing term in $A(x + \Delta x)^n - Ax^n$ when it is expressed in powers of Δx. Just such an expression is provided by the binomial expansion, Equation (1.41), which states that

$$(x + y)^n = \sum_{k=0}^{k=n} {}^nC_k x^{n-k} y^k,$$

or, in this case, replacing y by Δx,

$$(x + \Delta x)^n = \sum_{k=0}^{k=n} {}^nC_k x^{n-k} (\Delta x)^k = x^n + {}^nC_1 x^{n-1} \Delta x + {}^nC_2 x^{n-2} (\Delta x)^2 + \cdots$$

The terms in the expansion that have not been written explicitly all contain third or higher powers of Δx. Now, as recorded in Section 1.4.1, ${}^nC_1 = n$ for any n. So, for any n, the calculation of the derivative of $f(x) = Ax^n$ proceeds as follows

$$f'(x) = \lim_{\Delta x \to 0} \frac{f(x + \Delta x) - f(x)}{\Delta x}$$

$$= \lim_{\Delta x \to 0} \frac{A(x + \Delta x)^n - Ax^n}{\Delta x}$$

$$= \lim_{\Delta x \to 0} \frac{Ax^n + Anx^{n-1}\Delta x + A({}^nC_2)x^{n-2}(\Delta x)^2 + \cdots - Ax^n}{\Delta x}$$

$$= \lim_{\Delta x \to 0} \frac{Anx^{n-1}\Delta x + A({}^nC_2)x^{n-2}(\Delta x)^2 + \cdots}{\Delta x}$$

$$= \lim_{\Delta x \to 0} \left[Anx^{n-1} + A({}^nC_2)x^{n-2}\Delta x + \cdots \right].$$

As Δx tends to zero, $Anx^{n-1} + A({}^nC_2)x^{n-2}\Delta x + \cdots$ tends towards Anx^{n-1}. Hence, we have

$$f'(x) = nAx^{n-1}$$

as the first derivative of $f(x) = Ax^n$. ◂

Though it is not required for the above example, we can deduce that the second derivative of $f(x) = Ax^n$ is $f''(x) = n(n-1)Ax^{n-2}$ and, continuing in this way, that its nth derivative

Differential calculus

is $f^{(n)}(x) = n(n-1)(n-2) \cdots (2)(1) A x^{n-n} = n!A$; all higher derivatives than the nth are zero.

Derivatives of other functions can be obtained in the same way. The derivatives of some simple functions are listed below (note that a is a constant):[3]

$$\frac{d}{dx}(x^n) = nx^{n-1}, \quad \frac{d}{dx}(e^{ax}) = ae^{ax}, \quad \frac{d}{dx}(\ln ax) = \frac{1}{x},$$

$$\frac{d}{dx}(\sin ax) = a\cos ax, \quad \frac{d}{dx}(\cos ax) = -a\sin ax, \quad \frac{d}{dx}(\sec ax) = a\sec ax \tan ax,$$

$$\frac{d}{dx}(\tan ax) = a\sec^2 ax, \quad \frac{d}{dx}(\operatorname{cosec} ax) = -a\operatorname{cosec} ax \cot ax,$$

$$\frac{d}{dx}(\cot ax) = -a\operatorname{cosec}^2 ax, \quad \frac{d}{dx}\left(\sin^{-1}\frac{x}{a}\right) = \frac{1}{\sqrt{a^2 - x^2}},$$

$$\frac{d}{dx}\left(\cos^{-1}\frac{x}{a}\right) = \frac{-1}{\sqrt{a^2 - x^2}}, \quad \frac{d}{dx}\left(\tan^{-1}\frac{x}{a}\right) = \frac{a}{a^2 + x^2}.$$

Differentiation from first principles emphasises the definition of a derivative as the gradient of a function. However, for most practical purposes, returning to the definition (3.1) is time consuming and does not aid our understanding. Instead, as mentioned above, we employ a number of techniques, which use the derivatives listed above as 'building blocks', to evaluate the derivatives of more complicated functions than hitherto encountered. Sections 3.1.2–3.2 develop the methods required.

3.1.2 Differentiation of products

As a first example of the differentiation of a more complicated function, we consider finding the derivative of a function $f(x)$ that can be written as the product of two other functions of x, namely $f(x) = u(x)v(x)$. For example, if $f(x) = x^3 \sin x$ then we might take $u(x) = x^3$ and $v(x) = \sin x$. Clearly the separation is not unique; for the given example, an alternative break-up could be $u(x) = x^2$, $v(x) = x \sin x$, or, even more bizarrely, $u(x) = x^4 \tan x$, $v(x) = x^{-1} \cos x$. All the alternatives would give the same answer for the derivative in the end, but most would only increase the effort involved, rather than reduce it; keeping it as simple as possible is almost invariably the best policy.

The purpose of the separation is to split the function into two (or more) parts, of which we know the derivatives (or at least we can evaluate these derivatives more easily than that of the whole). We would gain little, however, if we did not know the relationship between the derivative of f and those of u and v. Fortunately, they are very simply related, as we will now show.

Since $f(x)$ is written as the product $u(x)v(x)$, it follows that

$$f(x + \Delta x) - f(x) = u(x + \Delta x)v(x + \Delta x) - u(x)v(x)$$
$$= u(x + \Delta x)[v(x + \Delta x) - v(x)] + [u(x + \Delta x) - u(x)]v(x),$$

---

[3] Prove the second result by setting $y = e^{ax}$, i.e. $\ln y = ax$, and using the expansion of $e^{a\Delta x}$. Then evaluate $d(\ln y)/dy$ to obtain the third result.

3.1 Differentiation

where we have both added and subtracted $u(x + \Delta x)v(x)$. Now, from definition (3.1) of a derivative,

$$\frac{df}{dx} = \lim_{\Delta x \to 0} \frac{f(x + \Delta x) - f(x)}{\Delta x}$$

$$= \lim_{\Delta x \to 0} \left\{ u(x + \Delta x) \left[\frac{v(x + \Delta x) - v(x)}{\Delta x} \right] + \left[\frac{u(x + \Delta x) - u(x)}{\Delta x} \right] v(x) \right\}.$$

In the limit $\Delta x \to 0$, the factors in square brackets become dv/dx and du/dx (by the definitions of these quantities) and $u(x + \Delta x)$ simply becomes $u(x)$. Consequently we obtain

$$\frac{df}{dx} = \frac{d}{dx}[u(x)v(x)] = u(x)\frac{dv(x)}{dx} + \frac{du(x)}{dx}v(x). \tag{3.7}$$

In primed notation and without writing the argument x explicitly, (3.7) is stated concisely as

$$f' = (uv)' = uv' + u'v. \tag{3.8}$$

This is a general result, obtained without making any assumptions about the specific forms f, u and v, other than that $f(x) = u(x)v(x)$. In words, the result reads as follows. *The derivative of the product of two functions is equal to the first function times the derivative of the second plus the second function times the derivative of the first.*

Example Find the derivative with respect to x of $f(x) = x^3 \sin x$.

Using the product rule, (3.7),

$$\frac{d}{dx}(x^3 \sin x) = x^3 \frac{d}{dx}(\sin x) + \frac{d}{dx}(x^3) \sin x$$

$$= x^3 \cos x + 3x^2 \sin x.$$

The obvious division of $f(x)$ into $u(x) = x^3$ and $v(x) = \sin x$ was used here before applying the product rule.[4]

As a further, quixotic but perhaps reassuring, example, consider differentiating $f(x) = x^n$, treating it as the product of two powers of x, namely x^{n-p} and x^p for some p. The calculation would proceed as follows:

$$\frac{df}{dx} = \frac{dx^n}{dx} = \frac{d}{dx}(x^{n-p} x^p)$$

$$= x^{n-p} \frac{dx^p}{dx} + x^p \frac{dx^{n-p}}{dx}$$

$$= x^{n-p} p x^{p-1} + x^p (n-p) x^{n-p-1}$$

$$= p x^{n-1} + (n-p) x^{n-1}$$

$$= n x^{n-1}.$$

---

[4] Repeat the calculation with $u(x) = x^2$ and $v(x) = x \sin x$, showing that the same answer is obtained.

Differential calculus

The final answer is as expected, and, of course, could have been found by inspection. But, since the same result is obtained whatever the value of p, carrying through the calculation does provide some further justification for the statement that the correct derivative is obtained however the original function is broken up into two factors.

The product rule may readily be extended to the product of three or more functions. Considering the function

$$f(x) = u(x)v(x)w(x) \tag{3.9}$$

and using (3.7), we obtain, again omitting the argument,

$$\frac{df}{dx} = u\frac{d}{dx}(vw) + \frac{du}{dx}vw.$$

Using (3.7) a second time to expand the first term on the RHS gives the complete result

$$\frac{d}{dx}(uvw) = uv\frac{dw}{dx} + u\frac{dv}{dx}w + \frac{du}{dx}vw \tag{3.10}$$

or

$$(uvw)' = uvw' + uv'w + u'vw. \tag{3.11}$$

It is readily apparent that this can be extended to products containing any number n of factors; the expression for the derivative will then consist of n terms with the prime appearing in successive terms on each of the n factors in turn. This is probably the easiest way to recall the product rule.

3.1.3 The chain rule

Products are just one type of complicated function that we may be required to differentiate. A second is the function of a function, e.g. $f(x) = (3 + x^2)^3 = [u(x)]^3$, where $u(x) = 3 + x^2$. If Δf, Δu and Δx are small finite quantities, it follows that

$$\frac{\Delta f}{\Delta x} = \frac{\Delta f}{\Delta u}\frac{\Delta u}{\Delta x}.$$

As the quantities become infinitesimally small we obtain

$$\frac{df}{dx} = \frac{df}{du}\frac{du}{dx}. \tag{3.12}$$

This is the *chain rule*, which we must apply when differentiating a function of a function.

Example Find the derivative with respect to x of $f(x) = (3 + x^2)^3$.

Rewriting the function as $f(x) = u^3$, where $u(x) = 3 + x^2$, and applying (3.12) we find

$$\frac{df}{dx} = 3u^2\frac{du}{dx} = 3u^2\frac{d}{dx}(3 + x^2) = 3u^2 \times 2x = 6x(3 + x^2)^2.$$

Of course, the same result could have been obtained by expanding $f(x)$ as a polynomial using the binomial theorem and then differentiating term-by-term.[5] ◀

---

[5] Do this, showing that $f(x) = x^6 + 9x^4 + 27x^2 + 27$ and that $6x(3 + x^2)^2$ expands to the derivative of this polynomial.

3.1 Differentiation

Similarly, the derivative with respect to x of $f(x) = 1/v(x)$ may be obtained by rewriting the function as $f(x) = v^{-1}$ and applying (3.12):

$$\frac{df}{dx} = -v^{-2}\frac{dv}{dx} = -\frac{1}{v^2}\frac{dv}{dx}. \tag{3.13}$$

The chain rule is also useful for calculating the derivative of a function f with respect to x when both x and f are written in terms of a further variable (or parameter), say t.

Example Find the derivative with respect to x of $f(t) = 2at$, where $x = at^2$.

We could of course substitute for t and then differentiate f as a function of x, but in this case it is quicker to use

$$\frac{df}{dx} = \frac{df}{dt}\frac{dt}{dx} = 2a\frac{1}{2at} = \frac{1}{t},$$

where we have used the result

$$\frac{dt}{dx} = \left(\frac{dx}{dt}\right)^{-1}.$$

as given in Equation (3.4).[6] ◀

3.1.4 Differentiation of quotients

Applying (3.7) for the derivative of a product to a function $f(x) = u(x)[1/v(x)]$, we may obtain the derivative of the quotient of two factors. Thus

$$f' = \left(\frac{u}{v}\right)' = u\left(\frac{1}{v}\right)' + u'\left(\frac{1}{v}\right) = u\left(-\frac{v'}{v^2}\right) + \frac{u'}{v},$$

where (3.13) has been used to evaluate $(1/v)'$. This can now be rearranged into the more convenient and memorisable form

$$f' = \left(\frac{u}{v}\right)' = \frac{vu' - uv'}{v^2}. \tag{3.14}$$

This can be expressed in words as *the derivative of a quotient is equal to the bottom times the derivative of the top minus the top times the derivative of the bottom, all over the bottom squared.*

Example Find the derivative with respect to x of $f(x) = \sin x / x$.

Using (3.14) with $u(x) = \sin x$, $v(x) = x$ and hence $u'(x) = \cos x$, $v'(x) = 1$, we find

$$f'(x) = \frac{x \cos x - \sin x}{x^2} = \frac{\cos x}{x} - \frac{\sin x}{x^2}.$$

At first sight, it might seem that both $f(x)$ and its derivative are infinite at $x = 0$. However, series expansions of all the sinusoids involved will show that this is not so; further the derivative of the series expansion for $f(x)$ is equal to the series expansion of the closed form derived for $f'(x)$. ◀

---

[6] Show that for an ellipse parameterised by $x = a\cos\phi$, $y = b\sin\phi$, the derivative $dy/dx = -(b^2 x)/(a^2 y)$.

As a more complicated example, one that involves all three of the methods so far discussed, consider obtaining the derivative with respect to x of

$$f(x) = \frac{x}{(x^2 + a^2)^{1/2}}.$$

This is a quotient, and so, from (3.14), we will need to find the derivatives of both the numerator and the denominator for substitution into the general quotient formula. The numerator, x, is simple enough; it has a derivative equal to 1. The denominator, $(x^2 + a^2)^{1/2}$, is more complicated; it is a function ($u^{1/2}$) of a function ($u = x^2 + a^2$). By the chain rule, its derivative is

$$\frac{d(x^2 + a^2)^{1/2}}{dx} = \frac{du^{1/2}}{du} \frac{du}{dx} = \tfrac{1}{2} u^{-1/2} 2x = x(x^2 + a^2)^{-1/2}.$$

We can now substitute these derivatives into (3.14) to yield

$$\frac{df}{dx} = \frac{(x^2 + a^2)^{1/2}(1) - (x)[x(x^2 + a^2)^{-1/2}]}{[(x^2 + a^2)^{1/2}]^2}.$$

This answer is correct but very ungainly, having both positive and negative exponents in the numerator. To tidy it up, let us multiply both numerator and denominator by $(x^2 + a^2)^{1/2}$ and finally obtain

$$\frac{df}{dx} = \frac{(x^2 + a^2) - x^2}{(x^2 + a^2)(x^2 + a^2)^{1/2}} = \frac{a^2}{(x^2 + a^2)^{3/2}}.$$

An alternative calculation of the same derivative, based on setting $x = a \tan \theta$, is the subject of Problem 3.7 at the end of this chapter.

3.1.5 Implicit differentiation

So far we have only differentiated functions written in the form $y = f(x)$. However, we may not always be presented with a relationship in this simple form. As an example consider the relation $x^3 - 3xy + y^3 = 2$. In this case it is not possible to rearrange the equation to give y as an explicit function of x. Nevertheless, by differentiating term by term with respect to x (*implicit differentiation*), we can find the derivative dy/dx.

For this method of obtaining derivatives, it is important to recognise that two types of procedures are involved. When a factor in one of the terms (or the whole term) is explicitly given as a function of x, then the differentiation proceeds directly, with Ax^n yielding nAx^{n-1}, $\cos x$ yielding $-\sin x$, etc. However, when a factor or the whole term is expressed in terms of y, the chain rule must be invoked; if the factor is $h(y)$, then its derivative with respect to x is obtained by first differentiating $h(y)$ with respect to y and then multiplying the result by dy/dx, i.e. the derivative is $dh(y)/dy \times dy/dx$. This is how the sought-after derivative dy/dx is introduced into the calculation. The following example illustrates this principle.

3.1 Differentiation

Example Find dy/dx if $x^3 - 3xy + y^3 = 2$.

Differentiating each term in the equation with respect to x we obtain

$$\frac{d}{dx}(x^3) - \frac{d}{dx}(3xy) + \frac{d}{dx}(y^3) = \frac{d}{dx}(2),$$

$$\Rightarrow \quad 3x^2 - \left(3x\frac{dy}{dx} + 3y\right) + 3y^2\frac{dy}{dx} = 0,$$

where the derivative of $3xy$ has been found using the product rule. Hence, rearranging for dy/dx,

$$\frac{dy}{dx} = \frac{y - x^2}{y^2 - x}.$$

Note that dy/dx is a function of both x and y and cannot be expressed as a function of x only.[7] A similar calculation can be found in Problem 3.8 at the end of this chapter. ◀

3.1.6 Logarithmic differentiation

In circumstances in which the variable with respect to which we are differentiating is an exponent, taking logarithms and then differentiating implicitly is the simplest way to find the derivative.

Example Find the derivative with respect to x of $y = b^x$.

To find the required derivative we first take logarithms and then differentiate implicitly. We will need the result, taken from the selection given on p. 106, that the derivative of $\ln(ax)$ is $1/x$ for any[8] constant a. On taking logarithms and then differentiating with respect to x, we obtain

$$\ln y = \ln b^x = x \ln b \quad \Rightarrow \quad \frac{1}{y}\frac{dy}{dx} = \ln b.$$

Now, rearranging and substituting for y, we find that

$$\frac{dy}{dx} = y \ln b = b^x \ln b.$$

This result is true for general positive values of b, but if we take the particular case of b equal to e, the base of natural logarithms, we obtain a well-known property of the function e^x:

$$\frac{d(e^x)}{dx} = e^x \ln e = e^x 1 = e^x,$$

i.e. the first derivative, and consequently *all* of its derivatives, are equal to the original function. ◀

---

[7] Obtain the same result as that given in the previous footnote by implicitly differentiating the equation of the corresponding ellipse.

[8] Since $\ln(ax)$ can be written as $\ln x + \ln a$, and $\ln a$ is itself a constant and therefore has a zero derivative, the derivative of $\ln(ax)$ does not depend upon a.

Differential calculus

EXERCISES 3.1

1. Find from first principles the derivatives of $\sin x$ and $\cos x$ and hence write down a general expression for the nth derivative of $f(x) = A \sin x + B \cos x$, distinguishing between the cases of n even and n odd.

2. Find the derivative of $f(x) = \sin ax / \cos^2 ax$, treating $f(x)$ as the product of $\tan ax$ and $\sec ax$.

3. Write down by inspection the derivative of $f(x) = x \sin ax \, e^{-bx}$.

4. Use the chain rule to find the (first) derivatives of the following, simplifying your answers as far as possible:

   (a) $\cos(\pi - x)$,   (b) $\exp(a^2 - x^2)$,   (c) $(1 + x^2)^3 - (1 - x^2)^3$.

5. By expressing them as quotients involving $\sin ax$ and $\cos ax$, verify the derivatives of $\tan ax$ and $\cot ax$ quoted in the list on p. 106.

6. A closed curve in the x–y plane is defined by $a^2(a^2 - y^2) = (2x^2 - a^2)^2$. Find the derivative dy/dx by each of the following methods.
   (a) Find an expression for y and use direct differentiation.
   (b) Parameterise the equation using $x = a \cos \phi$, $y = a \sin 2\phi$.
   (c) Use implicit differentiation.

   Show that each of your answers is equivalent to $\dfrac{2(a^2 - 2x^2)}{a\sqrt{a^2 - x^2}}$.

7. Calculate the second derivative of $y = x^x$, showing that it is $x^x(1 + \ln x)^2 + x^{x-1}$.

3.2 Leibnitz's theorem

We have discussed already how to find the derivative of a product of two or more functions. We now consider *Leibnitz's theorem*, which gives the corresponding results for the higher derivatives of products.

Consider again the function $f(x) = u(x)v(x)$. We know from the product rule that $f' = uv' + u'v$. Using the rule once more for each of the products we obtain

$$f'' = (uv'' + u'v') + (u'v' + u''v)$$
$$= uv'' + 2u'v' + u''v.$$

Similarly, differentiating twice more gives

$$f''' = uv''' + 3u'v'' + 3u''v' + u'''v,$$
$$f^{(4)} = uv^{(4)} + 4u'v''' + 6u''v'' + 4u'''v' + u^{(4)}v.$$

3.2 Leibnitz's theorem

The pattern emerging is clear and strongly suggests that the results generalise to

$$f^{(n)} = \sum_{r=0}^{n} \frac{n!}{r!(n-r)!} u^{(r)} v^{(n-r)} = \sum_{r=0}^{n} {}^nC_r u^{(r)} v^{(n-r)}, \qquad (3.15)$$

where the fraction $n!/[r!(n-r)!]$ is identified with the binomial coefficient ${}^nC_r$ (see Section 1.4). So as to keep the main text of these introductory chapters as free from detailed mathematical manipulation as possible, the *proof* that this is so has been placed in Appendix C; however, it does not require any topic that has not already been introduced, and could be studied at this point.

It will be noticed that, in each term of the summation, the sum of the orders of the derivatives that are multiplied together always add up to n; one is r and the other is $n-r$. So, for example, the fifth derivative of the general product uv will contain the zeroth derivative of u multiplied by the fifth of v, the first derivative of u multiplied by the fourth of v, ..., the third derivative of u multiplied by the second of v, ..., the fifth derivative of u multiplied by the zeroth of v. Remembering this general pattern makes it easier to write down the expansion without error.[9]

We continue with a straightforward worked example, in which all the required derivatives of the component functions can be obtained immediately, and attention can be focused on the structure of the calculation.

Example Find the third derivative of the function $f(x) = x^3 \sin x$.

When treating $f(x)$ as a product, $f = uv$, we should make the obvious choice of taking $u(x)$ as x^3 and $v(x)$ as $\sin x$. Since we seek the third derivative of f, we will need up to the third derivatives of each of u and v. They are, calculated successively, $u' = 3x^2$, $u'' = 6x$, $u''' = 6$ and $v' = \cos x$, $v'' = -\sin x$, $v''' = -\cos x$. Now, substituting in (3.15) with $n = 3$ we have that

$$f'''(x) = {}^3C_0 uv''' + {}^3C_1 u'v'' + {}^3C_2 u''v' + {}^3C_3 u'''v$$
$$= uv''' + 3u'v'' + 3u''v' + u'''v$$
$$= x^3(-\cos x) + 3(3x^2)(-\sin x) + 3(6x)\cos x + 6\sin x$$
$$= 3(2 - 3x^2)\sin x + x(18 - x^2)\cos x.$$

The same function was differentiated once in the worked example in Section 3.1.2; the reader may care to differentiate that result twice more and show that the above expression is obtained, as it must be. ◄

EXERCISES 3.2

1. Use Leibnitz's theorem to find the third derivative of $x^5 e^{-ax}$ and the fifth derivative of $x^3 e^{-ax}$.

---

[9] How many terms would you expect in the expression for the seventh derivative of a general product uv? And if v has the form $v(x) = 3x^5$?

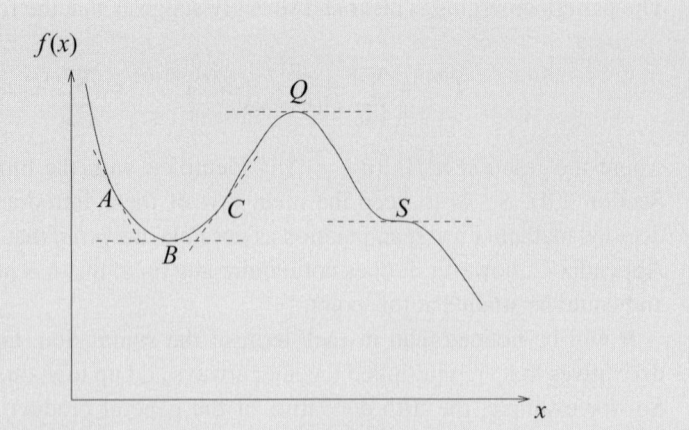

Figure 3.2 A graph of a function, $f(x)$, showing how differentiation corresponds to finding the gradient of the function at a particular point. Points B, Q and S are stationary points (see text).

2. Use Leibnitz's theorem to write down the third derivative of $f(x) = \sin x \cos x$. Express this result in terms of sinusoids of $2x$, and reconcile it with the fact that $f(x)$ could be written as $\frac{1}{2}\sin 2x$.

3.3 Special points of a function

We have interpreted the derivative of a function as the gradient of the function at the relevant point (Figure 3.1). As already discussed in a preliminary way on p. 57, if the gradient is zero for some particular value of x then a graph of the function has a horizontal tangent there. More formally, the function is said to have a *stationary point* at that value of x.

Stationary points may be divided into three categories, and an example of each is shown in Figure 3.2. Point B is said to be a *minimum* since the function *increases* in value in both directions away from it. Point Q is said to be a *maximum* since the function *decreases* in both directions away from it. Note that B is not the overall minimum value of the function and Q is not the overall maximum; rather, they are a local minimum and a local maximum. Maxima and minima are known collectively as *turning points*.

The third type of stationary point is the *stationary point of inflection*, S. In this case the function falls in the positive x-direction and rises in the negative x-direction so that S is neither a maximum nor a minimum. Nevertheless, the gradient of the function is zero at S, i.e. the graph of the function is flat there, and this justifies our calling it a stationary point. Of course, a point at which the gradient of the function is zero but the function rises in the positive x-direction and falls in the negative x-direction is also a stationary point of inflection.

The above distinction between the three types of stationary point has been made rather descriptively. However, it is possible to define and distinguish stationary points mathematically. From their definition as points of zero gradient, all stationary points of a function $f(x)$ must be characterised by $df/dx = 0$. In the case of the minimum, B, the

3.3 Special points of a function

slope, i.e. df/dx, changes from negative at A to positive at C through zero at B. Thus, df/dx is increasing and so the second derivative $d^2 f/dx^2$ must be positive. Conversely, at the maximum, Q, we must have that $d^2 f/dx^2$ is negative.[10]

It is less obvious, but intuitively reasonable, that $d^2 f/dx^2$ is zero at S. This may be inferred from the following observations. To the left of S the curve is concave upwards, so that df/dx is increasing with x and hence $d^2 f/dx^2 > 0$. To the right of S, however, the curve is concave downwards, so that df/dx is decreasing with x and hence $d^2 f/dx^2 < 0$.

In summary, at a stationary point $df/dx = 0$ and

(i) for a minimum, $d^2 f/dx^2 > 0$,
(ii) for a maximum, $d^2 f/dx^2 < 0$,
(iii) for a stationary point of inflection, $d^2 f/dx^2 = 0$ and $d^2 f/dx^2$ changes sign through the point.

It should be added that in the case of a stationary point of inflection (df/dx and $d^2 f/dx^2$ both zero), if it happens that $d^3 f/dx^3$ is also zero, then $d^2 f/dx^2$ does not necessarily change sign through the point. The actual rule is that if the first non-vanishing derivative of $f(x)$ at a stationary point is $f^{(n)}$, then the point is a maximum or minimum if n is even, but is a stationary point of inflection if n is odd. As examples that can be seen from a simple sketches: $f(x) = x^4$, which has a first non-vanishing derivative of $f^{(4)} = 24$ at $x = 0$, has a minimum there; $f(x) = x^5$, whose first non-vanishing derivative at the same point is $f^{(5)} = 120$, exhibits a stationary point of inflection. These general results may all be deduced from the Taylor expansion of the function about the stationary point (see Equation (6.18)), but are not proved here.

Example Find the positions and natures of the stationary points of the function

$$f(x) = 2x^3 - 3x^2 - 36x + 2.$$

The first criterion for a stationary point is that $df/dx = 0$, and hence we set

$$\frac{df}{dx} = 6x^2 - 6x - 36 = 0,$$

from which we obtain

$$(x - 3)(x + 2) = 0.$$

Hence the stationary points are at $x = 3$ and $x = -2$. To determine the nature of the stationary point we must evaluate $d^2 f/dx^2$:

$$\frac{d^2 f}{dx^2} = 12x - 6.$$

Now, we examine each stationary point in turn. For $x = 3$, $d^2 f/dx^2 = 30$. Since this is positive, we conclude that $x = 3$ is a minimum. Similarly, for $x = -2$, $d^2 f/dx^2 = -30$ and so $x = -2$ is a maximum.[11] ◀

---

[10] Give a formal proof that if a function $f(x) = \sin x$ or $f(x) = \cos x$ has a turning point at $x = x_0$, then that turning point is a maximum if $f(x_0)$ is positive and a minimum if $f(x_0)$ is negative.
[11] How many real zeros does $f(x)$ have?

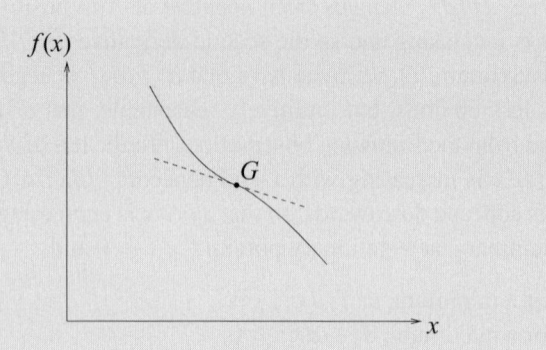

Figure 3.3 The graph of a function $f(x)$ that has a general point of inflection at the point G.

So far we have concentrated on stationary points, which are defined to have $df/dx = 0$. We have found that at a stationary point of inflection d^2f/dx^2 is also zero and changes sign. This naturally leads us to consider points at which d^2f/dx^2 is zero and changes sign but at which df/dx is *not*, in general, zero. Such points are called *general points of inflection* or simply *points of inflection*. Clearly, a stationary point of inflection is a special case for which df/dx is also zero. At a general point of inflection the graph of the function changes from being concave upwards to concave downwards (or vice versa), but the tangent to the curve at this point need not be horizontal. A typical example of a general point of inflection is shown in Figure 3.3.

The determination of the stationary points of a function, together with the identification of its zeros, infinities and possible asymptotes, is usually sufficient to enable a graph of the function showing most of its significant features to be sketched. This general topic is taken up again in Section 3.6 towards the end of this chapter, and some examples for the reader to try are included in the problems.

EXERCISES 3.3

1. Find the positions and natures of the stationary points of the following functions:

   (a) $2x^2 - 3x + 3$, (b) $2x^3 + 9x^2 - 60x + 12$, (c) $2x^3 + 9x^2 + 60x + 12$, (d) x^7.

2. For each of the functions in the previous exercise, determine the positions and natures of any points of inflection they possess.

3.4 Curvature of a function

In the previous section we saw that at a point of inflection of the function $f(x)$ the second derivative d^2f/dx^2 changes sign and passes through zero. The corresponding

3.4 Curvature of a function

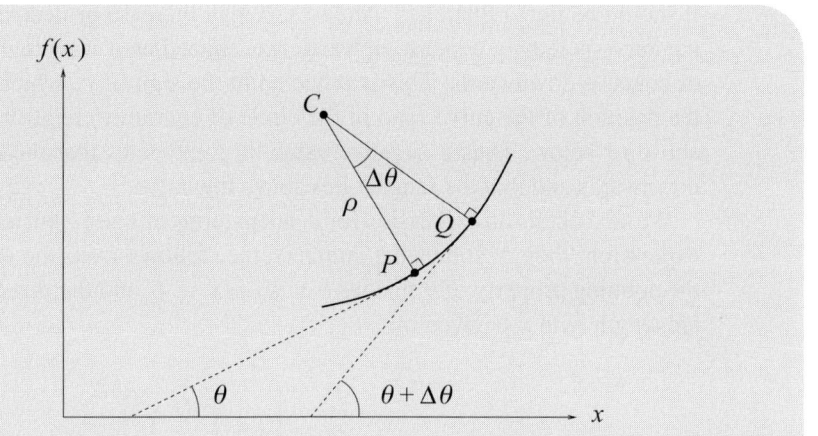

Figure 3.4 Two neighbouring tangents to the curve $f(x)$ whose slopes differ by $\Delta\theta$. The angular separation of the corresponding radii of the circle of curvature is also $\Delta\theta$.

graph of f shows an inversion of its curvature at the point of inflection. We now develop a more quantitative measure of the curvature of a function (or its graph), which is applicable at general points and not just in the neighbourhood of a point of inflection.

As in Figure 3.1, let θ be the angle made with the x-axis by the tangent at a point P on the curve $f = f(x)$, with $\tan\theta = df/dx$ evaluated at P. Now consider also the tangent at a neighbouring point Q on the curve, and suppose that it makes an angle $\theta + \Delta\theta$ with the x-axis, as illustrated in Figure 3.4.

It follows that the corresponding normals at P and Q, which are perpendicular to the respective tangents, also intersect at an angle $\Delta\theta$. Furthermore, their point of intersection, C in the figure, will be the position of the centre of a circle that approximates the arc PQ, at least to the extent of having the same tangents at the extremities of the arc. This circle is called the *circle of curvature*.

For a finite arc PQ, the lengths of CP and CQ will not, in general, be equal, as they would be if $f = f(x)$ *were* in fact the equation of a circle. But, as Q is allowed to tend to P, i.e. as $\Delta\theta \to 0$, they do become equal, their common value being ρ, the radius of the circle, known as the *radius of curvature*. It follows immediately that the curve and the circle of curvature have a common tangent at P and lie on the same side of it. The reciprocal of the radius of curvature, ρ^{-1}, defines the *curvature* of the function $f(x)$ at the point P.

The radius of curvature can be defined more mathematically as follows. The length Δs of arc PQ is approximately equal to $\rho\Delta\theta$ and, in the limit $\Delta\theta \to 0$, this relationship defines ρ as

$$\rho = \lim_{\Delta\theta \to 0} \frac{\Delta s}{\Delta\theta} = \frac{ds}{d\theta}. \tag{3.16}$$

Differential calculus

It should be noted that, as s increases, θ may increase or decrease according to whether the curve is locally concave upwards [i.e. shaped as if it were near a minimum in $f(x)$] or concave downwards. This is reflected in the sign of ρ, which therefore also indicates the position of the curve (and of the circle of curvature) relative to the common tangent, above or below. Thus a negative value of ρ indicates that the curve is locally concave downwards and that the tangent lies above the curve.

We next obtain an expression for ρ, not in terms of s and θ but in terms of x and $f(x)$. The expression, though somewhat cumbersome, follows from the defining Equation (3.16), the defining property of θ that $\tan\theta = df/dx \equiv f'$ and the fact that the rate of change of arc length with x is given by

$$\frac{ds}{dx} = \left[1 + \left(\frac{df}{dx}\right)^2\right]^{1/2}. \tag{3.17}$$

This last result, simply quoted here, is proved more formally in Section 4.8.

From the chain rule (3.12) it follows that

$$\rho = \frac{ds}{d\theta} = \frac{ds}{dx}\frac{dx}{d\theta}. \tag{3.18}$$

Differentiating both sides of $\tan\theta = df/dx$ with respect to x gives

$$\sec^2\theta \frac{d\theta}{dx} = \frac{d^2f}{dx^2} \equiv f'',$$

from which, using $\sec^2\theta = 1 + \tan^2\theta = 1 + (f')^2$, we can obtain $dx/d\theta$ as

$$\frac{dx}{d\theta} = \frac{1 + \tan^2\theta}{f''} = \frac{1 + (f')^2}{f''}. \tag{3.19}$$

Substituting (3.17) and (3.19) into (3.18) then yields the final expression for ρ:

$$\rho = [1 + (f')^2]^{1/2}\frac{1 + (f')^2}{f''} = \frac{[1 + (f')^2]^{3/2}}{f''}. \tag{3.20}$$

It should be noted that the quantity in brackets is always positive and that its three-halves root is also taken as positive. The sign of ρ is thus solely determined by that of d^2f/dx^2, in line with our previous discussion relating the sign to whether the curve is concave or convex upwards. If, as happens at a point of inflection, d^2f/dx^2 is zero, then ρ is formally infinite and the curvature of $f(x)$ is zero. As d^2f/dx^2 changes sign on passing through zero, both the local tangent and the circle of curvature change from their initial positions to the opposite side of the curve.

3.4 Curvature of a function

Example Show that the radius of curvature at the point (x, y) on the ellipse

$$\frac{x^2}{a^2} + \frac{y^2}{b^2} = 1$$

has magnitude $(a^4 y^2 + b^4 x^2)^{3/2}/(a^4 b^4)$ and the opposite sign to y. Check the special case $b = a$, for which the ellipse becomes a circle.

Implicit differentiation of the equation of the ellipse with respect to x gives

$$\frac{2x}{a^2} + \frac{2y}{b^2}\frac{dy}{dx} = 0$$

and so enables us to extract an expression for dy/dx as

$$\frac{dy}{dx} = -\frac{b^2 x}{a^2 y}.$$

A second differentiation, using (3.14), then yields

$$\frac{d^2 y}{dx^2} = -\frac{b^2}{a^2}\left(\frac{y - xy'}{y^2}\right) = -\frac{b^2}{a^2 y^2}\left(y + x\frac{b^2 x}{a^2 y}\right) = -\frac{b^4}{a^2 y^3}\left(\frac{y^2}{b^2} + \frac{x^2}{a^2}\right) = -\frac{b^4}{a^2 y^3},$$

where, to obtain the final equality, we have used the fact that (x, y) lies on the ellipse. We note that $d^2 y/dx^2$, and hence ρ, has the opposite sign to y^3 and hence to y. Substituting in (3.20) gives for the magnitude of the radius of curvature

$$|\rho| = \left|\frac{[1 + b^4 x^2/(a^4 y^2)]^{3/2}}{-b^4/(a^2 y^3)}\right| = \frac{(a^4 y^2 + b^4 x^2)^{3/2}}{a^4 b^4}.$$

For the special case $b = a$, $|\rho|$ reduces to $a^{-2}(y^2 + x^2)^{3/2}$ and, since $x^2 + y^2 = a^2$, this in turn gives $|\rho| = a$, as expected. ◀

Our discussion in this section has been restricted to curves that lie in one plane; a treatment of curvature in three dimensions is beyond the scope of the present book. Examples of the application of curvature to the bending of loaded beams and to particle orbits under the influence of a central force can be found in the problems at the end of Chapter 14.

EXERCISES 3.4

1. What is the radius of curvature of the graph of $f(x) = x^3 + 2x^2 + 5x + 6$ at the point $(-1, 2)$?

2. Show that the curvatures of the graph of $f(x) = 2x^3 + 5x^2 + 4x + 12$ at its two stationary points have equal magnitudes but opposite signs. What is the value of $f(x)$ at the point at which it has no curvature?

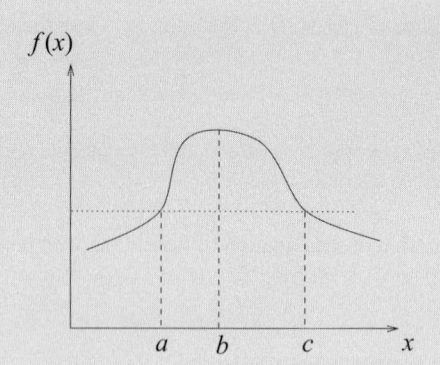

Figure 3.5 The graph of a function $f(x)$, showing that if $f(a) = f(c)$ then at at least one point between $x = a$ and $x = c$ the graph has zero gradient.

3.5 Theorems of differentiation

3.5.1 Rolle's theorem

The essential content of Rolle's theorem was discussed and virtually proved on p. 57. Put in a somewhat more mathematically precise way, it states that if a function $f(x)$ is continuous in the range $a \leq x \leq c$, is differentiable in the range $a < x < c$ and satisfies $f(a) = f(c)$, then for at least one point $x = b$, where $a < b < c$, $f'(b) = 0$ (see Figure 3.5). Thus, Rolle's theorem states that for a well-behaved (continuous and differentiable) function that has the same value at two points, either there is at least one stationary point between those points or the function is a constant between them.[12] The validity of the theorem is apparent from the figure and further analytic proof will not be given. The theorem is used in the derivation of the mean value theorem, which we now discuss.

3.5.2 Mean value theorem

The mean value theorem (Figure 3.6) states that if a function $f(x)$ is continuous in the range $a \leq x \leq c$ and differentiable in the range $a < x < c$ then

$$f'(b) = \frac{f(c) - f(a)}{c - a}, \tag{3.21}$$

for at least one value b where $a < b < c$. Thus, the mean value theorem states that for a well-behaved function the gradient of the line joining two points on the curve is equal to the slope of the tangent to the curve for at least one intervening point.

---

[12] Show that if $f(x) = (4x + 7)/(x^2 + 4x + 3)$, then $f(-2) = f(2) = 1$ and verify that $f'(x) = -g(x)/(x^2 + 4x + 3)^2$, where $g(x) = 4x^2 + 14x + 16$. Rolle's theorem indicates that $g(x) = 0$ for some x in $-2 < x < 2$, but, since $14^2 < 4 \times 4 \times 16$, $g(x)$ has no real zeros. Explain the apparent contradiction.

3.5 Theorems of differentiation

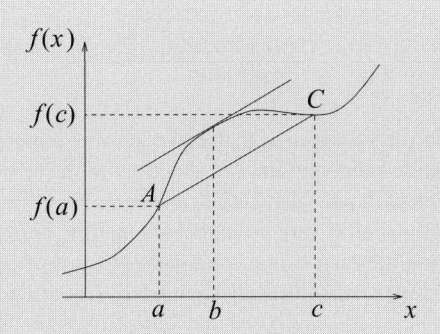

Figure 3.6 The graph of a function $f(x)$; at some point $x = b$ it has the same gradient as the line AC.

The proof of the mean value theorem follows from an analysis of Figure 3.6, as follows. The equation of the line AC is

$$g(x) = f(a) + (x - a)\frac{f(c) - f(a)}{c - a},$$

as can be checked by noting that $g(x)$ is linear in x, i.e. has the general form $y = mx + k$, and that $g(a) = f(a)$ whilst $g(c) = f(a) + f(c) - f(a) = f(c)$. Hence, the difference between the curve and the line is the function

$$h(x) = f(x) - g(x) = f(x) - f(a) - (x - a)\frac{f(c) - f(a)}{c - a}.$$

Since the curve and the line intersect at A and C, $h(x) = 0$ at both of these points. Hence, by an application of Rolle's theorem to $h(x)$, we know that $h'(x) = 0$ for at least one point $x = b$ between A and C. Differentiating our expression for $h(x)$ with respect to x (remembering that a, c, $f(a)$ and $f(c)$ are all constants), we obtain

$$h'(x) = f'(x) - \frac{f(c) - f(a)}{c - a}.$$

It follows that at $x = b$, where $h'(x) = 0$,

$$f'(b) = \frac{f(c) - f(a)}{c - a},$$

as given in the initial statement of the mean value theorem.

3.5.3 Applications of Rolle's theorem and the mean value theorem

Since the validity of Rolle's theorem is intuitively obvious, given the conditions imposed on $f(x)$, it will not be surprising that the problems that can be solved by applications of the theorem alone are relatively simple ones. Nevertheless, we will illustrate it with the following example.

122 Differential calculus

Example What semi-quantitative results can be deduced by applying Rolle's theorem to the following functions $f(x)$, with a and c chosen so that $f(a) = f(c) = 0$? (i) $\sin x$, (ii) $\cos x$, (iii) $x^2 - 3x + 2$, (iv) $x^2 + 7x + 3$, (v) $2x^3 - 9x^2 - 24x + k$.

(i) If the consecutive values of x that make $\sin x = 0$ are $\alpha_1, \alpha_2, \ldots$ (actually $x = n\pi$, for any integer n) then Rolle's theorem implies that the derivative of $\sin x$, namely $\cos x$, has at least one zero lying between each pair of values α_i and α_{i+1}.

(ii) In an exactly similar way, we conclude that the derivative of $\cos x$, namely $-\sin x$, has at least one zero lying between consecutive pairs of zeros of $\cos x$. These two results taken together (but neither separately) imply the well-known property of $\sin x$ and $\cos x$ that they have interleaving zeros.

(iii) For $f(x) = x^2 - 3x + 2$, $f(a) = f(c) = 0$ if a and c are taken as the roots of $f(x) = 0$. Either by factorisation of $f(x)$, or by using the standard formula for the roots of a quadratic equation, the required values are $a = 1$ and $c = 2$. Rolle's theorem then implies that $f'(x) = 2x - 3 = 0$ has a solution $x = b$ with b in the range $1 < b < 2$. This is obviously so, since $b = 3/2$.

(iv) With $f(x) = x^2 + 7x + 3$, the theorem tells us that if there are two roots of $x^2 + 7x + 3 = 0$ then they have the root of $f'(x) = 2x + 7 = 0$ lying between them. Thus, if there are any (real) roots of $x^2 + 7x + 3 = 0$ then they lie one on either side of $x = -7/2$. The actual roots are $(-7 \pm \sqrt{37})/2$.

(v) If $f(x) = 2x^3 - 9x^2 - 24x + k$ then $f'(x) = 0$ is the equation $6x^2 - 18x - 24 = 0$, which has solutions $x = -1$ and $x = 4$. Consequently, if α_1 and α_2 are two different roots of $f(x) = 0$ then at least one of -1 and 4 must lie in the open interval α_1 to α_2. If, as is the case for a certain range of values of k, $f(x) = 0$ has three roots, α_1, α_2 and α_3, then $\alpha_1 < -1 < \alpha_2 < 4 < \alpha_3$.

In each case, as might be expected, the application of Rolle's theorem does no more than focus attention on particular ranges of values; it does not yield precise answers. ◀

We now turn to the somewhat less obvious deductions that can be made using the mean value theorem. Direct verification of the theorem is straightforward when it is applied to simple functions. For example, if $f(x) = x^2$, it states that there is a value b in the interval $a < b < c$ such that

$$c^2 - a^2 = f(c) - f(a) = (c - a)f'(b) = (c - a)2b.$$

This is clearly so, since $b = (a + c)/2$ satisfies the relevant criteria.[13]

As a slightly more complicated example we may consider a cubic equation, say $f(x) = x^3 + 2x^2 + 4x - 6 = 0$, between two specified values of x, say 1 and 2. In this case we need to verify that there is a value of x lying in the range $1 < x < 2$ that satisfies

$$18 - 1 = f(2) - f(1) = (2 - 1)f'(x) = 1(3x^2 + 4x + 4).$$

This is easily done, either by evaluating $3x^2 + 4x + 4 - 17$ at $x = 1$ and at $x = 2$ and checking that the values have opposite signs or by solving $3x^2 + 4x + 4 - 17 = 0$ and showing that one of the roots lies in the stated interval. The actual root is $(-2 + \sqrt{43})/3 = 1.519$.

The following applications of the mean value theorem establish some general inequalities for two common functions.

---

[13] Show that the value of x at which the tangent to the curve $y(x) = x^3$ is parallel to the chord joining (a, a^3) to (c, c^3) is $x = (c^2 + ac + a^2)^{1/2}/\sqrt{3}$ and verify that this value is consistent with Rolle's theorem.

3.5 Theorems of differentiation

Example Determine inequalities satisfied by $\ln x$ and $\sin x$ for suitable ranges of the real variable x.

Since for positive values of its argument the derivative of $\ln x$ is x^{-1}, the mean value theorem gives us

$$\frac{\ln c - \ln a}{c - a} = \frac{1}{b}$$

for some b in $0 < a < b < c$. Further, since $a < b < c$ implies that $c^{-1} < b^{-1} < a^{-1}$, we have

$$\frac{1}{c} < \frac{\ln c - \ln a}{c - a} < \frac{1}{a},$$

or, multiplying through by $c - a$ and writing $c/a = x$ where $x > 1$,

$$1 - \frac{1}{x} < \ln x < x - 1.$$

This is not a particularly useful set of constraints on $\ln x$ for most values of x, as a little numerical substitution will show. However, when x is only just greater than 1 a more useful set of inequalities can be found. Let $x = 1 + \delta$, then for positive δ we have

$$1 - \frac{1}{1+\delta} < \ln(1+\delta) < \delta \quad \text{or} \quad \frac{\delta}{1+\delta} < \ln(1+\delta) < \delta.$$

The validity of this double inequality can be checked by substituting the Maclaurin series for $\ln(1 + \delta)$, given in Section 6.6.3, into it.

Applying the mean value theorem to $\sin x$ shows that

$$\frac{\sin c - \sin a}{c - a} = \cos b$$

for some b lying between a and c. If a and c are restricted to lie in the range $0 \leq a < c \leq \pi$, in which the cosine function is monotonically decreasing (i.e. there are no turning points), we can deduce that

$$\cos c < \frac{\sin c - \sin a}{c - a} < \cos a.$$

For the particular case $a = 0$ this reduces to

$$\cos c < \frac{\sin c}{c} < 1.$$

If we further restrict c to lie in $0 < c < \pi/2$, so that $\cos c$ is always positive, we can also deduce, by dividing through by $\cos c$, that[14]

$$1 < \frac{\tan c}{c} < \sec c.$$

Combining these two results gives

$$\cos c < \frac{\sin c}{c} < 1 < \frac{\tan c}{c} < \sec c,$$

a quadruple set of inequalities that can be confirmed by examining the Maclaurin expansions to order c^2 of each of the functions. If each is expressed as $1 + \alpha c^2$, then the successive values of α are $-\frac{1}{2}, -\frac{1}{6}, 0, \frac{1}{3}$ and $\frac{1}{2}$. ◀

---

[14] Carry this further by applying the mean value theorem to $y(x) = \cos x$, with $0 < x < \pi/2$, and conclude that $c < \tan c < 2c$ for $0 < c < \pi/4$.

EXERCISES 3.5

1. A particular function, known as the Legendre polynomial $P_4(x)$, has the form $P_4(x) = \frac{1}{8}(35x^4 - 30x^2 + 3)$. Verify by explicit calculation that, between the two positive zeros of $P_4(x)$, its derivative $P_4'(x)$ has a zero.

2. By applying the mean value theorem to the function $\tan x$, with $0 < x < \pi/4$, show that $\cos^{-1}(\sqrt{\pi}/2)$ lies within the same range.

3. Locate the points on the curve $y = x^3$ at which the tangents are parallel to the line $y = 13x - 12$. Determine the points at which the line and curve intersect and verify that the mean value theorem is satisfied in each case.

4. For the curve $f(x) = e^x$, let $X + y$ be the value of x at which the tangent to the curve is parallel to the straight line that intersects the curve at $x = X$ and $x = X + z$. Show that y is independent of X and find an expression for it in terms of z.

3.6 Graphs

We conclude this chapter with a brief discussion of graph sketching, which, as mentioned in the introduction to Chapter 2, is an important skill for the practising physical scientist. This applies whether the graphs are used to get a general impression of how a function behaves, to obtain an approximation to the solution of an algebraic equation or to determine a suitable method and starting point for the accurate solution of such an equation.

Modern computers and calculators offer a lot of help in this direction and many provide built-in graphing facilities, but such machines only do what they are programmed to do, often on screens that are too small for the reading of even approximate values, and, in any case, are not designed to deduce the salient points of the graph. Hence the need, in practice, for the physical scientist to be able to sketch or plot appropriate graphs as they are required.

In order to make accurate enough sketches for any of these purposes, it is important to be able to recognise the main features that will be possessed by the graph of a function $f(x)$, without having to make detailed calculations for many values of x. There is no automatic way to do this, but we give in the next subsection a list of aspects that should be investigated in a routine way. When the information gleaned from them has been incorporated into the graph, it is usually possible to draw a complete and qualitatively correct curve. The general procedures will be illustrated with a number of worked examples, each involving some of the aspects.

One type of function $f(x)$ that has to be sketched quite frequently is the ratio of two polynomials in x. The actual variable that normally appears in control engineering applications is s, since the function in question is usually the 'transfer function' of a linear electrical or mechanical system; this is the ratio of the Laplace transform of the output of the system to that of its input. Laplace transforms are defined, and their more elementary properties are discussed, in Chapter 13. In more advanced texts, Laplace transforms normally constitute a major, or the only, topic.

3.6 Graphs

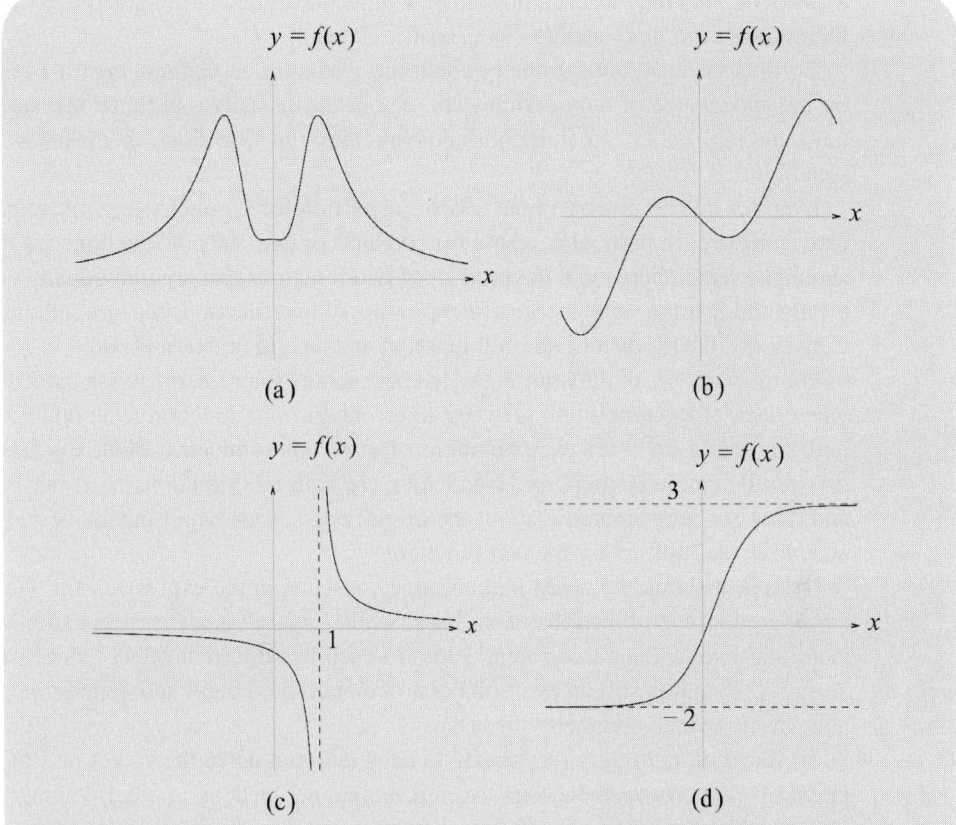

Figure 3.7 Graphs of functions $f(x)$ that show (a) symmetry about the y-axis, (b) antisymmetry about the y-axis, (c) a vertical asymptote and (d) two horizontal asymptotes $y_0 = 3$ and $y_0 = -2$.

The transforms can be used to convert differential equations into algebraic ones, with linear differential operators with respect to time being replaced by polynomial functions of an 'imaginary frequency' s; for our current purposes we only need to know that the transfer function takes the form $G(s) = P(s)/Q(s)$, where $P(x)$ and $Q(x)$ are polynomials in x. Several of our worked examples and test problems are concerned with such polynomial ratios, which are known more technically as *rational functions* when subjected to the proviso that $G(x)$ is undefined at values of x that make $Q(x) = 0$.

3.6.1 General considerations

The general headings under which the salient features of a graph of $y = f(x)$ may fall can, in most cases, be summarised as follows.

(i) *Symmetry* about the y-axis (or an *even function* of x). There is an obvious saving in calculational labour if the graph of a function has symmetry about the y-axis [see Figure 3.7(a)], or, indeed, about any axis. Symmetry about the y-axis is present if x

appears in $f(x)$ only as even powers of x or as the argument of named functions that are themselves even in x, such as $\cos x$, $\cosh x$ and $\exp(-x^2)$.

Correspondingly, antisymmetry about the y-axis (or an *odd function* of x) is indicated by the appearance of only odd powers of x or intrinsically odd functions, such as $\sin x$, $\tan x$ and $\tan^{-1} x$. For a continuous antisymmetric function $f(x)$, we must have $f(0) = 0$ [see Figure 3.7(b)].

Even for this relatively simple task of inspection for symmetry or antisymmetry, some care is needed. In particular, where the products or quotients of functions are involved, it should be remembered that the product of two functions that are individually even is also even – and that the same is true of the product of two functions that are individually odd. Conversely, the product of an even function and an odd function is odd.

For the purposes of illustration, we list here some examples, involving named functions, where lack of thought might give rise to errors: the sine function is an odd function, but both $\sin^2 x$ and $\sin x^2$ are even functions of x; $e^{-|x|}$ is symmetric about $x = 0$ but e^{-x} has no overall symmetry property; x and $\sin x$ are both odd functions of x, but both $x \sin x$ and $(\sin x)/x$ are symmetric about the origin; $\cos x$ is an even function of x, but $x \cos x$ and $\sin 2x = 2 \sin x \cos x$ are odd functions.

Further, it should be noted that additive constants in the expression for $f(x)$ must be considered as even functions of x. Thus, whilst $x^2 + x^6$ is an even function and $x + x^5$ is an odd one, adding a constant, 3 say, to each has different effects; the first becomes $3 + x^2 + x^6$ and is still an even function of x, but the second new function, $3 + x + x^5$, now has no overall symmetry property.

(ii) *Zeros* of $f(x)$. It is very useful to be able to pin down the values of x at which the graph of $f(x)$ crosses the x-axis. Even if this cannot be done in precise numerical terms, it is valuable information to know how many zeros $f(x)$ has in any particular range. This was considered at some length in Section 2.1, and the reader should refer back to that section, if the need arises.

(iii) *Particular values* of $f(x)$. Complementary to finding the zeros of $f(x)$ is the process of calculating $f(x)$ at those values of x at which it is especially simple to do so. Every case is different, but $f(0)$ and $f(1)$ are usually worth considering if $f(x)$ contains polynomials, as are $f(0)$, $f(\pi/2)$ and $f(\pi)$ if it involves sinusoidal functions.

(iv) *Vertical asymptotes* (or $f(x) \to \infty$). If $f(x)$ contains a function that 'goes to ∞' for a finite value of x, such as $\tan x$ whenever $x = (n + \frac{1}{2})\pi$ for integer n, or it contains a denominator that has the value zero for one or more values of x, the graph of f increases without limit as one of these points, say x_0, is approached.[15] The vertical line $x = x_0$ is described as a *vertical asymptote* and the graph of f approaches such an asymptote arbitrarily closely, without ever intersecting it. Just about the simplest example is given by $f(x) = A/(x - x_0)$ for, say, $A > 0$. Here, as $x \to x_0$ from below, the graph of $f(x)$ becomes very large and negative, but, if $x \to x_0$ from above, the graph takes on arbitrarily large positive values; in brief, $f(x)$ has an infinite discontinuity at $x = x_0$ [see Figure 3.7(c)].

---

[15] One instructor, having carefully explained that $\lim_{x \to 8}(x - 8)^{-1} = \infty$, set a written test asking for $\lim_{x \to \pi}(x - \pi)^{-1}$ and received $\lim_{x \to \pi}(x - \pi)^{-1} = ¤$ as the answer. Comment.

3.6 Graphs

One might add that when the plot of $f(x)$ contains a vertical asymptote, the physical system from which f was derived usually shows, at best, an abrupt change of behaviour as x passes through x_0, or, at worst, catastrophic failure!

(v) *Horizontal asymptotes* (or $x \to \infty$). Determining the behaviour of the graph of $f(x)$ for large values of $|x|$ is an important aspect of graph sketching. In many cases $f(x) \to 0$ as $x \to \pm\infty$, thus making the x-axis a horizontal line which f approaches, but never quite reaches; in other cases a non-zero value for y is one which is approached but never reached; all such lines are called *horizontal asymptotes* and are characterised by $y = y_0$, where y_0 is a constant that may be zero. Figure 3.7(d) shows a function that has two different horizontal asymptotes, one with y_0 positive for $x \to +\infty$ and the other, with a negative y_0, for large negative x.

Horizontal asymptotes with $y_0 \neq 0$ occur most frequently when $f(x)$ is the ratio of two polynomials of the same degree n; the value of y_0 is then the ratio of the coefficients multiplying x^n in the numerator and denominator. It is normally very helpful to be able to express $f(x)$ in the form $y_0 + g(x)$, where $g(x) \to 0$ as $x \to \pm\infty$. If this can be done, it establishes whether the horizontal asymptote is approached from above or from below. The first worked example below illustrates this procedure.

(vi) *Other asymptotes*. Vertical and horizontal asymptotes are not the only ones possible, as has already been noted in connection with equation (2.26) describing the asymptotes of the hyperbola in standard form. One of the worked examples in the next subsection provides further illustrations.

(vii) *Stationary points*. The procedure for locating and finding the natures of stationary points was discussed at length in Section 3.3 and will not be repeated here. For practical graph sketching, it is normally sufficient to establish the x_i that make $f'(x_i) = 0$ and not have to continue by investigating $f''(x_i)$, as the natures of the stationary points are usually clear from other features of the graph and knowledge of the values of $f(x_i)$.

3.6.2 Worked examples

This subsection consists of a series of worked examples whose solutions are presented with comments, most of which would normally be only in the student's mind and would not appear on paper.

Example Make a sketch graph of the function

$$f(x) = \frac{x^2 - 3}{x^2 - 4}$$

and determine whether there are any values that $f(x)$ cannot take.

One immediate observation is that the form of $f(x)$ contains only x^2 and constants, both of which are even functions; it follows that the graph will have symmetry about the y-axis. Equally obviously, f will have zeros at $x = \pm\sqrt{3}$ and vertical asymptotes at $x = \pm 2$.

Differential calculus

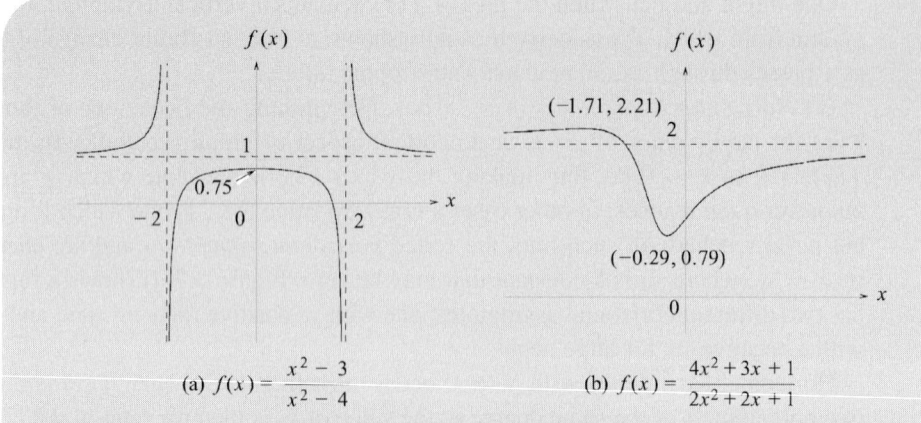

Figure 3.8 Graphs of the functions analysed in the first two worked examples in subsection 3.6.2.

(a) $f(x) = \dfrac{x^2 - 3}{x^2 - 4}$

(b) $f(x) = \dfrac{4x^2 + 3x + 1}{2x^2 + 2x + 1}$

It is clear that for both $x \to +\infty$ and $x \to -\infty$ the limiting value of f is 1, but to see on which side of this horizontal asymptote the graph lies we write $f(x)$ as

$$1 + g(x) = f(x) = \frac{x^2 - 3}{x^2 - 4} = 1 + \frac{1}{x^2 - 4}.$$

Thus, although $g(x) = 1/(x^2 - 4)$ tends to zero as $x \to \pm\infty$, it is always positive for $|x| > 2$. It follows that, at both extremities, the graph approaches its horizontal asymptote $y = 1$ from above.

Since we have symmetry about $x = 0$, we need consider what happens near a vertical asymptote only for $x = x_0 = 2$. In a form with a factored denominator – there is no need to factor a numerator unless the zeros of a rational polynomial are being sought – the given $f(x)$ becomes

$$f(x) = \frac{x^2 - 3}{(x + 2)(x - 2)}.$$

Near $x = 2$, this behaves like[16] the function $1/[4(x - 2)]$ and so is large and negative for x just less than 2, and large and positive for x just greater than 2. This links up with $f(x)$ approaching $y = 1$ from above for large positive x.

Of the 'routine enquiries', we are now left only with the question of stationary points. Since f is of the form u/v, f' will be zero when $vu' - uv' = 0$. In this case:

$$(x^2 - 4)2x - (x^2 - 3)2x = 0 \quad \Rightarrow \quad 2x(x^2 - 4 - x^2 + 3) = 0 \quad \Rightarrow \quad x = 0.$$

This result means that there is only one stationary point and that occurs at $x = 0$; the value of $f(x)$ there is $(-3)/(-4) = \frac{3}{4}$. Given that $f(\sqrt{3}) = f(-\sqrt{3}) = 0$ and f is large and negative for $|x|$ just less than 2, it is clear that the stationary point at $x = 0$ is a local maximum.

As $f(x) > 1$ for $|x| > 2$ and $f(x) < \frac{3}{4}$ for $|x| < 2$, it follows that $f(x)$ can never take values in the range $\frac{3}{4} < x < 1$.

A graph of $f(x)$ incorporating all of the features deduced above is sketched in Figure 3.8(a). ◀

---

[16] Note the strong connection with the third method of determining the coefficients of a partial fraction expansion, as described on p. 78.

3.6 Graphs

Our second worked example is also one that analyses a rational function.

Example Determine the range of values that

$$f(x) = \frac{4x^2 + 3x + 1}{2x^2 + 2x + 1}$$

can take and make a sketch graph of it.

The numerator and denominator both contain even and odd powers of x and there is no obvious simple change of origin, $x \to x' = x - \alpha$, that would make both either totally odd or totally even in x', and so the solution has no particular symmetry about any vertical line $x = \alpha$.

Both the numerator and denominator are quadratic polynomials in x, but each has '$b^2 < 4ac$' (explicitly, $3^2 < 16$ and $2^2 < 8$) and so neither has any real zeros. Consequently, f has neither zeros nor vertical asymptotes. It *does* have the obvious horizontal asymptote $y = 2$ and will approach it for both $x \to +\infty$ and $x \to -\infty$. To find out how it does so, we rewrite f as

$$f(x) = \frac{4x^2 + 3x + 1}{2x^2 + 2x + 1} = \frac{2(2x^2 + 2x + 1) - x - 1}{2x^2 + 2x + 1} = 2 - \frac{x + 1}{2x^2 + 2x + 1}.$$

This shows that when $x \to +\infty$ the asymptote is approached from below; conversely, it is approached from above when x is large and negative. Continuity then implies that the graph of f must cross the asymptote $y = 2$ at least once, and the above form for $f(x)$ shows that it does so only once, when $x + 1 = 0$, i.e. at $x = -1$.

What we have already learned about how the asymptote is approached and how often it is crossed, taken together with the fact that f has no discontinuities, implies that f must have (at least) one maximum stationary point to the left of $x = -1$ and (at least) one minimum with $x > -1$. In view of the question posed, actual maximum and minimum values are needed, and so, treating f as $f = u/v$ we set $vu' - uv' = 0$:

$$f' = (2x^2 + 2x + 1)(8x + 3) - (4x^2 + 3x + 1)(4x + 2) = 0,$$
$$16x^3 + 22x^2 + 14x + 3 - (16x^3 + 20x^2 + 10x + 2) = 0,$$
$$2x^2 + 4x + 1 = 0,$$
$$\Rightarrow \quad x = \frac{-4 \pm \sqrt{16 - 8}}{4} = -1 \pm \tfrac{1}{2}\sqrt{2} = -0.293 \text{ or } -1.707.$$

Straightforward substitution gives $f(-0.293) = 0.793$ and $f(-1.707) = 2.207$. Since $f'(x) = 0$ is a quadratic equation, it has only two solutions and so there are no further maxima or minima.

These conclusions, together with the trivial but useful result $f(0) = 1$, enable the graph shown in Figure 3.8(b) to be drawn and allow us to conclude that $f(x)$ is confined to the range $0.79 < f(x) < 2.21$. ◀

Our third worked example, which involves sinusoidal functions, is designed to show that it is worthwhile making simple analytical checks before drawing final conclusions from a graph.

Differential calculus

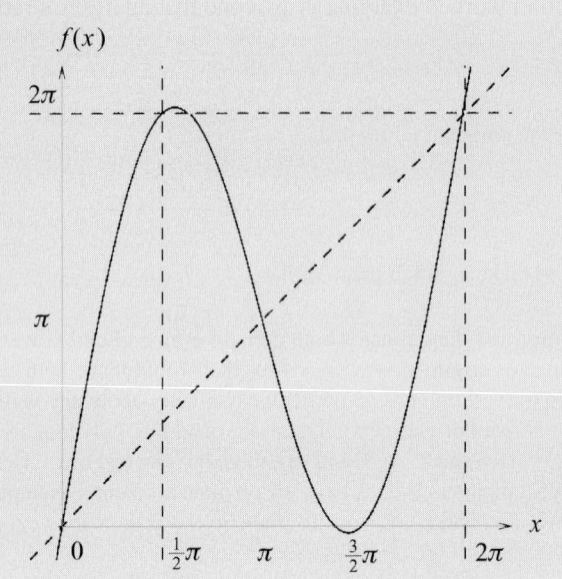

Figure 3.9 Graph of the function $f(x) = x + \frac{3\pi}{2} \sin x$ for $0 \leq x \leq 2\pi$.

Example By sketching its graph, find the values of $f(x) = x + \frac{3}{2}\pi \sin x$ at its turning points in the range $-2\pi \leq x \leq 2\pi$.

We first note that both x and $\sin x$ are odd functions of x and that, consequently, so is f. This means that we do not need to make a sketch for the region $-2\pi \leq x \leq 0$, as everything in that region can be deduced from the sketch for positive values of x.

It is clear that a large-range graph of f would consist of oscillations of fixed amplitude about the line $y = x$ and that ultimately f would become indefinitely large in modulus. We are concerned only with the region $0 \leq x \leq 2\pi$ and, within that region, some values that are simple to calculate are

$$f(0) = 0, \quad f\left(\frac{\pi}{2}\right) = 2\pi, \quad f(\pi) = \pi, \quad f\left(\frac{3\pi}{2}\right) = 0, \quad f(2\pi) = 2\pi.$$

It is tempting at this point to plot these points and to draw a smooth curve through them, one that has a maximum of 2π at $x = \pi/2$ and a minimum of zero at $x = 3\pi/2$. This graph would be virtually indistinguishable from the correct graph shown in Figure 3.9. For many purposes it would be good enough, but, for the actual question posed, it is as well to make a simple calculus check on these conclusions.

The first derivative of $f(x)$ is $f'(x) = 1 + (3\pi/2)\cos x$. At a stationary point this derivative should be zero, but, in fact, it has value 1 at both $x = \pi/2$ and $x = 3\pi/2$; thus, neither can be a stationary point. The actual values of x that make the derivative zero, and hence give the true positions of the turning points, satisfy the equation $\cos x = -2/3\pi$. One such point x_1 lies in the second quadrant and the other x_2 lies in the third; thus both turning points lie strictly within the range $x = \pi/2$ to $x = 3\pi/2$, excluding its end points.

Since $f(\pi/2) = 2\pi$ but $x = \pi/2$ is not the position of the maximum, the actual value of the nearby maximum must be greater than 2π; similarly the minimum at $x = x_2$, near to, but not coincident with, $x = 3\pi/2$, must have a negative value associated with it. The actual values have

3.6 Graphs

no special significance – other than answering the question! The turning points occur at $x_1 = 1.785$ and $x_2 = 4.499$, and the corresponding values of the function are

$$f(x_1) = x_1 + \frac{3\pi}{2}\left(1 - \frac{4}{9\pi^2}\right)^{1/2} = 6.390,$$

$$f(x_2) = x_2 - \frac{3\pi}{2}\left(1 - \frac{4}{9\pi^2}\right)^{1/2} = -0.107.$$

The value at the maximum is greater than 2π by only 1.7% and the negative value at the minimum is also small. For most purposes these small corrections could be ignored, but the actual use to which the results are to be put has to be the deciding factor. ◄

Our final worked example involves finding the sketch graphs of two functions whose equations differ only in the sign of one particular term; as will be seen, this can result in a significant qualitative difference between the forms of the two graphs. The example also illustrates the occurrence of asymptotes that are neither horizontal nor vertical.

Example Sketch the graphs of the functions

$$\text{(a) } f(x) = \frac{x^2 - 2x - 8}{x - 3}, \quad \text{(b) } g(x) = \frac{x^2 + 2x - 8}{x - 3}$$

and determine whether there are any values that the functions cannot take.

Neither function shows any symmetry properties, but both graphs will have a vertical asymptote, and hence an infinite discontinuity, at $x = 3$. The zeros of the functions are easily obtained by factorising their numerators as (a) $(x - 4)(x + 2)$ and (b) $(x - 2)(x + 4)$, with pairs of zeros at (a) 4 and -2 and (b) 2 and -4, respectively. It is also clear from this factorisation that when x is just greater than 3, with $x - 3$ positive, $f(x) = (x - 4)(x + 2)/(x - 3)$ will be large and negative; conversely, $g(x) = (x - 2)(x + 4)/(x - 3)$ will be large and positive.

To determine whether there are any non-vertical asymptotes, we expand f and g in partial fractions (see Section 2.3). In each case, the degree of the numerator is greater than that of the denominator by 1, and the ratio of the coefficients of the leading powers in numerator and denominator is also 1. Consequently, the general form for both expansions will be

$$h(x) = x + A + \frac{B}{x - 3}.$$

Multiplying each equation through by $x - 3$ gives[17]

$$\text{(a) } x^2 - 2x - 8 = x^2 - 3x + Ax - 3A + B \quad \Rightarrow \quad A = 1, \ B = -5,$$
$$\text{(b) } x^2 + 2x - 8 = x^2 - 3x + Ax - 3A + B \quad \Rightarrow \quad A = 5, \ B = 7.$$

Thus, expressed in terms of increasingly negative powers of x, the two functions take the forms

$$f(x) = x + 1 - \frac{5}{x - 3},$$

with an asymptote $y = x + 1$ approached from below as $x \to \infty$,

---

[17] Obtain the two values for B by inspection, using one of the methods available for determining partial fraction coefficients.

Differential calculus

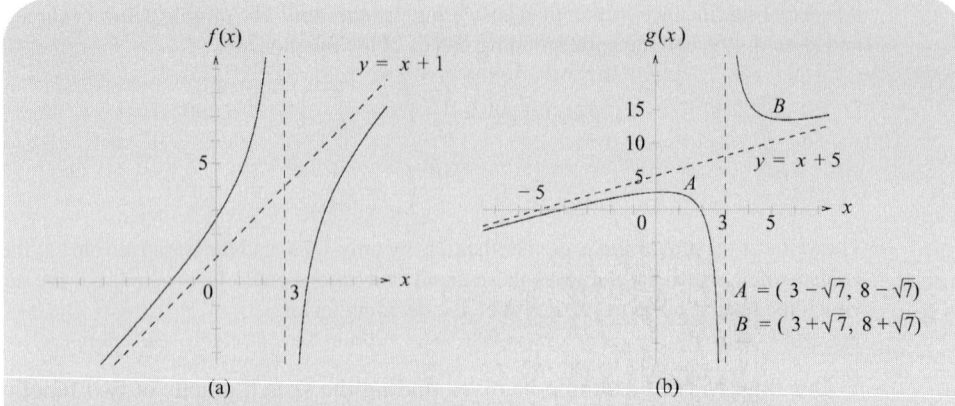

Figure 3.10 Graphs of the functions (a) $f(x) = (x^2 - 2x - 8)/(x - 3)$ and (b) $g(x) = (x^2 + 2x - 8)/(x - 3)$.

$$g(x) = x + 5 + \frac{7}{x - 3},$$

with an asymptote $y = x + 5$ approached from above as $x \to \infty$.

The two asymptotes are parallel, but have different intercepts with the y-axis. The functions themselves intercept that axis, i.e. when $x = 0$, at the common value $f(0) = g(0) = \frac{8}{3}$.

We now examine the derivatives of f and g to establish whether either has stationary points.

(a) For $f(x)$, setting $f'(x) = 0$ gives

$$0 = f'(x) = \frac{(x-3)(2x-2) - (x^2 - 2x - 8)}{(x-3)^2} \quad \Rightarrow \quad x^2 - 6x + 14 = 0.$$

Since $(-6)^2 < 4 \times 1 \times 14$ this quadratic equation has no real solutions, and we conclude that $f(x)$ has no turning points.

(b) For $g(x)$, setting $g'(x) = 0$ gives

$$0 = g'(x) = \frac{(x-3)(2x+2) - (x^2 + 2x - 8)}{(x-3)^2} \quad \Rightarrow \quad x^2 - 6x + 2 = 0.$$

Since $(-6)^2 > 4 \times 1 \times 2$ this quadratic equation does have real solutions, $x = 3 \pm \sqrt{7}$ and we conclude that $g(x)$ has turning points at these two values of x. The corresponding values of $g(x)$ are $8 \pm 2\sqrt{7}$ (see Problem 2.6).

Finally, collecting together what has been established:

(a) Vertical asymptote at $x_0 = 3$, with $f(x)$ large and negative just to the right of it; asymptotic to line $y = x + 1$ approached from below as $x \to \infty$ and from above as $x \to -\infty$; $f(-2) = f(4) = 0$; $f(0) = \frac{8}{3}$; no turning points;

(b) Vertical asymptote at $x_0 = 3$, with $g(x)$ large and positive just to the right of it; asymptotic to line $y = x + 5$ approached from above as $x \to \infty$ and from below as $x \to -\infty$; $f(-4) = f(2) = 0$; $f(0) = \frac{8}{3}$; turning points at $(3 - \sqrt{7}, 8 - 2\sqrt{7})$ and $(3 + \sqrt{7}, 8 + 2\sqrt{7})$.

With all this information to be fitted, the sketch graphs shown in (a) and (b) of Figure 3.10 have little room for manoeuvre; certainly, the main features of each are well established. The difference between the two graphs is quite striking and it will be apparent that, whilst $f(x)$ can take any real value, $g(x)$ cannot take values between $8 - 2\sqrt{7}$ and $8 + 2\sqrt{7}$. ◀

Summary

Since graph sketching consists essentially of the application of topics drawn from several sections in this and other chapters, it is impractical to provide short, straightforward exercises to test particular points. Consequently, no set of exercises is provided for this section, though the reader is referred to the final three of the more substantial problems that follow the summary.

SUMMARY

1. *Definitions*

$$\frac{df(x)}{dx} \equiv f^{(1)} \equiv f'(x) \equiv \lim_{\Delta x \to 0} \frac{f(x + \Delta x) - f(x)}{\Delta x},$$

$$f^{(n+1)}(x) \equiv \lim_{\Delta x \to 0} \frac{f^{(n)}(x + \Delta x) - f^{(n)}(x)}{\Delta x}.$$

2. *Standard derivatives*
   - $(e^{\alpha x})' = \alpha e^{\alpha x}$, $\quad (\ln \alpha x)' = \dfrac{1}{x}$, $\quad (a^x)' = a^x \ln a$.
   - For the derivatives of powers, sinusoidal and inverse sinusoidal functions, see p. 106.
   - For the derivatives of hyperbolic and inverse hyperbolic functions, see p. 204.

3. *Derivatives of compound functions*
   If $u, v, \ldots, w$ are all functions of x, then
   - $(uv)' = u'v + uv'$.
   - $(uv \ldots w)' = u'v \ldots w + uv' \ldots w + \cdots + uv \ldots w'$.
   - $\left(\dfrac{u}{v}\right)' = \dfrac{vu' - uv'}{v^2}$.
   - If $f = uv$, then $f^{(n)} = \sum_{r=0}^{n} {}^nC_r u^{(r)} v^{(n-r)}$ (Leibnitz).

4. *Change of variable*

$$\frac{dx}{df} = \left(\frac{df}{dx}\right)^{-1}, \qquad \frac{df}{dx} = \frac{df}{du}\frac{du}{dx} \quad \text{(chain rule)}.$$

5. *Stationary points*
   - For a stationary point of $f(x)$, $f'(x) = 0$ and for
     - maximum $\quad f'' < 0$,
     - minimum $\quad f'' > 0$,
     - point of inflection $\quad f'' = 0$ and changes sign through the point.

- If $a \le x \le c$, then for some b in $a < b < c$,
$$\frac{f(c) - f(a)}{c - a} = f'(b) \quad \text{(mean value theorem)}.$$
Rolle's theorem is a special case of this in which $f(c) = f(a)$ and $f'(b) = 0$.

6. *Radius of curvature* of $f(x)$
$$\rho = \frac{[1 + (f')^2]^{3/2}}{f''}.$$

7. *Graphs*
Aspects that may help in sketching a graph: symmetry or antisymmetry about the x- or y-axis; zeros; particular simply calculated values; vertical and horizontal asymptotes; other asymptotes; stationary points.

PROBLEMS

3.1. Obtain the following derivatives from first principles:
(a) the first derivative of $3x + 4$;
(b) the first, second and third derivatives of $x^2 + x$;
(c) the first derivative of $\sin 3x$.

3.2. Find from first principles the first derivative of $(x + 3)^2$ and compare your answer with that obtained using the chain rule.

3.3. Find the first derivatives of
(a) $x^2 \exp x$, (b) $2 \sin x \cos x$, (c) $\sin 2x$, (d) $x \sin ax$,
(e) $(\exp ax)(\sin ax) \tan^{-1} ax$, (f) $\ln(x^a + x^{-a})$,
(g) $\ln(a^x + a^{-x})$, (h) x^x.

3.4. Find the first derivatives of
(a) $x/(a + x)^2$, (b) $x/(1 - x)^{1/2}$, (c) $\tan^2 x$, as $\sin^2 x / \cos^2 x$,
(d) $(3x^2 + 2x + 1)/(8x^2 - 4x + 2)$.

3.5. Use result (3.13) to find the first derivatives of
(a) $(2x + 3)^{-3}$, (b) $\sec^2 x$, (c) $\operatorname{cosech}^3 3x$, (d) $1/\ln x$, (e) $1/[\sin^{-1}(x/a)]$.

3.6. Show that the function $y(x) = \exp(-|x|)$ defined by
$$y(x) = \begin{cases} \exp x & \text{for } x < 0, \\ 1 & \text{for } x = 0, \\ \exp(-x) & \text{for } x > 0, \end{cases}$$

Problems

is *not* differentiable at $x = 0$. Consider the limiting process for both $\Delta x > 0$ and $\Delta x < 0$.

3.7. Find the first derivative of

$$f(x) = \frac{x}{(x^2 + a^2)^{1/2}}$$

by making the substitution $x = a \tan \theta$. Show that $f(x) = g(\theta) = \sin \theta$ and then use the chain rule to obtain the derivative.

3.8. The equation of a particular curve is

$$x(x^2 + y^2) = 2y^3.$$

Show that the tangent to the curve at the point (2, 2) has unit slope. Excluding the origin, are there any points on the curve at which the tangent has (i) zero slope and (ii) infinite slope?

3.9. Find dy/dx if $x = (t-2)/(t+2)$ and $y = 2t/(t+1)$ for $-\infty < t < \infty$. Show that it is always non-negative and make use of this result in sketching the curve of y as a function of x.

3.10. If $2y + \sin y + 5 = x^4 + 4x^3 + 2\pi$, show that $dy/dx = 16$ when $x = 1$.

3.11. Find the second derivative of $y(x) = \cos[(\pi/2) - ax]$. Now set $a = 1$ and verify that the result is the same as that obtained by first setting $a = 1$ and simplifying $y(x)$ before differentiating.

3.12. Find the positions and natures of the stationary points of the following functions:
(a) $x^3 - 3x + 3$; (b) $x^3 - 3x^2 + 3x$; (c) $x^3 + 3x + 3$;
(d) $\sin ax$ with $a \neq 0$; (e) $x^5 + x^3$; (f) $x^5 - x^3$.

3.13. Show by differentiation and substitution that the differential equation

$$4x^2 \frac{d^2y}{dx^2} - 4x \frac{dy}{dx} + (4x^2 + 3)y = 0$$

has a solution of the form $y(x) = x^n \sin x$, and find the value of n.

3.14. By determining the turning point(s) of the function $(\ln x)/x$, show, without any numerical calculation, that $e^x > x^e$ for any positive x.

3.15. Show that the lowest value taken by the function $3x^4 + 4x^3 - 12x^2 + 6$ is -26.

3.16. A cone is formed by cutting a sector out of a circular sheet of paper and abutting the two straight edges of what is left. Show that the volume of the cone is a maximum when the angle of the sector removed is $\frac{1}{3}(6 - \sqrt{24})\pi$.

Differential calculus

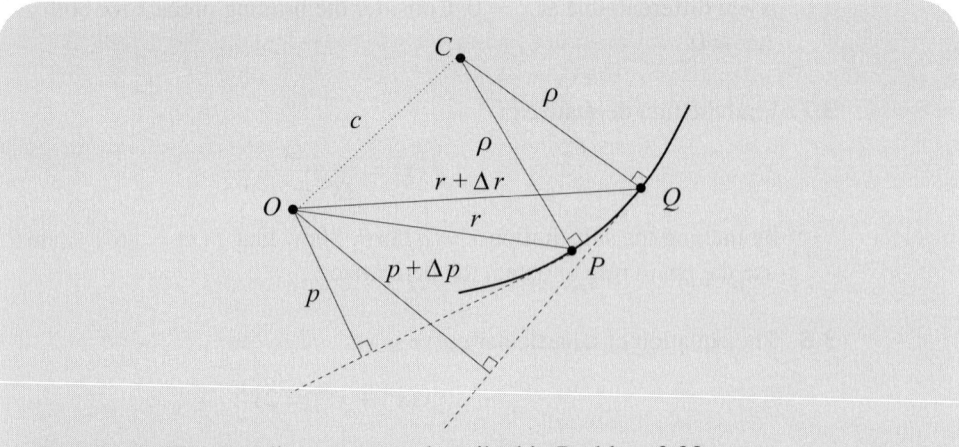

Figure 3.11 The coordinate system described in Problem 3.22.

3.17. Show that $y(x) = xa^{2x} \exp x^2$ has no stationary points other than $x = 0$, if $\exp(-\sqrt{2}) < a < \exp(\sqrt{2})$.

3.18. The curve $4y^3 = a^2(x + 3y)$ can be parameterised as $x = a \cos 3\theta$, $y = a \cos \theta$.
  (a) Obtain expressions for dy/dx (i) by implicit differentiation and (ii) in parameterised form. Verify that they are equivalent.
  (b) Show that the only point of inflection occurs at the origin. Is it a stationary point of inflection?
  (c) Use the information gained in (a) and (b) to sketch the curve, paying particular attention to its shape near the points $(-a, a/2)$ and $(a, -a/2)$ and to its slope at the 'end points' (a, a) and $(-a, -a)$.

3.19. The parametric equations for the motion of a charged particle released from rest in electric and magnetic fields at right angles to each other take the forms

$$x = a(\theta - \sin \theta), \qquad y = a(1 - \cos \theta).$$

Show that the tangent to the curve has slope $\cot(\theta/2)$. Use this result at a few calculated values of x and y to sketch the form of the particle's trajectory.

3.20. Show that the maximum curvature on the catenary $y(x) = a \cosh(x/a)$ is $1/a$. You will need some of the results about hyperbolic functions stated in Section 5.7.6.

3.21. The curve whose equation is $x^{2/3} + y^{2/3} = a^{2/3}$ for positive x and y and which is completed by its symmetric reflections in both axes is known as an astroid. Sketch it and show that its radius of curvature in the first quadrant is $3(axy)^{1/3}$.

3.22. A two-dimensional coordinate system useful for orbit problems is the tangential-polar coordinate system (Figure 3.11). In this system a curve is defined

Problems

by r, the distance from a fixed point O to a general point P of the curve, and p, the perpendicular distance from O to the tangent to the curve at P. By proceeding as indicated below, show that the radius of curvature, ρ, at P can be written in the form $\rho = r\, dr/dp$.

Consider two neighbouring points, P and Q, on the curve. The normals to the curve through those points meet at C, with (in the limit $Q \to P$) $CP = CQ = \rho$. Apply the cosine rule to triangles OPC and OQC to obtain two expressions for c^2, one in terms of r and p and the other in terms of $r + \Delta r$ and $p + \Delta p$. By equating them and letting $Q \to P$ deduce the stated result.

3.23. Use Leibnitz's theorem to find
   (a) the second derivative of $\cos x \sin 2x$,
   (b) the third derivative of $\sin x \ln x$,
   (c) the fourth derivative of $(2x^3 + 3x^2 + x + 2)e^{2x}$.

3.24. If $y = \exp(-x^2)$, show that $dy/dx = -2xy$ and hence, by applying Leibnitz's theorem, prove that for $n \geq 1$
$$y^{(n+1)} + 2xy^{(n)} + 2ny^{(n-1)} = 0.$$

3.25. Use the properties of functions at their turning points to do the following:
   (a) By considering its properties near $x = 1$, show that $f(x) = 5x^4 - 11x^3 + 26x^2 - 44x + 24$ takes negative values for some range of x.
   (b) Show that $f(x) = \tan x - x$ cannot be negative for $0 \leq x < \pi/2$, and deduce that $g(x) = x^{-1} \sin x$ decreases monotonically in the same range.

3.26. Determine what can be learned from applying Rolle's theorem to the following functions $f(x)$: (a) e^x; (b) $x^2 + 6x$; (c) $2x^2 + 3x + 1$; (d) $2x^2 + 3x + 2$; (e) $2x^3 - 21x^2 + 60x + k$. (f) If $k = -45$ in (e), show that $x = 3$ is one root of $f(x) = 0$, find the other roots and verify that the conclusions from (e) are satisfied.

3.27. By applying Rolle's theorem to $x^n \sin nx$, where n is an arbitrary positive integer, show that $\tan nx + x = 0$ has a solution α_1 with $0 < \alpha_1 < \pi/n$. Apply the theorem a second time to obtain the nonsensical result that there is a real α_2 in $0 < \alpha_2 < \pi/n$ such that $\cos^2(n\alpha_2) = -n$. Explain why this incorrect result arises.

3.28. Use the mean value theorem to establish bounds in the following cases.
   (a) For $-\ln(1-y)$, by considering $\ln x$ in the range $0 < 1 - y < x < 1$.
   (b) For $e^y - 1$, by considering $e^x - 1$ in the range $0 < x < y$.

3.29. For the function $y(x) = x^2 \exp(-x)$ obtain a simple relationship between y and dy/dx and then, by applying Leibnitz's theorem, prove that
$$xy^{(n+1)} + (n + x - 2)y^{(n)} + ny^{(n-1)} = 0.$$

Differential calculus

3.30. Use Rolle's theorem to deduce that, if the equation $f(x) = 0$ has a repeated root x_1, then x_1 is also a root of the equation $f'(x) = 0$.
   (a) Apply this result to the 'standard' quadratic equation $ax^2 + bx + c = 0$, to show that a necessary condition for equal roots is $b^2 = 4ac$.
   (b) Find all the roots of $f(x) = x^3 + 4x^2 - 3x - 18 = 0$, given that one of them is a repeated root.
   (c) The equation $f(x) = x^4 + 4x^3 + 7x^2 + 6x + 2 = 0$ has a repeated integer root. How many real roots does it have altogether?

3.31. Show that the curve $x^3 + y^3 - 12x - 8y - 16 = 0$ touches the x-axis.

3.32. Find a transformation $x \to x' = x - \alpha$ that takes

$$f(x) = \frac{2x^2 + 4x - 4}{x^2 + 2x - 3}$$

into a function $g(x')$ that has a definite symmetry property. By relating $g(x')$ to one of the functions used in the worked examples in the text, sketch the graph of $f(x)$ with a minimum of additional investigation.

3.33. Investigate the properties of the following functions and in each case make a sketchgraph incorporating the features you have identified.
   (a) $f(x) = (x^2 + 4x + 2)/[x(x + 2)]$.
   (b) $f(x) = [x(x^2 + 2x + 2)]/(x + 2)$.
   (c) $f(x) = 1 - e^{-x/3}(\frac{1}{6}\sin 2x + \cos 2x)$.

3.34. By finding their stationary points and examining their general forms, determine the range of values that each of the following functions $y(x)$ can take. In each case make a sketchgraph incorporating the features you have identified.
   (a) $y(x) = (x - 1)/(x^2 + 2x + 6)$.
   (b) $y(x) = 1/(4 + 3x - x^2)$.
   (c) $y(x) = (8 \sin x)/(15 + 8 \tan^2 x)$.

HINTS AND ANSWERS

3.1. (a) 3; (b) $2x + 1, 2, 0$; (c) $3 \cos 3x$.

3.3. Use the product rule in (a), (b), (d) and (e) [3 factors]; use the chain rule in (c), (f) and (g); use logarithmic differentiation in (g) and (h).
   (a) $(x^2 + 2x) \exp x$; (b) $2(\cos^2 x - \sin^2 x) = 2 \cos 2x$;
   (c) $2 \cos 2x$; (d) $\sin ax + ax \cos ax$;
   (e) $(a \exp ax)[(\sin ax + \cos ax) \tan^{-1} ax + (\sin ax)(1 + a^2 x^2)^{-1}]$;
   (f) $[a(x^a - x^{-a})]/[x(x^a + x^{-a})]$; (g) $[(a^x - a^{-x}) \ln a]/(a^x + a^{-x})$;
   (h) $(1 + \ln x)x^x$.

Hints and answers

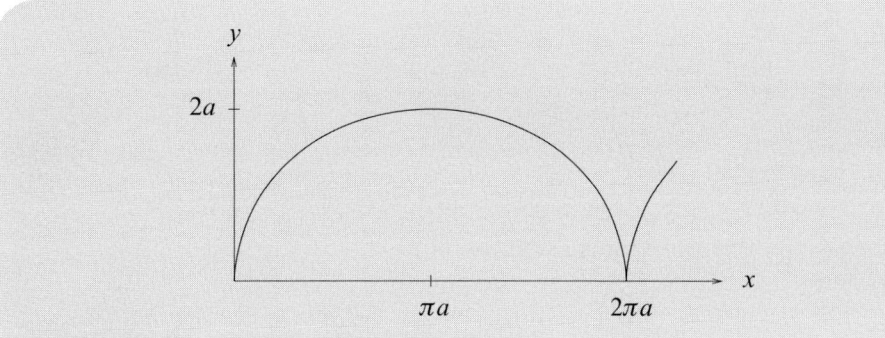

Figure 3.12 The solution to Problem 3.19.

3.5. (a) $-6(2x+3)^{-4}$; (b) $2\sec^2 x \tan x$; (c) $-9\operatorname{cosech}^3 3x \coth 3x$;
(d) $-x^{-1}(\ln x)^{-2}$; (e) $-(a^2-x^2)^{-1/2}[\sin^{-1}(x/a)]^{-2}$.

3.7. $d\theta/dx = a^{-1}\sec^{-2}\theta$. $df/dx = a^2(x^2+a^2)^{-3/2}$.

3.9. Calculate dy/dt and dx/dt and divide one by the other. $(t+2)^2/[2(t+1)^2]$.
Alternatively, eliminate t and find dy/dx by implicit differentiation.

3.11. $-\sin x$ in both cases.

3.13. The required conditions are $8n - 4 = 0$ and $4n^2 - 8n + 3 = 0$; both are satisfied by $n = \frac{1}{2}$.

3.15. The stationary points are the zeros of $12x^3 + 12x^2 - 24x$. The lowest stationary value is -26 at $x = -2$; other stationary values are 6 at $x = 0$ and 1 at $x = 1$.

3.17. Use logarithmic differentiation. Set $dy/dx = 0$, obtaining $2x^2 + 2x \ln a + 1 = 0$.

3.19. See Figure 3.12.

3.21. $\dfrac{dy}{dx} = -\left(\dfrac{y}{x}\right)^{1/3}$; $\dfrac{d^2y}{dx^2} = \dfrac{a^{2/3}}{3x^{4/3}y^{1/3}}$.

3.23. (a) $2(2 - 9\cos^2 x)\sin x$; (b) $(2x^{-3} - 3x^{-1})\sin x - (3x^{-2} + \ln x)\cos x$;
(c) $8(4x^3 + 30x^2 + 62x + 38)e^{2x}$.

3.25. (a) $f(1) = 0$ whilst $f'(1) \neq 0$ and so $f(x)$ must be negative in some region with $x = 1$ as an endpoint.
(b) $f'(x) = \tan^2 x > 0$ and $f(0) = 0$; $g'(x) = (-\cos x)(\tan x - x)/x^2$, which is never positive in the range.

3.27. The false result arises because $\tan nx$ is not differentiable at $x = \pi/(2n)$, which lies in the range $0 < x < \pi/n$, and so the conditions for applying Rolle's theorem are not satisfied.

3.29. The relationship is $x\,dy/dx = (2-x)y$.

3.31. By implicit differentiation, $y'(x) = (3x^2 - 12)/(8 - 3y^2)$, giving $y'(\pm 2) = 0$.
Since $y(2) = 4$ and $y(-2) = 0$, the curve touches the x-axis at the point $(-2, 0)$.

Figure 3.13 The solutions to Problem 3.33.

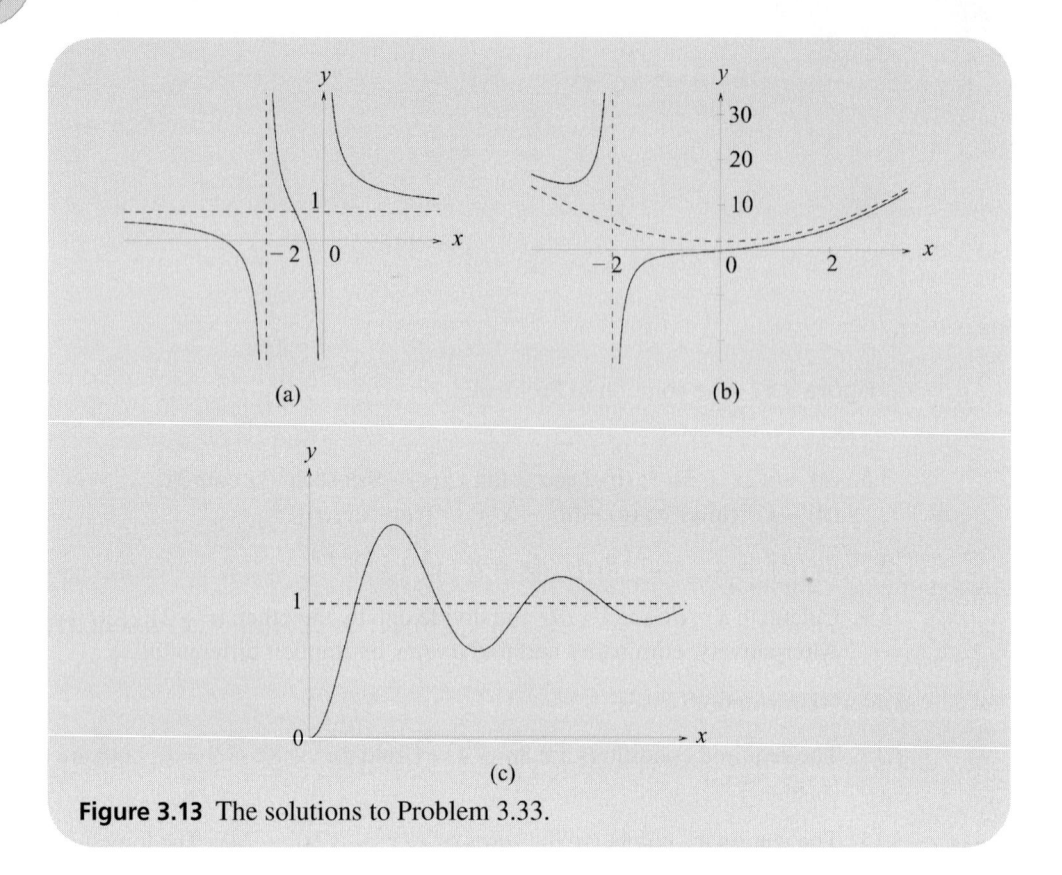

3.33. See Figure 3.13.
 (a) Zeros at $-2 \pm \sqrt{2}$; no turning points; write as $1 + [(2x + 2)/x(x + 2)]$ to determine asymptotes.
 (b) Write as $x^2 + 2 - 4(x + 2)^{-1}$; vertical asymptote at $x = -2$; asymptotic to $x^2 + 2$ as $x \to \infty$; one turning point at $x \approx -2.85$.
 (c) $f(x) \to 1$ as $x \to \infty$, with damped oscillation about that value; $f(0) = 1 - e^0[0 + 1] = 0$; $f(x)$ first crosses $f = 1$ when $\tan 2x = -6$, i.e. $x \approx 0.87$.

4 Integral calculus

As indicated at the start of the previous chapter, the differential calculus and its complement, the integral calculus, together form the most widely used tool for the analysis of physical systems. The link that connects the two is that they both deal with the effects of vanishingly small changes in related quantities; one seeks to obtain the ratio of two such changes, the other uses such a ratio to calculate the variation in one of the quantities resulting from a change in the other.

Any change in the value of any one property (or variable) of a physical system almost always results in the values of some or all of its other properties being altered; in general, the size of each consequential change depends upon the current values of all of the variables. As a result, during a finite change in any one of the values, that of x say, those associated with all of the other variables are continuously changing, making computation of the final situation difficult, if not impossible. The solution to this difficulty is provided by the integral calculus, which allows only vanishingly small changes, and, after any such change in one variable, brings all the other values 'up to date' (by infinitesimal amounts) before allowing any further change. By notionally carrying through an infinitely large number of these infinitesimally small changes, the effect of one or more finite changes can be calculated and the correct final values for all variables determined. This procedure has its basic mathematical representation in the formal definition of an integral, as given in Section 4.1.1.

4.1 Integration

The notion of an integral as the area under a curve will be familiar to the reader, and has already been used in Appendix A in connection with the definition of a natural logarithm. In Figure 4.1, in which the solid line is the plot of a function $f(x)$, the shaded area represents the quantity denoted by

$$I = \int_a^b f(x)\,dx. \tag{4.1}$$

This expression is known as the *definite integral* of $f(x)$ between the *lower limit* $x = a$ and the *upper limit* $x = b$, and $f(x)$ is called the *integrand*. In this context, it should be noted that the definite integral I does *not* depend upon x, which is known as the variable of integration or *dummy variable*. If x were replaced *throughout* by a new variable of integration, u say, then the value of I would not change in any way; I does depend on a, b

Integral calculus

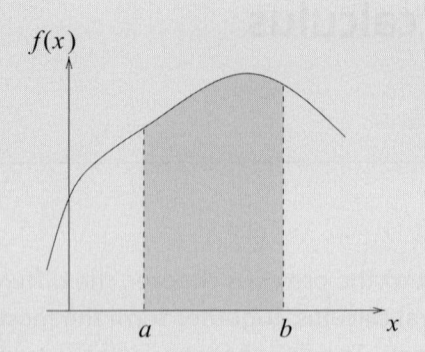

Figure 4.1 An integral as the area under a curve.

and the form of $f(x)$, but not upon x. However, despite having just emphasised this point, we must draw the reader's attention to the remarks immediately following Equation (4.11).

4.1.1 Integration from first principles

The definition of an integral as the area under a curve is not a formal definition, but one that can be readily visualised. The formal definition[1] of I involves subdividing the finite interval $a \leq x \leq b$ into a large number of subintervals, by defining intermediate points ξ_i such that $a = \xi_0 < \xi_1 < \xi_2 < \cdots < \xi_n = b$, and then forming the sum

$$S = \sum_{i=1}^{n} f(x_i)(\xi_i - \xi_{i-1}), \qquad (4.2)$$

where x_i is an arbitrary point that lies in the range $\xi_{i-1} \leq x_i \leq \xi_i$ (see Figure 4.2). If now n is allowed to tend to infinity in any way whatsoever, subject only to the restriction that the length of every subinterval ξ_{i-1} to ξ_i tends to zero, then S might, or might not, tend to a unique limit, I. If it does, then the definite integral of $f(x)$ between a and b is defined as having the value I. If no unique limit exists the integral is undefined. For continuous functions and a finite interval $a \leq x \leq b$ the existence of a unique limit is assured and the integral is guaranteed to exist.

Example

Evaluate from first principles the integral $I = \int_0^b x^2 \, dx$.

We first approximate the area under the curve $y = x^2$ between 0 and b by n rectangles of equal width h. If we take the value at the lower end of each subinterval (in the limit of an infinite number of subintervals we could equally well have chosen the value at the upper end) to give the height of the corresponding rectangle, then the area of the kth rectangle will be $(kh)^2 h = k^2 h^3$. The total area is thus

$$A = \sum_{k=0}^{n-1} k^2 h^3 = (h^3)\tfrac{1}{6}n(n-1)(2n-1),$$

---

[1] This definition defines the Riemann integral. Other, more abstract, procedures for integration are possible and enable the integration of more esoteric functions; but for integrations arising from physical situations, Riemann integration is adequate.

4.1 Integration

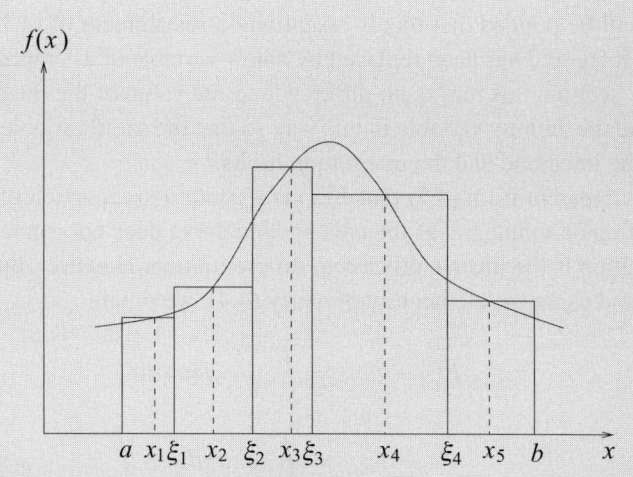

Figure 4.2 The evaluation of a definite integral by subdividing the interval $a \leq x \leq b$ into subintervals.

where we have used the expression for the sum of the squares of the natural numbers as given by Equation (2.40). Now $h = b/n$ and so

$$A = \left(\frac{b^3}{n^3}\right)\frac{n}{6}(n-1)(2n-1) = \frac{b^3}{6}\left(1 - \frac{1}{n}\right)\left(2 - \frac{1}{n}\right).$$

As $n \to \infty$, $A \to b^3/3$, which is thus the value I of the integral.

Some straightforward properties of definite integrals that are almost self-evident are as follows:

$$\int_a^b 0\,dx = 0, \qquad \int_a^a f(x)\,dx = 0, \tag{4.3}$$

$$\int_a^c f(x)\,dx = \int_a^b f(x)\,dx + \int_b^c f(x)\,dx, \tag{4.4}$$

$$\int_a^b [f(x) + g(x)]\,dx = \int_a^b f(x)\,dx + \int_a^b g(x)\,dx. \tag{4.5}$$

Combining (4.3) and (4.4) with c set equal to a shows that

$$\int_a^b f(x)\,dx = -\int_b^a f(x)\,dx. \tag{4.6}$$

4.1.2 Integration as the inverse of differentiation

The definite integral has been defined as the area under a curve between two fixed limits. Let us now consider the integral

$$F(x) = \int_a^x f(u)\,du \tag{4.7}$$

in which the lower limit a remains fixed but the upper limit x is now variable.

It will be noticed that this is essentially a restatement of (4.1), but that the variable x in the integrand has been replaced by a new variable u; as emphasised in the introduction to this section, this makes no difference to the value of the integral. It is conventional to rename the dummy variable in this way so that the same variable name does not appear in both the integrand and the integration limits.[2]

It is apparent from (4.7) that $F(x)$ is a continuous function of x, but at first glance the definition of an integral as the area under a curve does not connect with our assertion that integration is the inverse process to differentiation. However, by considering the integral (4.7) and using the elementary property (4.4), we obtain

$$F(x + \Delta x) = \int_a^{x+\Delta x} f(u)\,du$$

$$= \int_a^x f(u)\,du + \int_x^{x+\Delta x} f(u)\,du$$

$$= F(x) + \int_x^{x+\Delta x} f(u)\,du.$$

Rearranging and dividing through by Δx yields

$$\frac{F(x + \Delta x) - F(x)}{\Delta x} = \frac{1}{\Delta x}\int_x^{x+\Delta x} f(u)\,du.$$

Letting $\Delta x \to 0$ and using (3.1) we find that, by definition, the LHS becomes dF/dx, whereas the RHS becomes $f(x)$. The latter conclusion follows because when the integral range Δx is small the value of the integral on the RHS is approximately $f(x)\Delta x$ [since $f(u)$ is essentially constant throughout the range and has value $f(x)$], and in the limit $\Delta x \to 0$ no approximation is involved. Thus

$$\frac{dF(x)}{dx} = f(x), \tag{4.8}$$

or, substituting for $F(x)$ from (4.7),

$$\frac{d}{dx}\left[\int_a^x f(u)\,du\right] = f(x). \tag{4.9}$$

Pictorially, we can interpret Equation (4.9) as saying that, if we were to steadily increase the value of b in Figure 4.1, the rate at which the shaded area in the figure would be increasing at any point would vary as the value of $f(b)$ at that point; for an increase of db in b the shaded area would increase by $f(b)\,db$.

From the last two equations it is clear that integration can be considered as the inverse of differentiation. However, we see from the above analysis that the lower limit a is arbitrary and so differentiation does not have a *unique* inverse. Any function $F(x)$ obeying (4.8)

---

[2] When we come to study the integration of functions of more than one variable in Chapter 7, we will see that it *is* possible to have the same variable appearing in both the integrand and the integration limits – however, even here, the variable will be a fixed parameter so far as the integration is concerned, and it will be distinct from the dummy variable of integration.

4.1 Integration

is called an *indefinite integral* of $f(x)$, though any two such functions can only differ by at most an arbitrary additive constant. Since the lower limit is arbitrary, it is usual to write

$$F(x) = \int^x f(u)\,du \tag{4.10}$$

and explicitly include the arbitrary constant only when evaluating $F(x)$. The evaluation is conventionally written in the form

$$\int f(x)\,dx = F(x) + c \tag{4.11}$$

where c is called the *constant of integration*. It will be noticed that, in the absence of any integration limits, we use the same symbol for the arguments of both f and F. This can be confusing, but is sufficiently common practice that the reader needs to become familiar with it.

We also note that the definite integral of $f(x)$ between the fixed limits $x = a$ and $x = b$ can be written in terms of $F(x)$. From (4.7) we have

$$\int_a^b f(x)\,dx = \int_{x_0}^b f(x)\,dx - \int_{x_0}^a f(x)\,dx$$
$$= F(b) - F(a), \tag{4.12}$$

where x_0 is *any* third fixed point. Using the notation $F'(x) = dF/dx$, we may rewrite (4.8) as $F'(x) = f(x)$, and so express (4.12) as

$$\int_a^b F'(x)\,dx = F(b) - F(a) \equiv [F]_a^b.$$

The final identity symbol $\equiv$ defines the meaning of the square bracket notation. It is a generally accepted convention that $[g(x)]_a^b$ is a shorthand way of writing $g(b) - g(a)$, i.e. the difference between the value of whatever function is contained between the brackets when it is evaluated at $x = b$, and the value of the same function evaluated at $x = a$. For example $[x^3]_2^3 = 27 - 8 = 19$.

In contrast to differentiation, where repeated applications of the product rule and/or the chain rule will always give the required derivative, it is not always possible to find the integral of an arbitrary function. Indeed, in most real physical problems exact integration cannot be performed and we have to revert to numerical approximations. Despite this cautionary note, it is in fact possible to integrate many simple functions, and the following sections introduce the most common types.

EXERCISES 4.1

1. Evaluate

(a) $\left[\frac{1}{4}x^4\right]_1^3$, (b) $\left[\sin^2\theta\right]_0^{\pi/2}$, (c) $\left[x^2\right]_{-2}^2$, (d) $[\cos\theta]_{-\pi/2}^{\pi/2}$.

2. Given that $\int_1^x \dfrac{1}{u}\,du = \ln x$, evaluate $\int_{e^{-1}}^e \dfrac{1}{v}\,dv$.

4.2 Integration methods

Many of the techniques discussed in this section will probably be familiar to the reader and so are summarised largely by example.

4.2.1 Integration by inspection

The simplest method of integrating a function is by inspection. Some of the more elementary functions have well-known integrals that should be remembered. The reader will notice that these integrals are precisely the inverses of the derivatives found near the end of Section 3.1.1. A few are presented below, using the form given in (4.11):

$$\int a\,dx = ax + c, \qquad \int ax^n\,dx = \frac{ax^{n+1}}{n+1} + c,$$

$$\int e^{ax}\,dx = \frac{e^{ax}}{a} + c, \qquad \int \frac{a}{x}\,dx = a\ln x + c,$$

$$\int a\cos bx\,dx = \frac{a\sin bx}{b} + c, \qquad \int a\sin bx\,dx = \frac{-a\cos bx}{b} + c,$$

$$\int a\tan bx\,dx = \frac{-a\ln(\cos bx)}{b} + c, \qquad \int a\cos bx\,\sin^n bx\,dx = \frac{a\sin^{n+1} bx}{b(n+1)} + c,$$

$$\int \frac{a}{a^2 + x^2}\,dx = \tan^{-1}\left(\frac{x}{a}\right) + c, \qquad \int a\sin bx\,\cos^n bx\,dx = \frac{-a\cos^{n+1} bx}{b(n+1)} + c,$$

$$\int \frac{-1}{\sqrt{a^2 - x^2}}\,dx = \cos^{-1}\left(\frac{x}{a}\right) + c, \qquad \int \frac{1}{\sqrt{a^2 - x^2}}\,dx = \sin^{-1}\left(\frac{x}{a}\right) + c,$$

where the integrals that depend on n are valid for all $n \neq -1$ and where a and b are constants. In the two final results $|x| \leq a$.[3]

4.2.2 Integration of sinusoidal functions

Integrals of the type $\int \sin^n x\,dx$ and $\int \cos^n x\,dx$ may be found by using trigonometric expansions. Two methods are applicable, one for odd n and the other for even n.

In the first of these, when n is odd, use is made of the fact that, to within a possible minus sign, the pair of functions $\cos x$ and $\sin x$ are each the derivative of the other. This means that if one factor is separated off, the remaining ones form an even power of the original sinusoid, which can then be converted, using $\cos^2 x + \sin^2 x = 1$, to a sum of terms that contain only even powers of the other sinusoid, together with constants. When in this form, each term in the integrand is in the form of a power of a sinusoid multiplied by the derivative of that same sinusoid, and so can be integrated immediately. In the illustrative example that follows, one factor of $\sin x$ is used as (minus) the derivative of a sum of even powers of $\cos x$.

---

[3] Reconcile the stated form of the first of these results with what might have been expected from the second, namely that the indefinite integral of $-(a^2 - x^2)^{-1/2}$ is $-\sin^{-1}(x/a) + c$.

4.2 Integration methods

Example

Evaluate the integral $I = \int \sin^5 x\, dx$.

Following the approach outlined above, we rewrite the integral as a product of $\sin x$ and an even power of $\sin x$, and then use the relation $\sin^2 x = 1 - \cos^2 x$:

$$\begin{aligned} I &= \int \sin^4 x \sin x\, dx \\ &= \int (1 - \cos^2 x)^2 \sin x\, dx \\ &= \int (1 - 2\cos^2 x + \cos^4 x) \sin x\, dx \\ &= \int (\sin x - 2 \sin x \cos^2 x + \sin x \cos^4 x)\, dx \\ &= -\cos x + \tfrac{2}{3} \cos^3 x - \tfrac{1}{5} \cos^5 x + c, \end{aligned}$$

where the integration has been carried out using the results of Section 4.2.1. If the integrand had been of the form $\cos^{2m+1} x$, we would have separated off a single $\cos x$ and then expressed $\cos^{2m} x$ in terms of even powers of $\sin x$, again giving a sum of terms each of which could be integrated by inspection. ◀

The second method, used for integrating even powers of sinusoids, depends on rewriting the square of that sinusoid in terms of $\cos 2x$ and thus halving the power to which any sinusoidal function is raised. If the integrand still contains squares (or higher even powers) of sinusoids, the process is repeated. This reduction procedure comes at the 'price' of introducing multiples of x as the arguments of the sinusoids, but, for subsequent integration, this presents no added difficulty.

Example

Evaluate the integral $I = \int \cos^4 x\, dx$.

Rewriting the integral as a power of $\cos^2 x$ and then using the double-angle formula $\cos^2 x = \tfrac{1}{2}(1 + \cos 2x)$ yields

$$\begin{aligned} I &= \int (\cos^2 x)^2\, dx = \int \left(\frac{1 + \cos 2x}{2} \right)^2 dx \\ &= \int \tfrac{1}{4}(1 + 2\cos 2x + \cos^2 2x)\, dx. \end{aligned}$$

Using the double-angle formula again, we may write $\cos^2 2x = \tfrac{1}{2}(1 + \cos 4x)$, and hence

$$\begin{aligned} I &= \int \left[\tfrac{1}{4} + \tfrac{1}{2} \cos 2x + \tfrac{1}{8}(1 + \cos 4x) \right] dx \\ &= \tfrac{1}{4} x + \tfrac{1}{4} \sin 2x + \tfrac{1}{8} x + \tfrac{1}{32} \sin 4x + c \\ &= \tfrac{3}{8} x + \tfrac{1}{4} \sin 2x + \tfrac{1}{32} \sin 4x + c. \end{aligned}$$

Integral calculus

If the original integrand had been of the form $\sin^{2m} x$, we would have used the relationship $\sin^2 x = \frac{1}{2}(1 - \cos 2x)$, and so reduced it to one containing up to the mth power of $\cos 2x$. This could then be handled either by a repeat of the current procedure or by the method described in the previous worked example (depending upon whether m is even or odd).[4] ◄

4.2.3 Logarithmic integration

Integrals for which the integrand may be written as a fraction in which the numerator is proportional to the derivative of the denominator may be evaluated using

$$\int \frac{Af'(x)}{f(x)} dx = A \ln f(x) + c. \tag{4.13}$$

This follows directly from the differentiation of a logarithm as a function of a function (see Section 3.1.3).

Example Evaluate the integral

$$I = \int \frac{6x^2 + 2\cos x}{x^3 + \sin x} dx.$$

We note first that the numerator can be factorised to give $2(3x^2 + \cos x)$, and then that the quantity in parentheses is the derivative of the denominator. Hence

$$I = 2 \int \frac{3x^2 + \cos x}{x^3 + \sin x} dx = 2 \ln(x^3 + \sin x) + c,$$

where we have used (4.13) with $f(x) = x^3 + \sin x$. ◄

Sometimes the rearrangement needed to express the integrand in a form suitable for logarithmic integration is a bit more subtle, as is illustrated by the following two examples.[5]

$$\int \cot ax \, dx = \int \frac{\cos ax}{\sin ax} dx = \frac{1}{a} \ln(\sin ax) + c,$$

$$\int \frac{1}{1 + e^{-kx}} dx = \int \frac{e^{kx}}{e^{kx} + 1} dx = \frac{1}{k} \ln(e^{kx} + 1) + c.$$

4.2.4 Integration using partial fractions

The method of partial fractions was discussed at some length in Section 2.3, but in essence consists of the manipulation of a fraction (here the integrand) in such a way that it can be written as the sum of two or more simpler fractions. In that discussion it was shown that

---

[4] Note that the constant term in the integrand in its final form gives the average value of the power of the sinusoid around a complete cycle. Thus $\cos^4 x$ has an average value of $\frac{3}{8}$, whilst $\sin^5 x$ has zero average value. An important result, worth memorising, is that both $\cos^2 x$ and $\sin^2 x$ have an average value of $\frac{1}{2}$.
[5] Use a similar method to that of the first to show that the indefinite integral of $\tan ax$ is $a^{-1} \ln(\sec ax) + c$.

4.2 Integration methods

each term in the partial fraction expansion of a rational fraction is of one of three types; each of these types can be integrated directly in a standard way. More specifically:

$$\frac{1}{x-a} \quad \text{integrates to} \quad \ln(x-a) + c,$$

$$\frac{1}{(x-a)^n} \quad \text{integrates to} \quad \frac{-1}{n-1} \frac{1}{(x-a)^{n-1}} + c, \quad n \neq 1,$$

$$\frac{Ax+B}{x^2+a^2} \quad \text{integrates to} \quad \frac{A}{2} \ln(x^2 + a^2) + \frac{B}{a} \tan^{-1} \frac{x}{a} + c.$$

We illustrate the method with a simple example.

Example Evaluate the integral

$$I = \int \frac{1}{x^2 + x} \, dx.$$

The denominator factorises as $x(x+1)$, and so we separate the integrand into two partial fractions and integrate each directly:

$$I = \int \left(\frac{1}{x} - \frac{1}{x+1} \right) dx = \ln x - \ln(x+1) + c = \ln \left(\frac{x}{x+1} \right) + c.$$

In this case, both terms were of the first type listed above and so each gave rise to a logarithm.[6] ◀

4.2.5 Integration by substitution

Sometimes it is possible to make a substitution of variables that turns a complicated integral into a simpler one, which can then be integrated by a standard method. There are many useful substitutions, but knowing which to use is a matter of experience. We now present a few examples of particularly useful ones.

Example Evaluate the integral

$$I = \int \frac{1}{\sqrt{1-x^2}} \, dx.$$

Making the substitution $x = \sin u$, we note that $dx = \cos u \, du$, and hence

$$I = \int \frac{1}{\sqrt{1-\sin^2 u}} \cos u \, du = \int \frac{1}{\sqrt{\cos^2 u}} \cos u \, du = \int du = u + c.$$

Now substituting back for u,

$$I = \sin^{-1} x + c.$$

This corresponds to one of the results given in Section 4.2.1. ◀

---

[6] Factorise $f(x) = x^4 + 2x^3 + 5x^2 + 8x + 4$ and so write down the general form of the integral of $[f(x)]^{-1}$, leaving multiplicative constants unevaluated.

Integral calculus

As a general guide, if an integrand in the form of a fraction contains a term $\sqrt{a^2 - x^2}$ in its denominator, then some progress can usually be made by making the substitution $x = a \sin u$. The reason for this is that dx then becomes $a \cos u$ and this $\cos u$ in the numerator cancels with the square root in the denominator, since the latter has also become $a \cos u$. This assumes that $|x| \leq a$ throughout the integration region (as it must be if the integrand is to remain real).

If a square root is of the form $\sqrt{x^2 - a^2}$, with $|x| \geq a$, then the appropriate substitution is $x = a \cosh u$, where $\cosh u$ is a hyperbolic cosine. The hyperbolic functions are introduced and discussed in Chapter 5, where it is shown that, with this substitution, both dx and the square root become $a \sinh u$.[7] If the square root is in the denominator, the two terms cancel, but, even if it is in the numerator, the explicit square root has been removed. Correspondingly, and based on the same relationship (see footnote), square roots of the form $\sqrt{x^2 + a^2}$ are dealt with by substituting $x = a \sinh u$, with both dx and the square root becoming $a \cosh u$.

Another particular example of integration by substitution is afforded by integrals of the form

$$I = \int \frac{1}{a + b \cos x} dx \quad \text{or} \quad I = \int \frac{1}{a + b \sin x} dx. \tag{4.14}$$

In these cases, making the substitution $t = \tan(x/2)$ yields integrals that can be solved more easily than the originals. Formulae expressing $\sin x$ and $\cos x$ in terms of t were derived in Equations (1.73) and (1.74) (see p. 31), but before we can use them we must relate dx to dt as follows.

Since

$$\frac{dt}{dx} = \frac{1}{2} \sec^2 \frac{x}{2} = \frac{1}{2}\left(1 + \tan^2 \frac{x}{2}\right) = \frac{1 + t^2}{2},$$

the required relationship is

$$dx = \frac{2}{1 + t^2} dt. \tag{4.15}$$

We now have all the elements needed to change integrals of the form (4.14) into integrals of rational polynomials, as is illustrated by the following example.

Example Evaluate the integral

$$I = \int \frac{2}{1 + 3 \cos x} dx.$$

Rewriting $\cos x$ in terms of t, as in Equation (1.74), and using (4.15) yields

$$I = \int \frac{2}{1 + 3\left[(1 - t^2)(1 + t^2)^{-1}\right]} \left(\frac{2}{1 + t^2}\right) dt$$

$$= \int \frac{2(1 + t^2)}{1 + t^2 + 3(1 - t^2)} \left(\frac{2}{1 + t^2}\right) dt$$

---

[7] The analogue for hyperbolic functions of $\cos^2 x + \sin^2 x = 1$ (for sinusoids) is $\cosh^2 x - \sinh^2 x = 1$.

4.2 Integration methods

$$= \int \frac{2}{2-t^2} dt = \int \frac{2}{(\sqrt{2}-t)(\sqrt{2}+t)} dt$$

$$= \int \frac{1}{\sqrt{2}} \left(\frac{1}{\sqrt{2}-t} + \frac{1}{\sqrt{2}+t} \right) dt$$

$$= -\frac{1}{\sqrt{2}} \ln(\sqrt{2}-t) + \frac{1}{\sqrt{2}} \ln(\sqrt{2}+t) + c$$

$$= \frac{1}{\sqrt{2}} \ln \left[\frac{\sqrt{2}+\tan(x/2)}{\sqrt{2}-\tan(x/2)} \right] + c.$$

In the final line we resubstituted for t in terms of x, the original variable.[8]

Integrals of a similar form to (4.14), but involving $\sin 2x$, $\cos 2x$, $\tan 2x$, $\sin^2 x$, $\cos^2 x$ or $\tan^2 x$, rather than $\cos x$ and $\sin x$, should be evaluated by using the substitution $t = \tan x$.[9] In this case

$$\sin x = \frac{t}{\sqrt{1+t^2}}, \quad \cos x = \frac{1}{\sqrt{1+t^2}} \quad \text{and} \quad dx = \frac{dt}{1+t^2}. \quad (4.16)$$

A final example of the evaluation of integrals using substitution is the method of completing the square (cf. Section 2.1). This method can be used where a quadratic expression in the variable of integration x appears in the integrand; a change to a new variable y that is a linear function of x, i.e. $y = ax + b$, reduces the quadratic expression to one containing no linear term in y, but without introducing any complication into that for dx, which simply becomes $a\,dy$. The following illustrates this procedure.

Example Evaluate the integral

$$I = \int \frac{1}{x^2 + 4x + 7} dx.$$

We can write the integral in the form

$$I = \int \frac{1}{(x+2)^2 + 3} dx.$$

Substituting $y = x + 2$, we have that $dy = dx$ and hence that

$$I = \int \frac{1}{y^2 + 3} dy.$$

---

[8] Show that the value of x at which this final integral $\to \infty$ is the same as that at which the original integrand $\to \infty$.
[9] Demonstrate, using the substitutions given by (4.16), that the integral of $\sin 2x$ is given, within an integration constant, by $\cos^2 x$, which, again within a constant, is equal to $\frac{1}{2}\cos 2x$.

Integral calculus

Comparison with the table of standard integrals (see Section 4.2.1) then shows that

$$I = \frac{1}{\sqrt{3}} \tan^{-1}\left(\frac{y}{\sqrt{3}}\right) + c = \frac{1}{\sqrt{3}} \tan^{-1}\left(\frac{x+2}{\sqrt{3}}\right) + c.$$

If, after completing the square, the denominator had been of the form $y^2 - b^2$, rather than $y^2 + b^2$, then the integral would have been of the general form $(2b)^{-1} \ln[(y-b)/(y+b)]$, rather than an inverse tangent. ◀

EXERCISES 4.2

1. Find the indefinite integrals of the following integrands, first rearranging them into standard form where necessary:

   (a) $\dfrac{a}{a-x}$, (b) $\dfrac{x}{a-x}$, (c) $\dfrac{3}{4+(x+1)^2}$, (d) $\dfrac{2}{x^2+4x+8}$, (e) $\cos 2x \sin x$.

2. Evaluate the definite integrals $\displaystyle\int_0^{\pi/4} \cos^5 x \, dx$ and $\displaystyle\int_0^{\pi/4} \sin^4 x \, dx$.

3. Find the indefinite integrals of

   (a) $\dfrac{x}{a^2 - x^2}$, (b) $\dfrac{x(x+3)}{x^3 + 3x^2 + x + 3}$, (c) $\dfrac{1}{x^3 + 3x^2 + x + 3}$.

4. Find the derivative of $\ln(\tan x)$ and hence evaluate $\displaystyle\int \operatorname{cosec} 2x \, dx$.

5. Evaluate

   (a) $\displaystyle\int x\sqrt{a^2 - x^2}\, dx$, (b) $\displaystyle\int \dfrac{1}{x^2\sqrt{a^2 - x^2}}\, dx$,

   (c) $\displaystyle\int \dfrac{1}{2 - \sin x}\, dx$, (d) $\displaystyle\int_0^{\pi/2} \dfrac{1}{1 + \sin 2x}\, dx$.

4.3 Integration by parts

Integration by parts is the integration analogy of product differentiation. The principle is to break down a complicated function into two functions, at least one of which can be integrated by inspection. In fact, the method relies on the result for the derivative of a product. Recalling from (3.7) that

$$\frac{d}{dx}(uv) = u\frac{dv}{dx} + \frac{du}{dx}v,$$

where u and v are functions of x, we now integrate to find

$$uv = \int u\frac{dv}{dx}\,dx + \int \frac{du}{dx}v\,dx.$$

4.3 Integration by parts

Rearranging into the standard form for integration by parts gives

$$\int u \frac{dv}{dx} dx = uv - \int \frac{du}{dx} v \, dx. \qquad (4.17)$$

Integration by parts is often remembered for practical purposes in the form *the integral of a product of two functions is equal to {the first times the integral of the second} minus the integral of {the derivative of the first times the integral of the second}*. Here, u is 'the first' and dv/dx is 'the second'; clearly the integral v of 'the second' must be determinable by inspection.

Example

Evaluate the integral $I = \int x \sin x \, dx$.

In the notation given above, we identify x with u and $\sin x$ with dv/dx. Hence $v = -\cos x$ and $du/dx = 1$ and so using (4.17)

$$I = x(-\cos x) - \int (1)(-\cos x) \, dx = -x \cos x + \sin x + c.$$

Since both x and $\sin x$ can be both integrated and differentiated by inspection, there was a decision to be made about which would be u and which would be dv/dx in the general formula. The actual choice of x as u was dictated by the fact that x 'gets simpler' when it is differentiated and more complicated when integrated, whereas $\sin x$ 'stays about the same' in terms of complexity under either operation. ◀

The separation of the functions is not always so apparent, as is illustrated by the following example.

Example

Evaluate the integral $I = \int x^3 e^{-x^2} \, dx$.

Firstly we rewrite the integral as

$$I = \int x^2 \left(x e^{-x^2} \right) dx.$$

Now, using the notation given above, we identify x^2 with u and xe^{-x^2} with dv/dx. Hence $v = -\frac{1}{2} e^{-x^2}$ and $du/dx = 2x$, and so

$$I = -\frac{1}{2} x^2 e^{-x^2} - \int (-x) e^{-x^2} \, dx = -\frac{1}{2} x^2 e^{-x^2} - \frac{1}{2} e^{-x^2} + c.$$

Here, there was very little real choice for u. It had to be whatever was left over from x^3 after something proportional to the derivative $(-2x)$ of the exponent $(-x^2)$ of the exponential had been taken from it as a factor; without the derivative of the exponent being included in what was taken as dv/dx it would not be possible to carry out even one stage of integration. ◀

A trick that is useful when the integral of the given integrand is not known, but its derivative is, is to take '1' and the integrand as the two factors in a product, which is then integrated by parts. This is illustrated by the following example.

Integral calculus

Example

Evaluate the integral $I = \int \ln x \, dx$.

Firstly we rewrite the integral as

$$I = \int (\ln x) 1 \, dx.$$

Now, using the notation above, we identify $\ln x$ with u and 1 with dv/dx. Hence we have $v = x$ and $du/dx = 1/x$, and so

$$I = (\ln x)(x) - \int \left(\frac{1}{x}\right) x \, dx = x \ln x - x + c.$$

When this method is used, the '1' must always be identified with dv/dx, and the original integrand with u.[10]

It is sometimes necessary to integrate by parts more than once. In doing so, we may occasionally encounter a multiple of the original integral I. In such cases we can obtain a linear algebraic equation for I that can be solved to obtain its value.

Example

Evaluate the integral $I = \int e^{ax} \cos bx \, dx$.

Integrating by parts, taking e^{ax} as the first function,[11] we find

$$I = e^{ax}\left(\frac{\sin bx}{b}\right) - \int ae^{ax}\left(\frac{\sin bx}{b}\right) dx,$$

where, for convenience, we have omitted the constant of integration. Integrating by parts a second time,

$$I = e^{ax}\left(\frac{\sin bx}{b}\right) - ae^{ax}\left(\frac{-\cos bx}{b^2}\right) + \int a^2 e^{ax}\left(\frac{-\cos bx}{b^2}\right) dx.$$

Notice that the integral on the RHS is just $-a^2/b^2$ times the original integral I. Thus

$$I = e^{ax}\left(\frac{1}{b}\sin bx + \frac{a}{b^2}\cos bx\right) - \frac{a^2}{b^2}I.$$

Rearranging this expression to obtain I explicitly and including the constant of integration we find

$$I = \frac{e^{ax}}{a^2 + b^2}(b \sin bx + a \cos bx) + c. \quad (4.18)$$

Another method of evaluating this integral, using the exponential of a complex number, is given in Section 5.6.

---

[10] Use the same technique to show that the integral of $\sin^{-1} x$ between 0 and 1 has the value $\frac{1}{2}\pi - 1$.
[11] For this particular integral it does not matter whether $\cos bx$ or e^{ax} is taken as u, since both generate multiples of themselves after two integrations or differentiations. The reader should check this by taking $\cos bx$ as u, the opposite choice to that made in the text.

4.4 Reduction formulae

EXERCISES 4.3

1. Evaluate the following indefinite integrals:

   (a) $\int (3x^2 + 2x)e^{-x}\, dx$, (b) $\int x^5 e^{-x^3}\, dx$, (c) $\int \tan^{-1} x\, dx$.

2. Evaluate $\int \sin x \sin 2x\, dx$ by (a) rewriting the integrand and (b) using repeated integration by parts, showing that the two methods produce the same result.

4.4 Reduction formulae

Integration using reduction formulae is a process that involves first evaluating a simple integral and then, in stages, using it to find a more complicated one. In general structure, the procedure follows the same lines as proof by induction (see Section 2.4.1), in that the only direct calculations are for simple cases, with more complicated ones being tackled by indirect methods. Reduction formulae also share with induction the feature that a positive integer parameter characterises the various stages.

In practice, calculations using reduction formulae usually start with a form that is expressed in terms of a general integer parameter n and then aim to relate that to one with a lower value of n; that relationship is the *reduction formula*. The formula is then applied repeatedly until the original integral is related to one in which n has a very low value, usually 0 or 1, but occasionally 2 or 3. Finally, that low-n integral is evaluated directly. The method can be illustrated by the following worked example.

Example Using integration by parts, find a relationship between I_n and I_{n-1} where

$$I_n = \int_0^1 (1 - x^3)^n\, dx$$

and n is any positive integer. Hence evaluate $I_2 = \int_0^1 (1 - x^3)^2\, dx$.

Writing the integrand as a product and separating the integral into two we find

$$I_n = \int_0^1 (1 - x^3)(1 - x^3)^{n-1}\, dx$$

$$= \int_0^1 (1 - x^3)^{n-1}\, dx - \int_0^1 x^3(1 - x^3)^{n-1}\, dx.$$

The first term on the RHS is clearly I_{n-1} and so, writing the integrand in the second term on the RHS as a product,

$$I_n = I_{n-1} - \int_0^1 (x) x^2 (1 - x^3)^{n-1}\, dx.$$

Integral calculus

Integrating by parts we find

$$I_n = I_{n-1} + \left[\frac{x}{3n}(1-x^3)^n\right]_0^1 - \int_0^1 \frac{1}{3n}(1-x^3)^n \, dx$$

$$= I_{n-1} + 0 - \frac{1}{3n} I_n,$$

which on rearranging gives

$$I_n = \frac{3n}{3n+1} I_{n-1}.$$

We now have a reduction formula connecting successive integrals. Hence, if we can evaluate I_0, we can find I_1, I_2 etc. Evaluating I_0 is trivial:

$$I_0 = \int_0^1 (1-x^3)^0 \, dx = \int_0^1 dx = [x]_0^1 = 1.$$

Hence

$$I_1 = \frac{(3 \times 1)}{(3 \times 1) + 1} \times 1 = \frac{3}{4}, \qquad I_2 = \frac{(3 \times 2)}{(3 \times 2) + 1} \times \frac{3}{4} = \frac{9}{14}.$$

Although the first few I_n could be evaluated by direct expansion of the integrand, using the binomial theorem, followed by direct integration of terms like x^r, this becomes tedious for integrals containing higher values of n; these are therefore best evaluated using the reduction formula, which gives

$$I_n = \frac{3^n \, n!}{\prod_{r=0}^{n}(3r+1)}$$

as the general result.

EXERCISE 4.4

1. Using the identity $\sec^2 x = 1 + \tan^2 x$, find a reduction formula for $I_n = \int \tan^n x \, dx$, where n is a non-negative integer. Hence write down a general expression for I_n, distinguishing between n even and n odd. Evaluate the definite integrals

$$\int_0^{\pi/4} \tan^4 x \, dx \qquad \text{and} \qquad \int_0^{\pi/4} \tan^5 x \, dx.$$

4.5 Infinite and improper integrals

The definition of an integral given previously does not allow for cases in which either of the limits of integration is infinite (an *infinite integral*) or for cases in which $f(x)$ is infinite in some part of the range (an *improper integral*), e.g. $f(x) = (2-x)^{-1/4}$ near the point $x = 2$. Nevertheless, modification of the definition of an integral gives infinite and improper integrals each a meaning.

4.5 Infinite and improper integrals

In the case of an integral $I = \int_a^b f(x)\,dx$, the infinite integral, in which b tends to ∞, is defined by

$$I = \int_a^\infty f(x)\,dx = \lim_{b \to \infty} \int_a^b f(x)\,dx = \lim_{b \to \infty} F(b) - F(a).$$

As previously, $F(x)$ is the indefinite integral of $f(x)$ and $\lim_{b \to \infty} F(b)$ means the limit (or value) that $F(b)$ approaches as $b \to \infty$; it is evaluated *after* calculating the integral. The formal concept of a limit will be introduced in Chapter 6, but for the present purposes an intuitive interpretation will be sufficient.

Of course it may happen that as $b \to \infty$ the indefinite integral $F(b)$ does not tend to a limit; in that case, the infinite integral is not defined. To take an example that has already been discussed in Appendix A, the integral of x^{-1} between a and b is given by

$$\int_a^b \frac{1}{x}\,dx = \ln b - \ln a.$$

As $b \to \infty$ so does $\ln b$, albeit very slowly. Consequently, $\ln b$ does not approach any limit and as a result the infinite integral of x^{-1} is undefined. Our later, more precise, treatment of limits will not change this conclusion. It will be seen that the existence or otherwise of an infinite integral has to be tested against the definition on a case-by-case basis.

Integrals for which the two limits are $\pm\infty$ are first evaluated between limits of $\pm b$, and then b is allowed to approach ∞. This can result in either outcome. For example:

$$\int_{-\infty}^\infty x^3\,dx = \lim_{b \to \infty} \left[\frac{x^4}{4}\right]_{-b}^b = \lim_{b \to \infty} [\tfrac{1}{4}(b^4 - b^4)] = \lim_{b \to \infty} 0 = 0$$

has a well-defined limit and the integral is defined, but

$$\int_{-\infty}^\infty e^x\,dx = \lim_{b \to \infty} \left[e^x\right]_{-b}^b = \lim_{b \to \infty} (e^b - e^{-b}) = \infty - 0 = \infty$$

does not and the integral is undefined.

As a non-trivial example of a defined infinite integral, consider the following.

Example Evaluate the integral

$$I = \int_0^\infty \frac{x}{(x^2 + a^2)^2}\,dx.$$

Integrating, we find $F(x) = -\tfrac{1}{2}(x^2 + a^2)^{-1} + c$ and so

$$I = \lim_{b \to \infty} \left[\frac{-1}{2(b^2 + a^2)}\right] - \left(\frac{-1}{2a^2}\right) = \frac{1}{2a^2}.$$

The value of the constant of integration c is immaterial, as it always is for integrals between definite limits, here 0 and b.

Integral calculus

To define improper integrals, we adopt the approach of excluding the unbounded range from the integral, next performing the integration, and then letting the length of the excluded range tend to zero. If the value of the integral tends to a definite limit as the excluded length tends to zero, then that limit is defined as the value of the improper integral. Expressed in more mathematical terms, if the integrand $f(x)$ is infinite at $x = c$ (say), with $a \leq c \leq b$, then

$$\int_a^b f(x)\,dx = \lim_{\delta \to 0} \int_a^{c-\delta} f(x)\,dx + \lim_{\epsilon \to 0} \int_{c+\epsilon}^b f(x)\,dx,$$

provided both limits exist.[12] The following example illustrates the procedure.

Example

Evaluate the integral $I = \int_0^2 (2-x)^{-1/4}\,dx$.

Since the integrand becomes infinite at $x = 2$ and this is in the integration range, we initially cut out the interval from $x = 2 - \epsilon$ to $x = 2$. Then, integrating directly gives

$$I = \left[-\tfrac{4}{3}(2-x)^{3/4}\right]_0^{2-\epsilon}.$$

We now test whether this tends to a finite limit as ϵ is allowed to tend to zero.

$$I = \lim_{\epsilon \to 0}\left[-\tfrac{4}{3}\epsilon^{3/4}\right] + \tfrac{4}{3} 2^{3/4} = \left(\tfrac{4}{3}\right) 2^{3/4}.$$

It clearly does, and that limit is therefore the value of I. Notice that the result does not, and must not, depend upon any particular value of ϵ. ◀

EXERCISES 4.5

1. Determine whether the following infinite integrals exist and, where they do, evaluate them:

(a) $\int_0^\infty x e^{-\lambda x^2}\,dx,$ (b) $\int_0^\infty \dfrac{x}{a^2 + x^2}\,dx,$

(c) $\int_{-\infty}^\infty \dfrac{x}{a^2 + x^2}\,dx,$ (d) $\int_0^\infty \sin x\,dx.$

2. Determine whether the following improper integrals exist and, where they do, evaluate them:

(a) $\int_0^{\pi/2} \tan\theta\,d\theta,$ (b) $\int_{-\pi/2}^{\pi/2} \tan\theta\,d\theta,$

(c) $\int_0^1 \dfrac{x^2}{(1-x^3)^{3/4}}\,dx,$ (d) $\int_0^1 \dfrac{1}{(3x-1)^2}\,dx.$

---

[12] If a common quantity, h say, is used instead of both δ and ϵ, and the limit exists, then the limit is called the *principal value* of the integral.

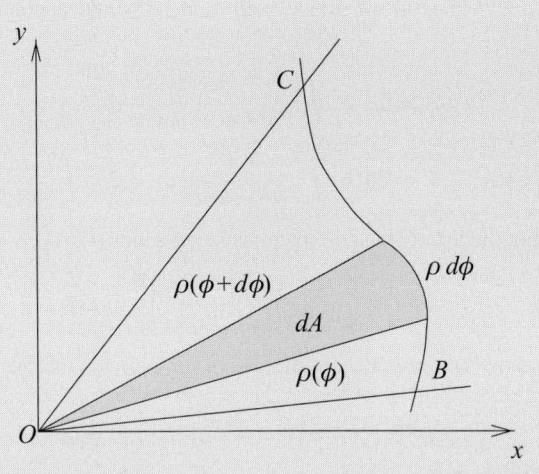

Figure 4.3 Finding the area of a sector OBC defined by the curve $\rho(\phi)$ and the radii OB, OC, at angles to the x-axis ϕ_1, ϕ_2 respectively.

4.6 Integration in plane polar coordinates

As described in Section 2.2.4, a curve is defined in plane polar coordinates ρ, ϕ by its distance ρ from the origin as a function of the angle ϕ between the line joining a point on the curve to the origin and the x-axis, i.e. $\rho = \rho(\phi)$. The size of an element of area is given in the same coordinate system by $dA = \tfrac{1}{2}\rho^2 \, d\phi$, as is illustrated in Figure 4.3. The total area enclosed by the curve in the sector defined by angles ϕ_1 and ϕ_2 is therefore given by

$$A = \int_{\phi_1}^{\phi_2} \tfrac{1}{2}\rho^2 \, d\phi. \tag{4.19}$$

One immediate, but hardly novel, deduction from this is that the area of a circle of radius a, is given by

$$A = \int_0^{2\pi} \tfrac{1}{2}a^2 \, d\phi = \left[\tfrac{1}{2}a^2 \phi\right]_0^{2\pi} = \pi a^2.$$

A more substantial calculation is provided by the following example.

Example The equation in polar coordinates of an ellipse with semi-axes a and b is

$$\frac{1}{\rho^2} = \frac{\cos^2 \phi}{a^2} + \frac{\sin^2 \phi}{b^2}.$$

Find the area A of the ellipse.

Using (4.19) and symmetry, we have

$$A = \frac{1}{2} \int_0^{2\pi} \frac{a^2 b^2}{b^2 \cos^2 \phi + a^2 \sin^2 \phi} \, d\phi = 2a^2 b^2 \int_0^{\pi/2} \frac{1}{b^2 \cos^2 \phi + a^2 \sin^2 \phi} \, d\phi.$$

To evaluate this integral we divide both numerator and denominator by $\cos^2 \phi$ and then write $\tan \phi = t$:[13]

$$A = 2a^2 b^2 \int_0^{\pi/2} \frac{\sec^2 \phi}{b^2 + a^2 \tan^2 \phi} d\phi$$

$$= 2a^2 b^2 \int_0^\infty \frac{1}{b^2 + a^2 t^2} dt = 2b^2 \int_0^\infty \frac{1}{(b/a)^2 + t^2} dt.$$

Finally, using the list of standard integrals (see Section 4.2.1),

$$A = 2b^2 \left[\frac{1}{(b/a)} \tan^{-1} \frac{t}{(b/a)} \right]_0^\infty = 2ab \left(\frac{\pi}{2} - 0 \right) = \pi ab.$$

Of course, if we let $a = b$, the familiar result for a circle is recovered.

EXERCISE 4.6

1. Using symmetry to avoid any ambiguity concerned with 'negative ρ-values', find the total areas of (a) the lemniscate of Bernoulli, $\rho^2 = a^2 \cos 2\phi$, and (b) the particular limaçon $\rho = \frac{1}{2}a(3 + 2\cos\phi)$.

4.7 Integral inequalities

Consider the functions $f(x)$, $\phi_1(x)$ and $\phi_2(x)$ such that $\phi_1(x) \leq f(x) \leq \phi_2(x)$ for all x in the range $a \leq x \leq b$. It immediately follows that

$$\int_a^b \phi_1(x)\, dx \leq \int_a^b f(x)\, dx \leq \int_a^b \phi_2(x)\, dx, \tag{4.20}$$

which gives us a way of putting bounds on the value of an integral that is difficult to evaluate explicitly.

Example Show that the value of the integral

$$I = \int_0^1 \frac{1}{(1 + x^2 + x^3)^{1/2}}\, dx$$

lies between 0.810 and 0.882.

What makes this integral difficult to evaluate is the x^3 term in the denominator. If it were absent, we would have an integrand of the form $(a^2 + x^2)^{-1/2}$; this could be handled in closed form, as we indicated in Section 4.2.5. We therefore need to put bounds on x^3, with the bounds expressed in terms of functions that can be managed. We note that for x in the range $0 \leq x \leq 1$, the double inequality $0 \leq x^3 \leq x^2$ holds. Hence

$$(1 + x^2)^{1/2} \leq (1 + x^2 + x^3)^{1/2} \leq (1 + 2x^2)^{1/2},$$

---

[13] Verify that the direct substitution of the relationships given in (4.16) gives the same t-integral.

and so
$$\frac{1}{(1+x^2)^{1/2}} \geq \frac{1}{(1+x^2+x^3)^{1/2}} \geq \frac{1}{(1+2x^2)^{1/2}}.$$

Consequently,
$$\int_0^1 \frac{1}{(1+x^2)^{1/2}}\,dx \geq \int_0^1 \frac{1}{(1+x^2+x^3)^{1/2}}\,dx \geq \int_0^1 \frac{1}{(1+2x^2)^{1/2}}\,dx.$$

We have not yet found the integral of $(a^2+x^2)^{-1/2}$; it can be expressed as $\ln(x+\sqrt{a^2+x^2})$ or, in terms of inverse hyperbolic functions (see Chapter 5), as $\sinh^{-1}(x/a)$. The first of these can be verified by direct differentiation. Taking this result for granted at this stage, we have

$$\left[\ln(x+\sqrt{1+x^2})\right]_0^1 \geq I \geq \left[\tfrac{1}{\sqrt{2}}\ln\left(x+\sqrt{\tfrac{1}{2}+x^2}\right)\right]_0^1$$

$$0.8814 \geq I \geq 0.8105$$

$$0.882 \geq I \geq 0.810.$$

In the last line the calculated values have been rounded to three significant figures, one rounded up and the other rounded down so that the proved inequality cannot be unknowingly made invalid. ◂

EXERCISE 4.7

1. Noting that, for $0 \leq x \leq \pi/2$, the double inequality $2x/\pi \leq \sin x \leq x$ holds, find to 3 s.f. limits for the value of $I = \int_0^{\pi/2} \frac{1}{1+\sin^2 x}\,dx$. Using an appropriate substitution, evaluate I exactly and so verify the validity of the limits.

4.8 Applications of integration

In this section we give brief outlines of some standard procedures that involve the use of integration. They typically form only a part of a larger calculation, and each would not normally be specified in any more detail than that given by the corresponding subsection heading.

4.8.1 Mean value of a function

The mean value m of a function between two limits a and b is defined by

$$m = \frac{1}{b-a} \int_a^b f(x)\,dx. \qquad (4.21)$$

The mean value may be thought of as the height of the rectangle that has the same area (over the same interval) as the area under the curve $f(x)$. This is illustrated in Figure 4.4.

Integral calculus

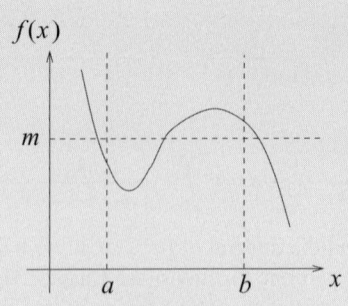

Figure 4.4 The mean value m of a function.

Example Find the mean value m of the function $f(x) = x^2$ between the limits $x = 2$ and $x = 4$.

Using (4.21),

$$m = \frac{1}{4-2} \int_2^4 x^2 \, dx = \frac{1}{2} \left[\frac{x^3}{3}\right]_2^4 = \frac{1}{2}\left(\frac{4^3}{3} - \frac{2^3}{3}\right) = \frac{28}{3}.$$

As expected, because x^2 increases more rapidly than x, this result is (slightly) more than the square of the mid-value of x over the given range; that would give $3^2 = 9$. ◀

4.8.2 Finding the length of a curve

Finding the area between a curve and certain straight lines provides one example of the use of integration, as we saw in Section 4.6. Another is that of finding the length of a curve. If a curve is defined by $y = f(x)$ then the distance along the curve, Δs, that corresponds to small changes Δx and Δy in x and y is given by

$$\Delta s \approx \sqrt{(\Delta x)^2 + (\Delta y)^2}; \qquad (4.22)$$

this follows directly from Pythagoras' theorem (see Figure 4.5). Dividing (4.22) through by Δx and letting $\Delta x \to 0$ we obtain[14]

$$\frac{ds}{dx} = \sqrt{1 + \left(\frac{dy}{dx}\right)^2}. \qquad (4.23)$$

Clearly the total length s of the curve between the points $x = a$ and $x = b$ is then given by integrating both sides of the equation:

$$s = \int_a^b \sqrt{1 + \left(\frac{dy}{dx}\right)^2} \, dx. \qquad (4.24)$$

The following provides a simple example of the use of this method.

---

[14] Instead of considering small changes Δx and Δy and letting these tend to zero, we could have derived (4.23) by considering infinitesimal changes dx and dy from the start. After writing $(ds)^2 = (dx)^2 + (dy)^2$, (4.23) may be deduced by using the formal device of dividing through by dx. Although not mathematically rigorous, this method is often used and generally leads to the correct result.

4.8 Applications of integration

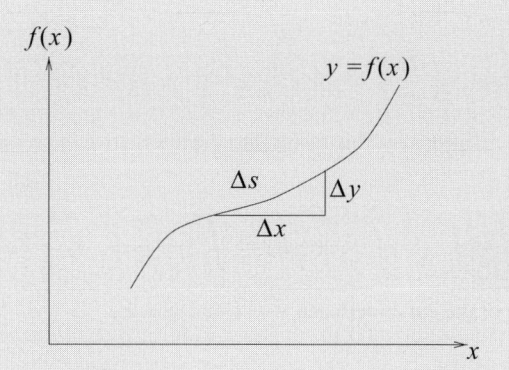

Figure 4.5 The distance moved along a curve, Δs, corresponding to the small changes Δx and Δy.

Example Find the length of the curve $y = x^{3/2}$ from $x = 0$ to $x = 2$.

Using (4.24) and noting that $dy/dx = \frac{3}{2}\sqrt{x}$, the length s of the curve is given by

$$s = \int_0^2 \sqrt{1 + \tfrac{9}{4}x}\, dx$$

$$= \left[\tfrac{2}{3}\left(\tfrac{4}{9}\right)\left(1 + \tfrac{9}{4}x\right)^{3/2}\right]_0^2 = \tfrac{8}{27}\left[\left(1 + \tfrac{9}{4}x\right)^{3/2}\right]_0^2$$

$$= \tfrac{8}{27}\left[\left(\tfrac{11}{2}\right)^{3/2} - 1\right].$$

For a more general power curve $y = x^n$ ($n > 0$), the integration would not be so straightforward; $n = 3/2$ gives a linear function of x under the square root sign and so makes the integration elementary. ◀

Although less often done, it is equally valid to divide (4.22) through by Δy and let $\Delta y \to 0$ and so obtain

$$\frac{ds}{dy} = \sqrt{1 + \left(\frac{dx}{dy}\right)^2} \quad \text{leading to} \quad s = \int_c^d \sqrt{1 + \left(\frac{dx}{dy}\right)^2}\, dy, \qquad (4.25)$$

where c and d are the y-values marking the beginning and end of the curve. If the extremes of the curve are given in this form, then this can be the best way to proceed. The hyperbolic functions $\cosh x$ and $\sinh x$ are not introduced formally until the next chapter, but we can use them expressed in exponential form to provide a worked example.

Example Find the length of the curve given by $y(x) = \tfrac{1}{2}(e^x + e^{-x})$ between the points at which $y = 1$ and $y = Y$, where $Y > 1$.

The $y = 1$ end of the curve clearly corresponds to $x = 0$, but solving $Y = \tfrac{1}{2}(e^x + e^{-x})$ for x is somewhat more complicated (though it can be done; see Section 5.7.5) and so we make y the variable of integration. We need $(dx/dy)^2$ but, given the equation of the curve, it is easier to calculate

$(dy/dx)^2$ as

$$\left(\frac{dy}{dx}\right)^2 = \left[\tfrac{1}{2}(e^x - e^{-x})\right]^2 = \left[\tfrac{1}{2}(e^x + e^{-x})\right]^2 - 1 = y^2 - 1.$$

Inserting the reciprocal of this result into the alternative expression for s and using the limits for y (not x) gives

$$s = \int_1^Y \sqrt{1 + \frac{1}{y^2 - 1}}\, dy = \int_1^Y \frac{y}{\sqrt{y^2 - 1}}\, dy = \left[\sqrt{y^2 - 1}\right]_1^Y = \sqrt{Y^2 - 1}$$

as the length of the curve between $y = 1$ and $y = Y$. ◀

In the other two-dimensional coordinate system we have met so far, namely plane polar coordinates, the corresponding expression for the length of a curve is

$$ds = \sqrt{(d\rho)^2 + (\rho\, d\phi)^2},$$

leading to

$$s = \int_{\rho_1}^{\rho_2} \sqrt{1 + \rho^2 \left(\frac{d\phi}{d\rho}\right)^2}\, d\rho \quad \text{or} \quad s = \int_{\phi_1}^{\phi_2} \sqrt{\rho^2 + \left(\frac{d\rho}{d\phi}\right)^2}\, d\phi \qquad (4.26)$$

For the simple spiral given by $\rho = b\phi$, the two equivalent expressions for the length of the spiral up to the point where it has completed one 'orbit' are

$$s = \int_0^{2\pi b} \sqrt{1 + \frac{\rho^2}{b^2}}\, d\rho \quad \text{or} \quad s = b \int_0^{2\pi} \sqrt{\phi^2 + 1}\, d\phi.$$

These integrals could be evaluated in terms of the hyperbolic functions that are studied in the next chapter, but doing so would add little to the main point of this subsection.[15]

4.8.3 Surfaces of revolution

Whilst it is easy to give an expression for the curved surface area of a uniform circular cylinder – $2\pi rh$ for a cylinder of constant radius r and length h – it is less straightforward if the radius of the cylinder varies along its length. In the former case, we could imagine the surface cut and rolled out into a flat rectangle with sides of $2\pi r$ and h, but, for the latter, the unrolled surface would not have a shape for which a readily available expression gives its area.

To make it into a set of calculable areas, we could cut the cylinder surface, perpendicularly to the axis of symmetry, into narrow strips; the length of any particular strip would be 2π times the local radius r of the cylinder. If the strip were the result of two cuts made a distance Δx apart, measured along the axis of the cylinder, then the width of the strip when laid out flat would be Δs, with Δs as given by Equation (4.22).[16] The area of the strip would therefore be $2\pi r \Delta s$ and the area of the surface would be the sum of all

---

[15] Extend this formalism to three dimensions and show that the length of thread on a uniform machine screw that has radius a and pitch h is $(h^2 + 4\pi^2 a^2)^{1/2}$ per turn.
[16] Note that, in general, Δs will be larger than Δx, and that it will never be smaller.

4.8 Applications of integration

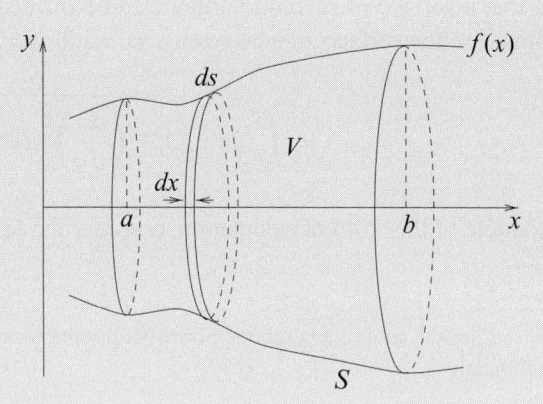

Figure 4.6 The surface and volume of revolution for the curve $y = f(x)$.

such quantities. This approach, in the limit that Δx becomes infinitesimal, is the basis of finding the area using integration. The following derivation is a much terser mathematical description of this procedure.

Consider the surface S formed by rotating the curve $y = f(x)$ about the x-axis (see Figure 4.6). The surface area of the 'collar' formed by rotating an element of the curve, ds, about the x-axis is $2\pi y \, ds$, and hence the total surface area is

$$S = \int_a^b 2\pi y \, ds.$$

Since, from (4.23), $ds = [1 + (dy/dx)^2]^{1/2} \, dx$, the total surface area between the planes $x = a$ and $x = b$ is

$$S = \int_a^b 2\pi y \sqrt{1 + \left(\frac{dy}{dx}\right)^2} \, dx. \qquad (4.27)$$

We now illustrate this result with a simple example.

Example Find the curved surface area of the cone formed by rotating about the x-axis the line $y = 2x$ between $x = 0$ and $x = h$.

Using (4.27), the surface area is given by

$$S = \int_0^h (2\pi) 2x \sqrt{1 + \left[\frac{d}{dx}(2x)\right]^2} \, dx$$

$$= \int_0^h 4\pi x \left(1 + 2^2\right)^{1/2} dx = \int_0^h 4\sqrt{5}\pi x \, dx$$

$$= \left[2\sqrt{5}\pi x^2\right]_0^h = 2\sqrt{5}\pi (h^2 - 0) = 2\sqrt{5}\pi h^2.$$

As it must be, this result is in agreement with the standard formula for the area of the curved surface of a cone, namely $S = \pi r \ell$, where r is the radius of its base (here $r = 2h$) and ℓ is its slope length, given in this case by $\ell = \sqrt{h^2 + (2h)^2} = \sqrt{5}h$.

◂

166　Integral calculus

We note that a surface of revolution may also be formed by rotating a line about the y-axis. In this case the surface area between $y = a$ and $y = b$ is

$$S = \int_a^b 2\pi x \sqrt{1 + \left(\frac{dx}{dy}\right)^2} \, dy. \tag{4.28}$$

As an example of this kind of calculation, consider the following problem.

Example Find the curved surface area of a parabolic bowl that has the form $x^2 = 4ay$, a base that is $4a$ in diameter, and height h.

Most of the calculation consists of algebraic manipulation, but we do need to find dx/dy. This is easily done by differentiating $x^2 = 4ay$ and obtaining $2x(dx/dy) = 4a$. As the bowl has a base of radius $2a$, its profile corresponds to the section of the parabola between $y = (2a)^2/4a = a$ and $y = a + h$. Substitution into (4.28) gives

$$S = \int_a^{a+h} 2\pi x \sqrt{1 + \left(\frac{2a}{x}\right)^2} \, dy$$

$$= \int_a^{a+h} 2\pi \sqrt{x^2 + 4a^2} \, dy$$

$$= \int_a^{a+h} 2\pi \sqrt{4ay + 4a^2} \, dy$$

$$= \int_a^{a+h} 4\pi \sqrt{a} \sqrt{y + a} \, dy$$

$$= 4\pi \sqrt{a} \left[\tfrac{2}{3}(y+a)^{3/2}\right]_a^{a+h}$$

$$= \tfrac{8}{3}\pi \sqrt{a} \left[(2a+h)^{3/2} - (2a)^{3/2}\right].$$

It should be noted that to obtain the integrand in the third line entirely in terms of y we used the curve-defining equation to replace x^2, and that it is the y-limits, a and $a + h$, that are appropriate. ◀

4.8.4　Volumes of revolution

The volume V enclosed by rotating the portion of the curve $y = f(x)$ between $x = a$ and $x = b$ about the x-axis (see Figure 4.6) can also be found using integration.

We consider the complete volume as made up of a very large number (formally, an infinitely large number) of thin discs, each of thickness dx. The volume of the single disc that lies between between x and $x + dx$, and has radius $y(x)$, is given by $dV = \pi y^2 \, dx$; since the disc is vanishingly thin, we can ignore any variation of its radius within the disc – more formally, the contribution to the volume of this variation is second order in

4.8 Applications of integration

dx. To obtain the total volume enclosed by the rotating curve, we integrate this infinitesimal volume between $x = a$ and $x = b$:

$$V = \int_a^b \pi y^2 \, dx. \tag{4.29}$$

Our final worked example uses the same cone as the first example in the previous subsection.

Example Find the volume of the cone enclosed by the surface that is formed when the portion of the line $y = 2x$ between $x = 0$ and $x = h$ is rotated about the x-axis.

Using (4.29), the volume is given by

$$V = \int_0^h \pi (2x)^2 \, dx = \int_0^h 4\pi x^2 \, dx$$

$$= \left[\tfrac{4}{3}\pi x^3\right]_0^h = \tfrac{4}{3}\pi (h^3 - 0) = \tfrac{4}{3}\pi h^3.$$

Again agreement is obtained with a standard formula: for a cone, the volume $V = \tfrac{1}{3}\pi r^2 h$, with $r = 2h$ in the current case.

◀

As before, it is also possible to form a volume of revolution by rotating a curve about the y-axis. In this case,

$$V = \int_a^b \pi x^2 \, dy. \tag{4.30}$$

gives the volume enclosed between $y = a$ and $y = b$.[17]

EXERCISES 4.8

1. Find the mean values of the following functions over the ranges indicated:

   (a) x^3 in $[0, 2]$, (b) x^3 in $[-2, 2]$, (c) $\sin \theta$ in $[0, \pi]$,
   (d) $\tan^2 \theta$ in $[0, \pi/4]$, (e) $x^3 e^{-x}$ in $[0, \infty]$.

2. Show that the length $L(x)$ of the curve $y = \ln(\cos x)$, with $0 \le x \le \pi/2$, as measured from the origin $x = 0$, $y = 0$, is given by $L(x) = \int_0^x \sec u \, du$. Using the substitution $t = \tan u/2$, evaluate $L(x)$ as

$$L(x) = \ln \frac{1 + \tan(x/2)}{1 - \tan(x/2)}.$$

---

[17] Show that the capacity of the parabolic bowl discussed in the second worked example of the previous subsection is $2\pi ah(2a + h)$.

168 Integral calculus

3. Show that the (outside) surface area of a flat-bottomed, straight-sided tumbler, $4a$ high, that has a base diameter of $2a$ and a diameter of $3a$ at its widest part, is

$$\pi a^2 \left(1 + \frac{5\sqrt{65}}{4}\right).$$

4. By considering them as a surface and volume of revolution generated by the semi-circular arc $x^2 + y^2 = a^2$, establish the well-known formulae for the surface area and volume of a sphere.

5. Find the volume of the solid obtained by rotating the curve $y = x(1-x)$ for $0 \le x \le 1$ around the x-axis.

SUMMARY

1. *Elementary properties of integrals*

$$\int_a^b 0 \, dx = 0, \qquad \int_a^a f(x) \, dx = 0, \qquad \int_a^b f(x) \, dx = -\int_b^a f(x) \, dx,$$

$$\int_a^c f(x) \, dx = \int_a^b f(x) \, dx + \int_b^c f(x) \, dx,$$

$$\int_a^b [f(x) + g(x)] \, dx = \int_a^b f(x) \, dx + \int_a^b g(x) \, dx,$$

$$\frac{d}{dx} F(x) \equiv \frac{d}{dx} \left[\int_{x_0}^x f(u) \, du\right] = f(x).$$

2. *Standard integrals*
   - For the integrals of elementary functions, including exponentials and sinusoids, see p. 146.
   - Some particularly important cases for physical science:

$$\int \sin x \, dx = -\cos x + c, \qquad \int \cos x \, dx = \sin x + c,$$

$$\int \frac{1}{a^2 + x^2} \, dx = \frac{1}{a} \tan^{-1}\left(\frac{x}{a}\right) + c,$$

$$\int_0^{n\pi/2} \cos^2 x \, dx = \frac{n\pi}{4} = \int_0^{n\pi/2} \sin^2 x \, dx,$$

$$\int_{x_0}^{x_0 + (n\pi/\alpha)} \cos^2(\alpha x) \, dx = \frac{n\pi}{2\alpha} = \int_{x_0}^{x_0 + (n\pi/\alpha)} \sin^2(\alpha x) \, dx.$$

Summary

3. *Common substitutions*
   With $t = \tan\theta/2$,

$$\sin\theta = \frac{2t}{1+t^2}, \quad \cos\theta = \frac{1-t^2}{1+t^2}, \quad d\theta = \frac{2}{1+t^2}\,dt.$$

Integrand contains	Substitution	Differential
$\sqrt{a^2 - x^2}$	$x = a\sin u$	$dx = a\cos u\,du$
$\sqrt{a^2 + x^2}$	$x = a\sinh u$	$dx = a\cosh u\,du$,
$\sqrt{x^2 - a^2}$	$x = a\cosh u$	$dx = a\sinh u\,du$

4. *Integration by parts*

$$\int u\frac{dv}{dx}\,dx = uv - \int \frac{du}{dx}v\,dx$$

or

$$\int uw\,dx = u\int^x w\,dx' - \int \frac{du}{dx}\left(\int^x w\,dx'\right)dx.$$

It is sometimes helpful to use the second form with w as (a hidden) unity.

5. *Infinite and improper integrals*
   - $\displaystyle\int_a^\infty f(x)\,dx = \lim_{b\to\infty}\int_a^b f(x)\,dx = \lim_{b\to\infty} F(b) - F(a).$
   - If $\lim_{x\to c} f(x) = \infty$ with $a \leq c \leq b$, then

$$\int_a^b f(x)\,dx = \lim_{\delta\to 0}\int_a^{c-\delta} f(x)\,dx + \lim_{\epsilon\to 0}\int_{c+\epsilon}^b f(x)\,dx,$$

   provided both limits exist.

6. *Curve lengths, and areas and volumes of revolution*
   - Curve length

$$s = \int_a^b \sqrt{1 + \left(\frac{dy}{dx}\right)^2}\,dx \quad \text{or} \quad s = \int_c^d \sqrt{1 + \left(\frac{dx}{dy}\right)^2}\,dy.$$

   - Area of solid of revolution

$$S = 2\pi\int_a^b y\sqrt{1 + \left(\frac{dy}{dx}\right)^2}\,dx \quad \text{or} \quad S = 2\pi\int_c^d x\sqrt{1 + \left(\frac{dx}{dy}\right)^2}\,dy.$$

   - Volume of solid of revolution

$$V = \pi\int_a^b y^2\,dx \quad \text{or} \quad V = \pi\int_c^d x^2\,dy.$$

Integral calculus

PROBLEMS

4.1. Find, by inspection, the indefinite integrals of
(a) $7x^6$; (b) $e^{3x} + e^{-3x}$; (c) $\cot 3x$; (d) $\sin x \sin 2x$; (e) $\cos x \sin 2x$;
(f) $(a - 2x)^{-1}$; (g) $(4 + x^2)^{-1}$; (h) $(4 - x^2)^{-1/2}$; (i) $x(4 + x^2)^{-1}$.

4.2. Find the following indefinite integrals:
(a) $\int (4 + x^2)^{-1} dx$; (b) $\int (8 + 2x - x^2)^{-1/2} dx$ for $2 \leq x \leq 4$;
(c) $\int (1 + \sin\theta)^{-1} d\theta$; (d) $\int (x\sqrt{1-x})^{-1} dx$ for $0 < x \leq 1$.

4.3. Find the indefinite integrals J of the following ratios of polynomials:
(a) $(x + 3)/(x^2 + x - 2)$;
(b) $(x^3 + 5x^2 + 8x + 12)/(2x^2 + 10x + 12)$;
(c) $(3x^2 + 20x + 28)/(x^2 + 6x + 9)$;
(d) $x^3/(a^8 + x^8)$.

4.4. Express $x^2(ax + b)^{-1}$ as the sum of powers of x and another integrable term, and hence evaluate

$$\int_0^{b/a} \frac{x^2}{ax + b} dx.$$

4.5. Find the integral J of $(ax^2 + bx + c)^{-1}$, with $a \neq 0$, distinguishing between the cases (i) $b^2 > 4ac$, (ii) $b^2 < 4ac$ and (iii) $b^2 = 4ac$.

4.6. Use logarithmic integration to find the indefinite integrals J of the following:
(a) $\sin 2x/(1 + 4\sin^2 x)$;
(b) $e^x/(e^x - e^{-x})$;
(c) $(1 + x \ln x)/(x \ln x)$;
(d) $[x(x^n + a^n)]^{-1}$.

4.7. Find the derivative of $f(x) = (1 + \sin x)/\cos x$ and hence determine the indefinite integral J of $\sec x$.

4.8. Find the indefinite integrals, J, of the following functions involving sinusoids:
(a) $\cos^5 x - \cos^3 x$;
(b) $(1 - \cos x)/(1 + \cos x)$;
(c) $\cos x \sin x/(1 + \cos x)$;
(d) $\sec^2 x/(1 - \tan^2 x)$.

4.9. By making the substitution $x = a\cos^2\theta + b\sin^2\theta$, evaluate the definite integrals J between limits a and b ($> a$) of the following functions:
(a) $[(x - a)(b - x)]^{-1/2}$;
(b) $[(x - a)(b - x)]^{1/2}$;
(c) $[(x - a)/(b - x)]^{1/2}$.

Problems

4.10. Determine whether the following integrals exist and, where they do, evaluate them:

(a) $\int_0^\infty \exp(-\lambda x)\,dx$; (b) $\int_{-\infty}^\infty \frac{x}{(x^2+a^2)^2}\,dx$;

(c) $\int_1^\infty \frac{1}{x+1}\,dx$; (d) $\int_0^1 \frac{1}{x^2}\,dx$;

(e) $\int_0^{\pi/2} \cot\theta\,d\theta$; (f) $\int_0^1 \frac{x}{(1-x^2)^{1/2}}\,dx$.

4.11. Use integration by parts to evaluate the following:

(a) $\int_0^y x^2 \sin x\,dx$; (b) $\int_1^y x \ln x\,dx$;

(c) $\int_0^y \sin^{-1} x\,dx$; (d) $\int_1^y \ln(a^2+x^2)/x^2\,dx$.

4.12. Show, using the following methods, that the indefinite integral of $x^3/(x+1)^{1/2}$ is
$$J = \tfrac{2}{35}(5x^3 - 6x^2 + 8x - 16)(x+1)^{1/2} + c.$$
(a) Repeated integration by parts.
(b) Setting $x+1 = u^2$ and determining dJ/du as $(dJ/dx)(dx/du)$.

4.13. The gamma function $\Gamma(n)$ is defined for all $n > -1$ by
$$\Gamma(n+1) = \int_0^\infty x^n e^{-x}\,dx.$$
Find a recurrence relation connecting $\Gamma(n+1)$ and $\Gamma(n)$.
(a) Deduce (i) the value of $\Gamma(n+1)$ when n is a non-negative integer and (ii) the value of $\Gamma\left(\tfrac{7}{2}\right)$, given that $\Gamma\left(\tfrac{1}{2}\right) = \sqrt{\pi}$.
(b) Now, taking factorial m for *any* m to be defined by $m! = \Gamma(m+1)$, evaluate $\left(-\tfrac{3}{2}\right)!$.

4.14. Define $J(m, n)$, for non-negative integers m and n, by the integral
$$J(m, n) = \int_0^{\pi/2} \cos^m\theta \sin^n\theta\,d\theta.$$
(a) Evaluate $J(0, 0)$, $J(0, 1)$, $J(1, 0)$, $J(1, 1)$, $J(m, 1)$, $J(1, n)$.
(b) Using integration by parts, prove that, for m and n both > 1,
$$J(m, n) = \frac{m-1}{m+n} J(m-2, n) \quad\text{and}\quad J(m, n) = \frac{n-1}{m+n} J(m, n-2).$$
(c) Evaluate (i) $J(5, 3)$, (ii) $J(6, 5)$ and (iii) $J(4, 8)$.

4.15. By integrating by parts twice, prove that I_n, as defined in the first equality below for positive integers n, has the value given in the second equality:
$$I_n = \int_0^{\pi/2} \sin n\theta \cos\theta\,d\theta = \frac{n - \sin(n\pi/2)}{n^2 - 1}.$$

Integral calculus

4.16. Evaluate the following definite integrals:
(a) $\int_0^\infty x e^{-x} \, dx$; (b) $\int_0^1 \left[(x^3 + 1)/(x^4 + 4x + 1)\right] dx$;
(c) $\int_0^{\pi/2} [a + (a-1)\cos\theta]^{-1} \, d\theta$ with $a > \frac{1}{2}$; (d) $\int_{-\infty}^\infty (x^2 + 6x + 18)^{-1} \, dx$.

4.17. If J_r is the integral
$$\int_0^\infty x^r \exp(-x^2) \, dx$$
show that
(a) $J_{2r+1} = (r!)/2$,
(b) $J_{2r} = 2^{-r}(2r-1)(2r-3)\cdots(5)(3)(1) J_0$.

4.18. Find positive constants a, b such that $ax \leq \sin x \leq bx$ for $0 \leq x \leq \pi/2$. Use this inequality to find (to two significant figures) upper and lower bounds for the integral
$$I = \int_0^{\pi/2} (1 + \sin x)^{1/2} \, dx.$$
Use the substitution $t = \tan(x/2)$ to evaluate I exactly.

4.19. By noting that for $0 \leq \eta \leq 1$, $\eta^{1/2} \geq \eta^{3/4} \geq \eta$, prove that
$$\frac{2}{3} \leq \frac{1}{a^{5/2}} \int_0^a (a^2 - x^2)^{3/4} \, dx \leq \frac{\pi}{4}.$$

4.20. The official specifications for a rugby ball allow one that has a length of 300 mm and a smallest circumference of 600 mm. By treating it as an ellipsoid of revolution, find its volume.

4.21. A vase has curved sides that are generated by rotating the part of the curve $x = \frac{1}{2}a(e^{y/a} + e^{-y/a})$ that lies between $y = 0$ and $y = ha$ around the y-axis. Show that the area of the curved surface is $\pi a^2 [\frac{1}{4}(e^{2h} - e^{-2h}) + h]$.

4.22. Show that the total length of the astroid $x^{2/3} + y^{2/3} = a^{2/3}$, which can be parameterised as $x = a\cos^3\theta$, $y = a\sin^3\theta$, is $6a$.

4.23. By noting that $\sinh x < \frac{1}{2}e^x < \cosh x$, and that $1 + z^2 < (1+z)^2$ for $z > 0$, show that, for $x > 0$, the length L of the curve $y = \frac{1}{2}e^x$ measured from the origin satisfies the inequalities $\sinh x < L < x + \sinh x$.

4.24. The equation of a cardioid in plane polar coordinates is
$$\rho = a(1 - \sin\phi).$$
Sketch the curve and find (i) its area, (ii) its total length, (iii) the surface area of the solid formed by rotating the cardioid about its axis of symmetry and (iv) the volume of the same solid.

HINTS AND ANSWERS

4.1. (a) x^7; (b) $\frac{1}{3}(e^x - e^{-x})$; (c) $\frac{1}{3}\ln\sin 3x$; (d) consider $\sin 2x$ as $2\sin x \cos x$, $\frac{2}{3}\sin^3 x$; (e) $-\frac{2}{3}\cos^3 x$; (f) $-\frac{1}{2}\ln(a-x)$; (g) $\frac{1}{2}\tan^{-1}(x/2)$ (h) $\sin^{-1}(x/2)$; (i) $\frac{1}{2}\ln(4+x^2)$.

4.3. (a) Express in partial fractions; $J = \frac{1}{3}\ln[(x-1)^4/(x+2)] + c$.
(b) Divide the numerator by the denominator and express the remainder in partial fractions; $J = x^2/4 + 4\ln(x+2) - 3\ln(x+3) + c$.
(c) After division of the numerator by the denominator the remainder can be expressed as $2(x+3)^{-1} - 5(x+3)^{-2}$; $J = 3x + 2\ln(x+3) + 5(x+3)^{-1} + c$.
(d) Set $x^4 = u$; $J = (4a^4)^{-1}\tan^{-1}(x^4/a^4) + c$.

4.5. Writing $b^2 - 4ac$ as $\Delta^2 > 0$, or $4ac - b^2$ as $\Delta'^2 > 0$:
(i) $\Delta^{-1}\ln[(2ax+b-\Delta)/(2ax+b+\Delta)] + k$;
(ii) $2\Delta'^{-1}\tan^{-1}[(2ax+b)/\Delta'] + k$;
(iii) $-2(2ax+b)^{-1} + k$.

4.7. $f'(x) = (1 + \sin x)/\cos^2 x = f(x)\sec x$; $J = \ln(f(x)) + c = \ln(\sec x + \tan x) + c$.

4.9. Note that $dx = 2(b-a)\cos\theta\sin\theta\,d\theta$.
(a) π; (b) $\pi(b-a)^2/8$; (c) $\pi(b-a)/2$.

4.11. (a) $(2-y^2)\cos y + 2y\sin y - 2$; (b) $[(y^2 \ln y)/2] + [(1-y^2)/4]$;
(c) $y\sin^{-1} y + (1-y^2)^{1/2} - 1$;
(d) $\ln(a^2+1) - (1/y)\ln(a^2+y^2) + (2/a)[\tan^{-1}(y/a) - \tan^{-1}(1/a)]$.

4.13. $\Gamma(n+1) = n\Gamma(n)$; (a) (i) $n!$, (ii) $15\sqrt{\pi}/8$; (b) $-2\sqrt{\pi}$.

4.15. By integrating twice recover a multiple of I_n.

4.17. $J_{2r+1} = rJ_{2r-1}$ and $2J_{2r} = (2r-1)J_{2r-2}$.

4.19. Set $\eta = 1 - (x/a)^2$ throughout and $x = a\sin\theta$ in one of the bounds.

4.21. Note that $(u + u^{-1})^2 = (u - u^{-1})^2 + 4$.

4.23. $L = \int_0^x \left(1 + \frac{1}{4}\exp 2x\right)^{1/2} dx$.

5
Complex numbers and hyperbolic functions

This chapter is concerned with the representation and manipulation of complex numbers. Complex numbers pervade this book, underscoring their wide application in the mathematics of the physical sciences. Some elementary applications of complex numbers to the description of physical systems appear in later chapters, but only the basic tools are presented here.

5.1 The need for complex numbers

Although complex numbers occur in many branches of mathematics, they arise most directly out of solving polynomial equations. We examine a specific quadratic equation as an example.

Consider the quadratic equation

$$z^2 - 4z + 5 = 0. \tag{5.1}$$

Equation (5.1) has two solutions, z_1 and z_2, such that

$$(z - z_1)(z - z_2) = 0. \tag{5.2}$$

Using the familiar formula for the roots of a quadratic equation, (2.4), the solutions z_1 and z_2, written in brief as $z_{1,2}$, are

$$z_{1,2} = \frac{4 \pm \sqrt{(-4)^2 - 4(1 \times 5)}}{2}$$

$$= 2 \pm \frac{\sqrt{-4}}{2}. \tag{5.3}$$

Both solutions contain the square root of a negative number. However, it would not be true to say that there are no solutions to the quadratic equation. The *fundamental theorem of algebra* states that a quadratic equation will always have two solutions and these are in fact given by (5.3). The second term on the RHS of (5.3) is called an *imaginary* term since it contains the square root of a negative number; the first term is called a *real* term. The full solution is the sum of a real term and an imaginary term and is called a *complex number*. A plot of the function $f(z) = z^2 - 4z + 5$ is shown in Figure 5.1. It will be seen that the plot does not intersect the z-axis, corresponding to the fact that the equation $f(z) = 0$ has no purely real solutions.

The choice of the symbol z for the quadratic variable was not arbitrary; the conventional representation of a complex number is z, where z is the sum of a real part x and i times

5.1 The need for complex numbers

Figure 5.1 The function $f(z) = z^2 - 4z + 5$.

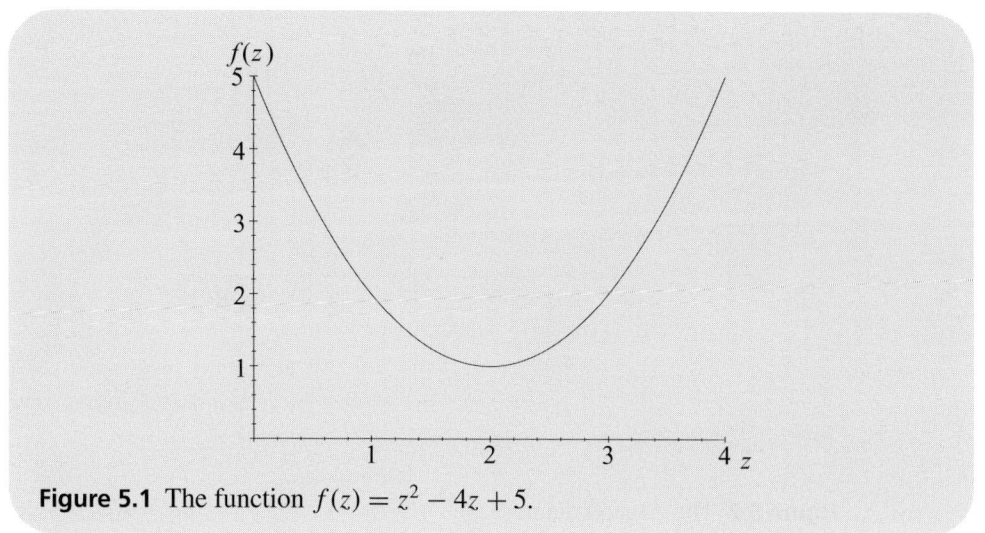

an imaginary part y, i.e.

$$z = x + iy,$$

where i is used to denote the square root of -1.[1] The real part x and the imaginary part y are usually denoted by Re z and Im z respectively. We note at this point that some physical scientists, engineers in particular, use j instead of i. However, for consistency, we will use i throughout this book.

In our particular example, $\sqrt{-4} = 2\sqrt{-1} = 2i$, and hence the two solutions of (5.1) are

$$z_{1,2} = 2 \pm \frac{2i}{2} = 2 \pm i.$$

Thus, here $x = 2$ and $y = \pm 1$.

For compactness a complex number is sometimes written in the form

$$z = (x, y),$$

where the components of z may be thought of as coordinates in an xy-plot. Such a plot is called an *Argand diagram* and is a common representation of complex numbers; an example is shown in Figure 5.2.

Our particular example of a quadratic equation may be readily generalised to polynomials whose highest power (degree) is greater than 2, e.g. cubic equations (degree 3), quartic equations (degree 4) and so on. For a general polynomial $f(z)$, of degree n, the fundamental theorem of algebra states that the equation $f(z) = 0$ will have exactly n solutions. We will examine cases of higher degree equations in Section 5.4.3.

---

[1] More strictly, we should say one of the square roots of -1. Since it is defined as a solution of the equation $z^2 + 1 = 0$, and this equation is quadratic, it follows that there must be (exactly) one other root to the equation; this is $z = -i$. Consequently, for a real and positive, $\sqrt{-a} = \pm i\sqrt{a}$.

Complex numbers and hyperbolic functions

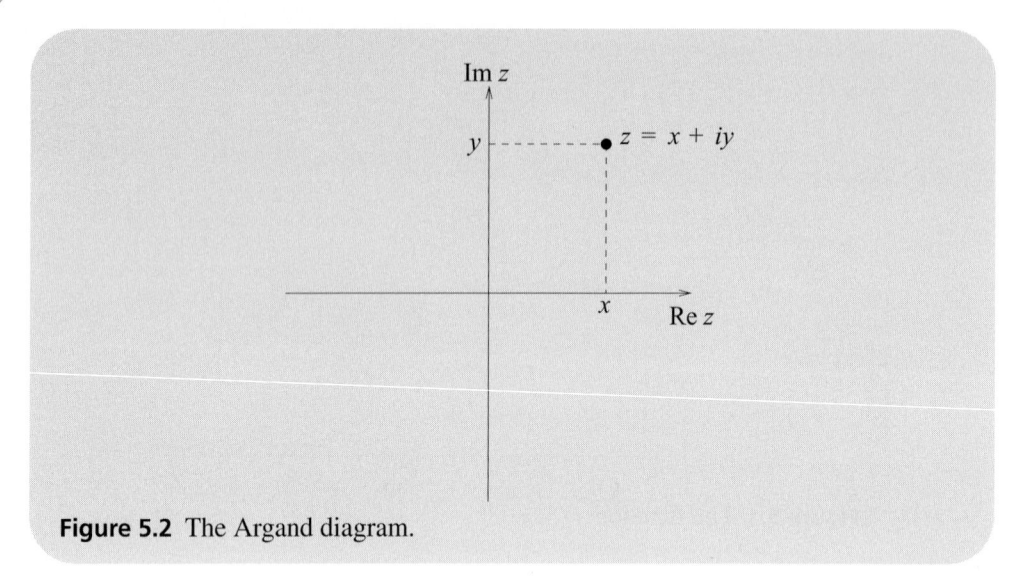

Figure 5.2 The Argand diagram.

The remainder of this chapter deals with: the algebra and manipulation of complex numbers; their polar representation, which has advantages in many circumstances; complex exponentials and logarithms; the use of complex numbers in finding the roots of polynomial equations; and hyperbolic functions.

EXERCISE 5.1

1. Plot the solutions of the following equations on an Argand diagram using the symbol a to label a solution of equation (a), b for equation (b), etc:

   (a) $z^2 - 5z + 6 = 0$, (b) $z^2 - 5z + 7 = 0$, (c) $z^2 + 5z + 7 = 0$,
   (d) $z^2 + 4 = 0$, (e) $z^2 + 4z + 4 = 0$, (f) $z^3 + z = 0$.

5.2 Manipulation of complex numbers

This section considers basic complex number manipulation. Some analogy may be drawn with vector manipulation (see Chapter 9), but this section stands alone as an introduction.

Before we move on to consider the ways in which complex numbers are combined, we discuss the procedure generally referred to as 'equating the real and imaginary parts', a procedure we will use many times. The phrase means that if we have an equation in which one or more of the terms could be complex, then we may equate the real terms on the LHS of the equation with the real terms on its RHS; similarly, the imaginary terms on the two sides can be equated (when doing so the factor i is normally omitted). In explicit form, if we have the equation

$$a + bi = c + di,$$

5.2 Manipulation of complex numbers

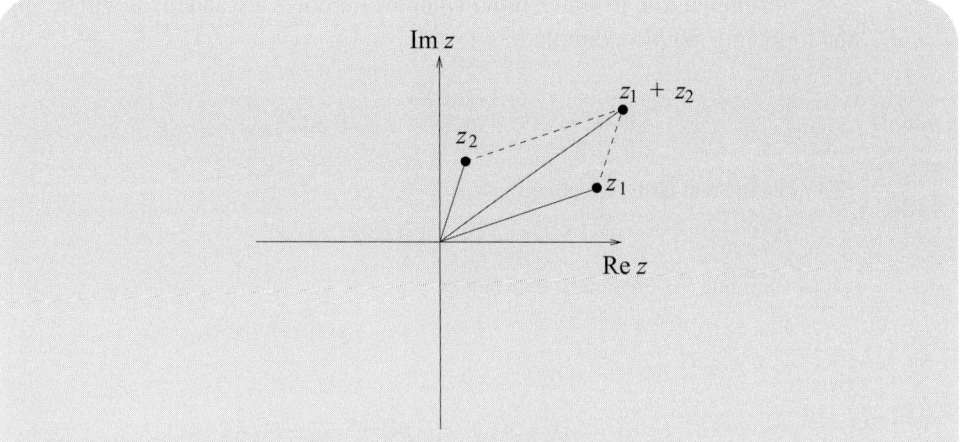

Figure 5.3 The addition of two complex numbers.

where a, b, c and d are real quantities or expressions, then we can conclude that $a = c$ and $b = d$, i.e. that the complex equation is really two separate equations. The justification for this conclusion is simple: the equation can be rearranged as $a - c = i(d - b)$, and if this equation is squared we obtain

$$(a - c)^2 = (-1)(d - b)^2 = -(d - b)^2.$$

Now since the square of *any* real quantity is either positive or zero, this equation equates a positive quantity to a negative one. This can only be so if both sides are zero, and so we conclude that $a = c$ and $d = b$.[2]

5.2.1 Addition and subtraction

The addition of two complex numbers, z_1 and z_2, generally gives another complex number. The real components and the imaginary components are added separately and in a like manner to the familiar addition of real numbers:

$$z_1 + z_2 = (x_1 + iy_1) + (x_2 + iy_2) = (x_1 + x_2) + i(y_1 + y_2),$$

or in component notation

$$z_1 + z_2 = (x_1, y_1) + (x_2, y_2) = (x_1 + x_2, y_1 + y_2).$$

The Argand representation of the addition of two complex numbers is shown in Figure 5.3.

By applying the commutativity and associativity of addition to the real and imaginary parts separately, we can show that the addition of complex numbers is itself commutative and associative, i.e.

$$z_1 + z_2 = z_2 + z_1,$$
$$z_1 + (z_2 + z_3) = (z_1 + z_2) + z_3.$$

---

[2] By trying to find some, show that the equation $x(x - 1) + ix(x + 4) = 6 - 3i$ has no real solutions for x.

Complex numbers and hyperbolic functions

Thus it is immaterial in which order complex numbers are added, as will be apparent from the following simple example.

Example Sum the complex numbers $z_1 = 1 + 2i$, $z_2 = 3 - 4i$ and $z_3 = -2 + i$.

Summing the real terms we obtain
$$1 + 3 - 2 = 2,$$
whilst summing the imaginary terms gives
$$2i - 4i + i = -i.$$

Hence
$$(1 + 2i) + (3 - 4i) + (-2 + i) = 2 - i$$
is the sum of the three individual complex numbers. Clearly, changing the order of the added numbers would not change the outcome. ◀

The subtraction of complex numbers is very similar to their addition and, as in the case of real numbers, if two identical complex numbers are subtracted then the result is zero. Multiplication of a complex number by a real number λ multiplies both the real and imaginary parts separately by λ. As a simple check, and an illustration of these points, the reader may wish to verify that for the three complex numbers z_1, z_2 and z_3 used in the above example:
$$z_1 + z_2 + 2z_3 = 0,$$
$$2z_1 - 3z_2 - z_3 = -5 + 15i.$$

5.2.2 Modulus and argument

The *modulus* (often referred to as the *magnitude*) of the complex number z is denoted by $|z|$ and is defined as
$$|z| = \sqrt{x^2 + y^2}. \tag{5.4}$$

Hence the modulus of the complex number is the distance between the corresponding point and the origin in the Argand diagram, as may be seen in Figure 5.4. The modulus of a complex number z is always positive or zero (never negative), and it is zero only when z is the zero complex number $0 + 0i$.

The argument of the complex number z is denoted by arg z and is defined as
$$\arg z = \tan^{-1}\left(\frac{y}{x}\right). \tag{5.5}$$

It can be seen that arg z is the angle[3] that the line joining the origin to z on the Argand diagram makes with the positive x-axis. The anticlockwise direction is taken to be positive

---
[3] In mathematics and most of physics, the argument is measured in radians, but in some engineering applications it is normal practice to refer to it as the *phase* of z and give its value in degrees.

5.2 Manipulation of complex numbers

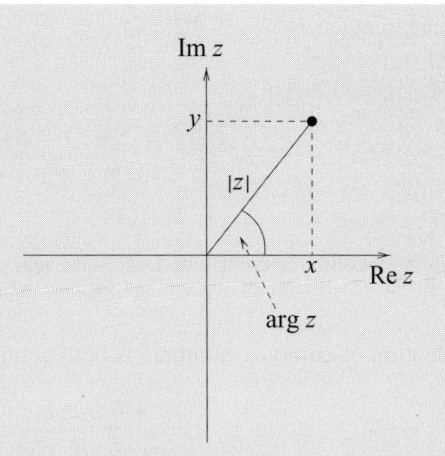

Figure 5.4 The modulus and argument of a complex number.

by convention. The angle arg z is shown in Figure 5.4. Account must be taken of the signs of x and y individually in determining in which quadrant arg z lies. Thus, for example, if x and y are both negative then arg z lies in the range $-\pi < \arg z < -\pi/2$ rather than in the first quadrant $(0 < \arg z < \pi/2)$, though both cases give the same value for the ratio of y to x.

Example Find the modulus and the argument of the complex number $z = 2 - 3i$.

Using (5.4), the modulus is given by

$$|z| = \sqrt{2^2 + (-3)^2} = \sqrt{13}.$$

Using (5.5), the argument is given by

$$\arg z = \tan^{-1}\left(-\tfrac{3}{2}\right).$$

The two angles whose tangents equal -1.5 are -0.9828 rad and 2.1588 rad. Since $x = 2$ and $y = -3$, z clearly lies in the fourth quadrant; therefore arg $z = -0.9828$ is the appropriate answer. ◂

5.2.3 Multiplication

Complex numbers may be multiplied together and, in general, give a complex number as the result. The product of two complex numbers z_1 and z_2 is found by multiplying them out in full and remembering that $i^2 = -1$, i.e.

$$\begin{aligned} z_1 z_2 &= (x_1 + iy_1)(x_2 + iy_2) \\ &= x_1 x_2 + ix_1 y_2 + iy_1 x_2 + i^2 y_1 y_2 \\ &= (x_1 x_2 - y_1 y_2) + i(x_1 y_2 + y_1 x_2). \end{aligned} \quad (5.6)$$

We next illustrate this general prescription with a concrete example.

Example Multiply the complex numbers $z_1 = 3 + 2i$ and $z_2 = -1 - 4i$.

By direct multiplication we find
$$\begin{aligned}z_1 z_2 &= (3 + 2i)(-1 - 4i) \\ &= -3 - 2i - 12i - 8i^2 \\ &= 5 - 14i.\end{aligned} \quad (5.7)$$

The term $-8i^2$ in the second line contributed $+8$ to the real part of the product. ◀

The multiplication of complex numbers is both commutative and associative, i.e.
$$z_1 z_2 = z_2 z_1, \quad (5.8)$$
$$(z_1 z_2) z_3 = z_1 (z_2 z_3). \quad (5.9)$$

The product of two complex numbers also has the simple properties
$$|z_1 z_2| = |z_1||z_2|, \quad (5.10)$$
$$\arg(z_1 z_2) = \arg z_1 + \arg z_2. \quad (5.11)$$

In words, the magnitude of a product is equal to the product of the magnitudes and the argument of a product is equal to the sum of the arguments. These relations can be proved most simply using the methods of Section 5.3.1, but they can be derived directly.[4]

Example Verify that (5.10) holds for the product of $z_1 = 3 + 2i$ and $z_2 = -1 - 4i$.

From (5.7), the modulus of $z_1 z_2$ is given by
$$|z_1 z_2| = |5 - 14i| = \sqrt{5^2 + (-14)^2} = \sqrt{221}.$$
For the individual factors, their moduli are
$$|z_1| = \sqrt{3^2 + 2^2} = \sqrt{13},$$
$$|z_2| = \sqrt{(-1)^2 + (-4)^2} = \sqrt{17}.$$
Substituting in both sides of (5.10),
$$|z_1||z_2| = \sqrt{13}\sqrt{17} = \sqrt{221} = |z_1 z_2|,$$
verifies that it is valid for this particular product (as it is, in fact, for all products). ◀

We now examine the effect on a complex number z of multiplying it by ± 1 and $\pm i$. These four multipliers have modulus unity and we can see immediately from (5.10) that multiplying z by another complex number of unit modulus gives a product with the same

---

[4] Prove the first of these directly by using (5.6) to compute $|z_1 z_2|^2$, simplifying the result, and then factorising it to show that it is equal to $|z_1|^2 |z_2|^2$. The second can be proved by dividing both the numerator and denominator of the expression for $\tan(\arg z_1 z_2)$ by $x_1 x_2$.

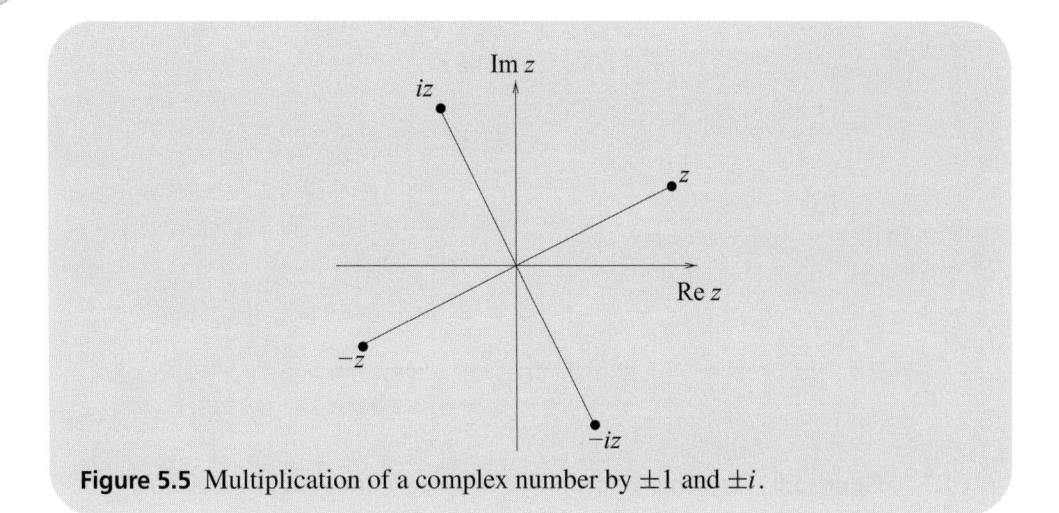

5.2 Manipulation of complex numbers

Figure 5.5 Multiplication of a complex number by ± 1 and $\pm i$.

modulus as z. We can also see from (5.11) that if we multiply z by a complex number then the argument of the product is the sum of the argument of z and the argument of the multiplier.

Taking each of the four multipliers in turn: multiplying z by unity (which has argument zero) leaves z unchanged in both modulus and argument, i.e. z is completely unaltered by the operation; multiplying by -1 (which has argument π) leads to rotation, through an angle π, of the line in the Argand diagram joining the origin to z; in a similar way, multiplication by i or $-i$ leads to corresponding rotations of $\pi/2$ or $-\pi/2$, respectively. These geometrical interpretations of multiplication are shown in Figure 5.5.

Example Using the geometrical interpretation of multiplication by i, find the product $i(1-i)$.

The complex number $1-i$ has argument $-\pi/4$ and modulus $\sqrt{2}$. Thus, using (5.10) and (5.11), its product with i has argument $+\pi/4$ and unchanged modulus $\sqrt{2}$. The complex number with modulus $\sqrt{2}$ and argument $+\pi/4$ is $1+i$ and so

$$i(1-i) = 1+i,$$

as is easily verified by direct multiplication. ◀

The process of the division for two complex numbers parallels that of multiplication, but, as it requires the notion of a complex conjugate (see the following subsection), discussion of it is postponed until Section 5.2.5.

5.2.4 Complex conjugate

If z has the convenient form $x+iy$ then its complex conjugate, denoted by z^*, may be found simply by changing the sign of the imaginary part, i.e. if $z = x+iy$ then $z^* = x-iy$. More generally, we may define the complex conjugate of z as the (complex)

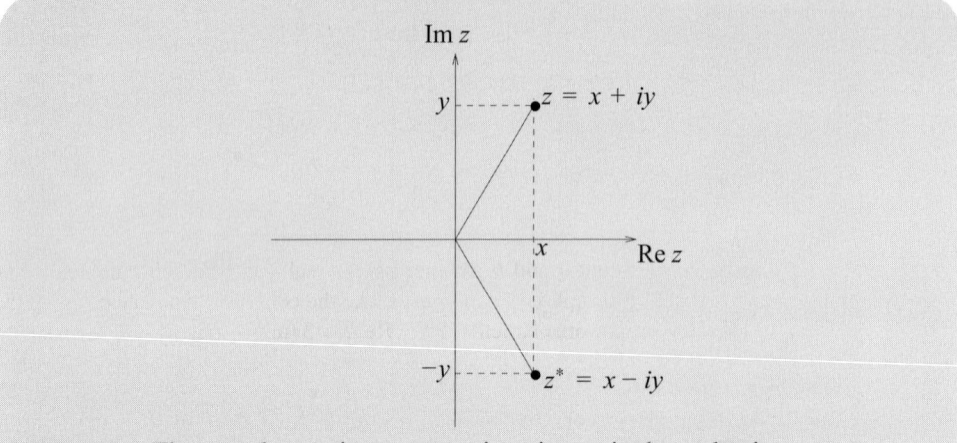

Figure 5.6 The complex conjugate as a mirror image in the real axis.

number that has the same magnitude (modulus) as z and when multiplied by z gives a real positive result, i.e. the product has no imaginary component and its real part is positive.[5]

If z happens to be purely real, then it is equal to its own complex conjugate; if it is purely imaginary, then it is equal to minus its complex conjugate. These latter two properties can be used as tests for purely real and purely imaginary quantities or expressions. A general complex number is neither real nor imaginary.

The following properties of the complex conjugate are easily proved and others may be derived from them. If $z = x + iy$ then

$$(z^*)^* = z, \tag{5.12}$$

$$z + z^* = 2\operatorname{Re} z = 2x, \tag{5.13}$$

$$z - z^* = 2i \operatorname{Im} z = 2iy, \tag{5.14}$$

In the case where z can be written in the form $x + iy$ it is easily verified, by direct multiplication of the components, that the product zz^* gives a real result:

$$zz^* = (x + iy)(x - iy) = x^2 - ixy + ixy - i^2 y^2 = x^2 + y^2 = |z|^2.$$

Not only is this result real, it is also equal to the square of the modulus of z, i.e.

$$zz^* = z^*z = |z|^2, \text{ which is real and } \geq 0. \tag{5.15}$$

Complex conjugation corresponds to a reflection of z in the real axis of the Argand diagram, as may be seen in Figure 5.6.

---

5 Suppose that $z^* = x_2 + iy_2$ is the complex conjugate of $z_1 = x_1 + iy_1$. Show that the requirements $|z^*| = |z_1|$ and $\operatorname{Im}(z_1 z^*) = 0$ together imply that $x_1^2 = x_2^2$ and $y_1^2 = y_2^2$. Show further that the requirement $\operatorname{Re}(z_1 z^*) > 0$ is violated if $x_2 = -x_1$. Hence deduce that $z^* = x_1 - iy_1$.

5.2 Manipulation of complex numbers

Example Find the complex conjugate of $z = a + 2i + 3ib$.

The complex number is written in the standard form

$$z = a + i(2 + 3b);$$

then, replacing i by $-i$, we obtain

$$z^* = a - i(2 + 3b).$$

We have assumed here that a and b are themselves real – the explicit numbers 2 and 3 clearly are! If a and b could be complex, then we must take the complex conjugate of every factor, giving $z^* = a^* - i(2 + 3b^*)$ or, more explicitly, $z^* = \text{Re}\,a - 3\,\text{Im}\,b - i(\text{Im}\,a + 2 + 3\,\text{Re}\,b)$. ◀

In some cases, however, it may not be simple to rearrange the expression for z into the standard form $x + iy$. Nevertheless, given two complex numbers, z_1 and z_2, it is straightforward to show that the complex conjugate of their sum (or difference) is equal to the sum (or difference) of their complex conjugates, i.e. $(z_1 \pm z_2)^* = z_1^* \pm z_2^*$. Similarly, it may be shown that the complex conjugate of the product (or quotient) of z_1 and z_2 is equal to the product (or quotient) of their complex conjugates, i.e. $(z_1 z_2)^* = z_1^* z_2^*$ and $(z_1/z_2)^* = z_1^*/z_2^*$.

Using these results, it can be deduced that, no matter how complicated the expression, its complex conjugate may *always* be found by replacing every i by $-i$. To apply this rule, however, we must always ensure that all complex parts are first written out in full, so that no i's are hidden. This is illustrated in the following example.

Example Find the complex conjugate of the complex number $z = w^{(3y+2ix)}$ where $w = x + 5i$.

Although we do not discuss complex powers until Section 5.5, the simple rule given above still enables us to find the complex conjugate of z.

In this case w itself contains real and imaginary components and so must be written out in full, i.e.

$$z = w^{3y+2ix} = (x + 5i)^{3y+2ix}.$$

Now we can replace each i by $-i$ to obtain

$$z^* = (x - 5i)^{3y-2ix}.$$

It can be shown that the product zz^* is real and this is done at the very end of Section 5.3. ◀

The quotient of $z = x + iy$ and its complex conjugate[6] is expressed in terms of x and y by

$$\frac{z}{z^*} = \left(\frac{x^2 - y^2}{x^2 + y^2}\right) + i\left(\frac{2xy}{x^2 + y^2}\right). \tag{5.16}$$

---

[6] Explain why this quotient must have unit modulus, and demonstrate by direct calculation that it has.

Complex numbers and hyperbolic functions

The derivation of this formula relies on the more general results of the following subsection.

5.2.5 Division

The procedure for the division of two complex numbers z_1 and z_2 bears some similarity to that for their multiplication. Writing the quotient in component form we obtain

$$\frac{z_1}{z_2} = \frac{x_1 + iy_1}{x_2 + iy_2}. \tag{5.17}$$

In order to separate the real and imaginary components of the quotient, we multiply both numerator and denominator by the complex conjugate of the denominator. By definition, this process will leave the denominator as a real quantity. Equation (5.17) gives

$$\frac{z_1}{z_2} = \frac{(x_1 + iy_1)(x_2 - iy_2)}{(x_2 + iy_2)(x_2 - iy_2)} = \frac{(x_1x_2 + y_1y_2) + i(x_2y_1 - x_1y_2)}{x_2^2 + y_2^2}$$

$$= \frac{x_1x_2 + y_1y_2}{x_2^2 + y_2^2} + i\frac{x_2y_1 - x_1y_2}{x_2^2 + y_2^2}.$$

Hence we have separated the quotient into real and imaginary components, as required.[7]

In the special case where $z_2 = z_1^*$, so that $x_2 = x_1$ and $y_2 = -y_1$, the general result reduces to (5.16).

Example Express z in the form $x + iy$, when

$$z = \frac{3 - 2i}{-1 + 4i}.$$

Multiplying numerator and denominator by the complex conjugate of the denominator we obtain

$$z = \frac{(3 - 2i)(-1 - 4i)}{(-1 + 4i)(-1 - 4i)} = \frac{-11 - 10i}{17}$$

$$= -\frac{11}{17} - \frac{10}{17}i.$$

We note that, as it must be, the modulus of the original fraction, $\sqrt{(13/17)}$, is equal to that of the final complex number, $\sqrt{(121 + 100)/(17)^2}$. ◀

In analogy to (5.10) and (5.11), which describe the multiplication of two complex numbers, the following relations apply to their division:

$$\left|\frac{z_1}{z_2}\right| = \frac{|z_1|}{|z_2|}, \tag{5.18}$$

$$\arg\left(\frac{z_1}{z_2}\right) = \arg z_1 - \arg z_2. \tag{5.19}$$

The proof of these relations is left until Section 5.3.1.

---

[7] Show that if $z = z_1/z_2 = z_3/|z_2|^2$, then $z^{-1} = z_3^*/|z_1|^2$.

5.3 Polar representation of complex numbers

EXERCISES 5.2

1. Do the equations

   (a) $z(z-2) = 1 - 2i$, (b) $z + 2 + i = z(z-i) + 4(i-1)$,
   (c) $z + 2 + i = z(z-i) + 4(1-i)$

   have any real solutions?

2. If $z_1 = 2 + 3i$ and $z_2 = 3 - i$, find the magnitudes and arguments of $z_1 + z_2$ and $z_1 - z_2$.

3. With z_1 and z_2 as in the previous exercise, find, by explicit calculation, the magnitudes and arguments of $z_1 z_2$ and z_1/z_2. Confirm that they are in agreement with results (5.10), (5.11) and (5.18), (5.19).

5.3 Polar representation of complex numbers

Although considering a complex number as the sum of a real and an imaginary part is often useful, sometimes the *polar representation* proves easier to manipulate. This makes use of the complex exponential function, which is defined by

$$e^z = \exp z \equiv 1 + z + \frac{z^2}{2!} + \frac{z^3}{3!} + \cdots \qquad (5.20)$$

Strictly speaking it is the function $\exp z$ that is defined by (5.20). The number e is the value of $\exp(1)$, i.e. it is just a number. However, it may be shown that e^z and $\exp z$ are equivalent (see Appendix A) when z is real and rational and mathematicians then *define* their equivalence for irrational and complex z. For the purposes of this book we will not concern ourselves further with this mathematical nicety but, rather, assume that (5.20) is valid for all z. We also note that, using (5.20), by multiplying together the appropriate series it can be shown that

$$e^{z_1} e^{z_2} = e^{z_1 + z_2}, \qquad (5.21)$$

which is analogous to the familiar result for exponentials of real numbers.[8]
From (5.20), it immediately follows that for $z = i\theta$, with θ real,

$$e^{i\theta} = 1 + i\theta - \frac{\theta^2}{2!} - \frac{i\theta^3}{3!} + \cdots \qquad (5.22)$$

$$= 1 - \frac{\theta^2}{2!} + \frac{\theta^4}{4!} - \cdots + i\left(\theta - \frac{\theta^3}{3!} + \frac{\theta^5}{5!} - \cdots\right) \qquad (5.23)$$

and hence that

$$e^{i\theta} = \cos\theta + i\sin\theta, \qquad (5.24)$$

---

[8] Show that $e^z \times e^{z^*}$ is real and that e^z/e^{z^*} has unit modulus.

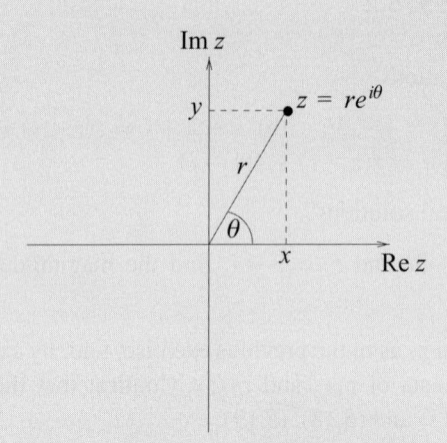

Figure 5.7 The polar representation of a complex number.

where the last equality follows from the series expansions of the sine and cosine functions (see Section 6.6.3 and Appendix B). This last relationship is called *Euler's equation*. It also follows from (5.24) that

$$e^{in\theta} = \cos n\theta + i \sin n\theta$$

for all n. From Euler's equation (5.24) and Figure 5.7 we deduce that

$$re^{i\theta} = r(\cos\theta + i\sin\theta)$$
$$= x + iy.$$

Thus a complex number may be represented in the polar form

$$z = re^{i\theta}. \tag{5.25}$$

Referring again to Figure 5.7, we can identify r with $|z|$ and θ with arg z. The simplicity of the representation of the modulus and argument is one of the main reasons for using the polar representation. The angle θ lies conventionally in the range $-\pi < \theta \leq \pi$, but, since rotation by θ is the same as rotation by $2n\pi + \theta$, where n is any integer,

$$re^{i\theta} \equiv re^{i(\theta+2n\pi)}.$$

The algebra of the polar representation is different from that of the real and imaginary component representation, though, of course, the results are identical. Some operations prove much easier in the polar representation, others much more complicated. The best representation for a particular problem must be determined by the manipulation required.

5.3 Polar representation of complex numbers

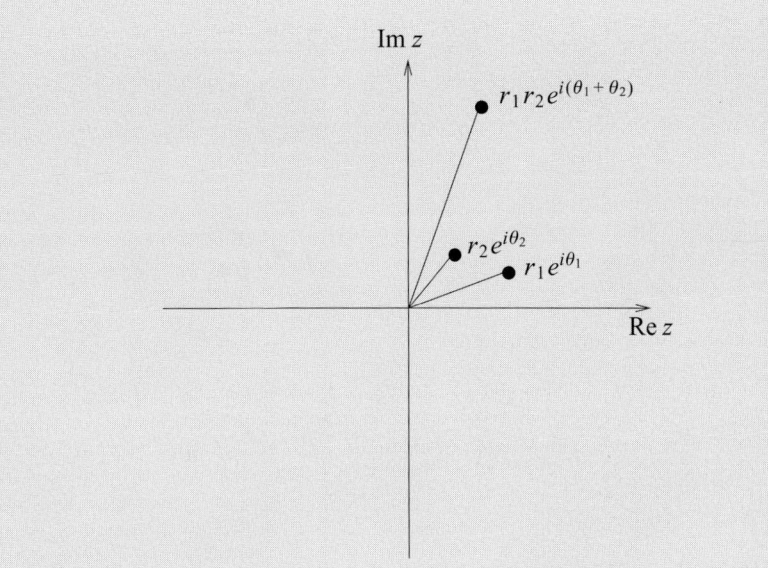

Figure 5.8 The multiplication of two complex numbers. In this case r_1 and r_2 are both greater than unity.

5.3.1 Multiplication and division in polar form

Multiplication and division in polar form are particularly simple. The product of $z_1 = r_1 e^{i\theta_1}$ and $z_2 = r_2 e^{i\theta_2}$ is given by

$$z_1 z_2 = r_1 e^{i\theta_1} r_2 e^{i\theta_2}$$
$$= r_1 r_2 e^{i(\theta_1 + \theta_2)}. \tag{5.26}$$

The relations $|z_1 z_2| = |z_1||z_2|$ and $\arg(z_1 z_2) = \arg z_1 + \arg z_2$ follow immediately. An example of the multiplication of two complex numbers is shown in Figure 5.8. Since no length scale is marked on the figure, it is not possible to check that the relationship between the moduli has been accurately represented,[9] but it can be seen that the angle between the line representing z_1 and the real z-axis is equal to that between the lines representing z_2 and $z_1 z_2$ – both should be equal to the argument of z_1.

Division is equally simple in polar form; the quotient of z_1 and z_2 is given by

$$\frac{z_1}{z_2} = \frac{r_1 e^{i\theta_1}}{r_2 e^{i\theta_2}} = \frac{r_1}{r_2} e^{i(\theta_1 - \theta_2)}. \tag{5.27}$$

The relations $|z_1/z_2| = |z_1|/|z_2|$ and $\arg(z_1/z_2) = \arg z_1 - \arg z_2$ are again immediately apparent. Complementing our previous statement about the product of two complex numbers (see Section 5.2.3), the above result can be put in the verbal form that the magnitude of a quotient is equal to the quotient of the magnitudes and the argument of a quotient is

---

[9] What could be done, by the interested reader, is to determine, on the assumption that the figure is to scale, the approximate radius the representation of the unit circle would have, if it were to be drawn.

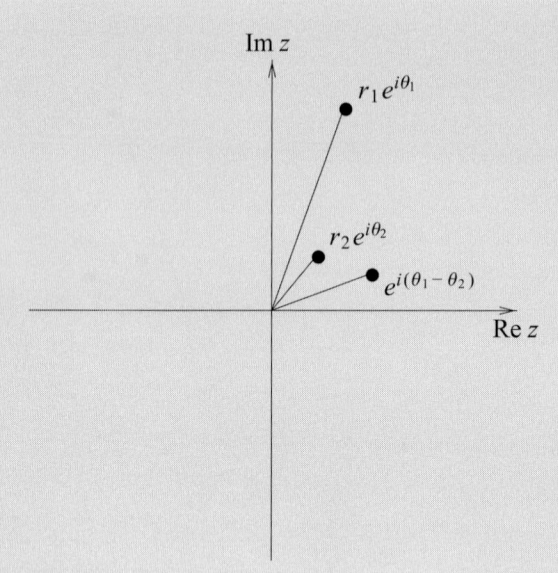

Figure 5.9 The division of two complex numbers. As in Figure 5.8, r_1 and r_2 are both greater than unity.

equal to the difference of the arguments. The division of two complex numbers in polar form is shown in Figure 5.9.

We are now in a position to prove the statement made near the end of Section 5.2.4 that the product of

$$z = w^{3y+2ix} = (x+5i)^{3y+2ix} \quad \text{and} \quad z^* = (w^*)^{3y-2ix} = (x-5i)^{3y-2ix}$$

is real. It should be remembered that x and y are themselves real, and so we can write the product zz^* as

$$zz^* = w^{3y}w^{2ix}(w^*)^{3y}(w^*)^{-2ix} = (|w|^2)^{3y}\left(\frac{w}{w^*}\right)^{2ix}.$$

The first factor in the final term on the RHS is clearly real and so we need to consider the second. Since w and w^* have equal magnitudes, w/w^* has the form $e^{2i\theta}$ where θ is the argument of w. Thus the second factor takes the form

$$(e^{2i\theta})^{2ix} = e^{-4\theta x}, \quad \text{which is real.}$$

Since both factors in the final term are real, so is their product, i.e. zz^* is real.

EXERCISES 5.3

1. Find the sum and product of $z_1 = 4 - 3i$ and $z_2 = 2e^{-i\pi/6}$, expressing your answers in both Cartesian and polar coordinates.

2. Show that the real and imaginary parts of z_1^7, where $z_1 = 4 - 3i$, are $-16\,124$ and $76\,443$, respectively. But do not attempt to do so by making six complex multiplications!

5.4 De Moivre's theorem

We now derive an extremely important theorem. Since $(e^{i\theta})^n = e^{in\theta}$, we have

$$(\cos\theta + i\sin\theta)^n = \cos n\theta + i\sin n\theta, \tag{5.28}$$

where the identity $e^{in\theta} = \cos n\theta + i\sin n\theta$ follows from the series definition of $e^{in\theta}$ [see (5.22)]. This result is called *de Moivre's theorem* and is often used in the manipulation of complex numbers. The theorem is valid for all n whether real, imaginary or complex.[10]

There are numerous applications of de Moivre's theorem, but this section examines just three: proofs of trigonometric identities; finding the nth roots of unity; and solving polynomial equations with complex roots.

5.4.1 Trigonometric identities

The use of de Moivre's theorem in finding trigonometric identities is best illustrated by examples. We consider first the problem of expressing a function of a multiple angle as a polynomial in the corresponding sinusoidal functions of a single angle.

Example Express $\sin 3\theta$ and $\cos 3\theta$ in terms of powers of $\cos\theta$ and $\sin\theta$.

Since we are considering a function of 3θ we use de Moivre's theorem with $n = 3$. Expanding the factor $(\cos\theta + i\sin\theta)^3$ using the binomial theorem, and recalling that $i^2 = -1$ and $i^3 = -i$, gives

$$\begin{aligned}\cos 3\theta + i\sin 3\theta &= (\cos\theta + i\sin\theta)^3 \\ &= \cos^3\theta + 3i\cos^2\theta\sin\theta + 3i^2\cos\theta\sin^2\theta + i^3\sin^3\theta \\ &= (\cos^3\theta - 3\cos\theta\sin^2\theta) + i(3\cos^2\theta\sin\theta - \sin^3\theta). \end{aligned} \tag{5.29}$$

We can equate the real and imaginary parts on the two sides of the equation separately. The real parts give

$$\begin{aligned}\cos 3\theta &= \cos^3\theta - 3\cos\theta\sin^2\theta \\ &= \cos^3\theta - 3\cos\theta(1 - \cos^2\theta) \\ &= 4\cos^3\theta - 3\cos\theta. \end{aligned} \tag{5.30}$$

Equating the imaginary parts yields

$$\begin{aligned}\sin 3\theta &= 3\cos^2\theta\sin\theta - \sin^3\theta \\ &= 3(1 - \sin^2\theta)\sin\theta - \sin^3\theta \\ &= 3\sin\theta - 4\sin^3\theta. \end{aligned}$$

In each case, in order to obtain the final form, we have used the identity $\cos^2\theta + \sin^2\theta = 1$. It will be noticed that $\cos 3\theta$, an even function of 3θ, and hence of θ, is expressed purely in terms of $\cos\theta$, an even function of θ; the fact it is raised to an odd power is irrelevant since $(+1)^3 = +1$. For $\sin 3\theta$, an odd function of θ, it does matter that, in each term of its expansion, $\sin\theta$, which is odd, is raised to an odd power; if it were raised to an even power the relevant term would be an even function of θ despite $\sin\theta$ being an odd function.[11]

◀

---

[10] Show that $(\cos\theta + i\sin\theta)^i$ is real and less than 1 if θ is real and positive.
[11] Which terms would you expect in the power expansion of $\sin 4\theta$? Try it out and show that it cannot be expressed as a *polynomial* in $\sin\theta$.

Complex numbers and hyperbolic functions

The general method can clearly be applied to finding power expansions of $\cos n\theta$ and $\sin n\theta$ for any positive integer n.

The converse process, that of expressing a power of $\cos\theta$ or $\sin\theta$ as a sum of terms containing the cosines and sines of multiple angles, uses the following two properties of $z = e^{i\theta}$:

$$z^n + \frac{1}{z^n} = 2\cos n\theta, \tag{5.31}$$

$$z^n - \frac{1}{z^n} = 2i\sin n\theta. \tag{5.32}$$

These equalities follow from simple applications of de Moivre's theorem combined with the respective oddness and evenness of the sine and cosine functions. For (5.31) we have

$$\begin{aligned} z^n + \frac{1}{z^n} &= (\cos\theta + i\sin\theta)^n + (\cos\theta + i\sin\theta)^{-n} \\ &= \cos n\theta + i\sin n\theta + \cos(-n\theta) + i\sin(-n\theta) \\ &= \cos n\theta + i\sin n\theta + \cos n\theta - i\sin n\theta \\ &= 2\cos n\theta, \end{aligned}$$

whilst (5.32) follows from

$$\begin{aligned} z^n - \frac{1}{z^n} &= (\cos\theta + i\sin\theta)^n - (\cos\theta + i\sin\theta)^{-n} \\ &= \cos n\theta + i\sin n\theta - \cos n\theta + i\sin n\theta \\ &= 2i\sin n\theta. \end{aligned}$$

For the particular case where $n = 1$,

$$z + \frac{1}{z} = e^{i\theta} + e^{-i\theta} = 2\cos\theta, \tag{5.33}$$

$$z - \frac{1}{z} = e^{i\theta} - e^{-i\theta} = 2i\sin\theta. \tag{5.34}$$

As expected, these relationships recover the series expansions for $\cos\theta$ and $\sin\theta$ when $e^{\pm i\theta}$ are expanded according to the definition of $\exp(x)$.

Example Find an expression for $\cos^3\theta$ in terms of $\cos 3\theta$ and $\cos\theta$.

Starting from (5.33), and using the binomial expansion, we obtain

$$\begin{aligned} \cos^3\theta &= \frac{1}{2^3}\left(z + \frac{1}{z}\right)^3 \\ &= \frac{1}{8}\left(z^3 + 3z + \frac{3}{z} + \frac{1}{z^3}\right) \end{aligned}$$

Because of the symmetry of the binomial coefficients, in that $^nC_r = {}^nC_{n-r}$, the coefficients of z^r and z^{-r} in such an expansion are bound to be equal in magnitude,[12,13] and they can be grouped to

---

[12] Convince yourself that the expansion of the nth power of $\sin\theta$, where n is an odd positive integer, could never contain a constant term and must consist entirely of sine functions of multiple angles.
[13] Show that the average value of $\cos^n\theta$ is zero if n is odd, but $2^{-n}n!/[(n/2)!]^2$ if n is even.

5.4 De Moivre's theorem

form the LHS of either (5.31) or (5.32). Making such a grouping in this case gives

$$\cos^3 \theta = \frac{1}{8}\left(z^3 + \frac{1}{z^3}\right) + \frac{3}{8}\left(z + \frac{1}{z}\right)$$
$$= \tfrac{1}{4}\cos 3\theta + \tfrac{3}{4}\cos \theta.$$

In line with our previous discussion, the expansion of this even function of θ consists entirely of multiple-angle functions that are also even functions of θ. ◀

This result happens to be a simple rearrangement of (5.30), but cases involving larger values of n are better handled using this direct method than by rearranging polynomial expansions of multiple-angle functions.

5.4.2 Finding the *n*th roots of unity

The equation $z^2 = 1$ has the familiar solutions $z = \pm 1$. However, now that we have introduced the concept of complex numbers we can solve the general equation $z^n = 1$. Recalling the fundamental theorem of algebra, we know that the equation has n solutions. In order to proceed, we recognise that the most general expression for 1, when it is considered as a complex number, is $e^{2ik\pi}$, where k is any integer, and rewrite the equation as

$$z^n = e^{2ik\pi}.$$

Now taking the *n*th root of each side of the equation we find

$$z = e^{2ik\pi/n}.$$

Hence, the solutions of $z^n = 1$ are

$$z_{1,2,\ldots,n} = 1,\ e^{2i\pi/n},\ \ldots,\ e^{2i(n-1)\pi/n},$$

corresponding to the values $0, 1, 2, \ldots, n-1$ for k. Larger integer values of k do not give new solutions, since the roots already listed are simply cyclically repeated for $k = n, n+1, n+2$, etc.

Example Find the solutions to the equation $z^3 = 1$.

By applying the above method we find

$$z = e^{2ik\pi/3}.$$

Hence the three solutions are $z_1 = e^{0i} = 1$, $z_2 = e^{2i\pi/3}$, $z_3 = e^{4i\pi/3}$. We note that, as expected, the next solution, for which $k = 3$, gives $z_4 = e^{6i\pi/3} = 1 = z_1$, so that there are only three separate solutions. ◀

Not surprisingly, given that $|z^3| = |z|^3$ from (5.10), all the roots of unity have unit modulus, i.e. they all lie on a circle in the Argand diagram of unit radius. The three roots are shown in Figure 5.10. Written in the form $z = x + iy$, the two complex roots are expressed as $z_2 = \tfrac{1}{2}(-1 + i\sqrt{3})$ and $z_3 = \tfrac{1}{2}(-1 - i\sqrt{3})$.

Complex numbers and hyperbolic functions

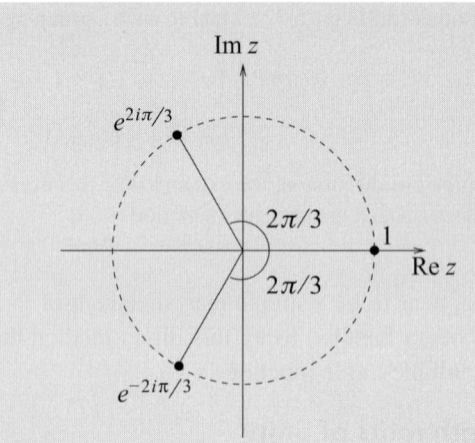

Figure 5.10 The solutions of $z^3 = 1$.

The cube roots of unity are often written as 1, ω and ω^2, with $\omega = e^{2i\pi/3}$. The properties $\omega^3 = 1$ and $1 + \omega + \omega^2 = 0$ are easily proved.[14]

5.4.3 Solving polynomial equations

A third application of de Moivre's theorem is to the solution of general polynomial equations. The methods used are very similar to those employed when finding the roots of real polynomial equations. Indeed, the first step in solving complex polynomial equations is to use those same methods to obtain equations of reduced degree that are satisfied by z, or powers of z. The complex roots may then be deduced, as is illustrated in the following example.

Example Solve the equation $f(z) = z^6 - z^5 + 4z^4 - 6z^3 + 2z^2 - 8z + 8 = 0$.

We first note that the sum of the coefficients in this sixth degree equation is zero; this means that $z = 1$ is one solution and that $z - 1$ is a factor of $f(z)$. Either by inspection or by writing $f(z) = (z - 1)(z^5 + a_4 z^4 + a_3 z^3 + a_2 z^2 + a_1 z + a_0)$ and equating coefficients as in Section 2.1.1, we find that

$$f(z) = (z - 1)(z^5 + 4z^3 - 2z^2 - 8) = 0.$$

The second term in parentheses can be factorised by inspection to give

$$f(z) = (z - 1)(z^3 - 2)(z^2 + 4) = 0.$$

Hence the roots are given by $z^3 = 2$, $z^2 = -4$ and $z = 1$, with the solutions to the quadratic equation given immediately by $z = \pm 2i$.

To find the complex cube roots, we first write the equation in the form

$$z^3 = 2 = 2e^{2ik\pi},$$

---

[14] This can be done either by direct substitution of the explicit forms or from the properties of the roots of a cubic equation as in Section 2.1.2.

5.4 De Moivre's theorem

where k is any integer. If we now take the cube root of both sides, we get

$$z = 2^{1/3} e^{2ik\pi/3}.$$

To avoid the duplication of solutions, we use the fact that $-\pi < \arg z \leq \pi$, which means taking only $k = 0$, $k = 1$ and $k = 2$, and find that the three solutions are

$$z_1 = 2^{1/3},$$
$$z_2 = 2^{1/3} e^{2\pi i/3} = 2^{1/3} \tfrac{1}{2}(-1 + \sqrt{3}i),$$
$$z_3 = 2^{1/3} e^{-2\pi i/3} = 2^{1/3} \tfrac{1}{2}(-1 - \sqrt{3}i).$$

The complex numbers z_1, z_2 and z_3, together with $z_4 = 2i$, $z_5 = -2i$ and $z_6 = 1$, are the solutions to the original polynomial equation.[15]

As expected from the fundamental theorem of algebra, we find that the total number of complex roots (six, in this case) is equal to the largest power of z in the polynomial. ◀

One generally useful result that can be established is that the roots of a polynomial with real coefficients occur in conjugate pairs (i.e. if z_1 is a root, then z_1^* is a second distinct root, unless z_1 is real). Most polynomial equations that arise from physical situations do have real coefficients as many coefficients are the direct results of physical measurements.

The proof of the assertion is as follows. Let the polynomial equation of which z is a root be

$$a_n z^n + a_{n-1} z^{n-1} + \cdots + a_1 z + a_0 = 0.$$

Taking the complex conjugate of this equation:

$$a_n^* (z^*)^n + a_{n-1}^* (z^*)^{n-1} + \cdots + a_1^* z^* + a_0^* = 0.$$

But the a_n are real, and so z^* satisfies

$$a_n (z^*)^n + a_{n-1} (z^*)^{n-1} + \cdots + a_1 z^* + a_0 = 0$$

and is therefore also a root of the original equation. An immediate corollary of this result is that any polynomial equation of odd degree with real coefficients has an odd number of real roots, and therefore at least one; this is the same conclusion as that reached less rigorously in the discussion following Equation (2.8).

EXERCISES 5.4

1. Find as polynomials in $\cos \theta$ and $\sin \theta$, respectively, expressions for $\cos 5\theta$ and $\sin 5\theta$.

2. Use the results of the previous exercise to deduce that

$$\cos\left(\frac{\pi}{10}\right) = \left(\frac{5 + \sqrt{5}}{8}\right)^{1/2} \quad \text{and} \quad \sin\left(\frac{\pi}{5}\right) = \left(\frac{5 - \sqrt{5}}{8}\right)^{1/2}.$$

---

[15] Check that the sum and product of these six roots have the values you expect.

3. Find all the distinct solutions of (a) $z^5 = 1$, and (b) $z^3 + 1 = 0$. Plot them all on a single Argand diagram.

4. By 'completing the cube' find all distinct roots of the equation
$$z^3 - 3z^2 + 3z - 9 = 0,$$
expressing them in terms of $\omega = e^{2i\pi/3}$. Verify that the product of the roots has its expected value.

5. Find, by factorising it, all the zeros of the function $f(z) = z^6 + 9z^4 - z^3 - 9z$ and plot them on an Argand diagram.

5.5 Complex logarithms and complex powers

The concept of a complex exponential has already been introduced in Section 5.3, where it was assumed that the definition of an exponential as a series was valid for complex numbers as well as for real numbers. Similarly, we can define the logarithm of a complex number and we can use complex numbers as exponents.

Let us denote the natural logarithm of a complex number z by $w = \operatorname{Ln} z$, where the notation Ln will be explained shortly. Thus, w must satisfy
$$z = e^w.$$
Using (5.21), we see that
$$z_1 z_2 = e^{w_1} e^{w_2} = e^{w_1 + w_2},$$
and taking logarithms of both sides we find
$$\operatorname{Ln}(z_1 z_2) = w_1 + w_2 = \operatorname{Ln} z_1 + \operatorname{Ln} z_2, \tag{5.35}$$
which shows that the familiar rule for the logarithm of the product of two real numbers also holds for complex numbers.

We may use (5.35) to investigate further the properties of Ln z. We have already noted that the argument of a complex number is multivalued, i.e. $\arg z = \theta + 2n\pi$, where n is any integer. Thus, in polar form, the complex number z should strictly be written as
$$z = re^{i(\theta + 2n\pi)}.$$
Taking the logarithm of both sides, and using (5.35), we find
$$\operatorname{Ln} z = \ln r + i(\theta + 2n\pi), \tag{5.36}$$
where $\ln r$ is the natural logarithm of the real positive quantity r and so is written normally. Thus from (5.36) we see that Ln z is itself multivalued. To avoid this multivalued behaviour it is conventional to define another function $\ln z$, the *principal value* of Ln z, which is obtained from Ln z by restricting the argument of z to lie in the range $-\pi <$

5.5 Complex logarithms and complex powers

$0 \leq \pi$. A straightforward illustration of these definitions is provided by the following example.

Example Evaluate $\text{Ln}(-i)$.

By rewriting $-i$ as a complex exponential, we find
$$\text{Ln}(-i) = \text{Ln}\left[e^{i(-\pi/2+2n\pi)}\right] = i(-\pi/2 + 2n\pi),$$
where n is any integer. Hence $\text{Ln}(-i) = \ldots, -5i\pi/2, -i\pi/2, 3i\pi/2, \ldots$ We note that $\ln(-i)$, the principal value of $\text{Ln}(-i)$, is given by $\ln(-i) = -i\pi/2$. ◂

If z and t are both complex numbers then the zth power of t is defined by
$$t^z = e^{z \text{Ln} t}.$$

Since $\text{Ln} t$ is multivalued, so too is this definition; its principal value is obtained by giving $\text{Ln} t$ its principal value.[16,17]

Example Simplify the expression $z = i^{-2i}$.

Firstly we take the logarithm of both sides of the equation and obtain
$$\text{Ln} z = -2i \, \text{Ln} i.$$
Now inverting the process, i.e. exponentiating both sides of the equation, we find
$$e^{\text{Ln} z} = z = e^{-2i \text{Ln} i}.$$
We can write $i = e^{i(\pi/2+2n\pi)}$, where n is any integer, and hence
$$\text{Ln} i = \text{Ln}\left[e^{i(\pi/2+2n\pi)}\right]$$
$$= i(\pi/2 + 2n\pi).$$
We can now simplify z to give
$$z = i^{-2i} = e^{-2i \times i(\pi/2+2n\pi)}$$
$$= e^{(\pi+4n\pi)},$$
which, perhaps surprisingly, is a real quantity rather than a complex one. ◂

The multivalued nature of complex powers and of the logarithms of complex numbers generally increase the difficulty of dealing with functions of a complex variable.[18]

---

[16] Consider whether the zth power of unity, 1^z, could have a non-unit modulus.
[17] On an Argand diagram show the approximate position(s) of the value(s) of the complex power $20^{i/(2\pi)}$.
[18] Although sometimes it can be used to advantage, e.g. some otherwise intractable integrals can be evaluated by exploiting the multivaluedness of their integrands.

Complex numbers and hyperbolic functions

However, the general topic of functions of a complex variable is beyond the scope of this book and we will not pursue the matter further.

EXERCISES 5.5

1. In which quadrant of the Argand diagram do all the values of $(3 + i)^{2i}$ lie?

2. Evaluate the multivalued functions $(1 + i)^i$ and $i^{(1+i)}$. In each case, describe qualitatively the disposition of the values on the Argand diagram.

5.6 Applications to differentiation and integration

We can use the exponential form of a complex number together with de Moivre's theorem (see Section 5.4) to simplify the differentiation of trigonometric functions, particularly when they are multiplied by linear exponential functions of the same variable.

> **Example** Find the derivative with respect to x of $e^{3x} \cos 4x$.
>
> We could differentiate this function straightforwardly using the product rule (see Section 3.1.2).[19] However, an alternative method in this case is to use a complex exponential. Let us consider the complex number
>
> $$z = e^{3x}(\cos 4x + i \sin 4x) = e^{3x} e^{4ix} = e^{(3+4i)x},$$
>
> where we have used de Moivre's theorem to rewrite the trigonometric functions as a complex exponential. This complex number has $e^{3x} \cos 4x$ as its real part. Now, differentiating z with respect to x we obtain
>
> $$\frac{dz}{dx} = (3 + 4i)e^{(3+4i)x} = (3 + 4i)e^{3x}(\cos 4x + i \sin 4x), \qquad (5.37)$$
>
> where we have again used de Moivre's theorem. Equating real parts we then find
>
> $$\frac{d}{dx}\left(e^{3x} \cos 4x\right) = e^{3x}(3 \cos 4x - 4 \sin 4x).$$
>
> By equating the imaginary parts of (5.37), we also obtain
>
> $$\frac{d}{dx}\left(e^{3x} \sin 4x\right) = e^{3x}(4 \cos 4x + 3 \sin 4x).$$
>
> Most procedures that replace real variables with complex ones produce a bonus of this kind, but many are of the disappointing $0 = 0$ form! ◂

In a similar way the complex exponential can be used to evaluate integrals containing both trigonometric and exponential functions.

---

[19] You should confirm that the same answers are obtained.

5.7 Hyperbolic functions

Example Evaluate the integral $I = \int e^{ax} \cos bx \, dx$.

Let us consider the integrand as the real part of the complex number

$$e^{ax}(\cos bx + i \sin bx) = e^{ax} e^{ibx} = e^{(a+ib)x},$$

where we use de Moivre's theorem to rewrite the trigonometric functions as a complex exponential. Integrating we find

$$\begin{aligned}
\int e^{(a+ib)x} \, dx &= \frac{e^{(a+ib)x}}{a+ib} + c \\
&= \frac{(a-ib)e^{(a+ib)x}}{(a-ib)(a+ib)} + c \\
&= \frac{e^{ax}}{a^2+b^2}\left(ae^{ibx} - ibe^{ibx}\right) + c,
\end{aligned} \quad (5.38)$$

where the constant of integration c is in general complex. Denoting this constant by $c = c_1 + ic_2$ and equating real parts in (5.38) we obtain

$$I = \int e^{ax} \cos bx \, dx = \frac{e^{ax}}{a^2+b^2}(a \cos bx + b \sin bx) + c_1,$$

which agrees with result (4.18) found using integration by parts. Equating imaginary parts in (5.38) we obtain, as a bonus,

$$J = \int e^{ax} \sin bx \, dx = \frac{e^{ax}}{a^2+b^2}(a \sin bx - b \cos bx) + c_2.$$

As would be expected, both results reduce to the normal integrals of $\cos x$ and $\sin x$ in the limit $a \to 0$. ◄

EXERCISE 5.6

1. Use the results that follow (5.38) to evaluate

$$\int_0^\infty e^{-2x} \sin 3x \, dx \quad \text{and} \quad \int_0^\infty e^{-2x} \cos 3x \, dx.$$

5.7 Hyperbolic functions

We have hitherto mentioned hyperbolic functions so many times that the reader may reasonably wonder whether many of the results in this current section have already been established by proclamation. Of course, that is not so, and we now proceed to justify what has previously been assumed or quoted.

The *hyperbolic functions* are the complex analogues of the trigonometric functions. The analogy may not be immediately apparent and their definitions may appear at first to be somewhat arbitrary. However, careful examination of their properties reveals the

Figure 5.11 Graphs of cosh x and sech x.

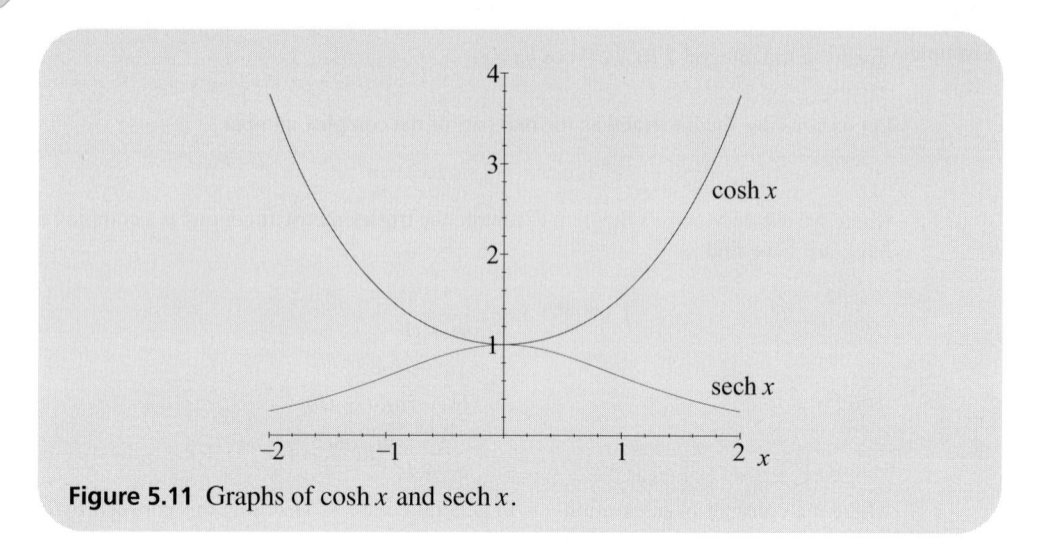

purpose of the definitions. For instance, their close relationship with the trigonometric functions, both in their identities and in their calculus, means that many of the familiar properties of trigonometric functions can also be applied to the hyperbolic functions. Further, hyperbolic functions occur regularly, and so giving them special names is a notational convenience.

5.7.1 Definitions

The two fundamental hyperbolic functions are cosh x and sinh x, which, as their names suggest, are the hyperbolic equivalents of cos x and sin x. They are defined by the following relations:

$$\cosh x = \tfrac{1}{2}(e^x + e^{-x}), \tag{5.39}$$
$$\sinh x = \tfrac{1}{2}(e^x - e^{-x}). \tag{5.40}$$

Note that cosh x is an even function of x and sinh x is an odd function. By analogy with the trigonometric functions, the remaining hyperbolic functions are

$$\tanh x = \frac{\sinh x}{\cosh x} = \frac{e^x - e^{-x}}{e^x + e^{-x}}, \tag{5.41}$$
$$\operatorname{sech} x = \frac{1}{\cosh x} = \frac{2}{e^x + e^{-x}}, \tag{5.42}$$
$$\operatorname{cosech} x = \frac{1}{\sinh x} = \frac{2}{e^x - e^{-x}}, \tag{5.43}$$
$$\coth x = \frac{1}{\tanh x} = \frac{e^x + e^{-x}}{e^x - e^{-x}}. \tag{5.44}$$

All the hyperbolic functions above have been defined in terms of the real variable x. However, this was simply so that they may be plotted (see Figures 5.11–5.13); the definitions are equally valid for any complex argument z.

5.7 Hyperbolic functions

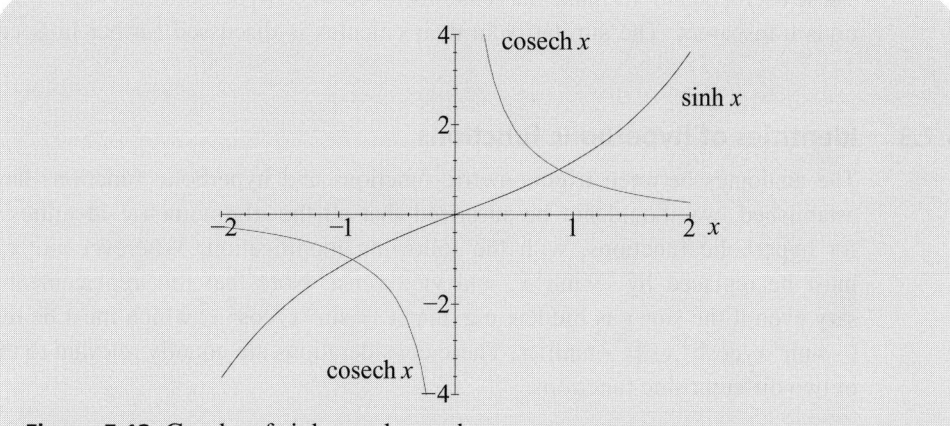

Figure 5.12 Graphs of sinh x and cosech x.

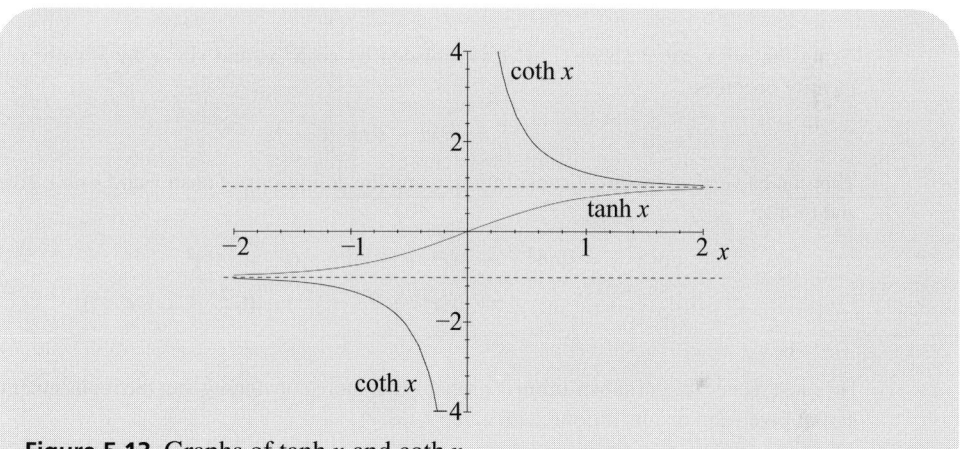

Figure 5.13 Graphs of tanh x and coth x.

5.7.2 Hyperbolic–trigonometric analogies

In the previous subsections we have alluded to the analogy between trigonometric and hyperbolic functions. Here, we discuss the close relationship between the two groups of functions.

Recalling (5.33) and (5.34), which express $\cos\theta$ and $\sin\theta$ in terms of exponential functions, we find, on setting θ as ix, that

$$\cos ix = \tfrac{1}{2}(e^{-x} + e^x) = \tfrac{1}{2}(e^x + e^{-x}),$$
$$\sin ix = \tfrac{1}{2i}(e^{-x} - e^x) = \tfrac{1}{2}i(e^x - e^{-x}).$$

Hence, by the definitions given in the previous subsection,

$$\cosh x = \cos ix, \tag{5.45}$$
$$i \sinh x = \sin ix, \tag{5.46}$$
$$\cos x = \cosh ix, \tag{5.47}$$
$$i \sin x = \sinh ix. \tag{5.48}$$

5.7.3 Identities of hyperbolic functions

The analogies between trigonometric functions and hyperbolic functions having been established, we should not be surprised that all the trigonometric identities also hold for hyperbolic functions, with the following modification. Wherever $\sin^2 x$ occurs it must be replaced by $-\sinh^2 x$, and vice versa. Note that this replacement is necessary even if the $\sin^2 x$ is hidden, e.g. $\tan^2 x = \sin^2 x / \cos^2 x$, which must be replaced by $(-\sinh^2 x / \cosh^2 x) = -\tanh^2 x$. These considerations are equally relevant to the product of two different sine functions.

Example Find the hyperbolic identity analogous to $\cos^2 x + \sin^2 x = 1$.

Using the rules stated above, $\cos^2 x$ is replaced by $\cosh^2 x$, and $\sin^2 x$ by $-\sinh^2 x$, and so the identity becomes

$$\cosh^2 x - \sinh^2 x = 1.$$

This can be verified by direct substitution, using the definitions of $\cosh x$ and $\sinh x$ given in (5.39) and (5.40):

$$\begin{aligned}\cosh^2 x - \sinh^2 x &= \tfrac{1}{4}(e^x + e^{-x})^2 - \tfrac{1}{4}(e^x - e^{-x})^2 \\ &= \tfrac{1}{4}[(e^{2x} + 2 + e^{-2x}) - [(e^{2x} - 2 + e^{-2x})] \\ &= 1.\end{aligned}$$

For most readers direct substitution is probably the more convincing, but the replacements indicated earlier give valid results if done with care. ◀

Some other identities that can be proved in a similar way are[20]

$$\text{sech}^2 x = 1 - \tanh^2 x, \tag{5.49}$$
$$\text{cosech}^2 x = \coth^2 x - 1, \tag{5.50}$$
$$\sinh 2x = 2 \sinh x \cosh x, \tag{5.51}$$
$$\cosh 2x = \cosh^2 x + \sinh^2 x. \tag{5.52}$$

5.7.4 Solving hyperbolic equations

When we are presented with a hyperbolic equation to solve, we may proceed by analogy with the solution of trigonometric equations. However, it is almost always easier to express the equation directly in terms of exponentials, as is done in the following example.

---

[20] As a check, write down what you expect to be the expression for $\tanh(u - v)$ in terms of $\tanh u$ and $\tanh v$, and then either prove, verify or correct it.

5.7 Hyperbolic functions

Example Solve the hyperbolic equation $\cosh x - 5 \sinh x - 5 = 0$.

Substituting the definitions of the hyperbolic functions we obtain
$$\tfrac{1}{2}(e^x + e^{-x}) - \tfrac{5}{2}(e^x - e^{-x}) - 5 = 0.$$
Rearranging, and then multiplying through by $-e^x$, gives in turn
$$-2e^x + 3e^{-x} - 5 = 0$$
and
$$2e^{2x} + 5e^x - 3 = 0.$$
Now we can factorise and solve:
$$(2e^x - 1)(e^x + 3) = 0.$$
Thus $e^x = 1/2$ or $e^x = -3$. Hence $x = -\ln 2$ or $x = \ln(-3)$. The interpretation of the logarithm of a negative number has been discussed in Section 5.5. ◀

For solving the general hyperbolic equation, of the form $a \cosh x + b \sinh x = c$ with $c > 0$, the original hyperbolic equation can be written as a quadratic equation in either e^x or e^{-x}; of course, the same solutions are found for x, whichever approach is adopted.[21]

5.7.5 Inverses of hyperbolic functions

Just like trigonometric functions, hyperbolic functions have inverses. If $y = \cosh x$ then $x = \cosh^{-1} y$, which serves as a definition of the inverse. By using the fundamental definitions of hyperbolic functions, we can find closed-form expressions for their inverses. This is best illustrated by example.

Example Find a closed-form expression for the inverse hyperbolic function $y = \sinh^{-1} x$, where y is real.

First we write x as a function of y, i.e.
$$y = \sinh^{-1} x \implies x = \sinh y.$$
Now, since $\cosh y = \tfrac{1}{2}(e^y + e^{-y})$ and $\sinh y = \tfrac{1}{2}(e^y - e^{-y})$,
$$e^y = \cosh y + \sinh y = \sqrt{1 + \sinh^2 y} + \sinh y = \sqrt{1 + x^2} + x,$$
and hence
$$y = \ln(\sqrt{1 + x^2} + x).$$
When substituting for $\cosh y$ we took the positive square root of $1 + \sinh^2 y$, since, for real y, the function $\cosh y$ is always positive. ◀

---

[21] Find explicit expressions for e^x and e^{-x}, where x is one of the solutions, and verify that their product is unity.

Complex numbers and hyperbolic functions

In a similar fashion it can be shown that

$$\cosh^{-1} x = \ln(x \pm \sqrt{x^2 - 1}),$$

the $\pm$ sign arising because $\sinh y = \pm\sqrt{\cosh^2 y - 1}$, and corresponding to the fact that for any given $x > 1$, there are two values,[22] $y = \pm\alpha$, that satisfy $\cosh y = x$. This is in contrast to the result of the worked example, since, for any given x (positive or negative), there is only one real value of y that satisfies $\sinh y = x$.

We finish this subsection with a second worked example that obtains an explicit expression for an inverse hyperbolic function.

Example Find a closed-form expression for the inverse hyperbolic function $y = \tanh^{-1} x$ for real y.

We note that x must lie in the range $-1 < x < 1$ and rearrange the equation to make x its subject:

$$y = \tanh^{-1} x \quad \Rightarrow \quad x = \tanh y.$$

Now, using the definition of $\tanh y$ and rearranging that, we find

$$x = \frac{e^y - e^{-y}}{e^y + e^{-y}} \quad \Rightarrow \quad (x+1)e^{-y} = (1-x)e^y.$$

Thus, it follows that

$$e^{2y} = \frac{1+x}{1-x} \quad \Rightarrow \quad e^y = \sqrt{\frac{1+x}{1-x}} \quad \Rightarrow \quad y = \ln\sqrt{\frac{1+x}{1-x}}.$$

Expressed purely in term of x, this becomes

$$\tanh^{-1} x = \frac{1}{2} \ln\left(\frac{1+x}{1-x}\right).$$

The final form of the answer is clearly consistent with the antisymmetry property $\tanh^{-1}(-x) = -\tanh^{-1}(x)$. ◀

Graphs of the inverse hyperbolic functions are shown in Figures 5.14–5.16.

5.7.6 Calculus of hyperbolic functions

Just as the identities of hyperbolic functions closely follow those of their trigonometric counterparts, so does their calculus. The derivatives of the two basic hyperbolic functions are given by

$$\frac{d}{dx}(\cosh x) = \sinh x, \tag{5.53}$$

$$\frac{d}{dx}(\sinh x) = \cosh x. \tag{5.54}$$

They may be deduced by considering the definitions (5.39), (5.40) as follows.

---

22 Show that $\ln(x - \sqrt{x^2 - 1}) = -\ln(x + \sqrt{x^2 - 1})$.

5.7 Hyperbolic functions

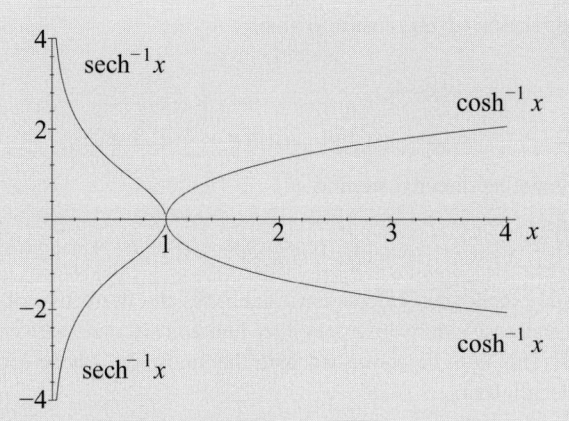

Figure 5.14 Graphs of $\cosh^{-1} x$ and $\text{sech}^{-1} x$.

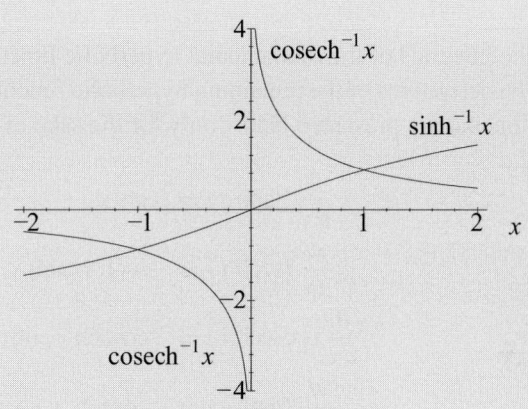

Figure 5.15 Graphs of $\sinh^{-1} x$ and $\text{cosech}^{-1} x$.

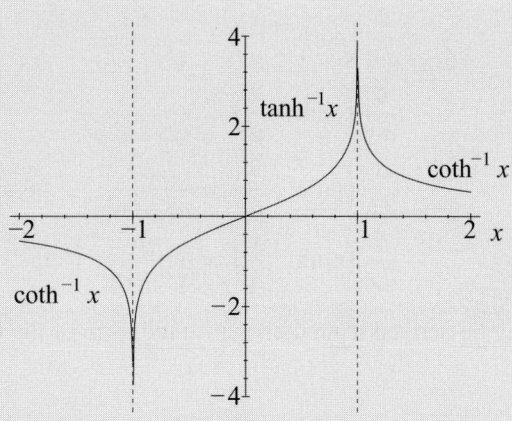

Figure 5.16 Graphs of $\tanh^{-1} x$ and $\coth^{-1} x$.

Complex numbers and hyperbolic functions

Example Verify the relation $(d/dx)\cosh x = \sinh x$.

Using the definition of $\cosh x$,
$$\cosh x = \tfrac{1}{2}(e^x + e^{-x}),$$
and differentiating directly, we find
$$\frac{d}{dx}(\cosh x) = \tfrac{1}{2}(e^x - e^{-x}) = \sinh x.$$

A completely analogous calculation establishes the derivative of $\sinh x$ as $\cosh x$. It should be noted that successive derivatives of either function alternate between the two, with no minus signs involved.[23] This is to be contrasted with the sinusoids, whose successive derivatives go through cycles of length four:
$$\ldots, \sin x, \cos x, -\sin x, -\cos x, \sin x, \ldots$$
and involve minus signs. ◀

Clearly the integrals of the fundamental hyperbolic functions are also defined by these relations. The derivatives of the remaining hyperbolic functions can be derived by product differentiation and are presented below only for the sake of completeness.

$$\frac{d}{dx}(\tanh x) = \operatorname{sech}^2 x, \tag{5.55}$$

$$\frac{d}{dx}(\operatorname{sech} x) = -\operatorname{sech} x \tanh x, \tag{5.56}$$

$$\frac{d}{dx}(\operatorname{cosech} x) = -\operatorname{cosech} x \coth x, \tag{5.57}$$

$$\frac{d}{dx}(\coth x) = -\operatorname{cosech}^2 x. \tag{5.58}$$

The inverse hyperbolic functions also have derivatives, which are given by the following:

$$\frac{d}{dx}\left(\cosh^{-1}\frac{x}{a}\right) = \frac{\pm 1}{\sqrt{x^2 - a^2}}, \tag{5.59}$$

$$\frac{d}{dx}\left(\sinh^{-1}\frac{x}{a}\right) = \frac{1}{\sqrt{x^2 + a^2}}, \tag{5.60}$$

$$\frac{d}{dx}\left(\tanh^{-1}\frac{x}{a}\right) = \frac{a}{a^2 - x^2}, \quad \text{for } x^2 < a^2, \tag{5.61}$$

$$\frac{d}{dx}\left(\coth^{-1}\frac{x}{a}\right) = \frac{-a}{x^2 - a^2}, \quad \text{for } x^2 > a^2. \tag{5.62}$$

These may be derived from the logarithmic form of the inverse (see Section 5.7.5).

---

[23] Differentiate both sides of the relationship $\sinh 2x = 2\sinh x \cosh x$ to obtain a formula for $\cosh 2x$, and then verify it by direct substitution from basic definitions.

Summary

Example Evaluate $(d/dx)\sinh^{-1} x$ using the logarithmic form of the inverse.

From the results of Section 5.7.5,

$$\frac{d}{dx}\left(\sinh^{-1} x\right) = \frac{d}{dx}\left[\ln\left(x + \sqrt{x^2 + 1}\right)\right]$$

$$= \frac{1}{x + \sqrt{x^2 + 1}}\left(1 + \frac{x}{\sqrt{x^2 + 1}}\right)$$

$$= \frac{1}{x + \sqrt{x^2 + 1}}\left(\frac{\sqrt{x^2 + 1} + x}{\sqrt{x^2 + 1}}\right)$$

$$= \frac{1}{\sqrt{x^2 + 1}}.$$

The same result can be obtained by writing $x = \sinh y$, differentiating with respect to y and identifying dy/dx as $(dx/dy)^{-1}$, with $\cosh y$ as $\sqrt{1 + x^2}$.

◀

EXERCISES 5.7

1. Evaluate the following: (a) $\tanh(i)$, (b) $\cosh(2 - i)$.

2. Prove results (5.51) and (5.52) by direct calculation from the definitions of the functions.

3. How many values does the expression $\sinh^{-1} 5 - \cosh^{-1} 5$ have? Find it/them.

4. For each of the following integrals, rearrange the polynomial in the denominator in such a way that it is the derivative of one of the standard inverse hyperbolic functions and so evaluate the integrals.

$$\text{(a) } \int_0^4 \frac{1}{\sqrt{x^2 - 4x + 8}}\,dx, \qquad \text{(b) } \int_4^8 \frac{1}{\sqrt{x^2 - 4x}}\,dx.$$

SUMMARY

1. *Real and imaginary parts*
   - $a + ib = c + id \;\Rightarrow\; a = c$ and $b = d$.
   - With $z = x + iy$, and $z^* = x - iy$,

   $$x = \text{Re } z = \frac{(z + z^*)}{2}, \qquad y = \text{Im } z = \frac{(z - z^*)}{2i}.$$

- In the Argand diagram, $z = re^{i\theta}$ and $z^* = re^{-i\theta}$ with
  (a) $|z| = r = \sqrt{x^2 + y^2} = \sqrt{zz^*}$,
  (b) $\arg z = \theta = \tan^{-1}\left(\dfrac{y}{x}\right)$, taking account of the signs of x and y,
  (c) $x = r\cos\theta$, $y = r\sin\theta$.

2. *Complex algebra*

   With $z_k = x_k + iy_k = r_k e^{i\theta_k}$,

   $$z_1 \pm z_2 = (x_1 \pm x_2) + i(y_1 \pm y_2),$$
   $$z_1 z_2 = (x_1 x_2 - y_1 y_2) + i(x_1 y_2 + y_1 x_2) = r_1 r_2 e^{i(\theta_1 + \theta_2)},$$
   $$|z_1 z_2| = |z_1||z_2|, \quad \arg z_1 z_2 = \arg z_1 + \arg z_2,$$
   $$\frac{z_1}{z_2} = \frac{r_1}{r_2} e^{i(\theta_1 - \theta_2)},$$
   $$\left|\frac{z_1}{z_2}\right| = \frac{|z_1|}{|z_2|}, \quad \arg \frac{z_1}{z_2} = \arg z_1 - \arg z_2.$$

3. *The unit circle*
   - $e^{i\theta} = \cos\theta + i\sin\theta$ (Euler's equation).
   - $(\cos\theta + i\sin\theta)^n = \cos n\theta + i\sin n\theta$ (de Moivre's theorem).
   - The nth roots of unity are $e^{2\pi i k/n}$ for $k = 0, 1, \ldots, n-1$.

4. *Hyperbolic functions*
   - $\cosh x = \tfrac{1}{2}(e^x + e^{-x})$, $\sinh x = \tfrac{1}{2}(e^x - e^{-x})$.
   - $\cos ix = \cosh x$, $\cosh ix = \cos x$, $\sin ix = i\sinh x$, $\sinh ix = i\sin x$.
   - $\cosh^2 x - \sinh^2 x = 1$.
   - $\sinh 2x = 2\sinh x \cosh x$, $\cosh 2x = \cosh^2 x + \sinh^2 x$.
   - $\cosh^{-1} x = \ln(x \pm \sqrt{x^2 - 1})$, $\sinh^{-1} x = \ln(x + \sqrt{x^2 + 1})$.
   - $\dfrac{d}{dx}(\cosh x) = \sinh x$, $\dfrac{d}{dx}(\sinh x) = \cosh x$.
   - $\dfrac{d}{dx}(\cosh^{-1}\dfrac{x}{a}) = \dfrac{\pm 1}{\sqrt{x^2 - a^2}}$, $\dfrac{d}{dx}(\sinh^{-1}\dfrac{x}{a}) = \dfrac{1}{\sqrt{x^2 + a^2}}$.

PROBLEMS

5.1. Express the following as single complex numbers.
(a) $(3 + 2i) + (-1 + i) - (5 + 2i)$, (b) $(3 + 2i)(4 - 3i)$,
(c) $(3 + 2i)/(4 - 3i)$.

Problems

5.2. Express the following as single complex numbers.
 (a) $(1+i)^2 + i(2-3i) - (2+i)^*$,
 (b) $(3+2i)^*/(4-3i)^2$,
 (c) $(3+2i)(1-2i) - (3-2i)(1+2i)$,
 (d) $\dfrac{3+2i}{4-5i} + \dfrac{4+5i}{3-2i}$.

5.3. Two complex numbers z and w are given by $z = 3+4i$ and $w = 2-i$. On an Argand diagram, plot
 (a) $z+w$, (b) $w-z$, (c) wz, (d) z/w,
 (e) $z^*w + w^*z$, (f) w^2, (g) $\ln z$, (h) $(1+z+w)^{1/2}$.

5.4. By considering the real and imaginary parts of the product $e^{i\theta} e^{i\phi}$ prove the standard formulae for $\cos(\theta + \phi)$ and $\sin(\theta + \phi)$.

5.5. By writing $\pi/12 = (\pi/3) - (\pi/4)$ and considering $e^{i\pi/12}$, evaluate $\cot(\pi/12)$.

5.6. Find the locus in the complex z-plane of points that satisfy the following equations.
 (a) $z - c = \rho \left(\dfrac{1+it}{1-it} \right)$, where c is complex, ρ is real and t is a real parameter that varies in the range $-\infty < t < \infty$.
 (b) $z = a + bt + ct^2$, in which t is a real parameter and a, b and c are complex numbers with b/c real.

5.7. Evaluate
 (a) $\mathrm{Re}(\exp 2iz)$, (b) $\mathrm{Im}(\cosh^2 z)$, (c) $(-1 + \sqrt{3}i)^{1/2}$,
 (d) $|\exp(i^{1/2})|$, (e) $\exp(i^3)$, (f) $\mathrm{Im}(2^{i+3})$, (g) i^i, (h) $\ln[(\sqrt{3}+i)^3]$.

5.8. Find the equations in terms of x and y of the sets of points in the Argand diagram that satisfy the following:
 (a) $\mathrm{Re}\, z^2 = \mathrm{Im}\, z^2$;
 (b) $(\mathrm{Im}\, z^2)/z^2 = -i$;
 (c) $\arg[z/(z-1)] = \pi/2$.

5.9. The two sets of points $z = a$, $z = b$, $z = c$, and $z = A$, $z = B$, $z = C$ are the corners of two similar triangles in the Argand diagram. Express in terms of $a, b, \ldots, C$
 (a) the equalities of corresponding angles and
 (b) the constant ratio of corresponding sides
in the two triangles.
By noting that any complex quantity can be expressed as
$$z = |z| \exp(i \arg z),$$
deduce that
$$a(B-C) + b(C-A) + c(A-B) = 0.$$

5.10. The most general type of transformation between one Argand diagram, in the z-plane, and another, in the Z-plane, that gives one and only one value of Z for each value of z (and conversely) is known as the *general bilinear transformation* and takes the form
$$z = \frac{aZ + b}{cZ + d}.$$
(a) Confirm that the transformation from the Z-plane to the z-plane is also a general bilinear transformation.
(b) Recalling that the equation of a circle can be written in the form
$$\left|\frac{z - z_1}{z - z_2}\right| = \lambda, \qquad \lambda \neq 1,$$
show that the general bilinear transformation transforms circles into circles (or straight lines). What is the condition that z_1, z_2 and λ must satisfy if the transformed circle is to be a straight line?

5.11. Sketch the parts of the Argand diagram in which
(a) Re $z^2 < 0$, $|z^{1/2}| \leq 2$;
(b) $0 \leq \arg z^* \leq \pi/2$;
(c) $|\exp z^3| \to 0$ as $|z| \to \infty$.
What is the area of the region in which all three sets of conditions are satisfied?

5.12. Denote the nth roots of unity by $1, \omega_n, \omega_n^2, \ldots, \omega_n^{n-1}$.
(a) Prove that
$$\text{(i)} \sum_{r=0}^{n-1} \omega_n^r = 0, \qquad \text{(ii)} \prod_{r=0}^{n-1} \omega_n^r = (-1)^{n+1}.$$
(b) Express $x^2 + y^2 + z^2 - yz - zx - xy$ as the product of two factors, each linear in x, y and z, with coefficients dependent on the third roots of unity (and those of the x terms arbitrarily taken as real).

5.13. Prove that $x^{2m+1} - a^{2m+1}$, where m is an integer ≥ 1, can be written as
$$x^{2m+1} - a^{2m+1} = (x - a) \prod_{r=1}^{m} \left[x^2 - 2ax \cos\left(\frac{2\pi r}{2m + 1}\right) + a^2\right].$$

5.14. The complex position vectors of two parallel interacting equal fluid vortices moving with their axes of rotation always perpendicular to the z-plane are z_1 and z_2. The equations governing their motions are
$$\frac{dz_1^*}{dt} = -\frac{i}{z_1 - z_2}, \qquad \frac{dz_2^*}{dt} = -\frac{i}{z_2 - z_1}.$$
Deduce that (a) $z_1 + z_2$, (b) $|z_1 - z_2|$ and (c) $|z_1|^2 + |z_2|^2$ are all constant in time, and hence describe the motion geometrically.

Problems

5.15. Solve the equation

$$z^7 - 4z^6 + 6z^5 - 6z^4 + 6z^3 - 12z^2 + 8z + 4 = 0,$$

(a) by examining the effect of setting $z^3 = 2$ and then
(b) by factorising and using the binomial expansion of $(z + a)^4$.
Plot the seven roots of the equation on an Argand plot, exemplifying that complex roots of a polynomial equation always occur in conjugate pairs if the polynomial has real coefficients.

5.16. The polynomial $f(z)$ is defined by

$$f(z) = z^5 - 6z^4 + 15z^3 - 34z^2 + 36z - 48.$$

(a) Show that the equation $f(z) = 0$ has roots of the form $z = \lambda i$, where λ is real, and hence factorise $f(z)$.
(b) Show further that the cubic factor of $f(z)$ can be written in the form $(z + a)^3 + b$, where a and b are real, and hence solve the equation $f(z) = 0$ completely.

5.17. The binomial expansion of $(1 + x)^n$, discussed in Chapter 1, can be written for a positive integer n as

$$(1 + x)^n = \sum_{r=0}^{n} {}^nC_r x^r,$$

where ${}^nC_r = n!/[r!(n-r)!]$.
(a) Use de Moivre's theorem to show that the sum

$$S_1(n) = {}^nC_0 - {}^nC_2 + {}^nC_4 - \cdots + (-1)^m \, {}^nC_{2m}, \qquad n - 1 \leq 2m \leq n,$$

has the value $2^{n/2} \cos(n\pi/4)$.
(b) Derive a similar result for the sum

$$S_2(n) = {}^nC_1 - {}^nC_3 + {}^nC_5 - \cdots + (-1)^m \, {}^nC_{2m+1}, \qquad n - 1 \leq 2m + 1 \leq n,$$

and verify it for the cases $n = 6, 7$ and 8.

5.18. By considering $(1 + \exp i\theta)^n$, prove that

$$\sum_{r=0}^{n} {}^nC_r \cos r\theta = 2^n \cos^n(\theta/2) \cos(n\theta/2),$$

$$\sum_{r=0}^{n} {}^nC_r \sin r\theta = 2^n \cos^n(\theta/2) \sin(n\theta/2),$$

where ${}^nC_r = n!/[r!(n-r)!]$.

Complex numbers and hyperbolic functions

5.19. Use de Moivre's theorem with $n = 4$ to prove that
$$\cos 4\theta = 8\cos^4\theta - 8\cos^2\theta + 1,$$
and deduce that
$$\cos\frac{\pi}{8} = \left(\frac{2+\sqrt{2}}{4}\right)^{1/2}.$$

5.20. Express $\sin^4\theta$ entirely in terms of the trigonometric functions of multiple angles and deduce that its average value over a complete cycle is $\frac{3}{8}$.

5.21. Use de Moivre's theorem to prove that
$$\tan 5\theta = \frac{t^5 - 10t^3 + 5t}{5t^4 - 10t^2 + 1},$$
where $t = \tan\theta$. Deduce the values of $\tan(n\pi/10)$ for $n = 1, 2, 3, 4$.

5.22. Prove the following results involving hyperbolic functions.
  (a) That
$$\cosh x - \cosh y = 2\sinh\left(\frac{x+y}{2}\right)\sinh\left(\frac{x-y}{2}\right).$$
  (b) That, if $y = \sinh^{-1} x$,
$$(x^2 + 1)\frac{d^2 y}{dx^2} + x\frac{dy}{dx} = 0.$$

5.23. Determine the conditions under which the equation
$$a\cosh x + b\sinh x = c, \qquad c > 0,$$
has zero, one or two real solutions for x. What is the solution if $a^2 = c^2 + b^2$?

5.24. Use the definitions and properties of hyperbolic functions to do the following:
  (a) Solve $\cosh x = \sinh x + 2\operatorname{sech} x$.
  (b) Show that the real solution x of $\tanh x = \operatorname{cosech} x$ can be written in the form $x = \ln(u + \sqrt{u})$. Find an explicit value for u.
  (c) Evaluate $\tanh x$ when x is the real solution of $\cosh 2x = 2\cosh x$.

5.25. Express $\sinh^4 x$ in terms of hyperbolic cosines of multiples of x, and hence find the real solutions of
$$2\cosh 4x - 8\cosh 2x + 5 = 0.$$

5.26. In the theory of special relativity, the relationship between the position and time coordinates of an event, as measured in two frames of reference that have parallel

x-axes, can be expressed in terms of hyperbolic functions. If the coordinates are x and t in one frame and x' and t' in the other, then the relationship takes the form

$$x' = x \cosh \phi - ct \sinh \phi,$$
$$ct' = -x \sinh \phi + ct \cosh \phi.$$

Express x and ct in terms of x', ct' and ϕ and show that

$$x^2 - (ct)^2 = (x')^2 - (ct')^2.$$

5.27. A closed barrel has as its curved surface the surface obtained by rotating about the x-axis the part of the curve

$$y = a[2 - \cosh(x/a)]$$

lying in the range $-b \leq x \leq b$, where $b < a \cosh^{-1} 2$. Show that the total surface area, A, of the barrel is given by

$$A = \pi a[9a - 8a \exp(-b/a) + a \exp(-2b/a) - 2b].$$

5.28. The principal value of the logarithmic function of a complex variable is defined to have its argument in the range $-\pi < \arg z \leq \pi$. By writing $z = \tan w$ in terms of exponentials show that

$$\tan^{-1} z = \frac{1}{2i} \ln \left(\frac{1+iz}{1-iz} \right).$$

Use this result to evaluate

$$\tan^{-1} \left(\frac{2\sqrt{3} - 3i}{7} \right).$$

HINTS AND ANSWERS

5.1. (a) $-3 + i$; (b) $18 - i$; (c) $\frac{1}{25}(6 + 17i)$.

5.3. (a) $5 + 3i$; (b) $-1 - 5i$; (c) $10 + 5i$; (d) $2/5 + 11i/5$; (e) 4; (f) $3 - 4i$; (g) $\ln 5 + i[\tan^{-1}(4/3) + 2n\pi]$; (h) $\pm(2.521 + 0.595i)$.

5.5. Use $\sin \pi/4 = \cos \pi/4 = 1/\sqrt{2}$, $\cos \pi/3 = 1/2$, $\sin \pi/3 = \sqrt{3}/2$. $\cot \pi/12 = 2 + \sqrt{3}$.

5.7. (a) $\exp(-2y) \cos 2x$; (b) $(\sin 2y \sinh 2x)/2$; (c) $\sqrt{2} \exp(\pi i/3)$ or $\sqrt{2} \exp(4\pi i/3)$; (d) $\exp(1/\sqrt{2})$ or $\exp(-1/\sqrt{2})$; (e) $0.540 - 0.841i$; (f) $8 \sin(\ln 2) = 5.11$; (g) $\exp(-\pi/2 - 2\pi n)$; (h) $\ln 8 + i(6n + 1/2)\pi$.

5.9. (a) $\arg[(b-a)/(c-a)] = \arg[(B-A)/(C-A)]$.
(b) $|(b-a)/(c-a)| = |(B-A)/(C-A)|$.

Complex numbers and hyperbolic functions

5.11. All three conditions are satisfied in $3\pi/2 \leq \theta \leq 7\pi/4$, $|z| \leq 4$; area $= 2\pi$.

5.13. Denoting $\exp[2\pi i/(2m+1)]$ by Ω, express $x^{2m+1} - a^{2m+1}$ as a product of factors like $(x - a\Omega^r)$ and then combine those containing Ω^r and Ω^{2m+1-r}. Use the fact that $\Omega^{2m+1} = 1$.

5.15. The roots are $2^{1/3} \exp(2\pi n i/3)$ for $n = 0, 1, 2$; $1 \pm 3^{1/4}$; $1 \pm 3^{1/4} i$.

5.17. Consider $(1+i)^n$. (b) $S_2(n) = 2^{n/2} \sin(n\pi/4)$. $S_2(6) = -8$, $S_2(7) = -8$, $S_2(8) = 0$.

5.19. Use the binomial expansion of $(\cos\theta + i\sin\theta)^4$.

5.21. Show that $\cos 5\theta = 16c^5 - 20c^3 + 5c$, where $c = \cos\theta$, and correspondingly for $\sin 5\theta$. Use $\cos^{-2}\theta = 1 + \tan^2\theta$. The four required values are
$[(5 - \sqrt{20})/5]^{1/2}$, $(5 - \sqrt{20})^{1/2}$, $[(5 + \sqrt{20})/5]^{1/2}$, $(5 + \sqrt{20})^{1/2}$.

5.23. Reality of the root(s) requires $c^2 + b^2 \geq a^2$ and $a + b > 0$. With these conditions, there are two roots if $a^2 > b^2$, but only one if $b^2 > a^2$.
For $a^2 = c^2 + b^2$, $x = \frac{1}{2} \ln[(a-b)/(a+b)]$.

5.25. Reduce the equation to $16 \sinh^4 x = 1$, yielding $x = \pm 0.481$.

5.27. Show that $ds = (\cosh x/a)\, dx$;
curved surface area $= \pi a^2 [8\sinh(b/a) - \sinh(2b/a)] - 2\pi ab$;
flat ends area $= 2\pi a^2 [4 - 4\cosh(b/a) + \cosh^2(b/a)]$.

6 Series and limits

Many examples exist in the physical sciences of situations where we are presented with a *sum of terms* to evaluate. As just two examples, there may be the need to add together the contributions from successive slits in a diffraction grating in order to find the total light intensity at a particular point, or to compute, for a particular site in a crystal, the electrostatic potential due to all the other ions in the crystal.

6.1 Series

A general series may have either a finite or infinite number of terms. In either case, the sum of the first N terms of a series (often called a partial sum) is written

$$S_N = u_1 + u_2 + u_3 + \cdots + u_N,$$

where the terms of the series u_n, $n = 1, 2, 3, \ldots, N$, are numbers that may in general be complex. If some or all of the terms are complex then, in general, S_N will also be complex, and we can write $S_N = X_N + iY_N$, where X_N and Y_N are the partial sums of the real and imaginary parts of each term separately and are therefore real. If a series has only N terms then the partial sum S_N is of course the sum of the series.

Sometimes we encounter series where each term depends on some variable, x, say. In this case the partial sum of the series will depend on the value assumed by x. For example, consider the infinite series

$$S(x) = 1 + x + \frac{x^2}{2!} + \frac{x^3}{3!} + \cdots$$

This is an example of a power series; these are discussed in more detail in Section 6.5. It is in fact the Maclaurin expansion of $\exp x$ (see Section 6.6.3 and Appendix A). Therefore, $S(x) = \exp x$ and, of course, its value varies according to the value of the variable x. A series might just as easily depend on a complex variable z.

A general, random sequence of numbers can be described as a series and a sum of the terms found. However, for cases of practical interest, there will usually be some sort of pattern in the form of the u_n – typically that u_n is a function of n – and hence a relationship between successive terms.[1] For example, if the nth term of a series is given by

$$u_n = \frac{1}{2^n},$$

---

[1] Write u_n as a function of n for the series $S(x) = \exp x$ and state the relationship between successive terms. Do the same for $S(x) = e^{-x}$.

for $n = 1, 2, 3, \ldots, N$, then the sum of the first N terms will be

$$S_N = \sum_{n=1}^{N} u_n = \frac{1}{2} + \frac{1}{4} + \frac{1}{8} + \cdots + \frac{1}{2^N}. \tag{6.1}$$

It is clear that the sum of a finite number of terms is always finite, provided that each term is itself finite. It is often of practical interest, however, to consider the sum of a series with an infinite number of finite terms. The sum of an infinite number of terms is best defined by first considering the partial sum of the first N terms, S_N. If the value of the partial sum S_N tends to a finite limit, S, as N tends to infinity, then the series is said to converge and its sum is given by the limit S. In other words, the sum of an infinite series is given by

$$S = \lim_{N \to \infty} S_N,$$

provided the limit exists.[2] For complex infinite series, if S_N approaches a limit $S = X + iY$ as $N \to \infty$, this means that $X_N \to X$ and $Y_N \to Y$ separately, i.e. the real and imaginary parts of the series are each convergent series with sums X and Y respectively.

However, not all infinite series have finite sums. As $N \to \infty$, the value of the partial sum S_N may diverge: it may approach $+\infty$ or $-\infty$, or oscillate finitely or infinitely. Moreover, for a series where each term depends on some variable, its convergence can depend on the value assumed by the variable. Whether an infinite series converges, diverges or oscillates has important implications when describing physical systems. Methods for determining whether a series converges are discussed in Section 6.3.

EXERCISES 6.1

1. Write down, in terms of x and n, an expression for the nth term of each of the following series: (a) $e^{-x^2/2}$, (b) $\sin \sqrt{x}$, (c) the annual increases in value of a capital sum A invested at $x\%$ p.a., with interest paid at the end of each year and then added to the capital.

2. Show that the series u_n given by

$$u_n = \frac{(1 + \sqrt{5})^n - (1 - \sqrt{5})^n}{2^n \sqrt{5}} \quad \text{for } n = 0, 1, 2, 3, \ldots$$

has the property $u_{n+1} = u_n + u_{n-1}$ with $u_0 = 0$ and $u_1 = 1$. *Hint*: note that $(1 \pm \sqrt{5})^2 = 2(3 \pm \sqrt{5})$.

This series is the Fibonacci series $0, 1, 1, 2, 3, 5, 8, 13, \ldots$, well known for the property that the ratio of successive terms approaches the 'golden mean', which is said to describe the most aesthetically pleasing proportions for the sides of a rectangle, e.g. the ideal picture frame.

---

[2] A more formal statement would be 'given *any* quantity $\epsilon > 0$, there exists an S and an N_0, the latter of which may depend upon ϵ, such that $|S_N - S| < \epsilon$ for *all* N greater than N_0'.

6.2 Summation of series

It is often necessary to find the sum of a finite series or a convergent infinite series. We now describe arithmetic, geometric and arithmetico-geometric series, which are particularly common and for which the sums are easily found. Other methods that can sometimes be used to sum more complicated series are discussed later.

6.2.1 Arithmetic series

An *arithmetic series* has the characteristic that the difference between successive terms is constant. The sum of a general arithmetic series is written

$$S_N = a + (a+d) + (a+2d) + \cdots + [a + (N-1)d] = \sum_{n=0}^{N-1}(a+nd).$$

If an infinite number of such terms were added, the series would ultimately increase indefinitely (or decrease indefinitely if d were negative); that is to say, it would diverge. Contrariwise, for a finite number of terms, however large that number, the series will not diverge, but have a finite value.

In order to get a compact, closed-form expression for such a value, we note that if we pair up the rth term, $a + (r-1)d$, and the $(N+1-r)$th term, which has value $a + (N-r)d$, their sum is $2a + (N-1)d$, i.e. independent of r. To make use of this, we rewrite the series in the opposite order and add this term by term to the original expression for S_N. This gives

$$2S_N = N[2a + (N-1)d] \quad \Rightarrow \quad S_N = \frac{N}{2}(\text{first term} + \text{last term}). \quad (6.2)$$

This can be thought of as N times the average value of a term in the series. The following example illustrates the method.

Example Sum the integers between 1 and 1000 inclusive.

This is an arithmetic series with $a = 1$, $d = 1$ and $N = 1000$. Therefore, using (6.2) we find

$$S_N = \frac{1000}{2}(1 + 1000) = 500\,500.$$

This can be checked directly – but only with considerable effort.[3]

6.2.2 Geometric series

Equation (6.1) is a particular example of a *geometric series*, which has the characteristic that the ratio of successive terms is a constant (one-half in this case). The sum of a general geometric series is written

$$S_N = a + ar + ar^2 + \cdots + ar^{N-1} = \sum_{n=0}^{N-1} ar^n,$$

---
[3] How many terms of the series 4, 5, 6, 7, ... must be added together to produce a total of 2695?

Series and limits

where a is a constant and r is the constant ratio of successive terms, the *common ratio*. The sum may be evaluated by considering S_N and rS_N; the expressions for these are

$$S_N = a + ar + ar^2 + ar^3 + \cdots + ar^{N-1},$$
$$rS_N = ar + ar^2 + ar^3 + ar^4 + \cdots + ar^N.$$

If we now subtract the second equation from the first, nearly all of the terms on the RHS cancel in pairs, leaving just the first term of the first equation and the final term of the second:

$$(1-r)S_N = a - ar^N.$$

Hence the expression for the sum of the first N terms is[4]

$$S_N = \frac{a(1-r^N)}{1-r}. \qquad (6.3)$$

For a series with an infinite number of terms and $|r| < 1$, we have $\lim_{N \to \infty} r^N = 0$, and the sum tends to the limit

$$S = \sum_{n=0}^{\infty} ar^n = \frac{a}{1-r}. \qquad (6.4)$$

In (6.1), $r = \frac{1}{2}$, $a = \frac{1}{2}$, and so $S = 1$. For $|r| \geq 1$, however, the series either diverges or oscillates.

Our illustrative example is based on the path of a bouncing ball.

Example Consider a ball that is dropped from a height of 27 m and on each bounce retains only a third of its kinetic energy; thus after one bounce it will return to a height of 9 m, after two bounces to 3 m, and so on. Find the total distance travelled between the first bounce and the Mth bounce.

The total distance travelled between the first bounce and the Mth bounce is given by the sum of $M - 1$ terms:

$$S_{M-1} = 2(9 + 3 + 1 + \cdots) = 2 \sum_{m=0}^{M-2} \frac{9}{3^m}$$

for $M > 1$, where the factor 2 is included to allow for both the upward and the downward journey. Inside the parentheses we clearly have a geometric series with first term 9 and common ratio 1/3 and hence the distance is given by (6.3), i.e.

$$S_{M-1} = 2 \times \frac{9\left[1 - \left(\frac{1}{3}\right)^{M-1}\right]}{1 - \frac{1}{3}} = 27\left[1 - \left(\frac{1}{3}\right)^{M-1}\right],$$

where the number of terms N in (6.3) has been replaced by $M - 1$.[5]

◀

---

[4] Note that the Nth term is ar^{N-1} and contains the $(N-1)$th power of r, not the Nth power.
[5] For the more general case in which a ball dropped from a height h retains a fraction r of its kinetic energy on bouncing, show that if the total distance travelled by the ball is αh then $r = (\alpha - 1)/(\alpha + 1)$.

6.2 Summation of series

6.2.3 Arithmetico-geometric series

An arithmetico-geometric series, as its name suggests, is a combined arithmetic and geometric series. It has the general form

$$S_N = a + (a+d)r + (a+2d)r^2 + \cdots + [a + (N-1)d]\,r^{N-1} = \sum_{n=0}^{N-1}(a+nd)r^n,$$

and can be summed, in a similar way to a pure geometric series, by multiplying by r and subtracting the result from the original series to obtain

$$(1-r)S_N = a + rd + r^2 d + \cdots + r^{N-1}d - [a+(N-1)d]\,r^N.$$

We now recognise that all the terms on the RHS, apart from the first and last, form a geometric series with first term rd and common ratio r. So, separating off the first and last terms and using expression (6.3) for the sum of a geometric series on the others, we find, after dividing through by $1-r$, that

$$S_N = \frac{a - [a+(N-1)d]\,r^N}{1-r} + \frac{rd(1-r^{N-1})}{(1-r)^2}. \qquad (6.5)$$

For an infinite series with $|r| < 1$, $\lim_{N\to\infty} r^N = 0$ as in the previous subsection, and the sum tends to the limit

$$S = \sum_{n=0}^{\infty}(a+nd)r^n = \frac{a}{1-r} + \frac{rd}{(1-r)^2}. \qquad (6.6)$$

As for a geometric series, if $|r| \geq 1$ then the series either diverges or oscillates.[6]

Example Sum the series

$$S = 2 + \frac{5}{2} + \frac{8}{2^2} + \frac{11}{2^3} + \cdots$$

This is an infinite arithmetico-geometric series with $a = 2$, $d = 3$ and $r = 1/2$. Therefore, from (6.6), we obtain $S = 10$, the first term contributing 4 and the second term 6. ◀

6.2.4 The difference method

The difference method is sometimes useful for summing series that are more complicated than the examples discussed above. Let us consider the general series

$$\sum_{n=1}^{N} u_n = u_1 + u_2 + \cdots + u_N.$$

If the terms of the series, u_n, can be expressed in the form

$$u_n = f(n) - f(n-1)$$

---

[6] Show, for an infinite arithmetico-geometric series that has $a > 0$ and a zero sum, that d is either negative or greater than $2a$.

Series and limits

for some function $f(n)$ then its (partial) sum is given by

$$S_N = \sum_{n=1}^{N} u_n = f(N) - f(0).$$

This can be shown as follows. The sum is given by

$$S_N = u_1 + u_2 + \cdots + u_N$$

and since $u_n = f(n) - f(n-1)$, it may be rewritten

$$S_N = [f(1) - f(0)] + [f(2) - f(1)] + \cdots + [f(N) - f(N-1)].$$

By cancelling terms we see that

$$S_N = f(N) - f(0),$$

a result that is used in the following example.

Example Evaluate the sum

$$\sum_{n=1}^{N} \frac{1}{n(n+1)}.$$

It is not immediately clear how this summation is to be carried out, but if the difference method is to become applicable, the product $(1/n) \times [1/(n+1)]$ has to be reformulated into the difference of two terms of similar structure. This indicates the use of partial fractions, and either by inspection or by using any of the standard methods described in Section 2.3, we find

$$u_n = -\left(\frac{1}{n+1} - \frac{1}{n}\right).$$

This is of the required form, namely $u_n = f(n) - f(n-1)$, with $f(n) = -1/(n+1)$ in this case. Using the general result developed above shows that

$$S_N = f(N) - f(0) = -\frac{1}{N+1} + 1 = \frac{N}{N+1}.$$

is the sum of the first N terms of the series. ◄

The difference method may be easily extended to evaluate sums in which each term can be expressed in the form

$$u_n = f(n) - f(n-m), \tag{6.7}$$

where m is an integer. By writing out the sum to N terms with each term expressed in this form, and cancelling terms in pairs as before, we find

$$S_N = \sum_{k=1}^{m} f(N-k+1) - \sum_{k=1}^{m} f(1-k).$$

6.2 Summation of series

Example Evaluate the sum

$$\sum_{n=1}^{N} \frac{1}{n(n+2)}.$$

Using partial fractions, as in the previous worked example, we find that

$$u_n = -\left[\frac{1}{2(n+2)} - \frac{1}{2n}\right].$$

Hence $u_n = f(n) - f(n-2)$ with $f(n) = -1/[2(n+2)]$, and so the sum is given by

$$S_N = f(N) + f(N-1) - f(0) - f(-1)$$
$$= -\frac{1}{2(N+2)} - \frac{1}{2(N+1)} + \frac{1}{2(2)} + \frac{1}{2(1)}$$
$$= \frac{3}{4} - \frac{1}{2}\left(\frac{1}{N+2} + \frac{1}{N+1}\right).$$

Note that although the summation only starts at $n = 1$, the expression for the sum,[7] perhaps puzzlingly, includes $f(0)$ and $f(-1)$. This is because u_n involves $f(n-2)$, and so, for example, u_1 includes a term $f(-1)$. ◂

In fact the difference method is quite flexible and may be used to evaluate some sums for which each term cannot be expressed in the form (6.7). The method still relies, however, on being able to write u_n in terms of a single function such that most terms in the sum cancel, leaving only a few terms at the beginning and the end. This is best illustrated by an example.

Example Evaluate the sum

$$\sum_{n=1}^{N} \frac{1}{n(n+1)(n+2)}.$$

Using partial fractions we find

$$u_n = \frac{1}{2(n+2)} - \frac{1}{n+1} + \frac{1}{2n}.$$

Hence $u_n = f(n) - 2f(n-1) + f(n-2)$ with $f(n) = 1/[2(n+2)]$. If we write out the sum, expressing each term u_n in this form, we find that most terms cancel[8] and the sum is given by

$$S_N = f(N) - f(N-1) - f(0) + f(-1) = \frac{1}{4} + \frac{1}{2}\left(\frac{1}{N+2} - \frac{1}{N+1}\right).$$

Clearly, the sum to infinity of the series is $\frac{1}{4}$. ◂

---

[7] Identify the origins of the four terms that appear explicitly in this expression. To which values of n do they correspond?

[8] Show that for this to happen requires that if $u_n = a_0 f(n) + a_1 f(n-1) + \cdots + a_m f(n-m)$, then $a_0 + a_1 + \cdots + a_m = 0$.

6.2.5 Series involving natural numbers

Series consisting of the natural numbers 1, 2, 3, ..., or the square or cube of these numbers, occur frequently and deserve a special mention. Let us first consider the sum of the first N natural numbers,

$$S_N = 1 + 2 + 3 + \cdots + N = \sum_{n=1}^{N} n.$$

This is clearly an arithmetic series with first term $a = 1$ and common difference $d = 1$. Therefore, from (6.2), $S_N = \frac{1}{2}N(N+1)$.

Next, we consider the sum of the squares of the first N natural numbers:

$$S_N = 1^2 + 2^2 + 3^2 + \cdots + N^2 = \sum_{n=1}^{N} n^2;$$

this may be evaluated using the difference method. The nth term in the series is $u_n = n^2$, which we need to express in the form $f(n) - f(n-1)$ for some function $f(n)$. Consider the function[9]

$$f(n) = n(n+1)(2n+1) \quad \Rightarrow \quad f(n-1) = (n-1)n(2n-1).$$

For this function $f(n) - f(n-1) = 6n^2$, and so we can write

$$u_n = \tfrac{1}{6}[f(n) - f(n-1)].$$

Therefore, by the difference method,

$$S_N = \tfrac{1}{6}[f(N) - f(0)] = \tfrac{1}{6}N(N+1)(2N+1).$$

Finally, we calculate the sum of the cubes of the first N natural numbers,

$$S_N = 1^3 + 2^3 + 3^3 + \cdots + N^3 = \sum_{n=1}^{N} n^3,$$

again using the difference method. Consider the function

$$f(n) = [n(n+1)]^2 \quad \Rightarrow \quad f(n-1) = [(n-1)n]^2,$$

for which $f(n) - f(n-1) = 4n^3$. Therefore, we can write the general nth term of the series as

$$u_n = \tfrac{1}{4}[f(n) - f(n-1)],$$

and again using the difference method we find

$$S_N = \tfrac{1}{4}[f(N) - f(0)] = \tfrac{1}{4}N^2(N+1)^2.$$

Note that this is the square of the sum of the natural numbers, i.e.

$$\sum_{n=1}^{N} n^3 = \left(\sum_{n=1}^{N} n\right)^2.$$

---

[9] Presented like this, the function $f(n)$ seems to have been 'pulled out of a hat'. This is not strictly so; a reasoned approach to the form of f is developed in Problem 6.4 at the end of this chapter.

6.2 Summation of series

The following worked example shows how to utilise the results for the natural numbers when a series consists of terms which are, in essence, polynomials in n.

Example Sum the series
$$\sum_{n=1}^{N}(n+1)(n+3).$$

The nth term in this series is
$$u_n = (n+1)(n+3) = n^2 + 4n + 3,$$
and therefore we can write
$$\sum_{n=1}^{N}(n+1)(n+3) = \sum_{n=1}^{N}(n^2 + 4n + 3)$$
$$= \sum_{n=1}^{N} n^2 + 4 \sum_{n=1}^{N} n + \sum_{n=1}^{N} 3$$
$$= \tfrac{1}{6}N(N+1)(2N+1) + 4 \times \tfrac{1}{2}N(N+1) + 3N$$
$$= \tfrac{1}{6}N(2N^2 + 15N + 31).$$

Since all terms of the series are integers, it follows, as a corollary, that $f(N) = N(2N^2 + 15N + 31)$ must be divisible by 6 for all positive integers N.[10] ◀

6.2.6 Transformation of series

A complicated series may sometimes be summed by transforming it into a familiar series for which we already know the sum, perhaps a geometric series or the Maclaurin expansion of a simple function (see Section 6.6.3). Various techniques are useful, and deciding which one to use in any given case is a matter of experience. We now discuss a few of the more common methods.

The differentiation or integration of a series is often useful in transforming an apparently intractable series into a more familiar one. If we wish to differentiate or integrate a series that already depends on some variable then we may do so in a straightforward manner. As an example:

Example Sum the series
$$S(x) = \frac{x^4}{3(0!)} + \frac{x^5}{4(1!)} + \frac{x^6}{5(2!)} + \cdots$$

The appearance of a factor n in the denominator of a term containing x^{n+1} suggests that the power should be reduced to x^n, after which differentiating with respect to x would produce a cancelling factor of n in the numerator. So, dividing both sides by x we obtain
$$\frac{S(x)}{x} = \frac{x^3}{3(0!)} + \frac{x^4}{4(1!)} + \frac{x^5}{5(2!)} + \cdots,$$

---

[10] By writing $f(N)$ in the form $2N(N+1)(N+2) + aN(N+1) + bN$, where a and b are constants you should determine, verify directly that this is so.

which is then easily differentiated to give

$$\frac{d}{dx}\left[\frac{S(x)}{x}\right] = \frac{x^2}{0!} + \frac{x^3}{1!} + \frac{x^4}{2!} + \frac{x^5}{3!} + \cdots.$$

We now have terms which all have the form $x^{n+2}/n!$. Recalling that the Maclaurin expansion of $\exp x$, given in Section 6.6.3, consists entirely of terms of the form $x^n/n!$, we recognise that the RHS of the equation is equal to $x^2 \exp x$.

Having done so, we must recover $S(x)$ from the derivative, and so now integrate both sides to obtain

$$\frac{S(x)}{x} = \int x^2 \exp x \, dx.$$

Integrating the RHS by parts we find

$$\frac{S(x)}{x} = x^2 \exp x - 2x \exp x + 2 \exp x + c,$$

where the value of the constant of integration c can be fixed by the requirement that $S(x)/x = 0$ at $x = 0$. Thus we find that $c = -2$ and that the sum is given by

$$S(x) = x^3 \exp x - 2x^2 \exp x + 2x \exp x - 2x,$$

a closed form that could hardly have been determined by inspection. ◀

Often, however, we require the sum of a series that does not depend on a variable. In this case, in order that we may differentiate or integrate the series, we define a function of some variable x such that the value of this function is equal to the sum of the series for some particular value of x (usually at $x = 1$).

Example Sum the series

$$S = 1 + \frac{2}{2} + \frac{3}{2^2} + \frac{4}{2^3} + \cdots$$

Let us begin by defining the function

$$f(x) = 1 + 2x + 3x^2 + 4x^3 + \cdots,$$

so that the sum $S = f(1/2)$. Integrating this function we obtain

$$\int f(x) \, dx = x + x^2 + x^3 + \cdots,$$

which we recognise as an infinite geometric series with first term $a = x$ and common ratio $r = x$. Therefore, from (6.4), we find that the sum of this series is $x/(1 - x)$. In other words

$$\int f(x) \, dx = \frac{x}{1 - x},$$

6.2 Summation of series

from which it follows that $f(x)$ is given by

$$f(x) = \frac{d}{dx}\left(\frac{x}{1-x}\right) = \frac{1}{(1-x)^2}.$$

The sum of the original series is therefore $S = f(1/2) = 4$. Clearly, many similar series can be summed by appropriate choices for x.[11] ◂

Aside from differentiation and integration, an appropriate substitution can sometimes transform a series into a more familiar form. In particular, series with terms that contain trigonometric functions can often be summed by the use of complex exponentials, as in the following example.

Example Sum the series

$$S(\theta) = 1 + \cos\theta + \frac{\cos 2\theta}{2!} + \frac{\cos 3\theta}{3!} + \cdots$$

Rewriting each cosine term as the real part of a complex exponential, we obtain

$$S(\theta) = \text{Re}\left\{1 + \exp i\theta + \frac{\exp 2i\theta}{2!} + \frac{\exp 3i\theta}{3!} + \cdots\right\}$$

$$= \text{Re}\left\{1 + \exp i\theta + \frac{(\exp i\theta)^2}{2!} + \frac{(\exp i\theta)^3}{3!} + \cdots\right\}.$$

After this second manipulation, the terms in the curly brackets are just those of the Maclaurin expansion of $\exp x$, given in Section 6.6.3, but with x set equal to $\exp i\theta$. Thus we may write $S(\theta)$ as

$$S(\theta) = \text{Re}\,[\exp(\exp i\theta)] = \text{Re}\,[\exp(\cos\theta + i\sin\theta)]$$
$$= \text{Re}\,\{[\exp(\cos\theta)][\exp(i\sin\theta)]\} = [\exp(\cos\theta)]\text{Re}\,[\exp(i\sin\theta)]$$
$$= [\exp(\cos\theta)][\cos(\sin\theta)],$$

giving an explicit closed-form expression for the sum of the infinite series.[12]

It should be noted that this approach is crucially dependent on de Moivre's theorem that allows us to replace $\cos n\theta$ by $\text{Re}\,[(\exp i\theta)^n]$; when this is done, all terms become powers of the same expression, $\exp i\theta$. ◂

---

[11] By making the substitution $x = e^{-y}$, with $y > 0$, and subsequently re-indexing the summation, prove that

$$S = \sum_{s=1}^{\infty} s\,e^{-sy} = \frac{1}{4\sinh^2(y/2)}.$$

[12] Check that $S(\theta)$ has the value you expect when $\theta = 0$.

Series and limits

EXERCISES 6.2

1. Show that the sums of the first N terms of the following series are as given:
   (a) $10, 7, 4, 1, -2, \ldots$ $\frac{1}{2}N(23 - 3N)$,
   (b) $2, -1, \frac{1}{2}, -\frac{1}{4}, \frac{1}{8}, \ldots$ $\frac{4}{3}\left(1 - \frac{(-1)^N}{2^N}\right)$,
   (c) $1, -4 \times 2, 7 \times 4, -10 \times 8, 13 \times 16, \ldots$ $\frac{1}{3}[(1 - 3N)(-2)^N - 1]$.

2. Factorise $n^2 + 4n + 3$ and hence write the sum
$$S_N = \sum_{n=1}^{N} \frac{1}{n^2 + 4n + 3}$$
in such a way that the difference summation method can be applied to it. Carry out the summation and deduce that as $N \to \infty$, $S_N \to 5/12$.

3. Prove that
$$\sum_{n=1}^{N}(n+1)(n+2)(n+3) = \frac{N}{4}\left(N^3 + 10N^2 + 35N + 50\right).$$

4. By transforming the following infinite series, evaluate their sums in closed form:
   (a) $x^{1/2} - \dfrac{x^{3/2}}{2!} + \dfrac{x^{5/2}}{4!} - \dfrac{x^{7/2}}{6!} + \cdots$,
   (b) $\dfrac{1}{2 \times 3^2} + \dfrac{1}{4 \times 3^4} + \dfrac{1}{6 \times 3^6} + \dfrac{1}{8 \times 3^8} + \cdots$,
   (c) $\sin\theta - \dfrac{\sin 3\theta}{3!} + \dfrac{\sin 5\theta}{5!} - \dfrac{\sin 7\theta}{7!} + \cdots$.

   Hint: for part (c) you will need to use $\sin\phi = \dfrac{1}{2i}\left(e^{i\phi} - e^{-i\phi}\right)$, an identity that is valid for all ϕ, both real and complex.

6.3 Convergence of infinite series

Although the sums of some commonly occurring infinite series may be found, the sum of a general infinite series is usually difficult to calculate. Nevertheless, it is often useful to know whether the partial sum of such a series converges to a limit, even if that limit cannot be found explicitly. As mentioned at the end of Section 6.1, if we allow N to tend to infinity, the partial sum
$$S_N = \sum_{n=1}^{N} u_n$$
of a series may tend to a definite limit (i.e. to the sum S of the series), or increase or decrease without limit, or oscillate finitely or infinitely.

6.3 Convergence of infinite series

To investigate the convergence of any given series, it is useful to have available a number of tests and theorems of general applicability. We discuss them below; some we will merely state, since once they have been stated they become almost self-evident, but are no less useful for that.

6.3.1 Absolute and conditional convergence

Let us first consider some general points concerning the convergence, or otherwise, of an infinite series. In general, an infinite series $\sum u_n$ can have complex terms, and, even in the special case of a real series, the terms can be positive or negative; these variations make it difficult to devise and describe tests for convergence that apply to any series.

However, whatever the form of the original series, we can always construct another series, $\sum |u_n|$, in which each term is simply the modulus of the corresponding term in the original series; for a series that already consists only of positive real terms, the two series are the same. Since each term in any such new series is a positive real number, convergence testing becomes a more standard procedure. If the series $\sum |u_n|$ converges then the series $\sum u_n$ is said to be *absolutely convergent*.

Further, the convergence of $\sum |u_n|$ implies that of $\sum u_n$, though, in general, the sums will not be the same.[13] However, it is clear that the non-convergence of $\sum |u_n|$ does *not* imply the non-convergence of $\sum u_n$, as is clear intuitively from considering the series[14]

$$S_1 = \sum u_n = 1 - \frac{1}{2} + \frac{1}{3} - \frac{1}{4} + \cdots \to 0.6931,$$

and its derived series of absolute terms[15]

$$S_2 = \sum |u_n| = 1 + \frac{1}{2} + \frac{1}{3} + \frac{1}{4} + \cdots \to \infty.$$

For an absolutely convergent series, the terms may be reordered without affecting whether or not the series converges or what its sum is.

If the series $\sum |u_n|$ diverges but $\sum u_n$ converges, then $\sum u_n$ is said to be *conditionally convergent*. Any conditionally convergent series must contain infinitely many terms of both signs, since any finite number of terms of either particular sign cannot change whether or not the partial sum *ultimately* tends to a limit – though, of course, they will directly affect what any such limit is.

For a conditionally convergent series, rearranging the order of the terms can affect the behaviour of the sum and, hence, whether the series converges or diverges. In fact, a theorem due to Riemann shows that, by a suitable rearrangement, a conditionally convergent series may be made to converge to any arbitrary limit, or to diverge, or to oscillate finitely or infinitely! Of course, if the original series $\sum u_n$ consists only of positive real terms and converges, then automatically it is absolutely convergent.

---

[13] Consider the series given by $u_n = (-1)^n x^n/n!$ for real x. Show that the sum of the moduli of the terms is $e^{|x|}$. Under what circumstances is the sum of the original series equal to this?

[14] Sum a modest number of terms from each series using a calculator or computer and note the very different behaviours of the two sums as each new term is added.

[15] The proof that this series diverges is given on p. 231.

6.3.2 Convergence of a series containing only real positive terms

As discussed above, in order to test for the absolute convergence of a series $\sum u_n$, we first construct the corresponding series $\sum |u_n|$; this contains only real positive terms. Therefore, in this subsection we will restrict our attention to series of this type.

We discuss below some tests that may be used to investigate the convergence of such a series. Before doing so, however, we note the following *crucial consideration*. In all the tests for, or discussions of, the convergence of a series, it is not what happens in the first ten, or the first thousand, or the first million terms (or any other finite number of terms) that matters, but what happens *ultimately*.

Preliminary test

A necessary *but not sufficient* condition for a series of real positive terms $\sum u_n$ to be convergent is that the term u_n tends to zero as n tends to infinity, i.e. we require

$$\lim_{n \to \infty} u_n = 0.$$

If this condition is not satisfied then the series must diverge. Even if it is satisfied, however, the series may still diverge, and further testing is required.

Comparison test

The comparison test is the most basic test for convergence. Let us consider two series $\sum u_n$ and $\sum v_n$ and suppose that we *know* the latter to be convergent (by some earlier analysis, for example). Then, if each term u_n in the first series is less than or equal to the corresponding term v_n in the second series, for all n greater than some fixed number N (that will vary from series to series), then the original series $\sum u_n$ is also convergent. In other words, if $\sum v_n$ is convergent and there exists an N such that

$$u_n \leq v_n \qquad \text{for all } n > N,$$

then $\sum u_n$ converges.

However, if $\sum v_n$ diverges and $u_n \geq v_n$ for all n greater than some fixed number then $\sum u_n$ diverges. We now illustrate the comparison test with a worked example.

Example Determine whether the following series converges:

$$\sum_{n=1}^{\infty} \frac{1}{n!+1} = \frac{1}{2} + \frac{1}{3} + \frac{1}{7} + \frac{1}{25} + \cdots \tag{6.8}$$

Let us compare this series with the series

$$\sum_{n=0}^{\infty} \frac{1}{n!} = \frac{1}{0!} + \frac{1}{1!} + \frac{1}{2!} + \frac{1}{3!} + \cdots = 2 + \frac{1}{2!} + \frac{1}{3!} + \cdots, \tag{6.9}$$

which is merely the series obtained by setting $x = 1$ in the Maclaurin expansion of $\exp x$ (see Section 6.6.3 and Appendix A), i.e.

$$\exp(1) = e = 1 + \frac{1}{1!} + \frac{1}{2!} + \frac{1}{3!} + \cdots$$

6.3 Convergence of infinite series

Clearly this second series is convergent, since it consists of only positive terms and has a finite sum. Thus, since, for $n > 1$, each term $u_n = 1/(n! + 1)$ in the series (6.8) is less than the corresponding term $1/n!$ in (6.9),[16] we conclude from the comparison test that (6.8) is also convergent. ◂

D'Alembert's ratio test

The ratio test determines whether a series converges by comparing the relative magnitudes of successive terms. If we consider a series $\sum u_n$ and set

$$\rho = \lim_{n \to \infty} \left(\frac{u_{n+1}}{u_n} \right), \tag{6.10}$$

then if $\rho < 1$ the series is convergent; if $\rho > 1$ the series is divergent; if $\rho = 1$ then the behaviour of the series is undetermined by this test.

To prove this we observe that if the limit (6.10) is less than unity, i.e. $\rho < 1$, then we can find a value r in the range $\rho < r < 1$ and a value N such that

$$\frac{u_{n+1}}{u_n} < r,$$

for all $n > N$. Now the terms u_n of the series that follow u_N are

$$u_{N+1}, \quad u_{N+2}, \quad u_{N+3}, \quad \ldots,$$

and each of these is less than the corresponding term of

$$r u_N, \quad r^2 u_N, \quad r^3 u_N, \quad \ldots \tag{6.11}$$

However, the terms of (6.11) are those of a geometric series with a common ratio r that is less than unity. This geometric series consequently converges and therefore, by the comparison test discussed above, so must the original series $\sum u_n$. An analogous argument may be used to prove the divergent case when $\rho > 1$.

Example Determine whether the following series converges:

$$\sum_{n=0}^{\infty} \frac{1}{n!} = \frac{1}{0!} + \frac{1}{1!} + \frac{1}{2!} + \frac{1}{3!} + \cdots = 2 + \frac{1}{2!} + \frac{1}{3!} + \cdots$$

As mentioned in the previous example, this series may be obtained by setting $x = 1$ in the Maclaurin expansion of $\exp x$, and hence we know already that it converges and has the sum $\exp(1) = e$. Nevertheless, we may use the ratio test to confirm that it converges.

Using (6.10), we have

$$\rho = \lim_{n \to \infty} \left[\frac{n!}{(n+1)!} \right] = \lim_{n \to \infty} \left(\frac{1}{n+1} \right) = 0 \tag{6.12}$$

and since $\rho < 1$, the series converges, as expected.[17] ◂

---

[16] This is also true for the $n = 1$ terms, in that $\frac{1}{2} < 2$, but even it were not, it would not matter, since any finite number of terms can be ignored when testing for convergence.

[17] What does the ratio test indicate about the convergence of the series of which the following are the nth terms: (a) $x^{2n+1}/(2n+1)!$, (b) x^n/n, (c) $nx^{1/n}$, and x is real and >0 in all cases? Comment further on case (c).

Ratio comparison test

As its name suggests, the ratio comparison test is a combination of the ratio and comparison tests. Let us consider the two series $\sum u_n$ and $\sum v_n$ and assume that we know the latter to be convergent. It may be shown that if a value N can be chosen so that

$$\frac{u_{n+1}}{u_n} \leq \frac{v_{n+1}}{v_n}$$

for *all* n greater than N, then $\sum u_n$ is also convergent.
Similarly, if

$$\frac{u_{n+1}}{u_n} \geq \frac{v_{n+1}}{v_n}$$

for all sufficiently large n, and $\sum v_n$ diverges then $\sum u_n$ also diverges.

Example Determine whether the following series converges:

$$\sum_{n=1}^{\infty} \frac{1}{(n!)^2} = 1 + \frac{1}{2^2} + \frac{1}{6^2} + \cdots$$

In this case the ratio of successive terms, as n tends to infinity, is given by

$$R = \lim_{n \to \infty} \left[\frac{n!}{(n+1)!} \right]^2 = \lim_{n \to \infty} \left(\frac{1}{n+1} \right)^2,$$

which is less than the ratio seen in (6.12). Hence, by the ratio comparison test, the series converges. It is clear that this series could also be shown to be convergent using the more-direct ratio test. ◀

A somewhat more subtle example of this test can be found in the footnote.[18]

Quotient test

The quotient test may also be considered as a combination of the ratio and comparison tests. Let us again consider the two series $\sum u_n$ and $\sum v_n$, and define ρ as the limit

$$\rho = \lim_{n \to \infty} \left(\frac{u_n}{v_n} \right). \tag{6.13}$$

Then, it can be shown that:

(i) if $\rho \neq 0$ but is finite then $\sum u_n$ and $\sum v_n$ either both converge or both diverge;
(ii) if $\rho = 0$ and $\sum v_n$ converges then $\sum u_n$ converges;
(iii) if $\rho = \infty$ and $\sum v_n$ diverges then $\sum u_n$ diverges.

The following worked example provides a simple illustration and the footnote provides a slightly more complicated one.[19]

---

[18] Consider the series in which $u_n = n^k r^n$, where k is any fixed positive constant and $0 < r < 1$. By choosing an r_1, where $0 < r < r_1 < 1$, as the common ratio of a geometric series, show that $\sum u_n$ converges.
[19] By applying the quotient test to $\sum n^{-2}$ and the series in which the nth term is $u_n = n \ln[(n+1)/(n-1)] - 2$, determine whether the latter converges or diverges. You will need one of the Maclaurin series from Section 6.6.3.

6.3 Convergence of infinite series

Example Given that the series $\sum_{n=1}^{\infty} 1/n$ diverges, determine whether the following series converges:

$$\sum_{n=1}^{\infty} \frac{4n^2 - n - 3}{n^3 + 2n}. \tag{6.14}$$

If we set $u_n = (4n^2 - n - 3)/(n^3 + 2n)$ and $v_n = 1/n$ then the limit (6.13) becomes

$$\rho = \lim_{n \to \infty} \left[\frac{(4n^2 - n - 3)/(n^3 + 2n)}{1/n} \right] = \lim_{n \to \infty} \left[\frac{4n^3 - n^2 - 3n}{n^3 + 2n} \right] = 4.$$

Since ρ is finite but non-zero and we are given that $\sum v_n$ diverges, from (i) above $\sum u_n$ must also diverge. ◀

Integral test

The integral test is an extremely powerful means of investigating the convergence of a series $\sum u_n$. Suppose that there exists a function $f(x)$ which monotonically decreases for x greater than some fixed value x_0 and for which $f(n) = u_n$, i.e. the value of the function at integer values of x is equal to the corresponding term in the series under investigation. Then it can be shown that, if the limit of the integral

$$\lim_{N \to \infty} \int^N f(x)\,dx$$

exists, the series $\sum u_n$ is convergent. Otherwise the series diverges. Note that the integral defined here has no lower limit; the test is sometimes stated with a lower limit, equal to unity, for the integral, but this can lead to unnecessary difficulties.

Example Determine whether the following series converges:

$$\sum_{n=1}^{\infty} \frac{1}{(n - 3/2)^2} = 4 + 4 + \frac{4}{9} + \frac{4}{25} + \cdots$$

Let us consider the function $f(x) = (x - 3/2)^{-2}$. Clearly $f(n) = u_n$ and $f(x)$ monotonically decreases for $x > 3/2$. Applying the integral test, we consider

$$\lim_{N \to \infty} \int^N \frac{1}{(x - 3/2)^2}\,dx = \lim_{N \to \infty} \left(\frac{-1}{N - 3/2} \right) = 0.$$

Since the limit exists the series converges. Note, however, that if we had included a lower limit, equal to unity, in the integral then we would have run into problems, since the integrand diverges at $x = 3/2$. ◀

The integral test is also useful for examining the convergence of the Riemann zeta series. This is a special series that occurs regularly and is of the form

$$\sum_{n=1}^{\infty} \frac{1}{n^p}.$$

Series and limits

It converges for $p > 1$ and diverges if $p \leq 1$. These convergence criteria may be derived as follows.

Using the integral test, we consider

$$\lim_{N \to \infty} \int^N \frac{1}{x^p} dx = \lim_{N \to \infty} \left(\frac{N^{1-p}}{1-p} \right),$$

and it is obvious that the limit tends to zero for $p > 1$ and to ∞ for $p \leq 1$.[20]

Cauchy's root test

Cauchy's root test may be useful in testing for convergence, especially if the nth terms of the series contain an nth power. If we define the limit

$$\rho = \lim_{n \to \infty} (u_n)^{1/n},$$

then it may be proved that the series $\sum u_n$ converges if $\rho < 1$. If $\rho > 1$ then the series diverges. Its behaviour is undetermined if $\rho = 1$.

Example Determine whether the following series converges:

$$\sum_{n=1}^{\infty} \left(\frac{1}{n} \right)^n = 1 + \frac{1}{4} + \frac{1}{27} + \cdots$$

Using Cauchy's root test, we find

$$\rho = \lim_{n \to \infty} \left(\frac{1}{n} \right) = 0,$$

and hence the series converges. The footnote provides another simple example.[21]

Grouping terms

We now consider the Riemann zeta series $\sum n^{-p}$, mentioned above, with an alternative proof of its convergence that uses the method of grouping terms. In general there are better ways of determining convergence, but the grouping method may be used if it is not immediately obvious how to approach a problem by a better method.

First consider the case where $p > 1$, and group the terms in the series as follows:

$$S_N = \frac{1}{1^p} + \left(\frac{1}{2^p} + \frac{1}{3^p} \right) + \left(\frac{1}{4^p} + \cdots + \frac{1}{7^p} \right) + \cdots$$

Now we can see that each bracket (except the first, which can, of course, be ignored) of this series is less than the corresponding term of the geometric series

$$S_N = \frac{1}{1^p} + \frac{2}{2^p} + \frac{4}{4^p} + \cdots$$

---

[20] Note that for the particular case $p = 1$ the integral takes the form $\ln N$ (rather than the form given) and this $\to \infty$ as $N \to \infty$.

[21] Determine the range of $a\ (>0)$ for which $\sum (n+a)a^n e^{-n}$ is convergent.

6.3 Convergence of infinite series

This geometric series has common ratio $r = \left(\frac{1}{2}\right)^{p-1}$; since $p > 1$, it follows that $r < 1$ and that the geometric series converges. Then the comparison test shows that the Riemann zeta series also converges for $p > 1$.

The divergence of the Riemann zeta series for $p \leq 1$ can be seen by first considering the case $p = 1$. The series is

$$S_N = 1 + \frac{1}{2} + \frac{1}{3} + \frac{1}{4} + \cdots,$$

which does *not* converge, as may be seen by bracketing the terms of the series in groups in the following way:

$$S_N = \sum_{n=1}^{N} u_n = 1 + \left(\frac{1}{2}\right) + \left(\frac{1}{3} + \frac{1}{4}\right) + \left(\frac{1}{5} + \frac{1}{6} + \frac{1}{7} + \frac{1}{8}\right) + \cdots$$

The sum of the terms in each bracket is $\geq \frac{1}{2}$; and, since as many such groupings can be made as we wish, it is clear that S_N increases indefinitely as N is increased.

Now returning to the case of the Riemann zeta series for $p < 1$, we note that each term in the series is greater than the corresponding one in the series for which $p = 1$. In other words, $1/n^p > 1/n$ for $n > 1$, $p < 1$. The comparison test then shows us that the Riemann zeta series will diverge for all $p \leq 1$.

6.3.3 Alternating series test

The tests discussed in the last subsection have been concerned with determining whether the series of real positive terms $\sum |u_n|$ converges, and so whether $\sum u_n$ is absolutely convergent. In practical cases it is usually just as important to know whether a series is actually convergent, irrespective of whether or not it is absolutely convergent. As noted earlier, cases of convergence, without absolute convergence, can only involve series that contain an infinite number of both positive and negative terms. In what follows, we will concentrate on the convergence or divergence of series in which the positive and negative terms alternate, i.e. an *alternating series*.

An alternating series can be written as

$$\sum_{n=1}^{\infty} (-1)^{n+1} u_n = u_1 - u_2 + u_3 - u_4 + u_5 - \cdots,$$

with all $u_n \geq 0$. Such a series can be shown to converge provided (i) $u_n \to 0$ as $n \to \infty$ and (ii) $u_n < u_{n-1}$ for all $n > N$ for some finite N. If these conditions are not met then the series oscillates.[22]

To prove this, suppose for definiteness that N is odd and consider the series starting at u_N. The sum of its first $2m$ terms is

$$S_{2m} = (u_N - u_{N+1}) + (u_{N+2} - u_{N+3}) + \cdots + (u_{N+2m-2} - u_{N+2m-1}).$$

---

[22] Note that it is not sufficient to show that (ii) is not met for some particular value of N; to establish oscillation, it has to be shown that there is *no* finite N that meets condition (ii).

By condition (ii) above, all the parentheses are positive, and so S_{2m} increases as m increases. However, we can also write S_{2m} in the form

$$S_{2m} = u_N - (u_{N+1} - u_{N+2}) - \cdots - (u_{N+2m-3} - u_{N+2m-2}) - u_{N+2m-1},$$

and since each expression within any one pair of parentheses is positive, we must have $S_{2m} < u_N$. Thus, since the positive quantity S_{2m} is always less than u_N for all m and $u_n \to 0$ as $n \to \infty$, we must have that $S_{2m} \to 0$. This implies that the original alternating series converges. It is clear that an analogous proof can be constructed in the case where N is even.

Example Determine whether the following series converges:

$$\sum_{n=1}^{\infty}(-1)^{n+1}\frac{1}{n} = 1 - \frac{1}{2} + \frac{1}{3} - \cdots$$

This alternating series clearly satisfies conditions (i) and (ii)[23] above and hence converges. However, as shown previously by the method of grouping terms, the corresponding series with all positive terms is divergent, i.e. the series is convergent, but not absolutely convergent. ◀

EXERCISES 6.3

1. Use appropriate tests to determine whether the series of which the following are typical terms are convergent; in several cases a choice of tests is available.

   (a) $\dfrac{n^2+1}{4n(n+1)(n+2)}$,   (b) $\dfrac{\ln n}{n^{3/2}}$,   (c) $\dfrac{e^n}{(e^{2n}+1)^{1/2}}$,

   (d) $\dfrac{(n^2+1)^{1/2}}{(\sqrt{n}+1)^3(n^3+1)^{1/5}}$,   (e) $\dfrac{(-1)^n 3n^2}{(n+1)^2 \ln n}$,   (f) $\dfrac{1}{[\ln(n+1)]^{n/2}}$.

2. The series of which the following are typical terms all have terms of alternating signs. Determine for each whether it is conditionally convergent, absolutely convergent, divergent, or finitely or infinitely oscillating.

   (a) $\dfrac{(-1)^n}{n^{3/4}}$,   (b) $\dfrac{(-1)^n(n^3+1)}{n(n+1)(n+2)}$,   (c) $\dfrac{(-1)^n}{\sqrt{n(n+1)}}$,

   (d) $\dfrac{(-1)^n 3^n}{n}$,   (e) $\dfrac{(-1)^n(n+1)(n+2)}{n!}$,   (f) $(-1)^n \ln\left(\dfrac{n+1}{n}\right)$.

6.4 Operations with series

Simple operations with series are fairly intuitive, and we discuss them here only for completeness. The following points apply to both finite and infinite series, unless otherwise stated.

---

[23] In this particular case N could be taken as low as $N = 1$.

6.5 Power series

(i) If $\sum u_n = S$ then $\sum k u_n = kS$ where k is any constant.
(ii) If $\sum u_n = S$ and $\sum v_n = T$ then $\sum (u_n \pm v_n) = S \pm T$.
(iii) If $\sum u_n = S$ then $a + \sum u_n = a + S$. A simple extension of this trivial result shows that the removal or insertion of a finite number of terms anywhere in a series does not affect its convergence.
(iv) If the infinite series $\sum u_n$ and $\sum v_n$ are both absolutely convergent then the series $\sum w_n$, where
$$w_n = u_1 v_n + u_2 v_{n-1} + \cdots + u_n v_1,$$
is also absolutely convergent. The series $\sum w_n$ is called the *Cauchy product* of the two original series. Furthermore, if $\sum u_n$ converges to the sum S and $\sum v_n$ converges to the sum T then $\sum w_n$ converges to the sum ST.
(v) It is not true in general that term-by-term differentiation or integration of a series will result in a new series with the same convergence properties.[24]

EXERCISE 6.4

1. Verify the validity of point (iv) above in the case where $u_r = a^r$ and $v_r = b^{-r}$, with $a < 1 < b$, as follows.
   (i) Show that
   $$w_n = \frac{a}{b^n} \left(\frac{1 - (ab)^n}{1 - ab} \right).$$
   (ii) Evaluate $S = \sum_{n=1}^{\infty} u_n$ and $T = \sum_{n=1}^{\infty} v_n$.
   (iii) Evaluate $\sum_{n=1}^{\infty} w_n$ and show that it is equal to ST.

6.5 Power series

A power series has the form
$$P(x) = a_0 + a_1 x + a_2 x^2 + a_3 x^3 + \cdots,$$
where a_0, a_1, a_2, a_3 etc. are constants; such series occur regularly in physics and engineering. Because, for $|x| < 1$, the later terms in the series usually become very small, they can often be neglected, thus effectively converting an infinite series into a finite polynomial. For example, the series
$$P(x) = 1 + x + x^2 + x^3 + \cdots,$$

---
[24] Demonstrate this by considering the sum $S(x) = \sum_{n=1}^{\infty} n^{-1} \sin nx$. Write out the first few terms of (a) $S(x)$, (b) dS/dx and (c) $\int S(x)\,dx$ and determine the convergence or otherwise of each at $x = \pi/2$.

although in principle infinitely long, in practice may be simplified if x happens to have a value small compared with unity. To see this, note that $P(x)$ for $x = 0.1$ has the following values: 1, if just one term is taken into account; 1.1, for two terms; 1.11, for three terms; 1.111, for four terms, etc. If the quantity that it represents can only be measured with an accuracy of two decimal places, then all but the first three terms may be ignored, i.e. when $x = 0.1$ or less

$$P(x) = 1 + x + x^2 + O(x^3) \approx 1 + x + x^2.$$

This sort of approximation is often used to simplify equations into manageable forms. It may seem imprecise at first but is perfectly acceptable insofar as it matches the experimental accuracy that can be achieved.

The symbols O and $\approx$ used above need some further explanation. They are used to compare the behaviour of two functions when a variable upon which both functions depend tends to a particular limit, usually zero or infinity (and obvious from the context). For two functions $f(x)$ and $g(x)$, with g positive, the formal *definitions* of the above two symbols, and an additional one o, are as follows:

(i) If there exists a constant k such that $|f| \leq kg$ as the limit is approached then $f = O(g)$.[25]
(ii) If as the limit of x is approached f/g tends to a limit l, where $l \neq 0$, then $f \approx lg$.[26] The statement $f \approx g$ means that the ratio of the two sides tends to unity.
(iii) If the limit f/g is zero, this is denoted by $f = o(g)$.[27]

6.5.1 Convergence of power series

The convergence or otherwise of power series is a crucial consideration in practical terms. For example, if we are to use a power series as an approximation, it is clearly important that it tends to the precise answer as more and more terms of the approximation are taken. Consider the general power series

$$P(x) = a_0 + a_1 x + a_2 x^2 + \cdots$$

Using d'Alembert's ratio test (see Section 6.3.2), we see that $P(x)$ converges absolutely if

$$\rho = \lim_{n \to \infty} \left| \frac{a_{n+1}}{a_n} x \right| = |x| \lim_{n \to \infty} \left| \frac{a_{n+1}}{a_n} \right| < 1.$$

Thus the convergence of $P(x)$ depends upon the value of x, i.e. there is, in general, a range of values of x for which $P(x)$ converges, an *interval of convergence*. What that range is will depend on the limiting value of the ratio of successive coefficients. Note that at the limits of this range $\rho = 1$, and so the series may converge or diverge there. The convergence of the series at the end-points may be determined by substituting those values of x into the power series $P(x)$ and testing the resulting series using any applicable method (as discussed in Section 6.3).

---

[25] Show that $f(x) = \sinh x - \sin x - \frac{1}{3}x^3$ is $O(x^7)$ as $x \to 0$.
[26] Find l in the previous footnote if $g(x) = 2x^7$.
[27] This is included here for completeness; it is not used elsewhere in the book.

6.5 Power series

Example Determine the range of values of x for which the following power series converges:
$$P(x) = 1 + 2x + 4x^2 + 8x^3 + \cdots$$

The general term of this power series is $2^n x^n$, and so, using the interval-of-convergence method discussed above,

$$\rho = \lim_{n \to \infty} \left| \frac{2^{n+1} x^{n+1}}{2^n x^n} \right| = \lim_{n \to \infty} \left| \frac{2^{n+1}}{2^n} x \right| = |2x|.$$

For convergence this needs to be < 1, and so the power series will converge for $|x| < 1/2$. Examining the end-points of the interval separately, we find

$$P(1/2) = 1 + 1 + 1 + \cdots,$$
$$P(-1/2) = 1 - 1 + 1 - \cdots$$

Clearly $P(1/2)$ diverges, whereas $P(-1/2)$ oscillates. Therefore, $P(x)$ is not convergent at either end-point of the region but is convergent for $-1 < x < 1$. ◀

The convergence of power series may be extended to the case where the parameter z is complex. For the power series

$$P(z) = a_0 + a_1 z + a_2 z^2 + \cdots,$$

we find that $P(z)$ converges if

$$\rho = \lim_{n \to \infty} \left| \frac{a_{n+1}}{a_n} z \right| = |z| \lim_{n \to \infty} \left| \frac{a_{n+1}}{a_n} \right| < 1.$$

We therefore have a range in $|z|$ for which $P(z)$ converges, i.e. $P(z)$ converges for values of z lying within a circle in the Argand diagram (in this case centred on the origin of the Argand diagram). The radius of the circle is called the *radius of convergence*: if z lies inside the circle, the series will converge, whereas if z lies outside the circle, the series will diverge; if, though, z lies on the circle then the convergence must be tested using another method. Clearly the radius of convergence R is given by $1/R = \lim_{n \to \infty} |a_{n+1}/a_n|$.[28]

Example Determine the range of values of z for which the following complex power series converges:
$$P(z) = 1 - \frac{z}{2} + \frac{z^2}{4} - \frac{z^3}{8} + \cdots$$

We find that $\rho = |z/2|$, which shows that $P(z)$ converges for $|z| < 2$. Therefore, the circle of convergence in the Argand diagram is centred on the origin and has a radius $R = 2$. On this circle we must test the convergence by substituting the value of z into $P(z)$ and considering the resulting series. On the circle of convergence we can write $z = 2 \exp i\theta$. Substituting this into $P(z)$, we

---

[28] Note that it is only the terms actually present in the power series that must be considered. For example, if the series were $P(z) = a_0 + a_2 z^2 + a_4 z^4 + \cdots$, then it is the ratio $|a_{n+2}/a_n|$ that must be considered. If the limit of this ratio were ρ, then the radius of convergence R would be given by $1/R = \sqrt{\rho}$. Similarly, if only powers of z^3 were present, then $1/R^3 = \lim_{n \to \infty} |a_{n+3}/a_n|$.

obtain

$$P(z) = 1 - \frac{2\exp i\theta}{2} + \frac{4\exp 2i\theta}{4} - \cdots$$
$$= 1 - \exp i\theta + [\exp i\theta]^2 - \cdots,$$

which is a complex infinite geometric series with first term $a = 1$ and common ratio $r = -\exp i\theta$. Therefore, on the circle of convergence we have

$$P(z) = \frac{1}{1 + \exp i\theta}.$$

Unless $\theta = \pi$ this is a finite complex number, and so $P(z)$ converges at all points on the circle $|z| = 2$ except at $\theta = \pi$ (i.e. $z = -2$), where it diverges. Note that $P(z)$ is just the binomial expansion of $(1 + z/2)^{-1}$, for which it is obvious that $z = -2$ is a singular point. In general, for power series expansions of complex functions about a given point in the complex plane, the circle of convergence extends as far as the nearest singular point.[29] ◀

Note that the centre of the circle of convergence does not necessarily lie at the origin. For example, applying the ratio test to the complex power series

$$P(z) = 1 + \frac{z-1}{2} + \frac{(z-1)^2}{4} + \frac{(z-1)^3}{8} + \cdots,$$

we find that for it to converge we require $|(z-1)/2| < 1$. Thus the series converges for z lying within a circle of radius 2 centred on the point $(1, 0)$ in the Argand diagram.

6.5.2 Operations with power series

The following rules are useful when manipulating power series; they apply to power series in a real or complex variable.

(i) If two power series $P(x)$ and $Q(x)$ have regions of convergence that overlap to some extent then the series produced by taking the sum, the difference or the product of $P(x)$ and $Q(x)$ converges in the common region.

(ii) If two power series $P(x)$ and $Q(x)$ converge for all values of x, then one series may be substituted into the other to give a third series, which also converges for all values of x. For example, consider the power series expansions of $\sin x$ and e^x given below in Section 6.6.3,

$$\sin x = x - \frac{x^3}{3!} + \frac{x^5}{5!} - \frac{x^7}{7!} + \cdots,$$

$$e^x = 1 + x + \frac{x^2}{2!} + \frac{x^3}{3!} + \frac{x^4}{4!} + \cdots,$$

both of which converge for all values of x. Substituting the series for $\sin x$ into that for e^x we obtain[30]

$$e^{\sin x} = 1 + x + \frac{x^2}{2!} - \frac{3x^4}{4!} - \frac{8x^5}{5!} + \cdots,$$

which also converges for all values of x.

---

[29] Find the radius of convergence of the complex power series $1 - z^2/2 + z^4/4 - z^6/8 + \cdots$ and relate it to the singular point(s) of an appropriate function.

[30] As the reader may wish to verify. See also part (c) of Problem 6.26.

6.5 Power series

If, however, either of the power series $P(x)$ and $Q(x)$ has only a limited region of convergence, or if they both do, then considerable care must be taken when substituting one series into the other. For example, suppose $Q(x)$ converges for all x, but $P(x)$ only converges for x within a finite range. We may substitute $Q(x)$ into $P(x)$ to obtain $P(Q(x))$, but if the value of $Q(x)$ lies outside the region of convergence for $P(x)$, the resulting series $P(Q(x))$ will not converge – even if x lies in the regions of convergence of both P and Q.

(iii) If a power series $P(x)$ converges for a particular range of x then the series obtained by differentiating every term and the series obtained by integrating every term also converge in this range.

This is easily seen for the power series

$$P(x) = a_0 + a_1 x + a_2 x^2 + \cdots,$$

which converges if $|x| < \lim_{n \to \infty} |a_n/a_{n+1}| \equiv k$. The series obtained by differentiating $P(x)$ with respect to x is given by

$$\frac{dP}{dx} = a_1 + 2a_2 x + 3a_3 x^2 + \cdots$$

and converges if

$$|x| < \lim_{n \to \infty} \left| \frac{n a_n}{(n+1) a_{n+1}} \right| = k.$$

Similarly, the series obtained by integrating $P(x)$ term by term,

$$\int P(x)\, dx = a_0 x + \frac{a_1 x^2}{2} + \frac{a_2 x^3}{3} + \cdots,$$

converges if

$$|x| < \lim_{n \to \infty} \left| \frac{(n+2) a_n}{(n+1) a_{n+1}} \right| = k.$$

Our conclusions follow from the fact that the additional factors $n/(n+1)$ and $(n+2)/(n+1)$ make no difference to the values of the two limits. Each factor has a limit of unity, making the overall ratio limits the same as that, k, of the original undifferentiated (or unintegrated) power series.

So, series resulting from differentiation or integration have the same interval of convergence as the original series. However, even if the original series converges at either end-point of the interval, it is not necessarily the case that the derived series will do so. The new series must be tested separately at the end-points in order to determine whether it converges there. Note that although power series may be integrated or differentiated without altering their interval of convergence, this is not true for series in general.

It is also worth noting that differentiating or integrating a power series term by term within its interval of convergence is equivalent to differentiating or integrating the function it represents. For example, consider the power series expansion of $\sin x$,

$$\sin x = x - \frac{x^3}{3!} + \frac{x^5}{5!} - \frac{x^7}{7!} + \cdots, \tag{6.15}$$

which converges for all values of x. If we differentiate term by term, the series becomes
$$1 - \frac{x^2}{2!} + \frac{x^4}{4!} - \frac{x^6}{6!} + \cdots,$$
which is the series expansion of $\cos x$, as we expect.[31]

EXERCISES 6.5

1. Write the relationships between the following pairs of functions, $p(x)$ and $q(x)$, in O, o and $\approx$ notation, (a) as $x \to 0$, and (b) as $x \to \infty$.

$p(x)$	$q(x)$
x	$\sin x$
$x^2 + 3x + 4$	$3x^2 + 2x + 4$
$3 \sin x$	1
$\cosh x$	$\sinh x$

2. Find the ranges of convergence of the following real power series, determining whether or not the end-points are included:
(a) $x + 2x^2 + 3x^3 + 4x^4 + \cdots$,
(b) $x - \dfrac{x^3}{3} + \dfrac{x^5}{5} - \dfrac{x^7}{7} + \cdots$,
(c) $1 - \dfrac{x^2}{1 \times 2} + \dfrac{x^4}{2 \times 4} - \dfrac{x^6}{3 \times 8} + \cdots$

3. Determine the radius of convergence of the complex power series
$$1 - \frac{z^2}{3} + \frac{z^4}{9} - \frac{z^6}{27} + \cdots$$
At which point(s) on the circle of convergence does the series diverge?

4. Illustrate the caveat stated in point (ii) of Section 6.5.2 by taking, for real x, $P(x) = e^x$ and $Q(x) = \ln(1 + x)$ and finding the ranges in which the two series $P(Q(x))$ and $Q(P(x))$ are convergent. What are the corresponding regions if x is replaced by the complex variable z?

5. By integrating the power series expansion of $1/\sqrt{1 + x^2}$, show that the power series for $\sinh^{-1} x$ is given by
$$\sinh^{-1} x = \sum_{n=0}^{\infty} \frac{(-1)^n (2n)! \, x^{2n+1}}{(2n + 1) \, 4^n \, (n!)^2}.$$

6.6 Taylor series

Taylor's theorem provides a way of expressing a function as a power series in x, known as a *Taylor series*, but it can be applied only to those functions that are continuous and differentiable within the x-range of interest.

---

[31] What modification to this statement is needed if we integrate the power series for $\sin x$ term by term?

6.6 Taylor series

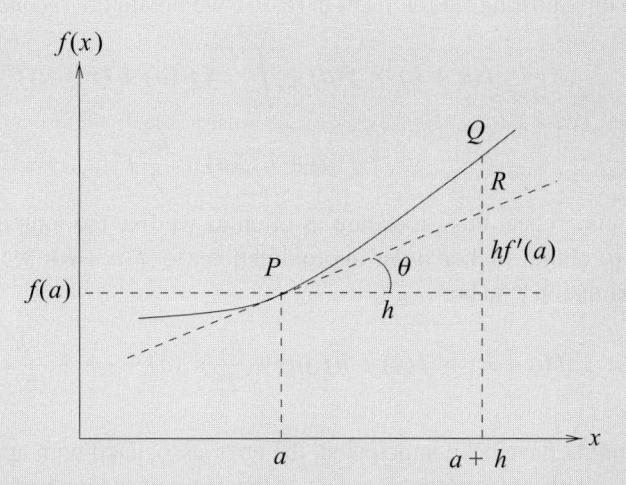

Figure 6.1 The first-order Taylor series approximation to a function $f(x)$. The slope of the function at P, i.e. $\tan\theta$, equals $f'(a)$. Thus the value of the function at Q, $f(a+h)$, is approximated by the ordinate of R, $f(a) + hf'(a)$.

6.6.1 Taylor's theorem

Suppose that we have a function $f(x)$ that we wish to express as a power series in $x - a$ about the point $x = a$. We shall assume that, in a given x-range, $f(x)$ is a continuous, single-valued function of x having continuous derivatives with respect to x, denoted by $f'(x)$, $f''(x)$ and so on, up to and including $f^{(n-1)}(x)$. We shall also assume that $f^{(n)}(x)$ exists in this range.

From the equation following (4.12) we may write

$$\int_a^{a+h} f'(x)\,dx = f(a+h) - f(a),$$

where a, $a + h$ are neighbouring values of x. Rearranging this equation, we may express the value of the function at $x = a + h$ in terms of its value at a by

$$f(a+h) = f(a) + \int_a^{a+h} f'(x)\,dx. \tag{6.16}$$

A *first approximation* for $f(a+h)$ may be obtained by substituting $f'(a)$ for $f'(x)$ in (6.16), to obtain

$$f(a+h) \approx f(a) + hf'(a).$$

This approximation is shown graphically in Figure 6.1. We may write this first approximation in terms of x and a as

$$f(x) \approx f(a) + (x-a)f'(a),$$

and, in a similar way,

$$f'(x) \approx f'(a) + (x-a)f''(a),$$
$$f''(x) \approx f''(a) + (x-a)f'''(a),$$

Series and limits

and so on. Substituting for $f'(x)$ in (6.16), we obtain the *second approximation*:

$$f(a+h) \approx f(a) + \int_a^{a+h} [f'(a) + (x-a)f''(a)] \, dx$$

$$\approx f(a) + hf'(a) + \frac{h^2}{2} f''(a).$$

We may repeat this procedure as often as we like (so long as the derivatives of $f(x)$ exist) to obtain higher order approximations to $f(a+h)$; we find the $(n-1)$th-order approximation[32] to be

$$f(a+h) \approx f(a) + hf'(a) + \frac{h^2}{2!} f''(a) + \cdots + \frac{h^{n-1}}{(n-1)!} f^{(n-1)}(a). \quad (6.17)$$

As might have been anticipated, the error associated with approximating $f(a+h)$ by this $(n-1)$th-order power series is of the order of the next term in the series. This error or *remainder* can be shown to be given by

$$R_n(h) = \frac{h^n}{n!} f^{(n)}(\xi),$$

for some ξ that lies in the range $[a, a+h]$. Taylor's theorem then states that we may write the *equality*

$$f(a+h) = f(a) + hf'(a) + \frac{h^2}{2!} f''(a) + \cdots + \frac{h^{(n-1)}}{(n-1)!} f^{(n-1)}(a) + R_n(h). \quad (6.18)$$

The theorem may also be written in a form suitable for finding $f(x)$ given the value of the function and its relevant derivatives at $x = a$, by substituting $x = a + h$ in the above expression. It then reads

$$f(x) = f(a) + (x-a)f'(a) + \frac{(x-a)^2}{2!} f''(a) + \cdots + \frac{(x-a)^{n-1}}{(n-1)!} f^{(n-1)}(a) + R_n(x),$$
$$(6.19)$$

where the remainder now takes the form

$$R_n(x) = \frac{(x-a)^n}{n!} f^{(n)}(\xi),$$

and ξ lies in the range $[a, x]$. Each of the formulae (6.18) and (6.19) gives us the *Taylor expansion* of the function about the point $x = a$. A special case occurs when $a = 0$. Such Taylor expansions, about $x = 0$, are called *Maclaurin series*.

Taylor's theorem is also valid without significant modification for functions of a complex variable. The extension of Taylor's theorem to functions of two real variables is given in Chapter 7.

For a function to be expressible as an infinite power series we require it to be infinitely differentiable and the remainder term R_n to tend to zero as n tends to infinity, i.e.

---

[32] The order of the approximation is simply the highest power of h in the series. Note, though, that the $(n-1)$th-order approximation contains n terms.

6.6 Taylor series

$\lim_{n\to\infty} R_n = 0$. In this case the infinite power series will represent the function within the interval of convergence of the series.

Example Expand $f(x) = \sin x$ as a Maclaurin series, i.e. about $x = 0$.

We must first verify that $\sin x$ may indeed be represented by an infinite power series. It is easily shown that the nth derivative of $f(x)$ is given by[33]

$$f^{(n)}(x) = \sin\left(x + \frac{n\pi}{2}\right).$$

Therefore, the remainder after expanding $f(x)$ as an $(n-1)$th-order polynomial about $x = 0$ is given by

$$R_n(x) = \frac{x^n}{n!} \sin\left(\xi + \frac{n\pi}{2}\right),$$

where ξ lies in the range $[0, x]$. Since the modulus of the sine term is always less than or equal to unity, we can write $|R_n(x)| < |x^n|/n!$. For any particular value of x, say $x = c$, $R_n(c) \to 0$ as $n \to \infty$. Hence $\lim_{n\to\infty} R_n(x) = 0$, and so $\sin x$ can be represented by an infinite Maclaurin series. Evaluating the function and its derivatives at $x = 0$ we obtain

$$f(0) = \sin 0 = 0,$$
$$f'(0) = \sin(\pi/2) = 1,$$
$$f''(0) = \sin \pi = 0,$$
$$f'''(0) = \sin(3\pi/2) = -1,$$

and so on. Therefore, the Maclaurin series expansion of $\sin x$ is given by

$$\sin x = x - \frac{x^3}{3!} + \frac{x^5}{5!} - \cdots$$

Note that, as expected, since $\sin x$ is an odd function, its power series expansion contains only odd powers of x. ◂

We may follow a similar procedure to obtain a Taylor series about an arbitrary point $x = a$, as in this second worked example.

Example Expand $f(x) = \cos x$ as a Taylor series about $x = \pi/3$.

As for $\sin x$, it is easily shown that each differentiation of $\cos x$ adds $\pi/2$ to its argument:

$$f^{(n)}(x) = \cos\left(x + \frac{n\pi}{2}\right).$$

Therefore, the remainder after expanding $f(x)$ as an $(n-1)$th-order polynomial about $x = \pi/3$ is given by

$$R_n(x) = \frac{(x - \pi/3)^n}{n!} \cos\left(\xi + \frac{n\pi}{2}\right),$$

---

[33] This can be verified by writing $\sin(x + n\pi/2)$ as $\sin x \cos(n\pi/2) + \cos x \sin(n\pi/2)$ and then considering the values of $\cos(n\pi/2)$ and $\sin(n\pi/2)$ for n even ($= 2m$) and n odd ($= 2m + 1$).

where ξ lies in the range $[\pi/3, x]$. The modulus of the cosine term is always less than or equal to unity, and so $|R_n(x)| < |(x - \pi/3)^n|/n!$. As in the previous example, $\lim_{n\to\infty} R_n(x) = 0$ for any particular value of x, and so $\cos x$ can be represented by an infinite Taylor series about $x = \pi/3$.

Evaluating the function and its derivatives at $x = \pi/3$ we obtain

$$f(\pi/3) = \cos(\pi/3) = 1/2,$$

$$f'(\pi/3) = \cos(5\pi/6) = -\sqrt{3}/2,$$

$$f''(\pi/3) = \cos(4\pi/3) = -1/2,$$

and so on. Thus the Taylor series expansion of $\cos x$ about $x = \pi/3$ is given by

$$\cos x = \frac{1}{2} - \frac{\sqrt{3}}{2}(x - \pi/3) - \frac{1}{2}\frac{(x - \pi/3)^2}{2!} + \cdots$$

It should be noted that, unlike the Taylor series for $\cos x$ about $x = 0$, this series contains both even and odd powers of the expansion variable; this is a reflection of the fact that the function does not possess the symmetry about a general value of x that it has about $x = 0$.[34] ◀

6.6.2 Approximation errors in Taylor series

In the previous subsection we saw how to represent a function $f(x)$ by an infinite power series that is exactly equal to $f(x)$ for all x within the interval of convergence of the series. However, in physical problems we usually do not want to have to sum an infinite number of terms, but prefer to use only a finite number of terms in the Taylor series to *approximate* the function in some given range of x. This being the case, it is desirable to know what is the maximum possible error associated with the approximation.

As given in (6.19), a function $f(x)$ can be represented by a finite $(n - 1)$th-order power series together with a remainder term such that

$$f(x) = f(a) + (x - a)f'(a) + \frac{(x - a)^2}{2!}f''(a) + \cdots + \frac{(x - a)^{n-1}}{(n - 1)!}f^{(n-1)}(a) + R_n(x),$$

where

$$R_n(x) = \frac{(x - a)^n}{n!}f^{(n)}(\xi)$$

and ξ lies in the range $[a, x]$. $R_n(x)$ is the remainder term, and represents the error in approximating $f(x)$ by the above $(n - 1)$th-order power series. Since the exact value of ξ that gives the correct value to $R_n(x)$ is not known, the best we can do is to find the maximum value of $|R_n(x)|$ in the range. This may be found by differentiating $R_n(x)$ with respect to ξ and equating the derivative to zero in the usual way for finding maxima.

---

[34] Which powers would you expect to be present in the following Taylor series: (a) $\cos^2 x - \sin^2 x$ about $x = 0$, (b) $\cos x$ about $x = \pi/2$, (c) $\sin 2x$ about $x = \pi/4$?

6.6 Taylor series

Example Expand $f(x) = \cos x$ as a Taylor series about $x = 0$ and find the error associated with using the approximation to evaluate $\cos(0.5)$ if only the first two non-vanishing terms are taken. (Note that the Taylor expansions of trigonometric functions are only valid for angles measured in radians.)

Evaluating the function and its derivatives at $x = 0$, we find

$$f(0) = \cos 0 = 1, \qquad f'(0) = -\sin 0 = 0,$$
$$f''(0) = -\cos 0 = -1, \quad f'''(0) = \sin 0 = 0.$$

So, for small $|x|$, we find from (6.19)

$$\cos x \approx 1 - \frac{x^2}{2}.$$

Note that since $\cos x$ is an even function about $x = 0$, its power series expansion contains only even powers of x. Therefore, in order to estimate the error in this approximation, we must consider the term in x^4, which is the next in the series. The required derivative is $f^{(4)}(x)$ and this is (by chance) equal to $\cos x$. Thus, adding in the remainder term $R_4(x)$, we find

$$\cos x = 1 - \frac{x^2}{2} + \frac{x^4}{4!} \cos \xi,$$

where ξ lies in the range $[0, x]$. Thus, the maximum possible error is $x^4/4!$, since $\cos \xi$ cannot exceed unity. If $x = 0.5$, taking just the first two terms yields $\cos(0.5) \approx 0.875\,00$ with a predicted error of less than $0.002\,60$. In fact, $\cos(0.5) = 0.877\,58$ to five decimal places. Thus, to this accuracy, the true error is $0.002\,58$, an error of about 0.3%.[35] ◂

6.6.3 Standard Maclaurin series

As it is often useful to have a readily available table of Maclaurin series for standard elementary functions, they are listed below.

$$\sin x = x - \frac{x^3}{3!} + \frac{x^5}{5!} - \frac{x^7}{7!} + \cdots \quad \text{for} \quad -\infty < x < \infty,$$

$$\cos x = 1 - \frac{x^2}{2!} + \frac{x^4}{4!} - \frac{x^6}{6!} + \cdots \quad \text{for} \quad -\infty < x < \infty,$$

$$\tan x = x + \frac{x^3}{3} + \frac{2x^5}{15} + \frac{17x^7}{315} + \cdots \quad \text{for} \quad -\pi/2 < x < \pi/2,$$

$$\tan^{-1} x = x - \frac{x^3}{3} + \frac{x^5}{5} - \frac{x^7}{7} + \cdots \quad \text{for} \quad -1 < x < 1,$$

$$e^x = 1 + x + \frac{x^2}{2!} + \frac{x^3}{3!} + \frac{x^4}{4!} + \cdots \quad \text{for} \quad -\infty < x < \infty,$$

$$\sinh x = x + \frac{x^3}{3!} + \frac{x^5}{5!} + \frac{x^7}{7!} + \cdots \quad \text{for} \quad -\infty < x < \infty,$$

---

[35] Show that including the quartic term reduces the error to about 0.0025% and that this is just less than the limit set by considering the term containing x^6.

Series and limits

$$\cosh x = 1 + \frac{x^2}{2!} + \frac{x^4}{4!} + \frac{x^6}{6!} + \cdots \quad \text{for } -\infty < x < \infty,$$

$$\ln(1+x) = x - \frac{x^2}{2} + \frac{x^3}{3} - \frac{x^4}{4} + \cdots \quad \text{for } -1 < x \leq 1,$$

$$(1+x)^n = 1 + nx + n(n-1)\frac{x^2}{2!} + n(n-1)(n-2)\frac{x^3}{3!} + \cdots \quad \text{for } -\infty < x < \infty.$$

These can all be derived by straightforward application of Taylor's theorem to the expansion of a function about $x = 0$.

EXERCISES 6.6

1. Find three-term approximations to $\sqrt[4]{82}$ using (a) a binomial expansion and (b) a Taylor series, showing that the two methods are equivalent. Evaluate the approximation and show that the error introduced by it is about 1 part in 10^7.

2. Find the maximum error that can occur if an $(n-1)$th-order Taylor series approximation is used to represent the function e^x in the range $1 < x < 1.2$. Use both the fourth-order approximation and a hand calculator to evaluate $e^{1.2}$ and show that the actual error lies within the expected range.

3. Write $f(x) = \exp(-x^2/2)$ as a Maclaurin series and estimate the error associated with an approximation to $e^{-1/2}$ that uses only the first three non-vanishing terms of the series. Compare it with the actual error and show that it is of the same order as the first omitted term in the expansion.

4. Find, up to terms in x^3, (a) a Taylor series for $\tan\left(\frac{\pi}{4} + x\right)$ and (b) a Maclaurin series for $f(x) = e^{-x}\cos x$.

6.7 Evaluation of limits

The idea of the limit of a function $f(x)$ as x approaches a value a is fairly intuitive, and we have used it several times already. However, there is a strict, more analytic, definition and this is given below.

In many cases the limit of the function as x approaches a limit point a will be simply the value $f(a)$, but sometimes this is not so. Firstly, the function may be undefined at $x = a$, as, for example, is the case when

$$f(x) = \frac{\sin x}{x},$$

which takes the value $0/0$ at $x = 0$. However, the limit as x approaches zero does exist for this function and can be evaluated; its value is unity, as is shown later.

6.7 Evaluation of limits

Another possibility is that, even if $f(x)$ is defined at $x = a$, its value may not be equal to the limiting value $\lim_{x \to a} f(x)$. This can occur for a discontinuous function at a point of discontinuity.

The strict definition of a limit is that *if $\lim_{x \to a} f(x)$ exists then, for any given number ϵ, however small, it must be possible to find numbers l and η such that $|f(x) - l| < \epsilon$ whenever $|x - a| < \eta$. The limit then has the value l.* In other words, as x becomes arbitrarily close to a, $f(x)$ becomes arbitrarily close to its limit, l (and stays there). If no such η can be found, then $f(x)$ does not tend to a limit as $x \to a$. To remove any ambiguity, it should be stated that, in general, the number η will depend on both ϵ and the form of $f(x)$.

The following observations are often useful for finding the limit of a function.

(i) A limit may be $\pm\infty$. For example as $x \to 0$, $1/x^2 \to \infty$.

(ii) A limit may be approached from below or above and the value may be different in each case. For example, consider the function $f(x) = \tan x$. As x tends to $\pi/2$ from below, $f(x) \to \infty$; but if the limit is approached from above, then $f(x) \to -\infty$. Another way of writing this is

$$\lim_{x \to \frac{\pi}{2}^-} \tan x = \infty, \qquad \lim_{x \to \frac{\pi}{2}^+} \tan x = -\infty.$$

(iii) It may ease the evaluation of a limit if the function under consideration is split into a sum, product or quotient. Provided that for each subunit so formed a limit exists, the rules for evaluating the original limit are as follows.

(a) $\lim_{x \to a} \{f(x) + g(x)\} = \lim_{x \to a} f(x) + \lim_{x \to a} g(x).$[36]

(b) $\lim_{x \to a} \{f(x)g(x)\} = \lim_{x \to a} f(x) \lim_{x \to a} g(x).$

(c) $\lim_{x \to a} \dfrac{f(x)}{g(x)} = \dfrac{\lim_{x \to a} f(x)}{\lim_{x \to a} g(x)},$

provided that the numerator and denominator are not both equal to zero or to infinity.

Illustrations using methods (a)–(c) are given in the following example.

Example Evaluate the limits

$$\lim_{x \to 1}(x^2 + 2x^3), \qquad \lim_{x \to 0}(x \cos x), \qquad \lim_{x \to \pi/2} \frac{\sin x}{x}.$$

Using (a) above,

$$\lim_{x \to 1}(x^2 + 2x^3) = \lim_{x \to 1} x^2 + \lim_{x \to 1} 2x^3 = 3.$$

Using (b),

$$\lim_{x \to 0}(x \cos x) = \lim_{x \to 0} x \lim_{x \to 0} \cos x = 0 \times 1 = 0.$$

---

[36] But comment on and correct the following

$$\lim_{x \to 0}\left[\frac{1}{x} - \frac{1}{x^2}\right] = \lim_{x \to 0}\frac{1}{x} - \lim_{x \to 0}\frac{1}{x^2} = \infty - \infty = 0.$$

Using (c),

$$\lim_{x \to \pi/2} \frac{\sin x}{x} = \frac{\lim_{x \to \pi/2} \sin x}{\lim_{x \to \pi/2} x} = \frac{1}{\pi/2} = \frac{2}{\pi}.$$

We note that, in the final example, we could not have found the limit as $x \to 0$ in this way, as the ratio would have become $0/0$. ◀

(iv) Limits of functions of x that contain exponents that themselves depend on x can often be found by taking logarithms.

Example Evaluate the limit

$$\lim_{x \to \infty} \left(1 - \frac{a^2}{x^2}\right)^{x^2}.$$

Let us define

$$y = \left(1 - \frac{a^2}{x^2}\right)^{x^2}$$

and consider the logarithm of the required limit, i.e.

$$\lim_{x \to \infty} \ln y = \lim_{x \to \infty} \left[x^2 \ln\left(1 - \frac{a^2}{x^2}\right)\right].$$

Using the Maclaurin series for $\ln(1 + x)$ given in Section 6.6.3, we can expand the logarithm as a series and obtain

$$\lim_{x \to \infty} \ln y = \lim_{x \to \infty} \left[x^2 \left(-\frac{a^2}{x^2} - \frac{a^4}{2x^4} + \cdots\right)\right] = -a^2.$$

Therefore, since $\lim_{x \to \infty} \ln y = -a^2$, it follows that $\lim_{x \to \infty} y = \exp(-a^2)$.[37] ◀

(v) L'Hôpital's rule may be used; it is an extension of (iii)(c) above. In cases in which both the numerator and denominator are zero, or both are infinite, subtler analysis is needed. Let us first consider $\lim_{x \to a} f(x)/g(x)$, where $f(a) = g(a) = 0$. To both overcome and make use of this difficulty, we expand both the numerator and the denominator as Taylor series:

$$\frac{f(x)}{g(x)} = \frac{f(a) + (x-a)f'(a) + [(x-a)^2/2!]f''(a) + \cdots}{g(a) + (x-a)g'(a) + [(x-a)^2/2!]g''(a) + \cdots}.$$

However, since $f(a) = g(a) = 0$, a factor of $x - a$ can be cancelled from every non-zero term to yield

$$\frac{f(x)}{g(x)} = \frac{f'(a) + [(x-a)/2!]f''(a) + \cdots}{g'(a) + [(x-a)/2!]g''(a) + \cdots}.$$

---

[37] Note the similarity of this result to the representation of the exponential function given in Equation (1.48).

6.7 Evaluation of limits

Therefore, in the limit $x \to a$ we find

$$\lim_{x \to a} \frac{f(x)}{g(x)} = \frac{f'(a)}{g'(a)},$$

provided $f'(a)$ and $g'(a)$ are not themselves both equal to zero.

If, however, $f'(a)$ and $g'(a)$ *are* both zero then the same process can be applied to the ratio $f'(x)/g'(x)$ to yield

$$\lim_{x \to a} \frac{f(x)}{g(x)} = \frac{f''(a)}{g''(a)},$$

provided that at least one of $f''(a)$ and $g''(a)$ is non-zero. If the original limit does exist, then it can be found by repeating the process as many times as is necessary for the ratio of corresponding nth derivatives not to be of the indeterminate form 0/0, i.e.

$$\lim_{x \to a} \frac{f(x)}{g(x)} = \frac{f^{(n)}(a)}{g^{(n)}(a)}.$$

The following example illustrates the process.

Example Evaluate the limit

$$\lim_{x \to 0} \frac{\sin x}{x}.$$

As noted earlier, if $x = 0$, both numerator and denominator are zero. Thus we need to apply l'Hôpital's rule: differentiating, we obtain

$$\lim_{x \to 0} \frac{\sin x}{x} = \lim_{x \to 0} \frac{\cos x}{1} = 1.$$

This is the result we would expect by dividing the Maclaurin series for $\sin x$ by x and then letting $x \to 0$. ◀

So far we have only considered the case where $f(a) = g(a) = 0$. For the case where $f(a) = g(a) = \infty$ we may still apply l'Hôpital's rule by writing

$$\lim_{x \to a} \frac{f(x)}{g(x)} = \lim_{x \to a} \frac{1/g(x)}{1/f(x)},$$

which is now of the form 0/0 at $x = a$.[38] Note also that l'Hôpital's rule is still valid for finding limits as $x \to \infty$, i.e. when $a = \infty$. This is easily shown by letting $y = 1/x$ as follows:

$$\lim_{x \to \infty} \frac{f(x)}{g(x)} = \lim_{y \to 0} \frac{f(1/y)}{g(1/y)}$$

$$= \lim_{y \to 0} \frac{-f'(1/y)/y^2}{-g'(1/y)/y^2}$$

$$= \lim_{y \to 0} \frac{f'(1/y)}{g'(1/y)}$$

$$= \lim_{x \to \infty} \frac{f'(x)}{g'(x)}.$$

---

[38] Find the limit of $\tan^n x / \sec^m x$ as $x \to \pi/2$, where n and m are positive integers.

Series and limits

In all of this section we have assumed that the functions involved are differentiable in an open interval that has the limit point as one of its end-points.

EXERCISES 6.7

1. Find the limits, where they exist, to which

$$f(x) = \frac{x^3 + 4x^2 - 7x - 10}{x^2 - 2x - 3}$$

tends as x tends to each of the following: (a) $-\infty$, (b) -5, (c) -1, (d) 0, (e) 1, (f) 2, (g) 3, (h) 5, (i) ∞.

2. Find the following limits. Try to select what is likely to be the most efficient method before starting to calculate.

(a) $\lim_{x \to 0} \dfrac{\tan^{-1}(x/a)}{\sin^{-1}(x/b)}$,

(b) $\lim_{x \to 0} \dfrac{x \cosh x - \sinh x}{\sinh x - x}$

(c) $\lim_{x \to \infty} \dfrac{\tan^{-1} x}{\tanh^{-1}[x/(x+1)]}$,

(d) $\lim_{x \to 0} \dfrac{x \sin x - x^2 \cos x}{(1 - \cos x)^2}$.

SUMMARY

1. *Finite and infinite series*

   Definitions: $S_N = \sum_{n=0}^{N-1} u_n$ and $S_\infty = \sum_{n=0}^{\infty} u_n$ with $|r| < 1$ where relevant.

Type	u_n	S_N	S_∞
Arithmetic	$a + nd$	$\frac{1}{2}N(u_0 + u_{N-1})$	∞
Geometric	ar^n	$\dfrac{a(1 - r^N)}{1 - r}$	$\dfrac{a}{1 - r}$
Arithmetico-geometric	$(a + nd)r^n$	see p. 217	$\dfrac{a}{1 - r} + \dfrac{rd}{(1 - r)^2}$

   - Difference method: if a function $f(n)$ can be found such that $u_n = f(n) - f(n-1)$, then $\sum_{n=1}^{N} u_n = f(N) - f(0)$.
   - Powers of the natural numbers:

   $$\sum_{n=1}^{N} n = \frac{1}{2}N(N+1), \qquad \sum_{n=1}^{N} n^2 = \frac{1}{6}N(N+1)(2N+1),$$

   $$\sum_{n=1}^{N} n^3 = \frac{1}{4}N^2(N+1)^2.$$

Summary

2. *Tests for the convergence of infinite series* $\sum u_n$

   In all tests only the *ultimate* behaviour matters; any finite number of terms can be disregarded. More symbolically, the criteria need only be satisfied for all $n > N$ where N can be as large as necessary, but must be finite.

   In all cases, a necessary (but not sufficient) requirement for convergence is that $\lim_{n \to \infty} u_n = 0$.

   - Alternating sign test: if successive terms alternate in sign and $|u_n| \to 0$ as $n \to \infty$, then $\sum u_n$ converges.
   - Integral test: if $f(n) = u_n$ when n is an integer and $\lim_{N \to \infty} \int^N f(x)\,dx$ exists, then $\sum u_n$ is convergent.
   - Other tests, based on quantitative comparisons, are given in the following table.

Test	Test quantity	Conclusion
Comparison	$u_n \leq v_n$	$\sum v_n$ conv. $\Rightarrow$ $\sum u_n$ conv.
Ratio	$\rho = \lim\limits_{n \to \infty} \left(\dfrac{u_{n+1}}{u_n}\right)$	$\begin{cases} < 1 & \text{the series converges,} \\ > 1 & \text{the series diverges,} \\ = 1 & \text{the test is inconclusive.} \end{cases}$
Ratio comparison	$\dfrac{u_{n+1}}{u_n} \leq \dfrac{v_{n+1}}{v_n}$	$\sum v_n$ conv. $\Rightarrow$ $\sum u_n$ conv.
Quotient	$\rho = \lim\limits_{n \to \infty} \left(\dfrac{u_n}{v_n}\right)$	$\neq 0$ and $\neq \infty$, then $\sum u_n$ and $\sum v_n$ converge or diverge together,
		$= 0$ and $\sum v_n$ converges, then so does $\sum u_n$,
		$= \infty$ and $\sum v_n$ diverges, then so does $\sum u_n$.
Cauchy's root	$\rho = \lim\limits_{n \to \infty} (u_n)^{1/n}$	$\begin{cases} < 1 & \text{the series converges,} \\ > 1 & \text{the series diverges,} \\ = 1 & \text{the test is inconclusive.} \end{cases}$

3. *Power series* $P(z) = \sum\limits_{n=0}^{\infty} a_n (z^m)^n$, with m usually equal to unity.

   - Radius of the circle of convergence: $R = \left(\lim\limits_{n \to \infty} \left|\dfrac{a_{n+1}}{a_n}\right|\right)^{-1/m}$.
   - Within its circle of convergence, a power series can be integrated or differentiated to produce another power series convergent in the same region.

- Taylor series for $f(x)$ about the point $x = a$;

$$f(x) = f(a) + (x-a)f'(a) + \frac{(x-a)^2}{2!} f''(a) +$$
$$\cdots + \frac{(x-a)^{n-1}}{(n-1)!} f^{(n-1)}(a) + \frac{(x-a)^n}{n!} f^{(n)}(\xi),$$

where $a \leq \xi \leq x$.
- Maclaurin series for common functions, see p. 243.

4. *Evaluation of limits* of $f(x)$ as $x \to a$.
   - The limits obtained when $x \to a^+$ and $x \to a^-$ are not necessarily equal.
   - The fractions $0/0$ and ∞/∞ are indeterminate.
   - L'Hôpital's rule for determining the limit as $x \to a$ of $f(x)/g(x)$ when an indeterminate form is encountered:

$$\lim_{x \to a} \frac{f(x)}{g(x)} = \lim_{x \to a} \frac{f^{(n)}(x)}{g^{(n)}(x)},$$

where n is the lowest value of m for which $f^{(m)}(a)/g^{(m)}(a)$ is not an indeterminate form.
   - If the indeterminate form $0 \times \infty$ is encountered, write it as $0/0$ or ∞/∞ by using the inverse of one of the factors involved.

PROBLEMS

6.1. Sum the even numbers between 1000 and 2000 inclusive.

6.2. If you invest £1000 on the first day of each year and interest is paid at 5% on your balance at the end of each year, how much money do you have after 25 years?

6.3. Prove that

$$\sum_1^N n(n+1)(n+2) = \tfrac{1}{4} N(N+1)(N+2)(N+3).$$

6.4. Based on the way that integration is defined and that, to within a small additive constant, $\int_1^N n^r \, dn \approx (r+1)^{-1} N^{r+1}$, we might expect that $\sum_1^N n \approx \tfrac{1}{2} N^2$; in fact, it is $\tfrac{1}{2} N(N+1)$. Similarly, we might expect that $\sum_1^N n^2 \approx \tfrac{1}{3} N^3$.
Assume that

$$S_N = \sum_{n=1}^N n^2 = \alpha N(N+a)(N+b),$$

Problems

with a and b constants and α a fraction between 0 and 1. By explicitly evaluating S_N for $N = 1, 2$ and 3, obtain three equations relating a, b and α and solve them.

Note that this is not a general *proof* of the form of S_N; it merely proposes a possible form. For a proof that establishes the validity of the proposal for all N, either the method of induction or that used in the main text has to be employed.

6.5. How does the convergence of the series

$$\sum_{n=r}^{\infty} \frac{(n-r)!}{n!}$$

depend on the integer r?

6.6. Show that for testing the convergence of the series

$$x + y + x^2 + y^2 + x^3 + y^3 + \cdots,$$

where $0 < x < y < 1$, the d'Alembert ratio test fails but the Cauchy root test is successful.

6.7. Find the sum S_N of the first N terms of the following series, and hence determine whether the series are convergent, divergent or oscillatory:

(a) $\sum_{n=1}^{\infty} \ln\left(\frac{n+1}{n}\right)$, (b) $\sum_{n=0}^{\infty} (-2)^n$, (c) $\sum_{n=1}^{\infty} \frac{(-1)^{n+1} n}{3^n}$.

For (c), adapt result (6.5) for an arithmetico-geometric series.

6.8. By grouping and rearranging terms of the absolutely convergent series

$$S = \sum_{n=1}^{\infty} \frac{1}{n^2}, \quad \text{show that} \quad S_o = \sum_{n \text{ odd}}^{\infty} \frac{1}{n^2} = \frac{3S}{4}.$$

6.9. Use the difference method to sum the series

$$\sum_{n=2}^{N} \frac{2n-1}{2n^2(n-1)^2}.$$

6.10. The $N + 1$ complex numbers ω_m are given by $\omega_m = \exp(2\pi i m/N)$, for $m = 0, 1, 2, \ldots, N$. Examine the cases $N = 1$ and $N = 2$ directly to check that your general formulae are still appropriate.
(a) Evaluate the following:

(i) $\sum_{m=0}^{N} \omega_m$, (ii) $\sum_{m=0}^{N} \omega_m^2$, (iii) $\sum_{m=0}^{N} \omega_m x^m$.

(b) Use these results to evaluate:

(i) $\sum_{m=0}^{N}\left[\cos\left(\frac{2\pi m}{N}\right) - \cos\left(\frac{4\pi m}{N}\right)\right]$, (ii) $\sum_{m=0}^{3} 2^m \sin\left(\frac{2\pi m}{3}\right)$.

6.11. Prove that

$$\cos\theta + \cos(\theta + \alpha) + \cdots + \cos(\theta + n\alpha) = \frac{\sin\frac{1}{2}(n+1)\alpha}{\sin\frac{1}{2}\alpha} \cos(\theta + \tfrac{1}{2}n\alpha).$$

6.12. Determine whether the following series converge (θ and p are positive real numbers):

(a) $\sum_{n=1}^{\infty} \frac{2\sin n\theta}{n(n+1)}$, (b) $\sum_{n=1}^{\infty} \frac{2}{n^2}$, (c) $\sum_{n=1}^{\infty} \frac{1}{2n^{1/2}}$,

(d) $\sum_{n=2}^{\infty} \frac{(-1)^n (n^2+1)^{1/2}}{n \ln n}$, (e) $\sum_{n=1}^{\infty} \frac{n^p}{n!}$.

6.13. Find the real values of x for which the following series are convergent:

(a) $\sum_{n=1}^{\infty} \frac{x^n}{n+1}$, (b) $\sum_{n=1}^{\infty} (\sin x)^n$, (c) $\sum_{n=1}^{\infty} n^x$, (d) $\sum_{n=1}^{\infty} e^{nx}$.

6.14. Determine whether the following series are convergent:

(a) $\sum_{n=1}^{\infty} \frac{n^{1/2}}{(n+1)^{1/2}}$, (b) $\sum_{n=1}^{\infty} \frac{n^2}{n!}$, (c) $\sum_{n=1}^{\infty} \frac{(\ln n)^n}{n^{n/2}}$, (d) $\sum_{n=1}^{\infty} \frac{n^n}{n!}$.

6.15. Determine whether the following series are absolutely convergent, convergent or oscillatory:

(a) $\sum_{n=1}^{\infty} \frac{(-1)^n}{n^{5/2}}$, (b) $\sum_{n=1}^{\infty} \frac{(-1)^n(2n+1)}{n}$, (c) $\sum_{n=0}^{\infty} \frac{(-1)^n |x|^n}{n!}$,

(d) $\sum_{n=0}^{\infty} \frac{(-1)^n}{n^2+3n+2}$, (e) $\sum_{n=1}^{\infty} \frac{(-1)^n 2^n}{n^{1/2}}$.

6.16. Obtain the positive values of x for which the following series converges:

$$\sum_{n=1}^{\infty} \frac{x^{n/2} e^{-n}}{n}.$$

Problems

6.17. Prove that

$$\sum_{n=2}^{\infty} \ln\left[\frac{n^r + (-1)^n}{n^r}\right]$$

is absolutely convergent for $r = 2$, but only conditionally convergent for $r = 1$.

6.18. An extension to the proof of the integral test (Section 6.3.2) shows that, if $f(x)$ is positive, continuous and monotonically decreasing, for $x \geq 1$, and the series $f(1) + f(2) + \cdots$ is convergent, then its sum does not exceed $f(1) + L$, where L is the integral

$$\int_1^{\infty} f(x)\,dx.$$

Use this result to show that the sum $\zeta(p)$ of the Riemann zeta series $\sum n^{-p}$, with $p > 1$, is not greater than $p/(p-1)$.

6.19. Demonstrate that rearranging the order of its terms can make a conditionally convergent series converge to a different limit by considering the series $\sum (-1)^{n+1} n^{-1} = \ln 2 = 0.693$. Rearrange the series as

$$S = \tfrac{1}{1} + \tfrac{1}{3} - \tfrac{1}{2} + \tfrac{1}{5} + \tfrac{1}{7} - \tfrac{1}{4} + \tfrac{1}{9} + \tfrac{1}{11} - \tfrac{1}{6} + \tfrac{1}{13} + \cdots$$

and group each set of three successive terms. Show that the series can then be written

$$\sum_{m=1}^{\infty} \frac{8m - 3}{2m(4m - 3)(4m - 1)},$$

which is convergent (by comparison with $\sum n^{-2}$) and contains only positive terms. Evaluate the first of these and hence deduce that S is not equal to $\ln 2$.

6.20. Illustrate result (iv) of Section 6.4, concerning Cauchy products, by considering the double summation

$$S = \sum_{n=1}^{\infty} \sum_{r=1}^{n} \frac{1}{r^2(n + 1 - r)^3}.$$

By examining the points in the nr-plane over which the double summation is to be carried out, show that S can be written as

$$S = \sum_{n=r}^{\infty} \sum_{r=1}^{\infty} \frac{1}{r^2(n + 1 - r)^3}.$$

Deduce that $S \leq 3$.

Series and limits

6.21. A Fabry–Pérot interferometer consists of two parallel heavily silvered glass plates; light enters normally to the plates and undergoes repeated reflections between them, with a small transmitted fraction emerging at each reflection. Find the intensity of the emerging wave, $|B|^2$, where

$$B = A(1-r)\sum_{n=0}^{\infty} r^n e^{in\phi},$$

with r and ϕ real.

6.22. Identify the series

$$\sum_{n=1}^{\infty} \frac{(-1)^{n+1} x^{2n}}{(2n-1)!},$$

and then, by integration and differentiation, deduce the values S of the following series:

(a) $\displaystyle\sum_{n=1}^{\infty} \frac{(-1)^{n+1} n^2}{(2n)!},$ (b) $\displaystyle\sum_{n=1}^{\infty} \frac{(-1)^{n+1} n}{(2n+1)!},$

(c) $\displaystyle\sum_{n=1}^{\infty} \frac{(-1)^{n+1} n \pi^{2n}}{4^n (2n-1)!},$ (d) $\displaystyle\sum_{n=0}^{\infty} \frac{(-1)^n (n+1)}{(2n)!}.$

For part (d), differentiate the result obtained in part (a) before x was given a particular value.

6.23. Starting from the Maclaurin series for $\cos x$, show that

$$(\cos x)^{-2} = 1 + x^2 + \frac{2x^4}{3} + \cdots$$

Deduce the first three terms in the Maclaurin series for $\tan x$.

6.24. Find the Maclaurin series for:

(a) $\ln\left(\dfrac{1+x}{1-x}\right),$ (b) $(x^2+4)^{-1},$ (c) $\sin^2 x.$

6.25. Writing the nth derivative of $f(x) = \sinh^{-1} x$ as

$$f^{(n)}(x) = \frac{P_n(x)}{(1+x^2)^{n-1/2}},$$

where $P_n(x)$ is a polynomial (of degree $n-1$), show that the $P_n(x)$ satisfy the recurrence relation

$$P_{n+1}(x) = (1+x^2) P_n'(x) - (2n-1) x P_n(x).$$

Hence generate the coefficients necessary to express $\sinh^{-1} x$ as a Maclaurin series up to terms in x^5.

Problems

6.26. Find the first three non-zero terms in the Maclaurin series for the following functions:

(a) $(x^2 + 9)^{-1/2}$, (b) $\ln[(2+x)^3]$, (c) $\exp(\sin x)$,
(d) $\ln(\cos x)$, (e) $\exp[-(x-a)^{-2}]$, (f) $\tan^{-1} x$.

6.27. By using the logarithmic series, prove that if a and b are positive and nearly equal then

$$\ln \frac{a}{b} \simeq \frac{2(a-b)}{a+b}.$$

Show that the error in this approximation is about $2(a-b)^3/[3(a+b)^3]$.

6.28. Determine whether the following functions $f(x)$ are (i) continuous and (ii) differentiable at $x = 0$:
(a) $f(x) = \exp(-|x|)$;
(b) $f(x) = (1 - \cos x)/x^2$ for $x \neq 0$, $f(0) = \frac{1}{2}$;
(c) $f(x) = x \sin(1/x)$ for $x \neq 0$, $f(0) = 0$;
(d) $f(x) = [4 - x^2]$, where $[y]$ denotes the integer part of y.

6.29. Find the limit as $x \to 0$ of $[\sqrt{1+x^m} - \sqrt{1-x^m}]/x^n$, in which m and n are positive integers.

6.30. Evaluate the following limits:

(a) $\lim_{x \to 0} \dfrac{\sin 3x}{\sinh x}$, (b) $\lim_{x \to 0} \dfrac{\tan x - \tanh x}{\sinh x - x}$,

(c) $\lim_{x \to 0} \dfrac{\tan x - x}{\cos x - 1}$, (d) $\lim_{x \to 0} \left(\dfrac{\csc x}{x^3} - \dfrac{\sinh x}{x^5} \right)$.

6.31. Find the limits of the following functions:

(a) $\dfrac{x^3 + x^2 - 5x - 2}{2x^3 - 7x^2 + 4x + 4}$, as $x \to 0$, $x \to \infty$ and $x \to 2$;

(b) $\dfrac{\sin x - x \cosh x}{\sinh x - x}$, as $x \to 0$;

(c) $\displaystyle\int_x^{\pi/2} \left(\dfrac{y \cos y - \sin y}{y^2} \right) dy$, as $x \to 0$.

6.32. Use Taylor expansions to three terms to find approximations to (a) $\sqrt[4]{17}$ and (b) $\sqrt[3]{26}$.

6.33. Using a first-order Taylor expansion about $x = x_0$, show that a better approximation than x_0 to the solution of the equation

$$f(x) = \sin x + \tan x = 2$$

is given by $x = x_0 + \delta$, where

$$\delta = \frac{2 - f(x_0)}{\cos x_0 + \sec^2 x_0}.$$

(a) Use this procedure twice to find the solution of $f(x) = 2$ to six significant figures, given that it is close to $x = 0.9$.

(b) Use the result in (a) and the substitution $y = \sin x$ to deduce, to the same degree of accuracy, one solution of the quartic equation

$$y^4 - 4y^3 + 4y^2 + 4y - 4 = 0.$$

6.34. Evaluate

$$\lim_{x \to 0} \left[\frac{1}{x^3} \left(\operatorname{cosec} x - \frac{1}{x} - \frac{x}{6} \right) \right].$$

6.35. In quantum theory, a system of oscillators, each of fundamental frequency ν and interacting at temperature T, has an average energy $\bar{E}$ given by

$$\bar{E} = \frac{\sum_{n=0}^{\infty} nh\nu e^{-nx}}{\sum_{n=0}^{\infty} e^{-nx}},$$

where $x = h\nu/kT$, h and k being the Planck and Boltzmann constants, respectively. Prove that both series converge, evaluate their sums and show that at high temperatures $\bar{E} \approx kT$, whilst at low temperatures $\bar{E} \approx h\nu \exp(-h\nu/kT)$.

6.36. In a very simple model of a crystal, point-like atomic ions are regularly spaced along an infinite one-dimensional row with spacing R. Alternate ions carry equal and opposite charges $\pm e$. The potential energy of the ith ion in the electric field due to another ion, the jth, is

$$\frac{q_i q_j}{4\pi \epsilon_0 r_{ij}},$$

where q_i, q_j are the charges on the ions and r_{ij} is the distance between them.
  Write down a series giving the total contribution V_i of the ith ion to the overall potential energy. Show that the series converges, and, if V_i is written as

$$V_i = \frac{\alpha e^2}{4\pi \epsilon_0 R},$$

find a closed-form expression for α, the Madelung constant for this (unrealistic) lattice.

6.37. One of the factors contributing to the high relative permittivity of water to static electric fields is the permanent electric dipole moment, p, of the water molecule. In an external field E the dipoles tend to line up with the field, but they do not do so completely because of thermal agitation corresponding to the temperature, T, of the water. A classical (non-quantum) calculation using the Boltzmann distribution shows that the average polarisability per molecule, α, is given by

$$\alpha = \frac{p}{E}(\coth x - x^{-1}),$$

where $x = pE/(kT)$ and k is the Boltzmann constant.

At ordinary temperatures, even with high field strengths (10^4 V m^{-1} or more), $x \ll 1$. By making suitable series expansions of the hyperbolic functions involved,[39] show that $\alpha = p^2/(3kT)$ to an accuracy of about one part in $15x^{-2}$.

6.38. In quantum theory, a certain method (the Born approximation) gives the (so-called) amplitude $f(\theta)$ for the scattering of a particle of mass m through an angle θ by a uniform potential well of depth V_0 and radius b (i.e. the potential energy of the particle is $-V_0$ within a sphere of radius b and zero elsewhere) as

$$f(\theta) = \frac{2mV_0}{\hbar^2 K^3}(\sin Kb - Kb \cos Kb).$$

Here $\hbar$ is the Planck constant divided by 2π, the energy of the particle is $\hbar^2 k^2/(2m)$ and K is $2k \sin(\theta/2)$.

Use l'Hôpital's rule to evaluate the amplitude at low energies, i.e. when k and hence K tend to zero, and so determine the low-energy total cross-section.

[Note: the differential cross-section is given by $|f(\theta)|^2$ and the total cross-section by the integral of this over all solid angles, i.e. $2\pi \int_0^\pi |f(\theta)|^2 \sin\theta \, d\theta$.]

HINTS AND ANSWERS

6.1. Write as $2(\sum_{n=1}^{1000} n - \sum_{n=1}^{499} n) = 751\,500$.

6.3. Write the general term as a cubic expression in n and then use the results derived in Section 6.2.5 to sum each power separately. Then factorise the resulting expression.

6.5. Divergent for $r \leq 1$; convergent for $r \geq 2$.

6.7. (a) $\ln(N+1)$, divergent; (b) $\frac{1}{3}[1-(-2)^n]$, oscillates infinitely; (c) add $\frac{1}{3}S_N$ to the S_N series; $\frac{3}{16}[1-(-3)^{-N}] + \frac{3}{4}N(-3)^{-N-1}$, convergent to $\frac{3}{16}$.

6.9. Write the nth term as the difference between two consecutive values of a partial-fraction function of n. The sum equals $\frac{1}{2}(1 - N^{-2})$.

6.11. Sum the geometric series with rth term $\exp[i(\theta + r\alpha)]$. Its real part is

$$\frac{\cos\theta - \cos[(n+1)\alpha + \theta] - \cos(\theta - \alpha) + \cos(\theta + n\alpha)}{4\sin^2(\alpha/2)},$$

which can be reduced to the given answer.

---

[39] Write $\coth x$ as $\cosh x/[x(1 + f(x))]$ and then use the binomial expansion of $[1 + f(x)]^{-1}$ up to and including the term containing $[f(x)]^2$.

Series and limits

6.13. (a) $-1 \le x < 1$; (b) all x except $x = (2n \pm 1)\pi/2$; (c) $x < -1$; (d) $x < 0$.

6.15. (a) Absolutely convergent; compare with Problem 6.6.12(b). (b) Oscillates finitely. (c) Absolutely convergent for all x. (d) Absolutely convergent; use partial fractions. (e) Oscillates infinitely.

6.17. Divide the series into two series, n odd and n even. For $r = 2$ both are absolutely convergent, by comparison with $\sum n^{-2}$. For $r = 1$ neither series is convergent, by comparison with $\sum n^{-1}$. However, the sum of the two is convergent, by the alternating sign test or by showing that the terms cancel in pairs.

6.19. The first term has value 0.833 and all other terms are positive.

6.21. $|A|^2(1-r)^2/(1+r^2-2r\cos\phi)$.

6.23. Use the binomial expansion and collect terms up to x^4. Integrate both sides of the displayed equation. $\tan x = x + x^3/3 + 2x^5/15 + \cdots$

6.25. For example, $P_5(x) = 24x^4 - 72x^2 + 9$. $\sinh^{-1} x = x - x^3/6 + 3x^5/40 - \cdots$

6.27. Set $a = D + \delta$ and $b = D - \delta$ and use the expansion for $\ln(1 \pm \delta/D)$.

6.29. The limit is 0 for $m > n$, 1 for $m = n$, and ∞ for $m < n$.

6.31. (a) $-\frac{1}{2}, \frac{1}{2}, \infty$; (b) -4; (c) $-1 + 2/\pi$.

6.33. (a) First approximation 0.886 452; second approximation 0.886 287. (b) Set $y = \sin x$ and re-express $f(x) = 2$ as a polynomial equation. $y = \sin(0.886\,287) = 0.774\,730$.

6.35. If $S(x) = \sum_{n=0}^{\infty} e^{-nx}$, then evaluate $S(x)$ and consider $dS(x)/dx$. $E = h\nu[\exp(h\nu/kT) - 1]^{-1}$.

6.37. The series expansion is $\dfrac{px}{E}\left(\dfrac{1}{3} - \dfrac{x^2}{45} + \cdots\right)$.

7 Partial differentiation

In Chapters 3 and 4 we discussed functions *f* of only one variable *x*, which were usually written *f* (*x*). Certain constants and parameters may also have appeared in the definition of *f*, e.g. *f* (*x*) = *ax* + 2 contains the constant 2 and the parameter *a*, but only *x* was considered as a variable and only the derivatives $f^{(n)}(x) = d^n f/dx^n$ were defined.

However, we can equally well consider functions that depend on more than one variable, e.g. the function $f(x, y) = x^2 + 3xy$, which depends on the two variables *x* and *y*. For any pair of values *x*, *y*, the function *f* (*x*, *y*) has a well-defined value, e.g. *f* (2, 3) = 22. This notion can clearly be extended to functions dependent on more than two variables. For the *n*-variable case, we write $f(x_1, x_2, \ldots, x_n)$ for a function that depends on the variables $x_1, x_2, \ldots, x_n$. When *n* = 2, x_1 and x_2 correspond to the variables *x* and *y* used above.

Functions of one variable, like *f* (*x*), can be represented by a graph on a plane sheet of paper, and it is apparent that functions of two variables can, with a little effort, be represented by a surface in three-dimensional space. Thus, we may also picture *f* (*x*, *y*) as describing the variation of height with position in a mountainous landscape. Functions of many variables, however, are usually very difficult to visualise, and so the preliminary discussion in this chapter will concentrate on functions of just two variables.

7.1 Definition of the partial derivative

It is clear that a function $f(x, y)$ of two variables will have a gradient in all directions in the *xy*-plane; in our mountainous landscape analogy, this would correspond to the initial steepness of the uphill or downhill slope encountered when setting off from any point to follow a particular compass direction. A general expression for this rate of change of *f* can be found and will be discussed in the next section. However, we first consider the simpler case of finding the rates of change of $f(x, y)$ in the positive *x*- and *y*-directions. These rates of change are called the *partial derivatives* with respect to *x* and *y* respectively, and they are extremely important in a wide range of physical applications.[1]

For a function of two variables $f(x, y)$ we may define the derivative with respect to *x*, for example, by saying that it is that for a one-variable function when *y* is held fixed and treated as a constant. To signify that a derivative is with respect to *x*, but at the same time

---

[1] In many, if not most, physical applications *x* and *y* are not distances and they may even represent physical parameters of different natures. For example, when describing the pressure of a mass of gas, *x* might be its volume and *y* its temperature.

Partial differentiation

to recognise that a derivative with respect to y also exists, the former is denoted by $\partial f / \partial x$ and is the *partial derivative of $f(x, y)$ with respect to x*. Similarly, the partial derivative of f with respect to y is denoted by $\partial f / \partial y$. Though each partial derivative is defined by differentiation with respect to one particular independent variable, both derivatives are, in general, functions of both variables.

To define formally the partial derivative of $f(x, y)$ with respect to x, we have

$$\frac{\partial f}{\partial x} = \lim_{\Delta x \to 0} \frac{f(x + \Delta x, y) - f(x, y)}{\Delta x}, \qquad (7.1)$$

provided that the limit exists. This is much the same as for the derivative of a one-variable function. The other partial derivative of $f(x, y)$ is similarly defined as a limit (provided it exists):

$$\frac{\partial f}{\partial y} = \lim_{\Delta y \to 0} \frac{f(x, y + \Delta y) - f(x, y)}{\Delta y}. \qquad (7.2)$$

It is common practice in connection with partial derivatives of functions involving more than one variable to indicate those variables that are held constant by writing them as subscripts to the derivative symbol. Thus, the partial derivatives defined in (7.1) and (7.2) would be written, respectively, as

$$\left(\frac{\partial f}{\partial x}\right)_y \quad \text{and} \quad \left(\frac{\partial f}{\partial y}\right)_x.$$

In this form, the subscript shows explicitly which variable is to be kept constant. A more compact notation for these partial derivatives is f_x and f_y, again respectively. However, it is extremely important when using partial derivatives to remember which variables are being held constant and it is wise to write out the partial derivative in explicit form if there is any possibility of confusion.

The extension of the definitions (7.1) and (7.2) to the general n-variable case is straightforward and can be written formally as

$$\frac{\partial f(x_1, x_2, \ldots, x_n)}{\partial x_i} = \lim_{\Delta x_i \to 0} \frac{[f(x_1, x_2, \ldots, x_i + \Delta x_i, \ldots, x_n) - f(x_1, x_2, \ldots, x_i, \ldots, x_n)]}{\Delta x_i},$$

provided that the limit exists.

Just as for one-variable functions, second (and higher) partial derivatives are defined in an analogous way. For a two-variable function $f(x, y)$ they are

$$\frac{\partial}{\partial x}\left(\frac{\partial f}{\partial x}\right) = \frac{\partial^2 f}{\partial x^2} = f_{xx}, \qquad \frac{\partial}{\partial y}\left(\frac{\partial f}{\partial y}\right) = \frac{\partial^2 f}{\partial y^2} = f_{yy},$$

$$\frac{\partial}{\partial x}\left(\frac{\partial f}{\partial y}\right) = \frac{\partial^2 f}{\partial x \partial y} = f_{xy}, \qquad \frac{\partial}{\partial y}\left(\frac{\partial f}{\partial x}\right) = \frac{\partial^2 f}{\partial y \partial x} = f_{yx}.$$

Although four second derivatives are defined in this way, only three of them are independent, since the relationship

$$\frac{\partial^2 f}{\partial x \partial y} = \frac{\partial^2 f}{\partial y \partial x}$$

7.2 The total differential and total derivative

is always obeyed, provided that these second partial derivatives are continuous at the point in question. This relationship often proves useful as a labour-saving device when evaluating second partial derivatives, as well as being the basis of several important identities relating to physical systems.[2] It can also be shown for a function of n variables, $f(x_1, x_2, \ldots, x_n)$, that, under the same conditions,

$$\frac{\partial^2 f}{\partial x_i \partial x_j} = \frac{\partial^2 f}{\partial x_j \partial x_i}$$

for any two values of i and j.[3] The following worked example illustrates the basic procedure.

Example Find the first and second partial derivatives of the function

$$f(x, y) = 2x^3 y^2 + y^3.$$

The first partial derivatives are

$$\frac{\partial f}{\partial x} = 6x^2 y^2, \qquad \frac{\partial f}{\partial y} = 4x^3 y + 3y^2.$$

It should be noted that when $\partial f/\partial x$ was calculated the term y^3 that appears in $f(x, y)$ was treated as a constant and so had a zero derivative. The second partial derivatives, calculated in the same way, are

$$\frac{\partial^2 f}{\partial x^2} = 12xy^2, \qquad \frac{\partial^2 f}{\partial y^2} = 4x^3 + 6y, \qquad \frac{\partial^2 f}{\partial x \partial y} = 12x^2 y, \qquad \frac{\partial^2 f}{\partial y \partial x} = 12x^2 y,$$

the last two being equal, as expected. ◂

EXERCISE 7.1

1. Find all first and second partial derivatives of

$$f(x, y) = x^3 + x(y^2 + a^2) \qquad \text{and} \qquad g(x, y) = (x + y)\sin x.$$

7.2 The total differential and total derivative

Having defined the (first) partial derivatives of a function $f(x, y)$, which give the rate of change of f along the positive x- and y-axes, we consider next the rate of change of $f(x, y)$ in an arbitrary direction. Suppose that we make simultaneous small changes Δx

---

[2] In particular, Maxwell's relationships describing the thermodynamics of gaseous systems; these are discussed in detail in Section 7.11.
[3] Convince yourself that a function of n variables has a total of $n(n + 1)/2$ independent second partial derivatives.

Partial differentiation

in x and Δy in y and that, as a result, f changes to $f + \Delta f$. Then we must have

$$\begin{aligned}\Delta f &= f(x + \Delta x, y + \Delta y) - f(x, y) \\ &= f(x + \Delta x, y + \Delta y) - f(x, y + \Delta y) + f(x, y + \Delta y) - f(x, y) \\ &= \left[\frac{f(x + \Delta x, y + \Delta y) - f(x, y + \Delta y)}{\Delta x}\right]\Delta x + \left[\frac{f(x, y + \Delta y) - f(x, y)}{\Delta y}\right]\Delta y.\end{aligned}$$

(7.3)

With regard to the last line, we note that the quantities in brackets are very similar to those involved in the definitions of partial derivatives (7.1) and (7.2). For them to be strictly equal to the partial derivatives, Δx and Δy would need to be infinitesimally small. But, even for finite (but not too large) Δx and Δy, the approximate formula

$$\Delta f \approx \frac{\partial f(x, y)}{\partial x}\Delta x + \frac{\partial f(x, y)}{\partial y}\Delta y \qquad (7.4)$$

can be obtained. It will be noticed that the first bracket in (7.3) actually approximates to $\partial f(x, y + \Delta y)/\partial x$ but that this has been replaced by $\partial f(x, y)/\partial x$ in (7.4). This approximation clearly has the same degree of validity as that which replaces the bracket by the partial derivative.

How valid an approximation (7.4) is to (7.3) depends not only on how small Δx and Δy are but also on the magnitudes of higher partial derivatives; this is discussed further in Section 7.7 in the context of Taylor series for functions of more than one variable. Nevertheless, by letting the small changes Δx and Δy in (7.4) become infinitesimal, we can define the *total differential* df of the function $f(x, y)$, without any approximation, as

$$df = \frac{\partial f}{\partial x}dx + \frac{\partial f}{\partial y}dy. \qquad (7.5)$$

The basic calculation is illustrated by the next worked example.

Example Find the total differential of the function $f(x, y) = y \exp(x + y)$.

Evaluating the first partial derivatives, we find

$$\frac{\partial f}{\partial x} = y \exp(x + y), \qquad \frac{\partial f}{\partial y} = \exp(x + y) + y \exp(x + y).$$

Applying (7.5), then gives

$$df = [y \exp(x + y)]dx + [(1 + y) \exp(x + y)]dy$$

as the total differential, i.e. the infinitesimal change in f that results from infinitesimal changes dx and dy in x and y respectively. ◀

Equation (7.5) can be extended to the case of a function of n variables, $f(x_1, x_2, \ldots, x_n)$:

$$df = \frac{\partial f}{\partial x_1}dx_1 + \frac{\partial f}{\partial x_2}dx_2 + \cdots + \frac{\partial f}{\partial x_n}dx_n. \qquad (7.6)$$

7.2 The total differential and total derivative

In some situations, despite the fact that several variables x_i, $i = 1, 2, \ldots, n$, appear to be involved, effectively only one of them is. This occurs if there are subsidiary relationships constraining all the x_i to have values dependent on the value of one of them, say x_1. These relationships may be represented by equations that are typically of the form[4]

$$x_i = x_i(x_1), \quad i = 2, 3, \ldots, n. \tag{7.7}$$

In principle f can then be expressed as a function of x_1 alone by substituting from (7.7) for $x_2, x_3, \ldots, x_n$, and then the *total derivative* (or simply the derivative) of f with respect to x_1 is obtained by ordinary differentiation.

Alternatively, (7.6) can be used to give

$$\frac{df}{dx_1} = \frac{\partial f}{\partial x_1} + \left(\frac{\partial f}{\partial x_2}\right)\frac{dx_2}{dx_1} + \cdots + \left(\frac{\partial f}{\partial x_n}\right)\frac{dx_n}{dx_1}, \tag{7.8}$$

either by notionally dividing all through by dx_1 or, more formally, by replacing dx_i by Δx_i and df by Δf, setting Δx_i equal to $(dx_i/dx_1)\Delta x_1$, dividing through by Δx_1 and finally letting $\Delta x_1 \to 0$.

It should be noted that the LHS of this equation is the total derivative df/dx_1, whilst the partial derivative $\partial f/\partial x_1$ forms only a part of the RHS. In evaluating this partial derivative account must be taken only of *explicit* appearances of x_1 in the function f, and *no* allowance must be made for the knowledge that changing x_1 necessarily changes $x_2, x_3, \ldots, x_n$. The contribution from these latter changes is precisely that of the remaining terms on the RHS of (7.8). Naturally, what has been shown using x_1 in the above argument applies equally well to any other of the x_i, with the appropriate consequent changes.

Example Find the total derivative of the two-variable function $f(x, y) = x^2 + 3xy$ with respect to x, given that $y = \sin^{-1} x$.

We can see immediately that

$$\frac{\partial f}{\partial x} = 2x + 3y, \quad \frac{\partial f}{\partial y} = 3x, \quad \frac{dy}{dx} = \frac{1}{(1 - x^2)^{1/2}}$$

and so, using (7.8) with $x_1 = x$ and $x_2 = y$,

$$\frac{df}{dx} = 2x + 3y + 3x \frac{1}{(1 - x^2)^{1/2}}$$

$$= 2x + 3\sin^{-1} x + \frac{3x}{(1 - x^2)^{1/2}}.$$

Obviously the same expression would have resulted if we had substituted for y from the start, but the above method often produces results with reduced calculation, particularly in more complicated examples. ◀

---

[4] The same symbol, here 'x_i', is used to represent both the function (with argument x_1) that generates x_i and the actual value of x_i; this can be confusing at first, but is common practice and needs to be recognised.

Partial differentiation

EXERCISE 7.2

1. Let $f(x, y) = y + x \sin y$. Find (a) the total differential if x and y are independent variables and (b) the total derivatives with respect to x and y if $y = \cos^{-1} x$. Simplify your answers as far as possible.

7.3 Exact and inexact differentials

In the previous section we discussed how to find the total differential of a function, i.e. its infinitesimal change in an arbitrary direction, in terms of its gradients $\partial f/\partial x$ and $\partial f/\partial y$ in the x- and y-directions [see (7.5)]. Sometimes, however, we wish to reverse the process and find the function f that differentiates to give a known differential. Usually, finding such functions relies on inspection and experience.

As an example, it is easy to see that the function whose differential is $df = x\,dy + y\,dx$ is simply $f(x, y) = xy + c$, where c is a constant. Differentials such as this, which integrate directly, are called *exact differentials*, whereas those that do not are *inexact differentials*. For example, $x\,dy + 3y\,dx$ is not the straightforward differential of any function (see the worked example below). Inexact differentials can be made exact, however, by multiplying through by a suitable function called an integrating factor. How to find such integrating factors in a deductive way is discussed in Section 14.2.3.[5]

Example Show that the differential $x\,dy + 3y\,dx$ is inexact.

On the one hand, since the multiplier of dx must be the partial derivative with respect to x of (a putative) $f(x, y)$, after integrating with respect to x we conclude that $f(x, y) = 3xy + g(y)$, where $g(y)$ is any function of y. On the other hand, if we integrate the multiplier of dy with respect to y we conclude that $f(x, y) = xy + h(x)$, where $h(x)$ is any function of x. These two deductions are inconsistent for any and every choice of $g(y)$ and $h(x)$. We therefore infer that there is no such $f(x, y)$ and that the differential is inexact. ◂

It is naturally of interest to investigate which properties of a differential make it exact. Consider the general differential containing two variables,

$$df = A(x, y)\,dx + B(x, y)\,dy.$$

If df is to be an exact differential we must have

$$\frac{\partial f}{\partial x} = A(x, y), \qquad \frac{\partial f}{\partial y} = B(x, y)$$

and, using the property $f_{xy} = f_{yx}$, we therefore require

$$\frac{\partial A}{\partial y} = \frac{\partial B}{\partial x}. \tag{7.9}$$

---

[5] But, as an interim measure, demonstrate that for $x\,dy + 3y\,dx$ a suitable integrating factor is x^2. What is the form of the corresponding $f(x, y)$?

7.3 Exact and inexact differentials

This is therefore a necessary condition for the differential to be exact; it is, in fact, also a sufficient condition.[6]

Example Using (7.9) show that $x\,dy + 3y\,dx$ is inexact.

In the above notation, $A(x, y) = 3y$ and $B(x, y) = x$ and so
$$\frac{\partial A}{\partial y} = 3, \qquad \frac{\partial B}{\partial x} = 1.$$

As these are not equal it follows that the differential cannot be exact. As it must be, this is the same conclusion as that reached in the previous worked example. ◀

Determining whether a differential containing many variables $x_1, x_2, \ldots, x_n$ is exact is a simple extension of the above. A differential containing many variables can be written in general form as
$$df = \sum_{i=1}^{n} g_i(x_1, x_2, \ldots, x_n)\,dx_i$$

and will be exact if
$$\frac{\partial g_i}{\partial x_j} = \frac{\partial g_j}{\partial x_i} \qquad \text{for all pairs } i, j. \tag{7.10}$$

There will be $\frac{1}{2}n(n-1)$ such relationships to be satisfied. In the next worked example $n = 3$.

Example Show that
$$(y + z)\,dx + x\,dy + x\,dz$$
is an exact differential.

In this case, $g_1(x, y, z) = y + z$, $g_2(x, y, z) = x$, $g_3(x, y, z) = x$ and hence $\partial g_1/\partial y = 1 = \partial g_2/\partial x$, $\partial g_3/\partial x = 1 = \partial g_1/\partial z$, $\partial g_2/\partial z = 0 = \partial g_3/\partial y$; therefore, from (7.10), the differential is exact. As mentioned above, it is sometimes possible to show that a differential is exact simply by finding, by inspection, the function from which it originates. In this example, it can be seen easily that $f(x, y, z) = x(y + z) + c$. ◀

EXERCISES 7.3

1. Determine whether or not the following differentials are exact:
   (a) $df = (y^2 + 3x^2)\,dx + 2xy\,dy$,
   (b) $df = (x^2 + 3y^2)\,dx + 2xy\,dy$,

---

[6] Justify this by showing that, provided (7.9) is satisfied, $f(x, y) = \int^x A(u, y)\,du$ has the appropriate total differential. Assume that you can differentiate with respect to y 'under the integral sign', i.e. that
$$\frac{\partial f}{\partial y} = \int^x \frac{\partial A(u, y)}{\partial y}\,du.$$

(c) $df = (\sin y - y \cos x) dx + (\sin x - x \cos y) dy$,

(d) $df = \dfrac{y^2}{x^2 + y^2} dx + \left(\tan^{-1} \dfrac{x}{y} - \dfrac{xy}{x^2 + y^2}\right) dy$.

2. Find the values of m and n that make $g(x, y) = x^m y^n$ an integrating factor for

$$df = (3x^2 y - 1) dx + \left(4x^3 - \dfrac{x}{y}\right) dy,$$

i.e. that make $d\phi = g\, df$ an exact differential. Determine the form of $\phi(x, y)$.

3. Test whether the following are exact differentials:
   (a) $df = (4x^3 + 2xy^2 - 2xz) dx + 2x^2 y\, dy - x^2 dz$,
   (b) $df = (4x^3 + y^3 - 3z^2) dx + 3xy^2 dy - 3xz^2 dz$,
   (c) $df = [y \cos(xy) + z \cos(xz)] dx + [x \cos(xy) - z \sin(yz)] dy$
   $\quad + [x \cos(xz) - y \sin(yz)] dz$.

7.4 Useful theorems of partial differentiation

So far our discussion has centred on a function $f(x, y)$ dependent on two variables, x and y. Equally, however, we could have expressed x as a function of f and y, or y as a function of f and x. To emphasise the point that all the variables are of equal standing, we now replace f by z. This does not imply that x, y and z are coordinate positions (though they might be). Since x is a function of y and z, it follows that

$$dx = \left(\dfrac{\partial x}{\partial y}\right)_z dy + \left(\dfrac{\partial x}{\partial z}\right)_y dz \qquad (7.11)$$

and similarly, since $y = y(x, z)$,

$$dy = \left(\dfrac{\partial y}{\partial x}\right)_z dx + \left(\dfrac{\partial y}{\partial z}\right)_x dz. \qquad (7.12)$$

We may now substitute (7.12) into (7.11) to obtain

$$dx = \left(\dfrac{\partial x}{\partial y}\right)_z \left(\dfrac{\partial y}{\partial x}\right)_z dx + \left[\left(\dfrac{\partial x}{\partial y}\right)_z \left(\dfrac{\partial y}{\partial z}\right)_x + \left(\dfrac{\partial x}{\partial z}\right)_y\right] dz. \qquad (7.13)$$

Now if we keep z constant, so that $dz = 0$, we obtain the *reciprocity relation*

$$\left(\dfrac{\partial x}{\partial y}\right)_z = \left(\dfrac{\partial y}{\partial x}\right)_z^{-1},$$

which is valid provided both partial derivatives exist and neither is equal to zero. Note, further, that this relationship only holds when the variable being kept constant, in this case z, is the same on both sides of the equation.

Alternatively we can put $dx = 0$ in (7.13). Then the contents of the square brackets also equal zero, and we obtain the *cyclic relation*

$$\left(\dfrac{\partial y}{\partial z}\right)_x \left(\dfrac{\partial z}{\partial x}\right)_y \left(\dfrac{\partial x}{\partial y}\right)_z = -1,$$

7.5 The chain rule

which holds unless any of the derivatives vanish.[7] In deriving this result we have used the reciprocity relation to replace $(\partial x/\partial z)_y^{-1}$ by $(\partial z/\partial x)_y$.

EXERCISE 7.4

1. One possible model equation connecting the pressure p, volume V and temperature T of a classical gas is

$$\left(p + \frac{a}{V^2}\right)(V - b) = RT.$$

Calculate $\left(\dfrac{\partial p}{\partial V}\right)_T$, $\left(\dfrac{\partial V}{\partial T}\right)_p$ and $\left(\dfrac{\partial T}{\partial p}\right)_V$ and verify that their product is -1, as stated above.

7.5 The chain rule

So far we have discussed the differentiation of a function $f(x, y)$ with respect to its variables x and y. We now consider the case where x and y are themselves functions of another variable, say u. If we wish to find the derivative df/du, we could simply substitute in $f(x, y)$ the expressions for $x(u)$ and $y(u)$ and then differentiate the resulting function of u. Such substitution will quickly give the desired answer in simple cases, but in more complicated examples it is easier to make use of the total differentials described in the previous section.

From Equation (7.5) the total differential of $f(x, y)$ is given by

$$df = \frac{\partial f}{\partial x}dx + \frac{\partial f}{\partial y}dy.$$

But we now note that by using the formal device of dividing through by du this immediately implies

$$\frac{df}{du} = \frac{\partial f}{\partial x}\frac{dx}{du} + \frac{\partial f}{\partial y}\frac{dy}{du}, \tag{7.14}$$

which is called the *chain rule* for partial differentiation. This expression provides a direct method for calculating the total derivative of f with respect to u and is particularly useful when an equation is expressed in a parametric form, as in the next worked example.

---

[7] Obtain expressions for the appropriate partial derivatives for the relationship $\tan z = y/x$ and verify that they satisfy the cyclic relation.

Example Given that $x(u) = 1 + au$ and $y(u) = bu^3$, find the rate of change of $f(x, y) = xe^{-y}$ with respect to u.

As discussed above, this problem could be addressed by substituting for x and y to obtain f as a function of u only, and then differentiating with respect to u. However, using (7.14) directly we obtain

$$\frac{df}{du} = (e^{-y})a + (-xe^{-y})3bu^2,$$

which on substituting for x and y gives

$$\frac{df}{du} = e^{-bu^3}(a - 3bu^2 - 3bau^3).$$

Note that, although it is defined in terms of two variables x and y, the function f is, in reality, one of u only. The derivative of f with respect to u is therefore a total derivative, not a partial one. ◀

Equation (7.14) is an example of the chain rule for a function of two variables each of which depends on a single variable. The chain rule may be extended to functions of many variables, each of which is itself a function of a variable u, i.e. $f(x_1, x_2, x_3, \ldots, x_n)$, with $x_i = x_i(u)$. In this case the chain rule gives

$$\frac{df}{du} = \sum_{i=1}^{n} \frac{\partial f}{\partial x_i} \frac{dx_i}{du} = \frac{\partial f}{\partial x_1} \frac{dx_1}{du} + \frac{\partial f}{\partial x_2} \frac{dx_2}{du} + \cdots + \frac{\partial f}{\partial x_n} \frac{dx_n}{du}. \tag{7.15}$$

EXERCISE 7.5

1. A twisted cubic curve is defined parametrically by $x = \lambda$, $y = \frac{3}{2}\lambda^2$, $z = \frac{3}{2}\lambda^3$. Show that the total derivative with respect to λ of s, the distance from the origin of a point on the curve, is

$$\frac{ds}{d\lambda} = \frac{4 + 18\lambda^2 + 27\lambda^4}{2(4 + 9\lambda^2 + 9\lambda^4)^{1/2}}.$$

7.6 Change of variables

It is sometimes necessary or desirable to make a change of variables during the course of an analysis, and consequently to have to change an equation expressed in one set of variables into an equation using another set. The same situation arises if a function f depends on one set of variables x_i, i.e. $f = f(x_1, x_2, \ldots, x_n)$, but the x_i are themselves functions of a further set of variables u_j and given by the equations

$$x_i = x_i(u_1, u_2, \ldots, u_m). \tag{7.16}$$

7.6 Change of variables

For each different value of i, x_i will be a different function of the u_j. In this case the chain rule (7.15) becomes

$$\frac{\partial f}{\partial u_j} = \sum_{i=1}^{n} \frac{\partial f}{\partial x_i} \frac{\partial x_i}{\partial u_j}, \quad j = 1, 2, \ldots, m, \tag{7.17}$$

and is said to express a *change of variables*. In general the number of variables in each set need not be equal, i.e. m need not equal n, but if both the x_i and the u_i are sets of independent variables then $m = n$.

The following worked example involves the transformation from one set of coordinates to another of a quantity written as $\nabla^2 \psi$, and usually read as 'del-squared ψ' or 'the Laplacian of ψ'. The differential operator ∇^2 is not formally introduced in this book until Section 11.6.2, but is one of the most commonly occurring operators in the equations of physical science.

Example Plane polar coordinates, ρ and ϕ, and Cartesian coordinates, x and y, are related by the expressions

$$x = \rho \cos \phi, \quad y = \rho \sin \phi,$$

as was shown in Figure 2.4. An arbitrary function $\psi(x, y)$ can be re-expressed as a function $\chi(\rho, \phi)$. Transform the expression

$$\frac{\partial^2 \psi}{\partial x^2} + \frac{\partial^2 \psi}{\partial y^2}$$

into one in ρ and ϕ.

We first note that $\rho^2 = x^2 + y^2$, $\phi = \tan^{-1}(y/x)$. We can now write down the four partial derivatives

$$\frac{\partial \rho}{\partial x} = \frac{x}{(x^2 + y^2)^{1/2}} = \cos \phi, \qquad \frac{\partial \phi}{\partial x} = \frac{-(y/x^2)}{1 + (y/x)^2} = -\frac{\sin \phi}{\rho},$$

$$\frac{\partial \rho}{\partial y} = \frac{y}{(x^2 + y^2)^{1/2}} = \sin \phi, \qquad \frac{\partial \phi}{\partial y} = \frac{1/x}{1 + (y/x)^2} = \frac{\cos \phi}{\rho}.$$

Thus, from (7.17), we may write

$$\frac{\partial}{\partial x} = \cos \phi \frac{\partial}{\partial \rho} - \frac{\sin \phi}{\rho} \frac{\partial}{\partial \phi}, \qquad \frac{\partial}{\partial y} = \sin \phi \frac{\partial}{\partial \rho} + \frac{\cos \phi}{\rho} \frac{\partial}{\partial \phi}.$$

Now it is only a matter of writing

$$\frac{\partial^2 \psi}{\partial x^2} = \frac{\partial}{\partial x}\left(\frac{\partial \psi}{\partial x}\right) = \frac{\partial}{\partial x}\left(\frac{\partial}{\partial x}\right)\psi$$

$$= \left(\cos \phi \frac{\partial}{\partial \rho} - \frac{\sin \phi}{\rho} \frac{\partial}{\partial \phi}\right)\left(\cos \phi \frac{\partial}{\partial \rho} - \frac{\sin \phi}{\rho} \frac{\partial}{\partial \phi}\right)\chi$$

$$= \left(\cos \phi \frac{\partial}{\partial \rho} - \frac{\sin \phi}{\rho} \frac{\partial}{\partial \phi}\right)\left(\cos \phi \frac{\partial \chi}{\partial \rho} - \frac{\sin \phi}{\rho} \frac{\partial \chi}{\partial \phi}\right)$$

$$= \cos^2 \phi \frac{\partial^2 \chi}{\partial \rho^2} + \frac{2 \cos \phi \sin \phi}{\rho^2} \frac{\partial \chi}{\partial \phi} - \frac{2 \cos \phi \sin \phi}{\rho} \frac{\partial^2 \chi}{\partial \phi \partial \rho} + \frac{\sin^2 \phi}{\rho} \frac{\partial \chi}{\partial \rho} + \frac{\sin^2 \phi}{\rho^2} \frac{\partial^2 \chi}{\partial \phi^2}$$

Partial differentiation

and a similar expression for $\partial^2\psi/\partial y^2$,

$$\frac{\partial^2\psi}{\partial y^2} = \left(\sin\phi\frac{\partial}{\partial\rho} + \frac{\cos\phi}{\rho}\frac{\partial}{\partial\phi}\right)\left(\sin\phi\frac{\partial}{\partial\rho} + \frac{\cos\phi}{\rho}\frac{\partial}{\partial\phi}\right)\chi$$

$$= \sin^2\phi\frac{\partial^2\chi}{\partial\rho^2} - \frac{2\cos\phi\sin\phi}{\rho^2}\frac{\partial\chi}{\partial\phi} + \frac{2\cos\phi\sin\phi}{\rho}\frac{\partial^2\chi}{\partial\phi\partial\rho}$$

$$+ \frac{\cos^2\phi}{\rho}\frac{\partial\chi}{\partial\rho} + \frac{\cos^2\phi}{\rho^2}\frac{\partial^2\chi}{\partial\phi^2}.$$

When these two expressions are added together the change of variables is complete and we obtain

$$\frac{\partial^2\psi}{\partial x^2} + \frac{\partial^2\psi}{\partial y^2} = \frac{\partial^2\chi}{\partial\rho^2} + \frac{1}{\rho}\frac{\partial\chi}{\partial\rho} + \frac{1}{\rho^2}\frac{\partial^2\chi}{\partial\phi^2}.$$

It should be remembered that, although for any pair of corresponding coordinates, (x, y) and (ρ, ϕ), $\psi(x, y) = \chi(\rho, \phi)$, the functional forms of ψ and χ will be quite different. ◀

EXERCISE 7.6

1. An arbitrary function $\phi(x, y)$ can be re-expressed as $\psi(u, v)$, where

$$u = x + iy \quad \text{and} \quad v = x - iy.$$

Use the chain rule to show that, under the change of variables,

$$\nabla^2\phi = \frac{\partial^2\phi}{\partial x^2} + \frac{\partial^2\phi}{\partial y^2} \quad \text{transforms to} \quad 4\frac{\partial^2\psi}{\partial u\partial v}.$$

7.7 Taylor's theorem for many-variable functions

We have already introduced Taylor's theorem for a function $f(x)$ of one variable, in Section 6.6. In an analogous way, the Taylor expansion of a function $f(x, y)$ of two variables is given by

$$f(x, y) = f(x_0, y_0) + \frac{\partial f}{\partial x}\Delta x + \frac{\partial f}{\partial y}\Delta y$$

$$+ \frac{1}{2!}\left[\frac{\partial^2 f}{\partial x^2}(\Delta x)^2 + 2\frac{\partial^2 f}{\partial x\partial y}\Delta x\Delta y + \frac{\partial^2 f}{\partial y^2}(\Delta y)^2\right] + \cdots, \quad (7.18)$$

where $\Delta x = x - x_0$ and $\Delta y = y - y_0$, and all the derivatives are to be evaluated at (x_0, y_0).

A straightforward worked example is the easiest way to see what this means in practice.

7.7 Taylor's theorem for many-variable functions

Example Find the Taylor expansion, up to quadratic terms in $x - 2$ and $y - 3$, of $f(x, y) = y \exp xy$ about the point $x = 2$, $y = 3$.

We first evaluate the required partial derivatives of the function, i.e.

$$\frac{\partial f}{\partial x} = y^2 \exp xy, \qquad \frac{\partial f}{\partial y} = \exp xy + xy \exp xy,$$

$$\frac{\partial^2 f}{\partial x^2} = y^3 \exp xy, \qquad \frac{\partial^2 f}{\partial y^2} = 2x \exp xy + x^2 y \exp xy,$$

$$\frac{\partial^2 f}{\partial x \partial y} = 2y \exp xy + xy^2 \exp xy.$$

These all need to be evaluated at the point $(2, 3)$, and then using (7.18), the Taylor expansion of a two-variable function, we find

$$f(x, y) \approx e^6 \{3 + 9(x - 2) + 7(y - 3) + (2!)^{-1}[27(x - 2)^2 + 48(x - 2)(y - 3) + 16(y - 3)^2]\}.$$

This could be expanded to read $f \approx e^6(234 - 117x - 89y + \frac{27}{2}x^2 + 24xy + 8y^2)$, but would generally be much more useful in its original form. ◀

It will be noticed that the terms in (7.18) containing first derivatives can be written as

$$\frac{\partial f}{\partial x} \Delta x + \frac{\partial f}{\partial y} \Delta y = \left(\Delta x \frac{\partial}{\partial x} + \Delta y \frac{\partial}{\partial y} \right) f(x, y),$$

where both sides of this relation should be evaluated at the point (x_0, y_0). Similarly, the terms in (7.18) containing second derivatives can be written as

$$\frac{1}{2!} \left[\frac{\partial^2 f}{\partial x^2} (\Delta x)^2 + 2 \frac{\partial^2 f}{\partial x \partial y} \Delta x \Delta y + \frac{\partial^2 f}{\partial y^2} (\Delta y)^2 \right] = \frac{1}{2!} \left(\Delta x \frac{\partial}{\partial x} + \Delta y \frac{\partial}{\partial y} \right)^2 f(x, y), \tag{7.19}$$

where it is understood that the partial derivatives resulting from squaring the expression in parentheses act only on $f(x, y)$ and its derivatives, not on Δx or Δy; again, both sides of (7.19) should be evaluated at (x_0, y_0). It can be shown that the higher order terms of the Taylor expansion of $f(x, y)$ can be written in an analogous way, and that we may write the full Taylor series as

$$f(x, y) = \sum_{n=0}^{\infty} \frac{1}{n!} \left[\left(\Delta x \frac{\partial}{\partial x} + \Delta y \frac{\partial}{\partial y} \right)^n f(x, y) \right]_{x_0, y_0}$$

where $\Delta x = x - x_0$ and $\Delta y = y - y_0$ and, as indicated, all the terms on the RHS are to be evaluated at (x_0, y_0).

Partial differentiation

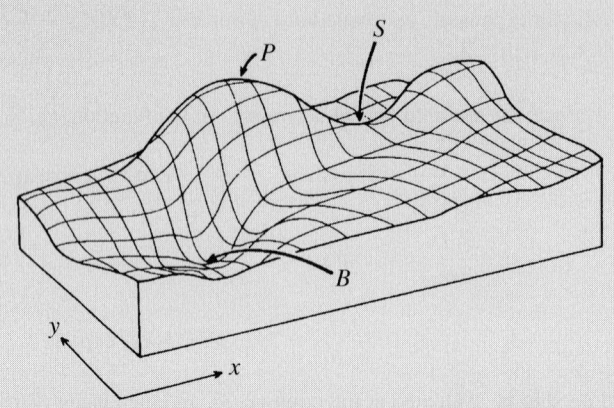

Figure 7.1 Stationary points of a function of two variables. A minimum occurs at B, a maximum at P and a saddle point at S.

EXERCISE 7.7

1. Show that a Taylor series approximation to the function $f(x, y) = (1 + x) \cos y \, e^{-x}$, likely to be accurate to better than 1% within a circle of radius 0.1 centred on the point $(1, \pi/2)$, is $e^{-1}(x - 3)(y - \pi/2)$.

7.8 Stationary values of two-variable functions

The idea of the *stationary points* of a function of just one variable has already been discussed in Section 3.3. We recall that the function $f(x)$ has a stationary point at $x = x_0$ if its gradient df/dx is zero at that point. A function may have any number of stationary points, and their nature, i.e. whether they are maxima, minima or stationary points of inflection, is determined by the value of the second derivative at the point. A stationary point is

(i) a minimum if $d^2 f/dx^2 > 0$;
(ii) a maximum if $d^2 f/dx^2 < 0$;
(iii) a stationary point of inflection if $d^2 f/dx^2 = 0$ and changes sign through the point.

We now consider the stationary points of functions of two variables; we will see that partial differential analysis is ideally suited to the determination of the positions and natures of such points. The methods developed here can be extended to functions of an arbitrary number of variables, but the analysis then requires techniques that are beyond the scope of this book; we will therefore restrict our attention to the two-variable case. Even in this case, the general situation is more complex than that for a function of one variable, as can be seen from Figure 7.1.

This figure shows part of a three-dimensional model of a function $f(x, y)$. At positions P and B there are a peak and a bowl respectively or, more mathematically, a local

7.8 Stationary values of two-variable functions

maximum and a local minimum. At position S the gradient in any direction is zero but the situation is complicated, since a section parallel to the plane $x = 0$ would show a maximum, but one parallel to the plane $y = 0$ would show a minimum. A point such as S is known as a *saddle point*. The orientation of the 'saddle' in the xy-plane is irrelevant; it is as shown in the figure solely for ease of discussion. For any saddle point the function increases in some directions away from the point but decreases in other directions.

For functions of two variables, such as the one shown, it should be clear that a necessary condition for a stationary point (maximum, minimum or saddle point) to occur is that

$$\frac{\partial f}{\partial x} = 0 \quad \text{and} \quad \frac{\partial f}{\partial y} = 0. \tag{7.20}$$

The vanishing of the partial derivatives in directions parallel to the axes is enough to ensure that, at that point, the partial derivative in any arbitrary direction is also zero. The latter can be considered as the superposition of two contributions, one along each axis; since both contributions are zero, so is the partial derivative in the arbitrary direction. This may be made more precise by considering the total differential

$$df = \frac{\partial f}{\partial x} dx + \frac{\partial f}{\partial y} dy.$$

Using (7.20) we see that, although the infinitesimal changes dx and dy can be chosen independently, the change in the value of the infinitesimal function df is always zero at a stationary point.

We now turn our attention to determining the nature of a stationary point of a function of two variables, i.e. whether it is a maximum, a minimum or a saddle point. By analogy with the one-variable case we see that $\partial^2 f/\partial x^2$ and $\partial^2 f/\partial y^2$ must both be positive for a minimum and both be negative for a maximum. However, these are not sufficient conditions, since they could also be obeyed at complicated saddle points. What is important for a minimum (or maximum) is that the second partial derivative must be positive (or negative) in *all* directions, not just in the x- and y-directions.

To establish just what constitutes sufficient conditions we first note that, since f is a function of two variables and $\partial f/\partial x = \partial f/\partial y = 0$, a Taylor expansion of the type (7.18) about the stationary point yields

$$f(x, y) - f(x_0, y_0) \approx \frac{1}{2!} \left[(\Delta x)^2 f_{xx} + 2 \Delta x \Delta y f_{xy} + (\Delta y)^2 f_{yy} \right],$$

where $\Delta x = x - x_0$ and $\Delta y = y - y_0$ and where the partial derivatives have been written in the more compact notation. Rearranging the contents of the bracket as the weighted sum of two squares, we find

$$f(x, y) - f(x_0, y_0) \approx \frac{1}{2} \left[f_{xx} \left(\Delta x + \frac{f_{xy} \Delta y}{f_{xx}} \right)^2 + (\Delta y)^2 \left(f_{yy} - \frac{f_{xy}^2}{f_{xx}} \right) \right]. \tag{7.21}$$

For a minimum, we require (7.21) to be positive for all Δx and Δy, and hence $f_{xx} > 0$ and $f_{yy} - (f_{xy}^2/f_{xx}) > 0$. Given the first constraint, i.e. that f_{xx} is positive, the second inequality can be written as

$$f_{xx} f_{yy} > f_{xy}^2. \tag{7.22}$$

Partial differentiation

For a maximum we require (7.21) to be negative for all Δx and Δy, and so firstly we require $f_{xx} < 0$. The second requirement is that $f_{yy} - (f_{xy}^2/f_{xx})$ is negative; since $f_{xx} < 0$, this becomes $f_{xx} f_{yy} > f_{xy}^2$ when multiplied through by f_{xx}. Thus both maxima and minima require the condition (7.22) to be satisfied. For minima and maxima, symmetry [or a reformulation of (7.21)] requires that f_{yy} obeys the same criteria as f_{xx}.

When (7.21) is negative (or zero) for some values of Δx and Δy but positive (or zero) for others, we have a saddle point. In this case (7.22) is not satisfied and $f_{xx} f_{yy} \leq f_{xy}^2$.

In summary, all stationary points have $f_x = f_y = 0$ and they may be classified further as

(i) minima if both f_{xx} and f_{yy} are positive *and* $f_{xy}^2 < f_{xx} f_{yy}$,
(ii) maxima if both f_{xx} and f_{yy} are negative *and* $f_{xy}^2 < f_{xx} f_{yy}$,
(iii) saddle points if f_{xx} and f_{yy} have opposite signs *or* if $f_{xy}^2 \geq f_{xx} f_{yy}$.

Note, however, that if $f_{xy}^2 = f_{xx} f_{yy}$ then it can be shown that there is one particular direction for which the difference between $f(x_0 + \Delta x, y_0 + \Delta y)$ and $f(x_0, y_0)$ is at least third order in Δx and Δy; in such situations further investigation is required. Moreover, if f_{xx}, f_{yy} and f_{xy} are all zero then the Taylor expansion has to be taken to a higher order. As simple examples, such extended investigations would show that the function $f(x, y) = x^4 + y^4$ has a minimum at the origin but that $g(x, y) = x^4 + y^3$ has a saddle point there.[8]

The following example shows some of these criteria in action.

Example Show that the function $f(x, y) = x^3 \exp(-x^2 - y^2)$ has a maximum at the point $(\sqrt{3/2}, 0)$, a minimum at $(-\sqrt{3/2}, 0)$ and a stationary point at the origin whose nature cannot be determined by the above procedures.

Setting the first two partial derivatives to zero to locate the stationary points, we find

$$\frac{\partial f}{\partial x} = (3x^2 - 2x^4) \exp(-x^2 - y^2) = 0, \tag{7.23}$$

$$\frac{\partial f}{\partial y} = -2yx^3 \exp(-x^2 - y^2) = 0. \tag{7.24}$$

For (7.24) to be satisfied we require $x = 0$ or $y = 0$ and for (7.23) to be satisfied we require $x = 0$ or $x = \pm\sqrt{3/2}$. Hence the stationary points are at $(0, 0)$, $(\sqrt{3/2}, 0)$ and $(-\sqrt{3/2}, 0)$. We now find the second partial derivatives:

$$f_{xx} = (4x^5 - 14x^3 + 6x) \exp(-x^2 - y^2),$$
$$f_{yy} = x^3(4y^2 - 2) \exp(-x^2 - y^2),$$
$$f_{xy} = 2x^2 y(2x^2 - 3) \exp(-x^2 - y^2).$$

---

[8] Make rough perspective sketches to show that these anticipated results are intuitively correct. Make a similar rough sketch for the function $h(x, y) = x^2 + 2xy$; you will probably find it helpful to apply criteria (i)–(iii).

7.8 Stationary values of two-variable functions

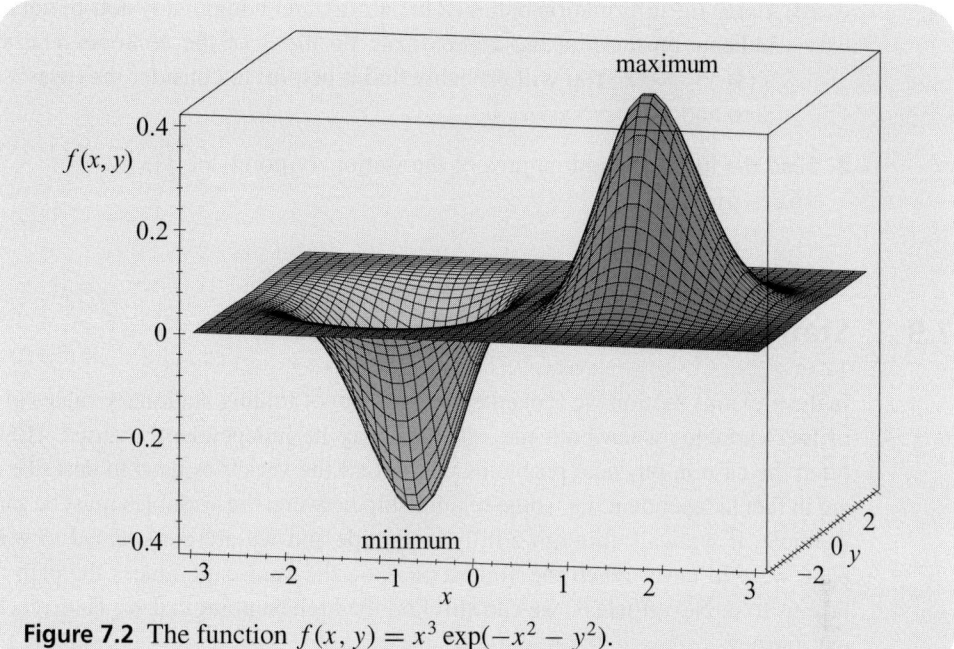

Figure 7.2 The function $f(x, y) = x^3 \exp(-x^2 - y^2)$.

We then substitute the pairs of values of x and y for each stationary point and find that at $(0, 0)$

$$f_{xx} = 0, \quad f_{yy} = 0, \quad f_{xy} = 0$$

and at $(\pm\sqrt{3/2}, 0)$

$$f_{xx} = \mp 6\sqrt{3/2} \exp(-3/2), \quad f_{yy} = \mp 3\sqrt{3/2} \exp(-3/2), \quad f_{xy} = 0.$$

Here the upper and lower signs in the values correspond to those in the coordinates, e.g. $f_{xx}(+\sqrt{3/2}, 0) = -6\sqrt{3/2} \exp(-3/2)$. Hence, applying criteria (i)–(iii) above, we find that $(0, 0)$ is an undetermined stationary point, $(\sqrt{3/2}, 0)$ is a maximum and $(-\sqrt{3/2}, 0)$ is a minimum. The function is shown in Figure 7.2. ◀

EXERCISES 7.8

1. The function $f(x, y)$ is given by $f(x, y) = x^4 - y^4 - 4x^2 + 4y^2$.
   (a) How many stationary points of each kind does $f(x, y)$ have, and where are they positioned?
   (b) Are there any stationary points whose nature cannot be determined by examining the first and second partial derivatives of f?
   (c) By writing $f(x, y)$ as $(x^2 - y^2)(x^2 + y^2 - 4)$, show on an x–y plot the contours $f(x, y) = 0$.

(d) Using the information obtained in part (a), and without any non-trivial calculation, indicate on the plot the approximate positions of the contours $f(x, y) = 2$ and $f(x, y) = -2$. You will probably find it helpful to consider the cases $x = 0$, $y \to \pm\infty$ and $x \to \pm\infty$, $y = 0$.

2. Find the locations and natures of the stationary points of $f(x, y) = x^3 - y^2 + xy - 16x + 6y + 10$.

7.9 Stationary values under constraints

In the previous section we looked at the problem of finding stationary values of a function of two variables when both the variables may be independently varied. However, it is often the case in physical problems that not all the variables used to describe a situation are in fact independent, i.e. some relationship between the variables must be satisfied. For example, if we walk through a hilly landscape and we are constrained to walk along a path, we will never reach the highest peak on the landscape unless the path happens to take us to it. Nevertheless, we can still find the highest point that we have reached during our journey.

We first discuss the case of a function of just two variables. The changes needed to accommodate more than two variables, in so far as taking account of constraints is concerned, are not significant and so we can later extend our results to more than two variables (to many more in the case of the final worked example in this section!). Let us consider finding the maximum value of the differentiable function $f(x, y)$ subject to the constraint $g(x, y) = c$, where c is a constant. In the above analogy, $f(x, y)$ might represent the height of the land above sea-level in some hilly region, whilst $g(x, y) = c$ is the equation of the path along which we walk.

We could, of course, use the constraint $g(x, y) = c$ to substitute for x or y in $f(x, y)$, thereby obtaining a new function of only one variable whose stationary points could be found using the methods discussed in Section 3.3. However, such a procedure can involve a lot of algebra and becomes very tedious for functions of more than two variables; further, even in the two-variable case, it may not be possible to manipulate $g(x, y) = c$ into an explicit expression for either x or y. A more direct method for solving such problems is the *method of Lagrange undetermined multipliers*, which we now discuss.

To maximise f we require

$$df = \frac{\partial f}{\partial x} dx + \frac{\partial f}{\partial y} dy = 0.$$

If dx and dy were independent, we could conclude $f_x = 0 = f_y$. However, here they are not independent, but constrained because g is constant:

$$dg = \frac{\partial g}{\partial x} dx + \frac{\partial g}{\partial y} dy = 0.$$

7.9 Stationary values under constraints

Multiplying dg by an as yet unknown number λ and adding it to df we obtain

$$d(f+\lambda g) = \left(\frac{\partial f}{\partial x} + \lambda \frac{\partial g}{\partial x}\right) dx + \left(\frac{\partial f}{\partial y} + \lambda \frac{\partial g}{\partial y}\right) dy = 0,$$

where λ is called a *Lagrange undetermined multiplier*. In this equation dx and dy are to be independent and arbitrary; we must therefore choose λ such that

$$\frac{\partial f}{\partial x} + \lambda \frac{\partial g}{\partial x} = 0, \qquad (7.25)$$

$$\frac{\partial f}{\partial y} + \lambda \frac{\partial g}{\partial y} = 0. \qquad (7.26)$$

These equations, together with the constraint $g(x, y) = c$, are sufficient, in principle at least, to find the three unknowns, i.e. λ and the values of x and y at the stationary point.
The following illustrates the method.

Example The temperature of a point (x, y) on a unit circle is given by $T(x, y) = 1 + xy$. Find the temperature of the two hottest points on the circle.

Since the only points eligible for consideration lie on the unit circle, we need to maximise $T(x, y)$ subject to the constraint $x^2 + y^2 = 1$. Applying (7.25) and (7.26) with $f(x, y) = T(x, y) = 1 + xy$ and $g(x, y) = x^2 + y^2$, we obtain

$$y + 2\lambda x = 0, \qquad (7.27)$$
$$x + 2\lambda y = 0. \qquad (7.28)$$

These results, together with the original constraint $x^2 + y^2 = 1$, provide three simultaneous equations that may be solved for λ, x and y.

From (7.27) and (7.28) we find $\lambda = \pm 1/2$, which in turn implies that $y = \mp x$. Remembering that $x^2 + y^2 = 1$, we find that

$$y = x \Rightarrow x = \pm\frac{1}{\sqrt{2}}, \quad y = \pm\frac{1}{\sqrt{2}},$$

$$y = -x \Rightarrow x = \mp\frac{1}{\sqrt{2}}, \quad y = \pm\frac{1}{\sqrt{2}}.$$

We have not yet determined which of these stationary points are maxima and which are minima. In this simple case, we need only substitute the four pairs of x- and y-values into $T(x, y) = 1 + xy$ to find that the maximum temperature on the unit circle is $T_{\max} = 3/2$ at the points $y = x = \pm 1/\sqrt{2}$.[9]

◀

The method of Lagrange multipliers can be used to find the stationary points of functions of more than two variables, subject to several constraints, provided that the number of constraints is smaller than the number of variables. For example, suppose that we wish

---

[9] Show that the same conclusion is reached by using the constraint to substitute for y in $T(x, y)$ and then treating the temperature as a function of a single variable.

Partial differentiation

to find the stationary points of $f(x, y, z)$ subject to the constraints $g(x, y, z) = c_1$ and $h(x, y, z) = c_2$, where c_1 and c_2 are constants. Because there are two constraints, we will need two Lagrange multipliers. Calling them λ and μ, we proceed as previously and obtain

$$\frac{\partial}{\partial x}(f + \lambda g + \mu h) = \frac{\partial f}{\partial x} + \lambda \frac{\partial g}{\partial x} + \mu \frac{\partial h}{\partial x} = 0,$$

$$\frac{\partial}{\partial y}(f + \lambda g + \mu h) = \frac{\partial f}{\partial y} + \lambda \frac{\partial g}{\partial y} + \mu \frac{\partial h}{\partial y} = 0, \quad (7.29)$$

$$\frac{\partial}{\partial z}(f + \lambda g + \mu h) = \frac{\partial f}{\partial z} + \lambda \frac{\partial g}{\partial z} + \mu \frac{\partial h}{\partial z} = 0.$$

We may now solve these three equations, together with the two constraints, to give λ, μ, x, y and z, as in part (ii) of the following example.

Example Find the stationary points of $f(x, y, z) = x^3 + y^3 + z^3$ subject to the following constraints:
(i) $g(x, y, z) = x^2 + y^2 + z^2 = 1$;
(ii) $g(x, y, z) = x^2 + y^2 + z^2 = 1$ and $h(x, y, z) = x + y + z = 0$.

Case (i). Since there is only one constraint in this case, we need only introduce a single Lagrange multiplier obtaining

$$\frac{\partial}{\partial x}(f + \lambda g) = 3x^2 + 2\lambda x = 0,$$

$$\frac{\partial}{\partial y}(f + \lambda g) = 3y^2 + 2\lambda y = 0, \quad (7.30)$$

$$\frac{\partial}{\partial z}(f + \lambda g) = 3z^2 + 2\lambda z = 0.$$

These equations are highly symmetrical and clearly have the solution $x = y = z = -2\lambda/3$. Using the constraint $x^2 + y^2 + z^2 = 1$ we find $\lambda = \pm\sqrt{3}/2$ and so stationary points occur at

$$x = y = z = \pm \frac{1}{\sqrt{3}}. \quad (7.31)$$

In solving the three equations (7.30) in this way, however, we have implicitly assumed that x, y and z are non-zero. However, it is clear from (7.30) that any of these values can equal zero, with the exception of the case $x = y = z = 0$ since this is prohibited by the constraint $x^2 + y^2 + z^2 = 1$. We must consider the other cases separately.

If $x = 0$, for example, the remaining three equations become

$$3y^2 + 2\lambda y = 0, \quad 3z^2 + 2\lambda z = 0, \quad y^2 + z^2 = 1.$$

Clearly, we require $\lambda \neq 0$, otherwise these equations are inconsistent. If neither y nor z is zero we find $y = -2\lambda/3 = z$ and from the third equation we require $y = z = \pm 1/\sqrt{2}$. If $y = 0$, however, then $z = \pm 1$ and, similarly, if $z = 0$ then $y = \pm 1$. Thus the stationary points having $x = 0$ are $(0, 0, \pm 1)$, $(0, \pm 1, 0)$ and $(0, \pm 1/\sqrt{2}, \pm 1/\sqrt{2})$. A similar procedure can be followed for the cases $y = 0$ and $z = 0$ respectively and, in addition to those already obtained, we find the stationary points $(\pm 1, 0, 0)$, $(\pm 1/\sqrt{2}, 0, \pm 1/\sqrt{2})$ and $(\pm 1/\sqrt{2}, \pm 1/\sqrt{2}, 0)$.

7.9 Stationary values under constraints

Case (ii). We now have two constraints and must therefore introduce two Lagrange multipliers, obtaining (cf. (7.29))

$$\frac{\partial}{\partial x}(f + \lambda g + \mu h) = 3x^2 + 2\lambda x + \mu = 0, \qquad (7.32)$$

$$\frac{\partial}{\partial y}(f + \lambda g + \mu h) = 3y^2 + 2\lambda y + \mu = 0, \qquad (7.33)$$

$$\frac{\partial}{\partial z}(f + \lambda g + \mu h) = 3z^2 + 2\lambda z + \mu = 0. \qquad (7.34)$$

These equations are again highly symmetrical and the simplest way to proceed is to subtract (7.33) from (7.32) to obtain

$$3(x^2 - y^2) + 2\lambda(x - y) = 0$$
$$\Rightarrow \quad 3(x + y)(x - y) + 2\lambda(x - y) = 0. \qquad (7.35)$$

This equation is clearly satisfied if $x = y$; then, from the second constraint, $x + y + z = 0$, we find $z = -2x$. Substituting these values into the first constraint, $x^2 + y^2 + z^2 = 1$, we obtain

$$x = \pm\frac{1}{\sqrt{6}}, \quad y = \pm\frac{1}{\sqrt{6}}, \quad z = \mp\frac{2}{\sqrt{6}}. \qquad (7.36)$$

Because of the high degree of symmetry amongst Equations (7.32)–(7.34), we may obtain by inspection two further relations analogous to (7.35), one containing the variables y, z and the other the variables x, z. Assuming $y = z$ in the first relation and $x = z$ in the second, we find the stationary points

$$x = \pm\frac{1}{\sqrt{6}}, \quad y = \mp\frac{2}{\sqrt{6}}, \quad z = \pm\frac{1}{\sqrt{6}} \qquad (7.37)$$

and

$$x = \mp\frac{2}{\sqrt{6}}, \quad y = \pm\frac{1}{\sqrt{6}}, \quad z = \pm\frac{1}{\sqrt{6}}. \qquad (7.38)$$

We note that in finding the stationary points (7.36)–(7.38) we did not need to evaluate the Lagrange multipliers λ and μ explicitly. This is not always the case, however, and in some problems it may be simpler to begin by finding the values of these multipliers.

Returning to (7.35) we must now consider whether there are any cases not yet discovered in which $x \neq y$ and the factor $(x - y)$ can be cancelled; the constrained stationary condition then becomes

$$3(x + y) + 2\lambda = 0. \qquad (7.39)$$

The stationary points already recorded in (7.37) and (7.38) have the property $x \neq y$, and they do indeed satisfy (7.39) for a common value for λ. They were, in fact, established by requiring $x = z$ and $y = z$ respectively and so, like (7.36), belong to cases in which two coordinates are equal.

Thus we need to consider only two further cases: $x = y = z$ and x, y and z are all different. The first is clearly prohibited by the constraint $x + y + z = 0$. For the second case, in which cancelling a term such as $(x - y)$ throughout cannot be equivalent to dividing by zero, (7.39), together with the analogous equations containing y, z and x, z respectively, must all be satisfied, i.e.

$$3(x + y) + 2\lambda = 0, \quad 3(y + z) + 2\lambda = 0, \quad 3(x + z) + 2\lambda = 0.$$

Adding these three equations together and using the constraint $x + y + z = 0$ we find $\lambda = 0$. However, for $\lambda = 0$ the equations are inconsistent for non-zero x, y and z.

All possibilities have now been examined and therefore all the stationary points have already been found; they are given by (7.36)–(7.38). ◀

Partial differentiation

The method may be extended to functions of any number n of variables subject to any smaller number m of constraints. This means that effectively there are $n - m$ independent variables and, as mentioned above, we could solve by substitution and then by the methods of the previous section. However, for large n this becomes cumbersome and the use of Lagrange undetermined multipliers is a useful simplification.[10]

Example A system contains a very large number N of particles, each of which can be in any of R energy levels with a corresponding energy E_i, $i = 1, 2, \ldots, R$. The number of particles in the ith level is n_i and the total energy of the system is a constant, E. Find the distribution of particles amongst the energy levels that maximises the expression

$$P = \frac{N!}{n_1! n_2! \cdots n_R!},$$

subject to the constraints that both the number of particles and the total energy remain constant, i.e.

$$g = N - \sum_{i=1}^{R} n_i = 0 \quad \text{and} \quad h = E - \sum_{i=1}^{R} n_i E_i = 0.$$

The way in which we proceed is as follows. In order to maximise P, we must minimise its denominator (since the numerator is fixed). Minimising the denominator is the same as minimising the logarithm of the denominator, i.e.

$$f = \ln(n_1! n_2! \cdots n_R!) = \ln(n_1!) + \ln(n_2!) + \cdots + \ln(n_R!).$$

Using Stirling's approximation, $\ln(n!) \approx n \ln n - n$, we find that

$$f = n_1 \ln n_1 + n_2 \ln n_2 + \cdots + n_R \ln n_R - (n_1 + n_2 + \cdots + n_R)$$

$$= \left(\sum_{i=1}^{R} n_i \ln n_i \right) - N.$$

It has been assumed here that, for the desired distribution, all the n_i are large. Thus, we now have a function f subject to two constraints, $g = 0$ and $h = 0$, and we can apply the Lagrange method, obtaining (cf. (7.29))

$$\frac{\partial f}{\partial n_1} + \lambda \frac{\partial g}{\partial n_1} + \mu \frac{\partial h}{\partial n_1} = 0,$$

$$\frac{\partial f}{\partial n_2} + \lambda \frac{\partial g}{\partial n_2} + \mu \frac{\partial h}{\partial n_2} = 0,$$

$$\vdots$$

$$\frac{\partial f}{\partial n_R} + \lambda \frac{\partial g}{\partial n_R} + \mu \frac{\partial h}{\partial n_R} = 0.$$

---

[10] In the following worked example, as applied to a normal macroscopic system, n is usually of the order of 10^{23} and some simplification is welcome!

7.9 Stationary values under constraints

Since all these equations are alike, we consider the general case

$$\frac{\partial f}{\partial n_k} + \lambda \frac{\partial g}{\partial n_k} + \mu \frac{\partial h}{\partial n_k} = 0,$$

for $k = 1, 2, \ldots, R$. Substituting the functions f, g and h into this relation we find

$$\frac{n_k}{n_k} + \ln n_k + \lambda(-1) + \mu(-E_k) = 0,$$

which can be rearranged to give

$$\ln n_k = \mu E_k + \lambda - 1,$$

and hence

$$n_k = C \exp \mu E_k.$$

We now have the general form for the distribution of particles amongst energy levels, but in order to determine the two constants μ and C, we recall that

$$\sum_{k=1}^{R} C \exp \mu E_k = N$$

and

$$\sum_{k=1}^{R} C E_k \exp \mu E_k = E.$$

This is known as the Boltzmann distribution and is a well-known result from statistical mechanics.[11]

◀

EXERCISES 7.9

1. An azimuthally symmetric hill of overall height h has a profile given in three-dimensional Cartesian coordinates by

$$z = \frac{ha^4}{(a^2 + x^2)(a^2 + y^2)}.$$

A footpath on the side of the hill has a projection onto the x–y plane given by

$$x^2 + y^2 + 2xy - bx + by + c = 0.$$

Show that the maximum height to which the path rises is $16a^4 b^4 h/(4a^2 b^2 + c^2)^2$.

2. By minimising $s^2 = x^2 + y^2$ subject to an appropriate constraint, determine the minimum distance from the origin of a point on the (rectangular) hyperbola

---

[11] Consider the particular case $R \to \infty$ and $E_k = (k-1)E_0$. Show that the (negative) value of μ is determined by the equation

$$E = \sum_{s=0}^{\infty} N(1 - e^{\mu E_0}) s E_0 e^{s\mu E_0}.$$

$(x - a)(y - a) = c^2$, where $a > 0$. Justify any choice of $\pm$ signs that you make in your solution.

7.10 Envelopes

As noted at the start of this chapter, many of the functions with which physicists, chemists and engineers have to deal contain, in addition to constants and one or more variables, quantities that are normally considered as parameters of the system under study. Such parameters may, for example, represent the capacitance of a capacitor, the length of a rod or the mass of a particle – quantities that are normally taken as fixed for any particular physical set-up. The corresponding variables may well be time, currents, charges, positions and velocities. However, the parameters *could* be varied, and in this section we study the effects of doing so; in particular we study how the form of dependence of one variable on another, typically $y = y(x)$, is affected when the value of a parameter is changed in a smooth and continuous way. In effect, we are making the parameter into an additional variable.

As a particular parameter, which we denote by α, is varied over its permitted range, the shape of the plot of y against x will change, usually, but not always, in a smooth and continuous way. For example, if the muzzle speed v of a shell fired from a gun is increased through a range of values then its height–distance trajectories will be a series of curves with a common starting point that are essentially just magnified copies of the original; furthermore, the curves do not cross each other. However, if the muzzle speed is kept constant but θ, the angle of elevation of the gun, is increased through a series of values, the corresponding trajectories do not vary in a monotonic way. When θ has been increased beyond $45°$ the trajectories then do cross some of the trajectories corresponding to $\theta < 45°$. All of the trajectories lie within a single curve that each individual trajectory touches at a different point. Such a curve is called the *envelope* to the set of trajectory solutions; it is to the study of such envelopes that this section is devoted.

For our general discussion of envelopes we will consider an equation of the form $f = f(x, y, \alpha) = 0$. A function of three Cartesian variables, $f = f(x, y, \alpha)$, is defined at all points in $xy\alpha$-space, whereas $f = f(x, y, \alpha) = 0$ is a *surface* in this space. A plane of constant α, which is parallel to the xy-plane, cuts such a surface in a curve. Thus, different values of the parameter α correspond to different curves, which can be plotted in the xy-plane. We now investigate how the *envelope equation* for such a family of curves is obtained.

Suppose $f(x, y, \alpha_1) = 0$ and $f(x, y, \alpha_1 + h) = 0$ are two neighbouring curves of a family for which the parameter α differs by a small amount h. Let them intersect at the point P with coordinates x, y, as shown in Figure 7.3. Then the envelope, indicated by the broken line in the figure, touches $f(x, y, \alpha_1) = 0$ at the point P_1, which is defined as the limiting position of P when α_1 is fixed but $h \to 0$. The full envelope is the curve traced out by P_1 as α_1 changes to generate successive members of the family of curves. Of course, for any finite h, $f(x, y, \alpha_1 + h) = 0$ is one of these curves and the envelope touches it at the point P_2.

7.10 Envelopes

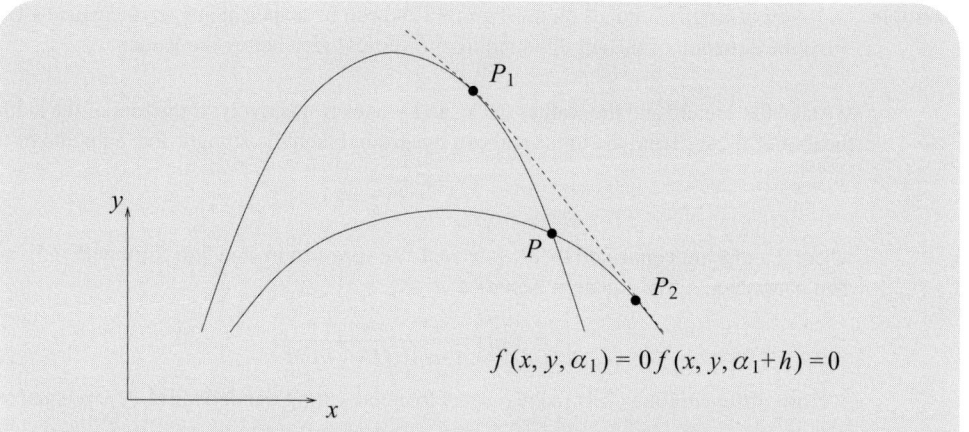

Figure 7.3 Two neighbouring curves in the xy-plane of the family $f(x, y, \alpha) = 0$ intersecting at P. For fixed α_1, the point P_1 is the limiting position of P as $h \to 0$. As α_1 is varied, P_1 delineates the envelope of the family (broken line).

We are now going to apply Rolle's theorem (see Section 3.5) with the parameter α as the independent variable and x and y fixed as constants. In this context, the two curves in Figure 7.3 can be thought of as the projections onto the xy-plane of the planar curves in which the *surface* $f = f(x, y, \alpha) = 0$ meets the planes $\alpha = \alpha_1$ and $\alpha = \alpha_1 + h$.

Along the normal to the page that passes through P, as α changes from α_1 to $\alpha_1 + h$ the value of $f = f(x, y, \alpha)$ will depart from zero, because the normal meets the surface $f = f(x, y, \alpha) = 0$ only at $\alpha = \alpha_1$ and at $\alpha = \alpha_1 + h$. However, at these end-points the values of $f = f(x, y, \alpha)$ will both be zero, and therefore equal. This allows us to apply Rolle's theorem and so to conclude that for some θ in the range $0 \leq \theta \leq 1$ the partial derivative $\partial f(x, y, \alpha_1 + \theta h)/\partial \alpha$ is zero. When h is made arbitrarily small, so that $P \to P_1$, the three defining equations reduce to two, which define the envelope point P_1:

$$f(x, y, \alpha_1) = 0 \quad \text{and} \quad \frac{\partial f(x, y, \alpha_1)}{\partial \alpha} = 0. \tag{7.40}$$

In (7.40), both the function and the gradient are evaluated at $\alpha = \alpha_1$. The equation of the envelope $g(x, y) = 0$ is found by eliminating α_1 between the two equations.[12]

As a simple example we will now solve the problem which when posed mathematically reads 'calculate the envelope appropriate to the family of straight lines in the xy-plane whose points of intersection with the coordinate axes are a fixed distance apart'. In more ordinary language, the problem is about a ladder leaning against a wall.

---

[12] Use these equations to show that the envelope of the family of circles, of which a typical one has radius a and is centred on $(2a, 0)$, is the pair of straight lines $y = \pm x/\sqrt{3}$. Confirm your conclusion using a sketch that shows a line through the origin that is tangent to a typical circle.

Example A ladder of length L stands on level ground and can be leaned at any angle against a vertical wall. Find the equation of the curve bounding the vertical area below the ladder.

We take the ground and the wall as the x- and y-axes respectively. If the foot of the ladder is a from the foot of the wall and the top is b above the ground then the straight-line equation of the ladder is

$$\frac{x}{a} + \frac{y}{b} = 1,$$

where a and b are connected by $a^2 + b^2 = L^2$. Expressed in standard form with only one independent parameter, a, the equation becomes

$$f(x, y, a) = \frac{x}{a} + \frac{y}{(L^2 - a^2)^{1/2}} - 1 = 0. \tag{7.41}$$

Now, differentiating (7.41) with respect to a and setting the derivative $\partial f/\partial a$ equal to zero gives

$$-\frac{x}{a^2} + \frac{ay}{(L^2 - a^2)^{3/2}} = 0;$$

from which it follows that

$$a = \frac{Lx^{1/3}}{(x^{2/3} + y^{2/3})^{1/2}} \quad \text{and} \quad (L^2 - a^2)^{1/2} = \frac{Ly^{1/3}}{(x^{2/3} + y^{2/3})^{1/2}}.$$

Eliminating a by substituting these values into (7.41) gives, for the equation of the envelope of all possible positions on the ladder,

$$x^{2/3} + y^{2/3} = L^{2/3}.$$

This is the equation of an astroid (mentioned in Problem 3.21), and, together with the wall and the ground, marks the boundary of the vertical area below the ladder. ◀

Other examples, drawn from both geometry and the physical sciences, are considered in the problems at the end of this chapter. The shell trajectory question discussed earlier in this section is solved there, but in the guise of a question about the water bell of an ornamental fountain.

EXERCISES 7.10

1. Show that all members of the family of parabolas

$$y^2 = a(x + \sqrt{a}),$$

with $a > 0$, are touched by the curve $27y^2 = 4x^3$.

2. (a) Prove that the envelope of the family of closed curves

$$f(x, y) = x^2 + y^2 - 2ay + 2a^2 - a = 0$$

is an ellipse with semi-axes of lengths $1/2$ and $1/\sqrt{2}$.
(b) By rearranging $f(x, y)$, determine and describe the form of a typical member of the family.

(c) Sketch a few family curves and the envelope and explain qualitatively why the two semi-axes are not both of length $1/2$. [A more quantitative understanding can be obtained by studying the behaviour of $a - \sqrt{a(1-a)}$.]

7.11 Thermodynamic relations

Thermodynamic relations provide a useful set of physical examples of partial differentiation. The relations we will derive are called *Maxwell's thermodynamic relations*. They express relationships between four thermodynamic quantities describing a unit mass of a substance. The quantities are the pressure P, the volume V, the thermodynamic temperature T and the entropy S of the substance. These four quantities are not independent; any two of them can be varied independently, but the other two are then determined.

The first law of thermodynamics may be expressed in total differential form [see (7.5)] as

$$dU = T\,dS - P\,dV, \tag{7.42}$$

where U is the internal energy of the substance. Essentially, this is a conservation of energy equation, but we will concern ourselves, not with the physics, but rather with the use of partial differentials to relate the four basic quantities discussed above. The method involves writing a total differential, dU say, in terms of the differentials of two variables, say X and Y, thus

$$dU = \left(\frac{\partial U}{\partial X}\right)_Y dX + \left(\frac{\partial U}{\partial Y}\right)_X dY, \tag{7.43}$$

and then using the relationship

$$\frac{\partial^2 U}{\partial X \partial Y} = \frac{\partial^2 U}{\partial Y \partial X}$$

to obtain one of the Maxwell relations. The variables X and Y are to be chosen from P, V, T and S.

Example

Show that $\left(\dfrac{\partial T}{\partial V}\right)_S = -\left(\dfrac{\partial P}{\partial S}\right)_V$.

Here the two variables that have to be held constant, in turn, happen to be those whose differentials appear on the RHS of (7.42). And so, taking X as S and Y as V in (7.43), we have

$$T\,dS - P\,dV = dU = \left(\frac{\partial U}{\partial S}\right)_V dS + \left(\frac{\partial U}{\partial V}\right)_S dV,$$

and find directly that

$$\left(\frac{\partial U}{\partial S}\right)_V = T \quad \text{and} \quad \left(\frac{\partial U}{\partial V}\right)_S = -P.$$

Partial differentiation

Differentiating the first expression with respect to V (whilst keeping S constant) and the second with respect to S (with constant V), and using

$$\frac{\partial^2 U}{\partial V \partial S} = \frac{\partial^2 U}{\partial S \partial V},$$

we find the Maxwell relation

$$\left(\frac{\partial T}{\partial V}\right)_S = -\left(\frac{\partial P}{\partial S}\right)_V,$$

as given in the question. ◀

A second Maxwell relation is derived in the next worked example.

Example Show that $(\partial S/\partial V)_T = (\partial P/\partial T)_V$.

Applying (7.43) to dS, with independent variables V and T, we find

$$dU = T\, dS - P\, dV = T\left[\left(\frac{\partial S}{\partial V}\right)_T dV + \left(\frac{\partial S}{\partial T}\right)_V dT\right] - P\, dV.$$

Similarly, applying (7.43) to dU we find

$$dU = \left(\frac{\partial U}{\partial V}\right)_T dV + \left(\frac{\partial U}{\partial T}\right)_V dT.$$

Thus, equating partial derivatives,

$$\left(\frac{\partial U}{\partial V}\right)_T = T\left(\frac{\partial S}{\partial V}\right)_T - P \quad \text{and} \quad \left(\frac{\partial U}{\partial T}\right)_V = T\left(\frac{\partial S}{\partial T}\right)_V.$$

But, since

$$\frac{\partial^2 U}{\partial T \partial V} = \frac{\partial^2 U}{\partial V \partial T}, \quad \text{i.e.} \quad \frac{\partial}{\partial T}\left(\frac{\partial U}{\partial V}\right)_T = \frac{\partial}{\partial V}\left(\frac{\partial U}{\partial T}\right)_V,$$

it follows that

$$\left(\frac{\partial S}{\partial V}\right)_T + T\frac{\partial^2 S}{\partial T \partial V} - \left(\frac{\partial P}{\partial T}\right)_V = \frac{\partial}{\partial V}\left[T\left(\frac{\partial S}{\partial T}\right)_V\right]_T = T\frac{\partial^2 S}{\partial V \partial T}.$$

Thus finally we obtain

$$\left(\frac{\partial S}{\partial V}\right)_T = \left(\frac{\partial P}{\partial T}\right)_V,$$

as a second Maxwell relation. ◀

The above derivation is rather cumbersome, however, and a useful device that can simplify the working is to define a new function, called a *thermodynamic potential*. The internal energy U discussed above is one example of a potential, but three others are commonly defined and they are described below.[13]

---

[13] For a valid thermodynamic potential, each term in the function must have the physical dimensions of energy and its value must depend only upon the current values of the variables it contains, and not, for example, on the system's previous history.

7.11 Thermodynamic relations

Example

Show that $\left(\dfrac{\partial S}{\partial V}\right)_T = \left(\dfrac{\partial P}{\partial T}\right)_V$ by considering the potential $U - ST$.

We first consider the differential $d(U - ST)$. From (7.5), we obtain

$$d(U - ST) = dU - SdT - TdS = -SdT - PdV$$

when use is made of (7.42). We rewrite $U - ST$ as F for convenience of notation; F is called the *Helmholtz potential*. Thus

$$dF = -SdT - PdV,$$

and it follows that

$$\left(\dfrac{\partial F}{\partial T}\right)_V = -S \quad \text{and} \quad \left(\dfrac{\partial F}{\partial V}\right)_T = -P.$$

Using these results together with

$$\dfrac{\partial^2 F}{\partial T \partial V} = \dfrac{\partial^2 F}{\partial V \partial T},$$

we can see immediately that

$$\left(\dfrac{\partial S}{\partial V}\right)_T = \left(\dfrac{\partial P}{\partial T}\right)_V,$$

which is the same Maxwell relation as before.

◀

Although the Helmholtz potential has other uses, in this context it has simply provided a means for a quick derivation of the Maxwell relation. The other Maxwell relations can be derived similarly by using two other potentials, the *enthalpy*, $H = U + PV$, and the *Gibbs free energy*, $G = U + PV - ST$ (see Problem 7.25).

EXERCISE 7.11

1. The entropy $S(H, T)$, the magnetisation $M(H, T)$ and the internal energy $U(H, T)$ of a magnetic salt placed in a magnetic field of strength H, at temperature T, are connected by the equation

$$TdS = dU - HdM. \qquad (*)$$

(a) Prove that $\left(\dfrac{\partial T}{\partial M}\right)_S = \left(\dfrac{\partial H}{\partial S}\right)_M.$

(b) Taking T and M as the independent variables, use $(*)$ to prove that

$$\left(\dfrac{\partial S}{\partial M}\right)_T = -\left(\dfrac{\partial H}{\partial T}\right)_M.$$

7.12 Differentiation of integrals

We conclude this chapter with a discussion of the differentiation of integrals. For functions of one variable, we have already seen in Equation (4.9) that it is meaningful to differentiate an indefinite integral with respect to its upper limit, though the result hardly ever provides new information. The situation is summarised by

$$F(x) = \int^x f(t)\,dt, \qquad \frac{\partial F(x)}{\partial x} = f(x).$$

However, if an integrand, or one or both of the limits between which a dummy variable x is integrated, are functions of a second variable y, then the result of the integration can be differentiated with respect to y if the need arises. For example, we may wish to evaluate the integral over x of a function $g(x, y)$ that contains y as a parameter and then chose the value of y that gives the integral $G(y)$ its minimum value. This would involve evaluating $dG(y)/dy$ and equating it to zero. We now show how these more general cases are handled, starting with integrands that contain a parameter.

Consider the indefinite integral of an integrand containing the parameter x, and the corresponding derivative of the integral with respect its upper limit:

$$F(x, t) = \int^t f(x, t')\,dt', \qquad \frac{\partial F(x, t)}{\partial t} = f(x, t).$$

Assuming that the second partial derivatives of $F(x, t)$ are continuous, we have

$$\frac{\partial^2 F(x, t)}{\partial t\, \partial x} = \frac{\partial^2 F(x, t)}{\partial x\, \partial t},$$

and so we can write

$$\frac{\partial}{\partial t}\left[\frac{\partial F(x, t)}{\partial x}\right] = \frac{\partial}{\partial x}\left[\frac{\partial F(x, t)}{\partial t}\right] = \frac{\partial f(x, t)}{\partial x}.$$

Reversing this equality, integrating with respect to t and then substituting the integral form for $F(x, t)$ gives

$$\int^t \frac{\partial f(x, t')}{\partial x}\,dt' = \frac{\partial F(x, t)}{\partial x} = \frac{\partial}{\partial x}\left[\int^t f(x, t')\,dt'\right]. \qquad (7.44)$$

In words, the integral of the derivative of the integrand with respect to a parameter is equal to the derivative of the integral with respect to the same parameter.[14]

The corresponding (but scarcely different) result for definite integrals follows from considering

$$I(x) = \int_{t=u}^{t=v} f(x, t)\,dt$$
$$= F(x, v) - F(x, u),$$

---

[14] Evaluate the integral of $e^{-\alpha t}$ between 0 and T and hence show that the integral of $t e^{-\alpha t}$ between the same limits is $\alpha^{-2}[1 - (1 + \alpha T)e^{-\alpha T}]$.

7.12 Differentiation of integrals

where u and v are constants. Differentiating this integral with respect to x, and using (7.44), we see that

$$\frac{dI(x)}{dx} = \frac{\partial F(x, v)}{\partial x} - \frac{\partial F(x, u)}{\partial x}$$
$$= \int^{v} \frac{\partial f(x, t)}{\partial x} dt - \int^{u} \frac{\partial f(x, t)}{\partial x} dt$$
$$= \int_{u}^{v} \frac{\partial f(x, t)}{\partial x} dt.$$

This is the *Leibnitz rule* for differentiating integrals, and states that for constant limits of integration the order of integration and differentiation can be reversed. This is the same result as that derived above for indefinite integrals.

In the more general case where the limits of the integral are themselves functions of x, it follows immediately that

$$I(x) = \int_{t=u(x)}^{t=v(x)} f(x, t) \, dt$$
$$= F(x, v(x)) - F(x, u(x)),$$

which yields the partial derivatives

$$\frac{\partial I}{\partial v} = f(x, v(x)), \qquad \frac{\partial I}{\partial u} = -f(x, u(x)).$$

Consequently,

$$\frac{dI}{dx} = \left(\frac{\partial I}{\partial v}\right) \frac{dv}{dx} + \left(\frac{\partial I}{\partial u}\right) \frac{du}{dx} + \frac{\partial I}{\partial x}$$
$$= f(x, v(x)) \frac{dv}{dx} - f(x, u(x)) \frac{du}{dx} + \frac{\partial}{\partial x} \int_{u(x)}^{v(x)} f(x, t) dt$$
$$= f(x, v(x)) \frac{dv}{dx} - f(x, u(x)) \frac{du}{dx} + \int_{u(x)}^{v(x)} \frac{\partial f(x, t)}{\partial x} dt, \qquad (7.45)$$

where the partial derivative with respect to x in the last term has been taken inside the integral sign using (7.44). This procedure is valid because $u(x)$ and $v(x)$ are being held constant in this term.

To illustrate this result, we give the following example.

Example Find the derivative with respect to x of the integral

$$I(x) = \int_{x}^{x^2} \frac{\sin xt}{t} \, dt.$$

Applying (7.45), we see that

$$\frac{dI}{dx} = \frac{\sin x^3}{x^2}(2x) - \frac{\sin x^2}{x}(1) + \int_{x}^{x^2} \frac{t \cos xt}{t} dt$$

$$= \frac{2\sin x^3}{x} - \frac{\sin x^2}{x} + \left[\frac{\sin xt}{x}\right]_x^{x^2}$$

$$= 3\frac{\sin x^3}{x} - 2\frac{\sin x^2}{x}$$

$$= \frac{1}{x}(3\sin x^3 - 2\sin x^2).$$

In this example all three possible sources contributed to the total derivative: the upper limit, the lower limit and a parameter in the integrand itself.

EXERCISES 7.12

1. Given that $\int_0^\infty e^{-\alpha t}\,dt = \alpha^{-1}$, prove, without any explicit integration, that

$$\int_0^\infty t^n e^{-\alpha t}\,dt = \frac{n!}{\alpha^{n+1}}.$$

2. By considering $\int_0^\pi \sin(xy)\,dx$, show that

$$\int_0^\pi [\sin(xy) + xy\cos(xy)]\,dx = \pi\sin\pi y.$$

3. The function $J(a)$ is defined by

$$J(a) = \int_0^{a^2} \tan^{-1}\left(\frac{x}{a^2}\right) dx.$$

(a) Show that $dJ/da = a\left(\frac{\pi}{2} - \ln 2\right)$.
(b) Determine the value of $J(0)$ by inspection.
(c) Deduce the value of $J(a)$.

SUMMARY

1. *Definitions and notation based on* $f = f(x, y)$
   - Partial derivative definition:

$$f_x \equiv \left(\frac{\partial f}{\partial x}\right)_y = \lim_{\Delta x \to 0} \frac{f(x+\Delta x, y) - f(x, y)}{\Delta x},$$

   i.e. y is held fixed.

Summary

- Second derivatives:

$$f_{xx} = \frac{\partial^2 f}{\partial x^2} = \frac{\partial}{\partial x}\left(\frac{\partial f}{\partial x}\right), \qquad f_{yy} = \frac{\partial^2 f}{\partial y^2} = \frac{\partial}{\partial y}\left(\frac{\partial f}{\partial y}\right),$$

$$f_{xy} = \frac{\partial}{\partial x}\left(\frac{\partial f}{\partial y}\right) = \frac{\partial^2 f}{\partial x \partial y} = \frac{\partial^2 f}{\partial y \partial x} = \frac{\partial}{\partial y}\left(\frac{\partial f}{\partial x}\right) = f_{yx}.$$

- Total differential: $df = \left(\frac{\partial f}{\partial x}\right)_y dx + \left(\frac{\partial f}{\partial y}\right)_x dy.$

- $\left(\frac{\partial x}{\partial y}\right)_f = \left(\frac{\partial y}{\partial x}\right)_f^{-1}$ and $\left(\frac{\partial y}{\partial f}\right)_x \left(\frac{\partial f}{\partial x}\right)_y \left(\frac{\partial x}{\partial y}\right)_f = -1.$

2. *Differentials*
   - If $df = A(x, y)\,dx + B(x, y)\,dy$, then df is exact $\Leftrightarrow \dfrac{\partial A}{\partial y} = \dfrac{\partial B}{\partial x}.$
   - Chain rule: if $x = x(u)$ and $y = y(u)$, then

   $$\frac{df}{du} = \left(\frac{\partial f}{\partial x}\right)_y \frac{dx}{du} + \left(\frac{\partial f}{\partial y}\right)_x \frac{dy}{du}.$$

   - Taylor's theorem for $f(x, y)$:

   $$f(x, y) = f(x_0, y_0) + \frac{\partial f}{\partial x}\Delta x + \frac{\partial f}{\partial y}\Delta y$$

   $$+ \frac{1}{2!}\left[\frac{\partial^2 f}{\partial x^2}(\Delta x)^2 + 2\frac{\partial^2 f}{\partial x \partial y}\Delta x \Delta y + \frac{\partial^2 f}{\partial y^2}(\Delta y)^2\right] + \cdots,$$

   where $\Delta x = x - x_0$ and $\Delta y = y - y_0$ and all derivatives are evaluated at (x_0, y_0).

3. *Stationary values* for $f(x, y)$
   - A necessary condition is $f_x = f_y = 0$, and then
     (i) minimum if both f_{xx} and f_{yy} are positive *and* $f_{xy}^2 < f_{xx}f_{yy}$,
     (ii) maximum if both f_{xx} and f_{yy} are negative *and* $f_{xy}^2 < f_{xx}f_{yy}$,
     (iii) saddle point if f_{xx} and f_{yy} have opposite signs *or* if $f_{xy}^2 \geq f_{xx}f_{yy}$.
   - Under a single constraint $g(x, y) = 0$, consider $h(x, y) = f(x, y) + \lambda g(x, y)$ and apply $h_x = 0, h_y = 0$, together with $g(x, y) = 0$, to solve for x, y and λ.
   - General procedure for $f(x_i)$ with $i = 1, 2, \ldots, N$ subject to constraints $g_j(x_i) = 0$ with $j = 1, 2, \ldots, M$ and $M < N$: form $h(x_i) = f(x_i) + \sum_j \lambda_j g_j(x_i)$ and then solve $\partial h/\partial x_i = 0$, together with $g_j(x_i) = 0$, for the $N + M$ quantities x_i and λ_j.

4. *Envelopes*
   The family of curves in the x–y plane given by $f(x, y, \alpha) = 0$, where α is a parameter, has an envelope given by eliminating α between the two equations

   $$f(x, y, \alpha) = 0 \quad \text{and} \quad \frac{\partial}{\partial \alpha} f(x, y, \alpha) = 0.$$

5. *Differentiating integrals* (Leibnitz's rule)
   - For fixed limits $\dfrac{d}{dx}\displaystyle\int_u^v f(x,t)\,dt = \int_u^v \dfrac{\partial f(x,t)}{\partial x}\,dt.$
   - For x-dependent limits $u = u(x)$, $v = v(x)$,
   $$\frac{d}{dx}\int_u^v f(x,t)\,dt = f(x,v(x))\frac{dv}{dx} - f(x,u(x))\frac{du}{dx} + \int_{u(x)}^{v(x)} \frac{\partial f(x,t)}{\partial x}\,dt.$$

PROBLEMS

7.1. Using the appropriate properties of ordinary derivatives, perform the following.
  (a) Find all the first partial derivatives of the following functions $f(x, y)$:
     (i) $x^2 y$, (ii) $x^2 + y^2 + 4$, (iii) $\sin(x/y)$, (iv) $\tan^{-1}(y/x)$,
     (v) $r(x, y, z) = (x^2 + y^2 + z^2)^{1/2}$.
  (b) For (i), (ii) and (v), find $\partial^2 f/\partial x^2$, $\partial^2 f/\partial y^2$ and $\partial^2 f/\partial x \partial y$.
  (c) For (iv) verify that $\partial^2 f/\partial x \partial y = \partial^2 f/\partial y \partial x$.

7.2. Determine which of the following are exact differentials:
  (a) $(3x + 2)y\,dx + x(x + 1)\,dy$;
  (b) $y \tan x\,dx + x \tan y\,dy$;
  (c) $y^2(\ln x + 1)\,dx + 2xy \ln x\,dy$;
  (d) $y^2(\ln x + 1)\,dy + 2xy \ln x\,dx$;
  (e) $[x/(x^2 + y^2)]\,dy - [y/(x^2 + y^2)]\,dx$.

7.3. Show that the differential
$$df = x^2\,dy - (y^2 + xy)\,dx$$
is not exact, but that $dg = (xy^2)^{-1}df$ is exact.

7.4. Show that
$$df = y(1 + x - x^2)\,dx + x(x + 1)\,dy$$
is not an exact differential.
  Find the differential equation that a function $g(x)$ must satisfy if $d\phi = g(x)df$ is to be an exact differential. Verify that $g(x) = e^{-x}$ is a solution of this equation and deduce the form of $\phi(x, y)$.

7.5. The equation $3y = z^3 + 3xz$ defines z implicitly as a function of x and y. Evaluate all three second partial derivatives of z with respect to x and/or y. Verify that z is a solution of
$$x\frac{\partial^2 z}{\partial y^2} + \frac{\partial^2 z}{\partial x^2} = 0.$$

Problems

7.6. A possible equation of state for a gas takes the form

$$PV = RT \exp\left(-\frac{\alpha}{VRT}\right),$$

in which α and R are constants. Calculate expressions for

$$\left(\frac{\partial P}{\partial V}\right)_T, \quad \left(\frac{\partial V}{\partial T}\right)_P, \quad \left(\frac{\partial T}{\partial P}\right)_V$$

and show that their product is -1, as stated in Section 7.4.

7.7. The function $G(t)$ is defined by

$$G(t) = F(x, y) = x^2 + y^2 + 3xy,$$

where $x(t) = at^2$ and $y(t) = 2at$. Use the chain rule to find the values of (x, y) at which $G(t)$ has stationary values as a function of t. Do any of them correspond to the stationary points of $F(x, y)$ as a function of x and y?

7.8. In the x–y plane, new coordinates s and t are defined by

$$s = \tfrac{1}{2}(x + y), \qquad t = \tfrac{1}{2}(x - y).$$

Transform the equation

$$\frac{\partial^2 \phi}{\partial x^2} - \frac{\partial^2 \phi}{\partial y^2} = 0$$

into the new coordinates and deduce that its general solution can be written

$$\phi(x, y) = f(x + y) + g(x - y),$$

where $f(u)$ and $g(v)$ are arbitrary functions of u and v, respectively.

7.9. The function $f(x, y)$ satisfies the differential equation

$$y\frac{\partial f}{\partial x} + x\frac{\partial f}{\partial y} = 0.$$

By changing to new variables $u = x^2 - y^2$ and $v = 2xy$, show that f is, in fact, a function of $x^2 - y^2$ only.

7.10. If $x = e^u \cos\theta$ and $y = e^u \sin\theta$, show that

$$\frac{\partial^2 \phi}{\partial u^2} + \frac{\partial^2 \phi}{\partial \theta^2} = (x^2 + y^2)\left(\frac{\partial^2 f}{\partial x^2} + \frac{\partial^2 f}{\partial y^2}\right),$$

where $f(x, y) = \phi(u, \theta)$.

7.11. Find and evaluate the maxima, minima and saddle points of the function

$$f(x, y) = xy(x^2 + y^2 - 1).$$

7.12. Show that
$$f(x, y) = x^3 - 12xy + 48x + by^2, \qquad b \neq 0,$$
has two, one or zero stationary points, according to whether $|b|$ is less than, equal to or greater than 3.

7.13. Locate the stationary points of the function
$$f(x, y) = (x^2 - 2y^2)\exp[-(x^2 + y^2)/a^2],$$
where a is a non-zero constant.
  Sketch the function along the x- and y-axes and hence identify the nature and values of the stationary points.

7.14. Find the stationary points of the function
$$f(x, y) = x^3 + xy^2 - 12x - y^2$$
and identify their natures.

7.15. Find the stationary values of
$$f(x, y) = 4x^2 + 4y^2 + x^4 - 6x^2y^2 + y^4$$
and classify them as maxima, minima or saddle points. Make a rough sketch of the contours of f in the quarter plane $x, y \geq 0$.

7.16. The temperature of a point (x, y, z) in or on the unit sphere is given by
$$T(x, y, z) = 1 + xy + yz.$$
By using the method of Lagrange multipliers, find the temperature of the hottest point on the sphere.

7.17. A rectangular parallelepiped has all eight vertices on the ellipsoid
$$x^2 + 3y^2 + 3z^2 = 1.$$
Using the symmetry of the parallelepiped about each of the planes $x = 0$, $y = 0$, $z = 0$, write down the surface area of the parallelepiped in terms of the coordinates of the vertex that lies in the octant $x, y, z \geq 0$. Hence find the maximum value of the surface area of such a parallelepiped.

7.18. Two horizontal corridors, $0 \leq x \leq a$ with $y \geq 0$, and $0 \leq y \leq b$ with $x \geq 0$, meet at right angles. Find the length L of the longest ladder (considered as a stick) that may be carried horizontally around the corner.

7.19. A barn is to be constructed with a uniform cross-sectional area A throughout its length. The cross-section is to be a rectangle of wall height h (fixed) and width w, surmounted by an isosceles triangular roof that makes an angle θ with the horizontal. The cost of construction is α per unit height of wall and β per unit

Problems

(slope) length of roof. Show that, irrespective of the values of α and β, to minimise costs w should be chosen to satisfy the equation

$$w^4 = 16A(A - wh)$$

and θ made such that $2\tan 2\theta = w/h$.

7.20. Show that the envelope of all concentric ellipses that have their axes along the x- and y-coordinate axes, and that have the sum of their semi-axes equal to a constant L, is the same curve (an astroid) as that found in the worked example in Section 7.10.

7.21. Find the area of the region covered by points on the lines

$$\frac{x}{a} + \frac{y}{b} = 1,$$

where the sum of any line's intercepts on the coordinate axes is fixed and equal to c.

7.22. Prove that the envelope of the circles whose diameters are those chords of a given circle that pass through a fixed point on its circumference is the cardioid

$$r = a(1 + \cos\theta).$$

Here a is the radius of the given circle and (r, θ) are the polar coordinates of the envelope. Take as the system parameter the angle ϕ between a chord and the polar axis from which θ is measured.

7.23. A water feature contains a spray head at water level at the centre of a round basin. The head is in the form of a small hemisphere perforated by many evenly distributed small holes, through which water spurts out at the same speed, v_0, in all directions.
(a) What is the shape of the 'water bell' so formed?
(b) What must be the minimum diameter of the bowl if no water is to be lost?

7.24. In order to make a focusing mirror that concentrates parallel axial rays to one spot (or conversely forms a parallel beam from a point source), a parabolic shape should be adopted. If a mirror that is part of a circular cylinder or sphere were used, the light would be spread out along a curve. This curve is known as a *caustic* and is the envelope of the rays reflected from the mirror. Denoting by θ the angle which a typical incident axial ray makes with the normal to the mirror at the place where it is reflected, the geometry of reflection (the angle of incidence equals the angle of reflection) is shown in Figure 7.4.

Show that a parametric specification of the caustic is

$$x = R\cos\theta\left(\tfrac{1}{2} + \sin^2\theta\right), \qquad y = R\sin^3\theta,$$

where R is the radius of curvature of the mirror. The curve is, in fact, part of an epicycloid.

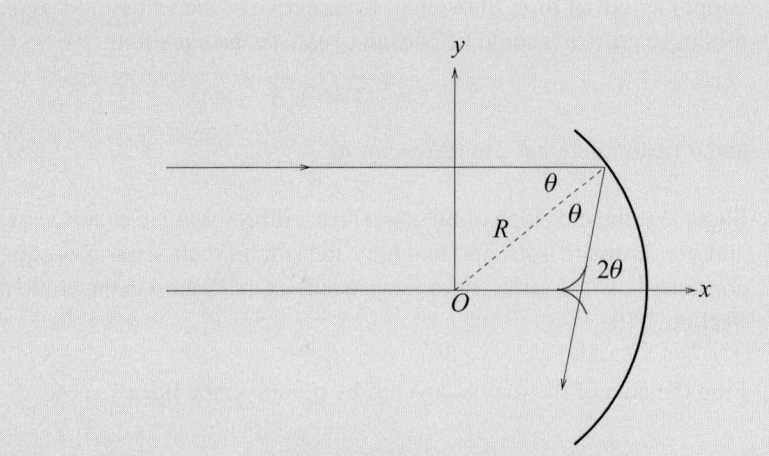

Figure 7.4 The reflecting mirror discussed in Problem 7.24.

7.25. By considering the differential
$$dG = d(U + PV - ST),$$
where G is the Gibbs free energy, P the pressure, V the volume, S the entropy and T the temperature of a system, and given further that the internal energy U satisfies
$$dU = T\,dS - P\,dV,$$
derive a Maxwell relation connecting $(\partial V/\partial T)_P$ and $(\partial S/\partial P)_T$.

7.26. Functions $P(V, T)$, $U(V, T)$ and $S(V, T)$ are related by
$$T\,dS = dU + P\,dV,$$
where the symbols have the same meaning as in the previous question. The pressure P is known from experiment to have the form
$$P = \frac{T^4}{3} + \frac{T}{V},$$
in appropriate units. If
$$U = \alpha V T^4 + \beta T,$$
where α and β are constants (or, at least, do not depend on T or V), deduce that α must have a specific value, but that β may have any value. Find the corresponding form of S.

7.27. As in the previous two problems on the thermodynamics of a simple gas, the quantity $dS = T^{-1}(dU + P\,dV)$ is an exact differential. Use this to prove that
$$\left(\frac{\partial U}{\partial V}\right)_T = T\left(\frac{\partial P}{\partial T}\right)_V - P.$$

Problems

In the van der Waals model of a gas, P obeys the equation

$$P = \frac{RT}{V-b} - \frac{a}{V^2},$$

where R, a and b are constants. Further, in the limit $V \to \infty$, the form of U becomes $U = cT$, where c is another constant. Find the complete expression for $U(V, T)$.

7.28. The entropy $S(H, T)$, the magnetisation $M(H, T)$ and the internal energy $U(H, T)$ of a magnetic salt placed in a magnetic field of strength H, at temperature T, are connected by the equation

$$T\, dS = dU - H\, dM.$$

By considering $d(U - TS - HM)$ prove that

$$\left(\frac{\partial M}{\partial T}\right)_H = \left(\frac{\partial S}{\partial H}\right)_T.$$

For a particular salt,

$$M(H, T) = M_0[1 - \exp(-\alpha H/T)].$$

Show that if, at a fixed temperature, the applied field is increased from zero to a strength such that the magnetisation of the salt is $\tfrac{3}{4}M_0$, then the salt's entropy *decreases* by an amount

$$\frac{M_0}{4\alpha}(3 - \ln 4).$$

7.29. Using the results of Section 7.12, evaluate the integral

$$I(y) = \int_0^\infty \frac{e^{-xy} \sin x}{x}\, dx.$$

Hence show that

$$J = \int_0^\infty \frac{\sin x}{x}\, dx = \frac{\pi}{2}.$$

7.30. The integral

$$\int_{-\infty}^\infty e^{-\alpha x^2}\, dx$$

has the value $(\pi/\alpha)^{1/2}$. Use this result to evaluate

$$J(n) = \int_{-\infty}^\infty x^{2n} e^{-x^2}\, dx,$$

where n is a positive integer. Express your answer in terms of factorials.

Partial differentiation

7.31. The function $f(x)$ is differentiable and $f(0) = 0$. A second function $g(y)$ is defined by

$$g(y) = \int_0^y \frac{f(x)\,dx}{\sqrt{y-x}}.$$

Prove that

$$\frac{dg}{dy} = \int_0^y \frac{df}{dx} \frac{dx}{\sqrt{y-x}}.$$

For the case $f(x) = x^n$, prove that

$$\frac{d^n g}{dy^n} = 2(n!)\sqrt{y}.$$

7.32. The functions $f(x,t)$ and $F(x)$ are defined by

$$f(x,t) = e^{-xt},$$
$$F(x) = \int_0^x f(x,t)\,dt.$$

Verify, by explicit calculation, that

$$\frac{dF}{dx} = f(x,x) + \int_0^x \frac{\partial f(x,t)}{\partial x}\,dt.$$

7.33. If

$$I(\alpha) = \int_0^1 \frac{x^\alpha - 1}{\ln x}\,dx, \qquad \alpha > -1,$$

what is the value of $I(0)$? Show that

$$\frac{d}{d\alpha} x^\alpha = x^\alpha \ln x$$

and deduce that

$$\frac{d}{d\alpha} I(\alpha) = \frac{1}{\alpha + 1}.$$

Hence prove that $I(\alpha) = \ln(1 + \alpha)$.

7.34. Find the derivative, with respect to x, of the integral

$$I(x) = \int_x^{3x} \exp xt\,dt.$$

7.35. The function $G(t, \xi)$ is defined for $0 \leq t \leq \pi$ by

$$G(t, \xi) = \begin{cases} -\cos t \sin \xi & \text{for } \xi \leq t, \\ -\sin t \cos \xi & \text{for } \xi > t. \end{cases}$$

Show that the function $x(t)$ defined by

$$x(t) = \int_0^\pi G(t,\xi) f(\xi)\, d\xi$$

satisfies the equation

$$\frac{d^2 x}{dt^2} + x = f(t),$$

where $f(t)$ can be *any* arbitrary (continuous) function. Show further that $x(0) = [dx/dt]_{t=\pi} = 0$, again for any $f(t)$, but that the *value* of $x(\pi)$ does depend upon the form of $f(t)$.

[The function $G(t, \xi)$ is an example of a Green's function, an important concept in the solution of differential equations.]

HINTS AND ANSWERS

7.1. (a) (i) $2xy$, x^2; (ii) $2x$, $2y$; (iii) $y^{-1}\cos(x/y)$, $(-x/y^2)\cos(x/y)$; (iv) $-y/(x^2+y^2)$, $x/(x^2+y^2)$; (v) x/r, y/r, z/r.
(b) (i) $2y$, 0, $2x$; (ii) 2, 2, 0; (v) $(y^2+z^2)r^{-3}$, $(x^2+z^2)r^{-3}$, $-xyr^{-3}$.
(c) Both second derivatives are equal to $(y^2-x^2)(x^2+y^2)^{-2}$.

7.3. $2x \ne -2y - x$. For g, both sides of Equation (7.9) equal y^{-2}.

7.5. $\partial^2 z/\partial x^2 = 2xz(z^2+x)^{-3}$, $\partial^2 z/\partial x\partial y = (z^2-x)(z^2+x)^{-3}$, $\partial^2 z/\partial y^2 = -2z(z^2+x)^{-3}$.

7.7. $(0,0)$, $(a/4, -a)$ and $(16a, -8a)$. Only the saddle point at $(0,0)$.

7.9. The transformed equation is $2(x^2+y^2)\partial f/\partial v = 0$; hence f does not depend on v.

7.11. Maxima, equal to $1/8$, at $\pm(1/2, -1/2)$, minima, equal to $-1/8$, at $\pm(1/2, 1/2)$, saddle points, equalling 0, at $(0,0)$, $(0, \pm 1)$, $(\pm 1, 0)$.

7.13. Maxima equal to $a^2 e^{-1}$ at $(\pm a, 0)$, minima equal to $-2a^2 e^{-1}$ at $(0, \pm a)$, saddle point equalling 0 at $(0,0)$.

7.15. Minimum at $(0,0)$; saddle points at $(\pm 1, \pm 1)$. To help with sketching the contours, determine the behaviour of $g(x) = f(x,x)$.

7.17. The Lagrange multiplier method gives $z = y = x/2$, for a maximal area of 4.

7.19. The cost always includes $2\alpha h$, which can therefore be ignored in the optimisation. With Lagrange multiplier λ, $\sin\theta = \lambda w/(4\beta)$ and $\beta \sec\theta - \frac{1}{2}\lambda w\tan\theta = \lambda h$, leading to the stated results.

7.21. The envelope of the lines $x/a + y/(c-a) - 1 = 0$, as a is varied, is $\sqrt{x} + \sqrt{y} = \sqrt{c}$. Area $= c^2/6$.

Partial differentiation

7.23. (a) Using $\alpha = \cot\theta$, where θ is the initial angle a jet makes with the vertical, the equation is $f(z, \rho, \alpha) = z - \rho\alpha + [g\rho^2(1+\alpha^2)/(2v_0^2)]$, and setting $\partial f/\partial \alpha = 0$ gives $\alpha = v_0^2/(g\rho)$. The water bell has a parabolic profile $z = v_0^2/(2g) - g\rho^2/(2v_0^2)$.
(b) Setting $z = 0$ gives the minimum diameter as $2v_0^2/g$.

7.25. Show that $(\partial G/\partial P)_T = V$ and $(\partial G/\partial T)_P = -S$. From each result obtain an expression for $\partial^2 G/\partial T \partial P$ and equate these, giving $(\partial V/\partial T)_P = -(\partial S/\partial P)_T$.

7.27. Find expressions for $(\partial S/\partial V)_T$ and $(\partial S/\partial T)_V$, and equate $\partial^2 S/\partial V \partial T$ with $\partial^2 S/\partial T \partial V$. $U(V, T) = cT - aV^{-1}$.

7.29. $dI/dy = -\text{Im}[\int_0^\infty \exp(-xy + ix)\,dx] = -1/(1+y^2)$. Integrate dI/dy from 0 to ∞. $I(\infty) = 0$ and $I(0) = J$.

7.31. Integrate the RHS of the equation by parts before differentiating with respect to y. Repeated application of the method establishes the result for all orders of derivative.

7.33. $I(0) = 0$; use Leibnitz's rule.

7.35. Write $x(t) = -\cos t \int_0^t \sin\xi\, f(\xi)\,d\xi - \sin t \int_t^\pi \cos\xi\, f(\xi)\,d\xi$ and differentiate each term as a product to obtain dx/dt. Obtain d^2x/dt^2 in a similar way. Note that integrals that have equal lower and upper limits have value zero. The value of $x(\pi)$ is $\int_0^\pi \sin\xi\, f(\xi)\,d\xi$.

8 Multiple integrals

Just as functions of several variables may be differentiated with respect to two or more of them, so may their integrals with respect to more than one variable be formed. The formal definitions of such multiple integrals are extensions of that given for a single variable in Chapter 3. In this chapter, we first discuss double and triple integrals and illustrate some of their applications. We then consider how to change the variables in multiple integrals and, finally, discuss some general properties of Jacobians.

8.1 Double integrals

For an integral involving two variables – a double integral – we have a function, $f(x, y)$ say, to be integrated with respect to x and y between certain limits. These limits can usually be represented by a closed curve C bounding a region R in the xy-plane. Following the discussion of single integrals given in Chapter 3, let us divide the region R into N subregions ΔR_p of area ΔA_p, $p = 1, 2, \ldots, N$, and let (x_p, y_p) be any point in subregion ΔR_p. Now consider the sum

$$S = \sum_{p=1}^{N} f(x_p, y_p) \Delta A_p,$$

and let $N \to \infty$ as each of the areas $\Delta A_p \to 0$. If the sum S tends to a unique limit, I, then this is called the *double integral of $f(x, y)$ over the region R* and is written

$$I = \int_R f(x, y) \, dA, \tag{8.1}$$

where dA stands for the element of area in the xy-plane. By choosing the subregions to be small rectangles each of area $\Delta A = \Delta x \Delta y$, and letting both Δx and $\Delta y \to 0$, we can also write the integral as

$$I = \int \int_R f(x, y) \, dx \, dy, \tag{8.2}$$

where we have written out the element of area explicitly as the product of the two coordinate differentials (see Figure 8.1).

Some authors use a single integration symbol whatever the dimension of the integral; others use as many symbols as the dimension. In different circumstances both have their advantages. We will adopt the convention used in (8.1) and (8.2), that as many integration symbols will be used as differentials *explicitly* written.

Multiple integrals

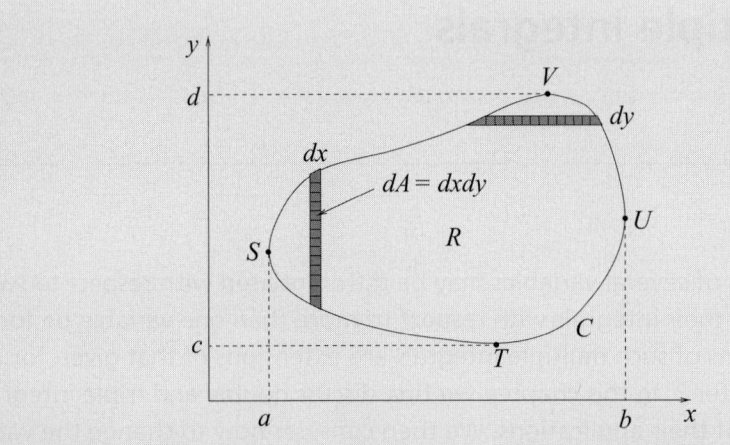

Figure 8.1 A simple curve C in the xy-plane, enclosing a region R.

The form (8.2) gives us a clue as to how we may proceed in the evaluation of a double integral. Referring to Figure 8.1, the limits on the integration may be written as an equation $c(x, y) = 0$ giving the boundary curve C. However, an explicit statement of the limits can be written in two distinct ways.

One way of evaluating the integral is first to sum up the contributions from the small rectangular elemental areas in a horizontal strip of width dy (as shown in the figure) and then to combine the contributions of these horizontal strips to cover the region R. In this case, we write

$$I = \int_{y=c}^{y=d} \left\{ \int_{x=x_1(y)}^{x=x_2(y)} f(x, y)\, dx \right\} dy, \tag{8.3}$$

where $x = x_1(y)$ and $x = x_2(y)$ are the equations of the curves TSV and TUV respectively. This expression indicates that first $f(x, y)$ is to be integrated with respect to x (treating y as a constant) between the values $x = x_1(y)$ and $x = x_2(y)$ and then the result, considered as a function of y, is to be integrated between the limits $y = c$ and $y = d$. Thus the double integral is evaluated by expressing it in terms of two single integrals called *iterated* (or *repeated*) integrals.

An alternative way of evaluating the integral, however, is first to sum up the contributions from the elemental rectangles arranged into *vertical* strips and then to combine these vertical strips to cover the region R. We then write

$$I = \int_{x=a}^{x=b} \left\{ \int_{y=y_1(x)}^{y=y_2(x)} f(x, y)\, dy \right\} dx, \tag{8.4}$$

where $y = y_1(x)$ and $y = y_2(x)$ are the equations of the curves STU and SVU respectively. In going to (8.4) from (8.3), we have essentially interchanged the order of integration.

In the discussion above we assumed that the curve C was such that any line parallel to either the x- or y-axis intersected C at most twice. In general, provided $f(x, y)$ is continuous everywhere in R and the boundary curve C has this simple shape, the same result is obtained irrespective of the order of integration. In cases where the region R has

8.1 Double integrals

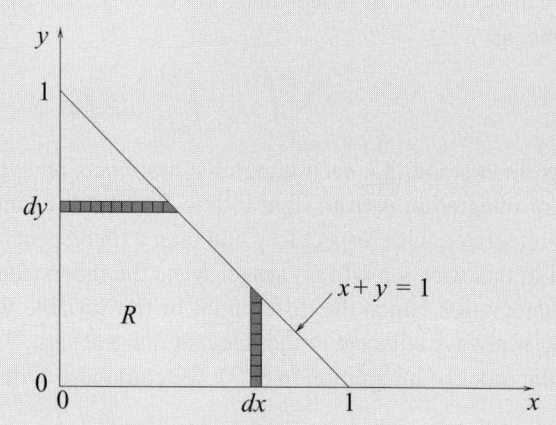

Figure 8.2 The triangular region whose sides are the axes $x = 0$, $y = 0$ and the line $x + y = 1$.

a more complicated shape, it can usually be subdivided into smaller simpler regions R_1, R_2 etc. that satisfy this criterion. The double integral over R is then merely the sum of the double integrals over the subregions. Our first worked example is an integration over a simple 'convex' region and no such subdivision is needed.

Example Evaluate the double integral

$$I = \int\int_R x^2 y \, dx \, dy,$$

where R is the triangular area bounded by the lines $x = 0$, $y = 0$ and $x + y = 1$. Reverse the order of integration and demonstrate that the same result is obtained.

The area of integration is shown in Figure 8.2. Suppose we choose to carry out the integration with respect to y first. With x fixed, the range of y is 0 to $1 - x$. We can therefore write

$$I = \int_{x=0}^{x=1} \left\{ \int_{y=0}^{y=1-x} x^2 y \, dy \right\} dx$$

$$= \int_{x=0}^{x=1} \left[\frac{x^2 y^2}{2} \right]_{y=0}^{y=1-x} dx = \int_0^1 \frac{x^2(1-x)^2}{2} dx = \frac{1}{60}.$$

Alternatively, we may choose to perform the integration with respect to x first. With y fixed, the range of x is 0 to $1 - y$, so we have

$$I = \int_{y=0}^{y=1} \left\{ \int_{x=0}^{x=1-y} x^2 y \, dx \right\} dy$$

$$= \int_{y=0}^{y=1} \left[\frac{x^3 y}{3} \right]_{x=0}^{x=1-y} dx = \int_0^1 \frac{(1-y)^3 y}{3} dy = \frac{1}{60}.$$

As expected, we obtain the same result irrespective of the order of integration.[1] ◀

---

[1] Note that $\int_0^1 x^n \, dx = 1/(n+1)$. For practice, write down directly the results of the final x and y integrations, each as the sum of a number of fractions. Verify that each sum does total to $1/60$.

Multiple integrals

We may avoid the use of braces in expressions such as (8.3) and (8.4) by writing (8.4), for example, as

$$I = \int_a^b dx \int_{y_1(x)}^{y_2(x)} dy\, f(x, y),$$

where it is understood that each integral symbol acts on everything to its right, and that the order of integration is from right to left. So, in this example, the integrand $f(x, y)$ is first to be integrated with respect to y and then with respect to x. With the double integral expressed in this way, we will no longer write the independent variables explicitly in the limits of integration, since the differential of the variable with respect to which we are integrating is always adjacent to the relevant integral sign.

Using the order of integration in (8.3), we could also write the double integral as

$$I = \int_c^d dy \int_{x_1(y)}^{x_2(y)} dx\, f(x, y).$$

Occasionally, however, interchange of the order of integration in a double integral is not permissible, as it yields a different result. For example, difficulties might arise if the region R were unbounded with some of the limits infinite, though in many cases involving infinite limits the same result is obtained whichever order of integration is used. Difficulties can also occur if the integrand $f(x, y)$ has any discontinuities in the region R or on its boundary C.

The above discussion for double integrals can easily be extended to triple integrals. Consider the function $f(x, y, z)$ defined in a closed three-dimensional region R. Proceeding as we did for double integrals, let us divide the region R into N subregions ΔR_p of volume ΔV_p, $p = 1, 2, \ldots, N$, and let (x_p, y_p, z_p) be any point in the subregion ΔR_p. Now we form the sum

$$S = \sum_{p=1}^{N} f(x_p, y_p, z_p) \Delta V_p,$$

and let $N \to \infty$ as each of the volumes $\Delta V_p \to 0$. If the sum S tends to a unique limit, I, then this is called the *triple integral of $f(x, y, z)$ over the region R* and is written

$$I = \int_R f(x, y, z)\, dV, \qquad (8.5)$$

where dV stands for the element of volume. By choosing the subregions to be small cuboids, each of volume $\Delta V = \Delta x \Delta y \Delta z$, and proceeding to the limit, we can also write the integral as

$$I = \int\int\int_R f(x, y, z)\, dx\, dy\, dz, \qquad (8.6)$$

where we have written out the element of volume explicitly as the product of the three coordinate differentials. Extending the notation used for double integrals, we may write triple integrals as three iterated integrals, for example,

$$I = \int_{x_1}^{x_2} dx \int_{y_1(x)}^{y_2(x)} dy \int_{z_1(x,y)}^{z_2(x,y)} dz\, f(x, y, z),$$

8.2 Applications of multiple integrals

where the limits on each of the integrals describe the values that x, y and z take on the boundary of the region R. As for double integrals, in most cases the order of integration does not affect the value of the integral.

We can extend these ideas to define multiple integrals of higher dimensionality in a similar way.

EXERCISES 8.1

1. Evaluate the integral $\int_R x^2 y^2 \, dA$ over the region R bounded by the circle $x^2 + y^2 = a^2$. You will need the results of Problem 4.14 to evaluate the final integral.

2. Show that, as expected, the value of the double integral

$$\int_R (x^2 + y^2) \sin x \cos y \, dA,$$

where R is the region $0 \leq x \leq \pi/2$, $0 \leq y \leq \pi/2$, is independent of the order of the integrations. [A number of careful integrations by parts will be needed to obtain the common value of $\frac{1}{4}\pi^2 + \pi - 4$.]

3. Evaluate the volume integral

$$I = \int z(x^2 + y^2) \, dV$$

over the parallelepiped $0 \leq x \leq a$, $0 \leq y \leq b$, $0 \leq z \leq c$.

4. Evaluate the integral

$$I = \int (x^2 + y^2) \, dV$$

over the volume bounded by the planes $x = 0$, $y = 0$, $z = 0$ and $x + y + z = 1$. Before each integration, simplify the integrand as far as possible.

8.2 Applications of multiple integrals

Multiple integrals have many uses in the physical sciences, since there are numerous physical quantities which can be written in terms of them. Many of these quantities have a physical reality that can be seen or measured in our ordinary three-dimensional space, such as the volume or mass of a body, but some of the most important are abstract quantities with no physical existence. For example, in quantum theory, most calculations of the possible and expected outcomes of physical measurements take the form of triple integrals over all space. We now discuss a few of the more common physical examples; for examples taken from quantum physics, see Problems 8.6 and 8.7 at the end of this chapter.

Multiple integrals

8.2.1 Areas and volumes

Multiple integrals are often used to find areas and volumes. For example, the integral

$$A = \int_R dA = \int\int_R dx\, dy$$

is simply equal to the area of the region R. Similarly, if we consider the surface $z = f(x, y)$ in three-dimensional Cartesian coordinates then the volume under this surface that stands vertically above the region R is given by the integral

$$V = \int_R z\, dA = \int\int_R f(x, y)\, dx\, dy,$$

where volumes above the xy-plane are counted as positive and those below as negative. To illustrate this approach, consider the following example.

Example Find the volume of the tetrahedron bounded by the three coordinate surfaces $x = 0$, $y = 0$ and $z = 0$ and the plane $x/a + y/b + z/c = 1$.

Referring to Figure 8.3, the elemental volume of the shaded region is given by $dV = z\, dx\, dy$, and we must integrate over the triangular region R in the xy-plane whose sides are $x = 0$, $y = 0$ and $y = b - bx/a$. The total volume of the tetrahedron is therefore given by

$$V = \int\int_R z\, dx\, dy = \int_0^a dx \int_0^{b-bx/a} dy\, c\left(1 - \frac{y}{b} - \frac{x}{a}\right)$$

$$= c \int_0^a dx \left[y - \frac{y^2}{2b} - \frac{xy}{a} \right]_{y=0}^{y=b-bx/a}$$

$$= c \int_0^a dx \left(\frac{bx^2}{2a^2} - \frac{bx}{a} + \frac{b}{2} \right)$$

$$= c \left[\frac{bx^3}{6a^2} - \frac{bx^2}{2a} + \frac{bx}{2} \right]_0^a = \frac{abc}{6}.$$

As expected, this result is symmetrical in a, b and c and $\to 0$ if any one of them does so.[2]

Alternatively, and a little more generally, we can write the volume of a three-dimensional region R as

$$V = \int_R dV = \int\int\int_R dx\, dy\, dz, \qquad (8.7)$$

where the only difficulty arises in setting the correct limits on each of the integrals. For the above example, writing the volume in this way corresponds to dividing the tetrahedron into elemental boxes of volume $dx\, dy\, dz$ (as shown in Figure 8.3); integration over z then adds up the boxes to form the shaded column in the figure. The limits of integration are $z = 0$ to $z = c(1 - y/b - x/a)$, and the total volume of the tetrahedron is given by

$$V = \int_0^a dx \int_0^{b-bx/a} dy \int_0^{c(1-y/b-x/a)} dz. \qquad (8.8)$$

---

[2] Deduce that a regular octahedron occupies less than a third of the volume of the smallest sphere that contains it.

8.2 Applications of multiple integrals

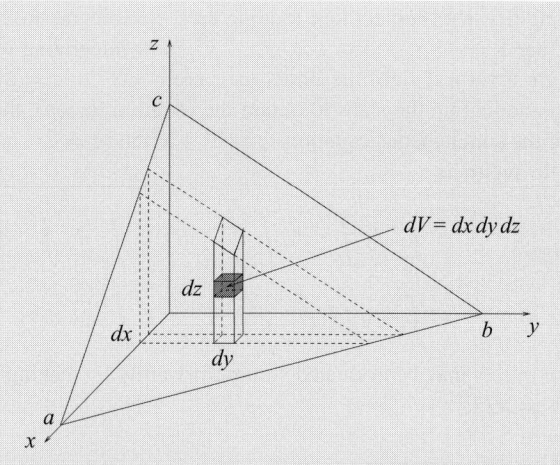

Figure 8.3 The tetrahedron bounded by the coordinate surfaces and the plane $x/a + y/b + z/c = 1$ is divided up into vertical slabs, the slabs into columns and the columns into small boxes.

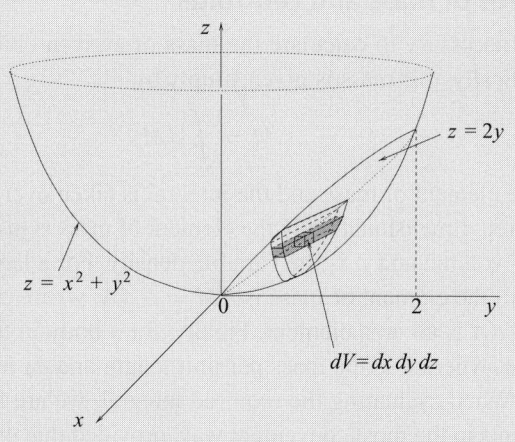

Figure 8.4 The region bounded by the paraboloid $z = x^2 + y^2$ and the plane $z = 2y$ is divided into vertical slabs, the slabs into horizontal strips and the strips into boxes.

After the initial z-integration, this calculation is exactly the same as the previous one and clearly gives the same result. However, it does show that the integrations could have been done in any order.

The approach specified by (8.7) is illustrated further in the following example.

Example Find the volume of the region bounded by the paraboloid $z = x^2 + y^2$ and the plane $z = 2y$.

The required region is shown in Figure 8.4. In order to write the volume of the region in the form (8.7), we must deduce the limits on each of the integrals. Since the integrations can be performed in any order, let us first divide the region into vertical slabs of thickness dy perpendicular to the y-axis and then, as shown in the figure, we cut each slab into horizontal strips of height dz and each strip into elemental boxes of volume $dV = dx\,dy\,dz$.

Multiple integrals

Integrating first with respect to x (adding up the elemental boxes to get a horizontal strip), the limits on x are $x = -\sqrt{z - y^2}$ to $x = \sqrt{z - y^2}$. Now integrating with respect to z (adding up the strips to form a vertical slab) the limits on z are $z = y^2$ to $z = 2y$. Finally, for integration with respect to y (adding up the slabs to obtain the required region), the limits are those given by the solutions of the simultaneous equations $z = 0^2 + y^2$ and $z = 2y$, namely $y = 0$ and $y = 2$. So the volume of the region is

$$V = \int_0^2 dy \int_{y^2}^{2y} dz \int_{-\sqrt{z-y^2}}^{\sqrt{z-y^2}} dx = \int_0^2 dy \int_{y^2}^{2y} dz \, 2\sqrt{z - y^2}$$

$$= \int_0^2 dy \, [\tfrac{4}{3}(z - y^2)^{3/2}]_{z=y^2}^{z=2y} = \int_0^2 dy \, \tfrac{4}{3}(2y - y^2)^{3/2}.$$

The integral over y may be evaluated straightforwardly by making the substitution $y = 1 + \sin u$, and gives $V = \pi/2$.[3] ◀

In general, when calculating the volume (area) of a region, the volume (area) elements need not be small boxes as in the previous example, but may be of any convenient shape. The latter is usually chosen so as to make the evaluation of the integral as simple as possible.

8.2.2 Masses, centres of mass and centroids

It is sometimes necessary to calculate the mass of a given object having a non-uniform density. Symbolically, this mass is given simply by

$$M = \int dM,$$

where dM is the element of mass and the integral is taken over the extent of the object.

For a solid three-dimensional body the element of mass is just $dM = \rho \, dV$, where dV is an element of volume and ρ is the variable density. For a laminar body (i.e. a uniform sheet of material) the element of mass is $dM = \sigma \, dA$, where σ is the mass per unit area of the body and dA is an area element. Finally, for a body in the form of a thin wire we have $dM = \lambda \, ds$, where λ is the mass per unit length and ds is an element of arc length along the wire. When evaluating the required integral, we are free to divide up the body into mass elements in the most convenient way, provided that over each mass element the density is approximately constant.

Example Find the mass of the tetrahedron bounded by the three coordinate surfaces and the plane $x/a + y/b + z/c = 1$, if its density is given by $\rho(x, y, z) = \rho_0(1 + x/a)$.

From (8.8), we can immediately write down the mass of the tetrahedron as

$$M = \int_R \rho_0 \left(1 + \frac{x}{a}\right) dV = \int_0^a dx \, \rho_0 \left(1 + \frac{x}{a}\right) \int_0^{b-bx/a} dy \int_0^{c(1-y/b-x/a)} dz,$$

---

[3] Show generally that integrals of the form $\int_a^b [(y-a)(b-y)]^{n/2} \, dy$ can be evaluated by setting $y = \tfrac{1}{2}(a+b) + \tfrac{1}{2}(b-a) \sin u$ and have the value $2[(b-a)/2]^{n+1} J(n+1, 0)$, where $J(n, m)$ are the integrals derived in Problem 4.14.

8.2 Applications of multiple integrals

where we have taken the density outside the integrations with respect to z and y since it depends only on x. Therefore, the integrations with respect to z and y proceed exactly as they did when finding the volume of the tetrahedron, and we have

$$M = c\rho_0 \int_0^a dx \left(1 + \frac{x}{a}\right) \left(\frac{bx^2}{2a^2} - \frac{bx}{a} + \frac{b}{2}\right). \tag{8.9}$$

We could have arrived at (8.9) more directly by dividing the tetrahedron into slabs of thickness dx perpendicular to the x-axis (see Figure 8.3), each of which is of constant density, since ρ depends on x alone. A triangular slab at position x has a height of $c(1 - x/a)$ and a base of length $b(1 - x/a)$. Its surface area is $\frac{1}{2}$base$\times$height, and therefore the slab has volume $dV = \frac{1}{2}c(1 - x/a)b(1 - x/a)\,dx$ and mass $dM = \rho\,dV = \rho_0(1 + x/a)\,dV$. Integrating over x we again obtain (8.9). This integral is easily evaluated[4] and gives $M = \frac{5}{24}abc\rho_0$. ◀

The coordinates of the centre of mass of a solid or laminar body may also be written in terms of multiple integrals. The centre of mass of a body has coordinates $\bar{x}$, $\bar{y}$, $\bar{z}$ given by the three equations

$$\bar{x}\int dM = \int x\,dM, \quad \bar{y}\int dM = \int y\,dM, \quad \bar{z}\int dM = \int z\,dM,$$

where again dM is an element of mass as described above, x, y, z are the coordinates of the centre of mass of the element dM and the integrals are taken over the entire body. Obviously, for any body that lies entirely in, or is symmetrical about, the xy-plane (say), we immediately have $\bar{z} = 0$. For completeness, we note that the three equations above can be written as the single vector equation (see Chapter 9)

$$\bar{\mathbf{r}} = \frac{1}{M}\int \mathbf{r}\,dM,$$

where $\bar{\mathbf{r}}$ is the position vector of the body's centre of mass with respect to the origin, $\mathbf{r}$ is the position vector of the centre of mass of the element dM and $M = \int dM$ is the total mass of the body. As previously, we may divide the body into the most convenient mass elements for evaluating the necessary integrals, provided each mass element is of constant density.

We further note that the coordinates of the *centroid* of a body are defined as those of its centre of mass if the body had uniform density.

Example Find the centre of mass of the solid hemisphere bounded by the surfaces $x^2 + y^2 + z^2 = a^2$ and the xy-plane, assuming that it has a uniform density ρ.

Referring to Figure 8.5, we know from symmetry that the centre of mass must lie on the z-axis. Let us divide the hemisphere into volume elements that are circular slabs of thickness dz parallel to the xy-plane. A slab at a height z has a radius of $\sqrt{a^2 - z^2}$, and hence the mass of the element is $dM = \rho\,dV = \rho\pi(a^2 - z^2)\,dz$. Integrating over z, we find that the z-coordinate of the centre of

---

[4] See footnote 1.

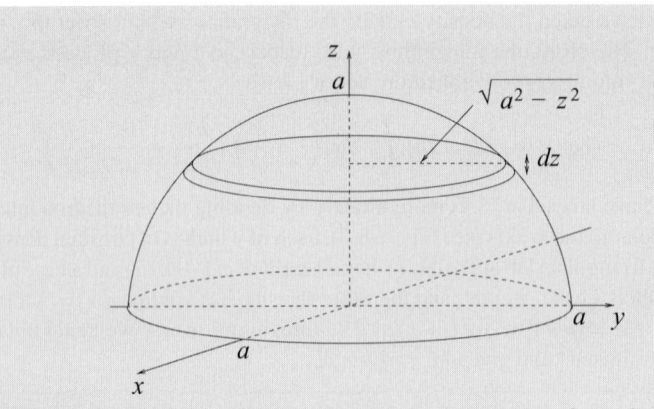

Figure 8.5 The solid hemisphere bounded by the surfaces $x^2 + y^2 + z^2 = a^2$ and the xy-plane.

mass of the hemisphere is given by

$$\bar{z} \int_0^a \rho\pi(a^2 - z^2)\,dz = \int_0^a z\rho\pi(a^2 - z^2)\,dz.$$

The integrals are easily evaluated and give $\bar{z} = 3a/8$. Since the hemisphere is of uniform density, this is also the position of its centroid.

8.2.3 Pappus's theorems

The theorems of Pappus[5] relate centroids to the volumes and areas of surfaces of revolution, as discussed in Chapter 3, and may be useful for finding one quantity given another that can be calculated more easily.

If a plane area is rotated about an axis that does not intersect it then the solid so generated is called a *volume of revolution*. Pappus's first theorem states that the volume of such a solid is given by the plane area A multiplied by the distance moved by its centroid (see Figure 8.6). This may be proved by considering the definition of the centroid of the plane area as the position of the centre of mass if the density is uniform, so that

$$\bar{y} = \frac{1}{A}\int y\,dA.$$

Now the volume generated by rotating the plane area about the x-axis is given by

$$V = \int 2\pi y\,dA = 2\pi\bar{y}A,$$

which is the area multiplied by the distance moved by the centroid.

Pappus's second theorem states that if a plane curve is rotated about a coplanar axis that does not intersect it then the area of the *surface of revolution* so generated is given by the length of the curve L multiplied by the distance moved by its centroid (see Figure 8.7).

---

[5] Which are about 17 centuries old.

8.2 Applications of multiple integrals

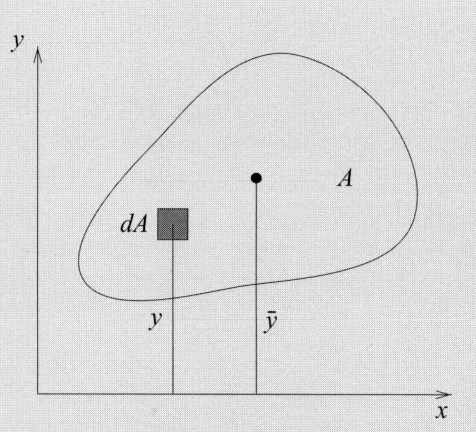

Figure 8.6 An area A in the xy-plane, which may be rotated about the x-axis to form a volume of revolution.

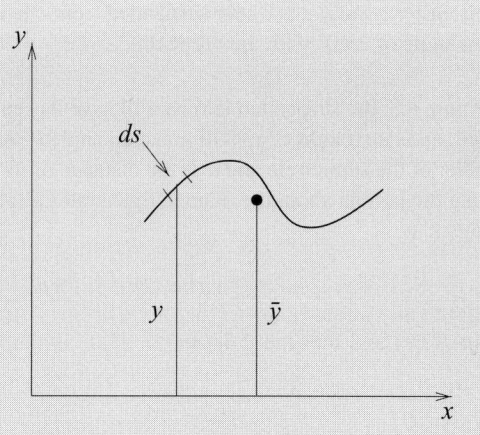

Figure 8.7 A curve in the xy-plane, which may be rotated about the x-axis to form a surface of revolution.

This may be proved in a similar manner to the first theorem by considering the definition of the centroid of a plane curve,

$$\bar{y} = \frac{1}{L} \int y \, ds,$$

and noting that the surface area generated is given by

$$S = \int 2\pi y \, ds = 2\pi \bar{y} L,$$

which is equal to the length of the curve multiplied by the distance moved by its centroid.[6]

---

[6] Note that the centroid of the curve will not, in general, lie on the curve.

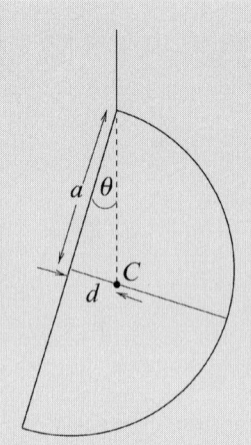

Figure 8.8 Suspending a semicircular lamina from one of its corners.

Example A semicircular uniform lamina is freely suspended from one of its corners. Show that its straight edge makes an angle of 23.0° with the vertical.

Referring to Figure 8.8, the suspended lamina will have its centre of gravity C vertically below the suspension point and its straight edge will make an angle $\theta = \tan^{-1}(d/a)$ with the vertical, where $2a$ is the diameter of the semicircle and d is the distance of its centre of mass from the diameter.

Since rotating the lamina about the diameter generates a sphere of volume $\frac{4}{3}\pi a^3$, Pappus's first theorem requires that

$$\tfrac{4}{3}\pi a^3 = 2\pi d \times \tfrac{1}{2}\pi a^2.$$

Hence $d = (4a)/(3\pi)$ and $\theta = \tan^{-1} 4/(3\pi) = 23.0°$ is the angle the diameter makes with the vertical.[7]

8.2.4 Moments of inertia

For problems in rotational mechanics it is often necessary to calculate the moment of inertia of a body about a given axis. This is defined by the multiple integral

$$I = \int l^2 \, dM,$$

where l is the distance of a mass element dM from the axis. We may again choose our mass elements so that they are as convenient as possible for evaluating the integral. In this case, however, we require that each part of any particular element, in addition to having an essentially constant density, is at approximately the same distance from the axis about which the moment of inertia is required.

---

[7] Show that if a semicircular uniform wire were suspended in the same way, the corresponding angle would be 32.5°.

8.2 Applications of multiple integrals

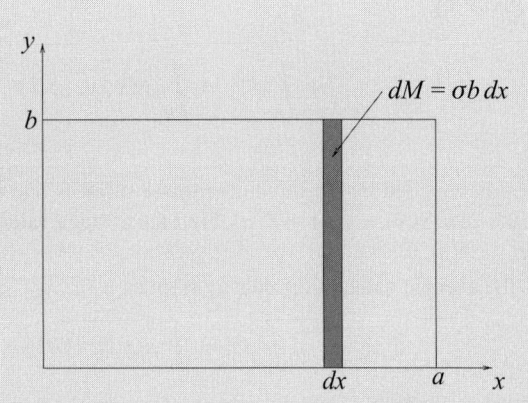

Figure 8.9 A uniform rectangular lamina of mass M with sides a and b can be divided into vertical strips.

Example Find the moment of inertia of a uniform rectangular lamina of mass M with sides a and b about one of the sides of length b.

Referring to Figure 8.9, we wish to calculate the moment of inertia about the y-axis. We therefore divide the rectangular lamina into elemental strips parallel to the y-axis of width dx. The mass of such a strip is $dM = \sigma b\, dx$, where σ is the mass per unit area of the lamina. The moment of inertia of a strip at a distance x from the y-axis is simply $dI = x^2\, dM = \sigma b x^2\, dx$. The total moment of inertia of the lamina about the y-axis is therefore[8]

$$I = \int_0^a \sigma b x^2\, dx = \frac{\sigma b a^3}{3}.$$

Since the total mass of the lamina is $M = \sigma a b$, we can write $I = \frac{1}{3}Ma^2$.

8.2.5 Mean values of functions

In Chapter 3 we discussed average values for functions of a single variable. This can be extended to functions of several variables. Let us consider, for example, a function $f(x, y)$ defined in some region R of the xy-plane. Then the average value $\bar{f}$ of the function is given by

$$\bar{f} \int_R dA = \int_R f(x, y)\, dA. \qquad (8.10)$$

This definition is easily extended to three (and higher) dimensions; if a function $f(x, y, z)$ is defined in some three-dimensional region of space R then the average value $\bar{f}$ of the

---

[8] What would the moment of inertia be about an axis parallel to the y-axis, but passing through the centre of the lamina rather than its edge?

function is given by

$$\bar{f} \int_R dV = \int_R f(x, y, z)\, dV. \tag{8.11}$$

Example A tetrahedron is bounded by the three coordinate surfaces and the plane $x/a + y/b + z/c = 1$ and has density $\rho(x, y, z) = \rho_0(1 + x/a)$. Find the average value of the density.

From (8.11), the average value of the density is given by

$$\bar{\rho} \int_R dV = \int_R \rho(x, y, z)\, dV.$$

Now the integral on the LHS is just the volume of the tetrahedron, which we found in Section 8.2.1 to be $V = \frac{1}{6}abc$, and the integral on the RHS is its mass $M = \frac{5}{24}abc\rho_0$, calculated in Section 8.2.2. Therefore $\bar{\rho} = M/V = \frac{5}{4}\rho_0$. ◀

EXERCISES 8.2

1. Find the area of the region bounded by the parabolas
$$y^2 = 4a(b-x) \quad \text{and} \quad y^2 = 4b(a+x), \quad \text{with } a, b > 0.$$

2. Write down a triple integral that gives the volume V of a circular dish that has straight sloping sides, is of depth c, and has lower and upper radii of a and b respectively. Carry out the first two integrations by inspection and hence show that $V = \frac{1}{3}\pi c(a^2 + ab + b^2)$.

3. A solid wooden block of uniform density has five plane faces. Its rectangular base has sides of lengths a and b and its apex, at height h, stands directly above one corner of the base; the block therefore has vertical planes for two of its faces. Taking these faces to be the planes $x = 0$ and $y = 0$, and the base to be the plane $z = 0$, show that the centre of mass of the block is the point $(\frac{3}{8}a, \frac{3}{8}b, \frac{1}{4}h)$.

4. A solid sphere of radius a has a density that varies linearly along one of its diameters, being $\rho_0(1 - \alpha)$ at one end of the diameter and $\rho_0(1 + \alpha)$ at the other. It stands freely, in its equilibrium position, on a table.
   (a) Taking the origin of coordinates at the centre of the sphere and the z-axis vertical, write down an expression giving the local density $\rho(z)$.
   (b) Find the mass of the sphere. Explain why it does not depend upon α.
   (c) Show that the centre of mass of the sphere is $\frac{1}{5}\alpha a$ below its centroid.

5. A curling stone (without its handle) is azimuthally symmetric and has a vertical cross-section consisting of a rectangle of length $2c$ and height $2a$, together with two semicircles, each of radius a, adjoining its vertical sides.

(a) Use a result from the worked example in Section 8.2.3 to show that the volume of the stone is

$$V = \frac{\pi a}{3}(6c^2 + 3\pi ac + 4a^2).$$

(b) Find an expression for its total surface area.

6. Calculate the moment of inertia of a uniform square lamina of side a and mass M (a) about an axis through its centre parallel to one of its sides and (b) about one of its main diagonals. Explain by reference to the note in Problem 8.10 why these two results must have the relationship they do; deduce the moment of inertia of the lamina about an axis perpendicular to its plane and passing through its centre.

7. Find the average value of $x^2 y$ over the semicircle $x^2 + y^2 \leq a^2$ with $y \geq 0$.

8. The function $f(r, \theta)$ is given in spherical polar coordinates by

$$f(r, \theta) = \frac{3\cos^2\theta - 1}{r} e^{-r/a}.$$

Find the average value of f^2 over the sphere $r \leq a$.

8.3 Change of variables in multiple integrals

It often happens that, either because of the form of the integrand involved or because of the boundary shape of the region of integration, it is desirable to express a multiple integral in terms of a new set of variables. We now consider how to do this.

8.3.1 Change of variables in double integrals

Let us begin by examining the change of variables in a double integral. Suppose that we require to change an integral

$$I = \int\int_R f(x, y)\, dx\, dy,$$

in terms of coordinates x and y, into one expressed in new coordinates u and v, given in terms of x and y by differentiable equations $u = u(x, y)$ and $v = v(x, y)$ with inverses $x = x(u, v)$ and $y = y(u, v)$. The region R in the xy-plane and the curve C that bounds it will become a new region R' and a new boundary C' in the uv-plane, and so we must change the limits of integration accordingly. Also, the function $f(x, y)$ becomes a new function $g(u, v)$ of the new coordinates.

Now the part of the integral that requires most consideration is the area element. In the xy-plane the element is the rectangular area $dA_{xy} = dx\, dy$ generated by constructing a grid of straight lines parallel to the x- and y-axes respectively. Our task is to determine the corresponding area element in the uv-coordinates. In general, the corresponding element dA_{uv} will not be the same shape as dA_{xy}, but this does not matter since all elements are infinitesimally small and the value of the integrand is considered constant over them; it is only their relative sizes that matter.

Multiple integrals

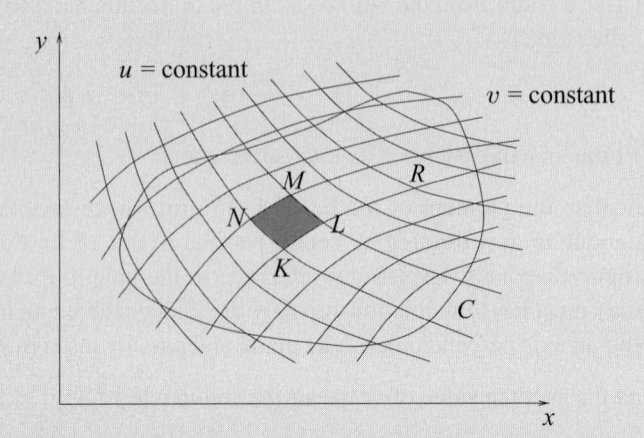

Figure 8.10 A region of integration R overlaid with a grid formed by the family of curves $u = $ constant and $v = $ constant. The parallelogram $KLMN$ defines the area element dA_{uv}.

Since the sides of the area element are infinitesimal, dA_{uv} will in general have the shape of a parallelogram. We can find the connection between dA_{xy} and dA_{uv} by considering the grid formed by the family of curves $u = $ constant and $v = $ constant, as shown in Figure 8.10. Since v is constant along the line element KL, the latter has components $(\partial x/\partial u)\, du$ and $(\partial y/\partial u)\, du$ in the directions of the x- and y-axes respectively. Similarly, since u is constant along the line element KN, the latter has corresponding components $(\partial x/\partial v)\, dv$ and $(\partial y/\partial v)\, dv$.

Anticipating the result for the area of a parallelogram given by Equations (9.28) and (9.33) in Chapter 9, we find that the area of the parallelogram $KLMN$ is

$$dA_{uv} = \left| \frac{\partial x}{\partial u}\, du\, \frac{\partial y}{\partial v}\, dv - \frac{\partial x}{\partial v}\, dv\, \frac{\partial y}{\partial u}\, du \right|$$

$$= \left| \frac{\partial x}{\partial u}\frac{\partial y}{\partial v} - \frac{\partial x}{\partial v}\frac{\partial y}{\partial u} \right| du\, dv.$$

The particular combination of partial differential coefficients appearing between the modulus signs in the final line is given a special name and called the *Jacobian* of x, y with respect to u, v. It is represented symbolically by a partial derivative that has two arguments in both the numerator and the denominator:

$$J = \frac{\partial(x, y)}{\partial(u, v)} \equiv \frac{\partial x}{\partial u}\frac{\partial y}{\partial v} - \frac{\partial x}{\partial v}\frac{\partial y}{\partial u}.$$

Clearly, the sign of J changes if either x and y, or u and v, are interchanged.

For our particular application to infinitesimal areas we have

$$dA_{uv} = \left| \frac{\partial(x, y)}{\partial(u, v)} \right| du\, dv.$$

8.3 Change of variables in multiple integrals

The reader acquainted with determinants will notice that the Jacobian can also be written as the 2 × 2 determinant

$$J = \frac{\partial(x, y)}{\partial(u, v)} = \begin{vmatrix} \dfrac{\partial x}{\partial u} & \dfrac{\partial y}{\partial u} \\ \dfrac{\partial x}{\partial v} & \dfrac{\partial y}{\partial v} \end{vmatrix}.$$

Such determinants can be evaluated using the methods of Chapter 10.

So, in summary, the relationship between the size of the area element generated by dx, dy and the size of the corresponding area element generated by du, dv is[9]

$$dx\,dy = \left|\frac{\partial(x, y)}{\partial(u, v)}\right| du\,dv. \tag{8.12}$$

This equality should be taken as meaning that, when transforming from coordinates x, y to coordinates u, v, the area element $dx\,dy$ should be replaced by the expression on the RHS of the above equality.[10] Of course, the Jacobian can, and in general will, vary over the region of integration. We may express the double integral in either coordinate system as

$$I = \iint_R f(x, y)\,dx\,dy = \iint_{R'} g(u, v) \left|\frac{\partial(x, y)}{\partial(u, v)}\right| du\,dv. \tag{8.13}$$

When evaluating the integral in the new coordinate system, it is usually advisable to sketch the region of integration R' in the uv-plane.

When evaluating double integrals, the decision as to whether or not to make a change to new variables u and v is a matter of experience, but it is seldom worth doing unless some, and preferably all, parts of the new boundary C' of R' can be described by setting either u or v equal to a constant. Usually, some parts of the boundary will correspond to constant values for u and other parts to constant values for v. In the following worked example it might seem that all boundaries except $\rho = a$ have disappeared, but this is not so, as is explained in the solution.

Example Evaluate the double integral

$$I = \iint_R \left(a + \sqrt{x^2 + y^2}\right) dx\,dy,$$

where R is the region bounded by the circle $x^2 + y^2 = a^2$.

In Cartesian coordinates, the integral may be written

$$I = \int_{-a}^{a} dx \int_{-\sqrt{a^2 - x^2}}^{\sqrt{a^2 - x^2}} dy \left(a + \sqrt{x^2 + y^2}\right),$$

---

[9] Note that the two sets of vertical bars used do not have the same meaning. Those in the definition of a Jacobian indicate a determinant, those placed around the Jacobian symbol indicate that its modulus must be taken.
[10] Note that u and v do not have to have the same physical dimensions as x and y, and can have different dimensions from each other. The Jacobian *automatically* has the dimensions needed to make the physical dimensions of the RHS of Equation (8.12) the same as those on the LHS.

and can be calculated directly. However, because of the circular boundary of the integration region, a change of variables to plane polar coordinates ρ, ϕ is indicated. The relationship between Cartesian and plane polar coordinates is given by $x = \rho \cos \phi$ and $y = \rho \sin \phi$. Using (8.13) we can therefore write

$$I = \int\int_{R'} (a+\rho) \left| \frac{\partial(x,y)}{\partial(\rho,\phi)} \right| d\rho\, d\phi,$$

where R' is the rectangular region in the $\rho\phi$-plane whose sides are $\rho = 0$, $\rho = a$, $\phi = 0$ and $\phi = 2\pi$. These are the four boundary values required to cover the circular region once and only once, from its centre to its perimeter, and for one complete turn around its centre.

The partial derivatives of x and y needed for the Jacobian can be read off immediately and we obtain

$$J = \frac{\partial(x,y)}{\partial(\rho,\phi)} = \begin{vmatrix} \cos\phi & \sin\phi \\ -\rho\sin\phi & \rho\cos\phi \end{vmatrix} = \rho(\cos^2\phi + \sin^2\phi) = \rho.$$

So the relationship between the area elements in Cartesian and in plane polar coordinates is

$$dx\, dy = \rho\, d\rho\, d\phi.$$

Therefore, when expressed in plane polar coordinates, the integral is given by

$$I = \int\int_{R'} (a+\rho)\rho\, d\rho\, d\phi$$

$$= \int_0^{2\pi} d\phi \int_0^a d\rho\, (a+\rho)\rho = 2\pi \left[\frac{a\rho^2}{2} + \frac{\rho^3}{3} \right]_0^a = \frac{5\pi a^3}{3}.$$

We note, in passing, that a 'dimensional check' on the calculated answer is satisfactory; if we think of x, y and a as 'lengths', then the original integral has the dimensions of L^3 – one power for each of dx, dy and the linear integrand, – and so has the final answer, as the only dimensional quantity it contains is a^3. ◀

8.3.2 Evaluation of the integral $I = \int_{-\infty}^{\infty} e^{-x^2}\, dx$

By making a judicious change of variables, it is sometimes possible to evaluate an integral that would be intractable otherwise. An important example of this method is provided by the evaluation of the integral

$$I = \int_{-\infty}^{\infty} e^{-x^2}\, dx.$$

This integral is central to the normalisation of the Gaussian (or normal) distribution that figures prominently in probability theory, statistical mechanics and kinetic theory.

Its value may be found by first constructing I^2, as follows:

$$I^2 = \int_{-\infty}^{\infty} e^{-x^2}\, dx \int_{-\infty}^{\infty} e^{-y^2}\, dy = \int_{-\infty}^{\infty} dx \int_{-\infty}^{\infty} dy\, e^{-(x^2+y^2)}$$

$$= \int\int_R e^{-(x^2+y^2)}\, dx\, dy,$$

where the region R is the whole xy-plane. The presence of the factor $x^2 + y^2$ then indicates that a change to plane polar coordinates would be beneficial. From the previous

8.3 Change of variables in multiple integrals

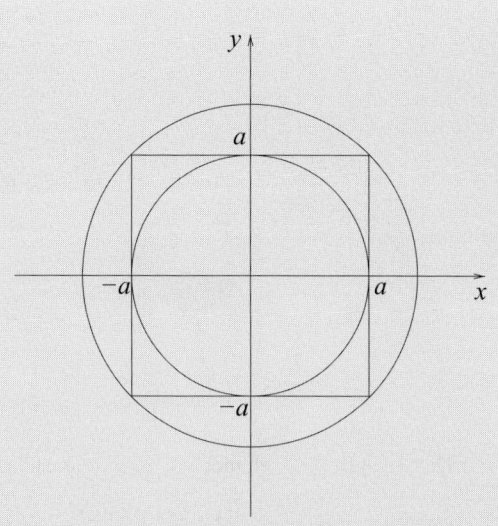

Figure 8.11 The regions used to illustrate the convergence properties of the integral $I(a) = \int_{-a}^{a} e^{-x^2} \, dx$ as $a \to \infty$.

worked example, we already know that the Jacobian for such a change is $J = \rho$ and that $dx\,dy = \rho\,d\rho\,d\phi$, and so making the change we find that

$$I^2 = \int\int_{R'} e^{-\rho^2} \rho\, d\rho\, d\phi = \int_0^{2\pi} d\phi \int_0^\infty d\rho\, \rho e^{-\rho^2} = 2\pi \left[-\tfrac{1}{2} e^{-\rho^2} \right]_0^\infty = \pi.$$

It follows that the original integral is given by $I = \sqrt{\pi}$, and that, since the integrand is an even function of x, the value of the integral from 0 to ∞ is simply $\sqrt{\pi}/2$.

We note, however, that, unlike in all the previous examples, the regions of integration R and R' are both infinite in extent (i.e. unbounded). It is therefore prudent to derive this result again using a more rigorous method; this we do by considering the same integral, but between finite limits $\pm a$, denoting it by $I(a)$:

$$I(a) = \int_{-a}^{a} e^{-x^2} \, dx.$$

Clearly $I = \lim_{a \to \infty} I(a)$. Now, using the same initial approach as previously, we have

$$I^2(a) = \int\int_R e^{-(x^2+y^2)} \, dx\, dy,$$

where R is the square of side $2a$ centred on the origin, and shown in Figure 8.11.

Since the integrand is everywhere positive, the value of the integral taken over the square R lies between two other values of the same integral, one taken over an inner circular region of radius a and the other taken over the region bounded by the outer circle of radius $\sqrt{2}a$; both circles are shown in the figure. Because of their circular boundaries, we may evaluate the integrals over the inner and outer circles explicitly by transforming to plane polar coordinates.[11]

---

[11] Since the regions involved are both finite, the doubts raised by the previous method do not need to be considered.

Multiple integrals

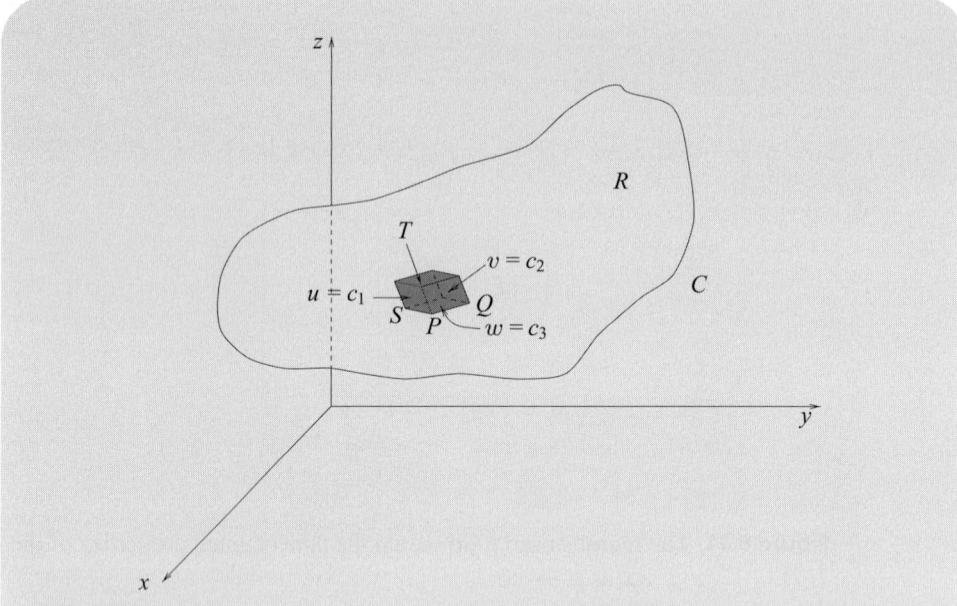

Figure 8.12 A three-dimensional region of integration R, showing an element of volume in u, v, w coordinates formed by the coordinate surfaces $u =$ constant, $v =$ constant, $w =$ constant.

The same integration as previously, but with finite upper limits $\rho = a$ and $\rho = \sqrt{2}a$, respectively, shows that

$$\pi\left(1 - e^{-a^2}\right) < I^2(a) < \pi\left(1 - e^{-2a^2}\right).$$

We may now take the limit $a \to \infty$, and, as both negative exponentials appearing above $\to 0$, we find that $\pi \leq \lim_{a \to \infty} I^2(a) \leq \pi$ and so conclude that $\lim_{a \to \infty} I^2(a) = \pi$. It follows that $I = \sqrt{\pi}$, as we found previously. Substituting $x = \sqrt{\alpha} y$ into this result shows that

$$\int_{-\infty}^{\infty} e^{-\alpha x^2} \, dx = \sqrt{\frac{\pi}{\alpha}}. \qquad (8.14)$$

As indicated earlier, this result has direct application in the study of probability, where it is used to give the correct normalisation of the normal (Gaussian) distribution.

8.3.3 Change of variables in triple integrals

A change of variable in a triple integral follows the same general lines as that for a double integral. Suppose we wish to change variables from x, y, z to u, v, w. In x, y, z coordinates the element of volume is a cuboid of sides dx, dy, dz and volume $dV_{xyz} = dx\,dy\,dz$. If, however, we divide up the total volume into infinitesimal elements by constructing a grid formed from the coordinate surfaces $u =$ constant, $v =$ constant and $w =$ constant, then the element of volume dV_{uvw} in the new coordinates will have the shape of a parallelepiped whose faces are the coordinate surfaces and whose edges are the curves formed by the intersections of these surfaces (see Figure 8.12). Along the line element PQ the

8.3 Change of variables in multiple integrals

coordinates v and w are constant, and so PQ has components $(\partial x/\partial u)\,du$, $(\partial y/\partial u)\,du$ and $(\partial z/\partial u)\,du$ in the directions of the x-, y- and z-axes respectively. The corresponding components of the line elements PS and ST are found by replacing u by v and w respectively.

The expression for the volume of a parallelepiped in terms of the components of its edges with respect to the x-, y- and z-axes is given on p. 347 of Chapter 9. Using this, we find that the element of volume in u, v, w coordinates is given by

$$dV_{uvw} = \left|\frac{\partial(x,y,z)}{\partial(u,v,w)}\right| du\,dv\,dw,$$

where the Jacobian of x, y, z with respect to u, v, w is a shorthand for a 3×3 determinant:

$$\frac{\partial(x,y,z)}{\partial(u,v,w)} \equiv \begin{vmatrix} \dfrac{\partial x}{\partial u} & \dfrac{\partial y}{\partial u} & \dfrac{\partial z}{\partial u} \\ \dfrac{\partial x}{\partial v} & \dfrac{\partial y}{\partial v} & \dfrac{\partial z}{\partial v} \\ \dfrac{\partial x}{\partial w} & \dfrac{\partial y}{\partial w} & \dfrac{\partial z}{\partial w} \end{vmatrix}.$$

So, in summary, the relationship between the elemental volumes in multiple integrals formulated in the two coordinate systems is given in Jacobian form by[12]

$$dx\,dy\,dz = \left|\frac{\partial(x,y,z)}{\partial(u,v,w)}\right| du\,dv\,dw,$$

and we can write a triple integral in either set of coordinates as

$$I = \iiint_R f(x,y,z)\,dx\,dy\,dz = \iiint_{R'} g(u,v,w) \left|\frac{\partial(x,y,z)}{\partial(u,v,w)}\right| du\,dv\,dw.$$

This is illustrated in the following example.

Example Find an expression for a volume element in spherical polar coordinates, and hence calculate the moment of inertia about a diameter of a uniform sphere of radius a and mass M.

Spherical polar coordinates r, θ, ϕ are defined by

$$x = r\sin\theta\cos\phi, \quad y = r\sin\theta\sin\phi, \quad z = r\cos\theta$$

(and are discussed fully in Chapter 11). The required Jacobian is therefore

$$J = \frac{\partial(x,y,z)}{\partial(r,\theta,\phi)} = \begin{vmatrix} \sin\theta\cos\phi & \sin\theta\sin\phi & \cos\theta \\ r\cos\theta\cos\phi & r\cos\theta\sin\phi & -r\sin\theta \\ -r\sin\theta\sin\phi & r\sin\theta\cos\phi & 0 \end{vmatrix}.$$

---

12 See footnote 9.

The determinant is most easily evaluated by expanding it with respect to the last column (see Chapter 10), which gives

$$J = \cos\theta[r^2 \sin\theta \cos\theta(\cos^2\phi + \sin^2\phi)] + r\sin\theta[r\sin^2\theta(\cos^2\phi + \sin^2\phi)]$$
$$= r^2 \sin\theta(\cos^2\theta + \sin^2\theta) = r^2 \sin\theta.$$

Therefore, the volume element in spherical polar coordinates is given by

$$dV = \frac{\partial(x, y, z)}{\partial(r, \theta, \phi)} dr\, d\theta\, d\phi = r^2 \sin\theta\, dr\, d\theta\, d\phi,$$

which agrees with the result given in Chapter 11.

If we place the sphere with its centre at the origin of an x, y, z coordinate system, then its moment of inertia about the z-axis (which is, of course, a diameter of the sphere) is

$$I = \int (x^2 + y^2)\, dM = \rho \int (x^2 + y^2)\, dV,$$

where the integral is taken over the sphere, and ρ is the (constant) density. Using spherical polar coordinates, we can write this as

$$I = \rho \int\int\int_V (r^2 \sin^2\theta)\, r^2 \sin\theta\, dr\, d\theta\, d\phi$$
$$= \rho \int_0^{2\pi} d\phi \int_0^\pi d\theta \sin^3\theta \int_0^a dr\, r^4$$
$$= \rho \times 2\pi \times \tfrac{4}{3} \times \tfrac{1}{5}a^5 = \tfrac{8}{15}\pi a^5 \rho.$$

Since the mass of the sphere is $M = \tfrac{4}{3}\pi a^3 \rho$, the moment of inertia can also be written as $I = \tfrac{2}{5}Ma^2$.

◀

8.3.4 General properties of Jacobians

Although we will not prove it, the general result for a change of coordinates in an n-dimensional integral from a set x_i to a set y_j (where i and j both run from 1 to n) is

$$dx_1\, dx_2 \cdots dx_n = \left|\frac{\partial(x_1, x_2, \ldots, x_n)}{\partial(y_1, y_2, \ldots, y_n)}\right| dy_1\, dy_2 \cdots dy_n,$$

where the n-dimensional Jacobian can be written as an $n \times n$ determinant (see Chapter 10) in an analogous way to the two- and three-dimensional cases.

For readers who already have sufficient familiarity with matrices (see Chapter 10) and their properties, a fairly compact proof of some useful general properties of Jacobians can be given as follows. Other readers should turn straight to the results (8.18) and (8.19) and return to the proof at some later time.

Consider three sets of variables x_i, y_i and z_i, with i running from 1 to n for each set. From the chain rule in partial differentiation [see (7.17)], we know that

$$\frac{\partial x_i}{\partial z_j} = \sum_{k=1}^n \frac{\partial x_i}{\partial y_k}\frac{\partial y_k}{\partial z_j}. \tag{8.15}$$

8.3 Change of variables in multiple integrals

Now let **A**, **B** and **C** be the matrices whose ijth elements are $\partial x_i/\partial y_j$, $\partial y_i/\partial z_j$ and $\partial x_i/\partial z_j$ respectively. We can then write (8.15) as the matrix product

$$c_{ij} = \sum_{k=1}^{n} a_{ik}b_{kj} \quad \text{or} \quad \mathbf{C} = \mathbf{AB}. \tag{8.16}$$

We may now use the general result for the determinant of the product of two matrices, namely $|\mathbf{AB}| = |\mathbf{A}||\mathbf{B}|$, and recall that the Jacobian

$$J_{xy} = \frac{\partial(x_1, \ldots, x_n)}{\partial(y_1, \ldots, y_n)} = |\mathbf{A}|, \tag{8.17}$$

and similarly for J_{yz} and J_{xz}. On taking the determinant of (8.16), we therefore obtain

$$J_{xz} = J_{xy}J_{yz}$$

or, in the usual notation,

$$\frac{\partial(x_1, \ldots, x_n)}{\partial(z_1, \ldots, z_n)} = \frac{\partial(x_1, \ldots, x_n)}{\partial(y_1, \ldots, y_n)} \frac{\partial(y_1, \ldots, y_n)}{\partial(z_1, \ldots, z_n)}. \tag{8.18}$$

As a special case, if the set z_i is taken to be identical to the set x_i, and the obvious result $J_{xx} = |I_n| = 1$ is used,[13] we obtain

$$J_{xy}J_{yx} = 1$$

or, in the usual notation,

$$\frac{\partial(x_1, \ldots, x_n)}{\partial(y_1, \ldots, y_n)} = \left[\frac{\partial(y_1, \ldots, y_n)}{\partial(x_1, \ldots, x_n)}\right]^{-1}. \tag{8.19}$$

The similarity between the properties of Jacobians and those of derivatives is apparent, and to some extent is suggested by the notation. We further note from (8.17) that since $|\mathbf{A}| = |\mathbf{A}^T|$, where $\mathbf{A}^T$ is the transpose of $\mathbf{A}$, we can interchange the rows and columns in the determinantal form of the Jacobian without changing its value.

EXERCISES 8.3

1. A uniform right circular cylinder has radius a and height $2h$.
   (a) Calculate its moment of inertia about its axis of symmetry.
   (b) Write down an integral expression for its moment of inertia about an axis that passes through the cylinder's centre and is perpendicular to the axis of symmetry. Convert this to an integral expressed in cylindrical polar coordinates and hence evaluate it.
   (c) How must its height be chosen if the cylinder is to have equal moments of inertia about three mutually perpendicular axes passing through its centre?
   [In fact, a cylinder of this particular height has the same moment of inertia about *any* axis passing through its centre – but you are not asked to prove this.]

---

[13] I_n is the $n \times n$ unit matrix with each entry on the leading diagonal equal to 1 and with all other entries equal to 0.

Multiple integrals

2. (a) Make a sketch of the family of parabolic curves given by $y^2 = 4u(u-x)$ and $y^2 = 4v(v+x)$, and identify the values of u and v that correspond to the x-axis.
   (b) Express x and y in terms of u and v and so calculate the Jacobian $\dfrac{\partial(x, y)}{\partial(u, v)}$.
   (c) Calculate the area bounded by the x-axis and the two parabolas $y^2 = 4a(a-x)$ and $y^2 = 4b(b+x)$.

3. Make a change of integration variable, $x \to u + \gamma$, for a choice of γ such that result (8.14) can be used, to evaluate

$$J(\beta) = \int_{-\infty}^{\infty} e^{-\alpha x^2 + \beta x} \, dx,$$

where α is real and > 0, and β is real. Show that whatever the sign of β, $J(\beta) \geq J(0)$ and indicate on a sketch graph why this should be so.

4. Using results stated or derived in the main text, determine the Jacobian $\dfrac{\partial(r, \theta, \phi)}{\partial(\rho, \phi, z)}$ for a change of coordinates from spherical to cylindrical polars, expressing your answer in terms of each set of coordinates.

SUMMARY

The value of a multiple integral is independent of the order in which the integrations are carried out, though the difficulty of finding it may not be.

1. *Areas, volumes and masses*

$$A = \int\int dx \, dy, \quad V = \int\int\int dx \, dy \, dz, \quad M = \int dM.$$

2. *Average values*
   - Centre of gravity: $\bar{x} = \dfrac{\int x \, dM}{\int dM}$, and similarly for $\bar{y}$ and $\bar{z}$.
   - Mean value of $f(x_i) = \dfrac{\int\int\cdots\int f(x_i) \, dx_1 \, dx_2 \ldots dx_n}{\int\int\cdots\int dx_1 \, dx_2 \ldots dx_n}$.

3. *Pappus's theorems*
   For volumes or areas of revolution formed by rotating a plane area or plane line segment, respectively, about an axis that *does not* intersect it.
   (i) The volume of revolution = plane area × the distance moved by its centroid.
   (ii) The area of revolution = segment length × the distance moved by its centroid.

4. *Change of variables*
   - The integral $I = \int\int \cdots \int_R f(x_1, x_2, \ldots, x_n)\, dx_1\, dx_2 \ldots dx_n$ can be written as
   $\int\int \cdots \int_{R'} g(y_1, y_2, \ldots, y_n)\, dy_1\, dy_2 \ldots dy_n$, where the volume elements are related by $dx_1\, dx_2 \ldots dx_n = |J_{xy}|\, dy_1\, dy_2 \ldots dy_n$, and the Jacobian J_{xy} is given by the determinant

   $$J_{xy} \equiv \frac{\partial(x_1, x_2, \ldots, x_n)}{\partial(y_1, y_2, \ldots, y_n)} \equiv \begin{vmatrix} \frac{\partial x_1}{\partial y_1} & \frac{\partial x_2}{\partial y_1} & \cdots & \frac{\partial x_n}{\partial y_1} \\ \frac{\partial x_1}{\partial y_2} & \frac{\partial x_2}{\partial y_2} & \cdots & \frac{\partial x_n}{\partial y_2} \\ \vdots & & \ddots & \vdots \\ \frac{\partial x_1}{\partial y_n} & \frac{\partial x_2}{\partial y_n} & \cdots & \frac{\partial x_n}{\partial y_n} \end{vmatrix}.$$

   - The rows and columns of J_{xy} can be interchanged without changing its value.
   - $J_{xy} J_{yx} = 1$ and $J_{xz} = J_{xy} J_{yz}$.

PROBLEMS

8.1. Identify the curved wedge bounded by the surfaces $y^2 = 4ax$, $x + z = a$ and $z = 0$, and hence calculate its volume V.

8.2. Evaluate the volume integral of $x^2 + y^2 + z^2$ over the rectangular parallelepiped bounded by the six surfaces $x = \pm a$, $y = \pm b$ and $z = \pm c$.

8.3. Find the volume integral of $x^2 y$ over the tetrahedral volume bounded by the planes $x = 0$, $y = 0$, $z = 0$, and $x + y + z = 1$.

8.4. Evaluate the surface integral of $f(x, y)$ over the rectangle $0 \leq x \leq a$, $0 \leq y \leq b$ for the functions

  (a) $f(x, y) = \dfrac{x}{x^2 + y^2}$,   (b) $f(x, y) = (b - y + x)^{-3/2}$.

8.5. Calculate the volume of an ellipsoid as follows:
  (a) Prove that the area of the ellipse

  $$\frac{x^2}{a^2} + \frac{y^2}{b^2} = 1$$

  is πab.

Multiple integrals

(b) Use this result to obtain an expression for the volume of a slice of thickness dz of the ellipsoid

$$\frac{x^2}{a^2} + \frac{y^2}{b^2} + \frac{z^2}{c^2} = 1.$$

Hence show that the volume of the ellipsoid is $4\pi abc/3$.

8.6. The function

$$\Psi(r) = A\left(2 - \frac{Zr}{a}\right) e^{-Zr/2a}$$

gives the form of the quantum-mechanical wavefunction representing the electron in a hydrogen-like atom of atomic number Z, when the electron is in its first allowed spherically symmetric excited state. Here r is the usual spherical polar coordinate, but, because of the spherical symmetry, the coordinates θ and ϕ do not appear explicitly in Ψ. Determine the value that A (assumed real) must have if the wavefunction is to be correctly normalised, i.e. if the volume integral of $|\Psi|^2$ over all space is to be equal to unity.

8.7. In quantum mechanics the electron in a hydrogen atom in some particular state is described by a wavefunction Ψ, which is such that $|\Psi|^2 \, dV$ is the probability of finding the electron in the infinitesimal volume dV. In spherical polar coordinates $\Psi = \Psi(r, \theta, \phi)$ and $dV = r^2 \sin\theta \, dr \, d\theta \, d\phi$. Two such states are described by

$$\Psi_1 = \left(\frac{1}{4\pi}\right)^{1/2} \left(\frac{1}{a_0}\right)^{3/2} 2e^{-r/a_0},$$

$$\Psi_2 = -\left(\frac{3}{8\pi}\right)^{1/2} \sin\theta \, e^{i\phi} \left(\frac{1}{2a_0}\right)^{3/2} \frac{re^{-r/2a_0}}{a_0\sqrt{3}}.$$

(a) Show that each Ψ_i is normalised, i.e. the integral over all space $\int |\Psi|^2 \, dV$ is equal to unity – physically, this means that the electron must be somewhere.

(b) The (so-called) dipole matrix element between the states 1 and 2 is given by the integral

$$p_x = \int \Psi_1^* qr \sin\theta \cos\phi \, \Psi_2 \, dV,$$

where q is the charge on the electron. Prove that p_x has the value $-2^7 q a_0/3^5$.

8.8. A planar figure is formed from uniform wire and consists of two equal semicircular arcs, each with its own closing diameter, joined so as to form a letter 'B'. The figure is freely suspended from its top left-hand corner. Show that the straight edge of the figure makes an angle θ with the vertical given by $\tan\theta = (2 + \pi)^{-1}$.

Problems

8.9. A certain torus has a circular vertical cross-section of radius a centred on a horizontal circle of radius c ($> a$).
   (a) Find the volume V and surface area A of the torus, and show that they can be written as
$$V = \frac{\pi^2}{4}(r_o^2 - r_i^2)(r_o - r_i), \qquad A = \pi^2(r_o^2 - r_i^2),$$
   where r_o and r_i are, respectively, the outer and inner radii of the torus.
   (b) Show that a vertical circular cylinder of radius c, coaxial with the torus, divides A in the ratio
$$\pi c + 2a \; : \; \pi c - 2a.$$

8.10. A thin uniform circular disc has mass M and radius a.
   (a) Prove that its moment of inertia about an axis perpendicular to its plane and passing through its centre is $\frac{1}{2}Ma^2$.
   (b) Prove that the moment of inertia of the same disc about a diameter is $\frac{1}{4}Ma^2$. This is an example of the general result for planar bodies that the moment of inertia of the body about an axis perpendicular to the plane is equal to the sum of the moments of inertia about two perpendicular axes lying in the plane; in an obvious notation
$$I_z = \int r^2 \, dm = \int (x^2 + y^2) \, dm = \int x^2 \, dm + \int y^2 \, dm = I_y + I_x.$$

8.11. In some applications in mechanics the moment of inertia of a body about a single point (as opposed to about an axis) is needed. The moment of inertia, I, about the origin of a uniform solid body of density ρ is given by the volume integral
$$I = \int_V (x^2 + y^2 + z^2) \rho \, dV.$$
Show that the moment of inertia of a right circular cylinder of radius a, length $2b$ and mass M about its centre is
$$M \left(\frac{a^2}{2} + \frac{b^2}{3} \right).$$

8.12. The shape of an axially symmetric hard-boiled egg, of uniform density ρ_0, is given in spherical polar coordinates by $r = a(2 - \cos\theta)$, where θ is measured from the axis of symmetry.
   (a) Prove that the mass M of the egg is $M = \frac{40}{3}\pi\rho_0 a^3$.
   (b) Prove that the egg's moment of inertia about its axis of symmetry is $\frac{342}{175}Ma^2$.

8.13. In spherical polar coordinates r, θ, ϕ the element of volume for a body that is symmetrical about the polar axis is $dV = 2\pi r^2 \sin\theta \, dr \, d\theta$, whilst its element of surface area is $2\pi r \sin\theta [(dr)^2 + r^2(d\theta)^2]^{1/2}$. A particular surface is defined by $r = 2a\cos\theta$, where a is a constant and $0 \leq \theta \leq \pi/2$. Find its total surface area and the volume it encloses, and hence identify the surface.

Multiple integrals

8.14. By expressing both the integrand and the surface element in spherical polar coordinates, show that the surface integral

$$\int \frac{x^2}{x^2 + y^2} \, dS$$

over the surface $x^2 + y^2 = z^2$, $0 \leq z \leq 1$, has the value $\pi/\sqrt{2}$.

8.15. By transforming to cylindrical polar coordinates, evaluate the integral

$$I = \int\int\int \ln(x^2 + y^2) \, dx \, dy \, dz$$

over the interior of the conical region $x^2 + y^2 \leq z^2$, $0 \leq z \leq 1$.

8.16. Sketch the two families of curves

$$y^2 = 4u(u - x), \qquad y^2 = 4v(v + x),$$

where u and v are parameters.

By transforming to the uv-plane, evaluate the integral of $y/(x^2 + y^2)^{1/2}$ over the part of the quadrant $x > 0$, $y > 0$ that is bounded by the lines $x = 0$, $y = 0$ and the curve $y^2 = 4a(a - x)$.

8.17. By making two successive simple changes of variables, evaluate

$$I = \int\int\int x^2 \, dx \, dy \, dz$$

over the ellipsoidal region

$$\frac{x^2}{a^2} + \frac{y^2}{b^2} + \frac{z^2}{c^2} \leq 1.$$

8.18. Sketch the domain of integration for the integral

$$I = \int_0^1 \int_{x=y}^{1/y} \frac{y^3}{x} \exp[y^2(x^2 + x^{-2})] \, dx \, dy$$

and characterise its boundaries in terms of new variables $u = xy$ and $v = y/x$. Show that the Jacobian for the change from (x, y) to (u, v) is equal to $(2v)^{-1}$, and hence evaluate I.

8.19. Sketch the part of the region $0 \leq x$, $0 \leq y \leq \pi/2$ that is bounded by the curves $x = 0$, $y = 0$, $\sinh x \cos y = 1$ and $\cosh x \sin y = 1$. By making a suitable change of variables, evaluate the integral

$$I = \int\int (\sinh^2 x + \cos^2 y) \sinh 2x \sin 2y \, dx \, dy$$

over the bounded subregion.

Hints and answers

8.20. Define a coordinate system u, v whose origin coincides with that of the usual x, y system and whose u-axis coincides with the x-axis, whilst the v-axis makes an angle α with it. By considering the integral $I = \int \exp(-r^2)\, dA$, where r is the radial distance from the origin, over the area defined by $0 \leq u < \infty, 0 \leq v < \infty$, prove that

$$\int_0^\infty \int_0^\infty \exp(-u^2 - v^2 - 2uv \cos\alpha)\, du\, dv = \frac{\alpha}{2\sin\alpha}.$$

8.21. As stated in Section 7.11, the first law of thermodynamics can be expressed as

$$dU = T\, dS - P\, dV.$$

By calculating and equating $\partial^2 U/\partial Y \partial X$ and $\partial^2 U/\partial X \partial Y$, where X and Y are an unspecified pair of variables (drawn from P, V, T and S), prove that

$$\frac{\partial(S, T)}{\partial(X, Y)} = \frac{\partial(V, P)}{\partial(X, Y)}.$$

Using the properties of Jacobians, deduce that

$$\frac{\partial(S, T)}{\partial(V, P)} = 1.$$

8.22. The distances of the variable point P, which has coordinates x, y, z, from the fixed points $(0, 0, 1)$ and $(0, 0, -1)$, are denoted by u and v respectively. New variables ξ, η, ϕ are defined by

$$\xi = \tfrac{1}{2}(u + v), \qquad \eta = \tfrac{1}{2}(u - v)$$

and ϕ is the angle between the plane $y = 0$ and the plane containing the three points. Prove that the Jacobian $\partial(\xi, \eta, \phi)/\partial(x, y, z)$ has the value $(\xi^2 - \eta^2)^{-1}$ and that

$$\iiint_{\text{all space}} \frac{(u - v)^2}{uv} \exp\left(-\frac{u + v}{2}\right) dx\, dy\, dz = \frac{16\pi}{3e}.$$

HINTS AND ANSWERS

8.1. For integration order z, y, x, the limits are $(0, a - x), (-\sqrt{4ax}, \sqrt{4ax})$ and $(0, a)$. For integration order y, x, z, the limits are $(-\sqrt{4ax}, \sqrt{4ax}), (0, a - z)$ and $(0, a)$.
$V = 16a^3/15$.

8.3. $1/360$.

8.5. (a) Evaluate $\int 2b[1 - (x/a)^2]^{1/2}\, dx$ by setting $x = a\cos\phi$;
(b) $dV = \pi \times a[1 - (z/c)^2]^{1/2} \times b[1 - (z/c)^2]^{1/2}\, dz$.

8.7. Write $\sin^3\theta$ as $(1 - \cos^2\theta)\sin\theta$ when integrating $|\Psi_2|^2$.

8.9. (a) $V = 2\pi c \times \pi a^2$ and $A = 2\pi a \times 2\pi c$. Setting $r_o = c + a$ and $r_i = c - a$ gives the stated results. (b) Show that the centre of gravity of either half is $2a/\pi$ from the cylinder.

8.11. Transform to cylindrical polar coordinates.

8.13. $4\pi a^2$; $4\pi a^3/3$; a sphere.

8.15. The volume element is $\rho\, d\phi\, d\rho\, dz$. The integrand for the final z-integration is given by $2\pi[(z^2 \ln z) - (z^2/2)]$; $I = -5\pi/9$.

8.17. Set $\xi = x/a$, $\eta = y/b$, $\zeta = z/c$ to map the ellipsoid onto the unit sphere, and then change from (ξ, η, ζ) coordinates to spherical polar coordinates; $I = 4\pi a^3 bc/15$.

8.19. Set $u = \sinh x \cos y$ and $v = \cosh x \sin y$; $J_{xy,uv} = (\sinh^2 x + \cos^2 y)^{-1}$ and the integrand reduces to $4uv$ over the region $0 \le u \le 1$, $0 \le v \le 1$; $I = 1$.

8.21. Terms such as $T\partial^2 S/\partial Y \partial X$ cancel in pairs. Use Equations (8.19) and (8.18).

9 Vector algebra

This chapter introduces space vectors and their manipulation. Firstly we deal with the description and algebra of vectors, then we consider how vectors may be used to describe lines, planes and spheres, and finally we look at the practical use of vectors in finding distances. The calculus of vectors will be developed in a later chapter; this chapter gives only some basic rules.

9.1 Scalars and vectors

The simplest kind of physical quantity is one that can be completely specified by its magnitude, a single number, together with the units in which it is measured. Such a quantity is called a *scalar*, and examples include temperature, time and density.

A *vector* is a quantity that requires both a magnitude (≥ 0) and a direction in space to specify it completely; we may think of it as an arrow in space. A familiar example is force, which has a magnitude (strength) measured in newtons and a direction of application. The large number of vectors that are used to describe the physical world include velocity, displacement, momentum and electric field. Vectors can also be used to describe quantities such as angular momentum and surface elements (a surface element has a magnitude, defined by its area, and a direction defined by the normal to its tangent plane); in such cases their definitions may seem somewhat arbitrary (though in fact they are standard) and not as physically intuitive as for vectors such as force. A vector is denoted by bold type, the convention of this book, or by underlining, the latter being much used in handwritten work.

This chapter considers basic vector algebra and illustrates just how powerful vector analysis can be. All the techniques are presented for three-dimensional space but most can be readily extended to more dimensions.

Throughout the book we will represent a vector in diagrams as a line together with an arrowhead. We will make no distinction between an arrowhead at the end of the line and one along the line's length but, rather, use that which gives the clearer diagram. Furthermore, even though we are considering three-dimensional vectors, we have to draw them in the plane of the paper. It should not be assumed that vectors drawn thus are coplanar, unless this is explicitly stated.

EXERCISE 9.1

1. For each of the following physical quantities, say whether it is a scalar or a vector, or is insufficiently specified to be uniquely classified: (a) velocity, (b) speed, (c) work,

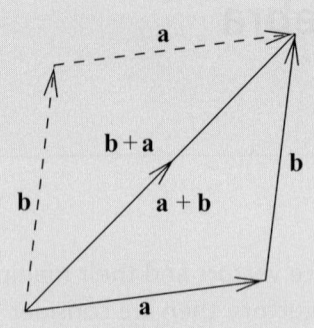

Figure 9.1 Addition of two vectors showing the commutation relation. We make no distinction between an arrowhead at the end of the line and one along the line's length, but rather use that which gives the clearer diagram.

(d) magnetic field, (e) fluid velocity component, (f) pressure, (g) electric current, (h) potential energy, (i) height, (j) gradient, (k) voltage, (l) surface charge density, (m) pressure gradient.

9.2 Addition, subtraction and multiplication of vectors

The *resultant* or *vector sum* of two displacement vectors is the displacement vector that results from performing first one and then the other displacement, as shown in Figure 9.1; this process is known as vector addition. However, the principle of addition has physical meaning for vector quantities other than displacements; for example, if two forces act on the same body then the resultant force acting on the body is the vector sum of the two. The addition of vectors only makes physical sense if they are of a like kind, for example if they are both forces acting in three dimensions. It may be seen from Figure 9.1 that vector addition is commutative, i.e.

$$\mathbf{a} + \mathbf{b} = \mathbf{b} + \mathbf{a}. \tag{9.1}$$

The generalisation of this procedure to the addition of three (or more) vectors is clear and leads to the associativity property of addition (see Figure 9.2), e.g.

$$\mathbf{a} + (\mathbf{b} + \mathbf{c}) = (\mathbf{a} + \mathbf{b}) + \mathbf{c}. \tag{9.2}$$

Thus, it is immaterial in what order any number of vectors are added; their resultant is always the same, whatever the order of addition.

The subtraction of two vectors is very similar to their addition (see Figure 9.3); that is,

$$\mathbf{a} - \mathbf{b} = \mathbf{a} + (-\mathbf{b})$$

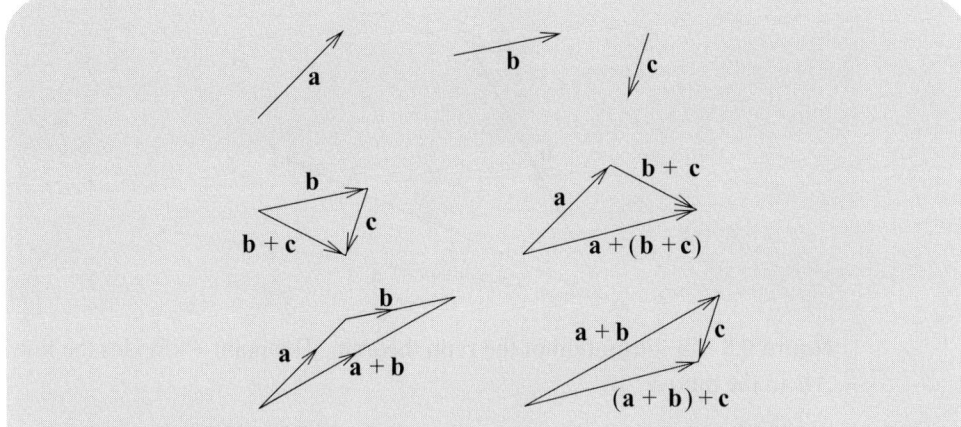

9.2 Addition, subtraction and multiplication of vectors

Figure 9.2 Addition of three vectors showing the associativity relation.

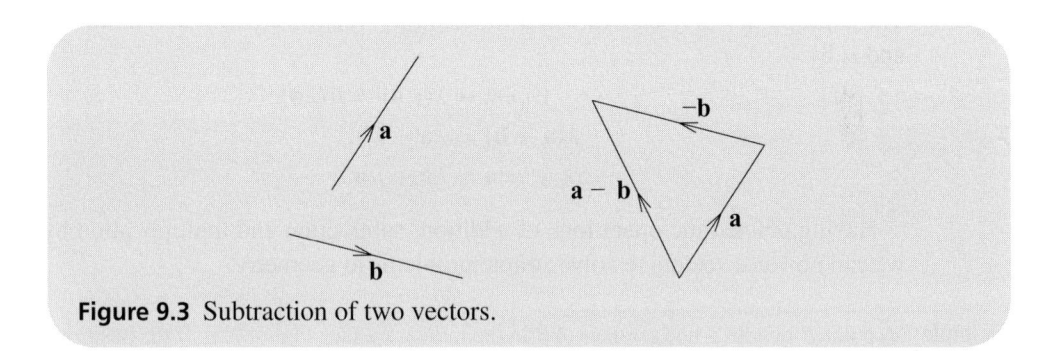

Figure 9.3 Subtraction of two vectors.

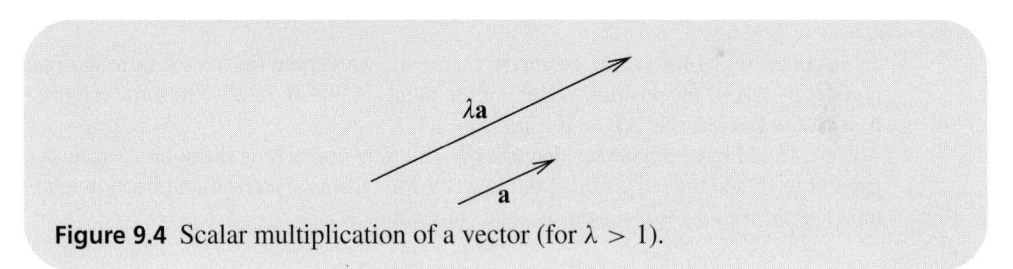

Figure 9.4 Scalar multiplication of a vector (for $\lambda > 1$).

where $-\mathbf{b}$ is a vector of equal magnitude but exactly opposite direction to vector $\mathbf{b}$. The subtraction of two equal vectors yields the zero vector, $\mathbf{0}$, which has zero magnitude and no associated direction.[1]

Multiplication of a vector by a scalar (not to be confused with the 'scalar product', to be discussed in Section 9.4.1) gives a vector in the same direction as the original but of a proportional magnitude. This can be seen in Figure 9.4. The scalar may be positive, negative or zero. It can also be complex in some applications. Clearly, when the scalar is negative we obtain a vector pointing in the opposite direction to the original vector.

---

[1] Show that if a body is in equilibrium under the action of n forces then the vector diagram representing the forces is a closed n-sided polygon.

Vector algebra

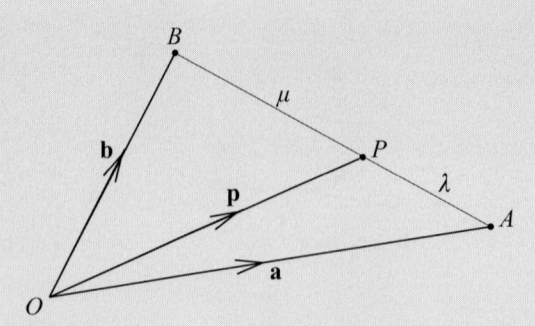

Figure 9.5 An illustration of the ratio theorem. The point P divides the line segment AB in the ratio $\lambda : \mu$.

Multiplication by a scalar is associative, commutative and distributive over addition. These properties may be expressed for arbitrary vectors $\mathbf{a}$ and $\mathbf{b}$ and arbitrary scalars λ and μ by

$$(\lambda\mu)\mathbf{a} = \lambda(\mu\mathbf{a}) = \mu(\lambda\mathbf{a}), \tag{9.3}$$
$$\lambda(\mathbf{a} + \mathbf{b}) = \lambda\mathbf{a} + \lambda\mathbf{b}, \tag{9.4}$$
$$(\lambda + \mu)\mathbf{a} = \lambda\mathbf{a} + \mu\mathbf{a}. \tag{9.5}$$

Having defined the operations of addition, subtraction and multiplication by a scalar, we can now use vectors to solve simple problems in geometry.

Example A point P divides a line segment AB in the ratio $\lambda : \mu$ (see Figure 9.5). If the position vectors of the points A and B are $\mathbf{a}$ and $\mathbf{b}$ respectively, find the position vector of the point P.

As is conventional for vector geometry problems, we denote the vector from the point A to the point B by $\mathbf{AB}$. If the position vectors of the points A and B, relative to some origin O, are $\mathbf{a}$ and $\mathbf{b}$, it should be clear that $\mathbf{AB} = \mathbf{b} - \mathbf{a}$.

Now, from Figure 9.5 we see that one possible way of reaching the point P from O is first to go from O to A, and then to go along the line AB for a distance equal to the fraction $\lambda/(\lambda + \mu)$ of its total length. We may express this in terms of vectors as

$$\begin{aligned}\mathbf{OP} = \mathbf{p} &= \mathbf{a} + \frac{\lambda}{\lambda + \mu}\mathbf{AB} \\ &= \mathbf{a} + \frac{\lambda}{\lambda + \mu}(\mathbf{b} - \mathbf{a}) \\ &= \left(1 - \frac{\lambda}{\lambda + \mu}\right)\mathbf{a} + \frac{\lambda}{\lambda + \mu}\mathbf{b} \\ &= \frac{\mu}{\lambda + \mu}\mathbf{a} + \frac{\lambda}{\lambda + \mu}\mathbf{b},\end{aligned} \tag{9.6}$$

which expresses the position vector of the point P in terms of those of A and B. We would, of course, obtain the same result by considering the path from O to B and then to P.[2] ◀

---

[2] Identify the points given by $\mathbf{p} = (\mu - \lambda)^{-1}(\mu\mathbf{a} - \lambda\mathbf{b})$. Consider both $\mu > \lambda$ and $\lambda > \mu$. If necessary, draw some simple examples on graph paper.

9.2 Addition, subtraction and multiplication of vectors

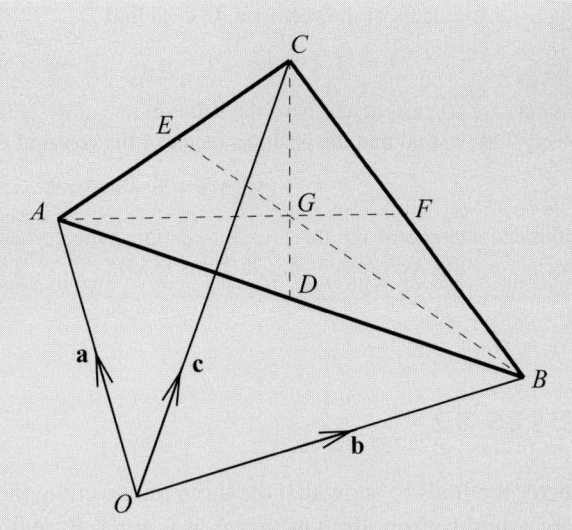

Figure 9.6 The centroid of a triangle. The triangle is defined by the points A, B and C that have position vectors $\mathbf{a}$, $\mathbf{b}$ and $\mathbf{c}$. The broken lines CD, BE, AF connect the vertices of the triangle to the mid-points of the opposite sides; these lines intersect at the centroid G of the triangle.

Result (9.6) is a version of the *ratio theorem* and we may use it in solving more complicated problems.

Example The vertices of triangle ABC have position vectors $\mathbf{a}$, $\mathbf{b}$ and $\mathbf{c}$ relative to some origin O (see Figure 9.6). Find the position vector of the centroid G of the triangle.

From Figure 9.6, the points D and E bisect the lines AB and AC respectively. Thus from the ratio theorem (9.6), with $\lambda = \mu = 1/2$, the position vectors of D and E relative to the origin are

$$\mathbf{d} = \tfrac{1}{2}\mathbf{a} + \tfrac{1}{2}\mathbf{b},$$
$$\mathbf{e} = \tfrac{1}{2}\mathbf{a} + \tfrac{1}{2}\mathbf{c}.$$

Using the ratio theorem again, we may write the position vector of a general point on the line CD that divides the line in the ratio $\lambda : (1 - \lambda)$ as

$$\mathbf{r} = (1 - \lambda)\mathbf{c} + \lambda\mathbf{d},$$
$$= (1 - \lambda)\mathbf{c} + \tfrac{1}{2}\lambda(\mathbf{a} + \mathbf{b}), \qquad (9.7)$$

where we have expressed $\mathbf{d}$ in terms of $\mathbf{a}$ and $\mathbf{b}$. Similarly, the position vector of a general point on the line BE can be expressed as

$$\mathbf{r} = (1 - \mu)\mathbf{b} + \mu\mathbf{e},$$
$$= (1 - \mu)\mathbf{b} + \tfrac{1}{2}\mu(\mathbf{a} + \mathbf{c}). \qquad (9.8)$$

Thus, at the intersection of the lines CD and BE we require, from (9.7) and (9.8),

$$(1 - \lambda)\mathbf{c} + \tfrac{1}{2}\lambda(\mathbf{a} + \mathbf{b}) = (1 - \mu)\mathbf{b} + \tfrac{1}{2}\mu(\mathbf{a} + \mathbf{c}).$$

Vector algebra

By equating the coefficients of the vectors **a**, **b**, **c** we find

$$\lambda = \mu, \qquad \tfrac{1}{2}\lambda = 1 - \mu, \qquad 1 - \lambda = \tfrac{1}{2}\mu.$$

These equations are consistent and have the solution $\lambda = \mu = 2/3$. Substituting these values into either (9.7) or (9.8) we find that the position vector of the centroid G is given by

$$\mathbf{g} = \tfrac{1}{3}(\mathbf{a} + \mathbf{b} + \mathbf{c}),$$

i.e. the arithmetic average of the three vectors defining the corners of the triangle.[3] Note that a change of origin for the vectors does not alter this result. ◀

EXERCISES 9.2

1. Use vector methods to show that the three lines joining the mid-points of the opposite edges of a tetrahedron are concurrent at a point P, and that each is the bisector of the other two. Take the origin as one of the corners of the tetrahedron, but give a prescription for the location of P for a general tetrahedron with vertices at **a**, **b**, **c** and **d**.

2. $ABCD$ is a parallelogram and M is the mid-point of AB. Use vector methods to show that DM and AC divide each other in the ratio $1:2$.

3. A triangle ABC has points D, E and F lying on sides BC, CA and AB respectively. If the lines AD, BE and CF are concurrent at G, then a part of Ceva's theorem states that

$$\frac{CD}{DB} \cdot \frac{BF}{FA} \cdot \frac{AE}{EC} = 1.$$

Prove this result by taking $\mathbf{d} = \lambda_1 \mathbf{b} + (1 - \lambda_1)\mathbf{c}$, $\mathbf{e} = \lambda_2 \mathbf{c} + (1 - \lambda_2)\mathbf{a}$, etc. and then writing $\mathbf{g} = \mu_1 \mathbf{a} + (1 - \mu_1)\mathbf{d}$, etc., where **x** is the vector position of point X.

Hint: Deduce the stated result by using the simultaneous equations that must be satisfied to write expressions for $\prod_{i=1}^{3} \lambda_i$ and $\prod_{i=1}^{3}(1 - \lambda_i)$, each entirely in terms of the μ_j.

9.3 Basis vectors, components and magnitudes

Given any three different vectors $\mathbf{e}_1$, $\mathbf{e}_2$ and $\mathbf{e}_3$, which do not all lie in a plane, it is possible, in a three-dimensional space, to write any other vector in terms of scalar multiples of them:

$$\mathbf{a} = a_1 \mathbf{e}_1 + a_2 \mathbf{e}_2 + a_3 \mathbf{e}_3. \tag{9.9}$$

The three vectors $\mathbf{e}_1$, $\mathbf{e}_2$ and $\mathbf{e}_3$ are said to form a *basis* (for the three-dimensional space); the scalars a_1, a_2 and a_3, which may be positive, negative or zero, are called the *components*

---

[3] Verify that G does lie on the line AF and that it divides it in the expected ratio.

9.3 Basis vectors, components and magnitudes

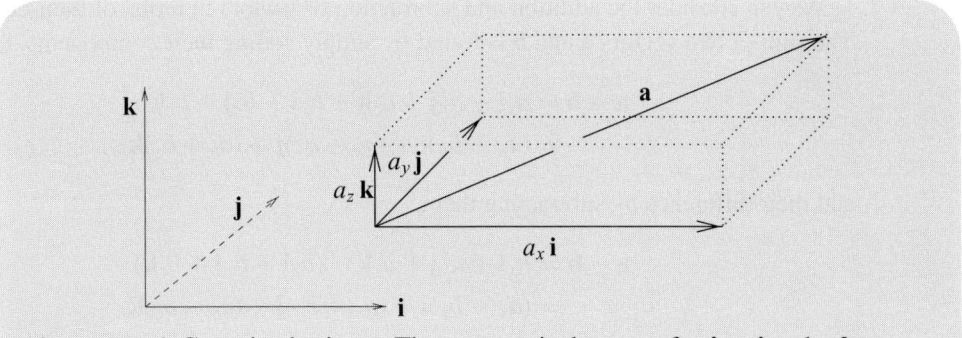

Figure 9.7 A Cartesian basis set. The vector **a** is the sum of $a_x\mathbf{i}$, $a_y\mathbf{j}$ and $a_z\mathbf{k}$.

of the vector **a** with respect to this basis. We say that the vector has been *resolved* into components.

Most often we will use basis vectors that are mutually perpendicular, for ease of manipulation, though this is not necessary. In general, a basis set must

(i) have as many basis vectors as the number of dimensions (in more formal language, the basis vectors must span the space) and
(ii) be such that no basis vector may be described as a (weighted) sum of the others, or, more formally, the basis vectors must be *linearly independent*. Putting this mathematically, in N dimensions, we require

$$c_1\mathbf{e}_1 + c_2\mathbf{e}_2 + \cdots + c_N\mathbf{e}_N \neq \mathbf{0},$$

for any set of coefficients $c_1, c_2, \ldots, c_N$, except the set $c_1 = c_2 = \cdots = c_N = 0$.

A more extended discussion of bases in general vector spaces is given in Section 10.1.1. In this chapter we will consider only vectors in three dimensions; algebraic extension to higher dimensionalities can be made in the obvious way, though visualisation becomes increasingly more difficult!

If we wish to label points in space using a Cartesian coordinate system (x, y, z), we may introduce the unit vectors **i**, **j** and **k**, which point along the positive x-, y- and z-axes respectively. A general vector **a** may then be written as a sum of three vectors, each parallel to a different coordinate axis:

$$\mathbf{a} = a_x\mathbf{i} + a_y\mathbf{j} + a_z\mathbf{k}. \qquad (9.10)$$

A vector in three-dimensional space thus requires three components to describe fully both its direction and its magnitude. As (9.10) and Figure 9.7 indicate, a general displacement in space may be thought of as the result of three successive displacements, one each along the x-, y- and z-directions. For brevity, the components of a vector **a** with respect to a particular coordinate system are sometimes written in the form (a_x, a_y, a_z). Note that the basis vectors **i**, **j** and **k** may themselves be represented by $(1, 0, 0)$, $(0, 1, 0)$ and $(0, 0, 1)$ respectively.

Vector algebra

We can consider the addition and subtraction of vectors in terms of their components. The sum of two vectors **a** and **b** is found by simply adding their components, i.e.

$$\mathbf{a} + \mathbf{b} = a_x\mathbf{i} + a_y\mathbf{j} + a_z\mathbf{k} + b_x\mathbf{i} + b_y\mathbf{j} + b_z\mathbf{k}$$
$$= (a_x + b_x)\mathbf{i} + (a_y + b_y)\mathbf{j} + (a_z + b_z)\mathbf{k}, \quad (9.11)$$

and their difference by subtracting them,

$$\mathbf{a} - \mathbf{b} = a_x\mathbf{i} + a_y\mathbf{j} + a_z\mathbf{k} - (b_x\mathbf{i} + b_y\mathbf{j} + b_z\mathbf{k})$$
$$= (a_x - b_x)\mathbf{i} + (a_y - b_y)\mathbf{j} + (a_z - b_z)\mathbf{k}, \quad (9.12)$$

as in the following example.

Example Two particles have velocities $\mathbf{v}_1 = \mathbf{i} + 3\mathbf{j} + 6\mathbf{k}$ and $\mathbf{v}_2 = \mathbf{i} - 2\mathbf{k}$ respectively. Find the velocity **u** of the second particle relative to the first.

The required relative velocity is given by

$$\mathbf{u} = \mathbf{v}_2 - \mathbf{v}_1 = (1-1)\mathbf{i} + (0-3)\mathbf{j} + (-2-6)\mathbf{k}$$
$$= -3\mathbf{j} - 8\mathbf{k}.$$

As expected, although both particles have non-zero speeds in the x direction, the two speeds are equal and so this component does not contribute to the particles' relative velocity. ◂

The magnitude of the vector **a** is denoted by |**a**| or a. In terms of its components in three-dimensional Cartesian coordinates, the magnitude of **a** is given by

$$a \equiv |\mathbf{a}| = \sqrt{a_x^2 + a_y^2 + a_z^2}. \quad (9.13)$$

Hence, the magnitude of a vector is a measure of its length. Such an analogy is useful for displacement vectors, but magnitude is better described, for example, by 'strength' for vectors such as force, or by 'speed' for velocity vectors.[4] For instance, in the previous example, the speed of the second particle relative to the first is given by

$$u = |\mathbf{u}| = \sqrt{(-3)^2 + (-8)^2} = \sqrt{73}.$$

A vector whose magnitude equals unity is called a *unit vector*. The unit vector in the direction **a** is usually notated **â** and may be evaluated as

$$\hat{\mathbf{a}} = \frac{\mathbf{a}}{|\mathbf{a}|}. \quad (9.14)$$

The unit vector is a useful concept because a vector written as $\lambda\hat{\mathbf{a}}$ then has magnitude λ and direction **â**. Thus magnitude and direction are explicitly separated.

---

[4] Note that, although **a** is a vector, its magnitude a is a (non-negative) scalar quantity.

9.4 Multiplication of two vectors

EXERCISES 9.3

1. Would any of the following sets of vectors, all expressed on the basis of **i**, **j** and **k**, be satisfactory as a basis for three-dimensional space? Explain your reasoning.
   (a) $(1, 2, -3), (-2, 3, 0), (-5, 4, 3)$,
   (b) $(1, 1, -1), (1, -1, 1), (-1, 1, 1)$,
   (c) $(0, 2, 2), (2, 0, 2), (2, 2, 0), (2, 2, 2)$,
   (d) $(1, 0, 1), (1, 0, -1), (-1, 0, 2)$,
   (e) $(1, 2, 3), (2, 3, 1), (3, 1, 2)$.

2. If, referred to the usual Cartesian basis, **i**, **j**, **k**,
$$\mathbf{a} = (4, -2, 1) \quad \text{and} \quad \mathbf{b} = (2, 1, -3),$$
find the magnitudes of the vectors $\mathbf{a} + \mathbf{b}$, $\mathbf{a} - \mathbf{b}$ and $3\mathbf{b} - \mathbf{a}$.

3. What would be the components of the vector $\mathbf{a} = 3\mathbf{i} - 2\mathbf{j} + \mathbf{k}$, if the vectors $\mathbf{f}_1 = \mathbf{j} + \mathbf{k}$, $\mathbf{f}_2 = \mathbf{k} + \mathbf{i}$ and $\mathbf{f}_3 = \mathbf{i} + \mathbf{j}$ were used as a basis? Are the $\mathbf{f}_i$ unit vectors? If so, say why; if not, convert them to unit vectors and re-express **a** in terms of the new basis.

9.4 Multiplication of two vectors

We have already considered multiplying a vector by a scalar. Now we consider the concept of multiplying one vector by another vector. It is not immediately obvious what the product of two vectors represents and, in fact, two different products are commonly defined, the *scalar product* and the *vector product*. As their names imply, the scalar product of two vectors is just a number, whereas the vector product is itself a vector. Although neither the scalar nor the vector product is what we might normally think of as a product, their use is widespread and several examples appear later in this book.

9.4.1 Scalar product

The scalar product (or dot product) of two vectors **a** and **b** is denoted by $\mathbf{a} \cdot \mathbf{b}$ (hence the name 'dot product') and is given by

$$\mathbf{a} \cdot \mathbf{b} \equiv |\mathbf{a}||\mathbf{b}| \cos \theta, \quad 0 \leq \theta \leq \pi, \quad (9.15)$$

where θ is the angle between the two vectors, placed 'tail to tail' or 'head to head'. Thus, the value of the scalar product $\mathbf{a} \cdot \mathbf{b}$ equals the magnitude of **a** multiplied by the projection of **b** onto **a** (see Figure 9.8).[5]

From (9.15) we see that the scalar product has the particularly useful property that

$$\mathbf{a} \cdot \mathbf{b} = 0 \quad (9.16)$$

is a necessary and sufficient condition for **a** to be perpendicular to **b** (unless either of them is zero). It should be noted in particular that the Cartesian basis vectors **i**, **j** and **k**, being

---

[5] Clearly, it could equally well be described as the magnitude of **b** multiplied by the projection of **a** onto **b**.

Vector algebra

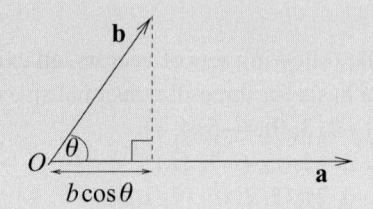

Figure 9.8 The projection of **b** onto the direction of **a** is $b\cos\theta$. The scalar product of **a** and **b** is $ab\cos\theta$.

mutually orthogonal unit vectors, satisfy the equations

$$\mathbf{i} \cdot \mathbf{i} = \mathbf{j} \cdot \mathbf{j} = \mathbf{k} \cdot \mathbf{k} = 1, \quad (9.17)$$
$$\mathbf{i} \cdot \mathbf{j} = \mathbf{j} \cdot \mathbf{k} = \mathbf{k} \cdot \mathbf{i} = 0. \quad (9.18)$$

Examples of scalar products arise naturally throughout physics and in particular in connection with energy. Perhaps the simplest is the work done $\mathbf{F} \cdot \mathbf{r}$ in moving the point of application of a constant force $\mathbf{F}$ through a displacement $\mathbf{r}$; notice that, as expected, if the displacement is perpendicular to the direction of the force then $\mathbf{F} \cdot \mathbf{r} = 0$ and no work is done. A second simple example is afforded by the potential energy $-\mathbf{m} \cdot \mathbf{B}$ of a magnetic dipole, represented in strength and orientation by a vector $\mathbf{m}$, placed in an external magnetic field $\mathbf{B}$.

As the name implies, the scalar product has a magnitude but no direction. The scalar product is commutative and distributive over addition:

$$\mathbf{a} \cdot \mathbf{b} = \mathbf{b} \cdot \mathbf{a} \quad (9.19)$$
$$\mathbf{a} \cdot (\mathbf{b} + \mathbf{c}) = \mathbf{a} \cdot \mathbf{b} + \mathbf{a} \cdot \mathbf{c}. \quad (9.20)$$

Our next example exploits the vector prescription for orthogonality, Equation (9.16), and thereby avoids the need for a complicated diagram.

Example Four non-coplanar points A, B, C, D are positioned such that the line AD is perpendicular to BC and BD is perpendicular to AC. Show that CD is perpendicular to AB.

Denote the four position vectors by **a**, **b**, **c**, **d**. As none of the three pairs of lines actually intersect, it would be difficult to indicate their orthogonality in the diagram we would normally draw. However, as already noted, the orthogonality can be expressed in vector form and we start by putting the fact that $AD \perp BC$ into this form:

$$(\mathbf{d} - \mathbf{a}) \cdot (\mathbf{c} - \mathbf{b}) = 0.$$

Similarly, since $BD \perp AC$,

$$(\mathbf{d} - \mathbf{b}) \cdot (\mathbf{c} - \mathbf{a}) = 0.$$

9.4 Multiplication of two vectors

Combining these two equations we find

$$(\mathbf{d} - \mathbf{a}) \cdot (\mathbf{c} - \mathbf{b}) = (\mathbf{d} - \mathbf{b}) \cdot (\mathbf{c} - \mathbf{a}),$$

which, on multiplying out the parentheses, gives

$$\mathbf{d} \cdot \mathbf{c} - \mathbf{a} \cdot \mathbf{c} - \mathbf{d} \cdot \mathbf{b} + \mathbf{a} \cdot \mathbf{b} = \mathbf{d} \cdot \mathbf{c} - \mathbf{b} \cdot \mathbf{c} - \mathbf{d} \cdot \mathbf{a} + \mathbf{b} \cdot \mathbf{a}.$$

Cancelling terms that appear on both sides and rearranging yields

$$\mathbf{d} \cdot \mathbf{b} - \mathbf{d} \cdot \mathbf{a} - \mathbf{c} \cdot \mathbf{b} + \mathbf{c} \cdot \mathbf{a} = 0,$$

which simplifies to give

$$(\mathbf{d} - \mathbf{c}) \cdot (\mathbf{b} - \mathbf{a}) = 0.$$

From (9.16), we see that this implies that CD is perpendicular to AB. ◂

If we introduce a set of basis vectors that are mutually orthogonal, such as $\mathbf{i}, \mathbf{j}, \mathbf{k}$, we can write the components of a vector $\mathbf{a}$, with respect to that basis, in terms of the scalar product of $\mathbf{a}$ with each of the basis vectors, i.e. $a_x = \mathbf{a} \cdot \mathbf{i}$, $a_y = \mathbf{a} \cdot \mathbf{j}$ and $a_z = \mathbf{a} \cdot \mathbf{k}$. In terms of their components a_x, a_y, a_z and b_x, b_y, b_z, the scalar product of vectors $\mathbf{a}$ and $\mathbf{b}$ is given by

$$\mathbf{a} \cdot \mathbf{b} = (a_x \mathbf{i} + a_y \mathbf{j} + a_z \mathbf{k}) \cdot (b_x \mathbf{i} + b_y \mathbf{j} + b_z \mathbf{k}) = a_x b_x + a_y b_y + a_z b_z, \quad (9.21)$$

where the cross terms such as $a_x \mathbf{i} \cdot b_y \mathbf{j}$ are zero because the basis vectors are mutually perpendicular; see Equation (9.18). It should be clear from (9.15) that the value of $\mathbf{a} \cdot \mathbf{b}$ has a geometrical definition and that this value is independent of the actual basis vectors used.

Example Find the angle between the vectors $\mathbf{a} = \mathbf{i} + 2\mathbf{j} + 3\mathbf{k}$ and $\mathbf{b} = 2\mathbf{i} + 3\mathbf{j} + 4\mathbf{k}$.

From (9.15) the cosine of the angle θ between $\mathbf{a}$ and $\mathbf{b}$ is given by

$$\cos \theta = \frac{\mathbf{a} \cdot \mathbf{b}}{|\mathbf{a}||\mathbf{b}|}.$$

From (9.21) the scalar product $\mathbf{a} \cdot \mathbf{b}$ has the value

$$\mathbf{a} \cdot \mathbf{b} = 1 \times 2 + 2 \times 3 + 3 \times 4 = 20,$$

and from (9.13) the lengths of the vectors are

$$|\mathbf{a}| = \sqrt{1^2 + 2^2 + 3^2} = \sqrt{14} \quad \text{and} \quad |\mathbf{b}| = \sqrt{2^2 + 3^2 + 4^2} = \sqrt{29}.$$

Thus,

$$\cos \theta = \frac{20}{\sqrt{14}\sqrt{29}} \approx 0.9926,$$

which implies that $\theta = 0.12$ rad. ◂

Vector algebra

We can see from the expressions (9.15) and (9.21) for the scalar product that if θ is the angle between **a** and **b** then

$$\cos\theta = \frac{a_x\,b_x}{a\,b} + \frac{a_y\,b_y}{a\,b} + \frac{a_z\,b_z}{a\,b}$$

where a_x/a, a_y/a and a_z/a are called the *direction cosines* of **a**, since they give the cosine of the angle made by **a** with each of the basis vectors. Similarly b_x/b, b_y/b and b_z/b are the direction cosines of **b**.

If we take the scalar product of any vector **a** with itself then clearly $\theta = 0$ and from (9.15) we have

$$\mathbf{a}\cdot\mathbf{a} = |\mathbf{a}|^2.$$

Thus the magnitude of **a** can be written in a coordinate-independent form as $|\mathbf{a}| = \sqrt{\mathbf{a}\cdot\mathbf{a}}$.

Finally, we note that the scalar product may be extended to vectors with complex components if it is redefined as

$$\mathbf{a}\cdot\mathbf{b} = a_x^* b_x + a_y^* b_y + a_z^* b_z,$$

where the asterisk represents the operation of complex conjugation. To accommodate this extension the commutation property (9.19) must be modified to read[6]

$$\mathbf{a}\cdot\mathbf{b} = (\mathbf{b}\cdot\mathbf{a})^*. \tag{9.22}$$

In particular it should be noted that $(\lambda\mathbf{a})\cdot\mathbf{b} = \lambda^*\mathbf{a}\cdot\mathbf{b}$, whereas $\mathbf{a}\cdot(\lambda\mathbf{b}) = \lambda\mathbf{a}\cdot\mathbf{b}$. However, the magnitude of a complex vector is still given by $|\mathbf{a}| = \sqrt{\mathbf{a}\cdot\mathbf{a}}$, since $\mathbf{a}\cdot\mathbf{a}$ is always real.

9.4.2 Vector product

The vector product (or cross product) of two vectors **a** and **b** is denoted by $\mathbf{a}\times\mathbf{b}$ and is defined to be a vector of magnitude $|\mathbf{a}||\mathbf{b}|\sin\theta$ in a direction perpendicular to both **a** and **b**:

$$|\mathbf{a}\times\mathbf{b}| = |\mathbf{a}||\mathbf{b}|\sin\theta.$$

The direction is found by 'rotating' **a** into **b** through the smallest possible angle. The sense of rotation is that of a right-handed screw that moves forward in the direction $\mathbf{a}\times\mathbf{b}$ (see Figure 9.9). Again, θ is the angle between the two vectors placed 'tail to tail' or 'head to head'. With this definition, **a**, **b** and $\mathbf{a}\times\mathbf{b}$ (in that order) form a right-handed set.

A more directly usable description of the relative directions in a vector product is provided by a right hand whose first two fingers and thumb are held to be as nearly mutually perpendicular as possible. If the first finger (the index or pointer finger) is pointed in the direction of the first vector and the second finger in the direction of the second vector, then the thumb gives the direction of the vector product.

---

[6] For the vectors $\mathbf{a} = (1+i, 2i, 3)$ and $\mathbf{b} = (2-i, 3+i, 1-i)$ calculate $\mathbf{a}\cdot\mathbf{b}$ and $\mathbf{b}\cdot\mathbf{a}$ and confirm the stated relationship. What are the magnitudes of **a** and **b**?

9.4 Multiplication of two vectors

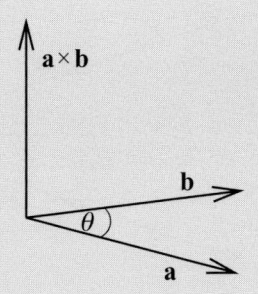

Figure 9.9 The vector product. The vectors **a**, **b** and **a** × **b** (in that order) form a right-handed set.

The vector product may (with a little work) be shown to be distributive over addition, but *anticommutative* and *non-associative*:[7]

$$(\mathbf{a} + \mathbf{b}) \times \mathbf{c} = (\mathbf{a} \times \mathbf{c}) + (\mathbf{b} \times \mathbf{c}), \tag{9.23}$$

$$\mathbf{b} \times \mathbf{a} = -(\mathbf{a} \times \mathbf{b}), \tag{9.24}$$

$$(\mathbf{a} \times \mathbf{b}) \times \mathbf{c} \neq \mathbf{a} \times (\mathbf{b} \times \mathbf{c}). \tag{9.25}$$

From its definition, we see that the vector product has the very useful property that if $\mathbf{a} \times \mathbf{b} = \mathbf{0}$ then **a** is parallel or antiparallel to **b** (unless either of them is zero). We also note that

$$\mathbf{a} \times \mathbf{a} = \mathbf{0}. \tag{9.26}$$

Example Show that if $\mathbf{a} = \mathbf{b} + \lambda \mathbf{c}$, for some scalar λ, then $\mathbf{a} \times \mathbf{c} = \mathbf{b} \times \mathbf{c}$.

From (9.23) we have

$$\mathbf{a} \times \mathbf{c} = (\mathbf{b} + \lambda \mathbf{c}) \times \mathbf{c} = \mathbf{b} \times \mathbf{c} + \lambda \mathbf{c} \times \mathbf{c}.$$

However, from (9.26), $\mathbf{c} \times \mathbf{c} = \mathbf{0}$ and so

$$\mathbf{a} \times \mathbf{c} = \mathbf{b} \times \mathbf{c}. \tag{9.27}$$

We note in passing that the fact that (9.27) is satisfied does *not* imply that $\mathbf{a} = \mathbf{b}$, as is clear from giving λ any non-zero value.[8] ◀

An example of the use of the vector product is that of finding the area, A, of a parallelogram with sides **a** and **b**, using the formula

$$A = |\mathbf{a} \times \mathbf{b}|. \tag{9.28}$$

---

[7] Make sketches to convince yourself that the vector on the LHS of (9.25) lies in the plane defined by vectors **a** and **b**, whilst that on the RHS lies in the plane defined by vectors **b** and **c**. For general vectors this establishes the inequality, but show that an exception arises if **a** and **c** are each orthogonal to **b**, but not to each other.
[8] State precisely what $\mathbf{a} \times \mathbf{c} = \mathbf{b} \times \mathbf{c}$ *does* imply.

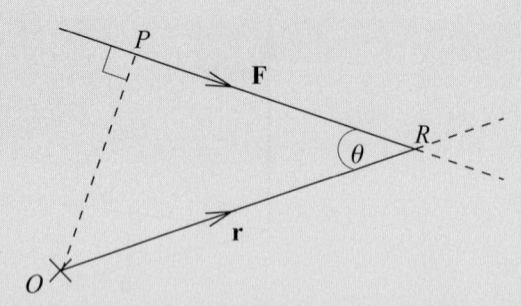

Figure 9.10 The moment of the force **F** about O is $\mathbf{r} \times \mathbf{F}$. The cross represents the direction of $\mathbf{r} \times \mathbf{F}$, which is perpendicularly into the plane of the paper.

Another example is afforded by considering a force **F** acting through a point R, whose vector position relative to the origin O is **r** (see Figure 9.10). Its *moment* or *torque* about O is the strength of the force times the perpendicular distance OP, which numerically is just $Fr\sin\theta$, i.e. the magnitude of $\mathbf{r} \times \mathbf{F}$. Furthermore, the sense of the moment is clockwise about an axis through O that points perpendicularly into the plane of the paper (the axis is represented by a cross in the figure). Thus the moment is completely represented by the vector $\mathbf{r} \times \mathbf{F}$, in both magnitude and spatial sense. It should be noted that the same vector product is obtained wherever the point R is chosen, so long as it lies on the line of action of **F**.

Similarly, if a solid body is rotating about some axis that passes through the origin, with an angular velocity ω, then we can describe this rotation by a vector $\boldsymbol{\omega}$ that has magnitude ω and points along the axis of rotation. The direction of $\boldsymbol{\omega}$ is the forward direction of a right-handed screw rotating in the same sense as the body. The velocity of any point in the body with position vector **r** is then given by $\mathbf{v} = \boldsymbol{\omega} \times \mathbf{r}$. Even if the axis of rotation does not pass through the origin, the rotation can still be represented by a vector with the appropriate components, though the velocity of points within the body is no longer given by this simple formula.[9]

Since the basis vectors **i**, **j**, **k** are mutually perpendicular unit vectors, forming a right-handed set, their vector products are easily seen to be

$$\mathbf{i} \times \mathbf{i} = \mathbf{j} \times \mathbf{j} = \mathbf{k} \times \mathbf{k} = \mathbf{0}, \qquad (9.29)$$

$$\mathbf{i} \times \mathbf{j} = -\mathbf{j} \times \mathbf{i} = \mathbf{k}, \qquad (9.30)$$

$$\mathbf{j} \times \mathbf{k} = -\mathbf{k} \times \mathbf{j} = \mathbf{i}, \qquad (9.31)$$

$$\mathbf{k} \times \mathbf{i} = -\mathbf{i} \times \mathbf{k} = \mathbf{j}. \qquad (9.32)$$

Using these relations, it is straightforward to show that the vector product of two general vectors **a** and **b** is given in terms of their components with respect to the basis set **i**, **j**, **k**,

---

[9] The Arctic Circle is at latitude 66.5°N. Taking the Sun as the origin of coordinates, the plane of the ecliptic as the x–y plane, and the x-axis as the line joining the Sun to the Earth at the winter solstice, obtain numerical values for the components of the vector representing the Earth's angular velocity about its own axis at (a) the winter solstice, (b) the summer solstice and (c) the spring equinox.

9.4 Multiplication of two vectors

by

$$\mathbf{a} \times \mathbf{b} = (a_y b_z - a_z b_y)\mathbf{i} + (a_z b_x - a_x b_z)\mathbf{j} + (a_x b_y - a_y b_x)\mathbf{k}. \qquad (9.33)$$

For the reader who is familiar with determinants (see Chapter 10), we record that this can also be written as[10]

$$\mathbf{a} \times \mathbf{b} = \begin{vmatrix} \mathbf{i} & \mathbf{j} & \mathbf{k} \\ a_x & a_y & a_z \\ b_x & b_y & b_z \end{vmatrix}.$$

That the cross product $\mathbf{a} \times \mathbf{b}$ is perpendicular to both $\mathbf{a}$ and $\mathbf{b}$ can be verified in component form by forming its dot products with each of the two vectors and showing that it is zero in both cases.

Example Find the area A of the parallelogram with sides $\mathbf{a} = \mathbf{i} + 2\mathbf{j} + 3\mathbf{k}$ and $\mathbf{b} = 4\mathbf{i} + 5\mathbf{j} + 6\mathbf{k}$.

The vector product $\mathbf{a} \times \mathbf{b}$ is given in component form by

$$\mathbf{a} \times \mathbf{b} = (2 \times 6 - 3 \times 5)\mathbf{i} + (3 \times 4 - 1 \times 6)\mathbf{j} + (1 \times 5 - 2 \times 4)\mathbf{k}$$
$$= -3\mathbf{i} + 6\mathbf{j} - 3\mathbf{k}.$$

Thus the area of the parallelogram is

$$A = |\mathbf{a} \times \mathbf{b}| = \sqrt{(-3)^2 + 6^2 + (-3)^2} = \sqrt{54}.$$

This result could also be obtained from a more geometric approach.[11]

◀

A useful formula that involves both scalar and vector products is Lagrange's identity (see Problem 9.9). It reads

$$(\mathbf{a} \times \mathbf{b}) \cdot (\mathbf{c} \times \mathbf{d}) \equiv (\mathbf{a} \cdot \mathbf{c})(\mathbf{b} \cdot \mathbf{d}) - (\mathbf{a} \cdot \mathbf{d})(\mathbf{b} \cdot \mathbf{c}). \qquad (9.34)$$

Its proof uses the properties of scalar triple products, as developed in the next section.

EXERCISES 9.4

1. For the vectors $(1, -3, 2)$ and $(3, 2, -1)$:
   (a) From their scalar product find the cosine of the angle θ between them.
   (b) From their vector product find the sine of θ.
   (c) Verify that the previous two results are consistent.

2. $\mathbf{a}$ and $\mathbf{b}$ are real non-zero vectors with $a = b$. Evaluate $(\mathbf{a} + \mathbf{b}) \cdot (\mathbf{a} - \mathbf{b})$ and interpret the result geometrically.

---

[10] Note that the anticommutative nature of the vector product is reflected in the antisymmetry of the determinantal form under row interchange.
[11] Starting from $A = ab \sin \theta_{ab}$, show that $A = \sqrt{a^2 b^2 - (\mathbf{a} \cdot \mathbf{b})^2}$. Evaluate A in the current case.

Vector algebra

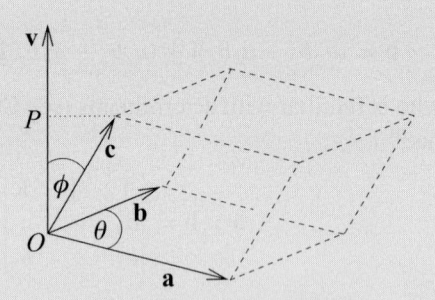

Figure 9.11 The scalar triple product gives the volume of a parallelepiped.

3. Which pair(s) of the following vectors are orthogonal?

$$\mathbf{a} = (1, 2, -4), \quad \mathbf{b} = (6, 1, 2), \quad \mathbf{c} = (4, -1, 1), \quad \mathbf{d} = (-2, 2, 5).$$

4. As judged by their scalar product, is the 'angle' between the vectors $(1 + i, i, -2)$ and $(-2i, 2, 1 - 2i)$ real? What size does it have?

5. Show that if $\mathbf{a} + \mathbf{b} + \mathbf{c} = \mathbf{0}$, then $\mathbf{a} \times \mathbf{b} = \mathbf{b} \times \mathbf{c} = \mathbf{c} \times \mathbf{a}$. Explain this result in geometric terms.

6. Calculate the area of the triangle whose vertices are at the points $(3, -1, 4), (2, 3, -1)$ and $(-3, 0, -2)$. Find the direction cosines of the normal to the plane of the triangle.

7. The angular momentum about the origin of a mass m moving with constant velocity $\mathbf{v}$ is given by $\mathbf{J} = \mathbf{r} \times m\mathbf{v}$. What are the magnitude and units of the angular momentum about the point $(2, 0, -1)$ m of a point body of mass 2.5 kg moving with constant velocity $(1, -1, 1)$ m s^{-1} along a path that passes through the point $(3, 4, -2)$ m.

9.5 Triple products

Now that we have defined the scalar and vector products, both of which involve two vectors, we can extend our discussion to define products of three vectors. Again, there are two possibilities, the *scalar triple product* and the *vector triple product*.

9.5.1 Scalar triple product

The scalar triple product is denoted and defined by

$$[\mathbf{a}, \mathbf{b}, \mathbf{c}] \equiv (\mathbf{a} \times \mathbf{b}) \cdot \mathbf{c}$$

and, as its name suggests, it is just a number. It is most simply interpreted as the volume of a parallelepiped whose edges are given by $\mathbf{a}$, $\mathbf{b}$ and $\mathbf{c}$ (see Figure 9.11). The vector $\mathbf{v} = \mathbf{a} \times \mathbf{b}$ is perpendicular to the base of the solid and has magnitude $v = ab \sin \theta$, i.e. the area of the base. Further, $\mathbf{v} \cdot \mathbf{c} = vc \cos \phi$. Thus, since $c \cos \phi = OP$ is the vertical height of the parallelepiped, it is clear that $(\mathbf{a} \times \mathbf{b}) \cdot \mathbf{c} =$ area of the base $\times$ perpendicular

9.5 Triple products

height = volume. It follows that, if the vectors **a**, **b** and **c** are coplanar, and so the parallelepiped has zero volume, then $(\mathbf{a} \times \mathbf{b}) \cdot \mathbf{c} = 0$.[12]

Expressed in terms of the components of each vector with respect to the Cartesian basis set **i**, **j**, **k** the scalar triple product is

$$(\mathbf{a} \times \mathbf{b}) \cdot \mathbf{c} = (a_y b_z - a_z b_y)c_x + (a_z b_x - a_x b_z)c_y + (a_x b_y - a_y b_x)c_z. \quad (9.35)$$

The RHS of this form can be algebraically rearranged as

$$a_x(b_y c_z - b_z c_y) + a_y(b_z c_x - b_x c_z) + a_z(b_x c_y - b_y c_x) = \mathbf{a} \cdot (\mathbf{b} \times \mathbf{c}), \quad (9.36)$$

proving that

$$(\mathbf{a} \times \mathbf{b}) \cdot \mathbf{c} = \mathbf{a} \cdot (\mathbf{b} \times \mathbf{c}).$$

This shows that the dot and cross symbols can be interchanged without changing the result (but, of course, the order of the vectors must not be changed). More generally, the scalar triple product is unchanged under cyclic permutation of the vectors **a**, **b**, **c**. Other permutations simply give the negative of the original scalar triple product. These results can be summarised by

$$[\mathbf{a}, \mathbf{b}, \mathbf{c}] = [\mathbf{b}, \mathbf{c}, \mathbf{a}] = [\mathbf{c}, \mathbf{a}, \mathbf{b}] = -[\mathbf{a}, \mathbf{c}, \mathbf{b}] = -[\mathbf{b}, \mathbf{a}, \mathbf{c}] = -[\mathbf{c}, \mathbf{b}, \mathbf{a}]. \quad (9.37)$$

Readers already familiar with determinants will note that the triple vector product can also be written in determinantal form:

$$\mathbf{a} \cdot (\mathbf{b} \times \mathbf{c}) = \begin{vmatrix} a_x & a_y & a_z \\ b_x & b_y & b_z \\ c_x & c_y & c_z \end{vmatrix}.$$

The formal study of determinants is taken up in the next chapter.

Example Find the volume V of the parallelepiped with sides $\mathbf{a} = \mathbf{i} + 2\mathbf{j} + 3\mathbf{k}$, $\mathbf{b} = 4\mathbf{i} + 5\mathbf{j} + 6\mathbf{k}$ and $\mathbf{c} = 7\mathbf{i} + 8\mathbf{j} + 10\mathbf{k}$.

We have already found, in Section 9.4.2, that $\mathbf{a} \times \mathbf{b} = -3\mathbf{i} + 6\mathbf{j} - 3\mathbf{k}$. Hence the volume of the parallelepiped is given by

$$V = |(\mathbf{a} \times \mathbf{b}) \cdot \mathbf{c}|$$
$$= |(-3\mathbf{i} + 6\mathbf{j} - 3\mathbf{k}) \cdot (7\mathbf{i} + 8\mathbf{j} + 10\mathbf{k})|$$
$$= |(-3)(7) + (6)(8) + (-3)(10)| = 3.$$

It would be a useful exercise, at this stage, for the reader to check that the same result is obtained using the form $V = |\mathbf{a} \cdot (\mathbf{b} \times \mathbf{c})|$. ◀

---

[12] Show this more algebraically by noting that if **a**, **b** and **c** are coplanar, then **c** can be written as $\mathbf{c} = \lambda \mathbf{a} + \mu \mathbf{b}$ for some λ and μ.

Vector algebra

9.5.2 Vector triple product

By the vector triple product of three vectors **a**, **b**, **c** we mean the vector $\mathbf{a} \times (\mathbf{b} \times \mathbf{c})$. As was indicated in footnote 7, $\mathbf{a} \times (\mathbf{b} \times \mathbf{c})$ is perpendicular to **a** and lies in the plane of **b** and **c** and so can be expressed in terms of them [see Equation (9.38) below]. We have already noted, in (9.25), that the vector triple product is not associative, i.e. $\mathbf{a} \times (\mathbf{b} \times \mathbf{c}) \neq (\mathbf{a} \times \mathbf{b}) \times \mathbf{c}$.

Two useful formulae involving the vector triple product are

$$\mathbf{a} \times (\mathbf{b} \times \mathbf{c}) = (\mathbf{a} \cdot \mathbf{c})\mathbf{b} - (\mathbf{a} \cdot \mathbf{b})\mathbf{c}, \tag{9.38}$$

$$(\mathbf{a} \times \mathbf{b}) \times \mathbf{c} = (\mathbf{a} \cdot \mathbf{c})\mathbf{b} - (\mathbf{b} \cdot \mathbf{c})\mathbf{a}, \tag{9.39}$$

which may be derived by writing each vector in component form (see Problem 9.8). It can also be shown[13] that for any three vectors **a**, **b**, **c**,

$$\mathbf{a} \times (\mathbf{b} \times \mathbf{c}) + \mathbf{b} \times (\mathbf{c} \times \mathbf{a}) + \mathbf{c} \times (\mathbf{a} \times \mathbf{b}) = \mathbf{0}.$$

EXERCISES 9.5

1. The force acting on a length $d\ell$ of a current-carrying wire in a magnetic field of induction **B** is $I\mathbf{B} \times d\boldsymbol{\ell}$, where I is the strength of the current. A straight wire of length 50 cm lies in the direction $3^{-1/2}(1, 1, 1)$ and carries a steady current of 2 A. It is placed in a magnetic field of induction 1.5 T acting in the direction $2^{-1/2}(0, 1, 1)$. Calculate the work needed to move the wire bodily by 15 cm in the direction $2^{-1/2}(-1, 0, 1)$. Ignore any induced back-e.m.f. effects.

2. Three non-coplanar vectors are $\mathbf{a} = (1, 0, 1)$, $\mathbf{b} = (-1, 1, 0)$ and $\mathbf{c} = (3, 4, 5)$.
   (a) Show that the volume of the parallelepiped with edges **a**, **b** and **c** is one half of that of the parallelepiped with edges $\mathbf{a} + \mathbf{b}$, $\mathbf{b} + \mathbf{c}$ and $\mathbf{c} + \mathbf{a}$.
   (b) Prove the same result more generally for *any* three non-coplanar vectors **a**, **b** and **c**.

3. For the vectors **a**, **b** and **c** as given in Exercise 2 above, find the angle between the vector triple products

$$\mathbf{a} \times (\mathbf{b} \times \mathbf{c}) \quad \text{and} \quad (\mathbf{a} \times \mathbf{b}) \times \mathbf{c}.$$

4. Prove that $[\mathbf{a} \times \mathbf{b}, \mathbf{a} \times \mathbf{c}, \mathbf{d}] = (\mathbf{a} \cdot \mathbf{d})[\mathbf{a}, \mathbf{b}, \mathbf{c}]$.

9.6 Equations of lines, planes and spheres

Now that we have described the basic algebra of vectors, we can apply the results to a variety of problems, the first of which is to find the equation of a line in vector form.

---

[13] Use (9.38) three times, and recall that the scalar product is commutative.

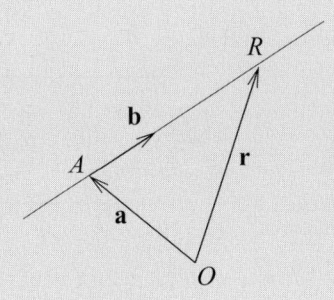

Figure 9.12 The equation of a line. The vector **b** is in the direction AR and $\lambda\mathbf{b}$ is the vector from A to R.

9.6.1 Equation of a line

Consider the line that passes through the fixed point A with position vector **a** and has direction **b** (see Figure 9.12). It is clear that the position vector **r** of a general point R on the line can be written as

$$\mathbf{r} = \mathbf{a} + \lambda\mathbf{b}, \tag{9.40}$$

since R can be reached by starting from O, going along the translation vector **a** to the point A on the line and then adding some multiple $\lambda\mathbf{b}$ of the vector **b**. Different values of λ give different points R on the line. The special case of $\mathbf{a} = \mathbf{0}$ gives $\mathbf{r} = \lambda\mathbf{b}$ and clearly represents a line through the origin in the direction of **b**.

Writing (9.40) in terms of its components, we see that the equation of the line can also be written in the form

$$\frac{x - a_x}{b_x} = \frac{y - a_y}{b_y} = \frac{z - a_z}{b_z} = \text{constant}. \tag{9.41}$$

Taking the vector product of (9.40) with **b** and remembering that $\mathbf{b} \times \mathbf{b} = \mathbf{0}$ gives

$$(\mathbf{r} - \mathbf{a}) \times \mathbf{b} = \mathbf{0}$$

as an alternative equation for the line.

We may also find the equation of the line that passes through two fixed points A and C with position vectors **a** and **c**. Since AC is given by $\mathbf{c} - \mathbf{a}$, the position vector of a general point on the line is

$$\mathbf{r} = \mathbf{a} + \lambda(\mathbf{c} - \mathbf{a}).$$

This equation can also be written as $\mathbf{r} = (1 - \lambda)\mathbf{a} + \lambda\mathbf{c} = \mu\mathbf{a} + (1 - \mu)\mathbf{c}$, showing that A and C are on equal footings.

9.6.2 Equation of a plane

The equation of a plane containing the point A with position vector **a** and perpendicular to a unit vector $\hat{\mathbf{n}}$ (see Figure 9.13) is

$$(\mathbf{r} - \mathbf{a}) \cdot \hat{\mathbf{n}} = 0. \tag{9.42}$$

Vector algebra

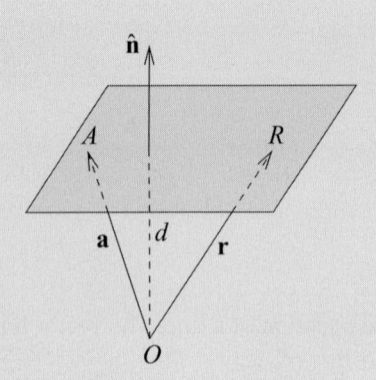

Figure 9.13 The equation of the plane is $(\mathbf{r} - \mathbf{a}) \cdot \hat{\mathbf{n}} = 0$.

This follows since the vector joining A to a general point R with position vector $\mathbf{r}$ is $\mathbf{r} - \mathbf{a}$; since $\mathbf{a}$ lies in the plane, $\mathbf{r}$ will also do so provided this vector, $\mathbf{r} - \mathbf{a}$, is perpendicular to the normal to the plane. Rewriting (9.42) as $\mathbf{r} \cdot \hat{\mathbf{n}} = \mathbf{a} \cdot \hat{\mathbf{n}}$, we see that the equation of the plane may also be expressed in the form $\mathbf{r} \cdot \hat{\mathbf{n}} = d$, or in component form as

$$lx + my + nz = d, \qquad (9.43)$$

where the unit[14] normal to the plane is $\hat{\mathbf{n}} = l\mathbf{i} + m\mathbf{j} + n\mathbf{k}$. The quantity $d = \mathbf{a} \cdot \hat{\mathbf{n}}$ is the component of $\mathbf{a}$ in the direction of $\hat{\mathbf{n}}$ and so is the perpendicular distance of the plane from the origin.

As well as being determined by one point it contains and the direction of its normal, a plane can also be defined by any three points that lie in it, provided they are not collinear. The equation of a plane containing the points $\mathbf{a}$, $\mathbf{b}$ and $\mathbf{c}$ is

$$\mathbf{r} = \mathbf{a} + \lambda(\mathbf{b} - \mathbf{a}) + \mu(\mathbf{c} - \mathbf{a}).$$

This is apparent because starting from the point $\mathbf{a}$ in the plane, all other points may be reached by moving a distance along each of two (non-parallel) directions in the plane. Two such directions are given by $\mathbf{b} - \mathbf{a}$ and $\mathbf{c} - \mathbf{a}$. It can be shown that the equation of this plane may also be written in the more symmetrical form

$$\mathbf{r} = \alpha\mathbf{a} + \beta\mathbf{b} + \gamma\mathbf{c},$$

where $\alpha + \beta + \gamma = 1$.[15] The following example exploits the implicit presence, in their equations, of the vector normals characterising two planes to find the line of intersection of the planes.

---

[14] With $l^2 + m^2 + n^2 = 1$.
[15] Take $\alpha = 1 - \lambda - \mu$, $\beta = \lambda$ and $\gamma = \mu$.

9.6 Equations of lines, planes and spheres

Example Find the direction of the line of intersection of the two planes $x + 3y - z = 5$ and $2x - 2y + 4z = 3$.

The two planes have normal vectors $\mathbf{n}_1 = \mathbf{i} + 3\mathbf{j} - \mathbf{k}$ and $\mathbf{n}_2 = 2\mathbf{i} - 2\mathbf{j} + 4\mathbf{k}$. It is clear that these are not parallel vectors and so the planes must intersect along some line. The direction $\mathbf{p}$ of this line must be parallel to both planes and hence perpendicular to both normals. Therefore

$$\begin{aligned}\mathbf{p} &= \mathbf{n}_1 \times \mathbf{n}_2 \\ &= [(3)(4) - (-2)(-1)]\mathbf{i} + [(-1)(2) - (1)(4)]\mathbf{j} + [(1)(-2) - (3)(2)]\mathbf{k} \\ &= 10\mathbf{i} - 6\mathbf{j} - 8\mathbf{k}.\end{aligned}$$

It is easily checked that the $\mathbf{p}$ so found has the correct properties by calculating $\mathbf{p} \cdot \mathbf{n}_1$ and $\mathbf{p} \cdot \mathbf{n}_2$ and showing that they are both zero.[16] ◀

9.6.3 Equation of a sphere

Clearly, the defining property of a sphere is that all points on it are equidistant from a fixed point in space and that the common distance is equal to the radius of the sphere. This is easily expressed in vector notation as

$$|\mathbf{r} - \mathbf{c}|^2 = (\mathbf{r} - \mathbf{c}) \cdot (\mathbf{r} - \mathbf{c}) = a^2, \tag{9.44}$$

where $\mathbf{c}$ is the position vector of the centre of the sphere and a is its radius.

The following example, involving the equations for planes, circles and spheres, is somewhat more complex than most of those worked through so far.

Example Find the radius ρ of the circle that is the intersection of the plane $\hat{\mathbf{n}} \cdot \mathbf{r} = p$ and the sphere of radius a centred on the point with position vector $\mathbf{c}$.

The equation of the sphere is

$$|\mathbf{r} - \mathbf{c}|^2 = a^2, \tag{9.45}$$

and that of the circle of intersection is

$$|\mathbf{r} - \mathbf{b}|^2 = \rho^2, \tag{9.46}$$

where $\mathbf{r}$ is restricted to lie in the plane and $\mathbf{b}$ is the position of the circle's centre.

As $\mathbf{b}$ lies on the plane whose normal is $\hat{\mathbf{n}}$, the vector $\mathbf{b} - \mathbf{c}$ must be parallel to $\hat{\mathbf{n}}$, i.e. $\mathbf{b} - \mathbf{c} = \lambda \hat{\mathbf{n}}$ for some λ. Further, by Pythagoras, we must have $\rho^2 + |\mathbf{b} - \mathbf{c}|^2 = a^2$. Thus $\lambda^2 = a^2 - \rho^2$.

---

[16] Show that a general point on the line of intersection can be written as $(3 + 5\lambda, \tfrac{1}{2} - 3\lambda, -\tfrac{1}{2} - 4\lambda)$.

Vector algebra

Writing $\mathbf{b} = \mathbf{c} + \sqrt{a^2 - \rho^2}\,\hat{\mathbf{n}}$ and substituting in (9.46) gives

$$r^2 - 2\mathbf{r} \cdot \left(\mathbf{c} + \sqrt{a^2 - \rho^2}\,\hat{\mathbf{n}}\right) + c^2 + 2(\mathbf{c} \cdot \hat{\mathbf{n}})\sqrt{a^2 - \rho^2} + a^2 - \rho^2 = \rho^2,$$

whilst, on expansion, (9.45) becomes

$$r^2 - 2\mathbf{r} \cdot \mathbf{c} + c^2 = a^2.$$

Subtracting these last two equations, using $\hat{\mathbf{n}} \cdot \mathbf{r} = p$ and simplifying yields

$$p - \mathbf{c} \cdot \hat{\mathbf{n}} = \sqrt{a^2 - \rho^2}.$$

On rearrangement, this gives ρ as $\sqrt{a^2 - (p - \mathbf{c} \cdot \hat{\mathbf{n}})^2}$, which places obvious geometrical constraints on the values a, $\mathbf{c}$, $\hat{\mathbf{n}}$ and p can take if a real intersection between the sphere and the plane is to occur. ◀

EXERCISES 9.6

1. Express each of the following equations for a line in the form used for the other one:

$$\mathbf{r} = \lambda(1, 1, 0) + (1 - \lambda)(2, -3, 1),$$
$$[\mathbf{r} - (2, -1, 2)] \times (2, 3, -1) = \mathbf{0}.$$

2. A plane whose normal is in the direction $(3, 1, -1)$ contains the point $(2, -2, -3)$. Which of the following points also lie on the plane?

(a) $(3, 0, 3)$, (b) $(1, 2, -2)$, (c) $(3, 3, 5)$, (d) $(-2, -3, -2)$.

3. A plane contains the points $(1, 4, 2)$, $(-2, 6, -2)$ and $(2, 0, 5)$. How close to the origin does it pass?

4. Show that the equation

$$[\mathbf{r}, \mathbf{b}, \mathbf{c}] + [\mathbf{r}, \mathbf{c}, \mathbf{a}] + [\mathbf{r}, \mathbf{a}, \mathbf{b}] = [\mathbf{a}, \mathbf{b}, \mathbf{c}]$$

is that of a plane. Verify that the plane in question is the one containing the points $\mathbf{a}$, $\mathbf{b}$ and $\mathbf{c}$.

5. Two spheres of radii a and b, centred on $\mathbf{A}$ and $\mathbf{B}$ respectively, intersect in a circle that lies in a plane P. Show that the origin is a distance d from P, where

$$d = \frac{(|\mathbf{A}|^2 - |\mathbf{B}|^2) - (a^2 - b^2)}{2|\mathbf{A} - \mathbf{B}|}.$$

6. An ellipse can be defined by the requirement that the sum of the two distances from a point $\mathbf{r}$ on the ellipse to each of the foci is a constant equal to $2a$. Taking the centre of

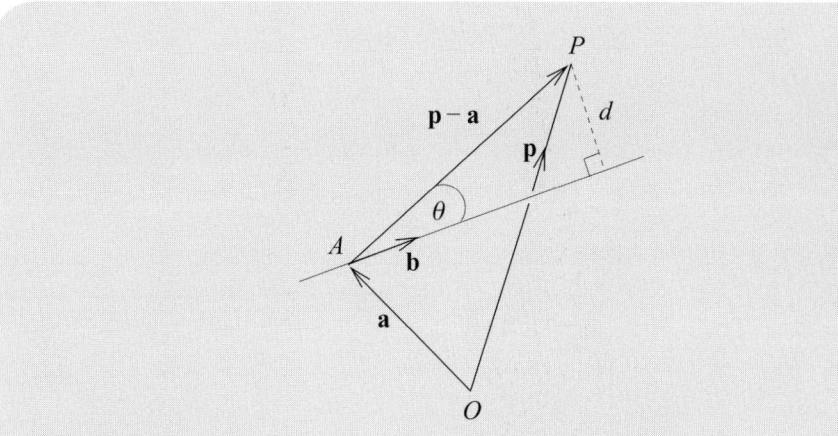

Figure 9.14 The minimum distance from a point to a line.

the ellipse as the origin O and the foci at $\pm\mathbf{f}$, express the requirement in vector form and hence show that the equation of the ellipse can be written as

$$a^4 - a^2(r^2 + f^2) + (\mathbf{r} \cdot \mathbf{f})^2 = 0.$$

9.7 Using vectors to find distances

This section deals with the practical application of vectors to finding distances. Some of these problems are extremely cumbersome in component form, but they all reduce to neat solutions when general vectors, with no explicit basis set, are used. These examples show the power of vectors in simplifying geometrical problems.

9.7.1 Distance from a point to a line

Figure 9.14 shows a line having direction $\mathbf{b}$ that passes through a point A whose position vector is $\mathbf{a}$. To find the *minimum distance* d of the line from a point P whose position vector is $\mathbf{p}$, we must solve the right-angled triangle shown. We see that $d = |\mathbf{p} - \mathbf{a}| \sin\theta$; so, from the definition of the vector product, it follows that[17]

$$d = |(\mathbf{p} - \mathbf{a}) \times \hat{\mathbf{b}}|.$$

It should be noted that it is $\hat{\mathbf{b}}$ (and not $\mathbf{b}$) that is required here, since the magnitude of $\mathbf{b}$ does not come into the expression for the minimum distance d; the appropriate value of $\sin\theta$ is generated by taking the vector product. The result is illustrated by the following example.

---

[17] Extend this result, using vector results so far obtained, to prove the following. If $\mathbf{a}$, $\mathbf{b}$ and $\mathbf{c}$ are three non-collinear points and d_c is defined as the minimum distance from $\mathbf{c}$ to the line passing through $\mathbf{a}$ and $\mathbf{b}$ (with d_a and d_b similarly defined) then $|\mathbf{c} - \mathbf{b}|d_a = |\mathbf{a} - \mathbf{c}|d_b = |\mathbf{b} - \mathbf{a}|d_c$. What is their common value? Interpret the result geometrically.

Vector algebra

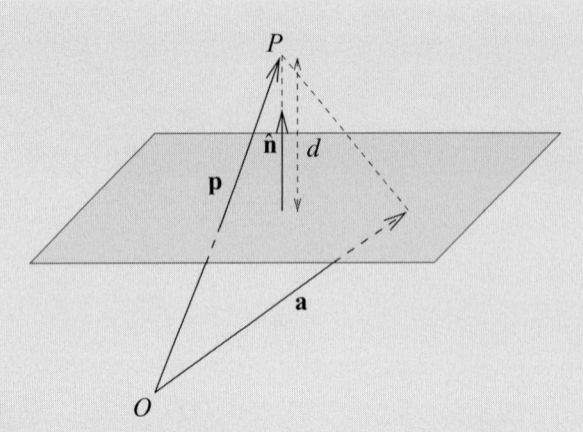

Figure 9.15 The minimum distance d from a point to a plane.

Example Find the minimum distance from the point P with coordinates $(1, 2, 1)$ to the line $\mathbf{r} = \mathbf{a} + \lambda \mathbf{b}$, where $\mathbf{a} = \mathbf{i} + \mathbf{j} + \mathbf{k}$ and $\mathbf{b} = 2\mathbf{i} - \mathbf{j} + 3\mathbf{k}$.

Comparison with (9.40) shows that the line passes through the point $(1, 1, 1)$ and has direction $2\mathbf{i} - \mathbf{j} + 3\mathbf{k}$. The unit vector in this direction is

$$\hat{\mathbf{b}} = \frac{1}{\sqrt{14}}(2\mathbf{i} - \mathbf{j} + 3\mathbf{k}).$$

The position vector of P is $\mathbf{p} = \mathbf{i} + 2\mathbf{j} + \mathbf{k}$ and we find

$$(\mathbf{p} - \mathbf{a}) \times \hat{\mathbf{b}} = \frac{1}{\sqrt{14}}\{[(1-1)\mathbf{i} + (2-1)\mathbf{j} + (1-1)\mathbf{k}] \times (2\mathbf{i} - \mathbf{j} + 3\mathbf{k})\}$$

$$= \frac{1}{\sqrt{14}}[\mathbf{j} \times (2\mathbf{i} - \mathbf{j} + 3\mathbf{k})]$$

$$= \frac{1}{\sqrt{14}}(3\mathbf{i} - 2\mathbf{k}).$$

Thus the minimum distance from the line to the point P is $d = \sqrt{(3^2 + 2^2)/14} = \sqrt{13/14}$. ◀

9.7.2 Distance from a point to a plane

The minimum distance d from a point P whose position vector is $\mathbf{p}$ to the plane defined by $(\mathbf{r} - \mathbf{a}) \cdot \hat{\mathbf{n}} = 0$ may be deduced by finding any vector from P to the plane and then determining its component in the normal direction. This is shown in Figure 9.15. Consider the vector $\mathbf{a} - \mathbf{p}$, which is a particular vector from P to the plane. Its component normal to the plane, and hence its distance from the plane, is given by

$$d = (\mathbf{a} - \mathbf{p}) \cdot \hat{\mathbf{n}}, \tag{9.47}$$

where the sign of d depends on which side of the plane P is situated.

9.7 Using vectors to find distances

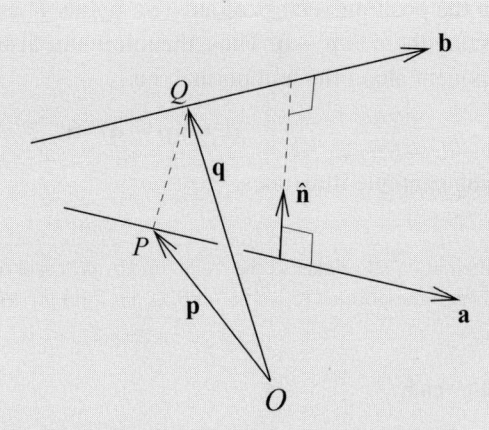

Figure 9.16 The minimum distance from one line to another.

Example Find the distance from the point P with coordinates $(1, 2, 3)$ to the plane that contains the points A, B and C having coordinates $(0, 1, 0)$, $(2, 3, 1)$ and $(5, 7, 2)$.

Let us denote the position vectors of the points A, B, C by $\mathbf{a}$, $\mathbf{b}$, $\mathbf{c}$. Two vectors in the plane are

$$\mathbf{b} - \mathbf{a} = 2\mathbf{i} + 2\mathbf{j} + \mathbf{k} \quad \text{and} \quad \mathbf{c} - \mathbf{a} = 5\mathbf{i} + 6\mathbf{j} + 2\mathbf{k},$$

and hence a vector normal to the plane is

$$\mathbf{n} = (2\mathbf{i} + 2\mathbf{j} + \mathbf{k}) \times (5\mathbf{i} + 6\mathbf{j} + 2\mathbf{k}) = -2\mathbf{i} + \mathbf{j} + 2\mathbf{k},$$

and its unit normal is

$$\hat{\mathbf{n}} = \frac{\mathbf{n}}{|\mathbf{n}|} = \tfrac{1}{3}(-2\mathbf{i} + \mathbf{j} + 2\mathbf{k}).$$

Denoting the position vector of P by $\mathbf{p}$, the minimum distance from the plane to P is given by

$$\begin{aligned} d &= (\mathbf{a} - \mathbf{p}) \cdot \hat{\mathbf{n}} \\ &= (-\mathbf{i} - \mathbf{j} - 3\mathbf{k}) \cdot \tfrac{1}{3}(-2\mathbf{i} + \mathbf{j} + 2\mathbf{k}) \\ &= \tfrac{2}{3} - \tfrac{1}{3} - 2 = -\tfrac{5}{3}. \end{aligned}$$

If we take P to be the origin O, then we find $d = \tfrac{1}{3}$, i.e. a positive quantity. It follows from this that the original point P with coordinates $(1, 2, 3)$, for which d was negative, is on the opposite side of the plane from the origin. ◀

9.7.3 Distance from a line to a line

Consider two lines in the directions $\mathbf{a}$ and $\mathbf{b}$, as shown in Figure 9.16. Since $\mathbf{a} \times \mathbf{b}$ is by definition perpendicular to both $\mathbf{a}$ and $\mathbf{b}$, the unit vector normal to both these lines is

$$\hat{\mathbf{n}} = \frac{\mathbf{a} \times \mathbf{b}}{|\mathbf{a} \times \mathbf{b}|}.$$

Vector algebra

If **p** and **q** are the position vectors of any two points P and Q on different lines then the vector connecting them is $\mathbf{p} - \mathbf{q}$. Thus, the minimum distance d between the lines is this vector's component along the unit normal, i.e.

$$d = |(\mathbf{p} - \mathbf{q}) \cdot \hat{\mathbf{n}}|,$$

as the following example illustrates.

Example A line is inclined at equal angles to the x-, y- and z-axes and passes through the origin. Another line passes through the points $(1, 2, 4)$ and $(0, 0, 1)$. Find the minimum distance between the two lines.

The first line is given by

$$\mathbf{r}_1 = \lambda(\mathbf{i} + \mathbf{j} + \mathbf{k})$$

and the second by

$$\mathbf{r}_2 = \mathbf{k} + \mu(\mathbf{i} + 2\mathbf{j} + 3\mathbf{k}),$$

with $\mu = 0$ corresponding to the point $(0, 0, 1)$ and $\mu = 1$ to $(1, 2, 4)$. Hence a vector normal to both lines is

$$\mathbf{n} = (\mathbf{i} + \mathbf{j} + \mathbf{k}) \times (\mathbf{i} + 2\mathbf{j} + 3\mathbf{k}) = \mathbf{i} - 2\mathbf{j} + \mathbf{k}$$

and the unit normal is

$$\hat{\mathbf{n}} = \frac{1}{\sqrt{6}}(\mathbf{i} - 2\mathbf{j} + \mathbf{k}).$$

A vector between the two lines is, for example, the one connecting the points $(0, 0, 0)$ and $(0, 0, 1)$, which is simply $\mathbf{k}$. Thus it follows that the minimum distance between the two lines is

$$d = \frac{1}{\sqrt{6}}|\mathbf{k} \cdot (\mathbf{i} - 2\mathbf{j} + \mathbf{k})| = \frac{1}{\sqrt{6}}.$$

This is sufficient for the question posed, but it is easy to verify that the same answer is obtained using a line joining *any* point $\mathbf{r}_1$ to *any* point $\mathbf{r}_2$.[18]

9.7.4 Distance from a line to a plane

Let us consider the line $\mathbf{r} = \mathbf{a} + \lambda \mathbf{b}$. This line will intersect any plane to which it is not parallel. Thus, if a plane has a normal $\hat{\mathbf{n}}$, the minimum distance from the line to the plane is zero unless

$$\mathbf{b} \cdot \hat{\mathbf{n}} = 0,$$

in which case the distance, d, will be

$$d = |(\mathbf{a} - \mathbf{r}) \cdot \hat{\mathbf{n}}|,$$

where $\mathbf{r}$ is any point in the plane.

---

[18] Verify this using $(\lambda, \lambda, \lambda)$ as $\mathbf{r}_1$ and $(\mu, 2\mu, 3\mu + 1)$ as $\mathbf{r}_2$.

9.8 Reciprocal vectors

Example A line is given by $\mathbf{r} = \mathbf{a} + \lambda \mathbf{b}$, where $\mathbf{a} = 5\mathbf{i} + 7\mathbf{j} + 9\mathbf{k}$ and $\mathbf{b} = 4\mathbf{i} + 5\mathbf{j} + 6\mathbf{k}$. Find the coordinates of the point P at which the line intersects the plane

$$x + 2y + 3z = 6.$$

A vector normal to the plane is

$$\mathbf{n} = \mathbf{i} + 2\mathbf{j} + 3\mathbf{k},$$

from which we find that $\mathbf{b} \cdot \mathbf{n} \neq 0$. Thus the line does indeed intersect the plane. To find the point of intersection we merely substitute the x-, y- and z-values of a general point on the line into the equation of the plane, obtaining

$$5 + 4\lambda + 2(7 + 5\lambda) + 3(9 + 6\lambda) = 6 \quad \Rightarrow \quad 46 + 32\lambda = 6.$$

This gives $\lambda = -\frac{5}{4}$, which we may substitute into the equation for the line to obtain $x = 5 - \frac{5}{4}(4) = 0$, $y = 7 - \frac{5}{4}(5) = \frac{3}{4}$ and $z = 9 - \frac{5}{4}(6) = \frac{3}{2}$. Thus the point of intersection is $(0, \frac{3}{4}, \frac{3}{2})$. ◀

EXERCISES 9.7

1. Find the minimum distance of the point $(3, -1, 2)$ from the line joining the points $(1, 1, -3)$ and $(2, -1, 1)$. Deduce the area of the triangle that has the three points as vertices.

2. Are the points $(3, -1, -4)$ and $(2, -3, -1)$ on the same or opposite sides of the plane $x + 2y - 2z = 12$?

3. Find the minimum distance between the line joining $(0, -2, 4)$ to $(-1, 3, 2)$ and the line in the direction $(1, 2, 1)$ that passes through $(3, 0, 4)$.

4. Find the point(s) at which the line $\mathbf{r} = \lambda(1, 2, 1) + (1 - \lambda)(2, -1, 3)$ meets the planes

    (a) $x + 2y - 2z = 12$,   (b) $4x + 2y + z = -10$.

9.8 Reciprocal vectors

The final section of this chapter introduces the concept of reciprocal vectors, which have particular uses in crystallography.

The two sets of vectors $\mathbf{a}, \mathbf{b}, \mathbf{c}$ and $\mathbf{a}', \mathbf{b}', \mathbf{c}'$ are called *reciprocal sets* if

$$\mathbf{a} \cdot \mathbf{a}' = \mathbf{b} \cdot \mathbf{b}' = \mathbf{c} \cdot \mathbf{c}' = 1 \tag{9.48}$$

and

$$\mathbf{a}' \cdot \mathbf{b} = \mathbf{a}' \cdot \mathbf{c} = \mathbf{b}' \cdot \mathbf{a} = \mathbf{b}' \cdot \mathbf{c} = \mathbf{c}' \cdot \mathbf{a} = \mathbf{c}' \cdot \mathbf{b} = 0. \tag{9.49}$$

Vector algebra

It can be verified (see Problem 9.19) that the reciprocal vectors of **a**, **b** and **c** are given by

$$\mathbf{a}' = \frac{\mathbf{b} \times \mathbf{c}}{\mathbf{a} \cdot (\mathbf{b} \times \mathbf{c})}, \qquad (9.50)$$

$$\mathbf{b}' = \frac{\mathbf{c} \times \mathbf{a}}{\mathbf{a} \cdot (\mathbf{b} \times \mathbf{c})}, \qquad (9.51)$$

$$\mathbf{c}' = \frac{\mathbf{a} \times \mathbf{b}}{\mathbf{a} \cdot (\mathbf{b} \times \mathbf{c})}, \qquad (9.52)$$

where $\mathbf{a} \cdot (\mathbf{b} \times \mathbf{c}) \neq 0$. In other words, reciprocal vectors only exist if **a**, **b** and **c** are not coplanar. Moreover, if **a**, **b** and **c** are mutually orthogonal unit vectors then $\mathbf{a}' = \mathbf{a}$, $\mathbf{b}' = \mathbf{b}$ and $\mathbf{c}' = \mathbf{c}$, so that the two systems of vectors are identical. As a straightforward example, consider the following.

Example Construct the reciprocal vectors of $\mathbf{a} = 2\mathbf{i}$, $\mathbf{b} = \mathbf{j} + \mathbf{k}$, $\mathbf{c} = \mathbf{i} + \mathbf{k}$.

First we evaluate the triple scalar product:

$$\mathbf{a} \cdot (\mathbf{b} \times \mathbf{c}) = 2\mathbf{i} \cdot [(\mathbf{j} + \mathbf{k}) \times (\mathbf{i} + \mathbf{k})]$$
$$= 2\mathbf{i} \cdot (\mathbf{i} + \mathbf{j} - \mathbf{k}) = 2.$$

This triple scalar product is not zero and so the three given vectors are not coplanar; thus reciprocal vectors will exist. Now we find them using prescriptions (9.50)–(9.52):

$$\mathbf{a}' = \tfrac{1}{2}(\mathbf{j} + \mathbf{k}) \times (\mathbf{i} + \mathbf{k}) = \tfrac{1}{2}(\mathbf{i} + \mathbf{j} - \mathbf{k}),$$
$$\mathbf{b}' = \tfrac{1}{2}(\mathbf{i} + \mathbf{k}) \times 2\mathbf{i} = \mathbf{j},$$
$$\mathbf{c}' = \tfrac{1}{2}(2\mathbf{i}) \times (\mathbf{j} + \mathbf{k}) = -\mathbf{j} + \mathbf{k}.$$

It is easily verified that these reciprocal vectors satisfy their defining properties, (9.48) and (9.49). ◀

We may also use the concept of reciprocal vectors to define the components of a vector **a** with respect to basis vectors $\mathbf{e}_1$, $\mathbf{e}_2$, $\mathbf{e}_3$ that are not mutually orthogonal. If the basis vectors are of unit length and mutually orthogonal, such as the Cartesian basis vectors **i**, **j**, **k**, then[19] the vector **a** can be written in the form

$$\mathbf{a} = (\mathbf{a} \cdot \mathbf{i})\mathbf{i} + (\mathbf{a} \cdot \mathbf{j})\mathbf{j} + (\mathbf{a} \cdot \mathbf{k})\mathbf{k}. \qquad (9.53)$$

In the more general case in which the basis is not orthonormal, this is no longer true. Nevertheless, we may write the components of **a** with respect to a non-orthonormal basis $\mathbf{e}_1$, $\mathbf{e}_2$, $\mathbf{e}_3$ in terms of its reciprocal basis vectors $\mathbf{e}'_1$, $\mathbf{e}'_2$, $\mathbf{e}'_3$, which are defined as in (9.50)–(9.52). If we let

$$\mathbf{a} = a_1 \mathbf{e}_1 + a_2 \mathbf{e}_2 + a_3 \mathbf{e}_3,$$

---

[19] See the text preceding (9.21).

Summary

then the scalar product $\mathbf{a} \cdot \mathbf{e}'_1$ is given by

$$\mathbf{a} \cdot \mathbf{e}'_1 = a_1 \mathbf{e}_1 \cdot \mathbf{e}'_1 + a_2 \mathbf{e}_2 \cdot \mathbf{e}'_1 + a_3 \mathbf{e}_3 \cdot \mathbf{e}'_1 = a_1,$$

where we have used the relations (9.49). Similarly, $a_2 = \mathbf{a} \cdot \mathbf{e}'_2$ and $a_3 = \mathbf{a} \cdot \mathbf{e}'_3$; so now

$$\mathbf{a} = (\mathbf{a} \cdot \mathbf{e}'_1)\mathbf{e}_1 + (\mathbf{a} \cdot \mathbf{e}'_2)\mathbf{e}_2 + (\mathbf{a} \cdot \mathbf{e}'_3)\mathbf{e}_3. \tag{9.54}$$

If the basis were orthonormal then, as noted earlier, $\mathbf{e}'_i = \mathbf{e}_i$ for each i and (9.54) is the same as formula (9.53) given above.

EXERCISES 9.8

1. Find the reciprocal vectors of the set

$$\mathbf{a} = (1, 1, 1), \quad \mathbf{b} = (1, -1, 1), \quad \mathbf{c} = (1, 1, -1).$$

What are the angles between the various pairs of (i) vectors and (ii) reciprocal vectors?

2. Use the reciprocal vectors found in the previous exercise to write the vector $\mathbf{x} = (4, -3, -2)$ with respect to a basis consisting of vectors $\mathbf{a}$, $\mathbf{b}$ and $\mathbf{c}$.

SUMMARY

1. *Vector algebra*
   - Addition, subtraction and scalar multiplication

   $$\mathbf{a} + \mathbf{b} = \mathbf{b} + \mathbf{a}, \quad \mathbf{a} + (\mathbf{b} + \mathbf{c}) = (\mathbf{a} + \mathbf{b}) + \mathbf{c},$$
   $$\mathbf{a} + (-\mathbf{a}) = \mathbf{0}, \quad \mathbf{a} - \mathbf{b} = \mathbf{a} + (-\mathbf{b}),$$
   $$\lambda(\mu \mathbf{a} + \nu \mathbf{b}) = \lambda\mu\mathbf{a} + \lambda\nu\mathbf{b}.$$

   - A unit vector in the direction of $\mathbf{a}$ is $\hat{\mathbf{a}} = \mathbf{a}/|\mathbf{a}|$, where $|\mathbf{a}|$ is the magnitude of $\mathbf{a}$.
   - The set of vectors $\{\mathbf{e}_i\}$ are linearly independent only if $\sum_i c_i \mathbf{e}_i = \mathbf{0}$ implies that $c_i = 0$ for all i.

2. *Scalar product*
   - Definition: scalar $s = \mathbf{a} \cdot \mathbf{b} = |\mathbf{a}||\mathbf{b}| \cos\theta = \mathbf{b} \cdot \mathbf{a}$ with $0 \leq \theta \leq \pi$.
   - $\mathbf{a} \cdot (\mathbf{b} + \mathbf{c}) = \mathbf{a} \cdot \mathbf{b} + \mathbf{a} \cdot \mathbf{c}$.
   - $\mathbf{a} \cdot \mathbf{a} = |\mathbf{a}|^2$.
   - *Warning*: if the vectors may have complex components, then $\mathbf{a} \cdot \mathbf{b} = (\mathbf{b} \cdot \mathbf{a})^*$ and $(\lambda \mathbf{a}) \cdot \mathbf{b} = \lambda^*(\mathbf{a} \cdot \mathbf{b})$.

Vector algebra

3. *Vector product*
   - Definition: vector $\mathbf{v} = \mathbf{a} \times \mathbf{b}$ with $\mathbf{a}$, $\mathbf{b}$ and $\mathbf{v}$ (in that order) forming a right-handed set.
   $$|\mathbf{v}| = |\mathbf{a}||\mathbf{b}| \sin\theta \text{ with } 0 \leq \theta \leq \pi.$$
   - Properties
   $$\mathbf{a} \times \mathbf{a} = \mathbf{0}, \quad \mathbf{b} \times \mathbf{a} = -(\mathbf{a} \times \mathbf{b}),$$
   $$(\mathbf{a} + \mathbf{b}) \times \mathbf{c} = (\mathbf{a} \times \mathbf{c}) + (\mathbf{b} \times \mathbf{c}),$$
   $$(\mathbf{a} \times \mathbf{b}) \times \mathbf{c} \neq \mathbf{a} \times (\mathbf{b} \times \mathbf{c}), \quad \text{(see below).}$$
   - In Cartesian components
   $$\mathbf{a} \times \mathbf{b} = (a_y b_z - a_z b_y)\mathbf{i} + (a_z b_x - a_x b_z)\mathbf{j} + (a_x b_y - a_y b_x)\mathbf{k}.$$

4. *Scalar triple product*
   - Definition: scalar $[\mathbf{a}, \mathbf{b}, \mathbf{c}] \equiv \mathbf{a} \cdot (\mathbf{b} \times \mathbf{c})$. The product $[\mathbf{a}, \mathbf{b}, \mathbf{c}]$ is equal to the volume of the parallelepiped with edges $\mathbf{a}$, $\mathbf{b}$ and $\mathbf{c}$. In Cartesian coordinates
   $$\mathbf{a} \cdot (\mathbf{b} \times \mathbf{c}) = a_x(b_y c_z - b_z c_y) + a_y(b_z c_x - b_x c_z) + a_z(b_x c_y - b_y c_x).$$
   - Properties
   $$[\mathbf{a}, \mathbf{b}, \mathbf{c}] = [\mathbf{b}, \mathbf{c}, \mathbf{a}] = [\mathbf{c}, \mathbf{a}, \mathbf{b}] = -[\mathbf{a}, \mathbf{c}, \mathbf{b}] = -[\mathbf{b}, \mathbf{a}, \mathbf{c}] = -[\mathbf{c}, \mathbf{b}, \mathbf{a}],$$
   $$(\mathbf{a} \times \mathbf{b}) \cdot (\mathbf{c} \times \mathbf{d}) = (\mathbf{a} \cdot \mathbf{c})(\mathbf{b} \cdot \mathbf{d}) - (\mathbf{a} \cdot \mathbf{d})(\mathbf{b} \cdot \mathbf{c}).$$

5. *Vector triple product*
   - Vector $\mathbf{a} \times (\mathbf{b} \times \mathbf{c})$ is perpendicular to $\mathbf{a}$ and lies in the plane defined by vectors $\mathbf{b}$ and $\mathbf{c}$.
   - Non-associativity
   $$\mathbf{a} \times (\mathbf{b} \times \mathbf{c}) = (\mathbf{a} \cdot \mathbf{c})\mathbf{b} - (\mathbf{a} \cdot \mathbf{b})\mathbf{c},$$
   $$(\mathbf{a} \times \mathbf{b}) \times \mathbf{c} = (\mathbf{a} \cdot \mathbf{c})\mathbf{b} - (\mathbf{b} \cdot \mathbf{c})\mathbf{a}.$$

6. *Lines, planes and spheres*
   - The point P that divides AB in the ratio $\lambda : \mu$ is given by
   $$\mathbf{p} = \frac{\mu}{\lambda + \mu}\mathbf{a} + \frac{\lambda}{\lambda + \mu}\mathbf{b}.$$
   - The centroid of the triangle ABC is given by $\mathbf{g} = \frac{1}{3}(\mathbf{a} + \mathbf{b} + \mathbf{c})$.
   - The line in the direction of $\mathbf{f}$ passing through the point A is
   $$\mathbf{r} = \mathbf{a} + \lambda \mathbf{f} \quad \text{or} \quad (\mathbf{r} - \mathbf{a}) \times \mathbf{f} = \mathbf{0}.$$
   - The line passing through A and C is $\mathbf{r} = \mathbf{a} + \lambda(\mathbf{c} - \mathbf{a})$.

- The plane with a normal in the direction of unit vector $\hat{\mathbf{n}}$ and containing the point A is

$$(\mathbf{r} - \mathbf{a}) \cdot \hat{\mathbf{n}} = 0 \quad \text{or} \quad \hat{\mathbf{n}} \cdot \mathbf{r} = p,$$

where p is the perpendicular distance from the origin to the plane.
- The plane containing points A, B and C is $\mathbf{r} = \alpha \mathbf{a} + \beta \mathbf{b} + \gamma \mathbf{c}$ with $\alpha + \beta + \gamma = 1$.
- The sphere with centre C and radius R is $(\mathbf{r} - \mathbf{c}) \cdot (\mathbf{r} - \mathbf{c}) = R^2$.

7. *Distances using vectors*
   - The distance of a point P from the line with direction $\mathbf{f}$ that passes through A is
   $d = |(\mathbf{a} - \mathbf{p}) \times \hat{\mathbf{f}}|$.
   - The distance of a point P from the plane with unit normal $\hat{\mathbf{n}}$ that contains A is
   $d = (\mathbf{a} - \mathbf{p}) \cdot \hat{\mathbf{n}}$, with the sign of d indicating which side of the plane P lies on.
   - The distance between the lines with directions $\mathbf{f}$ and $\mathbf{g}$, passing through the points A and B respectively, is

$$d = |(\mathbf{a} - \mathbf{b}) \cdot \hat{\mathbf{n}}|, \quad \text{where } \hat{\mathbf{n}} = \frac{\mathbf{f} \times \mathbf{g}}{|\mathbf{f} \times \mathbf{g}|}.$$

   - The distance between a line through A and a plane (to which it is parallel) with unit normal $\hat{\mathbf{n}}$ is $d = |(\mathbf{r} - \mathbf{a}) \cdot \hat{\mathbf{n}}|$, where $\mathbf{r}$ is any point on the plane.

8. *Reciprocal vectors* to the non-coplanar set $\{\mathbf{a}, \mathbf{b}, \mathbf{c}\}$

$$\mathbf{a}' = \frac{\mathbf{b} \times \mathbf{c}}{[\mathbf{a}, \mathbf{b}, \mathbf{c}]}, \quad \mathbf{b}' = \frac{\mathbf{c} \times \mathbf{a}}{[\mathbf{a}, \mathbf{b}, \mathbf{c}]}, \quad \mathbf{c}' = \frac{\mathbf{a} \times \mathbf{b}}{[\mathbf{a}, \mathbf{b}, \mathbf{c}]},$$

have the properties
   - $\mathbf{a} \cdot \mathbf{a}' = \mathbf{b} \cdot \mathbf{b}' = \mathbf{c} \cdot \mathbf{c}' = 1$.
   - $\mathbf{a}' \cdot \mathbf{b} = \mathbf{a}' \cdot \mathbf{c} = \mathbf{b}' \cdot \mathbf{a} = \mathbf{b}' \cdot \mathbf{c} = \mathbf{c}' \cdot \mathbf{a} = \mathbf{c}' \cdot \mathbf{b} = 0$.

PROBLEMS

9.1. Which of the following statements about general vectors $\mathbf{a}$, $\mathbf{b}$ and $\mathbf{c}$ are true?
(a) $\mathbf{c} \cdot (\mathbf{a} \times \mathbf{b}) = (\mathbf{b} \times \mathbf{a}) \cdot \mathbf{c}$;
(b) $\mathbf{a} \times (\mathbf{b} \times \mathbf{c}) = (\mathbf{a} \times \mathbf{b}) \times \mathbf{c}$;
(c) $\mathbf{a} \times (\mathbf{b} \times \mathbf{c}) = (\mathbf{a} \cdot \mathbf{c})\mathbf{b} - (\mathbf{a} \cdot \mathbf{b})\mathbf{c}$;
(d) $\mathbf{d} = \lambda \mathbf{a} + \mu \mathbf{b}$ implies $(\mathbf{a} \times \mathbf{b}) \cdot \mathbf{d} = 0$;
(e) $\mathbf{a} \times \mathbf{c} = \mathbf{b} \times \mathbf{c}$ implies $\mathbf{c} \cdot \mathbf{a} - \mathbf{c} \cdot \mathbf{b} = c|\mathbf{a} - \mathbf{b}|$;
(f) $(\mathbf{a} \times \mathbf{b}) \times (\mathbf{c} \times \mathbf{b}) = \mathbf{b}[\mathbf{b} \cdot (\mathbf{c} \times \mathbf{a})]$.

9.2. A unit cell of diamond is a cube of side A, with carbon atoms at each corner, at the centre of each face and, in addition, at positions displaced by $\frac{1}{4}A(\mathbf{i} + \mathbf{j} + \mathbf{k})$ from each of those already mentioned; $\mathbf{i}$, $\mathbf{j}$, $\mathbf{k}$ are unit vectors along the cube axes. One corner of the cube is taken as the origin of coordinates. What are the vectors joining the atom at $\frac{1}{4}A(\mathbf{i}+\mathbf{j}+\mathbf{k})$ to its four nearest neighbours? Determine the angle between the carbon bonds in diamond.

9.3. Identify the following surfaces:
(a) $|\mathbf{r}| = k$; (b) $\mathbf{r} \cdot \mathbf{u} = l$; (c) $\mathbf{r} \cdot \mathbf{u} = m|\mathbf{r}|$ for $-1 \leq m \leq +1$;
(d) $|\mathbf{r} - (\mathbf{r} \cdot \mathbf{u})\mathbf{u}| = n$.
Here k, l, m and n are fixed scalars and $\mathbf{u}$ is a fixed unit vector.

9.4. Find the angle between the position vectors to the points $(3, -4, 0)$ and $(-2, 1, 0)$ and find the direction cosines of a vector perpendicular to both.

9.5. A, B, C and D are the four corners, in order, of one face of a cube of side 2 units. The opposite face has corners E, F, G and H, with AE, BF, CG and DH as parallel edges of the cube. The centre O of the cube is taken as the origin and the x-, y- and z-axes are parallel to AD, AE and AB, respectively. Find the following:
(a) the angle between the face diagonal AF and the body diagonal AG;
(b) the equation of the plane through B that is parallel to the plane CGE;
(c) the perpendicular distance from the centre J of the face $BCGF$ to the plane OCG;
(d) the volume of the tetrahedron $JOCG$.

9.6. Use vector methods to prove that the lines joining the mid-points of the opposite edges of a tetrahedron $OABC$ meet at a point and that this point bisects each of the lines.

9.7. The edges OP, OQ and OR of a tetrahedron $OPQR$ are vectors $\mathbf{p}, \mathbf{q}$ and $\mathbf{r}$, respectively, where $\mathbf{p} = 2\mathbf{i} + 4\mathbf{j}$, $\mathbf{q} = 2\mathbf{i} - \mathbf{j} + 3\mathbf{k}$ and $\mathbf{r} = 4\mathbf{i} - 2\mathbf{j} + 5\mathbf{k}$. Show that OP is perpendicular to the plane containing OQR. Express the volume of the tetrahedron in terms of $\mathbf{p}, \mathbf{q}$ and $\mathbf{r}$ and hence calculate the volume.

9.8. Prove, by writing it out in component form, that

$$(\mathbf{a} \times \mathbf{b}) \times \mathbf{c} = (\mathbf{a} \cdot \mathbf{c})\mathbf{b} - (\mathbf{b} \cdot \mathbf{c})\mathbf{a}$$

and deduce the result, stated in equation (9.25), that the operation of forming the vector product is non-associative.

9.9. Prove Lagrange's identity, i.e.

$$(\mathbf{a} \times \mathbf{b}) \cdot (\mathbf{c} \times \mathbf{d}) = (\mathbf{a} \cdot \mathbf{c})(\mathbf{b} \cdot \mathbf{d}) - (\mathbf{a} \cdot \mathbf{d})(\mathbf{b} \cdot \mathbf{c}).$$

9.10. For four arbitrary vectors $\mathbf{a}, \mathbf{b}, \mathbf{c}$ and $\mathbf{d}$, evaluate

$$(\mathbf{a} \times \mathbf{b}) \times (\mathbf{c} \times \mathbf{d})$$

Problems

in two different ways and so prove that

$$a[b, c, d] - b[c, d, a] + c[d, a, b] - d[a, b, c] = 0.$$

Show that this reduces to the normal Cartesian representation of the vector d, i.e. $d_x i + d_y j + d_z k$, if a, b and c are taken as i, j and k, the Cartesian base vectors.

9.11. Show that the points $(1, 0, 1)$, $(1, 1, 0)$ and $(1, -3, 4)$ lie on a straight line. Give the equation of the line in the form

$$r = a + \lambda b.$$

9.12. The plane P_1 contains the points A, B and C, which have position vectors $a = -3i + 2j$, $b = 7i + 2j$ and $c = 2i + 3j + 2k$, respectively. Plane P_2 passes through A and is orthogonal to the line BC, whilst plane P_3 passes through B and is orthogonal to the line AC. Find the coordinates of r, the point of intersection of the three planes.

9.13. Two planes have non-parallel unit normals $\hat{n}$ and $\hat{m}$ and their closest distances from the origin are λ and μ, respectively. Find the vector equation of their line of intersection in the form $r = \nu p + a$.

9.14. Two fixed points, A and B, in three-dimensional space have position vectors a and b. Identify the plane P given by

$$(a - b) \cdot r = \tfrac{1}{2}(a^2 - b^2),$$

where a and b are the magnitudes of a and b.

Show also that the equation

$$(a - r) \cdot (b - r) = 0$$

describes a sphere S of radius $|a - b|/2$. Deduce that the intersection of P and S is also the intersection of two spheres, centred on A and B, and each of radius $|a - b|/\sqrt{2}$.

9.15. Let O, A, B and C be four points with position vectors 0, a, b and c, and denote by $g = \lambda a + \mu b + \nu c$ the position of the centre of the sphere on which they all lie.
(a) Prove that λ, μ and ν simultaneously satisfy

$$(a \cdot a)\lambda + (a \cdot b)\mu + (a \cdot c)\nu = \tfrac{1}{2}a^2$$

and two other similar equations.
(b) By making a change of origin, find the centre and radius of the sphere on which the points $p = 3i + j - 2k$, $q = 4i + 3j - 3k$, $r = 7i - 3k$ and $s = 6i + j - k$ all lie.

9.16. The vectors a, b and c are coplanar and related by

$$\lambda a + \mu b + \nu c = 0,$$

where λ, μ, ν are not all zero. Show that the condition for the points with position vectors $\alpha\mathbf{a}$, $\beta\mathbf{b}$ and $\gamma\mathbf{c}$ to be collinear is

$$\frac{\lambda}{\alpha} + \frac{\mu}{\beta} + \frac{\nu}{\gamma} = 0.$$

9.17. Using vector methods:
(a) Show that the line of intersection of the planes $x + 2y + 3z = 0$ and $3x + 2y + z = 0$ is equally inclined to the x- and z-axes and makes an angle $\cos^{-1}(-2/\sqrt{6})$ with the y-axis.
(b) Find the perpendicular distance between one corner of a unit cube and the major diagonal not passing through it.

9.18. Extend the derivation of Equation (9.47) to show that the volume of a tetrahedron whose vertices are at $\mathbf{a}$, $\mathbf{b}$, $\mathbf{c}$ and $\mathbf{d}$ is given by $\frac{1}{6}|[\mathbf{a} \cdot (\mathbf{b} \times \mathbf{c})] - [\mathbf{b} \cdot (\mathbf{c} \times \mathbf{d})] + [\mathbf{c} \cdot (\mathbf{d} \times \mathbf{a})] - [\mathbf{d} \cdot (\mathbf{a} \times \mathbf{b})]|$. Verify that this formula gives the correct result for the volume of the tetrahedron discussed in the worked example in Section 8.2.1.

9.19. The vectors $\mathbf{a}$, $\mathbf{b}$ and $\mathbf{c}$ are not coplanar. The vectors $\mathbf{a}'$, $\mathbf{b}'$ and $\mathbf{c}'$ are the associated reciprocal vectors. Verify that the expressions (9.50)–(9.52) define a set of reciprocal vectors $\mathbf{a}'$, $\mathbf{b}'$ and $\mathbf{c}'$ with the following properties:
(a) $\mathbf{a}' \cdot \mathbf{a} = \mathbf{b}' \cdot \mathbf{b} = \mathbf{c}' \cdot \mathbf{c} = 1$;
(b) $\mathbf{a}' \cdot \mathbf{b} = \mathbf{a}' \cdot \mathbf{c} = \mathbf{b}' \cdot \mathbf{a}$ etc. $= 0$;
(c) $[\mathbf{a}', \mathbf{b}', \mathbf{c}'] = 1/[\mathbf{a}, \mathbf{b}, \mathbf{c}]$;
(d) $\mathbf{a} = (\mathbf{b}' \times \mathbf{c}')/[\mathbf{a}', \mathbf{b}', \mathbf{c}']$.

9.20. Three non-coplanar vectors $\mathbf{a}$, $\mathbf{b}$ and $\mathbf{c}$, have as their respective reciprocal vectors the set $\mathbf{a}'$, $\mathbf{b}'$ and $\mathbf{c}'$. Show that the normal to the plane containing the points $k^{-1}\mathbf{a}$, $l^{-1}\mathbf{b}$ and $m^{-1}\mathbf{c}$ is in the direction of the vector $k\mathbf{a}' + l\mathbf{b}' + m\mathbf{c}'$.

9.21. In a crystal with a face-centred cubic structure, the basic cell can be taken as a cube of edge a with its centre at the origin of coordinates and its edges parallel to the Cartesian coordinate axes; atoms are sited at the eight corners and at the centre of each face. However, other basic cells are possible. One is the rhomboid shown in Figure 9.17, which has the three vectors $\mathbf{b}$, $\mathbf{c}$ and $\mathbf{d}$ as edges.
(a) Show that the volume of the rhomboid is one-quarter that of the cube.
(b) Show that the angles between pairs of edges of the rhomboid are 60° and that the corresponding angles between pairs of edges of the rhomboid defined by the reciprocal vectors to $\mathbf{b}$, $\mathbf{c}$, $\mathbf{d}$ are each 109.5°. (This rhomboid can be used as the basic cell of a body-centred cubic structure, more easily visualised as a cube with an atom at each corner and one at its centre.)
(c) In order to use the Bragg formula, $2d \sin\theta = n\lambda$, for the scattering of X-rays by a crystal, it is necessary to know the perpendicular distance d between successive planes of atoms; for a given crystal structure, d has a particular value for each set of planes considered. For the face-centred cubic structure

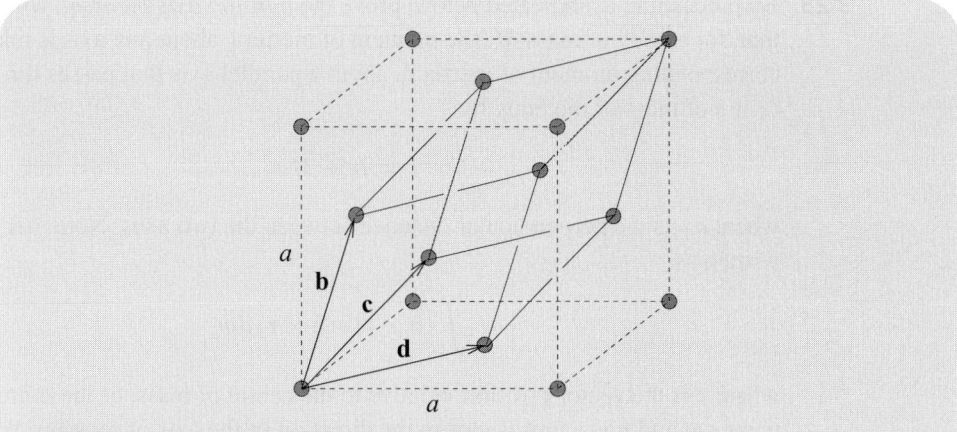

Figure 9.17 A face-centred cubic crystal.

find the distance between successive planes with normals in the $\mathbf{k}$, $\mathbf{i}+\mathbf{j}$ and $\mathbf{i}+\mathbf{j}+\mathbf{k}$ directions.

9.22. In Section 9.4.2 we showed how the moment or torque of a force about an axis could be represented by a vector in the direction of the axis. The magnitude of the vector gives the size of the moment and the sign of the vector gives the sense. Similar representations can be used for angular velocities and angular momenta.

(a) The magnitude of the angular momentum about the origin of a particle of mass m moving with velocity $\mathbf{v}$ on a path that is a perpendicular distance d from the origin is given by $m|\mathbf{v}|d$. Show that if $\mathbf{r}$ is the position of the particle then the vector $\mathbf{J} = \mathbf{r} \times m\mathbf{v}$ represents the angular momentum.

(b) Now consider a rigid collection of particles (or a solid body) rotating about an axis through the origin, the angular velocity of the collection being represented by $\boldsymbol{\omega}$.

(i) Show that the velocity of the ith particle is

$$\mathbf{v}_i = \boldsymbol{\omega} \times \mathbf{r}_i$$

and that the total angular momentum $\mathbf{J}$ is

$$\mathbf{J} = \sum_i m_i [r_i^2 \boldsymbol{\omega} - (\mathbf{r}_i \cdot \boldsymbol{\omega})\mathbf{r}_i].$$

(ii) Show further that the component of $\mathbf{J}$ along the axis of rotation can be written as $I\omega$, where I, the moment of inertia of the collection about the axis or rotation, is given by

$$I = \sum_i m_i \rho_i^2.$$

Interpret ρ_i geometrically.

(iii) Prove that the total kinetic energy of the particles is $\frac{1}{2}I\omega^2$.

Vector algebra

9.23. By proceeding as indicated below, prove the *parallel axis theorem*, which states that, for a body of mass M, the moment of inertia I about any axis is related to the corresponding moment of inertia I_0 about a parallel axis that passes through the centre of mass of the body by

$$I = I_0 + Ma_\perp^2,$$

where $a_\perp$ is the perpendicular distance between the two axes. Note that I_0 can be written as

$$\int (\hat{\mathbf{n}} \times \mathbf{r}) \cdot (\hat{\mathbf{n}} \times \mathbf{r}) \, dm,$$

where $\mathbf{r}$ is the vector position, relative to the centre of mass, of the infinitesimal mass dm and $\hat{\mathbf{n}}$ is a unit vector in the direction of the axis of rotation. Write a similar expression for I in which $\mathbf{r}$ is replaced by $\mathbf{r}' = \mathbf{r} - \mathbf{a}$, where $\mathbf{a}$ is the vector position of any point on the axis to which I refers. Use Lagrange's identity and the fact that $\int \mathbf{r} \, dm = \mathbf{0}$ (by the definition of the centre of mass) to establish the result.

9.24. Without carrying out any further integration, use the results of the previous problem, the worked example in Section 8.2.4 and Problem 8.10 to prove that the moment of inertia of a uniform rectangular lamina, of mass M and sides a and b, about an axis perpendicular to its plane and passing through the point $(\alpha a/2, \beta b/2)$, with $-1 \leq \alpha, \beta \leq 1$, is

$$\frac{M}{12}[a^2(1 + 3\alpha^2) + b^2(1 + 3\beta^2)].$$

9.25. Define a set of (non-orthogonal) base vectors $\mathbf{a} = \mathbf{j} + \mathbf{k}$, $\mathbf{b} = \mathbf{i} + \mathbf{k}$ and $\mathbf{c} = \mathbf{i} + \mathbf{j}$.
  (a) Establish their reciprocal vectors and hence express the vectors $\mathbf{p} = 3\mathbf{i} - 2\mathbf{j} + \mathbf{k}$, $\mathbf{q} = \mathbf{i} + 4\mathbf{j}$ and $\mathbf{r} = -2\mathbf{i} + \mathbf{j} + \mathbf{k}$ in terms of the base vectors $\mathbf{a}$, $\mathbf{b}$ and $\mathbf{c}$.
  (b) Verify that the scalar product $\mathbf{p} \cdot \mathbf{q}$ has the same value, -5, when evaluated using either set of components.

9.26. Systems that can be modelled as damped harmonic oscillators are widespread; pendulum clocks, car shock absorbers, tuning circuits in television sets and radios, and collective electron motions in plasmas and metals are just a few examples.

In all these cases, one or more variables describing the system obey(s) an equation of the form

$$\ddot{x} + 2\gamma \dot{x} + \omega_0^2 x = P \cos \omega t,$$

where $\dot{x} = dx/dt$, etc. and the inclusion of the factor 2 is conventional. In the steady state (i.e. after the effects of any initial displacement or velocity have been

Problems

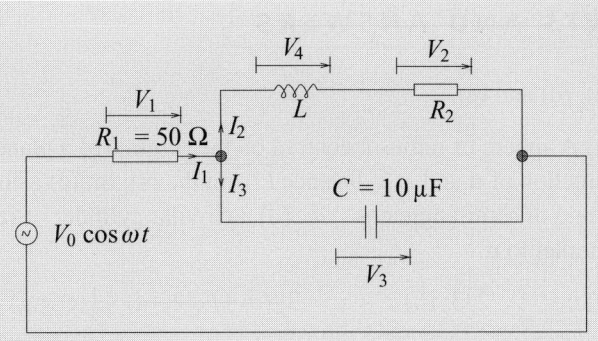

Figure 9.18 An oscillatory electric circuit. The power supply has angular frequency $\omega = 2\pi f = 400\pi$ s^{-1}.

damped out) the solution of the equation takes the form

$$x(t) = A\cos(\omega t + \phi).$$

By expressing each term in the form $B\cos(\omega t + \epsilon)$, and representing it by a vector of magnitude B making an angle ϵ with the x-axis, draw a closed vector diagram, at $t = 0$, say, that is equivalent to the equation.

(a) Convince yourself that whatever the value of ω (> 0) ϕ must be negative ($-\pi < \phi \leq 0$) and that

$$\phi = \tan^{-1}\left(\frac{-2\gamma\omega}{\omega_0^2 - \omega^2}\right).$$

(b) Obtain an expression for A in terms of P, ω_0 and ω.

9.27. According to alternating current theory, the currents and potential differences in the components of the circuit shown in Figure 9.18 are determined by Kirchhoff's laws and the relationships

$$I_1 = \frac{V_1}{R_1}, \quad I_2 = \frac{V_2}{R_2}, \quad I_3 = i\omega C V_3, \quad V_4 = i\omega L I_2.$$

The factor $i = \sqrt{-1}$ in the expression for I_3 indicates that the phase of I_3 is 90° ahead of V_3. Similarly, the phase of V_4 is 90° ahead of I_2.

Measurement shows that V_3 has an amplitude of $0.661 V_0$ and a phase of $+13.4°$ relative to that of the power supply. Taking $V_0 = 1$ V, and using a series of vector plots for potential differences and currents (they could all be on the same plot if suitable scales were chosen), determine all unknown currents and potential differences and find values for the inductance of L and the resistance of R_2.

[Scales of 1 cm = 0.1 V for potential differences and 1 cm = 1 mA for currents are convenient.]

HINTS AND ANSWERS

9.1. (c), (d) and (e).

9.3. (a) A sphere of radius k centred on the origin; (b) a plane with its normal in the direction of $\mathbf{u}$ and at a distance l from the origin; (c) a cone with its axis parallel to $\mathbf{u}$ and of semi-angle $\cos^{-1} m$; (d) a circular cylinder of radius n with its axis parallel to $\mathbf{u}$.

9.5. (a) $\cos^{-1}\sqrt{2/3}$; (b) $z - x = 2$; (c) $1/\sqrt{2}$; (d) $\frac{1}{3}\frac{1}{2}(\mathbf{c} \times \mathbf{g}) \cdot \mathbf{j} = \frac{1}{3}$.

9.7. Show that $\mathbf{q} \times \mathbf{r}$ is parallel to $\mathbf{p}$; volume $= \frac{1}{3}\left[\frac{1}{2}(\mathbf{q} \times \mathbf{r}) \cdot \mathbf{p}\right] = \frac{5}{3}$.

9.9. Note that $(\mathbf{a} \times \mathbf{b}) \cdot (\mathbf{c} \times \mathbf{d}) = \mathbf{d} \cdot [(\mathbf{a} \times \mathbf{b}) \times \mathbf{c}]$ and use the result for a triple vector product to expand the expression in square brackets.

9.11. Show that the position vectors of the points are linearly dependent; $\mathbf{r} = \mathbf{a} + \lambda\mathbf{b}$ where $\mathbf{a} = \mathbf{i} + \mathbf{k}$ and $\mathbf{b} = -\mathbf{j} + \mathbf{k}$.

9.13. Show that $\mathbf{p}$ must have the direction $\hat{\mathbf{n}} \times \hat{\mathbf{m}}$ and write $\mathbf{a}$ as $x\hat{\mathbf{n}} + y\hat{\mathbf{m}}$. By obtaining a pair of simultaneous equations for x and y, prove that $x = (\lambda - \mu\hat{\mathbf{n}} \cdot \hat{\mathbf{m}})/[1 - (\hat{\mathbf{n}} \cdot \hat{\mathbf{m}})^2]$ and that $y = (\mu - \lambda\hat{\mathbf{n}} \cdot \hat{\mathbf{m}})/[1 - (\hat{\mathbf{n}} \cdot \hat{\mathbf{m}})^2]$.

9.15. (a) Note that $|\mathbf{a} - \mathbf{g}|^2 = R^2 = |\mathbf{0} - \mathbf{g}|^2$, leading to $\mathbf{a} \cdot \mathbf{a} = 2\mathbf{a} \cdot \mathbf{g}$.
(b) Make $\mathbf{p}$ the new origin and solve the three simultaneous linear equations to obtain $\lambda = 5/18$, $\mu = 10/18$, $\nu = -3/18$, giving $\mathbf{g} = 2\mathbf{i} - \mathbf{k}$ and a sphere of radius $\sqrt{5}$ centred on $(5, 1, -3)$.

9.17. (a) Find two points on both planes, say $(0, 0, 0)$ and $(1, -2, 1)$, and hence determine the direction cosines of the line of intersection; (b) $(\frac{2}{3})^{1/2}$.

9.19. For (c) and (d), treat $(\mathbf{c} \times \mathbf{a}) \times (\mathbf{a} \times \mathbf{b})$ as a triple vector product with $\mathbf{c} \times \mathbf{a}$ as one of the three vectors.

9.21. (b) $\mathbf{b}' = a^{-1}(-\mathbf{i} + \mathbf{j} + \mathbf{k})$, $\mathbf{c}' = a^{-1}(\mathbf{i} - \mathbf{j} + \mathbf{k})$, $\mathbf{d}' = a^{-1}(\mathbf{i} + \mathbf{j} - \mathbf{k})$; (c) $a/2$ for direction $\mathbf{k}$; successive planes through $(0, 0, 0)$ and $(a/2, 0, a/2)$ give a spacing of $a/\sqrt{8}$ for direction $\mathbf{i} + \mathbf{j}$; successive planes through $(-a/2, 0, 0)$ and $(a/2, 0, 0)$ give a spacing of $a/\sqrt{3}$ for direction $\mathbf{i} + \mathbf{j} + \mathbf{k}$.

9.23. Note that $a^2 - (\hat{\mathbf{n}} \cdot \mathbf{a})^2 = a_\perp^2$.

9.25. $\mathbf{p} = -2\mathbf{a} + 3\mathbf{b}$, $\mathbf{q} = \frac{3}{2}\mathbf{a} - \frac{3}{2}\mathbf{b} + \frac{5}{2}\mathbf{c}$ and $\mathbf{r} = 2\mathbf{a} - \mathbf{b} - \mathbf{c}$. Remember that $\mathbf{a} \cdot \mathbf{a} = \mathbf{b} \cdot \mathbf{b} = \mathbf{c} \cdot \mathbf{c} = 2$ and $\mathbf{a} \cdot \mathbf{b} = \mathbf{a} \cdot \mathbf{c} = \mathbf{b} \cdot \mathbf{c} = 1$.

9.27. With currents in milliamps and potential differences in volts:
$I_1 = (7.76, -23.2°)$, $I_2 = (14.36, -50.8°)$, $I_3 = (8.30, 103.4°)$;
$V_1 = (0.388, -23.2°)$, $V_2 = (0.287, -50.8°)$, $V_4 = (0.596, 39.2°)$;
$L = 33$ mH, $R_2 = 20\,\Omega$.

10
Matrices and vector spaces

In Chapter 9 we defined a *vector* as a geometrical object which has both a magnitude and a direction and which may be thought of as an arrow fixed in our familiar three-dimensional space, a space which, if we need to, we define by reference to, say, the fixed stars. This geometrical definition of a vector is both useful and important since it is *independent* of any coordinate system with which we choose to label points in space.

In most specific applications, however, it is necessary at some stage to choose a coordinate system and to break down a vector into its *component vectors* in the directions of increasing coordinate values. Thus for a particular Cartesian coordinate system (for example) the component vectors of a vector **a** will be $a_x\mathbf{i}$, $a_y\mathbf{j}$ and $a_z\mathbf{k}$ and the complete vector will be

$$\mathbf{a} = a_x\mathbf{i} + a_y\mathbf{j} + a_z\mathbf{k}. \tag{10.1}$$

Although we have so far considered only real three-dimensional space, we may extend our notion of a vector to more abstract spaces, which in general can have an arbitrary number of dimensions N. We may still think of such a vector as an 'arrow' in this abstract space, so that it is again *independent* of any (N-dimensional) coordinate system with which we choose to label the space. As an example of such a space, which, though abstract, has very practical applications, we may consider the description of a mechanical or electrical system. If the state of a system is uniquely specified by assigning values to a set of N variables, which could include angles or currents, for example, then that state can be represented by a vector in an N-dimensional space, the vector having those values as its components.[1]

In this chapter we first discuss general *vector spaces* and their properties. We then go on to consider the transformation of one vector into another by a linear operator. This leads naturally to the concept of a *matrix*, a two-dimensional array of numbers. The general properties of matrices are then developed and lead to a discussion of how to use these properties to solve systems of linear equations. The chapter concludes with a study of more detailed properties associated with certain types of so-called 'square' matrices.

---

[1] This is an approach often used in control engineering.

Matrices and vector spaces

10.1 Vector spaces

A set of objects (vectors) **a**, **b**, **c**, ... is said to form a *linear vector space* V if:

(i) the set is closed under commutative and associative addition, so that

$$\mathbf{a} + \mathbf{b} = \mathbf{b} + \mathbf{a}, \tag{10.2}$$
$$(\mathbf{a} + \mathbf{b}) + \mathbf{c} = \mathbf{a} + (\mathbf{b} + \mathbf{c}), \tag{10.3}$$

with all of the vector sums belonging to the set;

(ii) the set is closed under multiplication by a scalar (any complex number) to form a new vector $\lambda \mathbf{a}$, the operation being both distributive and associative so that

$$\lambda(\mathbf{a} + \mathbf{b}) = \lambda \mathbf{a} + \lambda \mathbf{b}, \tag{10.4}$$
$$(\lambda + \mu)\mathbf{a} = \lambda \mathbf{a} + \mu \mathbf{a}, \tag{10.5}$$
$$\lambda(\mu \mathbf{a}) = (\lambda \mu)\mathbf{a}, \tag{10.6}$$

where λ and μ are arbitrary scalars;

(iii) there exists a *null vector* **0** such that $\mathbf{a} + \mathbf{0} = \mathbf{a}$ for all **a**;

(iv) multiplication by unity leaves any vector unchanged, i.e. $1 \times \mathbf{a} = \mathbf{a}$;

(v) all vectors have a corresponding *negative vector* $-\mathbf{a}$ such that $\mathbf{a} + (-\mathbf{a}) = \mathbf{0}$. It follows from (10.5) with $\lambda = 1$ and $\mu = -1$ that $-\mathbf{a}$ is the same vector as $(-1) \times \mathbf{a}$.

If all of the scalars are restricted to be real then we obtain a *real vector space* (an example of which is our familiar three-dimensional space); otherwise, in general, we obtain a *complex vector space*. It should be noted that it is common to use the terms 'vector space' and 'space', instead of the more formal 'linear vector space'.

The *span* of a set of vectors **a**, **b**, ..., **s** is defined as the set of all vectors that may be written as a linear sum of the original set, i.e. all vectors

$$\mathbf{x} = \alpha \mathbf{a} + \beta \mathbf{b} + \cdots + \sigma \mathbf{s} \tag{10.7}$$

that result from the infinite number of possible values of the (in general complex) scalars $\alpha, \beta, \ldots, \sigma$. If **x** in (10.7) is equal to **0** for some choice of $\alpha, \beta, \ldots, \sigma$ (not *all* zero), i.e. if

$$\alpha \mathbf{a} + \beta \mathbf{b} + \cdots + \sigma \mathbf{s} = \mathbf{0}, \tag{10.8}$$

then the set of vectors **a**, **b**, ..., **s** is said to be *linearly dependent*. In such a set at least one vector is redundant, since it can be expressed as a linear sum of the others. If, however, (10.8) is not satisfied by *any* set of coefficients (other than the trivial case in which all the coefficients are zero) then the vectors are *linearly independent*, and no vector in the set can be expressed as a linear sum of the others.

If, in a given vector space, there exist sets of N linearly independent vectors, but no set of $N + 1$ linearly independent vectors, then the vector space is said to be N-dimensional. In this chapter we will limit our discussion to vector spaces of finite dimensionality.

10.1.1 Basis vectors

If V is an N-dimensional vector space then *any* set of N linearly independent vectors $\mathbf{e}_1, \mathbf{e}_2, \ldots, \mathbf{e}_N$ forms a *basis* for V. If **x** is an arbitrary vector lying in V then it can be

10.1 Vector spaces

written as a linear sum of these basis vectors:

$$\mathbf{x} = x_1\mathbf{e}_1 + x_2\mathbf{e}_2 + \cdots + x_N\mathbf{e}_N = \sum_{i=1}^{N} x_i\mathbf{e}_i, \qquad (10.9)$$

for some set of coefficients x_i. Since any $\mathbf{x}$ lying in the span of V can be expressed in terms of the *basis* or *base vectors* $\mathbf{e}_i$, the latter are said to form a *complete* set.

The coefficients x_i are called the *components* of $\mathbf{x}$ with respect to the $\mathbf{e}_i$-basis. They are *unique*, since if both

$$\mathbf{x} = \sum_{i=1}^{N} x_i\mathbf{e}_i \quad \text{and} \quad \mathbf{x} = \sum_{i=1}^{N} y_i\mathbf{e}_i, \quad \text{then} \quad \sum_{i=1}^{N}(x_i - y_i)\mathbf{e}_i = \mathbf{0}. \qquad (10.10)$$

Since the $\mathbf{e}_i$ are linearly independent, each coefficient in the final equation in (10.10) must be individually zero and so $x_i = y_i$ for all $i = 1, 2, \ldots, N$.

It follows from this that *any* set of N linearly independent vectors can form a basis for an N-dimensional space.[2] If we choose a different set $\mathbf{e}'_i, i = 1, \ldots, N$ then we can write $\mathbf{x}$ as

$$\mathbf{x} = x'_1\mathbf{e}'_1 + x'_2\mathbf{e}'_2 + \cdots + x'_N\mathbf{e}'_N = \sum_{i=1}^{N} x'_i\mathbf{e}'_i, \qquad (10.11)$$

but this does not change the vector $\mathbf{x}$. The vector $\mathbf{x}$ (a geometrical entity) is independent of the basis – it is only the components of $\mathbf{x}$ that depend upon the basis.

10.1.2 The inner product

This subsection contains a working summary of the definition and properties of inner products; for a full discussion a more advanced text should be consulted. To describe how two vectors in a vector space 'multiply' (as opposed to add or subtract) we define their *inner product*, denoted in general by $\langle \mathbf{a}|\mathbf{b} \rangle$. This is a scalar function of vectors $\mathbf{a}$ and $\mathbf{b}$, though it is not necessarily real. Alternative notations for $\langle \mathbf{a}|\mathbf{b} \rangle$ are $(\mathbf{a}, \mathbf{b})$, or simply $\mathbf{a} \cdot \mathbf{b}$.

The scalar or dot product, $\mathbf{a} \cdot \mathbf{b} \equiv |\mathbf{a}||\mathbf{b}|\cos\theta$, of two vectors in real three-dimensional space (where θ is the angle between the vectors), was introduced in Chapter 9 and is an example of an inner product. In effect the notion of an inner product $\langle \mathbf{a}|\mathbf{b} \rangle$ is a generalisation of the dot product to more abstract vector spaces. The inner product has the following properties (in which, as usual, a superscript asterisk denotes complex conjugation):[3]

$$\langle \mathbf{a}|\mathbf{b} \rangle = \langle \mathbf{b}|\mathbf{a} \rangle^*, \qquad (10.12)$$

$$\langle \mathbf{a}|\lambda\mathbf{b} + \mu\mathbf{c} \rangle = \lambda \langle \mathbf{a}|\mathbf{b} \rangle + \mu \langle \mathbf{a}|\mathbf{c} \rangle, \qquad (10.13)$$

$$\langle \lambda\mathbf{a} + \mu\mathbf{b}|\mathbf{c} \rangle = \lambda^* \langle \mathbf{a}|\mathbf{c} \rangle + \mu^* \langle \mathbf{b}|\mathbf{c} \rangle, \qquad (10.14)$$

$$\langle \lambda\mathbf{a}|\mu\mathbf{b} \rangle = \lambda^*\mu \langle \mathbf{a}|\mathbf{b} \rangle. \qquad (10.15)$$

---

[2] All bases contain *exactly* N base vectors. A (putative) alternative base with M ($< N$) vectors would imply that there is no set of more than M linearly independent vectors – but the original base is just such a set, giving a contradiction. Equally, $M > N$ would imply the existence of a linearly independent set with more than N members – contradicting the specification for the original base set. Hence $M = N$.

[3] It is a useful exercise in close analysis to deduce properties (10.14) and (10.15), on a justified step-by-step basis, using only those given in (10.12) and (10.13) and the general properties of complex conjugation.

Matrices and vector spaces

Following the analogy with the dot product in three-dimensional real space, two vectors in a general vector space are defined to be *orthogonal* if $\langle a|b \rangle = 0$.

In the same way, the *norm* of a vector $\mathbf{a}$, defined by $||\mathbf{a}|| = \langle a|a \rangle^{1/2}$, is clearly a generalisation of the length or modulus $|\mathbf{a}|$ of a vector $\mathbf{a}$ in three-dimensional space. In a general vector space $\langle a|a \rangle$ can be positive or negative; however, we will be concerned only with spaces in which $\langle a|a \rangle \geq 0$ and which are therefore said to have a *positive semi-definite norm*. In such a space $\langle a|a \rangle = 0$ implies $\mathbf{a} = \mathbf{0}$.

It is usual when working with an N-dimensional vector space to use a basis $\hat{\mathbf{e}}_1, \hat{\mathbf{e}}_2, \ldots, \hat{\mathbf{e}}_N$ that has the desirable property of being *orthonormal* (the basis vectors are mutually orthogonal and each has unit norm), i.e. a basis that has the property

$$\langle \hat{\mathbf{e}}_i | \hat{\mathbf{e}}_j \rangle = \delta_{ij}. \tag{10.16}$$

Here δ_{ij} is the *Kronecker delta* symbol, defined by the properties

$$\delta_{ij} = \begin{cases} 1 & \text{for } i = j, \\ 0 & \text{for } i \neq j. \end{cases}$$

Using the above basis, any two vectors $\mathbf{a}$ and $\mathbf{b}$ can be written as

$$\mathbf{a} = \sum_{i=1}^{N} a_i \hat{\mathbf{e}}_i \quad \text{and} \quad \mathbf{b} = \sum_{i=1}^{N} b_i \hat{\mathbf{e}}_i.$$

Furthermore, *in such an orthonormal basis* we have, for any $\mathbf{a}$,

$$\langle \hat{\mathbf{e}}_j | \mathbf{a} \rangle = \sum_{i=1}^{N} \langle \hat{\mathbf{e}}_j | a_i \hat{\mathbf{e}}_i \rangle = \sum_{i=1}^{N} a_i \langle \hat{\mathbf{e}}_j | \hat{\mathbf{e}}_i \rangle = \sum_{i=1}^{N} a_i \delta_{ji} = a_j. \tag{10.17}$$

Thus the components of $\mathbf{a}$ are given by $a_i = \langle \hat{\mathbf{e}}_i | \mathbf{a} \rangle$. Note that this is *not* true unless the basis is orthonormal.

We can write the inner product of $\mathbf{a}$ and $\mathbf{b}$ in terms of their components in an orthonormal basis as

$$\langle a|b \rangle = \langle a_1 \hat{\mathbf{e}}_1 + a_2 \hat{\mathbf{e}}_2 + \cdots + a_N \hat{\mathbf{e}}_N | b_1 \hat{\mathbf{e}}_1 + b_2 \hat{\mathbf{e}}_2 + \cdots + b_N \hat{\mathbf{e}}_N \rangle$$

$$= \sum_{i=1}^{N} a_i^* b_i \langle \hat{\mathbf{e}}_i | \hat{\mathbf{e}}_i \rangle + \sum_{i=1}^{N} \sum_{j \neq i}^{N} a_i^* b_j \langle \hat{\mathbf{e}}_i | \hat{\mathbf{e}}_j \rangle$$

$$= \sum_{i=1}^{N} a_i^* b_i,$$

where the second equality follows from (10.15) and the third from (10.16) with all inner products in the first summation equal to unity and all those in the second (double) summation having zero value. This is clearly a generalisation of the expression (9.21) for the dot product of vectors in three-dimensional space.

The extension of the above results to the case where the base vectors $\mathbf{e}_1, \mathbf{e}_2, \ldots, \mathbf{e}_N$ are *not* orthonormal is more mathematically complicated, but fortunately will not be needed here.

10.1 Vector spaces

10.1.3 Some useful inequalities

For a set of objects (vectors) forming a linear vector space in which $\langle a|a \rangle \geq 0$ for all **a**, there are a number of inequalities that often prove useful. Here we list them without proofs.

(i) *Schwarz's inequality* states that

$$|\langle a|b \rangle| \leq ||a|| \, ||b||, \tag{10.18}$$

where the equality holds when **a** is a scalar multiple of **b**, i.e. when $\mathbf{a} = \lambda \mathbf{b}$. It is important here to distinguish between the *absolute value* of a scalar, $|\lambda|$, and the *norm* of a vector, $||\mathbf{a}||$.

(ii) The *triangle inequality* states that

$$||a + b|| \leq ||a|| + ||b|| \tag{10.19}$$

and is the intuitive analogue of the observation that the length of any one side of a triangle cannot be greater than the sum of the lengths of the other two sides.

(iii) *Bessel's inequality* states that if $\hat{\mathbf{e}}_i$, $i = 1, 2, \ldots, N$, form an orthonormal basis in an N-dimensional vector space, then

$$||a||^2 \geq \sum_i^M |\langle \hat{\mathbf{e}}_i | a \rangle|^2, \tag{10.20}$$

where the equality holds if $M = N$. If $M < N$ then inequality results, unless the basis vectors omitted all have $a_i = 0$. This is the analogue of $|\mathbf{x}|^2$ for a three-dimensional vector **v** being equal to the sum of the squares of all its components, and if any are omitted the sum may fall short of $|\mathbf{x}|^2$.

To these inequalities can be added one equality that sometimes proves useful. The *parallelogram equality* reads

$$||a + b||^2 + ||a - b||^2 = 2 \left(||a||^2 + ||b||^2 \right), \tag{10.21}$$

and may be proved straightforwardly from the properties of the inner product.

EXERCISES 10.1

1. Do the following form real linear vector spaces?
   (a) All first-degree polynomials with non-zero real coefficients.
   (b) All second-degree polynomials with real coefficients.
   (c) All numbers of the form $a + b\sqrt{p}$, with p a fixed prime and a and b real integers.
   (d) The natural logarithms of all positive real numbers.

2. Are the following sets of vectors linearly independent?
   (a) $(1, 1, 0), (1, 0, 1), (0, 1, 1)$.
   (b) $(2, -2, 2), (1, 2, -1), (2, -5, 4)$.
   (c) $(1, 1, 2, -2), (2, 3, 0, -1), (-1, 2, 1, 0)$.
   (d) $(1, 1, 2, -2), (3, 0, 3, -4), (-1, 2, 1, 0)$.

3. Find an orthonormal basis for three-dimensional Cartesian space that includes vectors in the directions

$$(2, -2, 1) \quad \text{and} \quad (3, 2, -2).$$

4. Evaluate the inner product $\langle a|b \rangle$ of the vectors

$$\mathbf{a} = (1 - i, i, 2) \quad \text{and} \quad \mathbf{b} = (1 + i, 1, 2i).$$

Find the norms of **a** and **b** and hence verify that Schwarz's inequality is satisfied.

10.2 Linear operators

We now discuss the action of *linear operators* on vectors in a vector space. A linear operator $\mathcal{A}$ associates with every vector **x** another vector

$$\mathbf{y} = \mathcal{A}\mathbf{x}$$

in such a way that, for two vectors **a** and **b**,

$$\mathcal{A}(\lambda \mathbf{a} + \mu \mathbf{b}) = \lambda \mathcal{A}\mathbf{a} + \mu \mathcal{A}\mathbf{b},$$

where λ, μ are scalars. We say that $\mathcal{A}$ 'operates' on **x** to give the vector **y**. We note that the action of $\mathcal{A}$ is *independent* of any basis or coordinate system and may be thought of as 'transforming' one geometrical entity (i.e. a vector) into another.

If we now introduce a basis $\mathbf{e}_i$, $i = 1, 2, \ldots, N$, into our vector space then the action of $\mathcal{A}$ on each of the basis vectors is to produce a linear combination of the latter; for the base vector $\mathbf{e}_j$ this may be written as

$$\mathcal{A}\mathbf{e}_j = \sum_{i=1}^{N} A_{ij}\mathbf{e}_i, \tag{10.22}$$

where A_{ij} is the ith component of the vector $\mathcal{A}\mathbf{e}_j$ in this basis; collectively the numbers A_{ij} are called the components of the linear operator in the $\mathbf{e}_i$-basis. *In this basis* we can express the relation $\mathbf{y} = \mathcal{A}\mathbf{x}$ in component form as

$$\mathbf{y} = \sum_{i=1}^{N} y_i \mathbf{e}_i = \mathcal{A}\left(\sum_{j=1}^{N} x_j \mathbf{e}_j\right) = \sum_{j=1}^{N} x_j \sum_{i=1}^{N} A_{ij}\mathbf{e}_i,$$

and hence, in purely component form, in this basis we have

$$y_i = \sum_{j=1}^{N} A_{ij} x_j. \tag{10.23}$$

If we had chosen a different basis $\mathbf{e}'_i$, in which the components of **x**, **y** and $\mathcal{A}$ are x'_i, y'_i and A'_{ij} respectively then the geometrical relationship $\mathbf{y} = \mathcal{A}\mathbf{x}$ would be represented in this

10.2 Linear operators

new basis by

$$y'_i = \sum_{j=1}^{N} A'_{ij} x'_j.$$

We have so far assumed that the vector **y** is in the same vector space as **x**. If, however, **y** belongs to a different vector space, which may in general be M-dimensional ($M \neq N$), then the above analysis needs a slight modification. By introducing a basis set $\mathbf{f}_i$, $i = 1, 2, \ldots, M$, into the vector space to which **y** belongs we may generalise (10.22) as

$$\mathcal{A}\mathbf{e}_j = \sum_{i=1}^{M} A_{ij} \mathbf{f}_i,$$

where the components A_{ij} of the linear operator $\mathcal{A}$ relate to both of the bases $\mathbf{e}_j$ and $\mathbf{f}_i$.

The basic properties of linear operators, arising from their definition, are summarised as follows. If **x** is a vector and $\mathcal{A}$ and $\mathcal{B}$ are two linear operators then

$$(\mathcal{A} + \mathcal{B})\mathbf{x} = \mathcal{A}\mathbf{x} + \mathcal{B}\mathbf{x},$$
$$(\lambda \mathcal{A})\mathbf{x} = \lambda(\mathcal{A}\mathbf{x}),$$
$$(\mathcal{A}\mathcal{B})\mathbf{x} = \mathcal{A}(\mathcal{B}\mathbf{x}),$$

where in the last equality we see that the action of two linear operators in succession is associative. However, the product of two general linear operators is not commutative, i.e. $\mathcal{A}\mathcal{B}\mathbf{x} \neq \mathcal{B}\mathcal{A}\mathbf{x}$ in general.[4]

In an obvious way we define the null (or zero) and identity operators by

$$\mathcal{O}\mathbf{x} = \mathbf{0} \quad \text{and} \quad \mathcal{I}\mathbf{x} = \mathbf{x},$$

for any vector **x** in our vector space. Two operators $\mathcal{A}$ and $\mathcal{B}$ are equal if $\mathcal{A}\mathbf{x} = \mathcal{B}\mathbf{x}$ for all vectors **x**. Finally, if there exists an operator $\mathcal{A}^{-1}$ such that

$$\mathcal{A}\mathcal{A}^{-1} = \mathcal{A}^{-1}\mathcal{A} = \mathcal{I}$$

then $\mathcal{A}^{-1}$ is the *inverse* of $\mathcal{A}$. Some linear operators do not possess an inverse and are called *singular*, whilst those operators that do have an inverse are termed *non-singular*.

EXERCISES 10.2

1. Are the following operators linear?
   (a) $\mathcal{A}\mathbf{x} = \mathbf{x} + \mathbf{d}$, where **d** is a fixed vector.
   (b) $\mathcal{A}\mathbf{x} = e^{i\theta}\mathbf{x}$, where θ is a fixed angle.
   (c) $\mathcal{A}\mathbf{x} = |\mathbf{x}|\,\mathbf{i}$, where **i** is a the unit vector in the x-direction.

---

[4] Consider a two-dimensional linear vector space in which a typical vector is $\mathbf{x} = (x_1, x_2)$, with linear operators $\mathcal{A}$, $\mathcal{B}$ and $\mathcal{C}$ defined by $\mathcal{A}\mathbf{x} = (2x_1 + x_2, x_2)$, $\mathcal{B}\mathbf{x} = (x_1, x_1 + 2x_2)$ and $\mathcal{C}\mathbf{x} = (x_1 - x_2, 2x_2)$. Show that, although $\mathcal{A}$ and $\mathcal{C}$ commute, $\mathcal{A}$ and $\mathcal{B}$ do not.

2. Are the following pairs of linear operations commutative?
   (a) $\mathcal{A} =$ a rotation of π about the x-axis,
       $\mathcal{B} =$ a rotation of π about the y-axis.
   (b) $\mathcal{A} =$ a rotation of $\pi/2$ about the x-axis,
       $\mathcal{B} =$ a rotation of $\pi/2$ about the y-axis.
   (c) $\mathcal{A} =$ a rotation of π about the x-axis,
       $\mathcal{B} =$ a rotation of $\pi/2$ about the y-axis.

10.3 Matrices

We have seen that in a particular basis $\mathbf{e}_i$ both vectors and linear operators can be described in terms of their components with respect to the basis. These components may be displayed as an array of numbers called a *matrix*. In general, if a linear operator $\mathcal{A}$ transforms vectors from an N-dimensional vector space, for which we choose a basis $\mathbf{e}_j$, $j = 1, 2, \ldots, N$, into vectors belonging to an M-dimensional vector space, with basis $\mathbf{f}_i$, $i = 1, 2, \ldots, M$, then we may represent the operator $\mathcal{A}$ by the matrix

$$\mathsf{A} = \begin{pmatrix} A_{11} & A_{12} & \cdots & A_{1N} \\ A_{21} & A_{22} & \cdots & A_{2N} \\ \vdots & \vdots & \ddots & \vdots \\ A_{M1} & A_{M2} & \cdots & A_{MN} \end{pmatrix}. \tag{10.24}$$

The *matrix elements* A_{ij} are the components of the linear operator with respect to the bases $\mathbf{e}_j$ and $\mathbf{f}_i$; the component A_{ij} of the linear operator appears in the ith row and jth column of the matrix. The array has M rows and N columns and is thus called an $M \times N$ matrix. If the dimensions of the two vector spaces are the same, i.e. $M = N$ (for example, if they are the same vector space) then we may represent $\mathcal{A}$ by an $N \times N$ or *square* matrix of *order* N. The component A_{ij}, which in general may be complex, is also commonly denoted by $(\mathsf{A})_{ij}$.

In a similar way we may denote a vector $\mathbf{x}$ in terms of its components x_i in a basis $\mathbf{e}_i$, $i = 1, 2, \ldots, N$, by the array

$$\mathsf{x} = \begin{pmatrix} x_1 \\ x_2 \\ \vdots \\ x_N \end{pmatrix},$$

which is a special case of (10.24) and is called a *column matrix* (or conventionally, and slightly confusingly, a *column vector* or even just a *vector* – strictly speaking the term 'vector' refers to the geometrical entity $\mathbf{x}$). The column matrix x can also be written as[5]

$$\mathsf{x} = (x_1 \quad x_2 \quad \cdots \quad x_N)^\mathrm{T},$$

which is the *transpose* of a *row matrix* (see Section 10.5).

---

[5] This alternative form is often used purely to save space in written or printed material.

10.4 Basic matrix algebra

We note that in a different basis $\mathbf{e}'_i$ the vector $\mathbf{x}$ would be represented by a *different* column matrix containing the components x'_i in the new basis, i.e.

$$\mathsf{x}' = \begin{pmatrix} x'_1 \\ x'_2 \\ \vdots \\ x'_N \end{pmatrix}.$$

Thus, we use x and x' to denote different column matrices which, in different bases $\mathbf{e}_i$ and $\mathbf{e}'_i$, represent the *same* vector $\mathbf{x}$. In many texts, however, this distinction is not made and $\mathbf{x}$ (rather than x) is equated to the corresponding column matrix; if we regard $\mathbf{x}$ as the geometrical entity, however, this can be misleading and so we explicitly make the distinction. A similar argument follows for linear operators; the same linear operator $\mathcal{A}$ is described in different bases by different matrices A and A', containing different matrix elements.

EXERCISE 10.3

1. The linear operator $\mathcal{A}$, which transforms vectors in a space with basis vectors $\mathbf{e}_j$ into one with basis vectors $\mathbf{f}_i$, is represented by the matrix

$$\mathsf{A} = \begin{pmatrix} 2 & 1 & -1 \\ 1 & 2 & -1 \end{pmatrix}.$$

Express the results of $\mathcal{A}$ acting on each of the $\mathbf{e}_j$ in terms of the $\mathbf{f}_i$. Either by inverting your answers, or by inspection, find a vector in the original linear vector space that is transformed into the zero vector.

10.4 Basic matrix algebra

The basic algebra of matrices may be deduced from the properties of the linear operators that they represent. In a given basis the action of two linear operators $\mathcal{A}$ and $\mathcal{B}$ on an arbitrary vector $\mathbf{x}$ (see towards the end of Section 10.2), when written in terms of components using (10.23), is given by

$$\sum_j (\mathsf{A} + \mathsf{B})_{ij} x_j = \sum_j A_{ij} x_j + \sum_j B_{ij} x_j,$$

$$\sum_j (\lambda \mathsf{A})_{ij} x_j = \lambda \sum_j A_{ij} x_j,$$

$$\sum_j (\mathsf{AB})_{ij} x_j = \sum_k A_{ik} (\mathsf{B}\mathsf{x})_k = \sum_j \sum_k A_{ik} B_{kj} x_j.$$

Matrices and vector spaces

Now, since **x** is arbitrary, we can immediately deduce the way in which matrices are added or multiplied, i.e.[6]

$$(\mathsf{A} + \mathsf{B})_{ij} = A_{ij} + B_{ij}, \qquad (10.25)$$

$$(\lambda \mathsf{A})_{ij} = \lambda A_{ij}, \qquad (10.26)$$

$$(\mathsf{AB})_{ij} = \sum_k A_{ik} B_{kj}. \qquad (10.27)$$

We note that a matrix element may, in general, be complex. We now discuss matrix addition and multiplication in more detail.

10.4.1 Matrix addition and multiplication by a scalar

From (10.25) we see that the sum of two matrices, $\mathsf{S} = \mathsf{A} + \mathsf{B}$, is the matrix whose elements are given by

$$S_{ij} = A_{ij} + B_{ij}$$

for every pair of subscripts i, j, with $i = 1, 2, \ldots, M$ and $j = 1, 2, \ldots, N$. For example, if A and B are 2×3 matrices then $\mathsf{S} = \mathsf{A} + \mathsf{B}$ is given by

$$\begin{pmatrix} S_{11} & S_{12} & S_{13} \\ S_{21} & S_{22} & S_{23} \end{pmatrix} = \begin{pmatrix} A_{11} & A_{12} & A_{13} \\ A_{21} & A_{22} & A_{23} \end{pmatrix} + \begin{pmatrix} B_{11} & B_{12} & B_{13} \\ B_{21} & B_{22} & B_{23} \end{pmatrix}$$

$$= \begin{pmatrix} A_{11} + B_{11} & A_{12} + B_{12} & A_{13} + B_{13} \\ A_{21} + B_{21} & A_{22} + B_{22} & A_{23} + B_{23} \end{pmatrix}. \qquad (10.28)$$

Clearly, for the sum of two matrices to have any meaning, the matrices must have the same dimensions, i.e. both be $M \times N$ matrices.

From definition (10.28) it follows that $\mathsf{A} + \mathsf{B} = \mathsf{B} + \mathsf{A}$ and that the sum of a number of matrices can be written unambiguously without bracketing, i.e. matrix addition is *commutative* and *associative*.

The difference of two matrices is defined by direct analogy with addition. The matrix $\mathsf{D} = \mathsf{A} - \mathsf{B}$ has elements

$$D_{ij} = A_{ij} - B_{ij}, \quad \text{for } i = 1, 2, \ldots, M, \ j = 1, 2, \ldots, N. \qquad (10.29)$$

From (10.26) the product of a matrix A with a scalar λ is the matrix with elements λA_{ij}, for example

$$\lambda \begin{pmatrix} A_{11} & A_{12} & A_{13} \\ A_{21} & A_{22} & A_{23} \end{pmatrix} = \begin{pmatrix} \lambda A_{11} & \lambda A_{12} & \lambda A_{13} \\ \lambda A_{21} & \lambda A_{22} & \lambda A_{23} \end{pmatrix}. \qquad (10.30)$$

Multiplication by a scalar is distributive and associative.

The following example illustrates these three elementary properties or definitions.

---

[6] Express the operators appearing in footnote 4 in matrix form and then use (10.27) to demonstrate their commutation or otherwise. Do operators $\mathcal{B}$ and $\mathcal{C}$ commute?

10.4 Basic matrix algebra

Example The matrices A, B and C are given by

$$A = \begin{pmatrix} 2 & -1 \\ 3 & 1 \end{pmatrix}, \quad B = \begin{pmatrix} 1 & 0 \\ 0 & -2 \end{pmatrix}, \quad C = \begin{pmatrix} -2 & 1 \\ -1 & 1 \end{pmatrix}.$$

Find the matrix $D = A + 2B - C$.

$$D = \begin{pmatrix} 2 & -1 \\ 3 & 1 \end{pmatrix} + 2\begin{pmatrix} 1 & 0 \\ 0 & -2 \end{pmatrix} - \begin{pmatrix} -2 & 1 \\ -1 & 1 \end{pmatrix}$$

$$= \begin{pmatrix} 2 + 2 \times 1 - (-2) & -1 + 2 \times 0 - 1 \\ 3 + 2 \times 0 - (-1) & 1 + 2 \times (-2) - 1 \end{pmatrix} = \begin{pmatrix} 6 & -2 \\ 4 & -4 \end{pmatrix}.$$

As a reminder, we note that for the question to have had any meaning, A, B and C all had to have the same dimensions, 2×2 in practice; the answer, D, is also 2×2. ◀

From the above considerations we see that the set of all, in general complex, $M \times N$ matrices (with fixed M and N) provide an example of a linear vector space – one whose elements have no obvious 'arrow-like' qualities.

The space is of dimension MN. One basis for it is the set of $M \times N$ matrices $\mathsf{E}^{(p,q)}$ with the property that $E_{ij}^{(p,q)} = 1$ if $i = p$ and $j = q$ whilst $E_{ij}^{(p,q)} = 0$ for all other values of i and j, i.e. each matrix has only one non-zero entry, and that equals unity. Here the pair (p, q) is simply a label that picks out a particular one of the matrices $\mathsf{E}^{(p,q)}$, the total number of which is MN.

10.4.2 Multiplication of matrices

Let us consider again the 'transformation' of one vector into another, $\mathbf{y} = \mathcal{A}\mathbf{x}$, which, from (10.23), may be described in terms of components with respect to a particular basis as

$$y_i = \sum_{j=1}^{N} A_{ij} x_j \quad \text{for } i = 1, 2, \ldots, M. \tag{10.31}$$

Writing this in matrix form as $\mathbf{y} = \mathsf{A}\mathbf{x}$ we have

$$\begin{pmatrix} y_1 \\ \boxed{y_2} \\ \vdots \\ y_M \end{pmatrix} = \begin{pmatrix} A_{11} & A_{12} & \ldots & A_{1N} \\ \boxed{A_{21}} & \boxed{A_{22}} & \ldots & \boxed{A_{2N}} \\ \vdots & \vdots & \ddots & \vdots \\ A_{M1} & A_{M2} & \ldots & A_{MN} \end{pmatrix} \begin{pmatrix} \boxed{x_1} \\ \boxed{x_2} \\ \vdots \\ \boxed{x_N} \end{pmatrix} \tag{10.32}$$

where we have highlighted with boxes the components used to calculate the element y_2: using (10.31) for $i = 2$,

$$y_2 = A_{21}x_1 + A_{22}x_2 + \cdots + A_{2N}x_N.$$

All the other components y_i are calculated similarly.

If, instead, we operate with $\mathcal{A}$ on a basis vector $\mathbf{e}_j$ having all components zero except for the jth, which equals unity, then we find

$$\mathcal{A}\mathbf{e}_j = \begin{pmatrix} A_{11} & A_{12} & \cdots & A_{1N} \\ A_{21} & A_{22} & \cdots & A_{2N} \\ \vdots & \vdots & \ddots & \vdots \\ A_{M1} & A_{M2} & \cdots & A_{MN} \end{pmatrix} \begin{pmatrix} 0 \\ 0 \\ \vdots \\ 1 \\ \vdots \\ 0 \end{pmatrix} = \begin{pmatrix} A_{1j} \\ A_{2j} \\ \vdots \\ A_{Mj} \end{pmatrix},$$

and so confirm our identification of the matrix element A_{ij} as the ith component of $\mathcal{A}\mathbf{e}_j$ in this basis.

From (10.27) we can extend our discussion to the product of two matrices $\mathsf{P} = \mathsf{AB}$, where P is the matrix of the quantities formed by the operation of the rows of A on the columns of B, treating each column of B in turn as the vector $\mathbf{x}$ represented in component form in (10.31). It is clear that, for this to be a meaningful definition, the number of columns in A must equal the number of rows in B. Thus the product AB of an $M \times N$ matrix A with an $N \times R$ matrix B is itself an $M \times R$ matrix P, where

$$P_{ij} = \sum_{k=1}^{N} A_{ik} B_{kj} \quad \text{for } i = 1, 2, \ldots, M, \quad j = 1, 2, \ldots, R.$$

For example, $\mathsf{P} = \mathsf{AB}$ may be written in matrix form

$$\begin{pmatrix} P_{11} & P_{12} \\ P_{21} & P_{22} \end{pmatrix} = \begin{pmatrix} A_{11} & A_{12} & A_{13} \\ A_{21} & A_{22} & A_{23} \end{pmatrix} \begin{pmatrix} B_{11} & B_{12} \\ B_{21} & B_{22} \\ B_{31} & B_{32} \end{pmatrix}$$

where

$$P_{11} = A_{11}B_{11} + A_{12}B_{21} + A_{13}B_{31},$$
$$P_{21} = A_{21}B_{11} + A_{22}B_{21} + A_{23}B_{31},$$
$$P_{12} = A_{11}B_{12} + A_{12}B_{22} + A_{13}B_{32},$$
$$P_{22} = A_{21}B_{12} + A_{22}B_{22} + A_{23}B_{32}.$$

Multiplication of more than two matrices follows naturally and is associative. So, for example,

$$\mathsf{A}(\mathsf{BC}) \equiv (\mathsf{AB})\mathsf{C}, \tag{10.33}$$

provided, of course, that all the products are defined.

As mentioned above, if A is an $M \times N$ matrix and B is an $N \times M$ matrix then two product matrices are possible, i.e.

$$\mathsf{P} = \mathsf{AB} \quad \text{and} \quad \mathsf{Q} = \mathsf{BA}.$$

These are clearly not the same, since P is an $M \times M$ matrix whilst Q is an $N \times N$ matrix. Thus, particular care must be taken to write matrix products in the intended order; $\mathsf{P} = \mathsf{AB}$

10.4 Basic matrix algebra

but $Q = BA$. We note in passing that A^2 means AA, A^3 means $A(AA) = (AA)A$, etc. Even if both A and B are square, in general

$$AB \neq BA, \tag{10.34}$$

i.e. the multiplication of matrices is not, in general, commutative. Consider the following.

Example Evaluate $P = AB$ and $Q = BA$ where

$$A = \begin{pmatrix} 3 & 2 & -1 \\ 0 & 3 & 2 \\ 1 & -3 & 4 \end{pmatrix}, \quad B = \begin{pmatrix} 2 & -2 & 3 \\ 1 & 1 & 0 \\ 3 & 2 & 1 \end{pmatrix}.$$

As we saw for the 2×2 case above, the element P_{ij} of the matrix $P = AB$ is found by mentally taking the 'scalar product' of the ith row of A with the jth column of B. For example, $P_{11} = 3 \times 2 + 2 \times 1 + (-1) \times 3 = 5$, $P_{12} = 3 \times (-2) + 2 \times 1 + (-1) \times 2 = -6$, etc. Thus

$$P = AB = \begin{pmatrix} 3 & 2 & -1 \\ 0 & 3 & 2 \\ 1 & -3 & 4 \end{pmatrix} \begin{pmatrix} 2 & -2 & 3 \\ 1 & 1 & 0 \\ 3 & 2 & 1 \end{pmatrix} = \begin{pmatrix} 5 & -6 & 8 \\ 9 & 7 & 2 \\ 11 & 3 & 7 \end{pmatrix},$$

and, similarly,

$$Q = BA = \begin{pmatrix} 2 & -2 & 3 \\ 1 & 1 & 0 \\ 3 & 2 & 1 \end{pmatrix} \begin{pmatrix} 3 & 2 & -1 \\ 0 & 3 & 2 \\ 1 & -3 & 4 \end{pmatrix} = \begin{pmatrix} 9 & -11 & 6 \\ 3 & 5 & 1 \\ 10 & 9 & 5 \end{pmatrix}.$$

These results illustrate that, in general, two matrices do not commute. ◀

The property that matrix multiplication is distributive over addition, i.e. that

$$(A + B)C = AC + BC \quad \text{and} \quad C(A + B) = CA + CB, \tag{10.35}$$

follows directly from its definition.[7]

10.4.3 The null and identity matrices

Both the null matrix and the identity matrix are frequently encountered, and we take this opportunity to introduce them briefly, leaving their uses until later. The *null* or *zero* matrix 0 has all elements equal to zero, and so its properties are

$$A0 = 0 = 0A,$$
$$A + 0 = 0 + A = A.$$

The *identity* matrix I has the property

$$AI = IA = A.$$

---

[7] But show that $(A + B)(A - B) = A^2 - B^2$ if, and only if, A and B commute.

Matrices and vector spaces

It is clear that, in order for the above products to be defined, the identity matrix must be square. The $N \times N$ identity matrix (often denoted by I_N) has the form

$$\mathsf{I}_N = \begin{pmatrix} 1 & 0 & \cdots & 0 \\ 0 & 1 & & \vdots \\ \vdots & & \ddots & 0 \\ 0 & \cdots & 0 & 1 \end{pmatrix}.$$

10.4.4 Functions of matrices

If a matrix A is *square* then, as mentioned above, one can define *powers* of A in a straightforward way. For example $\mathsf{A}^2 = \mathsf{AA}$, $\mathsf{A}^3 = \mathsf{AAA}$, or in the general case

$$\mathsf{A}^n = \mathsf{AA} \cdots \mathsf{A} \quad (n \text{ times}),$$

where n is a positive integer. Having defined powers of a square matrix A, we may construct *functions* of A of the form

$$\mathsf{S} = \sum_n a_n \mathsf{A}^n,$$

where the a_k are simple scalars and the number of terms in the summation may be finite or infinite. In the case where the sum has an infinite number of terms, the sum has meaning only if it converges. A common example of such a function is the *exponential* of a matrix, which is defined by

$$\exp \mathsf{A} = \sum_{n=0}^{\infty} \frac{\mathsf{A}^n}{n!}. \tag{10.36}$$

This definition can, in turn, be used to define other functions such as $\sin \mathsf{A}$ and $\cos \mathsf{A}$.[8]

EXERCISES 10.4

1. For the four matrices

$$\mathsf{A} = \begin{pmatrix} 3 & 1 & -3 \\ 1 & 2 & 0 \\ 0 & 3 & -1 \end{pmatrix}, \quad \mathsf{B} = \begin{pmatrix} 4 & 1 & -1 \\ -3 & 2 & 0 \end{pmatrix},$$

$$\mathsf{C} = \begin{pmatrix} 2 & 3 \\ -1 & 0 \\ 0 & 1 \end{pmatrix}, \quad \mathsf{D} = \begin{pmatrix} 1 & 2 & 2 \\ 2 & 1 & 2 \\ 2 & 2 & 1 \end{pmatrix},$$

which sums, differences and products are defined? Where they are, evaluate them.

2. For which matrix or matrices X, amongst those given in the previous exercise, is its cube defined? Calculate X^3 for one such matrix.

---

[8] For the 3×3 matrix A that has $A_{11} = A_{33} = 1$, $A_{22} = -1$ and all other $A_{ij} = 0$, show that the trace of $\exp i\mathsf{A}$, i.e. the sum of its diagonal elements, is equal to $3 \cos 1 + i \sin 1$.

10.5 The transpose and conjugates of a matrix

In the next few sections we will consider some of the quantities that characterise any given matrix and also some other matrices that can be derived from the original. A tabulation of these derived quantities and matrices is given in the end-of-chapter Summary. We start with the transposed matrix.

We have seen that the components of a linear operator in a given coordinate system can be written in the form of a matrix A. We will also find it useful, however, to consider the different (but clearly related) matrix formed by interchanging the rows and columns of A. The matrix is called the *transpose* of A and is denoted by A^T. It is obvious that if A is an $M \times N$ matrix then its transpose A^T is a $N \times M$ matrix.

Example Find the transpose of the matrix

$$\mathsf{A} = \begin{pmatrix} 3 & 1 & 2 \\ 0 & 4 & 1 \end{pmatrix}.$$

By interchanging the rows and columns of A we immediately obtain

$$\mathsf{A}^\mathsf{T} = \begin{pmatrix} 3 & 0 \\ 1 & 4 \\ 2 & 1 \end{pmatrix}.$$

As it must be, given that A is a 2×3 matrix, A^T is a 3×2 matrix. ◀

As mentioned in Section 10.3, the transpose of a column matrix is a row matrix and vice versa. An important use of column and row matrices is in the representation of the inner product of two real vectors in terms of their components in a given basis. This notion is discussed fully in the next section, where it is extended to complex vectors.

The transpose of the product of two matrices, $(\mathsf{AB})^\mathsf{T}$, is given by the product of their transposes taken in the reverse order, i.e.

$$(\mathsf{AB})^\mathsf{T} = \mathsf{B}^\mathsf{T} \mathsf{A}^\mathsf{T}. \tag{10.37}$$

This is proved as follows:

$$(\mathsf{AB})^\mathsf{T}_{ij} = (\mathsf{AB})_{ji} = \sum_k A_{jk} B_{ki}$$

$$= \sum_k (\mathsf{A}^\mathsf{T})_{kj} (\mathsf{B}^\mathsf{T})_{ik} = \sum_k (\mathsf{B}^\mathsf{T})_{ik} (\mathsf{A}^\mathsf{T})_{kj} = (\mathsf{B}^\mathsf{T} \mathsf{A}^\mathsf{T})_{ij},$$

and the proof can be extended to the product of several matrices to give[9]

$$(\mathsf{ABC} \cdots \mathsf{G})^\mathsf{T} = \mathsf{G}^\mathsf{T} \cdots \mathsf{C}^\mathsf{T} \mathsf{B}^\mathsf{T} \mathsf{A}^\mathsf{T}.$$

---

[9] Convince yourself that, even if $\mathsf{A}, \mathsf{B}, \mathsf{C}, \ldots, \mathsf{G}$ are not necessarily square matrices, but are compatible and the product $\mathsf{ABC} \cdots \mathsf{G}$ is meaningful, then their transposes are such that the product given on the RHS is also meaningful.

10.5.1 The complex and Hermitian conjugates

Two further matrices that can be derived from a given general $M \times N$ matrix are the *complex conjugate*, denoted by $\mathbf{A}^*$, and the *Hermitian conjugate*, denoted by $\mathbf{A}^\dagger$.

The complex conjugate of a matrix $\mathbf{A}$ is the matrix obtained by taking the complex conjugate of each of the elements of $\mathbf{A}$, i.e.

$$(\mathbf{A}^*)_{ij} = (A_{ij})^*.$$

Obviously if a matrix is *real* (i.e. it contains only real elements) then $\mathbf{A}^* = \mathbf{A}$.

Example Find the complex conjugate of the matrix

$$\mathbf{A} = \begin{pmatrix} 1 & 2 & 3i \\ 1+i & 1 & 0 \end{pmatrix}.$$

By taking the complex conjugate of each element in turn,

$$\mathbf{A}^* = \begin{pmatrix} 1 & 2 & -3i \\ 1-i & 1 & 0 \end{pmatrix},$$

the complex conjugate of the whole matrix is obtained immediately. ◀

The Hermitian conjugate, or *adjoint*, of a matrix $\mathbf{A}$ is the transpose of its complex conjugate, or equivalently, the complex conjugate of its transpose, i.e.

$$\mathbf{A}^\dagger = (\mathbf{A}^*)^\text{T} = (\mathbf{A}^\text{T})^*.$$

We note that if $\mathbf{A}$ is real (and so $\mathbf{A}^* = \mathbf{A}$) then $\mathbf{A}^\dagger = \mathbf{A}^\text{T}$, and taking the Hermitian conjugate is equivalent to taking the transpose. Following the previous line of argument for the transpose of the product of several matrices, the Hermitian conjugate of such a product can be shown to be given by

$$(\mathbf{AB} \cdots \mathbf{G})^\dagger = \mathbf{G}^\dagger \cdots \mathbf{B}^\dagger \mathbf{A}^\dagger. \tag{10.38}$$

Example Find the Hermitian conjugate of the matrix

$$\mathbf{A} = \begin{pmatrix} 1 & 2 & 3i \\ 1+i & 1 & 0 \end{pmatrix}.$$

Taking the complex conjugate of $\mathbf{A}$ from the previous example and then forming its transpose, we find

$$\mathbf{A}^\dagger = \begin{pmatrix} 1 & 1-i \\ 2 & 1 \\ -3i & 0 \end{pmatrix}.$$

We could obtain the same result, of course, by first taking the transpose of $\mathbf{A}$ and then forming its complex conjugate. ◀

10.6 The trace of a matrix

An important use of the Hermitian conjugate (or transpose in the real case) is in connection with the inner product of two vectors. Suppose that in a given orthonormal basis the vectors **a** and **b** may be represented by the column matrices

$$\mathbf{a} = \begin{pmatrix} a_1 \\ a_2 \\ \vdots \\ a_N \end{pmatrix} \quad \text{and} \quad \mathbf{b} = \begin{pmatrix} b_1 \\ b_2 \\ \vdots \\ b_N \end{pmatrix}. \tag{10.39}$$

Taking the Hermitian conjugate of **a**, to give a row matrix, and multiplying (on the right) by **b** we obtain

$$\mathbf{a}^\dagger \mathbf{b} = (a_1^* \ a_2^* \ \cdots \ a_N^*) \begin{pmatrix} b_1 \\ b_2 \\ \vdots \\ b_N \end{pmatrix} = \sum_{i=1}^N a_i^* b_i, \tag{10.40}$$

which is the expression for the inner product $\langle \mathbf{a}|\mathbf{b}\rangle$ in that basis.[10] The inner product could also be viewed as the 1×1 matrix obtained as the product of a $1 \times N$ matrix with an $N \times 1$ matrix. We note that for real vectors (10.40) reduces to $\mathbf{a}^T \mathbf{b} = \sum_{i=1}^N a_i b_i$.

EXERCISE 10.5

1. Write down the transpose, the complex conjugate and the Hermitian conjugate for each of the matrices

$$A = \begin{pmatrix} 1 & 0 & -1 \\ 1+i & i & 1-i \\ 2 & 2i & -i \end{pmatrix}, \quad B = \begin{pmatrix} 2 & i \\ i & 2 \\ 1 & -i \end{pmatrix}.$$

Verify by direct calculation that $(AB)^\dagger = B^\dagger A^\dagger$.

10.6 The trace of a matrix

For a given matrix A, in the previous two sections we have considered various other matrices that can be derived from it. However, sometimes one wishes to derive a single number from a matrix. The simplest example is the *trace* (or *spur*) of a square matrix, which is denoted by Tr A. This quantity is defined as the sum of the diagonal elements of the matrix,

$$\operatorname{Tr} A = A_{11} + A_{22} + \cdots + A_{NN} = \sum_{i=1}^N A_{ii}. \tag{10.41}$$

---

[10] It also follows that $\mathbf{a}^\dagger \mathbf{a} = \sum_{n=1}^N a_i^* a_i = \sum_{n=1}^N |a_i|^2$ is real for any vector **a**, whether or not it has complex components.

Matrices and vector spaces

At this point, the definition may seem arbitrary, but as will be seen in this section, as well as later in the chapter, the trace of a matrix has properties that characterise the linear operator it represents, and are independent of the basis chosen for that representation. It is clear that taking the trace is a linear operation so that, for example,

$$\text{Tr}(A \pm B) = \text{Tr}\, A \pm \text{Tr}\, B.$$

A very useful property of traces is that the trace of the product of two matrices is independent of the order of their multiplication; this results holds whether or not the matrices commute and is proved as follows:

$$\text{Tr}\, AB = \sum_{i=1}^{N}(AB)_{ii} = \sum_{i=1}^{N}\sum_{j=1}^{N} A_{ij}B_{ji} = \sum_{i=1}^{N}\sum_{j=1}^{N} B_{ji}A_{ij} = \sum_{j=1}^{N}(BA)_{jj} = \text{Tr}\, BA. \tag{10.42}$$

The result can be extended to the product of several matrices. For example, from (10.42), we immediately find

$$\text{Tr}\, ABC = \text{Tr}\, BCA = \text{Tr}\, CAB,$$

which shows that the trace of a multiple product is invariant under cyclic permutations of the matrices in the product. Other easily derived properties of the trace are, for example, $\text{Tr}\, A^T = \text{Tr}\, A$ and $\text{Tr}\, A^\dagger = (\text{Tr}\, A)^*$.

EXERCISE 10.6

1. For the three matrices

$$A = \begin{pmatrix} 1 & 0 \\ 0 & -1 \end{pmatrix}, \quad B = \begin{pmatrix} 0 & 1 \\ 1 & 0 \end{pmatrix}, \quad C = \begin{pmatrix} 0 & -i \\ i & 0 \end{pmatrix},$$

verify that $\text{Tr}\, ABC = \text{Tr}\, BCA = \text{Tr}\, CAB$. Show also that $\text{Tr}\, ABC \neq \text{Tr}\, BAC$.

10.7 The determinant of a matrix

For a given matrix A, the determinant det A (like the trace) is a single number (or algebraic expression) that depends upon the elements of A. Also like the trace, the determinant is defined only for *square* matrices. If, for example, A is a 3 × 3 matrix then its determinant, of *order* 3, is denoted by

$$\det A = |A| = \begin{vmatrix} A_{11} & A_{12} & A_{13} \\ A_{21} & A_{22} & A_{23} \\ A_{31} & A_{32} & A_{33} \end{vmatrix}, \tag{10.43}$$

i.e. the round or square brackets are replaced by vertical bars, similar to (large) modulus signs, but not to be confused with them.

In order to calculate the value of a general determinant of order n, we first define that of an order-1 determinant. We would not normally refer to an array with only one element

10.7 The determinant of a matrix

as a matrix, but formally it is a 1×1 matrix, and it is useful to think of it as such for the present purposes. The determinant of such a matrix is *defined* to be the value of its single entry. Notice that although when it is written in determinantal form it looks exactly like a modulus sign, $|a_{11}|$, it must not be treated as such, and, for example, a 1×1 matrix with a single entry -7 has determinant -7, not 7.

In order to define the determinant of an $n \times n$ matrix we will need to introduce the notions of the *minor* and the *cofactor* of an element of a matrix. We will then see that we can use the cofactors to write an order-3 determinant as the weighted sum of three order-2 determinants; these, in turn will each be formally expanded in terms of two order-1 determinants.[11]

The minor M_{ij} of the element A_{ij} of an $N \times N$ matrix $\mathbf{A}$ is the determinant of the $(N-1) \times (N-1)$ matrix obtained by removing all the elements of the ith row and jth column of $\mathbf{A}$; the associated cofactor, C_{ij}, is found by multiplying the minor by $(-1)^{i+j}$. The following example illustrates this.

Example Find the cofactor of the element A_{23} of the matrix

$$\mathbf{A} = \begin{pmatrix} A_{11} & A_{12} & A_{13} \\ A_{21} & A_{22} & A_{23} \\ A_{31} & A_{32} & A_{33} \end{pmatrix}.$$

Removing all the elements of the second row and third column of $\mathbf{A}$ and forming the determinant of the remaining terms gives the minor

$$M_{23} = \begin{vmatrix} A_{11} & A_{12} \\ A_{31} & A_{32} \end{vmatrix}.$$

Multiplying the minor by $(-1)^{2+3} = (-1)^5 = -1$ then gives

$$C_{23} = - \begin{vmatrix} A_{11} & A_{12} \\ A_{31} & A_{32} \end{vmatrix}$$

as the cofactor of A_{23}. ◂

We now define a determinant as *the sum of the products of the elements of any row or column and their corresponding cofactors*, e.g. $A_{21}C_{21} + A_{22}C_{22} + A_{23}C_{23}$ or $A_{13}C_{13} + A_{23}C_{23} + A_{33}C_{33}$. Such a sum is called a *Laplace expansion*. For example, in the first of these expansions, using the elements of the second row of the determinant defined by (10.43) and their corresponding cofactors, we write $|\mathbf{A}|$ as the Laplace expansion

$$|\mathbf{A}| = A_{21}(-1)^{(2+1)}M_{21} + A_{22}(-1)^{(2+2)}M_{22} + A_{23}(-1)^{(2+3)}M_{23}$$

$$= -A_{21}\begin{vmatrix} A_{12} & A_{13} \\ A_{32} & A_{33} \end{vmatrix} + A_{22}\begin{vmatrix} A_{11} & A_{13} \\ A_{31} & A_{33} \end{vmatrix} - A_{23}\begin{vmatrix} A_{11} & A_{12} \\ A_{31} & A_{32} \end{vmatrix}.$$

We will see later that the value of the determinant is independent of the row or column chosen. Of course, we have not yet determined the value of $|\mathbf{A}|$ but, rather, written it as

---

[11] Though in practice the values of order-2 determinants are nearly always computed directly by inspection.

Matrices and vector spaces

the weighted sum of three determinants of order 2. However, applying again the definition of a determinant, we can evaluate each of the order-2 determinants. As a typical example consider the first of these.

Example Evaluate the determinant

$$\begin{vmatrix} A_{12} & A_{13} \\ A_{32} & A_{33} \end{vmatrix}.$$

By considering the products of the elements of the first row in the determinant, and their corresponding cofactors (now order-1 determinants), we find

$$\begin{vmatrix} A_{12} & A_{13} \\ A_{32} & A_{33} \end{vmatrix} = A_{12}(-1)^{(1+1)}|A_{33}| + A_{13}(-1)^{(1+2)}|A_{32}|$$
$$= A_{12}A_{33} - A_{13}A_{32},$$

where the values of the order-1 determinants $|A_{33}|$ and $|A_{32}|$ are, as defined earlier, A_{33} and A_{32} respectively. It must be remembered that the determinant is *not* necessarily the same as the modulus, e.g. $\det(-2) = |-2| = -2$, not 2. ◂

We can now combine all the above results to show that the value of the determinant (10.43) is given by

$$|\mathsf{A}| = -A_{21}(A_{12}A_{33} - A_{13}A_{32}) + A_{22}(A_{11}A_{33} - A_{13}A_{31}) - A_{23}(A_{11}A_{32} - A_{12}A_{31})$$
(10.44)
$$= A_{11}(A_{22}A_{33} - A_{23}A_{32}) + A_{12}(A_{23}A_{31} - A_{21}A_{33}) + A_{13}(A_{21}A_{32} - A_{22}A_{31}),$$
(10.45)

where the final expression gives the form in which the determinant is usually remembered and is the form that is obtained immediately by considering the Laplace expansion using the first row of the determinant. The last equality, which essentially rearranges a Laplace expansion using the second row into one using the first row, supports our assertion that the value of the determinant is unaffected by which row or column is chosen for the expansion. An alternative, but equivalent, view is contained in the next example.

Example Suppose the rows of a real 3×3 matrix A are interpreted as the components, in a given basis, of three (three-component) vectors $\mathbf{a}$, $\mathbf{b}$ and $\mathbf{c}$. Show that the determinant of A can be written as

$$|\mathsf{A}| = \mathbf{a} \cdot (\mathbf{b} \times \mathbf{c}).$$

If the rows of A are written as the components in a given basis of three vectors $\mathbf{a}$, $\mathbf{b}$ and $\mathbf{c}$, we have from (10.45) that

$$|\mathsf{A}| = \begin{vmatrix} a_1 & a_2 & a_3 \\ b_1 & b_2 & b_3 \\ c_1 & c_2 & c_3 \end{vmatrix} = a_1(b_2c_3 - b_3c_2) + a_2(b_3c_1 - b_1c_3) + a_3(b_1c_2 - b_2c_1).$$

10.7 The determinant of a matrix

From expression (9.36) for the scalar triple product given in Section 9.5.1, it follows that we may write the determinant as

$$|A| = \mathbf{a} \cdot (\mathbf{b} \times \mathbf{c}). \tag{10.46}$$

In other words, $|A|$ is the volume of the parallelepiped defined by the vectors $\mathbf{a}$, $\mathbf{b}$ and $\mathbf{c}$. One could equally well interpret the *columns* of the matrix A as the components of three vectors, and result (10.46) would still hold.

This result provides a more memorable (and more meaningful) expression than (10.45) for the value of a 3×3 determinant. Indeed, using this geometrical interpretation, we see immediately that, if the vectors $\mathbf{a}_1$, $\mathbf{a}_2$, $\mathbf{a}_3$ are not linearly independent then the value of the determinant vanishes: $|A| = 0$.[12] ◀

The evaluation of determinants of order greater than 3 follows the same general method as that presented above, in that it relies on successively reducing the order of the determinant by writing it as a Laplace expansion. Thus, a determinant of order 4 is first written as a sum of four determinants of order 3, which are then evaluated using the above method. For higher order determinants, one cannot write down directly a simple geometrical expression for $|A|$ analogous to that given in (10.46). Nevertheless, it is still true that if the rows or columns of the $N \times N$ matrix A are interpreted as the components in a given basis of N (N-component) vectors $\mathbf{a}_1, \mathbf{a}_2, \ldots, \mathbf{a}_N$, then the determinant $|A|$ vanishes if these vectors are not all linearly independent.

10.7.1 Properties of determinants

A number of properties of determinants follow straightforwardly from the definition of det A; their use will often reduce the labour of evaluating a determinant. We present them here without specific proofs, though they all proved in Problem 10.37 using an alternative form for a determinant that is expressed in terms of the so-called Levi-Civita symbols. These are defined in Appendix D, in connection with a discussion of the summation convention (see Section 10.17).

(i) *Determinant of the transpose.* The transpose matrix A^T (which, we recall, is obtained by interchanging the rows and columns of A) has the same determinant as A itself, i.e.

$$|A^T| = |A|. \tag{10.47}$$

It follows that *any* theorem involving determinants established for the rows of A will apply to the columns as well, and vice versa.

(ii) *Determinant of the complex and Hermitian conjugate.* It is clear that the matrix A^* obtained by taking the complex conjugate of each element of A has the determinant $|A^*| = |A|^*$. Combining this result with (10.47), we find that

$$|A^\dagger| = |(A^*)^T| = |A^*| = |A|^*. \tag{10.48}$$

---

[12] Each can be expressed in terms of the other two; consequently, (i) they all lie in a plane and (ii) the parallelepiped they define has zero volume.

Matrices and vector spaces

(iii) *Interchanging two rows or two columns.* If two rows (columns) of A are interchanged, its determinant changes sign but is unaltered in magnitude.

(iv) *Removing factors.* If all the elements of a single row (column) of A have a common factor, λ, then this factor may be removed; the value of the determinant is given by the product of the remaining determinant and λ. Clearly this implies that if all the elements of any row (column) are zero then $|A| = 0$. It also follows that if every element of the $N \times N$ matrix A is multiplied by a constant factor λ then

$$|\lambda A| = \lambda^N |A|. \tag{10.49}$$

(v) *Identical rows or columns.* If any two rows (columns) of A are identical or are multiples of one another, then it can be shown that $|A| = 0$.

(vi) *Adding a constant multiple of one row (column) to another.* The determinant of a matrix is unchanged in value by adding to the elements of one row (column) any fixed multiple of the elements of another row (column).

(vii) *Determinant of a product.* If A and B are square matrices of the same order then

$$|AB| = |A||B| = |BA|. \tag{10.50}$$

A simple extension of this property gives, for example,

$$|AB \cdots G| = |A||B| \cdots |G| = |A||G| \cdots |B| = |A \cdots GB|,$$

which shows that the determinant is invariant under permutation of the matrices in a multiple product.

10.7.2 Evaluation of determinants

There is no explicit procedure for using the above results in the evaluation of any given determinant, and judging the quickest route to an answer is a matter of experience. A general guide is to try to reduce all terms but one in a row or column to zero and hence in effect to obtain a determinant of smaller size. The steps taken in evaluating the determinant in the example below are certainly not the fastest, but they have been chosen in order to illustrate the use of most of the properties listed above.

Example Evaluate the determinant

$$|A| = \begin{vmatrix} 1 & 0 & 2 & 3 \\ 0 & 1 & -2 & 1 \\ 3 & -3 & 4 & -2 \\ -2 & 1 & -2 & -1 \end{vmatrix}.$$

Taking a factor 2 out of the third column and then adding the second column to the third gives

$$|A| = 2 \begin{vmatrix} 1 & 0 & 1 & 3 \\ 0 & 1 & -1 & 1 \\ 3 & -3 & 2 & -2 \\ -2 & 1 & -1 & -1 \end{vmatrix} = 2 \begin{vmatrix} 1 & 0 & 1 & 3 \\ 0 & 1 & 0 & 1 \\ 3 & -3 & -1 & -2 \\ -2 & 1 & 0 & -1 \end{vmatrix}.$$

10.7 The determinant of a matrix

Subtracting the second column from the fourth gives

$$|A| = 2 \begin{vmatrix} 1 & 0 & 1 & 3 \\ 0 & 1 & 0 & 0 \\ 3 & -3 & -1 & 1 \\ -2 & 1 & 0 & -2 \end{vmatrix}.$$

We now note that the second row has only one non-zero element and so the determinant may conveniently be written as a Laplace expansion, i.e.

$$|A| = 2 \times 1 \times (-1)^{2+2} \begin{vmatrix} 1 & 1 & 3 \\ 3 & -1 & 1 \\ -2 & 0 & -2 \end{vmatrix} = 2 \begin{vmatrix} 4 & 0 & 4 \\ 3 & -1 & 1 \\ -2 & 0 & -2 \end{vmatrix},$$

where the last equality follows by adding the second row to the first. It can now be seen that the first row is minus twice the third, and so the value of the determinant is zero, by property (v) above. ◀

EXERCISES 10.7

1. Use the properties of determinants to show that the vectors

$$(2, -3, 1), \quad (-1, 4, 2), \quad (0, 1, 1),$$

   are coplanar.

2. For the matrices A, B and C defined in Exercise 10.6, prove by direct calculation that $|ABC| = |A||B||C|$. Show further that, although, as proved in that exercise, ABC and BAC do not have equal traces, they do have equal determinants.

3. Evaluate the determinant of

$$A = \begin{pmatrix} 1 & i & -1 \\ 0 & -1 & -i \\ 1 & i & 0 \end{pmatrix}$$

   by making a Laplace expansion (a) using the first row of A and (b) using the first column of A, and confirm that they yield the same value. Choose any other row or column and make the corresponding expansion, verifying that the same value for $|A|$ is obtained as previously.

4. By exploiting its general properties, evaluate the determinant of

$$A = \begin{pmatrix} 2 & 1 & -2 & 0 \\ -1 & 4 & 1 & 9 \\ 3 & -3 & -2 & -8 \\ -2 & -1 & 0 & -2 \end{pmatrix}$$

   as efficiently as you can.

10.8 The inverse of a matrix

Our first use of determinants will be in defining the *inverse* of a matrix. If we were dealing with ordinary numbers we would consider the relation $P = AB$ as equivalent to $B = P/A$, provided that $A \neq 0$. However, if A, B and P are matrices then this notation does not have an obvious meaning. What we really want to know is whether an explicit formula for B can be obtained in terms of A and P.

It will be shown that this is possible for those cases in which $|A| \neq 0$. A square matrix whose determinant is zero is called a *singular* matrix; otherwise it is *non-singular*. We will show that if A is non-singular we can define a matrix, denoted by A^{-1} and called the *inverse* of A, which has the property that if $AB = P$ then $B = A^{-1}P$. In words, B can be obtained by multiplying P from the left by A^{-1}. Analogously, if B is non-singular then, by multiplication from the right, $A = PB^{-1}$.

It is clear that

$$AI = A \quad \Rightarrow \quad I = A^{-1}A, \tag{10.51}$$

where I is the unit matrix, and so $A^{-1}A = I = AA^{-1}$.[13] These statements are equivalent to saying that if we first multiply a matrix, B say, by A and then multiply by the inverse A^{-1}, we end up with the matrix we started with, i.e.

$$A^{-1}AB = B. \tag{10.52}$$

This justifies our use of the term inverse. It is also clear that the inverse is only defined for square matrices.

So far we have only defined what we mean by the inverse of a matrix. Actually finding the inverse of a matrix A may be carried out in a number of ways. We will show that one method is to construct first the matrix C containing the cofactors of the elements of A, as discussed in Section 10.7. Then the required inverse A^{-1} can be found by forming the transpose of C and dividing by the determinant of A. Thus the elements of the inverse A^{-1} are given by

$$(A^{-1})_{ik} = \frac{(C)^T_{ik}}{|A|} = \frac{C_{ki}}{|A|}. \tag{10.53}$$

That this procedure does indeed result in the inverse may be seen by considering the components of $A^{-1}A$ with A^{-1} defined in this way, i.e.

$$(A^{-1}A)_{ij} = \sum_k (A^{-1})_{ik}(A)_{kj} = \sum_k \frac{C_{ki}}{|A|}A_{kj} = \frac{|A|}{|A|}\delta_{ij}. \tag{10.54}$$

The last equality in (10.54) relies on the property

$$\sum_k C_{ki}A_{kj} = |A|\delta_{ij}. \tag{10.55}$$

---

[13] It is not immediately obvious that $AA^{-1} = I$, since A^{-1} has only been defined as a left inverse. Prove that the left inverse is also a right inverse by defining A_R^{-1} by $AA_R^{-1} = I$ and then, by considering $A^{-1}AA_R^{-1}$, show that $A_R^{-1} = A^{-1}$.

10.8 The inverse of a matrix

This can be proved by considering the matrix $\mathbf{A}'$ obtained from the original matrix $\mathbf{A}$ when the ith column of $\mathbf{A}$ is replaced by one of the other columns, say the jth; as an equation, $A'_{ki} = A_{kj}$. With this construction, $\mathbf{A}'$ is a matrix with two identical columns and so has zero determinant. However, replacing the ith column by another does not change the cofactors C_{ki} of the elements in the ith column, which are therefore the same in $\mathbf{A}$ and $\mathbf{A}'$, i.e. $C_{ki} = C'_{ki}$ for all k. Recalling the Laplace expansion of a general determinant, i.e.

$$|\mathbf{A}| = \sum_k A_{ki} C_{ki},$$

we obtain for the case $i \neq j$ that

$$\sum_k A_{kj} C_{ki} = \sum_k A'_{ki} C'_{ki} = |\mathbf{A}'| = 0.$$

The Laplace expansion itself deals with the case $i = j$, and the two together establish result (10.55).

It is immediately obvious from (10.53) that the inverse of a matrix is not defined if the matrix is singular (i.e. if $|\mathbf{A}| = 0$).

Example Find the inverse of the matrix

$$\mathbf{A} = \begin{pmatrix} 2 & 4 & 3 \\ 1 & -2 & -2 \\ -3 & 3 & 2 \end{pmatrix}.$$

We first determine $|\mathbf{A}|$:

$$|\mathbf{A}| = 2[-2(2) - (-2)3] + 4[(-2)(-3) - (1)(2)] + 3[(1)(3) - (-2)(-3)]$$
$$= 11. \tag{10.56}$$

This is non-zero and so an inverse matrix can be constructed. To do this we need the matrix of the cofactors, $\mathbf{C}$, and hence $\mathbf{C}^T$. We find[14]

$$\mathbf{C} = \begin{pmatrix} 2 & 4 & -3 \\ 1 & 13 & -18 \\ -2 & 7 & -8 \end{pmatrix} \quad \text{and} \quad \mathbf{C}^T = \begin{pmatrix} 2 & 1 & -2 \\ 4 & 13 & 7 \\ -3 & -18 & -8 \end{pmatrix},$$

and hence

$$\mathbf{A}^{-1} = \frac{\mathbf{C}^T}{|\mathbf{A}|} = \frac{1}{11}\begin{pmatrix} 2 & 1 & -2 \\ 4 & 13 & 7 \\ -3 & -18 & -8 \end{pmatrix}. \tag{10.57}$$

This result can be checked (somewhat tediously) by computing $\mathbf{A}^{-1}\mathbf{A}$. ◂

For a 2×2 matrix, the inverse has a particularly simple form. If the matrix is

$$\mathbf{A} = \begin{pmatrix} A_{11} & A_{12} \\ A_{21} & A_{22} \end{pmatrix}$$

---

[14] The reader should calculate at least some of the cofactors for themselves, paying particular attention to the sign of each.

Matrices and vector spaces

then its determinant $|A|$ is given by $|A| = A_{11}A_{22} - A_{12}A_{21}$, and the matrix of cofactors is

$$C = \begin{pmatrix} A_{22} & -A_{21} \\ -A_{12} & A_{11} \end{pmatrix}.$$

Thus the inverse of A is given by

$$A^{-1} = \frac{C^T}{|A|} = \frac{1}{A_{11}A_{22} - A_{12}A_{21}} \begin{pmatrix} A_{22} & -A_{12} \\ -A_{21} & A_{11} \end{pmatrix}. \tag{10.58}$$

It can be seen that the transposed matrix of cofactors for a 2×2 matrix is the same as the matrix formed by swapping the elements on the leading diagonal (A_{11} and A_{22}) and changing the signs of the other two elements (A_{12} and A_{21}). This is completely general for a 2×2 matrix and is easy to remember.

The following are some further useful properties related to the inverse matrix and may be straightforwardly derived, as below.

(i) $(A^{-1})^{-1} = A$, (ii) $(A^T)^{-1} = (A^{-1})^T$, (iii) $(A^\dagger)^{-1} = (A^{-1})^\dagger$,

(iv) $(AB)^{-1} = B^{-1}A^{-1}$, (v) $(AB \cdots G)^{-1} = G^{-1} \cdots B^{-1}A^{-1}$.

Example Prove the properties (i)–(v) stated above.

We begin by writing down the fundamental expression defining the inverse of a non-singular square matrix A:

$$AA^{-1} = I = A^{-1}A. \tag{10.59}$$

Property (i). This follows immediately from the expression (10.59).
Property (ii). Taking the transpose of each expression in (10.59) gives

$$(AA^{-1})^T = I^T = (A^{-1}A)^T.$$

Using the result (10.37) for the transpose of a product of matrices and noting that $I^T = I$, we find

$$(A^{-1})^T A^T = I = A^T (A^{-1})^T.$$

However, from (10.59), this implies $(A^{-1})^T = (A^T)^{-1}$ and hence proves result (ii) above.

Property (iii). This may be proved in an analogous way to property (ii), by replacing the transposes in (ii) by Hermitian conjugates and using the result (10.38) for the Hermitian conjugate of a product of matrices.

Property (iv). Using (10.59), we may write

$$(AB)(AB)^{-1} = I = (AB)^{-1}(AB).$$

From the left-hand equality it follows, by multiplying on the left by A^{-1}, that

$$A^{-1}AB(AB)^{-1} = A^{-1}I \quad \text{and hence} \quad B(AB)^{-1} = A^{-1}.$$

Now multiplying on the left by B^{-1} gives

$$B^{-1}B(AB)^{-1} = B^{-1}A^{-1},$$

and hence the stated result.

10.9 The rank of a matrix

Property (v). Finally, result (iv) may be extended to case (v) in a straightforward manner. For example, using result (iv) twice we find

$$(ABC)^{-1} = (BC)^{-1}A^{-1} = C^{-1}B^{-1}A^{-1}.$$

Clearly, this can then be further extended to cover the product of any finite number of matrices. ◀

We conclude this section by noting that the determinant $|A^{-1}|$ of the inverse matrix can be expressed very simply in terms of the determinant $|A|$ of the matrix itself. Again we start with the fundamental expression (10.59). Then, using the property (10.50) for the determinant of a product, we find

$$|AA^{-1}| = |A||A^{-1}| = |I|.$$

It is straightforward to show by Laplace expansion that $|I| = 1$, and so we arrive at the useful result

$$|A^{-1}| = \frac{1}{|A|}. \tag{10.60}$$

EXERCISE 10.8

1. Where they exist, find the inverses of the following matrices:

$$\begin{pmatrix} 1-i & i \\ -i & 1+i \end{pmatrix}, \quad \begin{pmatrix} 4 & -2 & 10 \\ 3 & 1 & 5 \\ -3 & 1 & -7 \end{pmatrix}, \quad \begin{pmatrix} 1 & 3 & -2\sqrt{3} \\ 3 & 1 & 2\sqrt{3} \\ \sqrt{3} & -\sqrt{3} & -2 \end{pmatrix}.$$

10.9 The rank of a matrix

The *rank* of a general $M \times N$ matrix is an important concept, particularly in the solution of sets of simultaneous linear equations, as discussed in the next section, and we now consider it in some detail. Like the trace and determinant, the rank of matrix A is a single number (or algebraic expression) that depends on the elements of A. Unlike the trace and determinant, however, the rank of a matrix can be defined even when A is not square. As we shall see, there are two *equivalent* definitions of the rank of a general matrix.

Firstly, the rank of a matrix may be defined in terms of the *linear independence* of vectors. Suppose that the columns of an $M \times N$ matrix are interpreted as the components in a given basis of N (M-component) vectors $\mathbf{v}_1, \mathbf{v}_2, \ldots, \mathbf{v}_N$, as follows:

$$A = \begin{pmatrix} \uparrow & \uparrow & & \uparrow \\ \mathbf{v}_1 & \mathbf{v}_2 & \cdots & \mathbf{v}_N \\ \downarrow & \downarrow & & \downarrow \end{pmatrix}.$$

Then the *rank* of A, denoted by rank A or by $R(A)$, is defined as the number of *linearly independent* vectors in the set $\mathbf{v}_1, \mathbf{v}_2, \ldots, \mathbf{v}_N$, and equals the dimension of the vector space spanned by those vectors. Alternatively, we may consider the rows of A to contain the components in a given basis of the M (N-component) vectors $\mathbf{w}_1, \mathbf{w}_2, \ldots, \mathbf{w}_M$ as follows:

$$A = \begin{pmatrix} \leftarrow & \mathbf{w}_1 & \rightarrow \\ \leftarrow & \mathbf{w}_2 & \rightarrow \\ & \vdots & \\ \leftarrow & \mathbf{w}_M & \rightarrow \end{pmatrix}.$$

It may then be shown[15] that the rank of A is also equal to the number of linearly independent vectors in the set $\mathbf{w}_1, \mathbf{w}_2, \ldots, \mathbf{w}_M$. From this definition it is should be clear that the rank of A is unaffected by the exchange of two rows (or two columns) or by the multiplication of a row (or column) by a constant. Furthermore, suppose that a constant multiple of one row (column) is added to another row (column): for example, we might replace the row $\mathbf{w}_i$ by $\mathbf{w}_i + c\mathbf{w}_j$. This also has no effect on the number of linearly independent rows and so leaves the rank of A unchanged. We may use these properties to evaluate the rank of a given matrix.

A second (equivalent) definition of the rank of a matrix may be given and uses the concept of *submatrices*. A submatrix of A is any matrix that can be formed from the elements of A by ignoring one, or more than one, row or column. It may be shown that the rank of a general $M \times N$ matrix is equal to the size of the largest square submatrix of A whose determinant is non-zero. Therefore, if a matrix A has an $r \times r$ submatrix S with $|S| \neq 0$, but no $(r+1) \times (r+1)$ submatrix with non-zero determinant, then the rank of the matrix is r. From either definition it is clear that the rank of A is less than or equal to the smaller of M and N.[16]

Example Determine the rank of the matrix

$$A = \begin{pmatrix} 1 & 1 & 0 & -2 \\ 2 & 0 & 2 & 2 \\ 4 & 1 & 3 & 1 \end{pmatrix}.$$

The largest possible square submatrices of A must be of dimension 3×3. Clearly, A possesses four such submatrices, the determinants of which are given by

$$\begin{vmatrix} 1 & 1 & 0 \\ 2 & 0 & 2 \\ 4 & 1 & 3 \end{vmatrix} = 0, \qquad \begin{vmatrix} 1 & 1 & -2 \\ 2 & 0 & 2 \\ 4 & 1 & 1 \end{vmatrix} = 0,$$

$$\begin{vmatrix} 1 & 0 & -2 \\ 2 & 2 & 2 \\ 4 & 3 & 1 \end{vmatrix} = 0, \qquad \begin{vmatrix} 1 & 0 & -2 \\ 0 & 2 & 2 \\ 1 & 3 & 1 \end{vmatrix} = 0.$$

---

[15] For a fuller discussion, see, for example, C. D. Cantrell, *Modern Mathematical Methods for Physicists and Engineers* (Cambridge: Cambridge University Press, 2000), Chapter 6.
[16] State the rank of an $N \times N$ matrix all of whose entries are equal to the non-zero value λ. Justify your answer by separate references to (a) the independence of its columns and (b) the determinant of any arbitrary 2×2 submatrix.

10.10 Simultaneous linear equations

In each case the determinant may be evaluated in the way described in Section 10.7.1. The fact that the determinants of all four 3×3 submatrices are zero implies that the rank of A is less than 3.

The next largest square submatrices of A are of dimension 2×2. Consider, for example, the 2×2 submatrix formed by ignoring the third row and the third and fourth columns of A; this has determinant

$$\begin{vmatrix} 1 & 1 \\ 2 & 0 \end{vmatrix} = (1 \times 0) - (2 \times 1) = -2.$$

Since its determinant is non-zero, A is of rank 2 and we need not consider any other 2×2 submatrix. ◂

In the special case in which the matrix A is a *square* $N \times N$ matrix, by comparing either of the above definitions of rank with our discussion of determinants in Section 10.7, we see that $|A| = 0$ unless the rank of A is N. In other words, A is *singular* unless $R(A) = N$.

EXERCISE 10.9

1. Determine the ranks of the following matrices:

$$A = \begin{pmatrix} 1 & 2 & 3 \\ 1 & 2 & 3 \end{pmatrix}, \quad B = \begin{pmatrix} 1 & 1 \\ 2 & 2 \\ 3 & -3 \end{pmatrix}, \quad C = \begin{pmatrix} 2 & 3 & 0 \\ 1 & -4 & 2 \\ 2 & 1 & 3 \end{pmatrix},$$

$$D = \begin{pmatrix} 1 & -3 & 2 \\ 2 & 2 & -1 \\ 5 & 1 & 0 \end{pmatrix}, \quad E = \begin{pmatrix} 1 & 2 & 0 & -1 \\ 0 & -7 & 2 & 6 \\ 3 & -1 & 2 & 3 \end{pmatrix}.$$

10.10 Simultaneous linear equations

In physical applications we often encounter sets of simultaneous linear equations. In general we may have M equations in N unknowns $x_1, x_2, \ldots, x_N$ of the form

$$\begin{aligned} A_{11}x_1 + A_{12}x_2 + \cdots + A_{1N}x_N &= b_1, \\ A_{21}x_1 + A_{22}x_2 + \cdots + A_{2N}x_N &= b_2, \\ &\vdots \\ A_{M1}x_1 + A_{M2}x_2 + \cdots + A_{MN}x_N &= b_M, \end{aligned} \quad (10.61)$$

where the A_{ij} and b_i have known values. If all the b_i are zero then the system of equations is called *homogeneous*, otherwise it is *inhomogeneous*. Depending on the given values, this set of equations for the N unknowns $x_1, x_2, \ldots, x_N$ may have either a unique solution, no solution or infinitely many solutions. Matrix analysis may be used to distinguish between the possibilities.

Matrices and vector spaces

The set of equations may be expressed as a single matrix equation $Ax = b$, or, written out in full, as

$$\begin{pmatrix} A_{11} & A_{12} & \cdots & A_{1N} \\ A_{21} & A_{22} & \cdots & A_{2N} \\ \vdots & \vdots & \ddots & \vdots \\ A_{M1} & A_{M2} & \cdots & A_{MN} \end{pmatrix} \begin{pmatrix} x_1 \\ x_2 \\ \vdots \\ x_N \end{pmatrix} = \begin{pmatrix} b_1 \\ b_2 \\ \vdots \\ b_M \end{pmatrix}. \qquad (10.62)$$

A fourth way of writing the same equations is to interpret the columns of A as the components, in some basis, of N (M-component) vectors $v_1, v_2 \ldots, v_N$:

$$x_1 v_1 + x_2 v_2 + \cdots + x_N v_N = b. \qquad (10.63)$$

In passing, we recall that the number of linearly independent vectors is equal to r, the rank of A.

10.10.1 The number of solutions

The rank of A has far-reaching consequences for the existence of solutions to sets of simultaneous linear equations such as (10.61). As just mentioned, these equations may have *no solution*, a *unique solution* or *infinitely many solutions*. We now discuss these three cases in turn.

No solution

The system of equations possesses no solution unless, as expressed in Equation (10.63), **b** can be written as a linear combination of the columns of A; when it can, the $x_1, x_2, \ldots, x_N$ appearing in the combination give the solution. This in turn requires the set of vectors $b, v_1, v_2, \ldots, v_N$ to contain the same number of linearly independent vectors as the set $v_1, v_2, \ldots, v_N$. In terms of matrices, this is equivalent to the requirement that the matrix A and the *augmented matrix*

$$M = \begin{pmatrix} A_{11} & A_{12} & \cdots & A_{1N} & b_1 \\ A_{21} & A_{22} & \cdots & A_{2N} & b_1 \\ \vdots & & \ddots & & \vdots \\ A_{M1} & A_{M2} & \cdots & A_{MN} & b_M \end{pmatrix}$$

have the *same* rank r. If this condition is satisfied then the set of equations (10.61) will have either a unique solution or infinitely many solutions. If, however, A and M have different ranks, then there will be no solution.

A unique solution

If **b** can be expressed as in (10.63) and in addition $r = N$,[17] implying that the vectors $v_1, v_2, \ldots, v_N$ are linearly independent, then the equations have a *unique solution* $x_1, x_2, \ldots, x_N$. The uniqueness follows from the uniqueness of the expansion of any vector in the vector space for which the v_i form a basis [see Equation (10.10)].

---

[17] Note that M can be greater than N, but, if it is, then $M - N$ of the simultaneous equations must be expressible as linear combinations of the other N equations.

10.10 Simultaneous linear equations

Infinitely many solutions

If **b** can be expressed as in (10.63) but $r < N$, then only r of the vectors $\mathbf{v}_1, \mathbf{v}_2, \ldots, \mathbf{v}_N$ are linearly independent. We may therefore choose the coefficients of $n - r$ vectors in an arbitrary way, whilst still satisfying (10.63) for some set of coefficients $x_1, x_2, \ldots, x_N$; there are therefore *infinitely many solutions*.

We may use this result to investigate the special case of the solution of a *homogeneous* set of linear equations, for which $\mathbf{b} = \mathbf{0}$. Clearly the set *always* has the trivial solution $x_1 = x_2 = \ldots = x_N = 0$, and if $r = N$ this will be the only solution.

If $r < N$, however, there are infinitely many solutions; each will contain $N - r$ arbitrary components. In particular, we note that if $M < N$ (i.e. there are fewer equations than unknowns) then $r < N$ automatically. Hence a set of *homogeneous* linear equations with fewer equations than unknowns *always* has infinitely many solutions.

10.10.2 N simultaneous linear equations in N unknowns

A special case of (10.61) occurs when $M = N$. In this case the matrix **A** is *square* and we have the same number of equations as unknowns. Since **A** is square, the condition $r = N$ corresponds to $|\mathbf{A}| \neq 0$ and the matrix **A** is *non-singular*. The case $r < N$ corresponds to $|\mathbf{A}| = 0$, in which case **A** is *singular*.

As mentioned above, the equations will have a solution provided **b** can be written as in (10.63). If this is true then the equations will possess a unique solution when $|\mathbf{A}| \neq 0$ or infinitely many solutions when $|\mathbf{A}| = 0$. There exist several methods for obtaining the solution(s). Perhaps the most elementary method is *Gaussian elimination*; we will discuss this method first, and also address numerical subtleties such as equation interchange (pivoting). Following this, we will outline three further methods for solving a square set of simultaneous linear equations.

Gaussian elimination

This is probably one of the earliest techniques acquired by a student of algebra, namely the solving of simultaneous equations (initially only two in number) by the successive elimination of all the variables but one. This (known as *Gaussian elimination*) is achieved by using, at each stage, one of the equations to obtain an explicit expression for one of the remaining x_i in terms of the others and then substituting for that x_i in all other remaining equations. Eventually a single linear equation in just one of the unknowns is obtained. This is then solved and the result is resubstituted in previously derived equations (in reverse order) to establish values for all the x_i.

The method is probably very familiar to the reader, and so a specific example to illustrate this alone seems unnecessary. Instead, we will show how a calculation along such lines might be arranged so that the errors due to the inherent lack of precision in any calculating equipment do not become excessive. This can happen if the value of N is large and particularly (and we will merely state this) if the elements $A_{11}, A_{22}, \ldots, A_{NN}$ on the leading diagonal of the matrix of coefficients are small compared with the off-diagonal elements.

The process to be described is known as *Gaussian elimination with interchange*. The only, but essential, difference from straightforward elimination is that, before each variable

Matrices and vector spaces

x_i is eliminated, the equations are reordered to put the largest (in modulus) remaining coefficient of x_i on the leading diagonal.

We will take as an illustration a straightforward three-variable example, which can in fact be solved perfectly well without any interchange since, with simple numbers and only two eliminations to perform, rounding errors do not have a chance to build up. However, the important thing is that the reader should appreciate how this would apply in (say) a computer program for a 1000-variable case, perhaps with unforeseeable zeros or very small numbers appearing on the leading diagonal.

Example Solve the simultaneous equations

$$\begin{aligned}(a) \quad & x_1 & +6x_2 & -4x_3 = 8, \\ (b) \quad & 3x_1 & -20x_2 & +x_3 = 12, \\ (c) \quad & -x_1 & +3x_2 & +5x_3 = 3.\end{aligned} \quad (10.64)$$

Firstly, we interchange rows (a) and (b) to bring the term $3x_1$ onto the leading diagonal. In the following, we label the important equations (I), (II), (III), and the others alphabetically. A general (i.e. variable) label will be denoted by j.

$$\begin{aligned}(I) \quad & 3x_1 & -20x_2 & +x_3 = 12, \\ (d) \quad & x_1 & +6x_2 & -4x_3 = 8, \\ (e) \quad & -x_1 & +3x_2 & +5x_3 = 3.\end{aligned}$$

For (j) = (d) and (e), replace row (j) by

$$\text{row } (j) - \frac{a_{j1}}{3} \times \text{row } (I),$$

where a_{j1} is the coefficient of x_1 in row (j), to give the two equations

$$\begin{aligned}(II) \quad & \left(6 + \tfrac{20}{3}\right) x_2 & + \left(-4 - \tfrac{1}{3}\right) x_3 = 8 - \tfrac{12}{3}, \\ (f) \quad & \left(3 - \tfrac{20}{3}\right) x_2 & + \left(5 + \tfrac{1}{3}\right) x_3 = 3 + \tfrac{12}{3}.\end{aligned}$$

Now $|6 + \tfrac{20}{3}| > |3 - \tfrac{20}{3}|$ and so no interchange is required before the next elimination. To eliminate x_2, replace row (f) by

$$\text{row } (f) - \frac{\left(-\tfrac{11}{3}\right)}{\tfrac{38}{3}} \times \text{row } (II).$$

This gives

$$(III) \quad \left[\tfrac{16}{3} + \tfrac{11}{38} \times \tfrac{(-13)}{3}\right] x_3 = 7 + \tfrac{11}{38} \times 4.$$

Collecting together and tidying up the final equations, we have

$$\begin{aligned}(I) \quad & 3x_1 & -20x_2 & +x_3 = 12, \\ (II) \quad & & 38x_2 & -13x_3 = 12, \\ (III) \quad & & & x_3 = 2.\end{aligned}$$

Starting with (III) and working backwards, it is now a simple matter to obtain

$$x_1 = 10, \quad x_2 = 1, \quad x_3 = 2$$

as the complete solution of the simultaneous equations.

10.10 Simultaneous linear equations

Direct inversion

Since A is square it will possess an inverse, provided $|\mathsf{A}| \neq 0$. Thus, if A is non-singular, we immediately obtain

$$\mathsf{x} = \mathsf{A}^{-1}\mathsf{b} \tag{10.65}$$

as the unique solution to the set of equations. However, if $\mathsf{b} = \mathsf{0}$ then we see immediately that the set of equations possesses only the trivial solution $\mathsf{x} = \mathsf{0}$. The direct inversion method has the advantage that, once A^{-1} has been calculated, one may obtain the solutions x corresponding to different vectors $\mathsf{b}_1, \mathsf{b}_2, \ldots$ on the RHS, with little further work.

Example Show that the set of simultaneous equations

$$\begin{aligned} 2x_1 + 4x_2 + 3x_3 &= 4, \\ x_1 - 2x_2 - 2x_3 &= 0, \\ -3x_1 + 3x_2 + 2x_3 &= -7, \end{aligned} \tag{10.66}$$

has a unique solution, and find that solution.

The simultaneous equations can be represented by the matrix equation $\mathsf{Ax} = \mathsf{b}$, i.e.

$$\begin{pmatrix} 2 & 4 & 3 \\ 1 & -2 & -2 \\ -3 & 3 & 2 \end{pmatrix} \begin{pmatrix} x_1 \\ x_2 \\ x_3 \end{pmatrix} = \begin{pmatrix} 4 \\ 0 \\ -7 \end{pmatrix}.$$

As we have already shown that A^{-1} exists and have calculated it, see (10.57), it follows that $\mathsf{x} = \mathsf{A}^{-1}\mathsf{b}$ or, more explicitly, that

$$\begin{pmatrix} x_1 \\ x_2 \\ x_3 \end{pmatrix} = \frac{1}{11} \begin{pmatrix} 2 & 1 & -2 \\ 4 & 13 & 7 \\ -3 & -18 & -8 \end{pmatrix} \begin{pmatrix} 4 \\ 0 \\ -7 \end{pmatrix} = \begin{pmatrix} 2 \\ -3 \\ 4 \end{pmatrix}. \tag{10.67}$$

Thus the unique solution is $x_1 = 2$, $x_2 = -3$, $x_3 = 4$. ◀

LU decomposition

Although conceptually simple, finding the solution by calculating A^{-1} can be computationally demanding, especially when N is large. In fact, as we shall now show, it is not necessary to perform the full inversion of A in order to solve the simultaneous equations $\mathsf{Ax} = \mathsf{b}$. Rather, we can perform a *decomposition* of the matrix into the product of a square *lower triangular* matrix L and a square *upper triangular* matrix U, which are such that[18]

$$\mathsf{A} = \mathsf{LU}, \tag{10.68}$$

and then use the fact that triangular systems of equations can be solved very simply.

We must begin, therefore, by finding the matrices L and U such that (10.68) is satisfied. This may be achieved straightforwardly by writing out (10.68) in component form. For

---

[18] Lower and upper triangular matrices are not formally defined and discussed until Section 10.11.2, but relevant aspects of their general structure will be apparent from the way they are used here.

Matrices and vector spaces

illustration, let us consider the 3 × 3 case. It is, in fact, always possible, and convenient, to take the diagonal elements of L as unity, so we have

$$A = \begin{pmatrix} 1 & 0 & 0 \\ L_{21} & 1 & 0 \\ L_{31} & L_{32} & 1 \end{pmatrix} \begin{pmatrix} U_{11} & U_{12} & U_{13} \\ 0 & U_{22} & U_{23} \\ 0 & 0 & U_{33} \end{pmatrix}$$

$$= \begin{pmatrix} U_{11} & U_{12} & U_{13} \\ L_{21}U_{11} & L_{21}U_{12} + U_{22} & L_{21}U_{13} + U_{23} \\ L_{31}U_{11} & L_{31}U_{12} + L_{32}U_{22} & L_{31}U_{13} + L_{32}U_{23} + U_{33} \end{pmatrix}. \quad (10.69)$$

The nine unknown elements of L and U can now be determined by equating the nine elements of (10.69) to those of the 3 × 3 matrix A. This is done in the particular order illustrated in the example below.

Once the matrices L and U have been determined, one can use the decomposition to solve the set of equations $Ax = b$ in the following way. From (10.68), we have $LUx = b$, but this can be written as *two* triangular sets of equations

$$Ly = b \quad \text{and} \quad Ux = y,$$

where y is another column matrix to be determined. One may easily solve the first triangular set of equations for y, which is then substituted into the second set. The required solution x is then obtained readily from the second triangular set of equations. We note that, as with direct inversion, once the LU decomposition has been determined, one can solve for various RHS column matrices $b_1, b_2, \ldots$, with little extra work.

Example Use LU decomposition to solve the set of simultaneous equations (10.66).

We begin the determination of the matrices L and U by equating the elements of the matrix in (10.69) with those of the matrix

$$A = \begin{pmatrix} 2 & 4 & 3 \\ 1 & -2 & -2 \\ -3 & 3 & 2 \end{pmatrix}.$$

This is performed in the following order:

1st row: $U_{11} = 2,$ $U_{12} = 4,$ $U_{13} = 3$
1st column: $L_{21}U_{11} = 1,$ $L_{31}U_{11} = -3$ $\Rightarrow L_{21} = \frac{1}{2}, L_{31} = -\frac{3}{2}$
2nd row: $L_{21}U_{12} + U_{22} = -2$ $L_{21}U_{13} + U_{23} = -2$ $\Rightarrow U_{22} = -4, U_{23} = -\frac{7}{2}$
2nd column: $L_{31}U_{12} + L_{32}U_{22} = 3$ $\Rightarrow L_{32} = -\frac{9}{4}$
3rd row: $L_{31}U_{13} + L_{32}U_{23} + U_{33} = 2$ $\Rightarrow U_{33} = -\frac{11}{8}$

Thus we may write the matrix A as

$$A = LU = \begin{pmatrix} 1 & 0 & 0 \\ \frac{1}{2} & 1 & 0 \\ -\frac{3}{2} & -\frac{9}{4} & 1 \end{pmatrix} \begin{pmatrix} 2 & 4 & 3 \\ 0 & -4 & -\frac{7}{2} \\ 0 & 0 & -\frac{11}{8} \end{pmatrix}.$$

10.10 Simultaneous linear equations

We must now solve the set of equations $\mathsf{Ly} = \mathsf{b}$, which read

$$\begin{pmatrix} 1 & 0 & 0 \\ \frac{1}{2} & 1 & 0 \\ -\frac{3}{2} & -\frac{9}{4} & 1 \end{pmatrix} \begin{pmatrix} y_1 \\ y_2 \\ y_3 \end{pmatrix} = \begin{pmatrix} 4 \\ 0 \\ -7 \end{pmatrix}.$$

Since this set of equations is triangular, we quickly find

$$y_1 = 4, \quad y_2 = 0 - (\tfrac{1}{2})(4) = -2, \quad y_3 = -7 - (-\tfrac{3}{2})(4) - (-\tfrac{9}{4})(-2) = -\tfrac{11}{2}.$$

These values must then be substituted into the equations $\mathsf{Ux} = \mathsf{y}$, which read

$$\begin{pmatrix} 2 & 4 & 3 \\ 0 & -4 & -\frac{7}{2} \\ 0 & 0 & -\frac{11}{8} \end{pmatrix} \begin{pmatrix} x_1 \\ x_2 \\ x_3 \end{pmatrix} = \begin{pmatrix} 4 \\ -2 \\ -\frac{11}{2} \end{pmatrix}.$$

This set of equations is also triangular, and, starting with the final row, we find the solution (in the given order)

$$x_3 = 4, \quad x_2 = -3, \quad x_1 = 2,$$

which agrees with the result found above by direct inversion. ◀

We note, in passing, that one can calculate both the inverse and the determinant of A from its LU decomposition. To find the inverse A^{-1}, one solves the system of equations $\mathsf{Ax} = \mathsf{b}$ repeatedly for the N different RHS column matrices $\mathsf{b} = \mathsf{e}_i$, $i = 1, 2, \ldots, N$, where e_i is the column matrix with its ith element equal to unity and the others equal to zero. The solution x in each case gives the corresponding column of A^{-1}. Evaluation of the determinant $|\mathsf{A}|$ is much simpler. From (10.68), we have

$$|\mathsf{A}| = |\mathsf{LU}| = |\mathsf{L}||\mathsf{U}|. \tag{10.70}$$

Since L and U are triangular, however, we see from (10.75) that their determinants are equal to the products of their diagonal elements. Since $L_{ii} = 1$ for all i, we thus find

$$|\mathsf{A}| = U_{11} U_{22} \cdots U_{NN} = \prod_{i=1}^{N} U_{ii}.$$

As an illustration, in the above example we find $|\mathsf{A}| = (2)(-4)(-11/8) = 11$, which, as it must, agrees with our earlier calculation (10.56).

Finally, a related but slightly different decomposition is possible if matrix A is what is known as *positive semi-definite*. This latter concept is discussed more fully in Section 10.16 in connection with quadratic and Hermitian forms, but for our present purposes we take it as meaning that the scalar quantity $\mathsf{x}^\dagger \mathsf{A} \mathsf{x}$ is real and greater than or equal to zero for *all* column matrices x. An alternative prescription is that all of the eigenvectors (see Section 10.12) of A are non-negative.

Given this definition, if the matrix A is symmetric and positive semi-definite then we can decompose it as

$$\mathsf{A} = \mathsf{L}\mathsf{L}^\dagger, \qquad (10.71)$$

where L is a lower triangular matrix; this representation is known as a *Cholesky decomposition*.[19] We cannot set the diagonal elements of L equal to unity in this case, because we require the same number of independent elements in L as in A. The reason that the decomposition can only be applied to positive semi-definite matrices can be seen by considering the Hermitian form (or quadratic form in the real case)

$$\mathsf{x}^\dagger \mathsf{A} \mathsf{x} = \mathsf{x}^\dagger \mathsf{L}\mathsf{L}^\dagger \mathsf{x} = (\mathsf{L}^\dagger \mathsf{x})^\dagger (\mathsf{L}^\dagger \mathsf{x}).$$

Denoting the column matrix $\mathsf{L}^\dagger \mathsf{x}$ by y, we see that the last term on the RHS is $\mathsf{y}^\dagger \mathsf{y}$, which must be greater than or equal to zero. Thus, we require $\mathsf{x}^\dagger \mathsf{A} \mathsf{x} \geq 0$ for any arbitrary column matrix x.

As mentioned above, the requirement that a matrix be positive semi-definite is equivalent to demanding that all the eigenvalues of A are positive or zero. If one of the eigenvalues of A is zero, then, as will be shown in Equation (10.104), $|\mathsf{A}| = 0$ and A is *singular*. Thus, if A is a non-singular matrix, it must be *positive definite* (rather than just positive semi-definite) for a Cholesky decomposition (10.71) to be possible. In fact, in this case, the inability to find a matrix L that satisfies (10.71) implies that A cannot be positive definite.

The Cholesky decomposition can be used in a way analogous to that in which the LU decomposition was employed earlier, but we will not explore this aspect further. Some practice decompositions are included in the problems at the end of the chapter.

Cramer's rule

A further alternative method of solution is to use *Cramer's rule*, which also provides some insight into the nature of the solutions in the various cases. To illustrate this method let us consider a set of three equations in three unknowns,

$$A_{11}x_1 + A_{12}x_2 + A_{13}x_3 = b_1,$$
$$A_{21}x_1 + A_{22}x_2 + A_{23}x_3 = b_2, \qquad (10.72)$$
$$A_{31}x_1 + A_{32}x_2 + A_{33}x_3 = b_3,$$

which may be represented by the matrix equation $\mathsf{A}\mathsf{x} = \mathsf{b}$. We wish either to find the solution(s) x to these equations or to establish that there are no solutions. From result (vi) of Section 10.7.1, the determinant $|\mathsf{A}|$ is unchanged by adding to its first column the combination

$$\frac{x_2}{x_1} \times \text{(second column of } \mathsf{A}) + \frac{x_3}{x_1} \times \text{(third column of } \mathsf{A}).$$

---

[19] In the special case where A is real, the decomposition becomes $\mathsf{A} = \mathsf{L}\mathsf{L}^\mathsf{T}$.

10.10 Simultaneous linear equations

We thus obtain

$$|\mathsf{A}| = \begin{vmatrix} A_{11} & A_{12} & A_{13} \\ A_{21} & A_{22} & A_{23} \\ A_{31} & A_{32} & A_{33} \end{vmatrix} = \begin{vmatrix} A_{11} + (x_2/x_1)A_{12} + (x_3/x_1)A_{13} & A_{12} & A_{13} \\ A_{21} + (x_2/x_1)A_{22} + (x_3/x_1)A_{23} & A_{22} & A_{23} \\ A_{31} + (x_2/x_1)A_{32} + (x_3/x_1)A_{33} & A_{32} & A_{33} \end{vmatrix}.$$

Now the ith entry in the first column is simply b_i/x_1, with b_i as given by the original equations in (10.72). Therefore, substitution for the ith entry in the first column yields

$$|\mathsf{A}| = \frac{1}{x_1} \begin{vmatrix} b_1 & A_{12} & A_{13} \\ b_2 & A_{22} & A_{23} \\ b_3 & A_{32} & A_{33} \end{vmatrix} = \frac{1}{x_1} \Delta_1.$$

The determinant Δ_1 is known as a *Cramer determinant*. Similar manipulations of the second and third columns of $|\mathsf{A}|$ yield x_2 and x_3, and so the full set of results reads

$$x_1 = \frac{\Delta_1}{|\mathsf{A}|}, \quad x_2 = \frac{\Delta_2}{|\mathsf{A}|}, \quad x_3 = \frac{\Delta_3}{|\mathsf{A}|}, \quad (10.73)$$

where

$$\Delta_1 = \begin{vmatrix} b_1 & A_{12} & A_{13} \\ b_2 & A_{22} & A_{23} \\ b_3 & A_{32} & A_{33} \end{vmatrix}, \quad \Delta_2 = \begin{vmatrix} A_{11} & b_1 & A_{13} \\ A_{21} & b_2 & A_{23} \\ A_{31} & b_3 & A_{33} \end{vmatrix}, \quad \Delta_3 = \begin{vmatrix} A_{11} & A_{12} & b_1 \\ A_{21} & A_{22} & b_2 \\ A_{31} & A_{32} & b_3 \end{vmatrix}.$$

It can be seen that each Cramer determinant Δ_i is simply $|\mathsf{A}|$ but with column i replaced by the RHS of the original set of equations. If $|\mathsf{A}| \neq 0$ then (10.73) gives the unique solution. The proof given here appears to fail if any of the solutions x_i is zero, but it can be shown that result (10.73) is valid even in such a case. The following example uses Cramer's method to solve the same set of equations as used in the previous two worked examples.

Example Use Cramer's rule to solve the set of simultaneous equations (10.66).

Let us again represent these simultaneous equations by the matrix equation $\mathsf{Ax} = \mathsf{b}$, i.e.

$$\begin{pmatrix} 2 & 4 & 3 \\ 1 & -2 & -2 \\ -3 & 3 & 2 \end{pmatrix} \begin{pmatrix} x_1 \\ x_2 \\ x_3 \end{pmatrix} = \begin{pmatrix} 4 \\ 0 \\ -7 \end{pmatrix}.$$

From (10.56), the determinant of A is given by $|\mathsf{A}| = 11$. Following the discussion given above, the three Cramer determinants are

$$\Delta_1 = \begin{vmatrix} 4 & 4 & 3 \\ 0 & -2 & -2 \\ -7 & 3 & 2 \end{vmatrix}, \quad \Delta_2 = \begin{vmatrix} 2 & 4 & 3 \\ 1 & 0 & -2 \\ -3 & -7 & 2 \end{vmatrix}, \quad \Delta_3 = \begin{vmatrix} 2 & 4 & 4 \\ 1 & -2 & 0 \\ -3 & 3 & -7 \end{vmatrix}.$$

These may be evaluated using the properties of determinants listed in Section 10.7.1 and we find $\Delta_1 = 22$, $\Delta_2 = -33$ and $\Delta_3 = 44$. From (10.73) the solution to the equations (10.66) is given by

$$x_1 = \frac{22}{11} = 2, \quad x_2 = \frac{-33}{11} = -3, \quad x_3 = \frac{44}{11} = 4,$$

which agrees with the solution found in the previous example. ◀

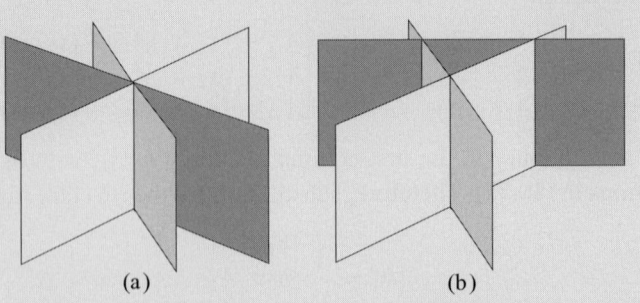

Figure 10.1 The two possible cases when A is of rank 2. In both cases all the normals lie in a horizontal plane but in (a) the planes all intersect on a single line (corresponding to an infinite number of solutions) whilst in (b) there are no common intersection points (no solutions).

10.10.3 A geometrical interpretation

A helpful view of what is happening when simultaneous equations are solved, is to consider each of the equations as representing a surface in an N-dimensional space. This is most easily visualised in three (or two) dimensions. So, for example, we think of each of the three equations (10.72) as representing a plane in three-dimensional Cartesian coordinates. Using result (9.43), the sets of components of the vectors normal to the planes are (A_{11}, A_{12}, A_{13}), (A_{21}, A_{22}, A_{23}) and (A_{31}, A_{32}, A_{33}), and, using (9.47), the perpendicular distances of the planes from the origin are given by

$$d_i = \frac{b_i}{\left(A_{i1}^2 + A_{i2}^2 + A_{i3}^2\right)^{1/2}} \quad \text{for } i = 1, 2, 3.$$

Finding the solution(s) to the simultaneous equations above corresponds to finding the point(s) of intersection of the planes.

If there is a unique solution the planes intersect at only a single point. This happens if their normals are linearly independent vectors. Since the rows of A represent the directions of these normals, this requirement is equivalent to $|A| \neq 0$. If $b = (0 \ 0 \ 0)^T = 0$ then all the planes pass through the origin and, since there is only a single solution to the equations, the origin is that (trivial) solution.

Let us now turn to the cases where $|A| = 0$. The simplest such case is that in which all three planes are parallel; this implies that the normals are all parallel and so A is of rank 1. Two possibilities exist:

(i) the planes are coincident, i.e. $d_1 = d_2 = d_3$, in which case there is an infinity of solutions;
(ii) the planes are not all coincident, i.e. $d_1 \neq d_2$ and/or $d_1 \neq d_3$ and/or $d_2 \neq d_3$, in which case there are no solutions.

It is apparent from (10.73) that case (i) occurs when all the Cramer determinants are zero and case (ii) occurs when at least one Cramer determinant is non-zero.

The most complicated cases with $|A| = 0$ are those in which the normals to the planes themselves lie in a plane but are not parallel. In this case A has rank 2. Again two possibilities exist, and these are shown in Figure 10.1. Just as in the rank-1 case, if all the

10.10 Simultaneous linear equations

Cramer determinants are zero then we get an infinity of solutions (this time on a line). Of course, in the special case in which $b = 0$ (and the system of equations is homogeneous), the planes all pass through the origin and so they must intersect on a line through it. If at least one of the Cramer determinants is non-zero, we get no solution.

These rules may be summarised as follows.

(i) $|A| \neq 0$, $b \neq 0$: the three planes intersect at a single point that is not the origin, and so there is only one solution, given by both (10.65) and (10.73).
(ii) $|A| \neq 0$, $b = 0$: the three planes intersect at the origin only and there is only the trivial solution $x = 0$.
(iii) $|A| = 0$, $b \neq 0$, Cramer determinants all zero: there is an infinity of solutions either on a line if A is rank 2, i.e. the cofactors are not all zero, or on a plane if A is rank 1, i.e. the cofactors are all zero.
(iv) $|A| = 0$, $b \neq 0$, Cramer determinants not all zero: no solutions.
(v) $|A| = 0$, $b = 0$: the three planes intersect on a line through the origin giving an infinity of solutions.

EXERCISES 10.10

1. Does the following set of equations have a non-trivial solution

$$x - y = \alpha$$
$$2y + 6z = \beta$$
$$3x + 2y + 3z = \gamma,$$

(a) when $(\alpha, \beta, \gamma) = (2, -7, -1)$ and (b) when $(\alpha, \beta, \gamma) = (3, -2, 10)$? If so, is the solution unique? [*Hint*: Matrix E in Exercise 10.9 is relevant.]

2. Do the following sets of equations have solutions? If so, how many?

(a)	(b)	(c)
$3x - 2y - z = 5,$	$3x - 2y - z = 5,$	$3x - 2y - z = 1,$
$x + 3y + 2z = 5,$	$x + 3y + 2z = 5,$	$x + 3y + 2z = -6,$
$x - 3y - 2z = -1.$	$5x + 4y + 3z = -1.$	$5x + 4y + 3z = -11.$

3. One of the sets of equations in the previous exercise has a unique solution. Find that solution using each of the following methods: (a) Gaussian elimination, (b) direct inversion of the relevant matrix A, (c) an LU decomposition and (d) Cramer's rule. Confirm that the determinant of A as deduced from method (c) agrees with that found in methods (b) and (d).

4. For each of the sets of equations in exercise 2 above, determine which of the conditions (i)–(v) listed in Section 10.10.3 is satisfied and sketch qualitatively (and/or describe) the relevant planes and their intersections in three-dimensional space.

10.11 Special types of square matrix

Having examined some of the properties and uses of matrices, and of other matrices derived from them, we now consider some sets of square matrices that are characterised by a common structure or property possessed by their members; a summarising table is given in the Summary. Matrices that are square, i.e. $N \times N$, appear in many physical applications, and some special forms of square matrix are of particular importance.

10.11.1 Diagonal matrices

The unit matrix, which we have already encountered, is an example of a *diagonal* matrix. Such matrices are characterised by having non-zero elements only on the *leading diagonal*, i.e. only elements A_{ij} with $i = j$ may be non-zero. For example,

$$\mathsf{A} = \begin{pmatrix} 1 & 0 & 0 \\ 0 & 2 & 0 \\ 0 & 0 & -3 \end{pmatrix}$$

is a 3×3 diagonal matrix. Such a matrix is often denoted by $\mathsf{A} = \text{diag}(1, 2, -3)$. By performing a Laplace expansion, it is easily shown that the determinant of an $N \times N$ diagonal matrix is equal to the product of the diagonal elements.[20] Thus, if the matrix has the form $\mathsf{A} = \text{diag}(A_{11}, A_{22}, \ldots, A_{NN})$ then

$$|\mathsf{A}| = A_{11} A_{22} \cdots A_{NN}. \qquad (10.74)$$

Moreover, it is also straightforward to show that the inverse of A is also a diagonal matrix given by

$$\mathsf{A}^{-1} = \text{diag}\left(\frac{1}{A_{11}}, \frac{1}{A_{22}}, \ldots, \frac{1}{A_{NN}}\right).$$

Finally, we note that if two matrices A and B are *both* diagonal then they have the useful property that their product is commutative:

$$\mathsf{AB} = \mathsf{BA}.$$

Thus the set of all $N \times N$ diagonal matrices form a commuting set under matrix multiplication. This property is *not* shared by square matrices in general.

10.11.2 Lower and upper triangular matrices

We have already encountered triangular matrices in connection with **LU** and Cholesky decompositions, but we include them here for the sake of completeness.

A square matrix A is called *lower triangular* if all the elements *above* the principal diagonal are zero. For example, the general form for a 3×3 lower triangular matrix is

$$\mathsf{A} = \begin{pmatrix} A_{11} & 0 & 0 \\ A_{21} & A_{22} & 0 \\ A_{31} & A_{32} & A_{33} \end{pmatrix},$$

---

[20] Using this notation write down the form of the *most general, non-zero, singular, traceless, diagonal* 3×3 matrix.

10.11 Special types of square matrix

where the elements A_{ij} may be zero or non-zero. Similarly, an *upper triangular* square matrix is one for which all the elements *below* the principal diagonal are zero. The general 3×3 form is thus

$$\mathsf{A} = \begin{pmatrix} A_{11} & A_{12} & A_{13} \\ 0 & A_{22} & A_{23} \\ 0 & 0 & A_{33} \end{pmatrix}.$$

By performing a Laplace expansion, it is straightforward to show that, in the general $N \times N$ case, the determinant of an upper or lower triangular matrix is equal to the product of its diagonal elements,

$$|\mathsf{A}| = A_{11} A_{22} \cdots A_{NN}. \tag{10.75}$$

Clearly property (10.74) of diagonal matrices is a special case of this more general result. Moreover, it may be shown that the inverse of a non-singular lower (upper) triangular matrix is also lower (upper) triangular.[21]

10.11.3 Symmetric and antisymmetric matrices

A square matrix A of order N with the property $\mathsf{A} = \mathsf{A}^{\mathrm{T}}$ is said to be *symmetric*. Similarly, a matrix for which $\mathsf{A} = -\mathsf{A}^{\mathrm{T}}$ is said to be *anti-* or *skew*-symmetric and its diagonal elements $a_{11}, a_{22}, \ldots, a_{NN}$ are necessarily zero. Moreover, if A is (anti-)symmetric then so too is its inverse A^{-1}. This is easily proved by noting that if $\mathsf{A} = \pm \mathsf{A}^{\mathrm{T}}$ then

$$(\mathsf{A}^{-1})^{\mathrm{T}} = (\mathsf{A}^{\mathrm{T}})^{-1} = \pm \mathsf{A}^{-1}.$$

Any $N \times N$ matrix A can be written as the sum of a symmetric and an antisymmetric matrix, since we may write

$$\mathsf{A} = \tfrac{1}{2}(\mathsf{A} + \mathsf{A}^{\mathrm{T}}) + \tfrac{1}{2}(\mathsf{A} - \mathsf{A}^{\mathrm{T}}) = \mathsf{B} + \mathsf{C},$$

where clearly $\mathsf{B} = \mathsf{B}^{\mathrm{T}}$ and $\mathsf{C} = -\mathsf{C}^{\mathrm{T}}$. The matrix B is therefore called the symmetric part of A and C is the antisymmetric part.

Example If A is an $N \times N$ antisymmetric matrix, show that $|\mathsf{A}| = 0$ if N is odd.

If A is antisymmetric then $\mathsf{A}^{\mathrm{T}} = -\mathsf{A}$. Using the properties of determinants (10.47) and (10.49), we have

$$|\mathsf{A}| = |\mathsf{A}^{\mathrm{T}}| = |-\mathsf{A}| = (-1)^N |\mathsf{A}|.$$

Thus, if N is odd then $|\mathsf{A}| = -|\mathsf{A}|$, which implies that $|\mathsf{A}| = 0$. ◀

---

[21] Determine where the following, clearly false, line of reasoning breaks down. Consider an upper triangular 3×3 matrix A which has unity for all its principal diagonal elements, $A_{12} = 0$, $A_{13} = a$ and $A_{23} = b$. It can be shown that $\mathsf{A} + \mathsf{A}^{-1} = 2\mathsf{I}$, and consequently (after multiplying through by A) we have $\mathsf{A}^2 - 2\mathsf{A} + \mathsf{I} = 0$. This can be written $(\mathsf{A} - \mathsf{I})(\mathsf{A} - \mathsf{I}) = (\mathsf{A} - \mathsf{I})^2 = 0$. Therefore $\mathsf{A} = \mathsf{I}$.

10.11.4 Orthogonal matrices

A non-singular matrix with the property that its transpose is also its inverse,

$$\mathsf{A}^\mathrm{T} = \mathsf{A}^{-1}, \tag{10.76}$$

is called an *orthogonal matrix*. It follows immediately that the inverse of an orthogonal matrix is also orthogonal, since

$$(\mathsf{A}^{-1})^\mathrm{T} = (\mathsf{A}^\mathrm{T})^{-1} = (\mathsf{A}^{-1})^{-1}.$$

Moreover, since for an orthogonal matrix $\mathsf{A}^\mathrm{T}\mathsf{A} = \mathsf{I}$, we have

$$|\mathsf{A}^\mathrm{T}\mathsf{A}| = |\mathsf{A}^\mathrm{T}||\mathsf{A}| = |\mathsf{A}|^2 = |\mathsf{I}| = 1.$$

Thus the determinant of an orthogonal matrix must be $|\mathsf{A}| = \pm 1$.

An orthogonal matrix[22] represents, in a particular basis, a linear operator that leaves the norms (lengths) of real vectors unchanged, as we will now show. Suppose that $\mathbf{y} = \mathcal{A}\mathbf{x}$ is represented in some coordinate system by the matrix equation $\mathbf{y} = \mathsf{A}\mathbf{x}$; then $\langle \mathbf{y}|\mathbf{y} \rangle$ is given in this coordinate system by

$$\mathbf{y}^\mathrm{T}\mathbf{y} = \mathbf{x}^\mathrm{T}\mathsf{A}^\mathrm{T}\mathsf{A}\mathbf{x} = \mathbf{x}^\mathrm{T}\mathbf{x}.$$

Hence $\langle \mathbf{y}|\mathbf{y} \rangle = \langle \mathbf{x}|\mathbf{x} \rangle$, showing that the action of a linear operator represented by an orthogonal matrix does not change the norm of a real vector.

10.11.5 Hermitian and anti-Hermitian matrices

An *Hermitian* matrix is one that satisfies $\mathsf{A} = \mathsf{A}^\dagger$, where $\mathsf{A}^\dagger$ is the Hermitian conjugate discussed in Section 10.5.1. Similarly, if $\mathsf{A}^\dagger = -\mathsf{A}$, then A is called *anti-Hermitian*. A real (anti-)symmetric matrix is a special case of an (anti-)Hermitian matrix, in which all the elements of the matrix are real. Also, if A is an (anti-)Hermitian matrix then so too is its inverse A^{-1}, since

$$(\mathsf{A}^{-1})^\dagger = (\mathsf{A}^\dagger)^{-1} = \pm \mathsf{A}^{-1}.$$

Any $N \times N$ matrix A can be written as the sum of an Hermitian matrix and an anti-Hermitian matrix, since

$$\mathsf{A} = \tfrac{1}{2}(\mathsf{A} + \mathsf{A}^\dagger) + \tfrac{1}{2}(\mathsf{A} - \mathsf{A}^\dagger) = \mathsf{B} + \mathsf{C},$$

where clearly $\mathsf{B} = \mathsf{B}^\dagger$ and $\mathsf{C} = -\mathsf{C}^\dagger$. The matrix B is called the Hermitian part of A and C is called the anti-Hermitian part.

10.11.6 Unitary matrices

A *unitary* matrix A is defined as one for which

$$\mathsf{A}^\dagger = \mathsf{A}^{-1}. \tag{10.77}$$

---

[22] A 2×2 matrix with both diagonal elements equal to $\cos\theta$ and off-diagonal elements $+\sin\theta$ and $-\sin\theta$ provides a practical example.

10.11 Special types of square matrix

Clearly, if A is real then $A^\dagger = A^T$, showing that a real orthogonal matrix is a special case of a unitary matrix, one in which all the elements are real.[23]

We note that the inverse A^{-1} of a unitary matrix is also unitary, since

$$(A^{-1})^\dagger = (A^\dagger)^{-1} = (A^{-1})^{-1}.$$

Moreover, since for a unitary matrix $A^\dagger A = I$, we have

$$|A^\dagger A| = |A^\dagger||A| = |A|^*|A| = |I| = 1.$$

Thus the determinant of a unitary matrix has unit modulus.

A unitary matrix represents, in a particular basis, a linear operator that leaves the norms (lengths) of complex vectors unchanged. If $y = \mathcal{A}x$ is represented in some coordinate system by the matrix equation $y = Ax$ then $\langle y|y \rangle$ is given in this coordinate system by

$$y^\dagger y = x^\dagger A^\dagger A x = x^\dagger x.$$

Hence $\langle y|y \rangle = \langle x|x \rangle$, showing that the action of the linear operator represented by a unitary matrix does not change the norm of a complex vector. The action of a unitary matrix on a complex column matrix thus parallels that of an orthogonal matrix acting on a real column matrix.

10.11.7 Normal matrices

A final important set of special matrices consists of the *normal* matrices, for which

$$AA^\dagger = A^\dagger A,$$

i.e. a normal matrix is one that commutes with its Hermitian conjugate.

We can easily show that Hermitian matrices and unitary matrices (or symmetric matrices and orthogonal matrices in the real case) are examples of normal matrices. For an Hermitian matrix, $A = A^\dagger$ and so

$$AA^\dagger = AA = A^\dagger A.$$

Similarly, for a unitary matrix, $A^{-1} = A^\dagger$ and so

$$AA^\dagger = AA^{-1} = I = A^{-1}A = A^\dagger A.$$

Finally, we note that if A is normal then so too is its inverse A^{-1}, since

$$A^{-1}(A^{-1})^\dagger = A^{-1}(A^\dagger)^{-1} = (A^\dagger A)^{-1} = (AA^\dagger)^{-1} = (A^\dagger)^{-1}A^{-1} = (A^{-1})^\dagger A^{-1}.$$

This broad class of matrices is formally important in the discussion of eigenvectors and eigenvalues (see the next section), as several general properties can be deduced purely on the basis that a matrix and its Hermitian conjugate commute. However, the corresponding general proofs tend to be more complicated than those treating only smaller classes of matrices and so, in the next sections, we have not pursued this broad approach.

---

[23] Three 2×2 matrices, S_x, S_y and S_z, are defined in Problem 10.10. Characterise each with respect to (a) reality, (b) symmetry, (c) Hermiticity, (d) orthogonality and (e) unitarity.

Matrices and vector spaces

EXERCISES 10.11

1. Characterise each of the following square matrices, in so far as their special properties are concerned:

$$A = \begin{pmatrix} 1 & 0 & 0 \\ 0 & -2 & 0 \\ 2 & 0 & 1 \end{pmatrix}, \quad B = \begin{pmatrix} 1 & i & -1 \\ -i & 0 & 1-i \\ -1 & 1+i & 1 \end{pmatrix}, \quad C = \begin{pmatrix} 1 & -1 & 2 \\ 1 & 3 & 0 \\ -2 & 0 & -4 \end{pmatrix},$$

$$D = \begin{pmatrix} 1 & e^{i\theta} \\ e^{-i\theta} & 1 \end{pmatrix}, \quad E = \begin{pmatrix} \cos\theta & \sin\theta \\ -\sin\theta & \cos\theta \end{pmatrix}, \quad F = \frac{1}{\sqrt{3}} \begin{pmatrix} 1 & -1-i \\ 1+i & i \end{pmatrix}.$$

2. Decompose the matrix

$$A = \begin{pmatrix} 5 & 4 & -1 \\ 2 & 3 & 0 \\ -1 & 6 & -2 \end{pmatrix}$$

into the sum of a traceless symmetric matrix, an antisymmetric matrix and a multiple of the unit matrix.

3. Prove that if U is unitary and $A \,(\neq I)$ is Hermitian, then $V = U^{-1}AU$ is Hermitian. Show further that if $A^2 = I$, then V is unitary. Give a simple possible 3×3 form for A.

10.12 Eigenvectors and eigenvalues

Suppose that a linear operator $\mathcal{A}$ transforms vectors $\mathbf{x}$ in an N-dimensional vector space into other vectors $\mathcal{A}\mathbf{x}$ in the same space. The possibility then arises that there exist vectors $\mathbf{x}$ each of which is transformed by $\mathcal{A}$ into a multiple of itself.[24] Such vectors would have to satisfy

$$\mathcal{A}\mathbf{x} = \lambda\mathbf{x}. \tag{10.78}$$

Any non-zero vector $\mathbf{x}$ that satisfies (10.78) for some value of λ is called an *eigenvector* of the linear operator $\mathcal{A}$, and λ is called the corresponding *eigenvalue*. As will be discussed below, in general the operator $\mathcal{A}$ has N independent eigenvectors $\mathbf{x}^i$, with eigenvalues λ_i. The λ_i are not necessarily all distinct.

If we choose a particular basis in the vector space, we can write (10.78) in terms of the components of $\mathcal{A}$ and $\mathbf{x}$ with respect to this basis as the matrix equation

$$A\mathbf{x} = \lambda\mathbf{x}, \tag{10.79}$$

where A is an $N \times N$ matrix. The column matrices $\mathbf{x}$ that satisfy (10.79) obviously represent the eigenvectors $\mathbf{x}$ of $\mathcal{A}$ in our chosen coordinate system. Conventionally, these

---

[24] That is, after the transformation the vector still 'points' in the same (or the directly opposite) direction in the vector space, even though it may have been changed in length.

10.12 Eigenvectors and eigenvalues

column matrices are also referred to as the *eigenvectors of the matrix* A.[25] Throughout this chapter we denote the ith eigenvector of a square matrix A by x^i and the corresponding eigenvalue by λ_i. This superscript notation for eigenvectors is used to avoid any confusion with components.

Clearly, if x is an eigenvector of A (with some eigenvalue λ) then any scalar multiple $\mu \mathsf{x}$ is also an eigenvector with the same eigenvalue; in other words, the factor by which the length of the vector is changed is independent of the original length. We therefore often use *normalised* eigenvectors, for which

$$\mathsf{x}^\dagger \mathsf{x} = 1$$

(note that $\mathsf{x}^\dagger \mathsf{x}$ corresponds to the inner product $\langle \mathsf{x} | \mathsf{x} \rangle$ in our basis). Any eigenvector x can be normalised by dividing all of its components by the scalar $(\mathsf{x}^\dagger \mathsf{x})^{1/2}$.

The problem of finding the eigenvalues and corresponding eigenvectors of a square matrix A plays an important role in many physical investigations. It is the standard basis for determining the normal modes of an oscillatory mechanical or electrical system, with applications ranging from the stability of bridges to the internal vibrations of molecules. It also provides the methodology for the particular formulation of quantum mechanics that is known as matrix mechanics.

We begin with an example that produces a simple deduction from the defining eigenvalue equation (10.79).

Example A non-singular matrix A has eigenvalues λ_i and eigenvectors x^i. Find the eigenvalues and eigenvectors of the inverse matrix A^{-1}.

The eigenvalues and eigenvectors of A satisfy

$$\mathsf{A}\mathsf{x}^i = \lambda_i \mathsf{x}^i.$$

Left-multiplying both sides of this equation by A^{-1}, we find

$$\mathsf{A}^{-1}\mathsf{A}\mathsf{x}^i = \lambda_i \mathsf{A}^{-1}\mathsf{x}^i.$$

Since $\mathsf{A}^{-1}\mathsf{A} = \mathsf{I}$, dividing through by λ_i and interchanging the two sides of the equation gives an eigenvalue equation for A^{-1}:

$$\mathsf{A}^{-1}\mathsf{x}^i = \frac{1}{\lambda_i} \mathsf{x}^i.$$

From this we see that each eigenvector x^i of A is also an eigenvector of A^{-1}, but that the corresponding eigenvalue is $1/\lambda_i$. As A and A^{-1} have the same dimensions, and hence the same number of independent eigenvectors, the two sets of eigenvectors are identical.[26,27] ◀

In the remainder of this section we will discuss some useful results concerning the eigenvectors and eigenvalues of certain special (though commonly occurring) square

---

[25] In this context, when referring to linear combinations of eigenvectors x we will normally use the term 'vector'.
[26] If any of the λ_i are repeated, then linear combinations of the corresponding x^i may have to be formed.
[27] Explain why, if one of the eigenvalues of A is 0, this does not imply that the inverse of A has an eigenvalue of ∞.

matrices. The results will be established for matrices whose elements may be complex; the corresponding properties for real matrices can be obtained as special cases.

10.12.1 Eigenvectors and eigenvalues of Hermitian and unitary matrices

We start by proving two powerful results about the eigenvalues and eigenfunctions of Hermitian matrices, namely:

(i) The eigenvalues of an Hermitian matrix are real.
(ii) The eigenvectors of an Hermitian matrix corresponding to different eigenvalues are orthogonal.

For the present we will assume that the eigenvalues of our Hermitian matrix A are distinct and later show what modifications are needed when they are not.

Consider two eigenvalues λ_i and λ_j and their corresponding eigenvectors satisfying

$$\mathsf{A}\mathsf{x}^i = \lambda_i \mathsf{x}^i, \tag{10.80}$$

$$\mathsf{A}\mathsf{x}^j = \lambda_j \mathsf{x}^j. \tag{10.81}$$

Taking the Hermitian conjugate of (10.80) we find $(\mathsf{x}^i)^\dagger \mathsf{A}^\dagger = \lambda_i^* (\mathsf{x}^i)^\dagger$. Multiplying this on the right by x^j gives

$$(\mathsf{x}^i)^\dagger \mathsf{A}^\dagger \mathsf{x}^j = \lambda_i^* (\mathsf{x}^i)^\dagger \mathsf{x}^j,$$

and similarly multiplying (10.81) through on the left by $(\mathsf{x}^i)^\dagger$ yields

$$(\mathsf{x}^i)^\dagger \mathsf{A} \mathsf{x}^j = \lambda_j (\mathsf{x}^i)^\dagger \mathsf{x}^j.$$

Then, since $\mathsf{A}^\dagger = \mathsf{A}$, the two LHSs are equal and on subtraction we obtain

$$0 = (\lambda_i^* - \lambda_j)(\mathsf{x}^i)^\dagger \mathsf{x}^j. \tag{10.82}$$

To prove result (i) we need only set $j = i$. Then (10.82) reads

$$0 = (\lambda_i^* - \lambda_i)(\mathsf{x}^i)^\dagger \mathsf{x}^i.$$

Now, since x is a non-zero vector, $(\mathsf{x}^i)^\dagger \mathsf{x}^i \neq 0$, implying that $\lambda_i^* = \lambda_i$, i.e. λ_i is real.

Result (ii) follows almost immediately because when $j \neq i$ in (10.82), and consequently $\lambda_j \neq \lambda_i = \lambda_i^*$, we must have $(\mathsf{x}^i)^\dagger \mathsf{x}^j = 0$, i.e. the relevant eigenvectors, x^i and x^j, are orthogonal.

We should also note at this point that if A is anti-Hermitian (rather than Hermitian) and $\mathsf{A}^\dagger = -\mathsf{A}$, then the bracket in (10.82) reads $(\lambda_i^* + \lambda_j)$ and when j is set equal to i we conclude that $\lambda_i^* = -\lambda_i$, i.e. λ_i is purely imaginary. The previous conclusion about the orthogonality of the eigenvectors is unaltered.

As a reminder, we also recall that real symmetric matrices are special cases of Hermitian matrices, and so they too have real eigenvalues and mutually orthogonal eigenvectors.

The importance of result (i) for Hermitian matrices will be apparent to any student of quantum mechanics. In quantum mechanics the eigenvalues of operators correspond to measured values of observable quantities, e.g. energy, angular momentum, parity and so on, and these clearly must be real. If we use Hermitian operators to formulate the

10.12 Eigenvectors and eigenvalues

theories of quantum mechanics, the above property guarantees physically meaningful results.

We now turn our attention to unitary matrices and prove, by very similar means to those just employed, that the eigenvalues of a unitary matrix necessarily have unit modulus. A unitary matrix satisfies $A^\dagger = A^{-1}$ or, equivalently, $A^\dagger A = I$.

Taking the Hermitian conjugate of (10.80) we have, as previously, that

$$(x^i)^\dagger A^\dagger = \lambda_i^* (x^i)^\dagger, \tag{10.83}$$

whilst from (10.81)

$$Ax^j = \lambda_j x^j. \tag{10.84}$$

Now, right-multiplying the LHS of (10.83) by the LHS of (10.84), and correspondingly for the two RHSs, gives

$$(x^i)^\dagger A^\dagger A x^j = \lambda_i^* (x^i)^\dagger \lambda_j x^j,$$
$$(x^i)^\dagger x^j = \lambda_i^* \lambda_j (x^i)^\dagger x^j,$$
$$\left[1 - \lambda_i^* \lambda_j\right] (x^i)^\dagger x^j = 0.$$

Finally, setting $j = i$, and again noting that x^i is a non-zero vector, shows that

$$1 - |\lambda_i|^2 = 0.$$

Thus, the eigenvalues of a unitary matrix have unit modulus. The proof of the orthogonality property of its eigenvectors is as for Hermitian matrices. For completeness, we also note that a real orthogonal matrix is a special case of a unitary matrix; it too has eigenvectors of unit modulus.[28]

If some of the eigenvalues of a matrix are equal and one eigenvalue corresponds to two or more different eigenvectors (i.e. no two are simple multiples of each other), that eigenvalue is said to be *degenerate*. In this case further justification of the orthogonality of the eigenvectors is needed. A process known as the Gram–Schmidt orthogonalisation procedure provides a proof of, and a means of achieving, orthogonality, in that it shows how to construct a mutually orthogonal set of linear combinations of those eigenvectors that correspond to the degenerate eigenvalue. In practice, however, the method is laborious and the example in Section 10.13 gives a less rigorous but considerably quicker way of achieving the same end.

10.12.2 Eigenvectors and eigenvalues of a general square matrix

When an $N \times N$ matrix does not qualify for the broad, but nevertheless restricted, class of normal matrices (see Section 10.11.7), there are no general properties that can be

---

[28] In fact, for a real orthogonal matrix, the only possible eigenvalues are $\lambda = \pm 1$. Show this by proving that they must satisfy $\lambda^2 = 1$.

ascribed to its eigenvalues and eigenvectors. In fact, in general it is not possible to find any orthogonal set of N eigenvectors or even to find *pairs* of orthogonal eigenvectors (except by chance in some cases). While its N non-orthogonal eigenvectors are usually linearly independent and hence form a basis for the N-dimensional vector space, even this is not necessarily so.

It may be shown (although we will not prove it) that any $N \times N$ matrix with *distinct* eigenvalues does have N linearly independent eigenvectors, which therefore do form a basis for the N-dimensional vector space. If a general square matrix has degenerate eigenvalues, however, then it may or may not have N linearly independent eigenvectors. A matrix whose eigenvectors are not linearly independent is said to be *defective*.

10.12.3 Simultaneous eigenvectors

We may now ask under what conditions two different Hermitian matrices can have a common set of eigenvectors. The result – that they do so if, and only if, they commute – has profound significance for the foundations of quantum mechanics.

To prove this important result let A and B be two $N \times N$ Hermitian matrices and x^i be the ith eigenvector of A corresponding to eigenvalue λ_i, i.e.

$$\mathsf{A}\mathsf{x}^i = \lambda_i \mathsf{x}^i \quad \text{for} \quad i = 1, 2, \ldots, N.$$

For the present we assume that the eigenvalues are all different.

(i) First suppose that A and B commute. Now consider

$$\mathsf{A}\mathsf{B}\mathsf{x}^i = \mathsf{B}\mathsf{A}\mathsf{x}^i = \mathsf{B}\lambda_i \mathsf{x}^i = \lambda_i \mathsf{B}\mathsf{x}^i,$$

where we have used the commutativity for the first equality and the eigenvector property for the second. It follows that $\mathsf{A}(\mathsf{B}\mathsf{x}^i) = \lambda_i(\mathsf{B}\mathsf{x}^i)$ and thus that $\mathsf{B}\mathsf{x}^i$ is an eigenvector of A corresponding to eigenvalue λ_i. But the eigenvector solutions of $(\mathsf{A} - \lambda_i \mathsf{I})\mathsf{x}^i = 0$ are unique to within a scale factor, and we therefore conclude that

$$\mathsf{B}\mathsf{x}^i = \mu_i \mathsf{x}^i$$

for some scale factor μ_i. However, this is just an eigenvector equation for B and shows that x^i is an eigenvector of B, in addition to being an eigenvector of A. By reversing the roles of A and B, it also follows that every eigenvector of B is an eigenvector of A. Thus the two sets of eigenvectors are identical.

(ii) Now suppose that A and B have all their eigenvectors in common, a typical one x^i satisfying both

$$\mathsf{A}\mathsf{x}^i = \lambda_i \mathsf{x}^i \quad \text{and} \quad \mathsf{B}\mathsf{x}^i = \mu_i \mathsf{x}^i.$$

As the eigenvectors span the N-dimensional vector space, any arbitrary vector x in the space can be written as a linear combination of the eigenvectors,

$$\mathsf{x} = \sum_{i=1}^{N} c_i \mathsf{x}^i.$$

10.12 Eigenvectors and eigenvalues

Now consider both

$$\mathsf{ABx} = \mathsf{AB}\sum_{i=1}^{N} c_i \mathsf{x}^i = \mathsf{A}\sum_{i=1}^{N} c_i \mu_i \mathsf{x}^i = \sum_{i=1}^{N} c_i \lambda_i \mu_i \mathsf{x}^i,$$

and

$$\mathsf{BAx} = \mathsf{BA}\sum_{i=1}^{N} c_i \mathsf{x}^i = \mathsf{B}\sum_{i=1}^{N} c_i \lambda_i \mathsf{x}^i = \sum_{i=1}^{N} c_i \mu_i \lambda_i \mathsf{x}^i.$$

It follows that ABx and BAx are the same for any arbitrary x and hence that

$$(\mathsf{AB} - \mathsf{BA})\mathsf{x} = 0$$

for all x. That is, A and B *commute*.

This completes the proof that a necessary and sufficient condition for two Hermitian matrices to have a set of eigenvectors in common is that they commute. It should be noted that if an eigenvalue of A, say, is degenerate then not all of its possible sets of eigenvectors will also constitute a set of eigenvectors of B. However, provided that by taking linear combinations one set of joint eigenvectors can be found, the proof is still valid and the result still holds.

When extended to the case of Hermitian differential operators (instead of matrices) and continuous eigenfunctions (instead of eigenvectors), the connection between commuting matrices and a set of common eigenvectors plays a fundamental role in the postulatory basis of quantum mechanics. It draws the distinction between commuting and non-commuting observables and sets limits on how much information about a system can be known, even in principle, at any one time.

EXERCISES 10.12

1. Determine which two of the matrices

$$\mathsf{A} = \begin{pmatrix} 10 & -3 \\ -3 & 2 \end{pmatrix}, \quad \mathsf{B} = \begin{pmatrix} 4 & 2 \\ 2 & -1 \end{pmatrix}, \quad \mathsf{C} = \begin{pmatrix} -5 & 3 \\ 3 & 3 \end{pmatrix}$$

commute and show that they have a common set of eigenvectors of the form $\mathsf{x}^1 = (\alpha, \beta)$ and $\mathsf{x}^2 = (\beta, -\alpha)$. Verify that the matrix that does not commute with either of the other two does not have x^1 and x^2 as its pair of eigenvectors. [In fact $\alpha : \beta = 1 : 3$, but try to deduce this rather than assume it.]

2. Confirm that the matrix F given in the first exercise of Section 10.11,

$$\mathsf{F} = \frac{1}{\sqrt{3}}\begin{pmatrix} 1 & -1-i \\ 1+i & i \end{pmatrix},$$

is unitary. Prove that its eigenvalues are given by

$$\lambda = \frac{1}{\sqrt{6}} \left(e^{i\pi/4} \pm \sqrt{5} e^{-i\pi/4} \right)$$

and so verify that **F**, like all unitary matrices, has eigenvalues of unit modulus.

10.13 Determination of eigenvalues and eigenvectors

The next step is to show how the eigenvalues and eigenvectors of a given $N \times N$ matrix **A** are found. To do this we refer to (10.79) and, by replacing **x** by **Ix** where **I** is the unit matrix of order N, rewrite it as

$$\mathsf{A}\mathsf{x} - \lambda \mathsf{I}\mathsf{x} = (\mathsf{A} - \lambda \mathsf{I})\mathsf{x} = 0. \tag{10.85}$$

The point of doing this is immediate, since (10.85) now has the form of a homogeneous set of simultaneous equations, the theory of which was developed in Section 10.10. What was proved there is that the equation $\mathsf{B}\mathsf{x} = 0$ only has a non-trivial solution **x** if $|\mathsf{B}| = 0$. Correspondingly, therefore, we must have in the present case that

$$|\mathsf{A} - \lambda \mathsf{I}| = 0, \tag{10.86}$$

if there are to be non-zero solutions **x** to (10.85).

Equation (10.86) is known as the *characteristic equation* for **A** and its LHS as the *characteristic* or *secular determinant* of **A**. The characteristic equation is a polynomial equation of degree N in the quantity λ. The N roots of this equation λ_i, $i = 1, 2, \ldots, N$, give the eigenvalues of **A**. Corresponding to each λ_i there will be a column vector x^i, which is the ith eigenvector of **A** and can be found by solving (10.85) for **x**.

Since λ only appears in (10.86) as part of an element on the leading diagonal of $\mathsf{A} - \lambda \mathsf{I}$, the λ^N- and λ^{N-1}-terms in the characteristic equation can only arise from the product of the N elements that lie on that diagonal. This means that, when (10.86) is written out as a polynomial equation in λ, the coefficient of $-\lambda^{N-1}$ in the equation will be simply $A_{11} + A_{22} + \cdots + A_{NN}$, whilst that of λ^N will be unity. As discussed in Section 10.6, the quantity $\sum_{i=1}^{N} A_{ii}$ is the *trace* of **A** and, from the ordinary theory of polynomial equations [see Equation (2.14)] will be equal to the sum of the roots of (10.86):

$$\sum_{i=1}^{N} \lambda_i = \operatorname{Tr} \mathsf{A}. \tag{10.87}$$

This can be used as one check that a computation of the eigenvalues λ_i has been done correctly. Unless Equation (10.87) is satisfied by a computed set of eigenvalues, they have not been calculated correctly. However, that Equation (10.87) is satisfied is a necessary, but not sufficient, condition for a correct computation. An alternative proof of (10.87) is given in Section 10.15. A straightforward example now follows.

10.13 Determination of eigenvalues and eigenvectors

Example Find the eigenvalues and normalised eigenvectors of the real symmetric matrix

$$A = \begin{pmatrix} 1 & 1 & 3 \\ 1 & 1 & -3 \\ 3 & -3 & -3 \end{pmatrix}.$$

Using (10.86),

$$\begin{vmatrix} 1-\lambda & 1 & 3 \\ 1 & 1-\lambda & -3 \\ 3 & -3 & -3-\lambda \end{vmatrix} = 0.$$

Expanding out this determinant gives[29]

$$(1-\lambda)[(1-\lambda)(-3-\lambda) - (-3)(-3)] + 1[(-3)(3) - 1(-3-\lambda)]$$
$$+ 3[1(-3) - (1-\lambda)(3)] = 0, \qquad (10.88)$$

which simplifies to give

$$(1-\lambda)(\lambda^2 + 2\lambda - 12) + (\lambda - 6) + 3(3\lambda - 6) = 0$$
$$\Rightarrow \quad (\lambda - 2)(\lambda - 3)(\lambda + 6) = 0.$$

Hence the roots of the characteristic equation, which are the eigenvalues of A, are $\lambda_1 = 2$, $\lambda_2 = 3$, $\lambda_3 = -6$. We note that, as expected,

$$\lambda_1 + \lambda_2 + \lambda_3 = -1 = 1 + 1 - 3 = A_{11} + A_{22} + A_{33} = \text{Tr } A.$$

For the first root, $\lambda_1 = 2$, a suitable eigenvector x^1, with elements x_1, x_2, x_3, must satisfy $Ax^1 = 2x^1$ or, equivalently,

$$x_1 + x_2 + 3x_3 = 2x_1,$$
$$x_1 + x_2 - 3x_3 = 2x_2, \qquad (10.89)$$
$$3x_1 - 3x_2 - 3x_3 = 2x_3.$$

These three equations are consistent (to ensure this was the purpose behind finding the particular values of λ) and yield $x_3 = 0$, $x_1 = x_2 = k$, where k is any non-zero number. A suitable eigenvector would thus be

$$x^1 = (k \quad k \quad 0)^T.$$

If we apply the normalisation condition, we require $k^2 + k^2 + 0^2 = 1$ or $k = 1/\sqrt{2}$. Hence

$$x^1 = \left(\frac{1}{\sqrt{2}} \quad \frac{1}{\sqrt{2}} \quad 0 \right)^T = \frac{1}{\sqrt{2}}(1 \quad 1 \quad 0)^T.$$

Repeating the last paragraph, but with the factor 2 on the RHS of (10.89) replaced successively by $\lambda_2 = 3$ and $\lambda_3 = -6$, gives

$$x^2 = \frac{1}{\sqrt{3}}(1 \quad -1 \quad 1)^T \quad \text{and} \quad x^3 = \frac{1}{\sqrt{6}}(1 \quad -1 \quad -2)^T$$

as two further normalised eigenvectors. ◂

---

[29] This 'head-on' method gives a cubic equation, the roots of which have to be obtained by inspection. Obtain the same result by (i) adding the second column to the first and (ii) taking out a common factor $(2 - \lambda)$. Then subtract the first row from the second and obtain a quadratic expression in λ that can be factorised by inspection.

Matrices and vector spaces

In the above example, the three values of λ are all different and A is a real symmetric matrix. Thus we expect, and it is easily checked, that the three eigenvectors are mutually orthogonal, i.e.

$$\left(x^1\right)^T x^2 = \left(x^1\right)^T x^3 = \left(x^2\right)^T x^3 = 0.$$

It will be apparent also that, as expected, the normalisation of the eigenvectors has no effect on their orthogonality.

Degenerate eigenvalues

We now return to the case of degenerate eigenvalues, i.e. those that have two or more associated eigenvectors. We have mentioned already that it is always possible to construct an orthogonal set of eigenvectors for a normal matrix; the following example, which exploits this natural or imposed mutual orthogonality, illustrates a heuristic method of finding such a set that is simpler than following the formal steps of the Gram–Schmidt process.

Example Construct an orthonormal set of eigenvectors for the matrix

$$A = \begin{pmatrix} 1 & 0 & 3 \\ 0 & -2 & 0 \\ 3 & 0 & 1 \end{pmatrix}.$$

We first determine the eigenvalues using $|A - \lambda I| = 0$:

$$0 = \begin{vmatrix} 1-\lambda & 0 & 3 \\ 0 & -2-\lambda & 0 \\ 3 & 0 & 1-\lambda \end{vmatrix} = -(1-\lambda)^2(2+\lambda) + 3(3)(2+\lambda)$$

$$= (4-\lambda)(\lambda+2)^2.$$

Thus $\lambda_1 = 4, \lambda_2 = -2 = \lambda_3$. The normalised eigenvector $x^1 = (x_1 \; x_2 \; x_3)^T$ corresponding to the unrepeated eigenvalue is found from

$$\begin{pmatrix} 1 & 0 & 3 \\ 0 & -2 & 0 \\ 3 & 0 & 1 \end{pmatrix} \begin{pmatrix} x_1 \\ x_2 \\ x_3 \end{pmatrix} = 4 \begin{pmatrix} x_1 \\ x_2 \\ x_3 \end{pmatrix} \quad \Rightarrow \quad x^1 = \frac{1}{\sqrt{2}} \begin{pmatrix} 1 \\ 0 \\ 1 \end{pmatrix}.$$

A general column vector that is orthogonal to x^1 is

$$x = (a \; b \; -a)^T, \tag{10.90}$$

and it is easily shown that

$$Ax = \begin{pmatrix} 1 & 0 & 3 \\ 0 & -2 & 0 \\ 3 & 0 & 1 \end{pmatrix} \begin{pmatrix} a \\ b \\ -a \end{pmatrix} = -2 \begin{pmatrix} a \\ b \\ -a \end{pmatrix} = -2x.$$

Thus x is a eigenvector of A with associated eigenvalue -2. It is clear, however, that there is an infinite set of eigenvectors x all possessing the required property; the geometrical analogue is that there are an infinite number of corresponding vectors x lying in the plane that has x^1 as its normal.

10.14 Change of basis and similarity transformations

We do require that the two remaining eigenvectors are orthogonal to one another, but this still leaves an infinite number of possibilities. For $\mathbf{x}^2$, therefore, let us choose a simple form of (10.90), suitably normalised, say,

$$\mathbf{x}^2 = (0 \quad 1 \quad 0)^T.$$

The third eigenvector is then specified (to within an arbitrary multiplicative constant) by the requirement that it must be orthogonal to $\mathbf{x}^1$ and $\mathbf{x}^2$; thus $\mathbf{x}^3$ may be found by evaluating the vector product of $\mathbf{x}^1$ and $\mathbf{x}^2$ and normalising the result. This gives

$$\mathbf{x}^3 = \frac{1}{\sqrt{2}} (-1 \quad 0 \quad 1)^T,$$

corresponding to $a = -1$ and $b = 0$, and completes the construction of an orthonormal set of eigenvectors.[30] ◀

EXERCISES 10.13

1. Find the eigenvalues and eigenvectors of the matrix

$$A = \begin{pmatrix} 2 & -1 & 0 \\ -1 & 2 & -1 \\ 0 & -1 & 2 \end{pmatrix}.$$

   Verify that its eigenvectors are mutually orthogonal.

2. Show that the matrix

$$A = \begin{pmatrix} 7 & 0 & 6 \\ 0 & -5 & 0 \\ 6 & 0 & -2 \end{pmatrix}$$

   has a degenerate eigenvalue. Construct an orthonormal set of eigenvectors for the matrix.

10.14 Change of basis and similarity transformations

Throughout this chapter we have considered the vector $\mathbf{x}$ as a geometrical quantity that is independent of any basis (or coordinate system). If we introduce a basis $\mathbf{e}_i$, $i = 1, 2, \ldots, N$, into our N-dimensional vector space then we may write

$$\mathbf{x} = x_1 \mathbf{e}_1 + x_2 \mathbf{e}_2 + \cdots + x_N \mathbf{e}_N,$$

and represent $\mathbf{x}$ in this basis by the column matrix

$$\mathbf{x} = (x_1 \quad x_2 \quad \cdots \quad x_n)^T,$$

---

[30] How would you find an orthonormal set if all three eigenvalues were equal?

having components x_i. We now consider how these components change as a result of a prescribed change of basis. Let us introduce a new basis $\mathbf{e}'_i$, $i = 1, 2, \ldots, N$, which is related to the old basis by

$$\mathbf{e}'_j = \sum_{i=1}^{N} S_{ij} \mathbf{e}_i, \tag{10.91}$$

the coefficient S_{ij} being the ith component of $\mathbf{e}'_j$ with respect to the old (unprimed) basis. For an arbitrary vector $\mathbf{x}$ it follows that

$$\mathbf{x} = \sum_{i=1}^{N} x_i \mathbf{e}_i = \sum_{j=1}^{N} x'_j \mathbf{e}'_j = \sum_{j=1}^{N} x'_j \sum_{i=1}^{N} S_{ij} \mathbf{e}_i.$$

From this we derive the relationship between the components of $\mathbf{x}$ in the two coordinate systems as

$$x_i = \sum_{j=1}^{N} S_{ij} x'_j,$$

which we can write in matrix form as

$$\mathsf{x} = \mathsf{S}\mathsf{x}' \tag{10.92}$$

where S is the *transformation matrix* associated with the change of basis.

Furthermore, since the vectors $\mathbf{e}'_j$ are linearly independent, the matrix S is non-singular and so possesses an inverse S^{-1}. Multiplying (10.92) on the left by S^{-1} we find

$$\mathsf{x}' = \mathsf{S}^{-1}\mathsf{x}, \tag{10.93}$$

which relates the components of $\mathbf{x}$ in the new basis to those in the old basis. Comparing (10.93) and (10.91) we note that the components of $\mathbf{x}$ transform inversely to the way in which the basis vectors $\mathbf{e}_i$ themselves transform. This has to be so, as the vector $\mathbf{x}$ itself must remain unchanged.

We may also find the transformation law for the components of a linear operator under the same change of basis. The operator equation $\mathbf{y} = \mathcal{A}\mathbf{x}$ (which is basis independent) can be written as a matrix equation in each of the two bases as

$$\mathsf{y} = \mathsf{A}\mathsf{x}, \qquad \mathsf{y}' = \mathsf{A}'\mathsf{x}'. \tag{10.94}$$

But, using (10.92) to change from the unprimed to the primed basis, we may rewrite the first equation as

$$\mathsf{S}\mathsf{y}' = \mathsf{A}\mathsf{S}\mathsf{x}' \quad \Rightarrow \quad \mathsf{y}' = \mathsf{S}^{-1}\mathsf{A}\mathsf{S}\mathsf{x}'.$$

Comparing this with the second equation in (10.94) we find that the components of the linear operator $\mathcal{A}$ transform as

$$\mathsf{A}' = \mathsf{S}^{-1}\mathsf{A}\mathsf{S}. \tag{10.95}$$

Equation (10.95) is an example of a *similarity transformation* – a transformation that can be particularly useful in the conversion of matrices into convenient forms for computation.

10.14 Change of basis and similarity transformations

Given a square matrix A, we may interpret it as representing a linear operator $\mathcal{A}$ in a given basis $\mathbf{e}_i$. From (10.95), however, we may also consider the matrix $\mathsf{A}' = \mathsf{S}^{-1}\mathsf{A}\mathsf{S}$, for any non-singular matrix S, as representing the same linear operator $\mathcal{A}$ but in a new basis $\mathbf{e}'_j$, related to the old basis by

$$\mathbf{e}'_j = \sum_i S_{ij} \mathbf{e}_i.$$

Therefore we would expect that any property of the matrix A that represents some (basis-independent) property of the linear operator $\mathcal{A}$ will also be a property of the matrix A'. We next list a number of such properties.

(i) If $\mathsf{A} = \mathsf{I}$ then $\mathsf{A}' = \mathsf{I}$, since, from (10.95),

$$\mathsf{A}' = \mathsf{S}^{-1}\mathsf{I}\mathsf{S} = \mathsf{S}^{-1}\mathsf{S} = \mathsf{I}. \tag{10.96}$$

(ii) The value of the determinant is unchanged:

$$|\mathsf{A}'| = |\mathsf{S}^{-1}\mathsf{A}\mathsf{S}| = |\mathsf{S}^{-1}||\mathsf{A}||\mathsf{S}| = |\mathsf{A}||\mathsf{S}^{-1}||\mathsf{S}| = |\mathsf{A}||\mathsf{S}^{-1}\mathsf{S}| = |\mathsf{A}|. \tag{10.97}$$

(iii) The characteristic determinant and hence the eigenvalues of A' are the same as those of A: from (10.86),

$$|\mathsf{A}' - \lambda \mathsf{I}| = |\mathsf{S}^{-1}\mathsf{A}\mathsf{S} - \lambda \mathsf{I}| = |\mathsf{S}^{-1}(\mathsf{A} - \lambda \mathsf{I})\mathsf{S}|$$
$$= |\mathsf{S}^{-1}||\mathsf{S}||\mathsf{A} - \lambda \mathsf{I}| = |\mathsf{A} - \lambda \mathsf{I}|. \tag{10.98}$$

(iv) The value of the trace is unchanged. This follows either from combining (10.87) and property (iii) above, or directly as follows:

$$\operatorname{Tr} \mathsf{A}' = \sum_i A'_{ii} = \sum_i \sum_j \sum_k (\mathsf{S}^{-1})_{ij} A_{jk} S_{ki}$$
$$= \sum_i \sum_j \sum_k S_{ki} (\mathsf{S}^{-1})_{ij} A_{jk} = \sum_j \sum_k \delta_{kj} A_{jk} = \sum_j A_{jj}$$
$$= \operatorname{Tr} \mathsf{A}. \tag{10.99}$$

An important class of similarity transformations is that for which S is a unitary matrix; in this case $\mathsf{A}' = \mathsf{S}^{-1}\mathsf{A}\mathsf{S} = \mathsf{S}^\dagger\mathsf{A}\mathsf{S}$. Unitary transformation matrices are particularly important for the following reason. If the original basis $\mathbf{e}_i$ is orthonormal and the transformation matrix S is unitary then

$$\langle \mathbf{e}'_i | \mathbf{e}'_j \rangle = \left\langle \sum_k S_{ki} \mathbf{e}_k \,\bigg|\, \sum_r S_{rj} \mathbf{e}_r \right\rangle$$
$$= \sum_k S^*_{ki} \sum_r S_{rj} \langle \mathbf{e}_k | \mathbf{e}_r \rangle$$
$$= \sum_k S^*_{ki} \sum_r S_{rj} \delta_{kr} = \sum_k S^*_{ki} S_{kj} = \sum_k S^\dagger_{ik} S_{kj} = (\mathsf{S}^\dagger \mathsf{S})_{ij} = \delta_{ij},$$

showing that the new basis is also orthonormal.

Furthermore, in addition to the properties of general similarity transformations, for unitary transformations the following hold.

(i) If A is Hermitian (anti-Hermitian) then A′ is Hermitian (anti-Hermitian), i.e. if $A^\dagger = \pm A$ then

$$(A')^\dagger = (S^\dagger A S)^\dagger = S^\dagger A^\dagger S = \pm S^\dagger A S = \pm A'. \quad (10.100)$$

(ii) If A is unitary (so that $A^\dagger = A^{-1}$) then A′ is unitary, since

$$(A')^\dagger A' = (S^\dagger A S)^\dagger (S^\dagger A S) = S^\dagger A^\dagger S S^\dagger A S = S^\dagger A^\dagger A S$$
$$= S^\dagger I S = I. \quad (10.101)$$

EXERCISE 10.14

1. Under a similarity transformation, matrices A and B transform to A′ and B′ respectively.
   (a) Show that if A and B commute, then so do A′ and B′.
   (b) Show that if $B = \exp A$, then $B' = \exp A'$.
   (c) If A and B do not commute, show that there is no similarity transformation such that A′ and B′ do commute.

10.15 Diagonalisation of matrices

Suppose that a linear operator $\mathcal{A}$ is represented in some basis e_i, $i = 1, 2, \ldots, N$, by the matrix A. Consider a new basis x^j given by

$$x^j = \sum_{i=1}^{N} S_{ij} e_i,$$

where the x^j are chosen to be the eigenvectors of the linear operator $\mathcal{A}$, i.e.

$$\mathcal{A} x^j = \lambda_j x^j. \quad (10.102)$$

In the new basis, $\mathcal{A}$ is represented by the matrix $A' = S^{-1} A S$, which has a particularly simple form, as we shall see shortly. The element S_{ij} of S is the ith component, in the old (unprimed) basis, of the jth eigenvector x^j of A, i.e. the columns of S are the eigenvectors of the matrix A:

$$S = \begin{pmatrix} \uparrow & \uparrow & & \uparrow \\ x^1 & x^2 & \cdots & x^N \\ \downarrow & \downarrow & & \downarrow \end{pmatrix},$$

10.15 Diagonalisation of matrices

that is, $S_{ij} = (\mathbf{x}^j)_i$. Therefore the ijth component of A' is given by

$$(\mathsf{S}^{-1}\mathsf{A}\mathsf{S})_{ij} = \sum_k \sum_l (\mathsf{S}^{-1})_{ik} A_{kl} S_{lj}$$

$$= \sum_k \sum_l (\mathsf{S}^{-1})_{ik} A_{kl} (\mathbf{x}^j)_l$$

$$= \sum_k (\mathsf{S}^{-1})_{ik} \lambda_j (\mathbf{x}^j)_k$$

$$= \sum_k \lambda_j (\mathsf{S}^{-1})_{ik} S_{kj} = \lambda_j \delta_{ij}.$$

So the matrix A' is diagonal with the eigenvalues of $\mathcal{A}$ as its diagonal elements, i.e.

$$\mathsf{A}' = \begin{pmatrix} \lambda_1 & 0 & \cdots & 0 \\ 0 & \lambda_2 & & \vdots \\ \vdots & & \ddots & 0 \\ 0 & \cdots & 0 & \lambda_N \end{pmatrix}.$$

Therefore, given a matrix A, if we construct the matrix S that has the eigenvectors of A as its columns then the matrix $\mathsf{A}' = \mathsf{S}^{-1}\mathsf{A}\mathsf{S}$ is diagonal and has the eigenvalues of A as its diagonal elements. Since we require S to be non-singular ($|\mathsf{S}| \neq 0$), the N eigenvectors of A must be linearly independent and form a basis for the N-dimensional vector space. It may be shown that *any matrix with distinct eigenvalues* can be diagonalised by this procedure. If, however, a general square matrix has degenerate eigenvalues then it may, or may not, have N linearly independent eigenvectors. If it does not then it *cannot* be diagonalised.

For normal matrices (which include Hermitian, anti-Hermitian and unitary matrices)[31] the N eigenvectors are indeed linearly independent. Moreover, when normalised, these eigenvectors form an *orthonormal* set (or can be made to do so). Therefore, the matrix S with these normalised eigenvectors as columns, i.e. whose elements are $S_{ij} = (\mathbf{x}^j)_i$, has the property

$$(\mathsf{S}^\dagger \mathsf{S})_{ij} = \sum_k (\mathsf{S}^\dagger)_{ik}(\mathsf{S})_{kj} = \sum_k S^*_{ki} S_{kj} = \sum_k (\mathbf{x}^i)^*_k (\mathbf{x}^j)_k = (\mathbf{x}^i)^\dagger \mathbf{x}^j = \delta_{ij}.$$

Hence S is unitary ($\mathsf{S}^{-1} = \mathsf{S}^\dagger$) and the original matrix A can be diagonalised by

$$\mathsf{A}' = \mathsf{S}^{-1}\mathsf{A}\mathsf{S} = \mathsf{S}^\dagger \mathsf{A}\mathsf{S}.$$

Therefore, any normal matrix A can be diagonalised by a similarity transformation using a *unitary* transformation matrix S.

---

[31] See Section 10.11.7.

Example Diagonalise the matrix

$$A = \begin{pmatrix} 1 & 0 & 3 \\ 0 & -2 & 0 \\ 3 & 0 & 1 \end{pmatrix}.$$

The matrix A is symmetric and so may be diagonalised by a transformation of the form $A' = S^\dagger AS$, where S has the normalised eigenvectors of A as its columns. We have already found these eigenvectors in Section 10.13, and so we can write straightaway

$$S = \frac{1}{\sqrt{2}} \begin{pmatrix} 1 & 0 & -1 \\ 0 & \sqrt{2} & 0 \\ 1 & 0 & 1 \end{pmatrix}.$$

We note that although the eigenvalues of A are degenerate, its three eigenvectors are linearly independent and so A can still be diagonalised. Thus, calculating $S^\dagger AS$ we obtain

$$S^\dagger AS = \frac{1}{2} \begin{pmatrix} 1 & 0 & 1 \\ 0 & \sqrt{2} & 0 \\ -1 & 0 & 1 \end{pmatrix} \begin{pmatrix} 1 & 0 & 3 \\ 0 & -2 & 0 \\ 3 & 0 & 1 \end{pmatrix} \begin{pmatrix} 1 & 0 & -1 \\ 0 & \sqrt{2} & 0 \\ 1 & 0 & 1 \end{pmatrix}$$

$$= \begin{pmatrix} 4 & 0 & 0 \\ 0 & -2 & 0 \\ 0 & 0 & -2 \end{pmatrix},$$

which is the required diagonal matrix, and has, as expected, the eigenvalues of A as its diagonal elements. ◀

If a matrix A is diagonalised by the similarity transformation $A' = S^{-1}AS$, so that $A' = \text{diag}(\lambda_1, \lambda_2, \ldots, \lambda_N)$, then we have immediately

$$\text{Tr } A' = \text{Tr } A = \sum_{i=1}^{N} \lambda_i, \qquad (10.103)$$

$$|A'| = |A| = \prod_{i=1}^{N} \lambda_i, \qquad (10.104)$$

since the eigenvalues of the matrix are unchanged by the transformation.

EXERCISE 10.15

1. Construct the orthogonal matrix S corresponding to the basis transformation that takes the matrix in exercise 2 of Section 10.13,

$$A = \begin{pmatrix} 7 & 0 & 6 \\ 0 & -5 & 0 \\ 6 & 0 & -2 \end{pmatrix},$$

into a diagonal matrix. Carry out the transformation and check that the resulting diagonal elements have the values you expect. Deduce the value of $|A|$ and confirm it by direct calculation.

10.16 Quadratic and Hermitian forms

Let us now introduce the concept of quadratic forms (and their complex analogues, Hermitian forms). A quadratic form Q is a scalar function of a real vector $\mathbf{x}$ given by

$$Q(\mathbf{x}) = \langle \mathbf{x} | \mathcal{A} \mathbf{x} \rangle, \tag{10.105}$$

for some real linear operator $\mathcal{A}$. In any given basis (coordinate system) we can write (10.105) in matrix form as

$$Q(\mathbf{x}) = \mathbf{x}^{\mathrm{T}} \mathsf{A} \mathbf{x}, \tag{10.106}$$

where A is a real matrix. In fact, as will be explained below, we need only consider the case where A is symmetric, i.e. $\mathsf{A} = \mathsf{A}^{\mathrm{T}}$. As an example in a three-dimensional space,

$$Q = \mathbf{x}^{\mathrm{T}} \mathsf{A} \mathbf{x} = \begin{pmatrix} x_1 & x_2 & x_3 \end{pmatrix} \begin{pmatrix} 1 & 1 & 3 \\ 1 & 1 & -3 \\ 3 & -3 & -3 \end{pmatrix} \begin{pmatrix} x_1 \\ x_2 \\ x_3 \end{pmatrix}$$

$$= x_1^2 + x_2^2 - 3x_3^2 + 2x_1 x_2 + 6x_1 x_3 - 6x_2 x_3. \tag{10.107}$$

It is reasonable to ask whether a quadratic form $Q = \mathbf{x}^{\mathrm{T}} \mathsf{M} \mathbf{x}$, where M is any (possibly non-symmetric) real square matrix, is a more general definition. That this is not the case may be seen by expressing M in terms of a symmetric matrix $\mathsf{A} = \frac{1}{2}(\mathsf{M} + \mathsf{M}^{\mathrm{T}})$ and an antisymmetric matrix $\mathsf{B} = \frac{1}{2}(\mathsf{M} - \mathsf{M}^{\mathrm{T}})$; clearly, M can be represented as $\mathsf{M} = \mathsf{A} + \mathsf{B}$. We then have

$$Q = \mathbf{x}^{\mathrm{T}} \mathsf{M} \mathbf{x} = \mathbf{x}^{\mathrm{T}} \mathsf{A} \mathbf{x} + \mathbf{x}^{\mathrm{T}} \mathsf{B} \mathbf{x}. \tag{10.108}$$

However, Q is a scalar quantity and so

$$Q = Q^{\mathrm{T}} = (\mathbf{x}^{\mathrm{T}} \mathsf{A} \mathbf{x})^{\mathrm{T}} + (\mathbf{x}^{\mathrm{T}} \mathsf{B} \mathbf{x})^{\mathrm{T}} = \mathbf{x}^{\mathrm{T}} \mathsf{A}^{\mathrm{T}} \mathbf{x} + \mathbf{x}^{\mathrm{T}} \mathsf{B}^{\mathrm{T}} \mathbf{x} = \mathbf{x}^{\mathrm{T}} \mathsf{A} \mathbf{x} - \mathbf{x}^{\mathrm{T}} \mathsf{B} \mathbf{x}. \tag{10.109}$$

Comparing (10.108) and (10.109) shows that $\mathbf{x}^{\mathrm{T}} \mathsf{B} \mathbf{x} = 0$, and hence $\mathbf{x}^{\mathrm{T}} \mathsf{M} \mathbf{x} = \mathbf{x}^{\mathrm{T}} \mathsf{A} \mathbf{x}$, i.e. Q is unchanged by considering only the symmetric part of M. Hence, with no loss of generality, we may assume $\mathsf{A} = \mathsf{A}^{\mathrm{T}}$ in (10.106).

From its definition (10.105), Q is clearly a basis- (i.e. coordinate-) independent quantity. Let us therefore consider a new basis related to the old one by an orthogonal transformation matrix S, the components in the two bases of any vector $\mathbf{x}$ being related [as in (10.92)] by $\mathbf{x} = \mathsf{S} \mathbf{x}'$ or, equivalently, by $\mathbf{x}' = \mathsf{S}^{-1} \mathbf{x} = \mathsf{S}^{\mathrm{T}} \mathbf{x}$. We then have

$$Q = \mathbf{x}^{\mathrm{T}} \mathsf{A} \mathbf{x} = (\mathbf{x}')^{\mathrm{T}} \mathsf{S}^{\mathrm{T}} \mathsf{A} \mathsf{S} \mathbf{x}' = (\mathbf{x}')^{\mathrm{T}} \mathsf{A}' \mathbf{x}',$$

where (as expected) the matrix describing the linear operator $\mathcal{A}$ in the new basis is given by $\mathsf{A}' = \mathsf{S}^{\mathrm{T}} \mathsf{A} \mathsf{S}$ (since $\mathsf{S}^{\mathrm{T}} = \mathsf{S}^{-1}$).[32] But, from the previous section, if we choose as S the matrix whose columns are the *normalised* eigenvectors of A then $\mathsf{A}' = \mathsf{S}^{\mathrm{T}} \mathsf{A} \mathsf{S}$ is diagonal with the eigenvalues of A as the diagonal elements. In the new basis

$$Q = \mathbf{x}^{\mathrm{T}} \mathsf{A} \mathbf{x} = (\mathbf{x}')^{\mathrm{T}} \Lambda \mathbf{x}' = \lambda_1 x_1'^2 + \lambda_2 x_2'^2 + \cdots + \lambda_N x_N'^2, \tag{10.110}$$

---

[32] Since A is symmetric, its normalised eigenvectors are orthogonal, or can be made so, and hence S is orthogonal with $\mathsf{S}^{-1} = \mathsf{S}^{\mathrm{T}}$.

where $\Lambda = \text{diag}(\lambda_1, \lambda_2, \ldots, \lambda_N)$ and the λ_i are the eigenvalues of A. It should be noted that Q contains no cross-terms of the form $x_1' x_2'$.

> **Example** Find an orthogonal transformation that takes the quadratic form (10.107) into the form
> $$\lambda_1 x_1'^2 + \lambda_2 x_2'^2 + \lambda_3 x_3'^2.$$
>
> The required transformation matrix S has the *normalised* eigenvectors of A as its columns. We have already found these in Section 10.13, and so we can write immediately
> $$\mathsf{S} = \frac{1}{\sqrt{6}} \begin{pmatrix} \sqrt{3} & \sqrt{2} & 1 \\ \sqrt{3} & -\sqrt{2} & -1 \\ 0 & \sqrt{2} & -2 \end{pmatrix},$$
> which is easily verified as being orthogonal. Since the eigenvalues of A are $\lambda = 2, 3$, and -6, the general result already proved shows that the transformation $\mathsf{x} = \mathsf{S}\mathsf{x}'$ will carry (10.107) into the form $2x_1'^2 + 3x_2'^2 - 6x_3'^2$. This may be verified most easily by writing out the inverse transformation $\mathsf{x}' = \mathsf{S}^{-1}\mathsf{x} = \mathsf{S}^T \mathsf{x}$ and substituting. The inverse equations are
> $$x_1' = (x_1 + x_2)/\sqrt{2},$$
> $$x_2' = (x_1 - x_2 + x_3)/\sqrt{3}, \quad (10.111)$$
> $$x_3' = (x_1 - x_2 - 2x_3)/\sqrt{6}.$$
> If these are substituted into the form $Q = 2x_1'^2 + 3x_2'^2 - 6x_3'^2$ then the original expression (10.107) is recovered. ◀

In the definition of Q it was assumed that the components x_1, x_2, x_3 and the matrix A were real. It is clear that in this case the quadratic form $Q = \mathsf{x}^T \mathsf{A} \mathsf{x}$ is real also. Another, rather more general, expression that is also real is the *Hermitian form*

$$H(\mathsf{x}) \equiv \mathsf{x}^\dagger \mathsf{A} \mathsf{x}, \quad (10.112)$$

where A is Hermitian (i.e. $\mathsf{A}^\dagger = \mathsf{A}$) and the components of x may now be complex. It is straightforward to show that H is real, since

$$H^* = (H^T)^* = \mathsf{x}^\dagger \mathsf{A}^\dagger \mathsf{x} = \mathsf{x}^\dagger \mathsf{A} \mathsf{x} = H.$$

With suitable generalisation, the properties of quadratic forms apply also to Hermitian forms, but to keep the presentation simple we will restrict our discussion to quadratic forms.

A special case of a quadratic (Hermitian) form is one for which $Q = \mathsf{x}^T \mathsf{A} \mathsf{x}$ is greater than zero for all column matrices x. By choosing as the basis the eigenvectors of A we have Q in the form

$$Q = \lambda_1 x_1^2 + \lambda_2 x_2^2 + \lambda_3 x_3^2.$$

The requirement that $Q > 0$ for all x means that all the eigenvalues λ_i of A must be positive. A symmetric (Hermitian) matrix A with this property is called *positive definite*.

10.16 Quadratic and Hermitian forms

If, instead, $Q \geq 0$ for all $\mathbf{x}$ then it is possible that some of the eigenvalues are zero, and A is called *positive semi-definite*.

10.16.1 The stationary properties of the eigenvectors

Consider a quadratic form, such as $Q(\mathbf{x}) = \langle \mathbf{x}|\mathcal{A}\mathbf{x}\rangle$, Equation (10.105), in a fixed basis. As the vector $\mathbf{x}$ is varied, through changes in its three components x_1, x_2 and x_3, the value of the quantity Q also varies. Because of the homogeneous form of Q we may restrict any investigation of these variations to vectors of unit length (since multiplying any vector $\mathbf{x}$ by any scalar k simply multiplies the value of Q by a factor k^2).

Of particular interest are any vectors $\mathbf{x}$ that make the value of the quadratic form a maximum or minimum. A necessary, but not sufficient, condition for this is that Q is stationary with respect to small variations $\Delta \mathbf{x}$ in $\mathbf{x}$, whilst $\langle \mathbf{x}|\mathbf{x}\rangle$ is maintained at a constant value (unity).

In the chosen basis the quadratic form is given by $Q = \mathbf{x}^T \mathsf{A} \mathbf{x}$ and, using Lagrange undetermined multipliers to incorporate the variational constraints, we are led to seek solutions of

$$\Delta[\mathbf{x}^T \mathsf{A} \mathbf{x} - \lambda(\mathbf{x}^T \mathbf{x} - 1)] = 0. \tag{10.113}$$

This may be used directly, together with the fact that $(\Delta \mathbf{x}^T) \mathsf{A} \mathbf{x} = \mathbf{x}^T \mathsf{A} \, \Delta \mathbf{x}$, since A is symmetric, to obtain

$$\mathsf{A} \mathbf{x} = \lambda \mathbf{x} \tag{10.114}$$

as the necessary condition that $\mathbf{x}$ must satisfy. If (10.114) is satisfied for some eigenvector $\mathbf{x}$ then the value of $Q(\mathbf{x})$ is given by

$$Q = \mathbf{x}^T \mathsf{A} \mathbf{x} = \mathbf{x}^T \lambda \mathbf{x} = \lambda. \tag{10.115}$$

However, if $\mathbf{x}$ and $\mathbf{y}$ are eigenvectors corresponding to different eigenvalues then they are (or can be chosen to be) orthogonal. Consequently, the expression $\mathbf{y}^T \mathsf{A} \mathbf{x}$ is necessarily zero, since

$$\mathbf{y}^T \mathsf{A} \mathbf{x} = \mathbf{y}^T \lambda \mathbf{x} = \lambda \mathbf{y}^T \mathbf{x} = 0. \tag{10.116}$$

Summarising, those column matrices $\mathbf{x}$ of unit magnitude that make the quadratic form Q stationary are eigenvectors of the matrix A, and the stationary value of Q is then equal to the corresponding eigenvalue. It is straightforward to see from the proof of (10.114) that, conversely, any eigenvector of A makes Q stationary.

Instead of maximising or minimising $Q = \mathbf{x}^T \mathsf{A} \mathbf{x}$ subject to the constraint $\mathbf{x}^T \mathbf{x} = 1$, an equivalent procedure is to extremise the function

$$\lambda(\mathbf{x}) = \frac{\mathbf{x}^T \mathsf{A} \mathbf{x}}{\mathbf{x}^T \mathbf{x}},$$

as we now show.

Example Show that if $\lambda(\mathbf{x})$ is stationary then $\mathbf{x}$ is an eigenvector of A and $\lambda(\mathbf{x})$ is equal to the corresponding eigenvalue.

We require $\Delta\lambda(\mathbf{x}) = 0$ with respect to small variations in $\mathbf{x}$. Now

$$\Delta\lambda = \frac{1}{(\mathbf{x}^T\mathbf{x})^2}\left[(\mathbf{x}^T\mathbf{x})\left(\Delta\mathbf{x}^T A\mathbf{x} + \mathbf{x}^T A\,\Delta\mathbf{x}\right) - \mathbf{x}^T A\mathbf{x}\left(\Delta\mathbf{x}^T\mathbf{x} + \mathbf{x}^T\Delta\mathbf{x}\right)\right]$$

$$= \frac{2\Delta\mathbf{x}^T A\mathbf{x}}{\mathbf{x}^T\mathbf{x}} - 2\left(\frac{\mathbf{x}^T A\mathbf{x}}{\mathbf{x}^T\mathbf{x}}\right)\frac{\Delta\mathbf{x}^T\mathbf{x}}{\mathbf{x}^T\mathbf{x}},$$

since $\mathbf{x}^T A\,\Delta\mathbf{x} = (\Delta\mathbf{x}^T)A\mathbf{x}$ and $\mathbf{x}^T\Delta\mathbf{x} = (\Delta\mathbf{x}^T)\mathbf{x}$. Thus

$$\Delta\lambda = \frac{2}{\mathbf{x}^T\mathbf{x}}\Delta\mathbf{x}^T[A\mathbf{x} - \lambda(\mathbf{x})\mathbf{x}].$$

If $\Delta\lambda$ is to be zero, we must have that $A\mathbf{x} = \lambda(\mathbf{x})\mathbf{x}$, i.e. $\mathbf{x}$ is an eigenvector of A with eigenvalue $\lambda(\mathbf{x})$. ◀

Thus the eigenvalues of a symmetric matrix A are the values of the function

$$\lambda(\mathbf{x}) = \frac{\mathbf{x}^T A\mathbf{x}}{\mathbf{x}^T\mathbf{x}}$$

at its stationary points. The eigenvectors of A lie along those directions in space for which the quadratic form $Q = \mathbf{x}^T A\mathbf{x}$ has stationary values, given a fixed magnitude for the vector $\mathbf{x}$. Similar results hold for Hermitian matrices.

10.16.2 Quadratic surfaces

The results of the previous subsection may be turned around to state that the surface given by

$$\mathbf{x}^T A\mathbf{x} = \text{constant} = 1 \text{ (say)}, \qquad (10.117)$$

and called a *quadratic surface*, has stationary values of its radius (i.e. origin–surface distance) in those directions that are along the eigenvectors of A. More specifically, in three dimensions the quadratic surface $\mathbf{x}^T A\mathbf{x} = 1$ has its principal axes along the three mutually perpendicular eigenvectors of A, and the squares of the corresponding principal radii are given by λ_i^{-1}, $i = 1, 2, 3$.

As well as having this stationary property of the radius, a *principal axis* is characterised by the fact that any section of the surface perpendicular to it has some degree of symmetry about it. If the eigenvalues corresponding to any two principal axes are degenerate then the quadratic surface has rotational symmetry about the third principal axis and the choice of a pair of axes perpendicular to that axis is not uniquely defined.

10.16 Quadratic and Hermitian forms

Example Find the shape of the quadratic surface

$$x_1^2 + x_2^2 - 3x_3^2 + 2x_1x_2 + 6x_1x_3 - 6x_2x_3 = 1.$$

If, instead of expressing the quadratic surface in terms of x_1, x_2, x_3, as in (10.107), we were to use the new variables x_1', x_2', x_3' defined in (10.111), for which the coordinate axes are along the three mutually perpendicular eigenvector directions $(1, 1, 0)$, $(1, -1, 1)$ and $(1, -1, -2)$, then the equation of the surface would take the form (see (10.110))

$$\frac{x_1'^2}{(1/\sqrt{2})^2} + \frac{x_2'^2}{(1/\sqrt{3})^2} - \frac{x_3'^2}{(1/\sqrt{6})^2} = 1.$$

Thus, for example, a section of the quadratic surface in the plane $x_3' = 0$, i.e. $x_1 - x_2 - 2x_3 = 0$, is an ellipse, with semi-axes $1/\sqrt{2}$ and $1/\sqrt{3}$. Similarly, a section in the plane $x_1' = x_1 + x_2 = 0$ is a hyperbola. ◀

Clearly the most general form of a quadratic surface, referred to its principal axes as coordinate axes, is

$$\pm\frac{x_1'^2}{a^2} \pm \frac{x_2'^2}{b^2} \pm \frac{x_3'^2}{c^2} = 1, \qquad (10.118)$$

where $\pm a^{-2}, \pm b^{-2}$ and $\pm c^{-2}$ are the three eigenvalues of the corresponding matrix **A**. For a real quadric surface, at least one of the signs on the LHS of (10.118) must be positive.

The simplest three-dimensional situation to visualise is that in which all of the eigenvalues are positive, since then the quadratic surface is an ellipsoid. If one eigenvalue is negative, as in the worked example, then the surface is ellipsoidal in some sections and hyperbolic in others, with the values of a, b and c and the value of the coordinate at which the section is taken determining where an ellipse terminates or a hyperbola begins.

The special case of one of the eigenvalues, λ_k say, being zero is worth mentioning. Then formally the corresponding principal radius becomes infinite or, more strictly, (10.118) becomes independent of x_k'. The corresponding quadratic surface is then a cylinder with its axis parallel to the x_k'-axis (i.e. in the direction of the corresponding eigenvector in the original coordinate system) and a cross-section given by (10.118) with no x_k' term; this will be an ellipse or hyperbola, depending on the relative sign of the other two eigenvalues.[33]

EXERCISES 10.16

1. Find a linear transformation of coordinates that takes the quadratic form

$$Q = 2x_1^2 + 2x_2^2 + 2x_3^2 - 4x_2x_3 - 14x_1x_3 - 4x_1x_2$$

into the form $9y_1^2 + \alpha y_2^2 + \beta y_3^2$, where the values of α and β are to be determined.

---

[33] What form do you expect the quadratic surface to take if two of the eigenvalues are zero (with the third one positive)?

432   Matrices and vector spaces

2. Determine, so far as possible by inspection, whether or not the following quadratic forms can take positive, zero and negative values. Confirm (or correct) your assessment by finding the eigenvalues of the relevant matrices. Remember that the x_i can take negative as well as positive values.
   (a) $Q = 41x_1^2 + 25x_2^2 + 34x_3^2 + 24x_1x_3$,
   (b) $Q = x_2^2 - 2x_1^2 - 2x_3^2 - 2x_2x_3 - x_1x_3 - 2x_1x_2$,
   (c) $Q = 48x_1x_3 - 18x_1^2 - 25x_2^2 - 32x_3^2$.

3. (a) By using Lagrange undetermined multipliers to find the stationary values of an appropriate quadratic form, subject to $x_1^2 + x_2^2 + x_3^2 = 1$, determine the eigenvalues of the matrix

$$A = \begin{pmatrix} 1 & 0 & 3 \\ 0 & -2 & 0 \\ 3 & 0 & 1 \end{pmatrix}.$$

   [This the same matrix as that used in Section 10.15.]
   (b) By reference to the solutions of the simultaneous equations generated, deduce that one of the eigenvalues is degenerate.
   (c) Obtain the appropriate eigenvectors (using a parameter for the degenerate case) and verify that $(x^i)^T A x^i / (x^i)^T x^i$ does have the value λ_i.

4. Using the results derived in exercise 2 above, determine and describe in geometric terms the quadric surfaces given, in each case, by $Q = 1$. Give the sizes of any relevant semi-axes and if there exists an axis of symmetry for the surface, determine its direction.

10.17 The summation convention

In this chapter we have often needed to take a sum over a number of terms which are all of the same general form, and differ only in the value of an indexing subscript. Such a summation has been indicated by a summation sign, $\sum$, with the range of the subscript written above and below the sign. This very explicit notation has been deliberately adopted for the purposes of introducing the general procedures. However, the reader will, after a time, doubtless have felt that much of the notation is superfluous, particularly when there have been multiple sums appearing in a single expression, each with its own explicit summation sign; the derivation of Equation (10.99) provides just such an example.

Such calculations can be significantly compacted, and in some cases simplified, if the Cartesian coordinates x, y and z are replaced symbolically by the indexed coordinates x_i, where i takes the values 1, 2 and 3, and the so-called *summation convention* is adopted. In this convention any *lower-case* alphabetic subscript that appears *exactly* twice in any term of an expression is understood to be summed over all the values that a subscript in that position can take (unless the contrary is specifically stated); there is no explicit summation sign.

The subscripted quantities may appear in the numerator and/or the denominator of a term in an expression. This naturally implies that any such pair of repeated subscripts

Summary

must occur only in subscript positions that have the same range of values. Sometimes the ranges of values have to be specified, but usually they are apparent from the context.

As a basic example, in this notation

$$P_{ij} = \sum_{k=1}^{N} A_{ik} B_{kj}$$

becomes

$$P_{ij} = A_{ik} B_{kj} \qquad \text{i.e. without the explicit summation sign.}$$

In order to use the convention, partial differentiation with respect to Cartesian coordinates x, y and z is denoted by the generic symbol $\partial/\partial x_i$; this facilitates a compact and efficient notation for the development of vector calculus identities. These are studied in Chapter 11, though, for the same reasons that matrix algebra was first presented here without using the convention, vector calculus is initially developed there without recourse to it.

Further discussion of the summation convention, together with additional examples of its use, form the content of Appendix D. Considerable care is needed when using the convention, but mastering it is well worthwhile, as it considerably shortens many matrix algebra and vector calculus calculations.

EXERCISE 10.17

1. This problem provides some more sophisticated uses of the summation convention for the reader who is interested; it is not an essential part of the text. Use the convention and Equation (D.4) in Appendix D to prove the vector identities below.
   (a) $\nabla \cdot (\mathbf{u} \times \mathbf{v}) = \mathbf{v} \cdot (\nabla \times \mathbf{u}) - \mathbf{u} \cdot (\nabla \times \mathbf{v})$.
   (b) $(\mathbf{u} \times \mathbf{v}) \times \mathbf{w} = (\mathbf{u} \cdot \mathbf{w})\mathbf{v} - (\mathbf{v} \cdot \mathbf{w})\mathbf{u}$.
   (c) $\nabla \times (\mathbf{u} \times \mathbf{v}) = (\nabla \cdot \mathbf{v})\mathbf{u} - (\nabla \cdot \mathbf{u})\mathbf{v} + (\mathbf{v} \cdot \nabla)\mathbf{u} - (\mathbf{u} \cdot \nabla)\mathbf{v}$.

SUMMARY

1. *Matrices and other quantities derived from an $M \times N$ matrix $\mathbf{A}$.*

Name	Symbol	How obtained	Notes
Trace	Tr $\mathbf{A}$	Sum the elements on the leading diagonal.	Needs $M = N$
2×2 determinant		$\begin{vmatrix} a_{11} & a_{12} \\ a_{21} & a_{22} \end{vmatrix} \equiv a_{11}a_{22} - a_{12}a_{21}$	Definition[34]

---

[34] The formal definition of a 1×1 determinant is the value of its single entry (including its sign); it is not to be confused with $|a_{11}|$ meaning the (positive) modulus of a_{11}.

Matrices and vector spaces

Name	Symbol	How obtained	Notes
Determinant	$|A|$	Make a Laplace expansion (Section 10.7) to reduce to a sum of 2×2 determinants.	Needs $M = N$
Rank	$R(A)$	The largest value of r for which A has an $r \times r$ submatrix with a non-zero determinant.	$R \leq \min\{M, N\}$
Transpose	A^T	Interchange rows and columns: $(A^T)_{ij} = A_{ji}$.	A^T is $N \times M$
Complex conjugate	A^*	Take the complex conjugate of each element: $(A^*)_{ij} = A^*_{ij}$.	A^* is $M \times N$
Hermitian conjugate	$A^\dagger$	Transpose the complex conjugate *or* complex conjugate the transpose: $(A^\dagger)_{ij} = A^*_{ji}$.	$A^\dagger$ is $N \times M$
Minor	M_{ij}	Evaluate the determinant of the $(N-1) \times (N-1)$ matrix formed by deleting the ith row and the jth column.	Needs $M = N$
Cofactor	C_{ij}	Multiply the minor M_{ij} by $(-1)^{i+j}$: $C_{ij} = (-1)^{i+j} M_{ij}$.	Needs $M = N$
Inverse	A^{-1}	Divide each element of the transpose C^T of the matrix of cofactors C by the determinant of A; $(A^{-1})_{ij} = C_{ji}/|A|$.	Needs $M = N$

2. *Matrix algebra*
   - $(A \pm B)_{ij} = A_{ij} \pm B_{ij}$.
   - $(\lambda A)_{ij} = \lambda A_{ij}$.
   - $(AB)_{ij} = \sum_k A_{ik} B_{kj}$.
   - $(A + B)C = AC + BC$ and $C(A + B) = CA + CB$.
   - $AB \neq BA$, in general.

3. *Special types of square matrices*

Type	Symbolic property	Descriptive property
Real	$A^* = A$	Every element is real.
Imaginary	$A^* = -A$	Every element is imaginary or zero.
Diagonal	$A_{ij} = 0$ for $i \neq j$	Every off-diagonal element is zero.

Summary

Type	Symbolic property	Descriptive property
Lower triangular	$A_{ij} = 0$ for $i < j$	Every element above the leading diagonal is zero.
Upper triangular	$A_{ij} = 0$ for $i > j$	Every element below the leading diagonal is zero.
Symmetric	$\mathbf{A}^T = \mathbf{A}$	The matrix is equal to its transpose; $A_{ij} = A_{ji}$.
Antisymmetric or skew-symmetric	$\mathbf{A}^T = -\mathbf{A}$	The matrix is equal to minus its transpose; $A_{ij} = -A_{ji}$. All of its diagonal elements must be zero.
Orthogonal	$\mathbf{A}^T = \mathbf{A}^{-1}$	The transpose is equal to the inverse.
Hermitian	$\mathbf{A}^\dagger = \mathbf{A}$	The matrix is equal to its Hermitian conjugate; $A_{ij} = A_{ji}^*$.
Anti-Hermitian	$\mathbf{A}^\dagger = -\mathbf{A}$	The matrix is equal to minus its Hermitian conjugate; $A_{ij} = -A_{ji}^*$.
Unitary	$\mathbf{A}^\dagger = \mathbf{A}^{-1}$	The Hermitian conjugate is equal to the inverse.
Normal	$\mathbf{A}^\dagger \mathbf{A} = \mathbf{A}\mathbf{A}^\dagger$	The matrix commutes with its Hermitian conjugate.
Singular	$|\mathbf{A}| = 0$	The matrix has zero determinant (and no inverse).
Non-singular	$|\mathbf{A}| \neq 0$	The matrix has a non-zero determinant (and an inverse).
Defective		The $N \times N$ matrix has fewer than N linearly independent eigenvectors.

- Normal matrices include real symmetric, orthogonal, Hermitian and unitary matrices.
- $|\mathbf{A}^T| = |\mathbf{A}|$; $|\mathbf{A}^{-1}| = |\mathbf{A}|^{-1}$.
- The determinant of an orthogonal matrix is equal to ± 1.

4. *Effects of matrix operations on matrix products.*

Name	Effect on matrix product	Notes
Trace	$\text{Tr}(\mathbf{AB}\ldots\mathbf{G}) = \text{Tr}(\mathbf{B}\ldots\mathbf{GA})$	The product matrix $\mathbf{AB}\ldots\mathbf{G}$ must be square, though the individual matrices need not be. However, they must be compatible.

Name	Effect on matrix product	Notes
Determinant	$\|AB\ldots G\| = \|A\|\|B\|\ldots\|G\|$	All matrices must be $N \times N$. Product is singular $\Leftrightarrow$ one or more of the individual matrices is singular.
Transpose	$(AB\ldots G)^T = G^T\ldots B^T A^T$	Matrices must be compatible but need not be square.
Complex conjugate	$(AB\ldots G)^* = A^* B^* \ldots G^*$	
Hermitian conjugate	$(AB\ldots G)^\dagger = G^\dagger \ldots B^\dagger A^\dagger$	Matrices must be compatible but need not be square.
Inverse	$(AB\ldots G)^{-1} = G^{-1}\ldots B^{-1}A^{-1}$	All matrices must be $N \times N$ and non-singular.

5. *Eigenvectors and eigenvalues*
   - The eigenvectors $\mathbf{x}^i$ and eigenvalues λ_i of a matrix $\mathbf{A}$ are defined by $\mathbf{A}\mathbf{x}^i = \lambda_i \mathbf{x}^i$.
   - The eigenvectors of a normal matrix corresponding to different eigenvalues are orthogonal.
   - The eigenvalues of an Hermitian (or real orthogonal) matrix are real.
   - The eigenvalues of a unitary matrix have unit modulus.
   - Two normal matrices commute $\Leftrightarrow$ they have a set of eigenvectors in common.
   - $\sum_i \lambda_i = \operatorname{Tr} \mathbf{A}, \quad \prod_i \lambda_i = \|\mathbf{A}\|$.
   - A square matrix is singular $\Leftrightarrow$ at least one of its eigenvalues is zero.

6. *To find the eigenvalues and eigenvectors of a square matrix $\mathbf{A}$ and diagonalise it.*
   (i) Solve the characteristic equation $\|\mathbf{A} - \lambda \mathbf{I}\| = 0$ for N values of λ, checking that $\sum_{i=1}^{N} \lambda_i = \operatorname{Tr} \mathbf{A}$.
   (ii) For each i, solve $\mathbf{A}\mathbf{x}^i = \lambda_i \mathbf{x}^i$ for $\mathbf{x}^i$.
   (iii) Construct the unitary matrix $\mathbf{S}$ whose columns are the normalised eigenvectors $\hat{\mathbf{x}}^i$ of $\mathbf{A}$.
   (iv) Then $\mathbf{A}' = \mathbf{S}^{-1}\mathbf{A}\mathbf{S} = \mathbf{S}^\dagger \mathbf{A}\mathbf{S}$ is diagonal, with diagonal elements λ_i $(i = 1, \ldots, N)$.

7. *Quadratic forms and surfaces for $N = 3$.*
   - The quadratic expression $Q(\mathbf{x}) = \mathbf{x}^T \mathbf{A}\mathbf{x}$, with $\mathbf{A}$ symmetric, can be put in the form $\sum_{n=1}^{N} \lambda_i (x'_i)^2$ using an real orthogonal change of basis $\mathbf{x} = \mathbf{S}\mathbf{x}'$, where $\mathbf{S}$ is as described in the previous section.
   - The equation $Q(\mathbf{x}) = 1$ represents a quadric surface whose principal axes lie in the directions of the eigenvectors $\mathbf{x}^i$ of $\mathbf{A}$ and have lengths $(\sqrt{|\lambda_i|})^{-1}$.

- If all the λ_i are positive the quadric surface is an ellipsoid; if one or two are negative, its cross-sections are a mixture of ellipses and hyperbolas. A zero eigenvalue gives rise to a 'cylinder' whose axis lies along the corresponding eigenvector direction.

8. *Simultaneous linear equations,* $\mathsf{A}\mathsf{x} = \mathsf{b}$.
   The Cramer determinant Δ_i is $|\mathsf{A}|$ but with the ith column of A replaced by the vector b.

$|\mathsf{A}|$	b	Δ_i	Number of solutions
$\neq 0$	$\neq 0$	–	One non-trivial, $\mathsf{x} = \mathsf{A}^{-1}\mathsf{b}$
	$= 0$	–	Only trivial $\mathsf{x} = 0$
$= 0$	$\neq 0$	all $\Delta_i = 0$	Infinite number
		at least one $\Delta_i \neq 0$	None
	$= 0$	–	Infinite number

Solution methods:
- Direct inversion, $\mathsf{x} = \mathsf{A}^{-1}\mathsf{b}$.
- *LU* decomposition: find a lower diagonal matrix L and an upper diagonal matrix U such that $\mathsf{A} = \mathsf{LU}$. Then solve, successively, $\mathsf{Ly} = \mathsf{b}$ and $\mathsf{Ux} = \mathsf{y}$ to obtain x.
- Cholesky decomposition: if A is symmetric and positive definite ($\mathsf{x}^\dagger \mathsf{A} \mathsf{x} > 0$ for all non-zero x) then find a lower diagonal matrix L such that $\mathsf{A} = \mathsf{LL}^\dagger$ and proceed as in *LU* decomposition.
- Cramer's rule: $x_i = \Delta_i/|\mathsf{A}|$.

PROBLEMS

10.1. Which of the following statements about linear vector spaces are true? Where a statement is false, give a counter-example to demonstrate this.
   (a) Non-singular $N \times N$ matrices form a vector space of dimension N^2.
   (b) Singular $N \times N$ matrices form a vector space of dimension N^2.
   (c) Complex numbers form a vector space of dimension 2.
   (d) Polynomial functions of x form an infinite-dimensional vector space.
   (e) Series $\{a_0, a_1, a_2, \ldots, a_N\}$ for which $\sum_{n=0}^{N} |a_n|^2 = 1$ form an N-dimensional vector space.
   (f) Absolutely convergent series form an infinite-dimensional vector space.
   (g) Convergent series with terms of alternating sign form an infinite-dimensional vector space.

10.2. Consider the matrices

$$\text{(a) } B = \begin{pmatrix} 0 & -i & i \\ i & 0 & -i \\ -i & i & 0 \end{pmatrix}, \quad \text{(b) } C = \frac{1}{\sqrt{8}} \begin{pmatrix} \sqrt{3} & -\sqrt{2} & -\sqrt{3} \\ 1 & \sqrt{6} & -1 \\ 2 & 0 & 2 \end{pmatrix}.$$

Are they (i) real, (ii) diagonal, (iii) symmetric, (iv) antisymmetric, (v) singular, (vi) orthogonal, (vii) Hermitian, (viii) anti-Hermitian, (ix) unitary, (x) normal?

10.3. By considering the matrices

$$A = \begin{pmatrix} 1 & 0 \\ 0 & 0 \end{pmatrix}, \quad B = \begin{pmatrix} 0 & 0 \\ 3 & 4 \end{pmatrix},$$

show that $AB = 0$ does *not* imply that either A or B is the zero matrix, but that it does imply that at least one of them is singular.

10.4. Evaluate the determinants

$$\text{(a) } \begin{vmatrix} a & h & g \\ h & b & f \\ g & f & c \end{vmatrix}, \quad \text{(b) } \begin{vmatrix} 1 & 0 & 2 & 3 \\ 0 & 1 & -2 & 1 \\ 3 & -3 & 4 & -2 \\ -2 & 1 & -2 & 1 \end{vmatrix}$$

and

$$\text{(c) } \begin{vmatrix} gc & ge & a+ge & gb+ge \\ 0 & b & b & b \\ c & e & e & b+e \\ a & b & b+f & b+d \end{vmatrix}.$$

10.5. Using the properties of determinants, solve with a minimum of calculation the following equations for x:

$$\text{(a) } \begin{vmatrix} x & a & a & 1 \\ a & x & b & 1 \\ a & b & x & 1 \\ a & b & c & 1 \end{vmatrix} = 0, \quad \text{(b) } \begin{vmatrix} x+2 & x+4 & x-3 \\ x+3 & x & x+5 \\ x-2 & x-1 & x+1 \end{vmatrix} = 0.$$

10.6. This problem considers a crystal whose unit cell has base vectors that are not necessarily mutually orthogonal.

(a) The basis vectors of the unit cell of a crystal, with the origin O at one corner, are denoted by $\mathbf{e}_1$, $\mathbf{e}_2$, $\mathbf{e}_3$. The matrix G has elements G_{ij}, where $G_{ij} = \mathbf{e}_i \cdot \mathbf{e}_j$ and H_{ij} are the elements of the matrix $H \equiv G^{-1}$. Show that the vectors $\mathbf{f}_i = \sum_j H_{ij} \mathbf{e}_j$ are the reciprocal vectors and that $H_{ij} = \mathbf{f}_i \cdot \mathbf{f}_j$.

(b) If the vectors $\mathbf{u}$ and $\mathbf{v}$ are given by

$$\mathbf{u} = \sum_i u_i \mathbf{e}_i, \quad \mathbf{v} = \sum_i v_i \mathbf{f}_i,$$

obtain expressions for $|\mathbf{u}|$, $|\mathbf{v}|$ and $\mathbf{u} \cdot \mathbf{v}$.

(c) If the basis vectors are each of length a and the angle between each pair is $\pi/3$, write down G and hence obtain H.
(d) Calculate (i) the length of the normal from O onto the plane containing the points $p^{-1}\mathbf{e}_1, q^{-1}\mathbf{e}_2, r^{-1}\mathbf{e}_3$ and (ii) the angle between this normal and $\mathbf{e}_1$.

10.7. Prove the following results involving Hermitian matrices.
(a) If A is Hermitian and U is unitary then $U^{-1}AU$ is Hermitian.
(b) If A is anti-Hermitian then iA is Hermitian.
(c) The product of two Hermitian matrices A and B is Hermitian if and only if A and B commute.
(d) If S is a real antisymmetric matrix then $A = (I - S)(I + S)^{-1}$ is orthogonal. If A is given by

$$A = \begin{pmatrix} \cos\theta & \sin\theta \\ -\sin\theta & \cos\theta \end{pmatrix}$$

then find the matrix S that is needed to express A in the above form.
(e) If K is skew-Hermitian, i.e. $K^\dagger = -K$, then $V = (I + K)(I - K)^{-1}$ is unitary.

10.8. A and B are real non-zero 3×3 matrices and satisfy the equation

$$(AB)^T + B^{-1}A = 0.$$

(a) Prove that if B is orthogonal then A is antisymmetric.
(b) Without assuming that B is orthogonal, prove that A is singular.

10.9. The *commutator* $[X, Y]$ of two matrices is defined by the equation

$$[X, Y] = XY - YX.$$

Two anticommuting matrices A and B satisfy

$$A^2 = I, \qquad B^2 = I, \qquad [A, B] = 2iC.$$

(a) Prove that $C^2 = I$ and that $[B, C] = 2iA$.
(b) Evaluate $[[[A, B], [B, C]], [A, B]]$.

10.10. The four matrices S_x, S_y, S_z and I are defined by

$$S_x = \begin{pmatrix} 0 & 1 \\ 1 & 0 \end{pmatrix}, \qquad S_y = \begin{pmatrix} 0 & -i \\ i & 0 \end{pmatrix},$$

$$S_z = \begin{pmatrix} 1 & 0 \\ 0 & -1 \end{pmatrix}, \qquad I = \begin{pmatrix} 1 & 0 \\ 0 & 1 \end{pmatrix},$$

where $i^2 = -1$. Show that $S_x^2 = I$ and $S_xS_y = iS_z$, and obtain similar results by permuting x, y and z. Given that $\mathbf{v}$ is a vector with Cartesian components

(v_x, v_y, v_z), the matrix $S(v)$ is defined as

$$S(v) = v_x S_x + v_y S_y + v_z S_z.$$

Prove that, for general non-zero vectors **a** and **b**,

$$S(a)S(b) = a \cdot b \, I + i \, S(a \times b).$$

Without further calculation, deduce that $S(a)$ and $S(b)$ commute if and only if **a** and **b** are parallel vectors.

10.11. A general triangle has angles α, β and γ and corresponding opposite sides a, b and c. Express the length of each side in terms of the lengths of the other two sides and the relevant cosines, writing the relationships in matrix and vector form, using the vectors having components a, b, c and $\cos\alpha, \cos\beta, \cos\gamma$. Invert the matrix and hence deduce the cosine-law expressions involving α, β and γ.

10.12. Given a matrix

$$A = \begin{pmatrix} 1 & \alpha & 0 \\ \beta & 1 & 0 \\ 0 & 0 & 1 \end{pmatrix},$$

where α and β are non-zero complex numbers, find its eigenvalues and eigenvectors. Find the respective conditions for (a) the eigenvalues to be real and (b) the eigenvectors to be orthogonal. Show that the conditions are jointly satisfied if and only if A is Hermitian.

10.13. Determine which of the matrices below are mutually commuting, and, for those that are, demonstrate that they have a complete set of eigenvectors in common:

$$A = \begin{pmatrix} 6 & -2 \\ -2 & 9 \end{pmatrix}, \quad B = \begin{pmatrix} 1 & 8 \\ 8 & -11 \end{pmatrix},$$

$$C = \begin{pmatrix} -9 & -10 \\ -10 & 5 \end{pmatrix}, \quad D = \begin{pmatrix} 14 & 2 \\ 2 & 11 \end{pmatrix}.$$

10.14. Do the following sets of equations have non-zero solutions? If so, find them.
(a) $3x + 2y + z = 0$, $\quad x - 3y + 2z = 0$, $\quad 2x + y + 3z = 0$.
(b) $2x = b(y+z)$, $\quad x = 2a(y-z)$, $\quad x = (6a-b)y - (6a+b)z$.

10.15. Solve the simultaneous equations

$$2x + 3y + z = 11,$$
$$x + y + z = 6,$$
$$5x - y + 10z = 34.$$

Problems

10.16. Solve the following simultaneous equations for x_1, x_2 and x_3, using matrix methods:

$$x_1 + 2x_2 + 3x_3 = 1,$$
$$3x_1 + 4x_2 + 5x_3 = 2,$$
$$x_1 + 3x_2 + 4x_3 = 3.$$

10.17. Show that the following equations have solutions only if $\eta = 1$ or 2, and find them in these cases:

$$x + y + z = 1,$$
$$x + 2y + 4z = \eta,$$
$$x + 4y + 10z = \eta^2.$$

10.18. Find the condition(s) on α such that the simultaneous equations

$$x_1 + \alpha x_2 = 1,$$
$$x_1 - x_2 + 3x_3 = -1,$$
$$2x_1 - 2x_2 + \alpha x_3 = -2$$

have (a) exactly one solution, (b) no solutions or (c) an infinite number of solutions; give all solutions where they exist.

10.19. Make an LU decomposition of the matrix

$$A = \begin{pmatrix} 3 & 6 & 9 \\ 1 & 0 & 5 \\ 2 & -2 & 16 \end{pmatrix}$$

and hence solve $Ax = b$, where (i) $b = (21 \ \ 9 \ \ 28)^T$, (ii) $b = (21 \ \ 7 \ \ 22)^T$.

10.20. Make an LU decomposition of the matrix

$$A = \begin{pmatrix} 2 & -3 & 1 & 3 \\ 1 & 4 & -3 & -3 \\ 5 & 3 & -1 & -1 \\ 3 & -6 & -3 & 1 \end{pmatrix}.$$

Hence solve $Ax = b$ for (i) $b = (-4 \ \ 1 \ \ 8 \ \ -5)^T$, (ii) $b = (-10 \ \ 0 \ -3 \ \ -24)^T$.

Deduce that $\det A = -160$ and confirm this by direct calculation.

10.21. Use the Cholesky decomposition method to determine whether the following matrices are positive definite. For each that is, determine the corresponding lower diagonal matrix L:

$$A = \begin{pmatrix} 2 & 1 & 3 \\ 1 & 3 & -1 \\ 3 & -1 & 1 \end{pmatrix}, \quad B = \begin{pmatrix} 5 & 0 & \sqrt{3} \\ 0 & 3 & 0 \\ \sqrt{3} & 0 & 3 \end{pmatrix}.$$

10.22. Find the eigenvalues and a set of eigenvectors of the matrix

$$\begin{pmatrix} 1 & 3 & -1 \\ 3 & 4 & -2 \\ -1 & -2 & 2 \end{pmatrix}.$$

Verify that its eigenvectors are mutually orthogonal.

10.23. Find three real orthogonal column matrices, each of which is a simultaneous eigenvector of

$$A = \begin{pmatrix} 0 & 0 & 1 \\ 0 & 1 & 0 \\ 1 & 0 & 0 \end{pmatrix} \quad \text{and} \quad B = \begin{pmatrix} 0 & 1 & 1 \\ 1 & 0 & 1 \\ 1 & 1 & 0 \end{pmatrix}.$$

10.24. Use the results of the first worked example in Section 10.13 to evaluate, without repeated matrix multiplication, the expression $A^6 x$, where $x = (2 \quad 4 \quad -1)^T$ and A is the matrix given in the example.

10.25. Given that A is a real symmetric matrix with normalised eigenvectors e^i, obtain the coefficients α_i involved when column matrix x, which is the solution of

$$Ax - \mu x = v, \qquad (*)$$

is expanded as $x = \sum_i \alpha_i e^i$. Here μ is a given constant and v is a given column matrix.
(a) Solve $(*)$ when

$$A = \begin{pmatrix} 2 & 1 & 0 \\ 1 & 2 & 0 \\ 0 & 0 & 3 \end{pmatrix},$$

$\mu = 2$ and $v = (1 \quad 2 \quad 3)^T$.
(b) Would $(*)$ have a solution if $\mu = 1$ and (i) $v = (1 \quad 2 \quad 3)^T$, (ii) $v = (2 \quad 2 \quad 3)^T$? Where it does, find it.

10.26. Demonstrate that the matrix

$$A = \begin{pmatrix} 2 & 0 & 0 \\ -6 & 4 & 4 \\ 3 & -1 & 0 \end{pmatrix}$$

is defective, i.e. does not have three linearly independent eigenvectors, by showing the following:

Problems

(a) its eigenvalues are degenerate and, in fact, all equal;
(b) any eigenvector has the form $(\mu \quad (3\mu - 2\nu) \quad \nu)^T$;
(c) if two pairs of values, μ_1, ν_1 and μ_2, ν_2, define two independent eigenvectors $\mathbf{v}_1$ and $\mathbf{v}_2$, then *any* third similarly defined eigenvector $\mathbf{v}_3$ can be written as a linear combination of $\mathbf{v}_1$ and $\mathbf{v}_2$, i.e.

$$\mathbf{v}_3 = a\mathbf{v}_1 + b\mathbf{v}_2,$$

where

$$a = \frac{\mu_3 \nu_2 - \mu_2 \nu_3}{\mu_1 \nu_2 - \mu_2 \nu_1} \quad \text{and} \quad b = \frac{\mu_1 \nu_3 - \mu_3 \nu_1}{\mu_1 \nu_2 - \mu_2 \nu_1}.$$

Illustrate (c) using the example $(\mu_1, \nu_1) = (1, 1)$, $(\mu_2, \nu_2) = (1, 2)$ and $(\mu_3, \nu_3) = (0, 1)$.

Show further that any matrix of the form

$$\begin{pmatrix} 2 & 0 & 0 \\ 6n - 6 & 4 - 2n & 4 - 4n \\ 3 - 3n & n - 1 & 2n \end{pmatrix}$$

is defective, with the same eigenvalues and eigenvectors as A.

10.27. By finding the eigenvectors of the Hermitian matrix

$$\mathsf{H} = \begin{pmatrix} 10 & 3i \\ -3i & 2 \end{pmatrix},$$

construct a unitary matrix U such that $\mathsf{U}^\dagger \mathsf{H} \mathsf{U} = \Lambda$, where Λ is a real diagonal matrix.

10.28. Use the stationary properties of quadratic forms to determine the maximum and minimum values taken by the expression

$$Q = 5x^2 + 4y^2 + 4z^2 + 2xz + 2xy$$

on the unit sphere, $x^2 + y^2 + z^2 = 1$. For what values of x, y, z do they occur?

10.29. Given that the matrix

$$\mathsf{A} = \begin{pmatrix} 2 & -1 & 0 \\ -1 & 2 & -1 \\ 0 & -1 & 2 \end{pmatrix}$$

has two eigenvectors of the form $(1 \quad y \quad 1)^T$, use the stationary property of the expression $J(\mathbf{x}) = \mathbf{x}^T \mathsf{A} \mathbf{x}/(\mathbf{x}^T \mathbf{x})$ to obtain the corresponding eigenvalues. Deduce the third eigenvalue.

10.30. Find the lengths of the semi-axes of the ellipse

$$73x^2 + 72xy + 52y^2 = 100$$

and determine their orientations.

10.31. The equation of a particular conic section is

$$Q \equiv 8x_1^2 + 8x_2^2 - 6x_1 x_2 = 110.$$

Determine the type of conic section this represents, the orientation of its principal axes and relevant lengths in the directions of these axes.

10.32. Show that the quadratic surface

$$5x^2 + 11y^2 + 5z^2 - 10yz + 2xz - 10xy = 4$$

is an ellipsoid with semi-axes of lengths 2, 1 and 0.5. Find the direction of its longest axis.

10.33. Find the direction of the axis of symmetry of the quadratic surface

$$7x^2 + 7y^2 + 7z^2 - 20yz - 20xz + 20xy = 3.$$

10.34. For the following matrices, find the eigenvalues and sufficient of the eigenvectors to be able to describe the quadratic surfaces associated with them:

(a) $\begin{pmatrix} 5 & 1 & -1 \\ 1 & 5 & 1 \\ -1 & 1 & 5 \end{pmatrix}$, (b) $\begin{pmatrix} 1 & 2 & 2 \\ 2 & 1 & 2 \\ 2 & 2 & 1 \end{pmatrix}$, (c) $\begin{pmatrix} 1 & 2 & 1 \\ 2 & 4 & 2 \\ 1 & 2 & 1 \end{pmatrix}$.

10.35. This problem demonstrates the reverse of the usual procedure of diagonalising a matrix.
(a) Rearrange the result $A' = S^{-1}AS$ of Section 10.15 to express the original matrix A in terms of the unitary matrix S and the diagonal matrix A'. Hence show how to construct a matrix A that has given eigenvalues and given (orthogonal) column matrices as its eigenvectors.
(b) Find the matrix that has as eigenvectors $(1\ \ 2\ \ 1)^T$, $(1\ \ -1\ \ 1)^T$ and $(1\ \ 0\ \ -1)^T$, with corresponding eigenvalues λ, μ and ν.
(c) Try a particular case, say $\lambda = 3$, $\mu = -2$ and $\nu = 1$, and verify by explicit solution that the matrix so found does have these eigenvalues.

10.36. Find an orthogonal transformation that takes the quadratic form

$$Q \equiv -x_1^2 - 2x_2^2 - x_3^2 + 8x_2 x_3 + 6x_1 x_3 + 8x_1 x_2$$

into the form

$$\mu_1 y_1^2 + \mu_2 y_2^2 - 4y_3^2,$$

and determine μ_1 and μ_2 (see Section 10.16).

10.37. The equation

$$|A|\epsilon_{lmn} = A_{li} A_{mj} A_{nk} \epsilon_{ijk}$$

is a more general form of the expression (10.45) for the determinant of a 3×3 matrix A. The latter could have been written as

$$|\mathsf{A}| = \epsilon_{ijk} A_{i1} A_{j2} A_{k3},$$

whilst the former removes the explicit mention of 1, 2, 3 at the expense of an additional Levi-Civita symbol. The definition can be readily extended to cover a general $N \times N$ matrix by using an ϵ with N subscripts; the symbol then has value -1 or $+1$ according to whether $i, j, k, \ldots, n$ is an odd or even permutation of $1, 2, 3, \ldots, N$, respectively.

Use the form given in (∗) to prove properties (i), (iii), (v), (vi) and (vii) of determinants stated in Section 10.7.1. Property (iv) is obvious by inspection. For definiteness take $N = 3$, but convince yourself that your methods of proof would be valid for any positive integer N.

HINTS AND ANSWERS

10.1. (a) False. $\mathsf{0}_N$, the $N \times N$ null matrix, is *not* non-singular.
(b) False. Consider the sum of $\begin{pmatrix} 1 & 0 \\ 0 & 0 \end{pmatrix}$ and $\begin{pmatrix} 0 & 0 \\ 0 & 1 \end{pmatrix}$.
(c) True.
(d) True.
(e) False. Consider $b_n = a_n + a_n$ for which $\sum_{n=0}^{N} |b_n|^2 = 4 \neq 1$, or note that there is no zero vector with unit norm.
(f) True.
(g) False. Consider the two series defined by

$$a_0 = \tfrac{1}{2}, \qquad a_n = 2(-\tfrac{1}{2})^n \text{ for } n \geq 1; \qquad b_n = -(-\tfrac{1}{2})^n \text{ for } n \geq 0.$$

The series that is the sum of $\{a_n\}$ and $\{b_n\}$ does not have alternating signs and so closure does not hold.

10.3. Use the property of the determinant of a matrix product.

10.5. (a) $x = a, b$ or c; (b) $x = -1$; the equation is linear in x.

10.7. (d) $\mathsf{S} = \begin{pmatrix} 0 & -\tan(\theta/2) \\ \tan(\theta/2) & 0 \end{pmatrix}$.
(e) Note that $(\mathsf{I} + \mathsf{K})(\mathsf{I} - \mathsf{K}) = \mathsf{I} - \mathsf{K}^2 = (\mathsf{I} - \mathsf{K})(\mathsf{I} + \mathsf{K})$.

10.9. (b) $32i\mathsf{A}$.

10.11. $a = b \cos \gamma + c \cos \beta$, and cyclic permutations; $a^2 = b^2 + c^2 - 2bc \cos \alpha$, and cyclic permutations.

10.13. C does not commute with the others; A, B and D have $(1 \ -2)^\mathsf{T}$ and $(2 \ 1)^\mathsf{T}$ as common eigenvectors.

10.15. $x = 3, y = 1, z = 2$.

Matrices and vector spaces

10.17. First show that A is singular. $\eta = 1$, $x = 1 + 2z$, $y = -3z$; $\eta = 2$, $x = 2z$, $y = 1 - 3z$.

10.19. $L = (1, 0, 0; \frac{1}{3}, 1, 0; \frac{2}{3}, 3, 1)$, $U = (3, 6, 9; 0, -2, 2; 0, 0, 4)$.
(i) $x = (-1 \quad 1 \quad 2)^T$. (ii) $x = (-3 \quad 2 \quad 2)^T$.

10.21. A is not positive definite as L_{33} is calculated to be $\sqrt{-6}$.
$B = LL^T$, where the non-zero elements of L are
$L_{11} = \sqrt{5}$, $L_{31} = \sqrt{3/5}$, $L_{22} = \sqrt{3}$, $L_{33} = \sqrt{12/5}$.

10.23. For A: $(1 \quad 0 \quad -1)^T$, $(1 \quad \alpha_1 \quad 1)^T$, $(1 \quad \alpha_2 \quad 1)^T$.
For B: $(1 \quad 1 \quad 1)^T$, $(\beta_1 \quad \gamma_1 \quad -\beta_1 - \gamma_1)^T$, $(\beta_2 \quad \gamma_2 \quad -\beta_2 - \gamma_2)^T$.
The α_i, β_i and γ_i are arbitrary.
Simultaneous and orthogonal: $(1 \quad 0 \quad -1)^T$, $(1 \quad 1 \quad 1)^T$, $(1 \quad -2 \quad 1)^T$.

10.25. $\alpha_j = (\mathbf{v} \cdot \mathbf{e}^{j*})/(\lambda_j - \mu)$, where λ_j is the eigenvalue corresponding to $\mathbf{e}^j$.
(a) $x = (2 \quad 1 \quad 3)^T$.
(b) Since μ is equal to one of A's eigenvalues λ_j, the equation only has a solution if $\mathbf{v} \cdot \mathbf{e}^{j*} = 0$; (i) no solution; (ii) $x = (1 \quad 1 \quad 3/2)^T$.

10.27. $U = (10)^{-1/2}(1, 3i; 3i, 1)$, $\Lambda = (1, 0; 0, 11)$.

10.29. $J = (2y^2 - 4y + 4)/(y^2 + 2)$ with stationary values at $y = \pm\sqrt{2}$ and corresponding eigenvalues $2 \mp \sqrt{2}$. From the trace property of A, the third eigenvalue equals 2.

10.31. Ellipse; $\theta = \pi/4$, $a = \sqrt{22}$; $\theta = 3\pi/4$, $b = \sqrt{10}$.

10.33. The direction of the eigenvector having the unrepeated eigenvalue is $(1, 1, -1)/\sqrt{3}$.

10.35. (a) $A = SA'S^\dagger$, where S is the matrix whose columns are the eigenvectors of the matrix A to be constructed, and $A' = \text{diag}(\lambda, \mu, \nu)$.
(b) $A = (\lambda + 2\mu + 3\nu, \, 2\lambda - 2\mu, \, \lambda + 2\mu - 3\nu; \, 2\lambda - 2\mu, \, 4\lambda + 2\mu, \, 2\lambda - 2\mu;$
$\lambda + 2\mu - 3\nu, \, 2\lambda - 2\mu, \, \lambda + 2\mu + 3\nu)$.
(c) $\frac{1}{3}(1, 5, -2; 5, 4, 5; -2, 5, 1)$.

10.37. This solution is fuller than most and to save even more additional text the summation convention (Appendix D) is used.
(i) We write the expression for $|A^T|$ using the given formalism, recalling that $(A^T)_{ij} = (A)_{ji}$. We then multiply both sides by ϵ_{lmn} and sum over l, m and n:

$$|A^T|\epsilon_{lmn} = A_{il} A_{jm} A_{kn} \epsilon_{ijk},$$
$$|A^T|\epsilon_{lmn}\epsilon_{lmn} = A_{il} A_{jm} A_{kn} \epsilon_{lmn} \epsilon_{ijk},$$
$$= |A|\epsilon_{ijk}\epsilon_{ijk},$$
$$|A^T| = |A|.$$

In the third line we have used the definition of $|A|$ (with the roles of the sets of dummy variables $\{i, j, k\}$ and $\{l, m, n\}$ interchanged), and in the fourth line we have cancelled the scalar quantity $\epsilon_{lmn}\epsilon_{lmn} = \epsilon_{ijk}\epsilon_{ijk}$; the value of this scalar is $N(N-1)$, but that is irrelevant here.

(iii) Every non-zero term on the RHS of $(*)$ contains any particular row index once and only once. The same can be said for the Levi-Civita symbol on the LHS. Thus, interchanging two rows is equivalent to interchanging two of the subscripts of ϵ_{lmn} and thereby reversing its sign. Consequently, the whole RHS changes sign and the magnitude of $|\mathbf{A}|$ remains the same, though its sign is changed.

(v) If, say, $A_{pi} = \lambda A_{pj}$, for some particular pair of values i and j and all p, then in the (multiple) summation on the RHS of $(*)$ each A_{nk} appears multiplied by (with no summation over i and j)

$$\epsilon_{ijk} A_{li} A_{mj} + \epsilon_{jik} A_{lj} A_{mi} = \epsilon_{ijk}\lambda A_{lj} A_{mj} + \epsilon_{jik} A_{lj}\lambda A_{mj} = 0,$$

since $\epsilon_{ijk} = -\epsilon_{jik}$. Consequently, grouped in this way, all pairs of terms contribute nothing to the sum and $|\mathbf{A}| = 0$.

(vi) Consider the matrix $\mathbf{B}$ whose m, jth element is defined by $B_{mj} = A_{mj} + \lambda A_{pj}$, where $p \neq m$. The only case that needs detailed analysis is when l, m and n are all different. Since $p \neq m$ it must be the same as either l or n; suppose that $p = l$. The determinant of $\mathbf{B}$ is given by

$$|\mathbf{B}|\epsilon_{lmn} = A_{li}(A_{mj} + \lambda A_{lj})A_{nk}\epsilon_{ijk}$$
$$= A_{li} A_{mj} A_{nk}\epsilon_{ijk} + \lambda A_{li} A_{lj} A_{nk}\epsilon_{ijk}$$
$$= |\mathbf{A}|\epsilon_{lmn} + \lambda 0,$$

where we have used the row equivalent of the intermediate result obtained for columns in (v). Thus we conclude that $|\mathbf{B}| = |\mathbf{A}|$.

(vii) If $\mathbf{X} = \mathbf{AB}$, then

$$|\mathbf{X}|\epsilon_{lmn} = A_{lx} B_{xi} A_{my} B_{yj} A_{nz} B_{zk}\epsilon_{ijk}.$$

Multiply both sides by ϵ_{lmn} and sum over l, m and n:

$$|\mathbf{X}|\epsilon_{lmn}\epsilon_{lmn} = \epsilon_{lmn} A_{lx} A_{my} A_{nz}\,\epsilon_{ijk} B_{xi} B_{yj} B_{zk}$$
$$= \epsilon_{xyz}|\mathbf{A}^\mathrm{T}|\epsilon_{xyz}|\mathbf{B}|,$$
$$\Rightarrow \quad |\mathbf{X}| = |\mathbf{A}^\mathrm{T}|\,|\mathbf{B}| = |\mathbf{A}|\,|\mathbf{B}|, \quad \text{using result (i)}.$$

To obtain the last line we have cancelled the non-zero scalar $\epsilon_{lmn}\epsilon_{lmn} = \epsilon_{xyz}\epsilon_{xyz}$ from both sides, as we did in the proof of result (i).

The extension to the product of any number of matrices is obvious. Replacing $\mathbf{B}$ by $\mathbf{CD}$ or by $\mathbf{DC}$ and applying the result just proved extends it to a product of three matrices. Extension to any higher number is done in the same way.

11
Vector calculus

In Chapter 9 we discussed the algebra of vectors and in Chapter 10 we considered how to transform one vector into another using a linear operator. In this chapter and the next we discuss the calculus of vectors, i.e. the differentiation and integration both of vectors describing particular bodies, such as the velocity of a particle, and of vector fields, in which a vector is defined as a function of the coordinates throughout some volume (one-, two- or three-dimensional). Since the aim of this chapter is to develop methods for handling multi-dimensional physical situations, we will assume throughout that the functions with which we have to deal have sufficiently amenable mathematical properties, in particular that they are continuous and differentiable.

11.1 Differentiation of vectors

Let us consider a vector $\mathbf{a}$ that is a function of a scalar variable u. By this we mean that with each value of u we associate a vector $\mathbf{a}(u)$. For example, in Cartesian coordinates $\mathbf{a}(u) = a_x(u)\mathbf{i} + a_y(u)\mathbf{j} + a_z(u)\mathbf{k}$, where $a_x(u)$, $a_y(u)$ and $a_z(u)$ are scalar functions of u and are the components of the vector $\mathbf{a}(u)$ in the x-, y- and z-directions respectively. We note that if $\mathbf{a}(u)$ is continuous at some point $u = u_0$ then this implies that each of the Cartesian components $a_x(u)$, $a_y(u)$ and $a_z(u)$ is also continuous there.

Let us consider the derivative of the vector function $\mathbf{a}(u)$ with respect to u. The derivative of a vector function is defined in a similar manner to the ordinary derivative of a scalar function $f(x)$ given in Chapter 3. The small change in the vector $\mathbf{a}(u)$ resulting from a small change Δu in the value of u is given by $\Delta \mathbf{a} = \mathbf{a}(u + \Delta u) - \mathbf{a}(u)$ (see Figure 11.1). The derivative of $\mathbf{a}(u)$ with respect to u is defined to be

$$\frac{d\mathbf{a}}{du} = \lim_{\Delta u \to 0} \frac{\mathbf{a}(u + \Delta u) - \mathbf{a}(u)}{\Delta u}, \tag{11.1}$$

assuming that the limit exists, in which case $\mathbf{a}(u)$ is said to be differentiable at that point. Note that $d\mathbf{a}/du$ is also a vector, which is not, in general, parallel to $\mathbf{a}(u)$. In Cartesian coordinates, the derivative of the vector $\mathbf{a}(u) = a_x\mathbf{i} + a_y\mathbf{j} + a_z\mathbf{k}$ is given by

$$\frac{d\mathbf{a}}{du} = \frac{da_x}{du}\mathbf{i} + \frac{da_y}{du}\mathbf{j} + \frac{da_z}{du}\mathbf{k}.$$

Perhaps the simplest application of the above is to finding the velocity and acceleration of a particle in classical mechanics. If the time-dependent position vector of the particle with respect to the origin in Cartesian coordinates is given by $\mathbf{r}(t) = x(t)\mathbf{i} + y(t)\mathbf{j} + z(t)\mathbf{k}$

11.1 Differentiation of vectors

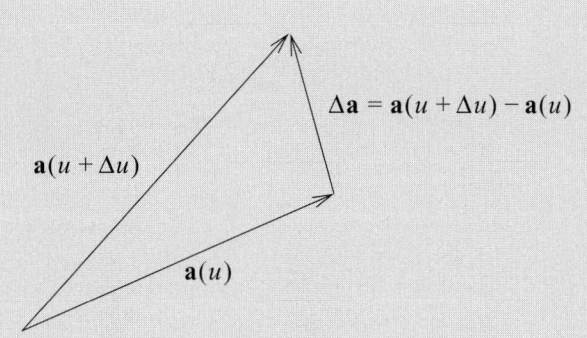

Figure 11.1 A small change in a vector $\mathbf{a}(u)$ resulting from a small change in u.

then the velocity of the particle is given by the vector

$$\mathbf{v}(t) = \frac{d\mathbf{r}}{dt} = \frac{dx}{dt}\mathbf{i} + \frac{dy}{dt}\mathbf{j} + \frac{dz}{dt}\mathbf{k}.$$

The direction of the velocity vector is along the tangent to the path $\mathbf{r}(t)$ at the instantaneous position of the particle, and its magnitude $|\mathbf{v}(t)|$ is equal to the speed of the particle. The acceleration of the particle is given in a similar manner by

$$\mathbf{a}(t) = \frac{d\mathbf{v}}{dt} = \frac{d^2x}{dt^2}\mathbf{i} + \frac{d^2y}{dt^2}\mathbf{j} + \frac{d^2z}{dt^2}\mathbf{k}.$$

These notions are illustrated in the following worked example.

Example The position vector of a particle at time t in Cartesian coordinates is given by $\mathbf{r}(t) = 2t^2\mathbf{i} + (3t - 2)\mathbf{j} + (3t^2 - 1)\mathbf{k}$. Find the speed of the particle at $t = 1$ and the component of its acceleration in the direction $\mathbf{s} = \mathbf{i} + 2\mathbf{j} + \mathbf{k}$.

The velocity and acceleration of the particle are given by

$$\mathbf{v}(t) = \frac{d\mathbf{r}}{dt} = 4t\mathbf{i} + 3\mathbf{j} + 6t\mathbf{k},$$

$$\mathbf{a}(t) = \frac{d\mathbf{v}}{dt} = 4\mathbf{i} + 6\mathbf{k}.$$

The speed of the particle at $t = 1$ is simply

$$|\mathbf{v}(1)| = \sqrt{4^2 + 3^2 + 6^2} = \sqrt{61}.$$

The acceleration of the particle is constant (i.e. independent of t) and its component in the direction $\mathbf{s}$ is given by

$$\mathbf{a} \cdot \hat{\mathbf{s}} = \frac{(4\mathbf{i} + 6\mathbf{k}) \cdot (\mathbf{i} + 2\mathbf{j} + \mathbf{k})}{\sqrt{1^2 + 2^2 + 1^2}} = \frac{5\sqrt{6}}{3}.$$

Note that the vector $\mathbf{s}$ had to be converted into the unit vector $\hat{\mathbf{s}}$, by dividing by its modulus, before it could be used to determine the component of $\mathbf{a}$ in its direction. ◀

Vector calculus

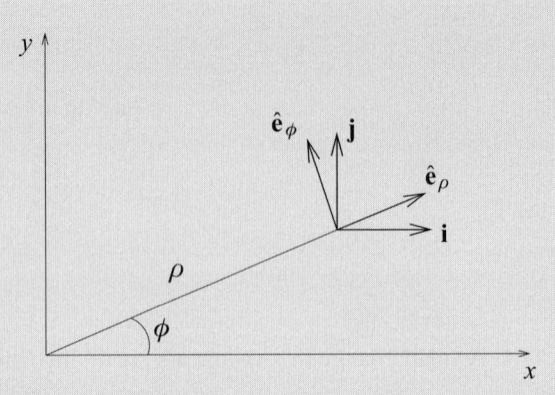

Figure 11.2 Unit basis vectors for two-dimensional Cartesian and plane polar coordinates.

In the case discussed above, **i**, **j** and **k** are fixed, time-independent basis vectors. This may not be true of basis vectors in general; when we are not using Cartesian coordinates the basis vectors themselves must also be differentiated. We discuss basis vectors for non-Cartesian coordinate systems in detail in Section 11.9. Nevertheless, as a simple example, let us now consider two-dimensional plane polar coordinates ρ, ϕ.

Referring to Figure 11.2, imagine holding ϕ fixed and moving radially outwards, i.e. in the direction of increasing ρ. Let us denote the unit vector in this direction by $\hat{\mathbf{e}}_\rho$. Similarly, imagine keeping ρ fixed and moving around a circle of fixed radius in the direction of increasing ϕ. Let us denote the unit vector tangent to the circle by $\hat{\mathbf{e}}_\phi$. The two vectors $\hat{\mathbf{e}}_\rho$ and $\hat{\mathbf{e}}_\phi$ are the basis vectors for this two-dimensional coordinate system, just as **i** and **j** are basis vectors for two-dimensional Cartesian coordinates. All these basis vectors are shown in Figure 11.2.

An important difference between the two sets of basis vectors is that, while **i** and **j** are constant in magnitude *and direction*, the vectors $\hat{\mathbf{e}}_\rho$ and $\hat{\mathbf{e}}_\phi$ have constant magnitudes but their directions change as ρ and ϕ vary. Therefore, when calculating the derivative of a vector written in polar coordinates we must also differentiate the basis vectors. One way of doing this is to express $\hat{\mathbf{e}}_\rho$ and $\hat{\mathbf{e}}_\phi$ in terms of **i** and **j**. From Figure 11.2, we see that

$$\hat{\mathbf{e}}_\rho = \cos\phi\,\mathbf{i} + \sin\phi\,\mathbf{j},$$
$$\hat{\mathbf{e}}_\phi = -\sin\phi\,\mathbf{i} + \cos\phi\,\mathbf{j}.$$

Since **i** and **j** are constant vectors, we find that the derivatives of the basis vectors $\hat{\mathbf{e}}_\rho$ and $\hat{\mathbf{e}}_\phi$ with respect to t are given by

$$\frac{d\hat{\mathbf{e}}_\rho}{dt} = -\sin\phi\frac{d\phi}{dt}\mathbf{i} + \cos\phi\frac{d\phi}{dt}\mathbf{j} = \dot\phi\,\hat{\mathbf{e}}_\phi, \tag{11.2}$$

$$\frac{d\hat{\mathbf{e}}_\phi}{dt} = -\cos\phi\frac{d\phi}{dt}\mathbf{i} - \sin\phi\frac{d\phi}{dt}\mathbf{j} = -\dot\phi\,\hat{\mathbf{e}}_\rho, \tag{11.3}$$

where the overdot is the conventional notation for differentiation with respect to time.

11.1 Differentiation of vectors

Example The position vector of a particle in plane polar coordinates is $\mathbf{r}(t) = \rho(t)\hat{\mathbf{e}}_\rho$. Find expressions for the velocity and acceleration of the particle in these coordinates.

By direct differentiation of a product or by using result (11.4) below, the velocity of the particle is given by

$$\mathbf{v}(t) = \dot{\mathbf{r}}(t) = \dot{\rho}\,\hat{\mathbf{e}}_\rho + \rho\,\dot{\hat{\mathbf{e}}}_\rho = \dot{\rho}\,\hat{\mathbf{e}}_\rho + \rho\dot{\phi}\,\hat{\mathbf{e}}_\phi,$$

where we have used (11.2). In a similar way its acceleration is given by

$$\begin{aligned}\mathbf{a}(t) &= \frac{d}{dt}(\dot{\rho}\,\hat{\mathbf{e}}_\rho + \rho\dot{\phi}\,\hat{\mathbf{e}}_\phi) \\ &= \ddot{\rho}\,\hat{\mathbf{e}}_\rho + \dot{\rho}\,\dot{\hat{\mathbf{e}}}_\rho + \dot{\rho}\dot{\phi}\,\hat{\mathbf{e}}_\phi + \rho\ddot{\phi}\,\hat{\mathbf{e}}_\phi + \rho\dot{\phi}\,\dot{\hat{\mathbf{e}}}_\phi \\ &= \ddot{\rho}\,\hat{\mathbf{e}}_\rho + \dot{\rho}(\dot{\phi}\,\hat{\mathbf{e}}_\phi) + \rho\dot{\phi}(-\dot{\phi}\,\hat{\mathbf{e}}_\rho) + \rho\ddot{\phi}\,\hat{\mathbf{e}}_\phi + \dot{\rho}\dot{\phi}\,\hat{\mathbf{e}}_\phi \\ &= (\ddot{\rho} - \rho\dot{\phi}^2)\,\hat{\mathbf{e}}_\rho + (\rho\ddot{\phi} + 2\dot{\rho}\dot{\phi})\,\hat{\mathbf{e}}_\phi.\end{aligned}$$

Here, (11.2) and (11.3) were used to go from the second line to the third.[1] ◀

11.1.1 Differentiation of composite vector expressions

In composite vector expressions each of the vectors or scalars involved may be a function of some scalar variable u, as we have seen. The derivatives of such expressions are easily found using the definition (11.1) and the rules of ordinary differential calculus. They may be summarised by the following, in which we assume that $\mathbf{a}$ and $\mathbf{b}$ are differentiable vector functions of a scalar u and that ϕ is a differentiable scalar function of u:

$$\frac{d}{du}(\phi \mathbf{a}) = \phi\frac{d\mathbf{a}}{du} + \frac{d\phi}{du}\mathbf{a}, \tag{11.4}$$

$$\frac{d}{du}(\mathbf{a} \cdot \mathbf{b}) = \mathbf{a} \cdot \frac{d\mathbf{b}}{du} + \frac{d\mathbf{a}}{du} \cdot \mathbf{b}, \tag{11.5}$$

$$\frac{d}{du}(\mathbf{a} \times \mathbf{b}) = \mathbf{a} \times \frac{d\mathbf{b}}{du} + \frac{d\mathbf{a}}{du} \times \mathbf{b}. \tag{11.6}$$

The order of the factors in the terms on the RHS of (11.6) is, of course, just as important as it is in the original vector product.

Example A particle of mass m with position vector $\mathbf{r}$ relative to some origin O experiences a force $\mathbf{F}$, which produces a torque (moment) $\mathbf{T} = \mathbf{r} \times \mathbf{F}$ about O. The angular momentum of the particle about O is given by $\mathbf{L} = \mathbf{r} \times m\mathbf{v}$, where $\mathbf{v}$ is the particle's velocity. Show that the rate of change of angular momentum is equal to the applied torque.

The rate of change of angular momentum is given by

$$\frac{d\mathbf{L}}{dt} = \frac{d}{dt}(\mathbf{r} \times m\mathbf{v}).$$

---

[1] Apply this analysis to the case of a body of mass m moving with constant angular velocity ω in a circle of radius R centred on the origin, showing that the force needed to sustain the motion has magnitude $mR\omega^2$ and is directed towards the origin.

Vector calculus

Using (11.6) we obtain

$$\frac{d\mathbf{L}}{dt} = \frac{d\mathbf{r}}{dt} \times m\mathbf{v} + \mathbf{r} \times \frac{d}{dt}(m\mathbf{v})$$
$$= \mathbf{v} \times m\mathbf{v} + \mathbf{r} \times \frac{d}{dt}(m\mathbf{v})$$
$$= \mathbf{0} + \mathbf{r} \times \mathbf{F} = \mathbf{T},$$

where in the last line we use Newton's second law, namely $\mathbf{F} = d(m\mathbf{v})/dt$.

◀

If a vector $\mathbf{a}$ is a function of a scalar variable s that is itself a function of u, so that $s = s(u)$, then the chain rule (see Section 3.1.3) gives

$$\frac{d\mathbf{a}(s)}{du} = \frac{ds}{du}\frac{d\mathbf{a}}{ds}. \tag{11.7}$$

The derivatives of more complicated vector expressions may be found by repeated application of the above equations.

One further useful result can be derived by considering the derivative

$$\frac{d}{du}(\mathbf{a} \cdot \mathbf{a}) = 2\mathbf{a} \cdot \frac{d\mathbf{a}}{du}.$$

Since $\mathbf{a} \cdot \mathbf{a} = a^2$, where $a = |\mathbf{a}|$, we see that

$$\mathbf{a} \cdot \frac{d\mathbf{a}}{du} = 0 \quad \text{if } a \text{ is constant.} \tag{11.8}$$

In other words, if a vector $\mathbf{a}(u)$ has a constant magnitude as u varies then it is perpendicular to the vector $d\mathbf{a}/du$.

11.1.2 Differential of a vector

As a final note on the differentiation of vectors, we can also define the *differential* of a vector, in a similar way to that of a scalar in ordinary differential calculus. In the definition of the vector derivative (11.1) we used the notion of a small change $\Delta\mathbf{a}$ in a vector $\mathbf{a}(u)$ resulting from a small change Δu in its argument. In the limit $\Delta u \to 0$, the change in $\mathbf{a}$ becomes infinitesimally small, and we denote it by the differential $d\mathbf{a}$. From (11.1) we see that the differential is given by

$$d\mathbf{a} = \frac{d\mathbf{a}}{du}du. \tag{11.9}$$

Note that the differential of a vector is also a vector. As an example, the infinitesimal change in the position vector of a particle in an infinitesimal time dt is

$$d\mathbf{r} = \frac{d\mathbf{r}}{dt}dt = \mathbf{v}\,dt,$$

where $\mathbf{v}$ is the particle's velocity.[2]

---

[2] In the same way, the infinitesimal change in velocity in an infinitesimal time dt is given by $d\mathbf{v} = \mathbf{a}\,dt$, where $\mathbf{a}$ is the particle's acceleration.

11.2 Integration of vectors

EXERCISES 11.1

1. A particle moves with velocity $\mathbf{v}(t)$ under the influence of an external force $\mathbf{F}(t)$. Show that the change in its kinetic energy between the times t_1 and t_2 is $\Delta K = \int_{t_1}^{t_2} \mathbf{v} \cdot \mathbf{F}\, dt$.

2. Find the derivative with respect to t of the scalar triple product $\mathbf{a}(t) \cdot (\mathbf{b}(t) \times \mathbf{c}(t))$. Use the permutation properties of a scalar triple product to arrange it in as symmetrical a form as possible, i.e. so that all three terms have the same general structure. Taking t to be time, give a geometrical interpretation of your answer.

3. A particle moves in such a way that its position at time t is $\mathbf{r}(t) = a \cos \omega t\, \mathbf{i} + b \sin \omega t\, \mathbf{j}$. Show that, at any instant, the particle's acceleration is directed towards the origin and has magnitude $\omega^2 r$. Describe the particle's trajectory geometrically.

11.2 Integration of vectors

The integration of a vector (or of an expression involving vectors that may itself be either a vector or scalar) with respect to a scalar u can be regarded as the inverse of differentiation. We must remember, however, that

(i) the integral has the same nature (vector or scalar) as the integrand;
(ii) the constant of integration for indefinite integrals must be of the same nature as the integral.

For example, if $\mathbf{a}(u) = d[\mathbf{A}(u)]/du$ then the indefinite integral of $\mathbf{a}(u)$ is given by

$$\int \mathbf{a}(u)\, du = \mathbf{A}(u) + \mathbf{b},$$

where $\mathbf{b}$ is a constant vector, of the same nature as $\mathbf{A}$. The definite integral of $\mathbf{a}(u)$ from $u = u_1$ to $u = u_2$ is given by

$$\int_{u_1}^{u_2} \mathbf{a}(u)\, du = \mathbf{A}(u_2) - \mathbf{A}(u_1).$$

Readers familiar with the physics of orbits will probably recognise the results of the next example as effectively the law of conservation of angular momentum under a central force and one of Keppler's laws of planetary motion.

Example A small particle of mass m orbits a much larger mass M centred at the origin O. According to Newton's law of gravitation, the position vector $\mathbf{r}$ of the small mass obeys the differential equation

$$m \frac{d^2 \mathbf{r}}{dt^2} = -\frac{GMm}{r^2} \hat{\mathbf{r}}.$$

Show that the vector $\mathbf{r} \times d\mathbf{r}/dt$ is a constant of the motion.

Forming the vector product of the differential equation with $\mathbf{r}$, we obtain

$$\mathbf{r} \times \frac{d^2\mathbf{r}}{dt^2} = -\frac{GM}{r^2}\mathbf{r} \times \hat{\mathbf{r}}.$$

Since $\mathbf{r}$ and $\hat{\mathbf{r}}$ are collinear, $\mathbf{r} \times \hat{\mathbf{r}} = \mathbf{0}$ and therefore we have

$$\mathbf{r} \times \frac{d^2\mathbf{r}}{dt^2} = \mathbf{0}. \tag{11.10}$$

However,

$$\frac{d}{dt}\left(\mathbf{r} \times \frac{d\mathbf{r}}{dt}\right) = \mathbf{r} \times \frac{d^2\mathbf{r}}{dt^2} + \frac{d\mathbf{r}}{dt} \times \frac{d\mathbf{r}}{dt} = \mathbf{0},$$

since the first term is zero by (11.10) and the second is zero because it is the vector product of two parallel (in this case identical) vectors. Integrating, we obtain the required result

$$\mathbf{r} \times \frac{d\mathbf{r}}{dt} = \mathbf{c}, \tag{11.11}$$

where $\mathbf{c}$ is a constant vector.

As a further point of interest we may note that in an infinitesimal time dt the change in the position vector of the small mass is $d\mathbf{r}$ and the element of area swept out by the position vector of the particle is simply $dA = \frac{1}{2}|\mathbf{r} \times d\mathbf{r}|$. Dividing both sides of this equation by dt, we conclude that

$$\frac{dA}{dt} = \frac{1}{2}\left|\mathbf{r} \times \frac{d\mathbf{r}}{dt}\right| = \frac{|\mathbf{c}|}{2},$$

and that the physical interpretation of the above result (11.11) is that the position vector $\mathbf{r}$ of the small mass sweeps out equal areas in equal times. This result is in fact valid for motion under any force that acts along the line joining the two particles. ◄

EXERCISES 11.2

1. Given that the integral $\int^t \mathbf{a}(u)\,du$, where $\mathbf{a}(u) = (u^2, 1+u, (1-u)^2)$, has the value $(1, 1, 1)$ at $t = 1$, find its value at $t = 2$.

2. If $\mathbf{a}(t) = (t^2, 1+t, (1-t)^2)$ and $\mathbf{b}(t) = (1-t, t^2, 1+t)$, evaluate the definite integrals (i) $\int_0^1 \mathbf{a} \cdot \mathbf{b}\,dt$ and (ii) $\int_0^1 \mathbf{a} \times \mathbf{b}\,dt$.

11.3 Vector functions of several arguments

The concept of the derivative of a vector is easily extended to cases where the vectors (or scalars) are functions of more than one independent scalar variable, $u_1, u_2, \ldots, u_n$. In this case, the results of Section 11.1.1 are still valid, except that the derivatives become partial derivatives $\partial \mathbf{a}/\partial u_i$ defined as in ordinary differential calculus. For example, in Cartesian coordinates,

$$\frac{\partial \mathbf{a}}{\partial u_r} = \frac{\partial a_x}{\partial u_r}\mathbf{i} + \frac{\partial a_y}{\partial u_r}\mathbf{j} + \frac{\partial a_z}{\partial u_r}\mathbf{k}.$$

11.4 Surfaces

In particular, (11.7) generalises to the chain rule of partial differentiation discussed in Section 7.5. If $\mathbf{a} = \mathbf{a}(u_1, u_2, \ldots, u_n)$ and each of the u_i is also a function $u_i(v_1, v_2, \ldots, v_n)$ of the variables v_i then, generalising (7.17),

$$\frac{\partial \mathbf{a}}{\partial v_i} = \frac{\partial \mathbf{a}}{\partial u_1}\frac{\partial u_1}{\partial v_i} + \frac{\partial \mathbf{a}}{\partial u_2}\frac{\partial u_2}{\partial v_i} + \cdots + \frac{\partial \mathbf{a}}{\partial u_n}\frac{\partial u_n}{\partial v_i} = \sum_{j=1}^{n} \frac{\partial \mathbf{a}}{\partial u_j}\frac{\partial u_j}{\partial v_i}. \tag{11.12}$$

A special case of this rule arises when $\mathbf{a}$ is an explicit function of some variable v, as well as of scalars $u_1, u_2, \ldots, u_n$ that are themselves functions of v; then we have

$$\frac{d\mathbf{a}}{dv} = \frac{\partial \mathbf{a}}{\partial v} + \sum_{j=1}^{n} \frac{\partial \mathbf{a}}{\partial u_j}\frac{\partial u_j}{\partial v}. \tag{11.13}$$

We may also extend the concept of the differential of a vector given in (11.9) to vectors dependent on several variables $u_1, u_2, \ldots, u_n$:

$$d\mathbf{a} = \frac{\partial \mathbf{a}}{\partial u_1} du_1 + \frac{\partial \mathbf{a}}{\partial u_2} du_2 + \cdots + \frac{\partial \mathbf{a}}{\partial u_n} du_n = \sum_{j=1}^{n} \frac{\partial \mathbf{a}}{\partial u_j} du_j. \tag{11.14}$$

As an example, the infinitesimal change in an electric field $\mathbf{E}$ in moving from a position $\mathbf{r}$ to a neighbouring one $\mathbf{r} + d\mathbf{r}$ is given by

$$d\mathbf{E} = \frac{\partial \mathbf{E}}{\partial x} dx + \frac{\partial \mathbf{E}}{\partial y} dy + \frac{\partial \mathbf{E}}{\partial z} dz. \tag{11.15}$$

EXERCISE 11.3

1. The vector field $\mathbf{a}$ is given in plane polar coordinates (ρ, ϕ) by $\mathbf{a} = (\rho \cos 2\phi, -\rho \sin 2\phi)$. Show that the vector $\partial \mathbf{a}/\partial x$, expressed in terms of Cartesian coordinates x and y, is $(x^2 + y^2)^{-3/2}(x^3 + 3xy^2, -2y^3)$. Show further that the y-component of this vector is $xy/(x^2 + y^2)$.

11.4 Surfaces

A surface S in space can be described by the vector $\mathbf{r}(u, v)$ joining the origin O of a coordinate system to a point on the surface (see Figure 11.3). As the parameters u and v vary, the end-point of the vector moves over the surface.

In Cartesian coordinates the surface is given by

$$\mathbf{r}(u, v) = x(u, v)\mathbf{i} + y(u, v)\mathbf{j} + z(u, v)\mathbf{k},$$

where $x = x(u, v)$, $y = y(u, v)$ and $z = z(u, v)$ are the parametric equations of the surface. We can also represent a surface by $z = f(x, y)$ or $g(x, y, z) = 0$. Either of these representations can be converted into the parametric form. For example, if $z = f(x, y)$ then by setting $u = x$ and $v = y$ the surface can be represented in parametric form by

$$\mathbf{r}(u, v) = u\mathbf{i} + v\mathbf{j} + f(u, v)\mathbf{k}.$$

Vector calculus

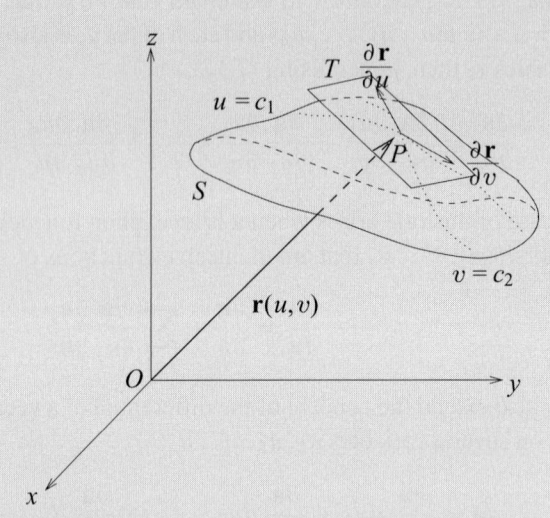

Figure 11.3 The tangent plane T to a surface S at a particular point P; $u = c_1$ and $v = c_2$ are the coordinate curves, shown by dotted lines, that pass through P. The broken line shows some particular parametric curve $\mathbf{r} = \mathbf{r}(\lambda)$ lying in the surface.

Any curve $\mathbf{r}(\lambda)$, where λ is a parameter, on the surface S can be represented by a pair of equations relating the parameters u and v, for example $u = f(\lambda)$ and $v = g(\lambda)$. A parametric representation of the curve can easily be found by straightforward substitution, i.e. $\mathbf{r}(\lambda) = \mathbf{r}(u(\lambda), v(\lambda))$. Using (11.12) for the case where the vector is a function of a single variable λ so that the LHS becomes a total derivative, the tangent to the curve $\mathbf{r}(\lambda)$ at any point is given by

$$\frac{d\mathbf{r}}{d\lambda} = \frac{\partial \mathbf{r}}{\partial u}\frac{du}{d\lambda} + \frac{\partial \mathbf{r}}{\partial v}\frac{dv}{d\lambda}. \tag{11.16}$$

The two curves $u =$ constant and $v =$ constant passing through any point P on S are called *coordinate curves*. For the curve $u =$ constant, for example, we have $du/d\lambda = 0$, and so from (11.16) its tangent vector is in the direction $\partial \mathbf{r}/\partial v$. Similarly, the tangent vector to the curve $v =$ constant is in the direction $\partial \mathbf{r}/\partial u$.

If the surface is smooth then at any point P on S the vectors $\partial \mathbf{r}/\partial u$ and $\partial \mathbf{r}/\partial v$ are linearly independent and define the *tangent plane* T at the point P (see Figure 11.3). A vector normal to the surface at P is given by

$$\mathbf{n} = \frac{\partial \mathbf{r}}{\partial u} \times \frac{\partial \mathbf{r}}{\partial v}. \tag{11.17}$$

In the neighbourhood of P, an infinitesimal vector displacement $d\mathbf{r}$ is written

$$d\mathbf{r} = \frac{\partial \mathbf{r}}{\partial u}du + \frac{\partial \mathbf{r}}{\partial v}dv.$$

11.4 Surfaces

The *element of area* at P, an infinitesimal parallelogram whose sides are the coordinate curves, has magnitude

$$dS = \left|\frac{\partial \mathbf{r}}{\partial u} du \times \frac{\partial \mathbf{r}}{\partial v} dv\right| = \left|\frac{\partial \mathbf{r}}{\partial u} \times \frac{\partial \mathbf{r}}{\partial v}\right| du\, dv = |\mathbf{n}|\, du\, dv. \qquad (11.18)$$

Thus the total area of the surface is

$$A = \int\int_R \left|\frac{\partial \mathbf{r}}{\partial u} \times \frac{\partial \mathbf{r}}{\partial v}\right| du\, dv = \int\int_R |\mathbf{n}|\, du\, dv, \qquad (11.19)$$

where R is the region in the uv-plane corresponding to the range of parameter values that define the surface.

In this generalised form, the various steps may initially be somewhat difficult to follow, and so we take as our worked example a familiar surface, namely that of a sphere, so that both the parameters and their roles in describing the surface will be well known to the reader.

Example Find the element of area on the surface of a sphere of radius a, and hence calculate the total surface area of the sphere.

We can represent a point $\mathbf{r}$ on the surface of the sphere in terms of the two parameters θ and ϕ:

$$\mathbf{r}(\theta, \phi) = a \sin\theta \cos\phi\, \mathbf{i} + a \sin\theta \sin\phi\, \mathbf{j} + a \cos\theta\, \mathbf{k},$$

where θ and ϕ are the polar and azimuthal angles respectively. At any point P, vectors tangent to the coordinate curves $\theta = $ constant and $\phi = $ constant are

$$\frac{\partial \mathbf{r}}{\partial \theta} = a \cos\theta \cos\phi\, \mathbf{i} + a \cos\theta \sin\phi\, \mathbf{j} - a \sin\theta\, \mathbf{k},$$

$$\frac{\partial \mathbf{r}}{\partial \phi} = -a \sin\theta \sin\phi\, \mathbf{i} + a \sin\theta \cos\phi\, \mathbf{j}.$$

A normal $\mathbf{n}$ to the surface at this point is then given by

$$\mathbf{n} = \frac{\partial \mathbf{r}}{\partial \theta} \times \frac{\partial \mathbf{r}}{\partial \phi} = \begin{vmatrix} \mathbf{i} & \mathbf{j} & \mathbf{k} \\ a \cos\theta \cos\phi & a \cos\theta \sin\phi & -a \sin\theta \\ -a \sin\theta \sin\phi & a \sin\theta \cos\phi & 0 \end{vmatrix}$$

$$= a^2 \sin\theta (\sin\theta \cos\phi\, \mathbf{i} + \sin\theta \sin\phi\, \mathbf{j} + \cos\theta\, \mathbf{k}),$$

which has a magnitude of $a^2 \sin\theta$. Therefore, the element of area at P is, from (11.18),

$$dS = a^2 \sin\theta\, d\theta\, d\phi,$$

and the total surface area of the sphere is given by

$$A = \int_0^\pi d\theta \int_0^{2\pi} d\phi\, a^2 \sin\theta = 4\pi a^2.$$

There are, of course, much simpler methods of proving this result![3]

---

[3] Use a similar method to show that the surface element on the paraboloid of revolution given in cylindrical polar coordinates by $\rho^2 = 4az$ is $dS = (2a)^{-1}(\rho^2 + 4a^2)^{1/2} d\rho\, d\phi$.

EXERCISE 11.4

1. A three-dimensional surface is given parametrically by

$$x = a\cosh\theta \cos\phi, \quad y = a\cosh\theta \sin\phi, \quad z = c\sinh\theta.$$

(a) Describe the geometric shapes of the coordinate surfaces.
(b) Show that the element of area corresponding to differentials $d\theta$, $d\phi$ is given by
$a\cosh\theta(c^2\cosh^2\theta + a^2\sinh^2\theta)^{1/2} \, d\theta \, d\phi$.
(c) A particular curve $\mathbf{r} = \mathbf{r}(\lambda)$ lying on the surface is defined by the relationship $\theta = \phi = \lambda$. Find, in terms of λ, the components of the tangent to the curve.

11.5 Scalar and vector fields

We now turn to the case where a particular scalar or vector quantity is defined not just at a point in space but continuously as a *field* throughout some region of space R (which is often the whole space). Although the concept of a field is valid for spaces with an arbitrary number of dimensions, in the remainder of this chapter we will restrict our attention to the familiar three-dimensional case. A *scalar field* $\phi(x, y, z)$ associates a scalar with each point in R, while a *vector field* $\mathbf{a}(x, y, z)$ associates a vector with each point. In what follows, we will assume that the variation in the scalar or vector field from point to point is both continuous and differentiable in R.

Simple examples of scalar fields include the pressure at each point in a fluid and the electrostatic potential at each point in space in the presence of an electric charge. Vector fields relating to the same physical systems are the velocity vector in a fluid (giving the local speed and direction of the flow) and the electric field.

With the study of continuously varying scalar and vector fields there arises the need to consider their derivatives and also the integration of field quantities along lines, over surfaces and throughout volumes in the field. We defer the discussion of line, surface and volume integrals until the next chapter, and in the remainder of this chapter we concentrate on the definitions of vector differential operators and their properties.

11.6 Vector operators

Certain differential operations may be performed on scalar and vector fields and have wide-ranging applications in the physical sciences. The most important operations are those of finding the *gradient* of a scalar field and the *divergence* and *curl* of a vector field. It is usual to define these operators from a strictly mathematical point of view, as we do below. In the following chapter, however, we will discuss their geometrical definitions, which rely on the concept of integrating vector quantities along lines and over surfaces.

Central to all these differential operations is the vector operator ∇, which is called *del* (or sometimes *nabla*) and in Cartesian coordinates is defined by

$$\nabla \equiv \mathbf{i}\frac{\partial}{\partial x} + \mathbf{j}\frac{\partial}{\partial y} + \mathbf{k}\frac{\partial}{\partial z}. \tag{11.20}$$

11.6 Vector operators

The form of this operator in non-Cartesian coordinate systems is discussed in Sections 11.8 and 11.9.

11.6.1 Gradient of a scalar field

The *gradient* of a scalar field $\phi(x, y, z)$ is defined by

$$\text{grad } \phi = \nabla\phi = \mathbf{i}\frac{\partial\phi}{\partial x} + \mathbf{j}\frac{\partial\phi}{\partial y} + \mathbf{k}\frac{\partial\phi}{\partial z}. \tag{11.21}$$

Clearly, $\nabla\phi$ is a vector field whose x-, y- and z-components are the first partial derivatives of $\phi(x, y, z)$ with respect to x, y and z respectively. Also note that the vector field $\nabla\phi$ should not be confused with $\phi\nabla$, which has components $(\phi\,\partial/\partial x, \phi\,\partial/\partial y, \phi\,\partial/\partial z)$, and is a vector operator.[4]

Example Find the gradient of the scalar field $\phi = xy^2z^3$.

From (11.21) the gradient of ϕ, obtained by differentiating with respect to x, y and z in turn, is given by

$$\nabla\phi = y^2z^3\mathbf{i} + 2xyz^3\mathbf{j} + 3xy^2z^2\mathbf{k}.$$

Note that each component can be a function of some or all of the coordinates. ◀

The gradient of a scalar field ϕ has some interesting geometrical properties. Let us first consider the problem of *calculating the rate of change of ϕ in some particular direction*. For an infinitesimal vector displacement $d\mathbf{r}$, forming its scalar product with $\nabla\phi$ we obtain

$$\nabla\phi \cdot d\mathbf{r} = \left(\mathbf{i}\frac{\partial\phi}{\partial x} + \mathbf{j}\frac{\partial\phi}{\partial y} + \mathbf{k}\frac{\partial\phi}{\partial z}\right) \cdot (\mathbf{i}\,dx + \mathbf{j}\,dy + \mathbf{k}\,dx)$$

$$= \frac{\partial\phi}{\partial x}dx + \frac{\partial\phi}{\partial y}dy + \frac{\partial\phi}{\partial z}dz$$

$$= d\phi, \tag{11.22}$$

which is the infinitesimal change in ϕ in going from position $\mathbf{r}$ to $\mathbf{r} + d\mathbf{r}$. In particular, if $\mathbf{r}$ depends on some parameter u such that $\mathbf{r}(u)$ defines a curve in space, then the total derivative of ϕ with respect to u along the curve is simply

$$\frac{d\phi}{du} = \nabla\phi \cdot \frac{d\mathbf{r}}{du}. \tag{11.23}$$

In the particular case where the parameter u is the arc length s along the curve, the total derivative of ϕ with respect to s along the curve is given by

$$\frac{d\phi}{ds} = \nabla\phi \cdot \hat{\mathbf{t}}, \tag{11.24}$$

where $\hat{\mathbf{t}}$ is the unit tangent to the curve at the given point.

---
[4] Distinguish between (i) $(\phi\nabla)\psi$, (ii) $(\nabla\phi)\psi$ and (iii) $\nabla(\phi\psi)$ and determine the x-component of each if $\phi(x, y, z) = x^2y^2z^2$ and $\psi(x, y, z) = x^2 + y^2 + z^2$.

Vector calculus

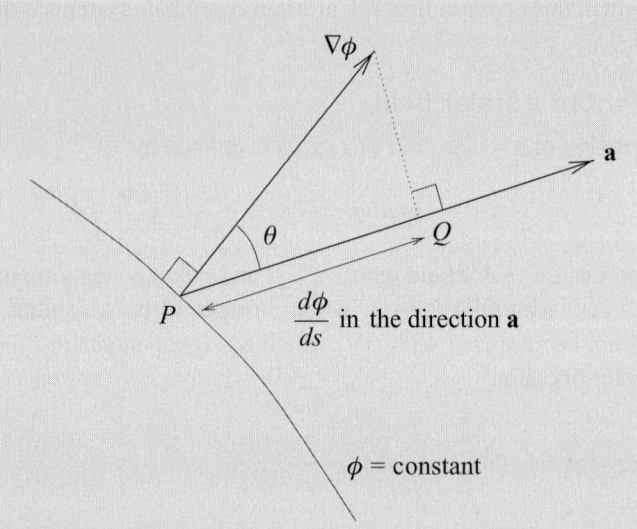

Figure 11.4 Geometrical properties of $\nabla\phi$. PQ gives the value of $d\phi/ds$ in the direction **a**.

In general, the rate of change of ϕ with respect to the distance s in a particular direction **a** is given by

$$\frac{d\phi}{ds} = \nabla\phi \cdot \hat{\mathbf{a}} \qquad (11.25)$$

and is called the directional derivative. Since $\hat{\mathbf{a}}$ is a unit vector we have

$$\frac{d\phi}{ds} = |\nabla\phi|\cos\theta$$

where θ is the angle between $\hat{\mathbf{a}}$ and $\nabla\phi$, as shown in Figure 11.4. Clearly, $\nabla\phi$ lies in the direction of the fastest increase in ϕ, and $|\nabla\phi|$ is the largest possible value of $d\phi/ds$. Similarly, the largest rate of decrease of ϕ is $d\phi/ds = -|\nabla\phi|$ in the direction of $-\nabla\phi$.

Example For the function $\phi = x^2y + yz$ at the point $(1, 2, -1)$, find its rate of change with distance in the direction $\mathbf{a} = \mathbf{i} + 2\mathbf{j} + 3\mathbf{k}$. At this same point, what is the greatest possible rate of change with distance and in which direction does it occur?

The gradient of ϕ is given by (11.21):

$$\nabla\phi = 2xy\mathbf{i} + (x^2 + z)\mathbf{j} + y\mathbf{k},$$
$$= 4\mathbf{i} + 2\mathbf{k} \quad \text{at the point } (1, 2, -1).$$

The unit vector in the direction of **a** is $\hat{\mathbf{a}} = \frac{1}{\sqrt{14}}(\mathbf{i} + 2\mathbf{j} + 3\mathbf{k})$, so the rate of change of ϕ with distance s in this direction is, using (11.25),

$$\frac{d\phi}{ds} = \nabla\phi \cdot \hat{\mathbf{a}} = \frac{1}{\sqrt{14}}(4 + 6) = \frac{10}{\sqrt{14}}.$$

From the above discussion, at the point $(1, 2, -1)$ the directional derivative $d\phi/ds$ will be greatest in the direction of $\nabla\phi = 4\mathbf{i} + 2\mathbf{k}$ and has the value $|\nabla\phi| = \sqrt{20}$ in this direction. ◂

11.6 Vector operators

We can extend the above analysis to find the rate of change of a vector field (rather than a scalar field as above) in a particular direction. The scalar differential operator $\hat{\mathbf{a}} \cdot \nabla$ can be shown to give the rate of change with distance in the direction $\hat{\mathbf{a}}$ of the quantity (vector or scalar) on which it acts. In Cartesian coordinates it may be written as

$$\hat{\mathbf{a}} \cdot \nabla = a_x \frac{\partial}{\partial x} + a_y \frac{\partial}{\partial y} + a_z \frac{\partial}{\partial z}. \tag{11.26}$$

Thus we can write the infinitesimal change in an electric field in moving from $\mathbf{r}$ to $\mathbf{r} + d\mathbf{r}$ given in (11.15) as $d\mathbf{E} = (d\mathbf{r} \cdot \nabla)\mathbf{E}$.

A second interesting geometrical property of $\nabla \phi$ may be found by considering the surface defined by $\phi(x, y, z) = c$, where c is some constant. If $\hat{\mathbf{t}}$ is a unit tangent to this surface at some point then clearly $d\phi/ds = 0$ in this direction and from (11.24) we have $\nabla \phi \cdot \hat{\mathbf{t}} = 0$. In other words, *$\nabla \phi$ is a vector normal to the surface $\phi(x, y, z) = c$ at every point*, as shown in Figure 11.4.[5]

If $\hat{\mathbf{n}}$ is a unit normal to the surface in the direction of increasing $\phi(x, y, z)$, then the gradient is sometimes written

$$\nabla \phi \equiv \frac{\partial \phi}{\partial n} \hat{\mathbf{n}}, \tag{11.27}$$

where $\partial \phi / \partial n \equiv |\nabla \phi|$ is the rate of change of ϕ in the direction $\hat{\mathbf{n}}$ and is called the *normal derivative*.

Example Find expressions for the equations of the tangent plane and the line normal to the surface $\phi(x, y, z) = c$ at the point P with coordinates x_0, y_0, z_0. Use the results to find the equations of the tangent plane and the line normal to the surface of the sphere $\phi = x^2 + y^2 + z^2 = a^2$ at the point $(0, 0, a)$.

A vector normal to the surface $\phi(x, y, z) = c$ at the point P is simply $\nabla \phi$ evaluated at that point; we denote it by $\mathbf{n}_0$. If $\mathbf{r}_0$ is the position vector of the point P relative to the origin and $\mathbf{r}$ is the position vector of any point on the tangent plane, then the vector equation of the tangent plane is, from (9.42),

$$(\mathbf{r} - \mathbf{r}_0) \cdot \mathbf{n}_0 = 0.$$

Similarly, if $\mathbf{r}$ is the position vector of any point on the straight line passing through P (with position vector $\mathbf{r}_0$) in the direction of the normal $\mathbf{n}_0$ then the vector equation of this line is, from Section 9.6.1,

$$(\mathbf{r} - \mathbf{r}_0) \times \mathbf{n}_0 = \mathbf{0}.$$

Now, applying the first of these two results to the surface of the sphere $\phi = x^2 + y^2 + z^2 = a^2$,

$$\nabla \phi = 2x\mathbf{i} + 2y\mathbf{j} + 2z\mathbf{k}$$
$$= 2a\mathbf{k} \quad \text{at the point } (0, 0, a).$$

Therefore, the equation of the tangent plane to the sphere at this point is

$$(\mathbf{r} - \mathbf{r}_0) \cdot 2a\mathbf{k} = 0.$$

---

[5] If $\phi(x, y, z) = Ar^{-n}$, with $A > 0$ and $r^2 = x^2 + y^2 + z^2$, identify the surfaces of constant ϕ and hence the direction of $\nabla \phi$. Confirm your conclusion by explicit calculation, working in Cartesian coordinates and using the chain rule to evaluate the required derivatives.

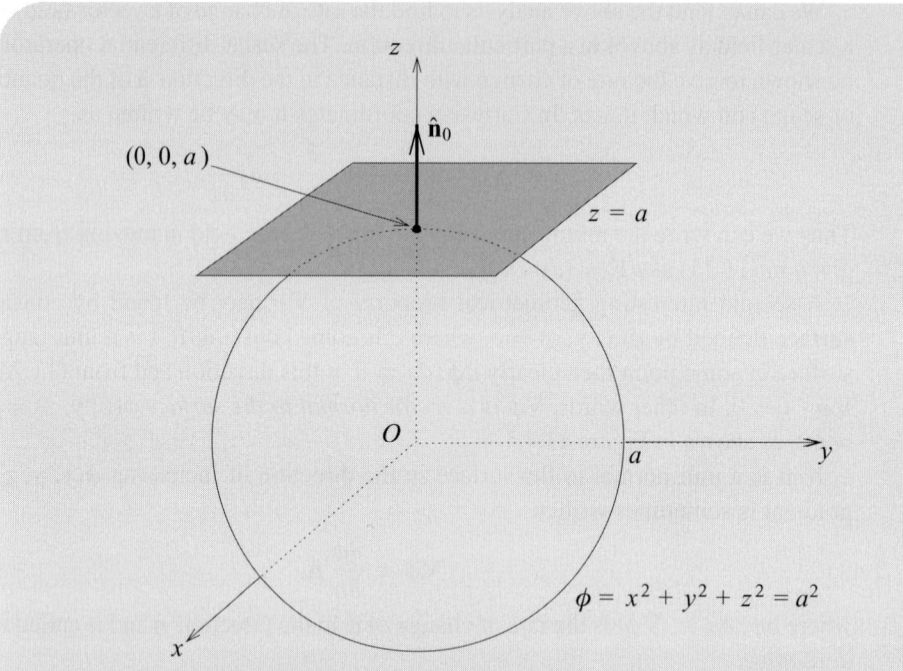

Figure 11.5 The tangent plane and the normal to the surface of the sphere $\phi = x^2 + y^2 + z^2 = a^2$ at the point $\mathbf{r}_0$ with coordinates $(0, 0, a)$.

This gives $2a(z - a) = 0$ or $z = a$, as expected. The equation of the line normal to the sphere at the point $(0, 0, a)$, given by the second result, is

$$(\mathbf{r} - \mathbf{r}_0) \times 2a\mathbf{k} = \mathbf{0},$$

which gives $2ay\mathbf{i} - 2ax\mathbf{j} = \mathbf{0}$ or $x = y = 0$, i.e. the z-axis, again as expected. Figure 11.5 shows the tangent plane and normal to the surface of the sphere at this point. ◀

Further properties of the gradient operation, which are analogous to those of the ordinary derivative, are listed in Section 11.7.1 and may be easily proved. In addition to these, we note that the gradient operation also obeys the chain rule as in ordinary differential calculus, i.e. if ϕ and ψ are scalar fields in some region R then[6]

$$\nabla [\phi(\psi)] = \frac{\partial \phi}{\partial \psi} \nabla \psi.$$

11.6.2 Divergence of a vector field

The *divergence* of a vector field $\mathbf{a}(x, y, z)$ is defined by

$$\operatorname{div} \mathbf{a} = \nabla \cdot \mathbf{a} = \frac{\partial a_x}{\partial x} + \frac{\partial a_y}{\partial y} + \frac{\partial a_z}{\partial z}, \tag{11.28}$$

---

[6] Evaluate both sides of this equation in the particular case that $\psi(x, y, z) = z(x^2 + y^2)$ and $\phi(\psi) = \psi^2$ and verify that they are the same function of x, y and z.

11.6 Vector operators

where a_x, a_y and a_z are the x-, y- and z-components of $\mathbf{a}$. Clearly, $\nabla \cdot \mathbf{a}$ is a scalar field. Any vector field $\mathbf{a}$ for which $\nabla \cdot \mathbf{a} = 0$ is said to be *solenoidal*.

Example Find the divergence of the vector field $\mathbf{a} = x^2y^2\mathbf{i} + y^2z^2\mathbf{j} + x^2z^2\mathbf{k}$.

From (11.28) the divergence of $\mathbf{a}$ is given by
$$\nabla \cdot \mathbf{a} = 2xy^2 + 2yz^2 + 2x^2z = 2(xy^2 + yz^2 + x^2z).$$

Although this expression contains three terms, they are simply added together and the expression is a scalar, not a vector.

The geometric definition of divergence and its physical meaning will be discussed in the next chapter. For the moment, we merely note that the divergence can be considered as a quantitative measure of how much a vector field diverges (spreads out) or converges at any given point. For example, if we consider the vector field $\mathbf{v}(x, y, z)$ describing the local velocity at any point in a fluid then $\nabla \cdot \mathbf{v}$ is equal to the net rate of outflow of fluid per unit volume, evaluated at a point (by letting a small volume at that point tend to zero).

Now if some vector field $\mathbf{a}$ is itself derived from a scalar field via $\mathbf{a} = \nabla \phi$ then $\nabla \cdot \mathbf{a}$ has the form $\nabla \cdot \nabla \phi$ or, as it is usually written, $\nabla^2 \phi$, where ∇^2 (del squared) is the scalar differential operator

$$\nabla^2 \equiv \frac{\partial^2}{\partial x^2} + \frac{\partial^2}{\partial y^2} + \frac{\partial^2}{\partial z^2}. \tag{11.29}$$

$\nabla^2 \phi$ is called the *Laplacian* of ϕ and appears in several important partial differential equations of mathematical physics.

Example Find the Laplacian of the scalar field $\phi = xy^2z^3$.

From (11.29) the Laplacian of ϕ is given by
$$\nabla^2 \phi = \frac{\partial^2 \phi}{\partial x^2} + \frac{\partial^2 \phi}{\partial y^2} + \frac{\partial^2 \phi}{\partial z^2} = 2xz^3 + 6xy^2z.$$

Note that, like the divergence of a vector, the Laplacian of a scalar is a single quantity (i.e. another scalar), even though the general expression for it may contain more than one term.

11.6.3 Curl of a vector field

The *curl* of a vector field $\mathbf{a}(x, y, z)$ is defined by

$$\operatorname{curl} \mathbf{a} = \nabla \times \mathbf{a} = \left(\frac{\partial a_z}{\partial y} - \frac{\partial a_y}{\partial z}\right)\mathbf{i} + \left(\frac{\partial a_x}{\partial z} - \frac{\partial a_z}{\partial x}\right)\mathbf{j} + \left(\frac{\partial a_y}{\partial x} - \frac{\partial a_x}{\partial y}\right)\mathbf{k},$$

where a_x, a_y and a_z are the x-, y- and z-components of $\mathbf{a}$; $\nabla \times \mathbf{a}$ is itself a vector field. The RHS can be written in a more memorable form as the determinant

$$\nabla \times \mathbf{a} = \begin{vmatrix} \mathbf{i} & \mathbf{j} & \mathbf{k} \\ \dfrac{\partial}{\partial x} & \dfrac{\partial}{\partial y} & \dfrac{\partial}{\partial z} \\ a_x & a_y & a_z \end{vmatrix}, \quad (11.30)$$

where it is understood that, on expanding the determinant, the partial derivatives in the second row act on the components of $\mathbf{a}$ in the third row.

Any vector field $\mathbf{a}$ for which $\nabla \times \mathbf{a} = \mathbf{0}$ is said to be *irrotational*; clearly, any vector field derived as the gradient of a scalar field has this characteristic throughout.

Example Find the curl of the vector field $\mathbf{a} = x^2 y^2 z^2 \mathbf{i} + y^2 z^2 \mathbf{j} + x^2 z^2 \mathbf{k}$.

The curl of $\mathbf{a}$ is given by

$$\nabla \times \mathbf{a} = \begin{vmatrix} \mathbf{i} & \mathbf{j} & \mathbf{k} \\ \dfrac{\partial}{\partial x} & \dfrac{\partial}{\partial y} & \dfrac{\partial}{\partial z} \\ x^2 y^2 z^2 & y^2 z^2 & x^2 z^2 \end{vmatrix}$$

$$= \left[\frac{\partial}{\partial y}(x^2 z^2) - \frac{\partial}{\partial z}(y^2 z^2)\right]\mathbf{i} + \left[\frac{\partial}{\partial z}(x^2 y^2 z^2) - \frac{\partial}{\partial x}(x^2 z^2)\right]\mathbf{j} + \left[\frac{\partial}{\partial x}(y^2 z^2) - \frac{\partial}{\partial y}(x^2 y^2 z^2)\right]\mathbf{k}$$

$$= -2\left[y^2 z \mathbf{i} + (x z^2 - x^2 y^2 z)\mathbf{j} + x^2 y z^2 \mathbf{k}\right].$$

As with any general vector, each of the components of the curl of a vector can depend on some or all of the coordinates.[7] ◂

For a vector field $\mathbf{v}(x, y, z)$ describing the local velocity at any point in a fluid, $\nabla \times \mathbf{v}$ is a measure of the angular velocity of the fluid in the neighbourhood of that point. If a small paddle wheel were placed at various points in the fluid then it would tend to rotate in regions where $\nabla \times \mathbf{v} \neq \mathbf{0}$, while it would not rotate in regions where $\nabla \times \mathbf{v} = \mathbf{0}$.

Another insight into the physical interpretation of the curl operator is gained by considering the vector field $\mathbf{v}$ describing the velocity at any point in a rigid body rotating about some axis with angular velocity $\boldsymbol{\omega}$. If $\mathbf{r}$ is the position vector of the point with respect to some origin on the axis of rotation then the velocity of the point is given by $\mathbf{v} = \boldsymbol{\omega} \times \mathbf{r}$. Without any loss of generality, we may take $\boldsymbol{\omega}$ to lie along the z-axis of our coordinate system, so that $\boldsymbol{\omega} = \omega \mathbf{k}$. The velocity field is then $\mathbf{v} = -\omega y \mathbf{i} + \omega x \mathbf{j}$. The curl of this vector field is easily found to be

$$\nabla \times \mathbf{v} = \begin{vmatrix} \mathbf{i} & \mathbf{j} & \mathbf{k} \\ \dfrac{\partial}{\partial x} & \dfrac{\partial}{\partial y} & \dfrac{\partial}{\partial z} \\ -\omega y & \omega x & 0 \end{vmatrix} = 2\omega \mathbf{k} = 2\boldsymbol{\omega}. \quad (11.31)$$

---

[7] For the field considered here, where is the field irrotational?

11.7 Vector operator formulae

Therefore, the curl of the velocity field is a vector equal to twice the angular velocity vector of the rigid body about its axis of rotation. We give a full geometrical discussion of the curl of a vector in the next chapter.

EXERCISES 11.6

1. Find the gradients of the following scalar fields:

    (a) xy^2z^3,　　(b) $x^2y^2 + y^2z^2 + z^2x^2$,　　(c) $xyze^{-(x^2+y^2)}$.

    What can be said about all of the resulting vector fields **a**? Demonstrate your answer explicitly in cases (a) and (b).

2. Where does the scalar field $\phi(x, y, z) = \exp[-(x^2 + y^2 + z^2)]$ have its greatest rate of change in the direction $(1, -2, 1)$, and what is that rate of change?

3. A scalar field ϕ is given by $\phi = z(x^2 + y^2)$ and a (so-called) twisted cubic curve is defined parametrically by $x = \lambda$, $y = \frac{3}{2}\lambda^2$, $z = \frac{3}{2}\lambda^3$.
    (a) Show (i) directly and (ii) by using (11.23) that the total derivative of ϕ with respect to λ along the curve is $\frac{3}{8}\lambda^4(20 + 63\lambda^2)$.
    (b) Find an expression for the arc length s along the curve and hence evaluate $d\phi/ds$ at the point $(2, 6, 12)$.

4. A function is said to be *harmonic* if its Laplacian vanishes. Which of the following two-dimensional functions are harmonic?

    (a) $x^2 + y^2$,　　(b) $x^2 - y^2$,　　(c) $2xy$,　　(d) $\dfrac{x}{x^2 + y^2}$.

5. The electrostatic potential associated with a uniform line charge on the negative z-axis is proportional to $\phi(x, y, z) = \ln(r + z)$, where $r^2 = x^2 + y^2 + z^2$. Noting that $\partial r/\partial x = x/r$, etc. and using the chain rule, show that (except where $z = -r$, i.e. except on the negative z-axis) ϕ satisfies Laplace's equation, $\nabla^2 \phi = 0$.

6. Which, if any, of the following Cartesian vector fields are irrotational?

    $$\mathbf{a} = (2x + z, 2y - z, x - y),$$
    $$\mathbf{b} = (2x + z, 2y - z, x + y),$$
    $$\mathbf{c} = (\sin z, -\sin z, (x - y)\cos z),$$
    $$\mathbf{d} = (-\cos z, \cos z, (x - y)\sin z),$$
    $$\mathbf{e} = (\cos z, -\cos z, (x - y)\sin z).$$

11.7 Vector operator formulae

In the same way as for ordinary vectors (Chapter 9), certain identities exist for vector operators. In addition, we must consider various relations involving the action of vector

Vector calculus

Table 11.1 *Vector operators acting on sums and products. The operator ∇ is defined in (11.20); ϕ and ψ are scalar fields, $\mathbf{a}$ and $\mathbf{b}$ are vector fields.*

$$\nabla(\phi + \psi) = \nabla\phi + \nabla\psi$$
$$\nabla \cdot (\mathbf{a} + \mathbf{b}) = \nabla \cdot \mathbf{a} + \nabla \cdot \mathbf{b}$$
$$\nabla \times (\mathbf{a} + \mathbf{b}) = \nabla \times \mathbf{a} + \nabla \times \mathbf{b}$$
$$\nabla(\phi\psi) = \phi\nabla\psi + \psi\nabla\phi$$
$$\nabla(\mathbf{a} \cdot \mathbf{b}) = \mathbf{a} \times (\nabla \times \mathbf{b}) + \mathbf{b} \times (\nabla \times \mathbf{a}) + (\mathbf{a} \cdot \nabla)\mathbf{b} + (\mathbf{b} \cdot \nabla)\mathbf{a}$$
$$\nabla \cdot (\phi\mathbf{a}) = \phi\nabla \cdot \mathbf{a} + \mathbf{a} \cdot \nabla\phi$$
$$\nabla \cdot (\mathbf{a} \times \mathbf{b}) = \mathbf{b} \cdot (\nabla \times \mathbf{a}) - \mathbf{a} \cdot (\nabla \times \mathbf{b})$$
$$\nabla \times (\phi\mathbf{a}) = \nabla\phi \times \mathbf{a} + \phi\nabla \times \mathbf{a}$$
$$\nabla \times (\mathbf{a} \times \mathbf{b}) = \mathbf{a}(\nabla \cdot \mathbf{b}) - \mathbf{b}(\nabla \cdot \mathbf{a}) + (\mathbf{b} \cdot \nabla)\mathbf{a} - (\mathbf{a} \cdot \nabla)\mathbf{b}$$

operators on sums and products of scalar and vector fields. Some of these relations have been mentioned earlier, but we list all the more important ones here for convenience. The validity of these relations may be verified by direct calculation; as indicated at the end of the previous chapter, in most cases, the quickest and most compact way of doing this is to use the notation and results discussed in Appendix D.

Although some of the following vector relations are expressed in Cartesian coordinates, it may be proved that they are all independent of the choice of coordinate system. This is to be expected, since grad, div and curl all have clear geometrical definitions, which are discussed more fully in the next chapter and which do not rely on any particular choice of coordinate system.

11.7.1 Vector operators acting on sums and products

Let ϕ and ψ be scalar fields and $\mathbf{a}$ and $\mathbf{b}$ be vector fields. Assuming these fields are differentiable, the action of grad, div and curl on various sums and products of them is presented in Table 11.1.

These relations can be proved by direct calculation. For example, the penultimate entry is proved as follows.

Example Show that

$$\nabla \times (\phi\mathbf{a}) = \nabla\phi \times \mathbf{a} + \phi\nabla \times \mathbf{a}.$$

The x-component of the LHS is

$$\frac{\partial}{\partial y}(\phi a_z) - \frac{\partial}{\partial z}(\phi a_y) = \phi\frac{\partial a_z}{\partial y} + \frac{\partial \phi}{\partial y}a_z - \phi\frac{\partial a_y}{\partial z} - \frac{\partial \phi}{\partial z}a_y,$$
$$= \phi\left(\frac{\partial a_z}{\partial y} - \frac{\partial a_y}{\partial z}\right) + \left(\frac{\partial \phi}{\partial y}a_z - \frac{\partial \phi}{\partial z}a_y\right),$$
$$= \phi(\nabla \times \mathbf{a})_x + (\nabla\phi \times \mathbf{a})_x,$$

where, for example, $(\nabla\phi \times \mathbf{a})_x$ denotes the x-component of the vector $\nabla\phi \times \mathbf{a}$. Incorporating the y- and z-components, which can be similarly found, we obtain the stated result. ◀

11.7 Vector operator formulae

An alternative proof using the methods of Appendix D and the summation convention is

$$[\nabla \times (\phi \mathbf{a})]_i = \epsilon_{ijk} \frac{\partial(\phi a_k)}{\partial x_j} = \epsilon_{ijk} \frac{\partial \phi}{\partial x_j} a_k + \epsilon_{ijk} \phi \frac{\partial a_k}{\partial x_j} = [\nabla \phi \times \mathbf{a}]_i + [\phi(\nabla \times \mathbf{a})]_i.$$

Some useful special cases of the relations in Table 11.1 are worth noting. If $\mathbf{r}$ is the position vector relative to some origin and $r = |\mathbf{r}|$, then

$$\nabla \phi(r) = \frac{d\phi}{dr} \hat{\mathbf{r}},$$

$$\nabla \cdot [\phi(r)\mathbf{r}] = 3\phi(r) + r \frac{d\phi(r)}{dr},$$

$$\nabla^2 \phi(r) = \frac{d^2\phi(r)}{dr^2} + \frac{2}{r} \frac{d\phi(r)}{dr},$$

$$\nabla \times [\phi(r)\mathbf{r}] = \mathbf{0}.$$

These results may be proved straightforwardly using Cartesian coordinates[8] but far more simply using spherical polar coordinates, which are discussed in Section 11.8.2. Particular cases of these results are

$$\nabla r = \hat{\mathbf{r}}, \quad \nabla \cdot \mathbf{r} = 3, \quad \nabla \times \mathbf{r} = \mathbf{0},$$

together with

$$\nabla \left(\frac{1}{r} \right) = -\frac{\hat{\mathbf{r}}}{r^2}.$$

11.7.2 Combinations of grad, div and curl

We now consider the action of two vector operators in succession on a scalar or vector field. We can immediately discard four of the nine obvious combinations of grad, div and curl, since they clearly do not make sense. If ϕ is a scalar field and $\mathbf{a}$ is a vector field, these four combinations are grad(grad ϕ), div(div $\mathbf{a}$), curl(div $\mathbf{a}$) and grad(curl $\mathbf{a}$). In each case the second (outer) vector operator is acting on the wrong type of field, i.e. scalar instead of vector or vice versa. In grad(grad ϕ), for example, grad acts on grad ϕ, which is a vector field, but we know that grad only acts on scalar fields (although in fact we can form the so-called *outer product* of the del operator with a vector to give a tensor, but that need not concern us here).

Of the five valid combinations of grad, div and curl, two are identically zero, namely[9]

$$\text{curl grad } \phi = \nabla \times \nabla \phi = \mathbf{0}, \tag{11.32}$$

$$\text{div curl } \mathbf{a} = \nabla \cdot (\nabla \times \mathbf{a}) = 0. \tag{11.33}$$

---

[8] Prove the second result using Cartesian coordinates. Use the chain rule and recall that $\partial r / \partial x = x/r$, etc.
[9] Prove these results by using the summation convention, showing that the two LHSs take the forms

$$\epsilon_{ijk} \frac{\partial^2 \phi}{\partial x_j \partial x_k} \quad \text{and} \quad \epsilon_{ijk} \frac{\partial^2 a_k}{\partial x_i \partial x_j},$$

and then considering the effects of interchanging j and k in the first case and i and j in the second.

Vector calculus

From (11.32), we see that, as noted earlier, if **a** is derived from the gradient of some scalar function such that $\mathbf{a} = \nabla \phi$ then it is necessarily irrotational ($\nabla \times \mathbf{a} = 0$). We also note that if **a** is an irrotational vector field then another irrotational vector field is $\mathbf{a} + \nabla \phi + \mathbf{c}$, where ϕ is any scalar field and **c** is any constant vector. This follows since

$$\nabla \times (\mathbf{a} + \nabla \phi + \mathbf{c}) = \nabla \times \mathbf{a} + \nabla \times \nabla \phi = \mathbf{0}.$$

Similarly, from (11.33) we may infer that if **b** is the curl of some vector field **a** such that $\mathbf{b} = \nabla \times \mathbf{a}$ then **b** is solenoidal ($\nabla \cdot \mathbf{b} = 0$). Obviously, if **b** is solenoidal and **c** is any constant vector then $\mathbf{b} + \mathbf{c}$ is also solenoidal.

The three remaining combinations of grad, div and curl are

$$\text{div grad } \phi = \nabla \cdot \nabla \phi = \nabla^2 \phi = \frac{\partial^2 \phi}{\partial x^2} + \frac{\partial^2 \phi}{\partial y^2} + \frac{\partial^2 \phi}{\partial z^2}, \tag{11.34}$$

$$\text{grad div } \mathbf{a} = \nabla(\nabla \cdot \mathbf{a}),$$
$$= \left(\frac{\partial^2 a_x}{\partial x^2} + \frac{\partial^2 a_y}{\partial x \partial y} + \frac{\partial^2 a_z}{\partial x \partial z} \right)\mathbf{i} + \left(\frac{\partial^2 a_x}{\partial y \partial x} + \frac{\partial^2 a_y}{\partial y^2} + \frac{\partial^2 a_z}{\partial y \partial z} \right)\mathbf{j}$$
$$+ \left(\frac{\partial^2 a_x}{\partial z \partial x} + \frac{\partial^2 a_y}{\partial z \partial y} + \frac{\partial^2 a_z}{\partial z^2} \right)\mathbf{k}, \tag{11.35}$$

$$\text{curl curl } \mathbf{a} = \nabla \times (\nabla \times \mathbf{a}) = \nabla(\nabla \cdot \mathbf{a}) - \nabla^2 \mathbf{a}, \tag{11.36}$$

where (11.34) and (11.35) are expressed in Cartesian coordinates. In (11.36), the term $\nabla^2 \mathbf{a}$ has the linear differential operator ∇^2 acting on a vector [as opposed to a scalar as in (11.34)], which of course consists of a sum of unit vectors multiplied by components. Two cases arise.

(i) If the unit vectors are constants (i.e. they are independent of the values of the coordinates) then the differential operator gives a non-zero contribution only when acting upon the components, the unit vectors being merely multipliers.
(ii) If the unit vectors vary as the values of the coordinates change (i.e. are not constant in direction throughout the whole space) then the derivatives of these vectors appear as contributions to $\nabla^2 \mathbf{a}$.

Cartesian coordinates are an example of the first case in which each component satisfies $(\nabla^2 \mathbf{a})_i = \nabla^2 a_i$. In this case (11.36) can be applied to each component separately:

$$[\nabla \times (\nabla \times \mathbf{a})]_i = [\nabla(\nabla \cdot \mathbf{a})]_i - \nabla^2 a_i. \tag{11.37}$$

However, cylindrical and spherical polar coordinates come in the second class. For them (11.36) is still true, but the further step to (11.37) cannot be made.

More complicated vector operator relations may be proved using combinations of the relations given above. The following example shows that the vector product of two vectors each of which has been derived as the gradient of a scalar is *always* solenoidal.

11.8 Cylindrical and spherical polar coordinates

Example Show that
$$\nabla \cdot (\nabla \phi \times \nabla \psi) = 0,$$
where ϕ and ψ are scalar fields.

From the previous section we have
$$\nabla \cdot (\mathbf{a} \times \mathbf{b}) = \mathbf{b} \cdot (\nabla \times \mathbf{a}) - \mathbf{a} \cdot (\nabla \times \mathbf{b}).$$
If we let $\mathbf{a} = \nabla \phi$ and $\mathbf{b} = \nabla \psi$ then we obtain
$$\nabla \cdot (\nabla \phi \times \nabla \psi) = \nabla \psi \cdot (\nabla \times \nabla \phi) - \nabla \phi \cdot (\nabla \times \nabla \psi) = 0, \qquad (11.38)$$
since $\nabla \times \nabla \phi = 0 = \nabla \times \nabla \psi$, from (11.32). ◀

EXERCISES 11.7

1. Prove by direct calculation that $\nabla \cdot (\phi \psi \mathbf{a}) = \phi \psi \nabla \cdot \mathbf{a} + \phi \mathbf{a} \cdot \nabla \psi + \psi \mathbf{a} \cdot \nabla \phi$.

2. Calculate the divergence of each of the following vector fields:
$$\mathbf{a} = (y^2 z^3, 2xyz^3, 3xy^2 z^2),$$
$$\mathbf{b} = (-xy, y^2 - z^2, -yz),$$
$$\mathbf{c} = (y^2 z^3 - xy, 4xyz^3 + y^2 - z^2, -xz^4 - yz).$$

   Which, if any, of the fields is solenoidal? What can be said about the vector field $\mathbf{c} - \mathbf{b}$? Relate it to the vector field $\mathbf{d} = (xyz^4, 0, \frac{1}{3} y^3 z^3)$.

3. For the Cartesian vector field $\mathbf{a} = (x^2(y+z), y^2(z+x), z^2(x+y))$, evaluate directly and separately (i) $\nabla^2 \mathbf{a}$, (ii) $\nabla(\nabla \cdot \mathbf{a})$ and (iii) $\nabla \times (\nabla \times \mathbf{a})$. Hence verify that both (11.36) and (11.37) are satisfied.

11.8 Cylindrical and spherical polar coordinates

The operators we have discussed in this chapter, i.e. grad, div, curl and ∇^2, have all been defined in terms of Cartesian coordinates, but for many physical situations other coordinate systems are more natural. For example, many systems, such as an isolated charge in space, have spherical symmetry and spherical polar coordinates would be the obvious choice. For axisymmetric systems, such as fluid flow in a pipe, cylindrical polar coordinates are the natural choice. The physical laws governing the behaviour of the systems are often expressed in terms of the vector operators we have been discussing, and so it is necessary to be able to express these operators in these other, non-Cartesian, coordinates. We first consider the two most common non-Cartesian coordinate systems, i.e. cylindrical and spherical polars, and then go on to discuss general curvilinear coordinates in the next section.

Vector calculus

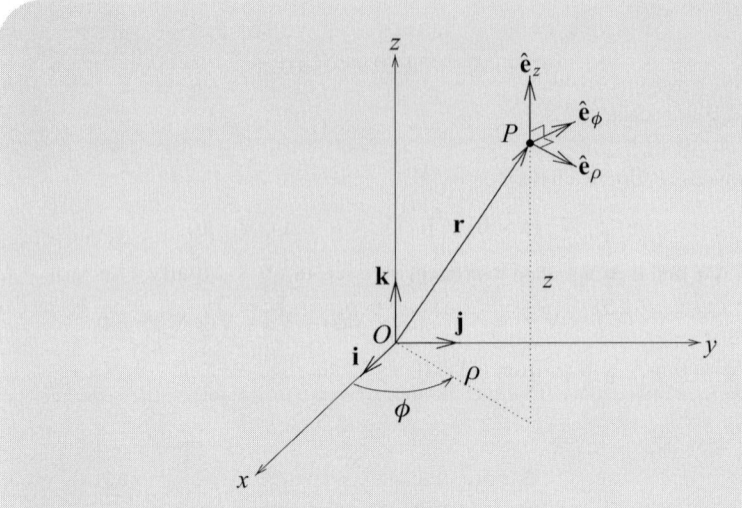

Figure 11.6 Cylindrical polar coordinates ρ, ϕ, z.

11.8.1 Cylindrical polar coordinates

As shown in Figure 11.6, the position of a point in space P having Cartesian coordinates x, y, z may be expressed in terms of cylindrical polar coordinates ρ, ϕ, z, where

$$x = \rho \cos\phi, \quad y = \rho \sin\phi, \quad z = z, \tag{11.39}$$

and $\rho \geq 0$, $0 \leq \phi < 2\pi$ and $-\infty < z < \infty$. The position vector of P may therefore be written

$$\mathbf{r} = \rho\cos\phi\,\mathbf{i} + \rho\sin\phi\,\mathbf{j} + z\,\mathbf{k}. \tag{11.40}$$

If we take the partial derivatives of $\mathbf{r}$ with respect to ρ, ϕ and z respectively then we obtain the three vectors

$$\mathbf{e}_\rho = \frac{\partial \mathbf{r}}{\partial \rho} = \cos\phi\,\mathbf{i} + \sin\phi\,\mathbf{j}, \tag{11.41}$$

$$\mathbf{e}_\phi = \frac{\partial \mathbf{r}}{\partial \phi} = -\rho\sin\phi\,\mathbf{i} + \rho\cos\phi\,\mathbf{j}, \tag{11.42}$$

$$\mathbf{e}_z = \frac{\partial \mathbf{r}}{\partial z} = \mathbf{k}. \tag{11.43}$$

These vectors lie in the directions of increasing ρ, ϕ and z respectively but are not all of unit length.[10] Although $\mathbf{e}_\rho$, $\mathbf{e}_\phi$ and $\mathbf{e}_z$ form a useful set of basis vectors in their own right (we will see in Section 11.9 that such a basis is sometimes the *most* useful), it is usual to work with the corresponding *unit* vectors, which are obtained by dividing each vector by

---

[10] $\mathbf{e}_\rho$ and $\mathbf{e}_z$ *are* of unit length, but $\mathbf{e}_\phi$ has length ρ, which, moreover, varies with the position of P.

11.8 Cylindrical and spherical polar coordinates

its modulus to give

$$\hat{\mathbf{e}}_\rho = \mathbf{e}_\rho = \cos\phi\,\mathbf{i} + \sin\phi\,\mathbf{j}, \tag{11.44}$$

$$\hat{\mathbf{e}}_\phi = \frac{1}{\rho}\mathbf{e}_\phi = -\sin\phi\,\mathbf{i} + \cos\phi\,\mathbf{j}, \tag{11.45}$$

$$\hat{\mathbf{e}}_z = \mathbf{e}_z = \mathbf{k}. \tag{11.46}$$

These three unit vectors, like the Cartesian unit vectors $\mathbf{i}$, $\mathbf{j}$ and $\mathbf{k}$, form an orthonormal triad[11] at each point in space, i.e. the basis vectors are mutually orthogonal and of unit length (see Figure 11.6). Unlike the fixed vectors $\mathbf{i}$, $\mathbf{j}$ and $\mathbf{k}$, however, $\hat{\mathbf{e}}_\rho$ and $\hat{\mathbf{e}}_\phi$ change direction as P moves.

The expression for a general infinitesimal vector displacement $d\mathbf{r}$ in the position of P is given, from (11.14), by

$$\begin{aligned} d\mathbf{r} &= \frac{\partial \mathbf{r}}{\partial \rho}\,d\rho + \frac{\partial \mathbf{r}}{\partial \phi}\,d\phi + \frac{\partial \mathbf{r}}{\partial z}\,dz \\ &= d\rho\,\mathbf{e}_\rho + d\phi\,\mathbf{e}_\phi + dz\,\mathbf{e}_z \\ &= d\rho\,\hat{\mathbf{e}}_\rho + \rho\,d\phi\,\hat{\mathbf{e}}_\phi + dz\,\hat{\mathbf{e}}_z. \end{aligned} \tag{11.47}$$

This expression illustrates an important difference between Cartesian and cylindrical polar coordinates (or non-Cartesian coordinates in general). In Cartesian coordinates, the distance moved in going from x to $x + dx$, with y and z held constant, is simply $ds = dx$. However, in cylindrical polars, if ϕ changes by $d\phi$, with ρ and z held constant, then the distance moved is *not* $d\phi$, but $ds = \rho\,d\phi$. Factors, such as the ρ in $\rho\,d\phi$, that multiply the coordinate differentials to give distances are known as *scale factors*. From (11.47), the scale factors for the ρ-, ϕ- and z-coordinates are therefore 1, ρ and 1 respectively.

The magnitude ds of the displacement $d\mathbf{r}$ is given in cylindrical polar coordinates by

$$(ds)^2 = d\mathbf{r}\cdot d\mathbf{r} = (d\rho)^2 + \rho^2(d\phi)^2 + (dz)^2,$$

where in the second equality we have used the fact that the basis vectors are orthonormal. We can also find the volume element in a cylindrical polar system (see Figure 11.7) by calculating the volume of the infinitesimal parallelepiped defined by the vectors $d\rho\,\hat{\mathbf{e}}_\rho$, $\rho\,d\phi\,\hat{\mathbf{e}}_\phi$ and $dz\,\hat{\mathbf{e}}_z$. As discussed in Section 9.5.1, this is given by the scalar triple product of the three vectors:

$$dV = |d\rho\,\hat{\mathbf{e}}_\rho \cdot (\rho\,d\phi\,\hat{\mathbf{e}}_\phi \times dz\,\hat{\mathbf{e}}_z)| = \rho\,d\rho\,d\phi\,dz,$$

which again uses the fact that the basis vectors are orthonormal. For a simple coordinate system such as cylindrical polars the expressions for $(ds)^2$ and dV are obvious from the geometry.

We will now express the vector operators discussed in this chapter in terms of cylindrical polar coordinates. Let us consider a vector field $\mathbf{a}(\rho, \phi, z)$ and a scalar field $\Phi(\rho, \phi, z)$, where we use Φ for the scalar field to avoid confusion with the azimuthal angle ϕ. We must

---

[11] Taken in the order given, ρ, ϕ, z, they form a right-handed set, as the reader should verify.

Vector calculus

Table 11.2 *Vector operators in cylindrical polar coordinates; Φ is a scalar field and $\mathbf{a}$ is a vector field.*

$$\nabla \Phi = \frac{\partial \Phi}{\partial \rho} \hat{\mathbf{e}}_\rho + \frac{1}{\rho}\frac{\partial \Phi}{\partial \phi} \hat{\mathbf{e}}_\phi + \frac{\partial \Phi}{\partial z} \hat{\mathbf{e}}_z$$

$$\nabla \cdot \mathbf{a} = \frac{1}{\rho}\frac{\partial}{\partial \rho}(\rho a_\rho) + \frac{1}{\rho}\frac{\partial a_\phi}{\partial \phi} + \frac{\partial a_z}{\partial z}$$

$$\nabla \times \mathbf{a} = \frac{1}{\rho}\begin{vmatrix} \hat{\mathbf{e}}_\rho & \rho\hat{\mathbf{e}}_\phi & \hat{\mathbf{e}}_z \\ \frac{\partial}{\partial \rho} & \frac{\partial}{\partial \phi} & \frac{\partial}{\partial z} \\ a_\rho & \rho a_\phi & a_z \end{vmatrix}$$

$$\nabla^2 \Phi = \frac{1}{\rho}\frac{\partial}{\partial \rho}\left(\rho \frac{\partial \Phi}{\partial \rho}\right) + \frac{1}{\rho^2}\frac{\partial^2 \Phi}{\partial \phi^2} + \frac{\partial^2 \Phi}{\partial z^2}$$

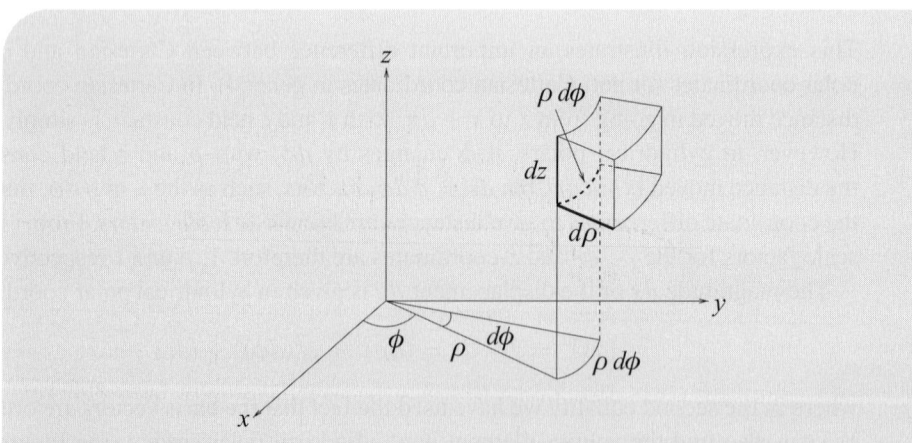

Figure 11.7 The element of volume in cylindrical polar coordinates is given by $\rho \, d\rho \, d\phi \, dz$.

first write the vector field in terms of the basis vectors of the cylindrical polar coordinate system, i.e.

$$\mathbf{a} = a_\rho \, \hat{\mathbf{e}}_\rho + a_\phi \, \hat{\mathbf{e}}_\phi + a_z \, \hat{\mathbf{e}}_z,$$

where a_ρ, a_ϕ and a_z are the components of $\mathbf{a}$ in the ρ-, ϕ- and z- directions respectively. The expressions for grad, div, curl and ∇^2 can then be calculated and are given in Table 11.2. Since the derivations of these expressions are rather complicated we leave them until our discussion of general curvilinear coordinates in the next section; the reader could well postpone examination of these formal proofs until some experience of using the expressions has been gained.

11.8 Cylindrical and spherical polar coordinates

Example Express the vector field $\mathbf{a} = yz\,\mathbf{i} - y\,\mathbf{j} + xz^2\,\mathbf{k}$ in cylindrical polar coordinates, and hence calculate its divergence. Show that the same result is obtained by evaluating the divergence in Cartesian coordinates.

The basis vectors of the cylindrical polar coordinate system are given in (11.44)–(11.46). Solving these equations simultaneously for $\mathbf{i}$, $\mathbf{j}$ and $\mathbf{k}$ we obtain

$$\mathbf{j} = \sin\phi\,\hat{\mathbf{e}}_\rho + \cos\phi\,\hat{\mathbf{e}}_\phi,$$
$$\mathbf{i} = \cos\phi\,\hat{\mathbf{e}}_\rho - \sin\phi\,\hat{\mathbf{e}}_\phi,$$
$$\mathbf{k} = \hat{\mathbf{e}}_z.$$

Substituting these relations and (11.39) into the expression for $\mathbf{a}$ we find

$$\mathbf{a} = z\rho\sin\phi\,(\cos\phi\,\hat{\mathbf{e}}_\rho - \sin\phi\,\hat{\mathbf{e}}_\phi) - \rho\sin\phi\,(\sin\phi\,\hat{\mathbf{e}}_\rho + \cos\phi\,\hat{\mathbf{e}}_\phi) + z^2\rho\cos\phi\,\hat{\mathbf{e}}_z$$
$$= (z\rho\sin\phi\cos\phi - \rho\sin^2\phi)\,\hat{\mathbf{e}}_\rho - (z\rho\sin^2\phi + \rho\sin\phi\cos\phi)\,\hat{\mathbf{e}}_\phi + z^2\rho\cos\phi\,\hat{\mathbf{e}}_z.$$

From this expression for $\mathbf{a}$, the individual components a_ρ, a_ϕ and a_z can be read off and substituted into the formula for $\nabla \cdot \mathbf{a}$ given in Table 11.2. When the partial differentiations indicated there are carried out,[12] the result is

$$\nabla \cdot \mathbf{a} = 2z\sin\phi\cos\phi - 2\sin^2\phi - 2z\sin\phi\cos\phi - \cos^2\phi + \sin^2\phi + 2z\rho\cos\phi$$
$$= 2z\rho\cos\phi - 1.$$

Alternatively, and much more quickly in this case, we can calculate the divergence directly in Cartesian coordinates. We obtain

$$\nabla \cdot \mathbf{a} = \frac{\partial a_x}{\partial x} + \frac{\partial a_y}{\partial y} + \frac{\partial a_z}{\partial z} = 0 + (-1) + 2xz = 2zx - 1,$$

which on substituting $x = \rho\cos\phi$ yields the same result as the calculation in cylindrical polars. ◀

Finally, we note that similar results can be obtained for (two-dimensional) polar coordinates in a plane by omitting the z-dependence. For example, $(ds)^2 = (d\rho)^2 + \rho^2(d\phi)^2$, while the element of volume is replaced by the element of area $dA = \rho\,d\rho\,d\phi$.

11.8.2 Spherical polar coordinates

As shown in Figure 11.8, the position of a point in space P, with Cartesian coordinates x, y, z, may be expressed in terms of spherical polar coordinates r, θ, ϕ, where

$$x = r\sin\theta\cos\phi, \quad y = r\sin\theta\sin\phi, \quad z = r\cos\theta, \tag{11.48}$$

and $r \geq 0, 0 \leq \theta \leq \pi$ and $0 \leq \phi < 2\pi$. The position vector of P may therefore be written as

$$\mathbf{r} = r\sin\theta\cos\phi\,\mathbf{i} + r\sin\theta\sin\phi\,\mathbf{j} + r\cos\theta\,\mathbf{k}.$$

If, in a similar manner to that used in the previous section for cylindrical polars, we find the partial derivatives of $\mathbf{r}$ with respect to r, θ and ϕ respectively and divide each of the

---

[12] Doing this for yourself gives useful practice.

Vector calculus

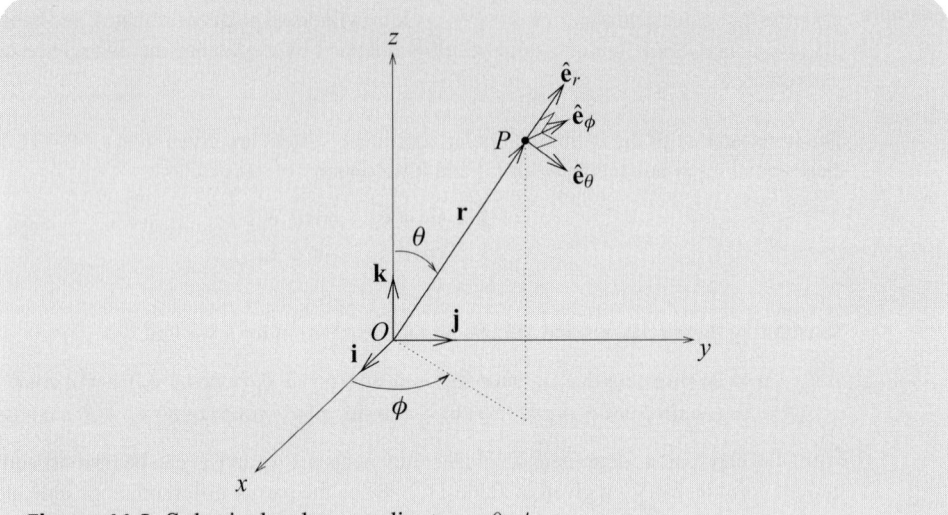

Figure 11.8 Spherical polar coordinates r, θ, ϕ.

resulting vectors by its modulus then we obtain the unit basis vectors

$$\hat{\mathbf{e}}_r = \sin\theta\cos\phi\,\mathbf{i} + \sin\theta\sin\phi\,\mathbf{j} + \cos\theta\,\mathbf{k},$$
$$\hat{\mathbf{e}}_\theta = \cos\theta\cos\phi\,\mathbf{i} + \cos\theta\sin\phi\,\mathbf{j} - \sin\theta\,\mathbf{k},$$
$$\hat{\mathbf{e}}_\phi = -\sin\phi\,\mathbf{i} + \cos\phi\,\mathbf{j}.$$

These unit vectors are in the directions of increasing r, θ and ϕ respectively and are the orthonormal basis set for spherical polar coordinates, as shown in Figure 11.8.

A general infinitesimal vector displacement in spherical polars is, from (11.14),

$$d\mathbf{r} = dr\,\hat{\mathbf{e}}_r + r\,d\theta\,\hat{\mathbf{e}}_\theta + r\sin\theta\,d\phi\,\hat{\mathbf{e}}_\phi; \qquad (11.49)$$

thus the scale factors for the r-, θ- and ϕ-coordinates are 1, r and $r\sin\theta$ respectively. The magnitude ds of the displacement $d\mathbf{r}$ is given by

$$(ds)^2 = d\mathbf{r}\cdot d\mathbf{r} = (dr)^2 + r^2(d\theta)^2 + r^2\sin^2\theta(d\phi)^2,$$

since the basis vectors form an orthonormal set. The element of volume in spherical polar coordinates (see Figure 11.9) is the volume of the infinitesimal parallelepiped defined by the vectors $dr\,\hat{\mathbf{e}}_r$, $r\,d\theta\,\hat{\mathbf{e}}_\theta$ and $r\sin\theta\,d\phi\,\hat{\mathbf{e}}_\phi$ and is given by

$$dV = \left|dr\,\hat{\mathbf{e}}_r \cdot (r\,d\theta\,\hat{\mathbf{e}}_\theta \times r\sin\theta\,d\phi\,\hat{\mathbf{e}}_\phi)\right| = r^2\sin\theta\,dr\,d\theta\,d\phi,$$

where again we use the fact that the basis vectors are orthonormal. The same expressions for $(ds)^2$ and dV could be obtained by visual examination of the geometry of the spherical-polar coordinate system.

We will now express the standard vector operators in spherical polar coordinates, using the same techniques as for cylindrical polar coordinates. We consider a scalar field $\Phi(r, \theta, \phi)$ and a vector field $\mathbf{a}(r, \theta, \phi)$. The latter may be written in terms of the basis

11.8 Cylindrical and spherical polar coordinates

Table 11.3 *Vector operators in spherical polar coordinates; Φ is a scalar field and $\mathbf{a}$ is a vector field.*

$$\nabla \Phi = \frac{\partial \Phi}{\partial r} \hat{\mathbf{e}}_r + \frac{1}{r} \frac{\partial \Phi}{\partial \theta} \hat{\mathbf{e}}_\theta + \frac{1}{r \sin \theta} \frac{\partial \Phi}{\partial \phi} \hat{\mathbf{e}}_\phi$$

$$\nabla \cdot \mathbf{a} = \frac{1}{r^2} \frac{\partial}{\partial r}(r^2 a_r) + \frac{1}{r \sin \theta} \frac{\partial}{\partial \theta}(\sin \theta \, a_\theta) + \frac{1}{r \sin \theta} \frac{\partial a_\phi}{\partial \phi}$$

$$\nabla \times \mathbf{a} = \frac{1}{r^2 \sin \theta} \begin{vmatrix} \hat{\mathbf{e}}_r & r \hat{\mathbf{e}}_\theta & r \sin \theta \, \hat{\mathbf{e}}_\phi \\ \frac{\partial}{\partial r} & \frac{\partial}{\partial \theta} & \frac{\partial}{\partial \phi} \\ a_r & r a_\theta & r \sin \theta \, a_\phi \end{vmatrix}$$

$$\nabla^2 \Phi = \frac{1}{r^2} \frac{\partial}{\partial r}\left(r^2 \frac{\partial \Phi}{\partial r}\right) + \frac{1}{r^2 \sin \theta} \frac{\partial}{\partial \theta}\left(\sin \theta \frac{\partial \Phi}{\partial \theta}\right) + \frac{1}{r^2 \sin^2 \theta} \frac{\partial^2 \Phi}{\partial \phi^2}$$

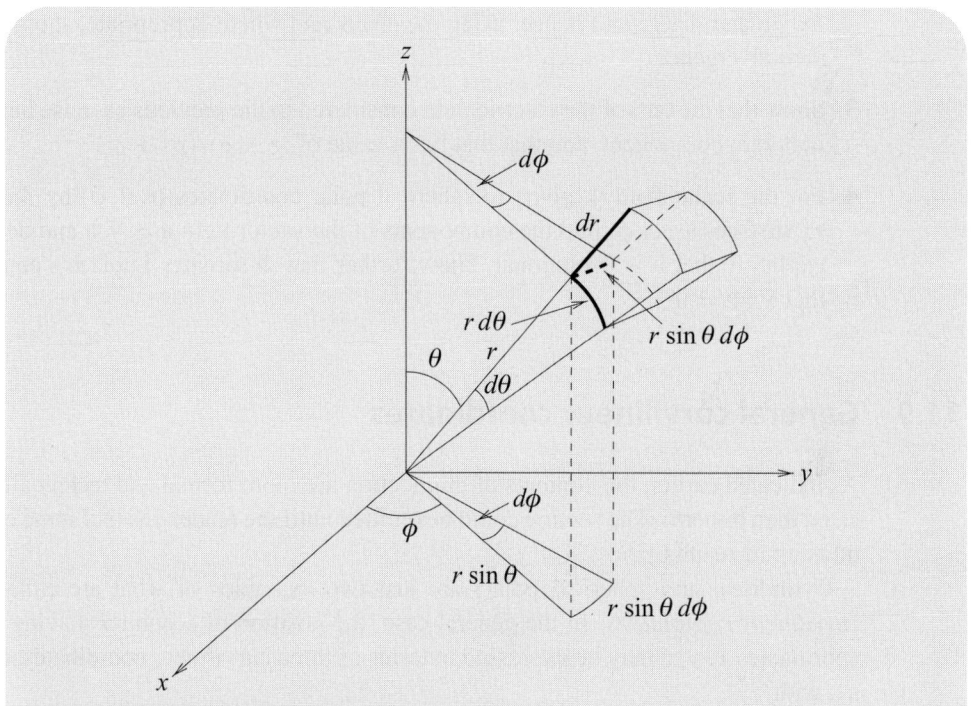

Figure 11.9 The element of volume in spherical polar coordinates is given by $r^2 \sin \theta \, dr \, d\theta \, d\phi$.

vectors of the spherical polar coordinate system as

$$\mathbf{a} = a_r \, \hat{\mathbf{e}}_r + a_\theta \, \hat{\mathbf{e}}_\theta + a_\phi \, \hat{\mathbf{e}}_\phi,$$

where a_r, a_θ and a_ϕ are the components of $\mathbf{a}$ in the r-, θ- and ϕ-directions respectively. The expressions for grad, div, curl and ∇^2 are given in Table 11.3. The derivations of these results are given in the next section.

Vector calculus

As a final note we mention that in the expression for $\nabla^2 \Phi$ given in Table 11.3 we can rewrite the first term on the RHS as follows:[13]

$$\frac{1}{r^2} \frac{\partial}{\partial r}\left(r^2 \frac{\partial \Phi}{\partial r}\right) = \frac{1}{r} \frac{\partial^2}{\partial r^2}(r\Phi).$$

This alternative expression can sometimes be useful in shortening calculations.

EXERCISES 11.8

1. Show that $\Phi(\rho, \phi, z) = A\rho^n \sin n\theta$ is a solution of Laplace's equation, $\nabla^2 \Phi = 0$, for arbitrary constant A and arbitrary integer n.

2. A certain axially symmetric electric field $\mathbf{E}$ is given in cylindrical polar coordinates (ρ, ϕ, z) by $\mathbf{E} = \left(\dfrac{\rho}{r^2 + zr}, 0, \dfrac{1}{r}\right)$, where $r^2 = \rho^2 + z^2$. Find simple expressions for $\partial r/\partial \rho$ and $\partial r/\partial z$ and hence, using the chain rule where appropriate, show that $\mathbf{E}$ has zero divergence.

3. Show that the curl of the electric field considered in the previous exercise has only one non-zero component, and that that has a value of $(z - \rho)/(\rho^2 + z^2)^{3/2}$.

4. For the scalar field Φ given in spherical polar coordinates (r, θ, ϕ) by $\Phi(r, \theta, \phi) = r^2 \sin\theta \cos\theta \cos\phi$, find the components of the vector field $\mathbf{a} = \nabla\Phi$ and demonstrate explicitly that it is irrotational. Show further that Φ satisfies Laplace's equation, i.e. that $\nabla^2 \Phi = 0$.

11.9 General curvilinear coordinates

As indicated earlier, the contents of this section are more formal and technically complicated than hitherto. The section could be omitted until the reader has had some experience of using its results.

Cylindrical and spherical polars are just two examples of what are called *general curvilinear coordinates*. In the general case, the position of a point P having Cartesian coordinates x, y, z may be expressed in terms of three curvilinear coordinates, u_1, u_2 and u_3, with

$$x = x(u_1, u_2, u_3), \quad y = y(u_1, u_2, u_3), \quad z = z(u_1, u_2, u_3).$$

Conversely, the u_i can be expressed in terms of x, y and z:

$$u_1 = u_1(x, y, z), \quad u_2 = u_2(x, y, z), \quad u_3 = u_3(x, y, z).$$

We assume that all these functions are continuous, differentiable and have a single-valued inverse, except perhaps at or on certain isolated points or lines, so that there is a one-to-one correspondence between the x, y, z and u_1, u_2, u_3 systems. The u_1-, u_2- and

---

[13] Show that both expressions are equal to $\partial^2 \Phi/\partial r^2 + (2/r) \partial \Phi/\partial r$.

11.9 General curvilinear coordinates

Figure 11.10 General curvilinear coordinates.

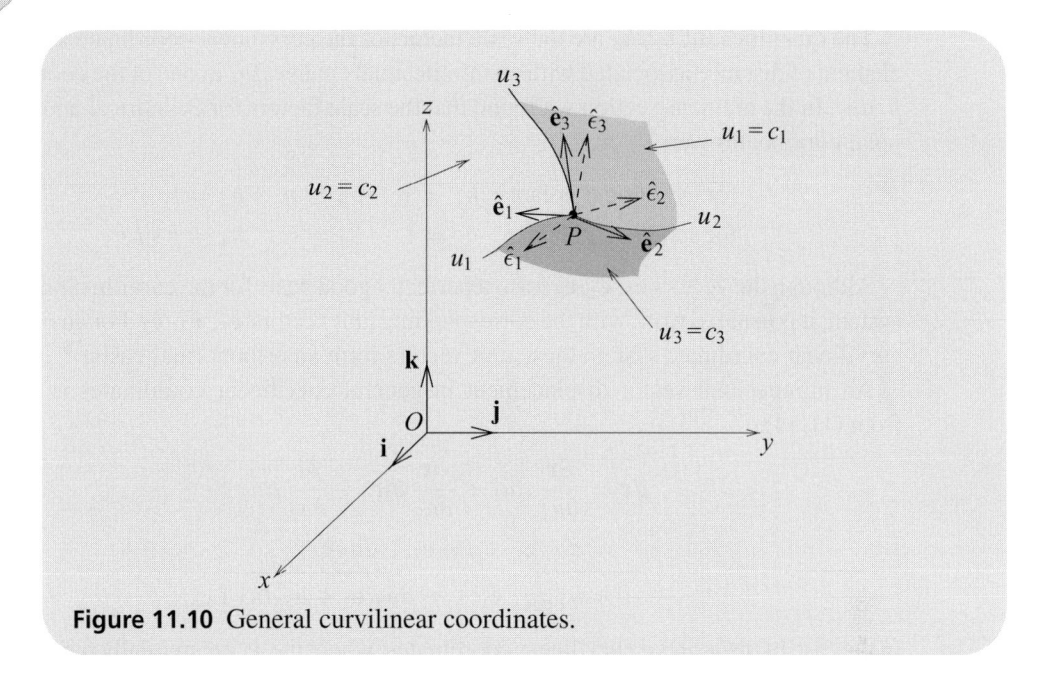

u_3-coordinate curves of a general curvilinear system are analogous to the x-, y- and z-axes of Cartesian coordinates. The surfaces $u_1 = c_1$, $u_2 = c_2$ and $u_3 = c_3$, where c_1, c_2, c_3 are constants, are called the *coordinate surfaces* and each pair of these surfaces has its intersection in a curve called a *coordinate curve* or *line* (see Figure 11.10).

As an example that is already familiar, in the spherical polar coordinate system the coordinate surfaces are spheres, cones and a 'sheaf' of half-planes containing the z-axis. The coordinate curves defined by the intersections of spheres and cones are circles, on which $u_3 = \phi$ identifies any particular point P; the curves determined by spheres and half-planes are semi-circular arcs (on which $u_2 = \theta$ defines P); the cones and half-planes meet in radial lines, on which the value of $u_1 = r$ picks out any particular point.

If at each point in space the three coordinate surfaces passing through the point meet at right angles then the curvilinear coordinate system is called *orthogonal*. In our example of spherical polars, the three coordinate surfaces passing through the point (R, Θ, Φ) are the sphere $r = R$, the circular cone $\theta = \Theta$ and the plane $\phi = \Phi$, which intersect at right angles at that point. Therefore, spherical polars form an orthogonal coordinate system (as do cylindrical polars[14]).

If $\mathbf{r}(u_1, u_2, u_3)$ is the position vector of the point P then $\mathbf{e}_1 = \partial \mathbf{r}/\partial u_1$ is a vector tangent to the u_1-curve at P (for which u_2 and u_3 are constants) in the direction of increasing u_1. Similarly, $\mathbf{e}_2 = \partial \mathbf{r}/\partial u_2$ and $\mathbf{e}_3 = \partial \mathbf{r}/\partial u_3$ are vectors tangent to the u_2- and u_3-curves at P in the direction of increasing u_2 and u_3 respectively. Denoting the lengths of these vectors by h_1, h_2 and h_3, the *unit* vectors in each of these directions are given by

$$\hat{\mathbf{e}}_1 = \frac{1}{h_1}\frac{\partial \mathbf{r}}{\partial u_1}, \quad \hat{\mathbf{e}}_2 = \frac{1}{h_2}\frac{\partial \mathbf{r}}{\partial u_2}, \quad \hat{\mathbf{e}}_3 = \frac{1}{h_3}\frac{\partial \mathbf{r}}{\partial u_3},$$

where $h_1 = |\partial \mathbf{r}/\partial u_1|$, $h_2 = |\partial \mathbf{r}/\partial u_2|$ and $h_3 = |\partial \mathbf{r}/\partial u_3|$.

---
[14] Identify the shapes of the coordinate surfaces for cylindrical polar coordinates.

Vector calculus

The quantities h_1, h_2, h_3 are the scale factors of the curvilinear coordinate system. The element of distance associated with an infinitesimal change du_i in one of the coordinates is $h_i\, du_i$. In the previous section we found that the scale factors for cylindrical and spherical polar coordinates were[15]

$$\text{for cylindrical polars} \quad h_\rho = 1, \quad h_\phi = \rho, \quad h_z = 1,$$
$$\text{for spherical polars} \quad h_r = 1, \quad h_\theta = r, \quad h_\phi = r\sin\theta.$$

Although the vectors $\mathbf{e}_1, \mathbf{e}_2, \mathbf{e}_3$ form a perfectly good basis for the curvilinear coordinate system, it is usual to work with the corresponding unit vectors $\hat{\mathbf{e}}_1, \hat{\mathbf{e}}_2, \hat{\mathbf{e}}_3$. For an orthogonal curvilinear coordinate system these unit vectors form an orthonormal basis.[16]

An infinitesimal vector displacement in general curvilinear coordinates is given by, from (11.14),

$$d\mathbf{r} = \frac{\partial \mathbf{r}}{\partial u_1}\, du_1 + \frac{\partial \mathbf{r}}{\partial u_2}\, du_2 + \frac{\partial \mathbf{r}}{\partial u_3}\, du_3 \tag{11.50}$$

$$= du_1\, \mathbf{e}_1 + du_2\, \mathbf{e}_2 + du_3\, \mathbf{e}_3 \tag{11.51}$$

$$= h_1\, du_1\, \hat{\mathbf{e}}_1 + h_2\, du_2\, \hat{\mathbf{e}}_2 + h_3\, du_3\, \hat{\mathbf{e}}_3. \tag{11.52}$$

In the case of *orthogonal* curvilinear coordinates, where the $\hat{\mathbf{e}}_i$ are mutually perpendicular, the element of arc length is given by[17]

$$(ds)^2 = d\mathbf{r} \cdot d\mathbf{r} = h_1^2 (du_1)^2 + h_2^2 (du_2)^2 + h_3^2 (du_3)^2. \tag{11.53}$$

The volume element for the coordinate system is the volume of the infinitesimal parallelepiped defined by the vectors $(\partial \mathbf{r}/\partial u_i)\, du_i = du_i\, \mathbf{e}_i = h_i\, du_i\, \hat{\mathbf{e}}_i$, for $i = 1, 2, 3$. For orthogonal coordinates this is given by

$$dV = |du_1\, \mathbf{e}_1 \cdot (du_2\, \mathbf{e}_2 \times du_3\, \mathbf{e}_3)|$$
$$= |h_1\, \hat{\mathbf{e}}_1 \cdot (h_2\, \hat{\mathbf{e}}_2 \times h_3\, \hat{\mathbf{e}}_3)|\, du_1\, du_2\, du_3$$
$$= h_1 h_2 h_3\, du_1\, du_2\, du_3.$$

Now, in addition to the set $\{\hat{\mathbf{e}}_i\}$, $i = 1, 2, 3$, there exists another set of three unit basis vectors at P. Since ∇u_1 is a vector normal to the surface $u_1 = c_1$, a unit vector in this direction is $\hat{\boldsymbol{\epsilon}}_1 = \nabla u_1 / |\nabla u_1|$. Similarly, $\hat{\boldsymbol{\epsilon}}_2 = \nabla u_2 / |\nabla u_2|$ and $\hat{\boldsymbol{\epsilon}}_3 = \nabla u_3 / |\nabla u_3|$ are unit vectors normal to the surfaces $u_2 = c_2$ and $u_3 = c_3$ respectively.

Therefore at each point P in a curvilinear coordinate system, there exist, in general, two sets of unit vectors: $\{\hat{\mathbf{e}}_i\}$, tangent to the coordinate curves, and $\{\hat{\boldsymbol{\epsilon}}_i\}$, normal to the coordinate surfaces. A vector $\mathbf{a}$ can be written in terms of either set of unit vectors:

$$\mathbf{a} = a_1\, \hat{\mathbf{e}}_1 + a_2\, \hat{\mathbf{e}}_2 + a_3\, \hat{\mathbf{e}}_3 = A_1 \hat{\boldsymbol{\epsilon}}_1 + A_2 \hat{\boldsymbol{\epsilon}}_2 + A_3 \hat{\boldsymbol{\epsilon}}_3,$$

where a_1, a_2, a_3 and A_1, A_2, A_3 are the components of $\mathbf{a}$ in the two systems.

---

[15] Formally, for Cartesian coordinates, $h_x = h_y = h_z = 1$.
[16] Though, in general, their directions in space depend upon where they are located – unlike the situation in a Cartesian system, where the directions are always the same.
[17] Remember that $\hat{\mathbf{e}}_i \cdot \hat{\mathbf{e}}_i = 1$ and that $\hat{\mathbf{e}}_i \cdot \hat{\mathbf{e}}_j = 0$ if $i \neq j$. Consequently, there are no cross-terms of the form $du_1 du_2$, say, in the expression for $(ds)^2$.

11.9 General curvilinear coordinates

However, it is intuitively the case, and it may be shown mathematically, that if the coordinate system is an orthogonal one, then the two bases are the same. In other words, in a system in which the coordinate surfaces passing through any one point meet at right angles, the direction in which the position vector moves when only u_1, say, is changed, is the same direction as that of the normal to the particular constant-u_1 surface that passes through the point.

Non-orthogonal coordinates are difficult to work with and beyond the scope of this book, and so from now on we will consider only orthogonal systems and not need to distinguish between the two sets of base vectors; for the rest of our discussion we will use the set $\{\hat{\mathbf{e}}_i\}$.

We next derive expressions for the standard vector operators in *orthogonal* curvilinear coordinates. The expressions for the vector operators in cylindrical and spherical polar coordinates given in Tables 11.2 and 11.3 respectively can be found from those derived below by inserting the appropriate expressions for the scale factors.

Gradient

The change $d\Phi$ in a scalar field Φ resulting from changes du_1, du_2, du_3 in the coordinates u_1, u_2, u_3 is given by, from (7.5),

$$d\Phi = \frac{\partial \Phi}{\partial u_1} du_1 + \frac{\partial \Phi}{\partial u_2} du_2 + \frac{\partial \Phi}{\partial u_3} du_3.$$

For orthogonal curvilinear coordinates u_1, u_2, u_3 we find from (11.52), and comparison with (11.22), that we can write this as

$$d\Phi = \nabla\Phi \cdot d\mathbf{r}, \tag{11.54}$$

where $\nabla\Phi$ is given by

$$\nabla\Phi = \frac{1}{h_1}\frac{\partial \Phi}{\partial u_1}\hat{\mathbf{e}}_1 + \frac{1}{h_2}\frac{\partial \Phi}{\partial u_2}\hat{\mathbf{e}}_2 + \frac{1}{h_3}\frac{\partial \Phi}{\partial u_3}\hat{\mathbf{e}}_3. \tag{11.55}$$

This implies that the del operator can be written

$$\nabla = \frac{\hat{\mathbf{e}}_1}{h_1}\frac{\partial}{\partial u_1} + \frac{\hat{\mathbf{e}}_2}{h_2}\frac{\partial}{\partial u_2} + \frac{\hat{\mathbf{e}}_3}{h_3}\frac{\partial}{\partial u_3}.$$

A particular result that we will need below is obtained by setting $\Phi = u_i$ in (11.55); this leads immediately to

$$\nabla u_i = \frac{\hat{\mathbf{e}}_i}{h_i} \quad \text{for } i = 1, 2, 3. \tag{11.56}$$

Divergence

In order to derive the expression for the divergence of a vector field in orthogonal curvilinear coordinates, we must first write the vector field in terms of the basis vectors of the coordinate system:

$$\mathbf{a} = a_1\,\hat{\mathbf{e}}_1 + a_2\,\hat{\mathbf{e}}_2 + a_3\,\hat{\mathbf{e}}_3.$$

Vector calculus

The divergence is then given by

$$\nabla \cdot \mathbf{a} = \frac{1}{h_1 h_2 h_3} \left[\frac{\partial}{\partial u_1}(h_2 h_3 a_1) + \frac{\partial}{\partial u_2}(h_3 h_1 a_2) + \frac{\partial}{\partial u_3}(h_1 h_2 a_3) \right], \quad (11.57)$$

as we now prove.

Example Prove the expression for $\nabla \cdot \mathbf{a}$ in orthogonal curvilinear coordinates.

Let us first consider the sub-expression $\nabla \cdot (a_1 \hat{\mathbf{e}}_1)$, the first term in the expression for $\mathbf{a}$. Using (11.56) twice, we can write $\hat{\mathbf{e}}_1 = \hat{\mathbf{e}}_2 \times \hat{\mathbf{e}}_3 = h_2 \nabla u_2 \times h_3 \nabla u_3$, and so

$$\nabla \cdot (a_1 \hat{\mathbf{e}}_1) = \nabla \cdot (a_1 h_2 h_3 \nabla u_2 \times \nabla u_3)$$
$$= \nabla(a_1 h_2 h_3) \cdot (\nabla u_2 \times \nabla u_3) + a_1 h_2 h_3 \nabla \cdot (\nabla u_2 \times \nabla u_3).$$

Here we have used the sixth vector identity in Table 11.1, with the product $a_1 h_2 h_3$ replacing ϕ and the vector product $\nabla u_2 \times \nabla u_3$ replacing the vector. However, both u_2 and u_3 are scalar fields (as well as being coordinates) and therefore, from (11.38), $\nabla \cdot (\nabla u_2 \times \nabla u_3) = 0$. So, using (11.56) again, we obtain

$$\nabla \cdot (a_1 \hat{\mathbf{e}}_1) = \nabla(a_1 h_2 h_3) \cdot \left(\frac{\hat{\mathbf{e}}_2}{h_2} \times \frac{\hat{\mathbf{e}}_3}{h_3} \right) = \nabla(a_1 h_2 h_3) \cdot \frac{\hat{\mathbf{e}}_1}{h_2 h_3}.$$

Using (11.55) to find the gradient of $a_1 h_2 h_3$, but retaining only the $\hat{\mathbf{e}}_1$ component (because of the scalar product with $\hat{\mathbf{e}}_1$ in the above equation), we find on substitution that

$$\nabla \cdot (a_1 \hat{\mathbf{e}}_1) = \frac{1}{h_1 h_2 h_3} \frac{\partial}{\partial u_1}(a_1 h_2 h_3).$$

Repeating the analysis for $\nabla \cdot (a_2 \hat{\mathbf{e}}_2)$ and $\nabla \cdot (a_3 \hat{\mathbf{e}}_3)$, and adding the results, we obtain the general expression for the divergence of $\mathbf{a}$ as stated in (11.57). ◀

Laplacian

In expression (11.57) for the divergence, we now replace $\mathbf{a}$ by $\nabla \Phi$ as given by (11.55), i.e. we set $a_i = h_i^{-1} \partial \Phi / \partial u_i$. When this is done we obtain

$$\nabla^2 \Phi = \frac{1}{h_1 h_2 h_3} \left[\frac{\partial}{\partial u_1}\left(\frac{h_2 h_3}{h_1} \frac{\partial \Phi}{\partial u_1} \right) + \frac{\partial}{\partial u_2}\left(\frac{h_3 h_1}{h_2} \frac{\partial \Phi}{\partial u_2} \right) + \frac{\partial}{\partial u_3}\left(\frac{h_1 h_2}{h_3} \frac{\partial \Phi}{\partial u_3} \right) \right],$$

which is the general expression for the Laplacian of Φ in orthogonal curvilinear coordinates.

Curl

The curl of a vector field $\mathbf{a} = a_1 \hat{\mathbf{e}}_1 + a_2 \hat{\mathbf{e}}_2 + a_3 \hat{\mathbf{e}}_3$ in orthogonal curvilinear coordinates is given by

$$\nabla \times \mathbf{a} = \frac{1}{h_1 h_2 h_3} \begin{vmatrix} h_1 \hat{\mathbf{e}}_1 & h_2 \hat{\mathbf{e}}_2 & h_3 \hat{\mathbf{e}}_3 \\ \dfrac{\partial}{\partial u_1} & \dfrac{\partial}{\partial u_2} & \dfrac{\partial}{\partial u_3} \\ h_1 a_1 & h_2 a_2 & h_3 a_3 \end{vmatrix}. \quad (11.58)$$

11.9 General curvilinear coordinates

Table 11.4 *Vector operators in orthogonal curvilinear coordinates u_1, u_2, u_3. Φ is a scalar field and $\mathbf{a}$ is a vector field.*

$$\nabla\Phi = \frac{1}{h_1}\frac{\partial\Phi}{\partial u_1}\hat{\mathbf{e}}_1 + \frac{1}{h_2}\frac{\partial\Phi}{\partial u_2}\hat{\mathbf{e}}_2 + \frac{1}{h_3}\frac{\partial\Phi}{\partial u_3}\hat{\mathbf{e}}_3$$

$$\nabla\cdot\mathbf{a} = \frac{1}{h_1 h_2 h_3}\left[\frac{\partial}{\partial u_1}(h_2 h_3 a_1) + \frac{\partial}{\partial u_2}(h_3 h_1 a_2) + \frac{\partial}{\partial u_3}(h_1 h_2 a_3)\right]$$

$$\nabla\times\mathbf{a} = \frac{1}{h_1 h_2 h_3}\begin{vmatrix} h_1\hat{\mathbf{e}}_1 & h_2\hat{\mathbf{e}}_2 & h_3\hat{\mathbf{e}}_3 \\ \dfrac{\partial}{\partial u_1} & \dfrac{\partial}{\partial u_2} & \dfrac{\partial}{\partial u_3} \\ h_1 a_1 & h_2 a_2 & h_3 a_3 \end{vmatrix}$$

$$\nabla^2\Phi = \frac{1}{h_1 h_2 h_3}\left[\frac{\partial}{\partial u_1}\left(\frac{h_2 h_3}{h_1}\frac{\partial\Phi}{\partial u_1}\right) + \frac{\partial}{\partial u_2}\left(\frac{h_3 h_1}{h_2}\frac{\partial\Phi}{\partial u_2}\right) + \frac{\partial}{\partial u_3}\left(\frac{h_1 h_2}{h_3}\frac{\partial\Phi}{\partial u_3}\right)\right]$$

The proof of this is similar to that for the divergence operator, and again we give it as a worked example.

Example Prove the expression for $\nabla\times\mathbf{a}$ in orthogonal curvilinear coordinates.

Let us first consider the sub-expression $\nabla\times(a_1\hat{\mathbf{e}}_1)$. Since $\hat{\mathbf{e}}_1 = h_1\nabla u_1$, we have, using the penultimate entry in Table 11.1, that

$$\nabla\times(a_1\hat{\mathbf{e}}_1) = \nabla\times(a_1 h_1 \nabla u_1)$$
$$= \nabla(a_1 h_1)\times\nabla u_1 + a_1 h_1(\nabla\times\nabla u_1).$$

But $\nabla\times\nabla u_1 = 0$, so we obtain

$$\nabla\times(a_1\hat{\mathbf{e}}_1) = \nabla(a_1 h_1)\times\frac{\hat{\mathbf{e}}_1}{h_1}.$$

Letting $\Phi = a_1 h_1$ in (11.55) and substituting into the above equation, we find

$$\nabla\times(a_1\hat{\mathbf{e}}_1) = \frac{\hat{\mathbf{e}}_2}{h_3 h_1}\frac{\partial}{\partial u_3}(a_1 h_1) - \frac{\hat{\mathbf{e}}_3}{h_1 h_2}\frac{\partial}{\partial u_2}(a_1 h_1).$$

Notice that, because of the cross product, it is the $\hat{\mathbf{e}}_3$ component of $\nabla(a_1 h_1)$ that produces the $\hat{\mathbf{e}}_2$ component of the above expression, and vice versa. The corresponding analysis of $\nabla\times(a_2\hat{\mathbf{e}}_2)$ produces terms in $\hat{\mathbf{e}}_3$ and $\hat{\mathbf{e}}_1$, whilst that of $\nabla\times(a_3\hat{\mathbf{e}}_3)$ produces terms in $\hat{\mathbf{e}}_1$ and $\hat{\mathbf{e}}_2$. When the three results are added together, the coefficients multiplying $\hat{\mathbf{e}}_1$, $\hat{\mathbf{e}}_2$ and $\hat{\mathbf{e}}_3$ are the same as those obtained by writing out (11.58) explicitly, thus proving the stated result. ◀

The general expressions for the vector operators in orthogonal curvilinear coordinates are shown for reference in Table 11.4. The explicit results for cylindrical and spherical polar coordinates, given in Tables 11.2 and 11.3 respectively, are obtained by substituting the appropriate set of scale factors in each case.

SUMMARY

1. *Derivatives of products*
   - $\dfrac{d}{du}(\phi \mathbf{a}) = \phi \dfrac{d\mathbf{a}}{du} + \dfrac{d\phi}{du}\mathbf{a},$
   - $\dfrac{d}{du}(\mathbf{a} \cdot \mathbf{b}) = \mathbf{a} \cdot \dfrac{d\mathbf{b}}{du} + \dfrac{d\mathbf{a}}{du} \cdot \mathbf{b},$
   - $\dfrac{d}{du}(\mathbf{a} \times \mathbf{b}) = \mathbf{a} \times \dfrac{d\mathbf{b}}{du} + \dfrac{d\mathbf{a}}{du} \times \mathbf{b}.$

2. *Integrals of vectors* with respect to scalars
   (i) The integral has the same nature (vector or scalar) as the integrand.
   (ii) The constant of integration for indefinite integrals must be of the same nature as the integral.

3. *For a surface* given by $\mathbf{r} = \mathbf{r}(u, v)$
   - If $u = u(\lambda)$ and $v = v(\lambda)$ is a curve $\mathbf{r} = \mathbf{r}(\lambda)$ lying in the surface, then the tangent vector to the curve is
   $$\mathbf{t} = \dfrac{d\mathbf{r}}{d\lambda} = \dfrac{\partial \mathbf{r}}{\partial u}\dfrac{du}{d\lambda} + \dfrac{\partial \mathbf{r}}{\partial v}\dfrac{dv}{d\lambda}.$$
   - A normal to the tangent plane and the size of an element of area are
   $$\mathbf{n} = \dfrac{\partial \mathbf{r}}{\partial u} \times \dfrac{\partial \mathbf{r}}{\partial v} \quad \text{and} \quad dS = |\mathbf{n}|\, du\, dv.$$

4. *Vector operators acting on fields*, and their Cartesian forms
   - Gradient of a scalar:
   $$\text{grad } \phi = \nabla \phi = \mathbf{i}\dfrac{\partial \phi}{\partial x} + \mathbf{j}\dfrac{\partial \phi}{\partial y} + \mathbf{k}\dfrac{\partial \phi}{\partial z} \quad \text{with} \quad \nabla \phi \cdot d\mathbf{r} = d\phi.$$
   - The rate of change in the direction of $\mathbf{a}$ is given by the operator
   $$\hat{\mathbf{a}} \cdot \nabla = \hat{a}_x \dfrac{\partial}{\partial x} + \hat{a}_y \dfrac{\partial}{\partial y} + \hat{a}_z \dfrac{\partial}{\partial z}.$$
   - Divergence of a vector:
   $$\text{div } \mathbf{a} = \nabla \cdot \mathbf{a} = \dfrac{\partial a_x}{\partial x} + \dfrac{\partial a_y}{\partial y} + \dfrac{\partial a_z}{\partial z}.$$
   - Curl of a vector:
   $$\text{curl } \mathbf{a} = \nabla \times \mathbf{a} = \begin{vmatrix} \mathbf{i} & \mathbf{j} & \mathbf{k} \\ \dfrac{\partial}{\partial x} & \dfrac{\partial}{\partial y} & \dfrac{\partial}{\partial z} \\ a_x & a_y & a_z \end{vmatrix}.$$
   - For the actions of the operators on sums and products, see Table 11.1 on p. 466.

5. *Combinations of vector operators*, and their Cartesian forms

Name	Symbolic form	Value or Cartesian form
del^2 ϕ	$\nabla^2 \phi = \nabla \cdot \nabla \phi$	$\dfrac{\partial^2 \phi}{\partial x^2} + \dfrac{\partial^2 \phi}{\partial y^2} + \dfrac{\partial^2 \phi}{\partial z^2}$
curl grad ϕ	$\nabla \times \nabla \phi$	$\mathbf{0}$
div curl $\mathbf{a}$	$\nabla \cdot (\nabla \times \mathbf{a})$	0
grad div $\mathbf{a}$	$\nabla(\nabla \cdot \mathbf{a})$	$\left(\dfrac{\partial^2 a_x}{\partial x^2} + \dfrac{\partial^2 a_y}{\partial x \partial y} + \dfrac{\partial^2 a_z}{\partial x \partial z}\right)\mathbf{i} +$ $+ (\cdots)\mathbf{j} + (\cdots)\mathbf{k}$
curl curl $\mathbf{a}$	$\nabla \times (\nabla \times \mathbf{a})$ $= \nabla(\nabla \cdot \mathbf{a}) - \nabla^2 \mathbf{a}$	$\left(\dfrac{\partial^2 a_y}{\partial x \partial y} + \dfrac{\partial^2 a_z}{\partial x \partial z} - \dfrac{\partial^2 a_x}{\partial y^2} - \dfrac{\partial^2 a_x}{\partial z^2}\right)\mathbf{i} +$ $+ (\cdots)\mathbf{j} + (\cdots)\mathbf{k}$

- In Cartesian coordinates, $(\nabla^2 \mathbf{a})_i = \nabla^2 a_i$.
- $\nabla \cdot (\nabla \phi \times \nabla \psi) = 0$ if ϕ and ψ are scalar fields.

6. *Vector operators in polar coordinates*
   - For cylindrical polars, see Table 11.2 on p. 472.
   - For spherical polars, see Table 11.3 on p. 475.

7. *General orthogonal curvilinear coordinates* with $\mathbf{r} = \mathbf{r}(u_1, u_2, u_3)$
   - Scale factors and unit vectors
   $$h_i = \left|\frac{\partial \mathbf{r}}{\partial u_i}\right|, \qquad \hat{\mathbf{e}}_i = \frac{1}{h_i}\frac{\partial \mathbf{r}}{\partial u_i}.$$
   - For cylindrical polars $h_1 = 1$, $h_2 = \rho$, $h_3 = 1$; for spherical polars $h_1 = 1$, $h_2 = r$, $h_3 = r \sin \theta$.
   - $(ds)^2 = d\mathbf{r} \cdot d\mathbf{r} = h_1^2 (du_1)^2 + h_2^2 (du_2)^2 + h_3^2 (du_3)^2$ with no cross-terms of the form $du_1\, du_2$.
   - $dV = h_1 h_2 h_3 \, du_1\, du_2\, du_3$.
   - $\nabla = \dfrac{\hat{\mathbf{e}}_1}{h_1}\dfrac{\partial}{\partial u_1} + \dfrac{\hat{\mathbf{e}}_2}{h_2}\dfrac{\partial}{\partial u_2} + \dfrac{\hat{\mathbf{e}}_3}{h_3}\dfrac{\partial}{\partial u_3}.$
   - For vector operators, see Table 11.4 on p. 481.

PROBLEMS

11.1. A particle's position at time $t \geq 0$ is given by $\mathbf{r} = \mathbf{r}(t)$ and remains within a bounded region.

(a) By considering the time derivative of $|\mathbf{r}|^2$, show mathematically the intuitive result that when the particle is furthest from the origin its position and velocity vectors are orthogonal.

(b) For $\mathbf{r}(t) = (2\mathbf{i} + 5t\,\mathbf{j})e^{-t}$, find the particle's maximum distance d from the origin and verify explicitly that result (a) is valid.

(c) At what time does the particle trajectory cross the line $\mathbf{r} = \lambda(\mathbf{i} + \mathbf{j})$, and how far from the origin is it when it does so? How does the trajectory approach its end-point?

11.2. In terms of spherical polar coordinates (r, θ, ϕ) the position vector $\mathbf{r}$ can be written on a Cartesian basis as

$$\mathbf{r} = r\sin\theta\cos\phi\,\mathbf{i} + r\sin\theta\sin\phi\,\mathbf{j} + r\cos\theta\,\mathbf{k}.$$

The corresponding unit basis vectors $\hat{\mathbf{e}}_r$, $\hat{\mathbf{e}}_\theta$ and $\hat{\mathbf{e}}_\phi$ are given explicitly in Section 11.8.2.

(a) Show that the derivatives of these basis vectors with respect to (time) t can be expressed by

$$\frac{d\hat{\mathbf{e}}_r}{dt} = \dot\theta\,\hat{\mathbf{e}}_\theta + \sin\theta\,\dot\phi\,\hat{\mathbf{e}}_\phi,$$

$$\frac{d\hat{\mathbf{e}}_\theta}{dt} = -\dot\theta\,\hat{\mathbf{e}}_r + \cos\theta\,\dot\phi\,\hat{\mathbf{e}}_\phi,$$

$$\frac{d\hat{\mathbf{e}}_\phi}{dt} = -\sin\theta\,\dot\phi\,\hat{\mathbf{e}}_r - \cos\theta\,\dot\phi\,\hat{\mathbf{e}}_\theta.$$

(b) A general point P has position vector $\mathbf{p}$ given in spherical polar coordinates by $\mathbf{p} = r\,\hat{\mathbf{e}}_r$. Find an expression for its velocity vector and show that its acceleration has components

$$\ddot r - r\dot\theta^2 - r\sin^2\theta\,\dot\phi^2, \quad r\ddot\theta + 2\dot r\dot\theta - r\sin\theta\cos\theta\,\dot\phi^2$$

$$\text{and} \quad 2(\dot r\sin\theta + r\cos\theta\,\dot\theta)\dot\phi + r\sin\theta\,\ddot\phi.$$

(c) Verify that this general result reduces to the ones expected when (i) ϕ is constant and (ii) $\theta = \pi/2$ with $\dot\theta = 0$.

11.3. Evaluate the integral

$$\int \left[\mathbf{a}(\dot{\mathbf{b}} \cdot \mathbf{a} + \mathbf{b} \cdot \dot{\mathbf{a}}) + \dot{\mathbf{a}}(\mathbf{b} \cdot \mathbf{a}) - 2(\dot{\mathbf{a}} \cdot \mathbf{a})\mathbf{b} - \dot{\mathbf{b}}|\mathbf{a}|^2\right] dt$$

in which $\dot{\mathbf{a}}$, $\dot{\mathbf{b}}$ are the derivatives of $\mathbf{a}$, $\mathbf{b}$ with respect to t.

11.4. At time $t = 0$, the vectors $\mathbf{E}$ and $\mathbf{B}$ are given by $\mathbf{E} = \mathbf{E}_0$ and $\mathbf{B} = \mathbf{B}_0$, where the unit vectors, $\mathbf{E}_0$ and $\mathbf{B}_0$ are fixed and orthogonal. The equations of motion are

$$\frac{d\mathbf{E}}{dt} = \mathbf{E}_0 + \mathbf{B} \times \mathbf{E}_0,$$

$$\frac{d\mathbf{B}}{dt} = \mathbf{B}_0 + \mathbf{E} \times \mathbf{B}_0.$$

Problems

Find **E** and **B** at a general time t, showing that after a long time the directions of **E** and **B** have almost interchanged.

11.5. The general equation of motion of a (non-relativistic) particle of mass m and charge q when it is placed in a region where there is a magnetic field **B** and an electric field **E** is

$$m\ddot{\mathbf{r}} = q(\mathbf{E} + \dot{\mathbf{r}} \times \mathbf{B});$$

here **r** is the position of the particle at time t and $\dot{\mathbf{r}} = d\mathbf{r}/dt$, etc. Write this as three separate equations in terms of the Cartesian components of the vectors involved.

For the simple case of crossed uniform fields $\mathbf{E} = E\mathbf{i}$, $\mathbf{B} = B\mathbf{j}$, in which the particle starts from the origin at $t = 0$ with $\dot{\mathbf{r}} = v_0\mathbf{k}$, find the equations of motion and show the following:
(a) if $v_0 = E/B$ then the particle continues its initial motion;
(b) if $v_0 = 0$ then the particle follows the space curve given in terms of the parameter ξ by

$$x = \frac{mE}{B^2 q}(1 - \cos\xi), \quad y = 0, \quad z = \frac{mE}{B^2 q}(\xi - \sin\xi).$$

Interpret this curve geometrically and relate ξ to t. Show that the total distance travelled by the particle after time t is given by

$$\frac{2E}{B} \int_0^t \left| \sin\frac{Bqt'}{2m} \right| dt'.$$

11.6. Use vector methods to find the maximum angle to the horizontal at which a stone may be thrown so as to ensure that it is always moving away from the thrower.

11.7. If two systems of coordinates with a common origin O are rotating with respect to each other, the measured accelerations differ in the two systems. Denoting by **r** and **r**′ position vectors in frames $OXYZ$ and $OX'Y'Z'$, respectively, the connection between the two is

$$\ddot{\mathbf{r}}' = \ddot{\mathbf{r}} + \dot{\boldsymbol{\omega}} \times \mathbf{r} + 2\boldsymbol{\omega} \times \dot{\mathbf{r}} + \boldsymbol{\omega} \times (\boldsymbol{\omega} \times \mathbf{r}),$$

where $\boldsymbol{\omega}$ is the angular velocity vector of the rotation of $OXYZ$ with respect to $OX'Y'Z'$ (taken as fixed). The third term on the RHS is known as the Coriolis acceleration, whilst the final term gives rise to a centrifugal force.

Consider the application of this result to the firing of a shell of mass m from a stationary ship on the steadily rotating earth, working to the first order in ω ($= 7.3 \times 10^{-5}$ rad s^{-1}). If the shell is fired with velocity **v** at time $t = 0$ and only reaches a height that is small compared with the radius of the earth, show that its acceleration, as recorded on the ship, is given approximately by

$$\ddot{\mathbf{r}} = \mathbf{g} - 2\boldsymbol{\omega} \times (\mathbf{v} + \mathbf{g}t),$$

where $m\mathbf{g}$ is the weight of the shell measured on the ship's deck.

The shell is fired at another stationary ship (a distance **s** away) and **v** is such that the shell would have hit its target had there been no Coriolis effect.

(a) Show that without the Coriolis effect the time of flight of the shell would have been $\tau = -2\mathbf{g} \cdot \mathbf{v}/g^2$.

(b) Show further that when the shell actually hits the sea it is off-target by approximately

$$\frac{2\tau}{g^2}[(\mathbf{g} \times \boldsymbol{\omega}) \cdot \mathbf{v}](\mathbf{g}\tau + \mathbf{v}) - (\boldsymbol{\omega} \times \mathbf{v})\tau^2 - \frac{1}{3}(\boldsymbol{\omega} \times \mathbf{g})\tau^3.$$

(c) Estimate the order of magnitude Δ of this miss for a shell for which the initial speed v is 300 m s^{-1}, firing close to its maximum range (**v** makes an angle of $\pi/4$ with the vertical) in a northerly direction, whilst the ship is stationed at latitude 45° North.

11.8. Find the areas of the given surfaces using parametric coordinates.

(a) Using the parameterisation $x = u \cos \phi$, $y = u \sin \phi$, $z = u \cot \Omega$, find the sloping surface area of a right circular cone of semi-angle Ω whose base has radius a. Verify that it is equal to $\frac{1}{2} \times$ perimeter of the base $\times$ slope height.

(b) Using the same parameterisation as in (a) for x and y, and an appropriate choice for z, find the surface area between the planes $z = 0$ and $z = Z$ of the paraboloid of revolution $z = \alpha(x^2 + y^2)$.

11.9. Parameterising the hyperboloid

$$\frac{x^2}{a^2} + \frac{y^2}{b^2} - \frac{z^2}{c^2} = 1$$

by $x = a \cos \theta \sec \phi$, $y = b \sin \theta \sec \phi$, $z = c \tan \phi$, show that an area element on its surface is

$$dS = \sec^2 \phi \left[c^2 \sec^2 \phi \left(b^2 \cos^2 \theta + a^2 \sin^2 \theta \right) + a^2 b^2 \tan^2 \phi \right]^{1/2} d\theta \, d\phi.$$

Use this formula to show that the area of the curved surface $x^2 + y^2 - z^2 = a^2$ between the planes $z = 0$ and $z = 2a$ is

$$\pi a^2 \left(6 + \frac{1}{\sqrt{2}} \sinh^{-1} 2\sqrt{2} \right).$$

11.10. For the function

$$z(x, y) = (x^2 - y^2)e^{-x^2 - y^2},$$

find the location(s) at which the steepest gradient occurs. What are the magnitude and direction of that gradient? The algebra involved is easier if plane polar coordinates are used.

11.11. Verify by direct calculation that

$$\nabla \cdot (\mathbf{a} \times \mathbf{b}) = \mathbf{b} \cdot (\nabla \times \mathbf{a}) - \mathbf{a} \cdot (\nabla \times \mathbf{b}).$$

Problems

11.12. In the following problems, **a**, **b** and **c** are vector fields.
(a) Simplify
$$\nabla \times \mathbf{a}(\nabla \cdot \mathbf{a}) + \mathbf{a} \times [\nabla \times (\nabla \times \mathbf{a})] + \mathbf{a} \times \nabla^2 \mathbf{a}.$$
(b) By explicitly writing out the terms in Cartesian coordinates, prove that
$$[\mathbf{c} \cdot (\mathbf{b} \cdot \nabla) - \mathbf{b} \cdot (\mathbf{c} \cdot \nabla)]\mathbf{a} = (\nabla \times \mathbf{a}) \cdot (\mathbf{b} \times \mathbf{c}).$$
(c) Prove that $\mathbf{a} \times (\nabla \times \mathbf{a}) = \nabla(\tfrac{1}{2}a^2) - (\mathbf{a} \cdot \nabla)\mathbf{a}$.

11.13. Evaluate the Laplacian of the function
$$\psi(x, y, z) = \frac{zx^2}{x^2 + y^2 + z^2}$$
(a) directly in Cartesian coordinates and (b) after changing to a spherical polar coordinate system. Verify that, as they must, the two methods give the same result.

11.14. Verify that (11.37) is valid for each component separately when **a** is the Cartesian vector $x^2 y\,\mathbf{i} + xyz\,\mathbf{j} + z^2 y\,\mathbf{k}$, by showing that each side of the equation is equal to $z\,\mathbf{i} + (2x + 2z)\,\mathbf{j} + x\,\mathbf{k}$.

11.15. The (Maxwell) relationship between a time-independent magnetic field **B** and the current density **J** (measured in SI units in A m^{-2}) producing it,
$$\nabla \times \mathbf{B} = \mu_0 \mathbf{J},$$
can be applied to a long cylinder of conducting ionised gas which, in cylindrical polar coordinates, occupies the region $\rho < a$.
(a) Show that a uniform current density $(0, C, 0)$ and a magnetic field $(0, 0, B)$, with B constant $(= B_0)$ for $\rho > a$ and $B = B(\rho)$ for $\rho < a$, are consistent with this equation. Given that $B(0) = 0$ and that **B** is continuous at $\rho = a$, obtain expressions for C and $B(\rho)$ in terms of B_0 and a.
(b) The magnetic field can be expressed as $\mathbf{B} = \nabla \times \mathbf{A}$, where **A** is known as the vector potential. Show that a suitable **A** that has only one non-vanishing component, $A_\phi(\rho)$, can be found, and obtain explicit expressions for $A_\phi(\rho)$ for both $\rho < a$ and $\rho > a$. Like **B**, the vector potential is continuous at $\rho = a$.
(c) The gas pressure $p(\rho)$ satisfies the hydrostatic equation $\nabla p = \mathbf{J} \times \mathbf{B}$ and vanishes at the outer wall of the cylinder. Find a general expression for p.

11.16. Evaluate the Laplacian of a vector field using two different coordinate systems as follows.
(a) For cylindrical polar coordinates ρ, ϕ, z, evaluate the derivatives of the three unit vectors with respect to each of the coordinates, showing that only $\partial \hat{\mathbf{e}}_\rho / \partial \phi$ and $\partial \hat{\mathbf{e}}_\phi / \partial \phi$ are non-zero.

(i) Hence evaluate $\nabla^2 \mathbf{a}$ when $\mathbf{a}$ is the vector $\hat{\mathbf{e}}_\rho$, i.e. a vector of unit magnitude everywhere directed radially outwards and expressed by $a_\rho = 1, a_\phi = a_z = 0$.
(ii) Note that it is trivially obvious that $\nabla \times \mathbf{a} = \mathbf{0}$ and hence that Equation (11.36) requires that $\nabla(\nabla \cdot \mathbf{a}) = \nabla^2 \mathbf{a}$.
(iii) Evaluate $\nabla(\nabla \cdot \mathbf{a})$ and show that the latter equation holds, but that

$$[\nabla(\nabla \cdot \mathbf{a})]_\rho \neq \nabla^2 a_\rho.$$

(b) Rework the same problem in Cartesian coordinates (where, as it happens, the algebra is more complicated).

11.17. Maxwell's equations for electromagnetism in free space (i.e. in the absence of charges, currents and dielectric or magnetic media) can be written

(i) $\nabla \cdot \mathbf{B} = 0$, (ii) $\nabla \cdot \mathbf{E} = 0$,
(iii) $\nabla \times \mathbf{E} + \dfrac{\partial \mathbf{B}}{\partial t} = \mathbf{0}$, (iv) $\nabla \times \mathbf{B} - \dfrac{1}{c^2}\dfrac{\partial \mathbf{E}}{\partial t} = \mathbf{0}$.

A vector $\mathbf{A}$ is defined by $\mathbf{B} = \nabla \times \mathbf{A}$, and a scalar ϕ by $\mathbf{E} = -\nabla \phi - \partial \mathbf{A}/\partial t$. Show that if the condition

(v) $\nabla \cdot \mathbf{A} + \dfrac{1}{c^2}\dfrac{\partial \phi}{\partial t} = 0$

is imposed (this is known as choosing the Lorentz gauge), then $\mathbf{A}$ and ϕ satisfy wave equations as follows:

(vi) $\nabla^2 \phi - \dfrac{1}{c^2}\dfrac{\partial^2 \phi}{\partial t^2} = 0$,

(vii) $\nabla^2 \mathbf{A} - \dfrac{1}{c^2}\dfrac{\partial^2 \mathbf{A}}{\partial t^2} = \mathbf{0}$.

The reader is invited to proceed as follows.
(a) Verify that the expressions for $\mathbf{B}$ and $\mathbf{E}$ in terms of $\mathbf{A}$ and ϕ are consistent with (i) and (iii).
(b) Substitute for $\mathbf{E}$ in (ii) and use the derivative with respect to time of (v) to eliminate $\mathbf{A}$ from the resulting expression. Hence obtain (vi).
(c) Substitute for $\mathbf{B}$ and $\mathbf{E}$ in (iv) in terms of $\mathbf{A}$ and ϕ. Then use the gradient of (v) to simplify the resulting equation and so obtain (vii).

11.18. For a description using spherical polar coordinates with axial symmetry, of the flow of a very viscous fluid, the components of the velocity field $\mathbf{u}$ are given in terms of the *stream function* ψ by

$$u_r = \dfrac{1}{r^2 \sin\theta}\dfrac{\partial \psi}{\partial \theta}, \qquad u_\theta = \dfrac{-1}{r \sin\theta}\dfrac{\partial \psi}{\partial r}.$$

Find an explicit expression for the differential operator E defined by

$$E\psi = -(r \sin\theta)(\nabla \times \mathbf{u})_\phi.$$

The stream function satisfies the equation of motion $E^2\psi = 0$ and, for the flow of a fluid past a sphere, takes the form $\psi(r, \theta) = f(r)\sin^2\theta$. Show that $f(r)$ satisfies the (ordinary) differential equation

$$r^4 f^{(4)} - 4r^2 f'' + 8rf' - 8f = 0.$$

11.19. Paraboloidal coordinates u, v, ϕ are defined in terms of Cartesian coordinates by

$$x = uv\cos\phi, \qquad y = uv\sin\phi, \qquad z = \tfrac{1}{2}(u^2 - v^2).$$

Identify the coordinate surfaces in the u, v, ϕ system. Verify that each coordinate surface ($u = $ constant, say) intersects every coordinate surface on which one of the other two coordinates (v, say) is constant. Show further that the system of coordinates is an orthogonal one and determine its scale factors. Prove that the u-component of $\nabla \times \mathbf{a}$ is given by

$$\frac{1}{(u^2+v^2)^{1/2}}\left(\frac{a_\phi}{v} + \frac{\partial a_\phi}{\partial v}\right) - \frac{1}{uv}\frac{\partial a_v}{\partial \phi}.$$

11.20. In a Cartesian system, A and B are the points $(0, 0, -1)$ and $(0, 0, 1)$ respectively. In a new coordinate system a general point P is given by (u_1, u_2, u_3) with $u_1 = \tfrac{1}{2}(r_1 + r_2)$, $u_2 = \tfrac{1}{2}(r_1 - r_2)$, $u_3 = \phi$; here r_1 and r_2 are the distances AP and BP and ϕ is the angle between the plane ABP and $y = 0$.
(a) Express z and the perpendicular distance ρ from P to the z-axis in terms of u_1, u_2, u_3.
(b) Evaluate $\partial x/\partial u_i, \partial y/\partial u_i, \partial z/\partial u_i$, for $i = 1, 2, 3$.
(c) Find the Cartesian components of $\hat{\mathbf{u}}_j$ and hence show that the new coordinates are mutually orthogonal. Evaluate the scale factors and the infinitesimal volume element in the new coordinate system.
(d) Determine and sketch the forms of the surfaces $u_i = $ constant.
(e) Find the most general function f of u_1 only that satisfies $\nabla^2 f = 0$.

11.21. Hyperbolic coordinates u, v, ϕ are defined in terms of Cartesian coordinates by

$$x = \cosh u \cos v \cos\phi, \qquad y = \cosh u \cos v \sin\phi, \qquad z = \sinh u \sin v.$$

Sketch the coordinate curves in the $\phi = 0$ plane, showing that far from the origin they become concentric circles and radial lines. In particular, identify the curves $u = 0$, $v = 0$, $v = \pi/2$ and $v = \pi$. Calculate the tangent vectors at a general point, show that they are mutually orthogonal and deduce that the appropriate scale factors are

$$h_u = h_v = (\cosh^2 u - \cos^2 v)^{1/2}, \qquad h_\phi = \cosh u \cos v.$$

Find the most general function $\psi(u)$ of u only that satisfies Laplace's equation $\nabla^2 \psi = 0$.

HINTS AND ANSWERS

11.1. (a) Note that $d(\mathbf{r} \cdot \mathbf{r})/dt = 2\mathbf{r} \cdot \dot{\mathbf{r}}$. (b) $d = |\mathbf{r}(\frac{4}{5})| = 2\sqrt{5}e^{-4/5} = 2.0095$.
(c) $t = 2/5$. 1.896. Tangentially to the y-axis.

11.3. Group the term so that they form the total derivatives of compound vector expressions. The integral has the value $\mathbf{a} \times (\mathbf{a} \times \mathbf{b}) + \mathbf{h}$.

11.5. For crossed uniform fields $\ddot{x} + (Bq/m)^2 x = q(E - Bv_0)/m$, $\ddot{y} = 0$,
$m\dot{z} = qBx + mv_0$;
(b) $\xi = Bqt/m$; the path is a cycloid in the plane $y = 0$; $ds = [(dx/dt)^2 + (dz/dt)^2]^{1/2} dt$.

11.7. $\mathbf{g} = \ddot{\mathbf{r}}' - \boldsymbol{\omega} \times (\boldsymbol{\omega} \times \mathbf{r})$, where $\ddot{\mathbf{r}}'$ is the shell's acceleration measured by an observer fixed in space. To first order in ω, the direction of $\mathbf{g}$ is radial, i.e. parallel to $\ddot{\mathbf{r}}'$.
(a) Note that $\mathbf{s}$ is orthogonal to $\mathbf{g}$.
(b) If the actual time of flight is T, use $(\mathbf{s} + \boldsymbol{\Delta}) \cdot \mathbf{g} = 0$ to show that
$$T \approx \tau(1 + 2g^{-2}(\mathbf{g} \times \boldsymbol{\omega}) \cdot \mathbf{v} + \cdots).$$
In the Coriolis terms, it is sufficient to put $T \approx \tau$.
(c) For this situation $(\mathbf{g} \times \boldsymbol{\omega}) \cdot \mathbf{v} = 0$ and $\boldsymbol{\omega} \times \mathbf{v} = \mathbf{0}$; $\tau \approx 43$ s and $\Delta = 10$–15 m to the East.

11.9. To integrate $\sec^2 \phi (\sec^2 \phi + \tan^2 \phi)^{1/2} d\phi$ put $\tan \phi = 2^{-1/2} \sinh \psi$.

11.11. Work in Cartesian coordinates, regrouping the terms obtained by evaluating the divergence on the LHS.

11.13. (a) $2z(x^2 + y^2 + z^2)^{-3}[(y^2 + z^2)(y^2 + z^2 - 3x^2) - 4x^4]$; (b) $2r^{-1} \cos\theta (1 - 5\sin^2\theta \cos^2\phi)$; both are equal to $2zr^{-4}(r^2 - 5x^2)$.

11.15. Use the formulae given in Table 11.2.
(a) $C = -B_0/(\mu_0 a)$; $B(\rho) = B_0 \rho/a$.
(b) $B_0 \rho^2/(3a)$ for $\rho < a$, and $B_0[\rho/2 - a^2/(6\rho)]$ for $\rho > a$.
(c) $[B_0^2/(2\mu_0)][1 - (\rho/a)^2]$.

11.17. Recall that $\nabla \times \nabla\phi = \mathbf{0}$ for any scalar ϕ and that $\partial/\partial t$ and ∇ act on different variables.

11.19. Two sets of paraboloids of revolution about the z-axis and the sheaf of planes containing the z-axis. For constant u, $-\infty < z < u^2/2$; for constant v, $-v^2/2 < z < \infty$. The scale factors are $h_u = h_v = (u^2 + v^2)^{1/2}$, $h_\phi = uv$.

11.21. The tangent vectors are as follows: for $u = 0$, the line joining $(1, 0, 0)$ and $(-1, 0, 0)$; for $v = 0$, the line joining $(1, 0, 0)$ and $(\infty, 0, 0)$; for $v = \pi/2$, the line $(0, 0, z)$; for $v = \pi$, the line joining $(-1, 0, 0)$ and $(-\infty, 0, 0)$.
$\psi(u) = 2\tan^{-1} e^u + c$, derived from $\partial[\cosh u (\partial \psi/\partial u)]/\partial u = 0$.

12

Line, surface and volume integrals

In Chapter 11 we encountered continuously varying scalar and vector fields and discussed the action of various differential operators on them. There is often a need to consider not only these differential operations, but also the integration of field quantities along lines, over surfaces and throughout volumes. In general the integrand may be scalar or vector in nature, but the evaluation of such integrals involves their reduction to one or more scalar integrals, which are then evaluated. In the case of surface and volume integrals this requires the evaluation of double and triple integrals (see Chapter 8).

12.1 Line integrals

In this section we discuss *line* or *path integrals*, in which some quantity related to the field is integrated between two given points in space, A and B, along a prescribed curve C that joins them. In general, we may encounter line integrals of the forms

$$\int_C \phi \, d\mathbf{r}, \quad \int_C \mathbf{a} \cdot d\mathbf{r}, \quad \int_C \mathbf{a} \times d\mathbf{r}, \tag{12.1}$$

where ϕ is a scalar field and $\mathbf{a}$ is a vector field. The three integrals themselves are respectively vector, scalar and vector in nature. As we will see below, in physical applications line integrals of the second type are by far the most common.

The formal definition of a line integral closely follows that of ordinary integrals and can be considered as the limit of a sum. We may divide the path C joining the points A and B into N small line elements $\Delta \mathbf{r}_p$, $p = 1, \ldots, N$. If (x_p, y_p, z_p) is any point on the line element $\Delta \mathbf{r}_p$ then the second type of line integral in (12.1), for example, is defined as

$$\int_C \mathbf{a} \cdot d\mathbf{r} = \lim_{N \to \infty} \sum_{p=1}^{N} \mathbf{a}(x_p, y_p, z_p) \cdot \Delta \mathbf{r}_p,$$

where it is assumed that all $|\Delta \mathbf{r}_p| \to 0$ as $N \to \infty$.

Each of the line integrals in (12.1) is evaluated over some curve C that may be either open (A and B being distinct points) or closed (the curve C forms a loop, so that A and B are coincident). In the case where C is closed, the line integral is written $\oint_C$ to indicate this. The curve may be given either parametrically by $\mathbf{r}(u) = x(u)\mathbf{i} + y(u)\mathbf{j} + z(u)\mathbf{k}$ or by means of simultaneous equations relating x, y, z for the given path (in Cartesian coordinates).

In general, the value of the line integral depends not only on the end-points A and B but also on the path C joining them. For a closed curve we must also specify the direction around the loop in which the integral is taken. It is usually taken to be such that a person

Line, surface and volume integrals

walking around the loop C in this direction always has the region R on their left; this is equivalent to traversing C in the anticlockwise direction (as viewed from above).

12.1.1 Evaluating line integrals

The method of evaluating a line integral is to reduce it to a set of scalar integrals. It is usual to work in Cartesian coordinates, in which case $d\mathbf{r} = dx\,\mathbf{i} + dy\,\mathbf{j} + dz\,\mathbf{k}$. The first type of line integral in (12.1) then becomes simply

$$\int_C \phi\, d\mathbf{r} = \mathbf{i} \int_C \phi(x, y, z)\,dx + \mathbf{j} \int_C \phi(x, y, z)\,dy + \mathbf{k} \int_C \phi(x, y, z)\,dz.$$

The three integrals on the RHS are ordinary scalar integrals that can be evaluated in the usual way once the path of integration C has been specified. Note that in the above we have used relations of the form

$$\int \phi\, \mathbf{i}\, dx = \mathbf{i} \int \phi\, dx,$$

which is allowable since the Cartesian unit vectors are of constant magnitude and direction and hence may be taken out of the integral. If we had been using a different coordinate system, such as spherical polars, then, as we saw in the previous chapter, the unit basis vectors would not be constant. In that case the basis vectors could not be factorised out of the integral.

The second and third line integrals in (12.1) can also be reduced to a set of scalar integrals by writing the vector field $\mathbf{a}$ in terms of its Cartesian components as $\mathbf{a} = a_x\mathbf{i} + a_y\mathbf{j} + a_z\mathbf{k}$, where a_x, a_y and a_z are each (in general) functions of x, y and z. The second line integral in (12.1), for example, can then be written as

$$\int_C \mathbf{a} \cdot d\mathbf{r} = \int_C (a_x\mathbf{i} + a_y\mathbf{j} + a_z\mathbf{k}) \cdot (dx\,\mathbf{i} + dy\,\mathbf{j} + dz\,\mathbf{k})$$

$$= \int_C (a_x\,dx + a_y\,dy + a_z\,dz)$$

$$= \int_C a_x\,dx + \int_C a_y\,dy + \int_C a_z\,dz. \qquad (12.2)$$

A similar procedure may be followed for the third type of line integral in (12.1), which involves a cross product.[1]

Line integrals have properties that are analogous to those of ordinary integrals. In particular, the following are useful properties (which we illustrate using the second form of line integral in (12.1) but which are valid for all three types).

(i) Reversing the path of integration changes the sign of the integral. If the path C along which the line integrals are evaluated has A and B as its end-points then

$$\int_A^B \mathbf{a} \cdot d\mathbf{r} = -\int_B^A \mathbf{a} \cdot d\mathbf{r}.$$

---

[1] Write out this integral explicitly in Cartesian coordinates.

12.1 Line integrals

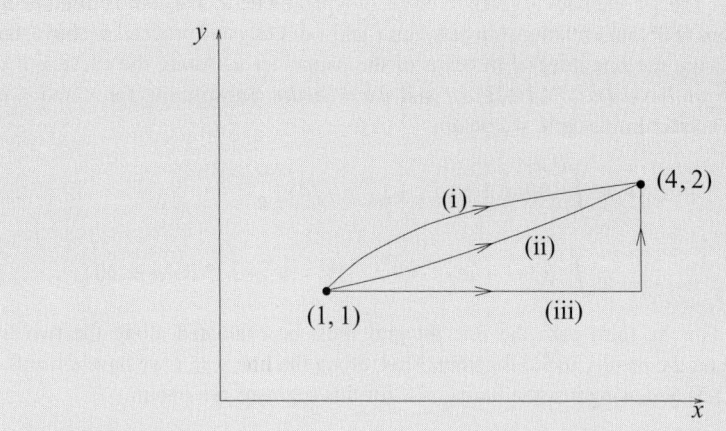

Figure 12.1 Different possible paths between the points (1, 1) and (4, 2).

This implies that if the path C is a loop then integrating around the loop in the opposite direction changes the sign of the integral.

(ii) If the path of integration is subdivided into smaller segments then the sum of the separate line integrals along each segment is equal to the line integral along the whole path. So, if P is any point on the path of integration that lies between the path's end-points A and B then

$$\int_A^B \mathbf{a} \cdot d\mathbf{r} = \int_A^P \mathbf{a} \cdot d\mathbf{r} + \int_P^B \mathbf{a} \cdot d\mathbf{r}.$$

Example Evaluate the line integral $I = \int_C \mathbf{a} \cdot d\mathbf{r}$, where $\mathbf{a} = (x+y)\mathbf{i} + (y-x)\mathbf{j}$, along each of the paths in the xy-plane shown in Figure 12.1, namely

(i) the parabola $y^2 = x$ from (1, 1) to (4, 2),
(ii) the curve $x = 2u^2 + u + 1$, $y = 1 + u^2$ from (1, 1) to (4, 2),
(iii) the line $y = 1$ from (1, 1) to (4, 1), followed by the line $x = 4$ from (4, 1) to (4, 2).

Since each of the paths lies entirely in the xy-plane, we have $d\mathbf{r} = dx\,\mathbf{i} + dy\,\mathbf{j}$. We can therefore write the line integral as

$$I = \int_C \mathbf{a} \cdot d\mathbf{r} = \int_C [(x+y)\,dx + (y-x)\,dy]. \tag{12.3}$$

We must now evaluate this line integral along each of the prescribed paths.

Case (i). Along the parabola $y^2 = x$ we have $2y\,dy = dx$. Substituting for x in (12.3) and using just the limits on y, we obtain

$$I = \int_{(1,1)}^{(4,2)} [(x+y)\,dx + (y-x)\,dy] = \int_1^2 [(y^2+y)2y + (y-y^2)]\,dy = 11\tfrac{1}{3}.$$

Note that we could just as easily have substituted for y and obtained an integral in x, which would have given the same result.[2]

---

2 Show that this is so.

Line, surface and volume integrals

Case (ii). The second path is given in terms of a parameter u. We could eliminate u between the two equations to obtain a relationship between x and y directly and proceed as above, but it is usually quicker to write the line integral in terms of the parameter u. Along the curve $x = 2u^2 + u + 1$, $y = 1 + u^2$ we have $dx = (4u + 1) du$ and $dy = 2u \, du$. Substituting for x and y in (12.3) and writing the correct limits on u, we obtain

$$I = \int_{(1,1)}^{(4,2)} [(x+y) \, dx + (y-x) \, dy]$$

$$= \int_0^1 [(3u^2 + u + 2)(4u+1) - (u^2 + u)2u] \, du = 10\tfrac{2}{3}.$$

Case (iii). For the third path the line integral must be evaluated along the two line segments separately and the results added together. First, along the line $y = 1$ we have $dy = 0$. Substituting this into (12.3) and using just the limits on x for this segment, we obtain

$$\int_{(1,1)}^{(4,1)} [(x+y) \, dx + (y-x) \, dy] = \int_1^4 (x+1) \, dx = 10\tfrac{1}{2}.$$

Next, along the line $x = 4$ we have $dx = 0$. Substituting this into (12.3) and using just the limits on y for this segment, we obtain

$$\int_{(4,1)}^{(4,2)} [(x+y) \, dx + (y-x) \, dy] = \int_1^2 (y-4) \, dy = -2\tfrac{1}{2}.$$

The value of the line integral along the whole path is just the sum of the values of the line integrals along each segment, and is given by $I = 10\tfrac{1}{2} - 2\tfrac{1}{2} = 8$. ◀

When calculating a line integral along some curve C, which is given in terms of x, y and z, we are sometimes faced with the problem that the curve C is such that x, y and z are not single-valued functions of one another over the entire length of the curve. This is a particular problem for closed loops in the xy-plane (and also for some open curves). In such cases the path may be subdivided into shorter line segments along which one coordinate is a single-valued function of the other two. The sum of the line integrals along these segments is then equal to the line integral along the entire curve C. A better solution, however, is to represent the curve in a parametric form $\mathbf{r}(u)$ that is valid for its entire length.

Example Evaluate the line integral $I = \oint_C x \, dy$, where C is the circle in the xy-plane defined by $x^2 + y^2 = a^2$, $z = 0$.

Adopting the usual convention mentioned above, the circle C is to be traversed in the anticlockwise direction. Taking the circle as a whole means x is not a single-valued function of y. We must therefore divide the path into two parts with $x = +\sqrt{a^2 - y^2}$ for the semicircle lying to the right of $x = 0$, and $x = -\sqrt{a^2 - y^2}$ for the semicircle lying to the left of $x = 0$. The required line

12.1 Line integrals

integral is then the sum of the integrals along the two semicircles. Substituting for x, and then setting $y = a \sin\theta$, it is given by

$$I = \oint_C x\,dy = \int_{-a}^{a} \sqrt{a^2 - y^2}\,dy + \int_{a}^{-a} \left(-\sqrt{a^2 - y^2}\right) dy$$

$$= 4\int_0^a \sqrt{a^2 - y^2}\,dy$$

$$= 4a^2 \int_0^{\pi/2} \sqrt{1 - \sin^2\theta}\,\cos\theta\,d\theta = 4a^2 \int_0^{\pi/2} \cos^2\theta\,d\theta = \pi a^2.$$

Alternatively, we can represent the entire circle parametrically, in terms of the azimuthal angle ϕ, so that $x = a\cos\phi$ and $y = a\sin\phi$ with ϕ running from 0 to 2π. The integral can now be evaluated over the whole circle at once. Noting that $dy = a\cos\phi\,d\phi$, we can rewrite the line integral completely in terms of the parameter ϕ and obtain

$$I = \oint_C x\,dy = a^2 \int_0^{2\pi} \cos^2\phi\,d\phi = \pi a^2.$$

The final evaluation used the fact that the integral with respect to ϕ of $\cos^2 \lambda\phi$ (or $\sin^2 \lambda\phi$) over any range that is $m\pi$ in length, is equal to $\frac{1}{2}$ times the length of the range if λm is an integer.[3] ◀

12.1.2 Physical examples of line integrals

There are many physical examples of line integrals, but perhaps the most common is the expression for the total work done by a force **F** when it moves its point of application from a point A to a point B along a given curve C. We allow the magnitude and direction of **F** to vary along the curve. Let the force act at a point **r** and consider a small displacement $d\mathbf{r}$ along the curve; then the small amount of work done is $dW = \mathbf{F} \cdot d\mathbf{r}$, as discussed in Section 9.4.1 (note that dW can be either positive or negative). Therefore, the total work done in traversing the path C is

$$W_C = \int_C \mathbf{F} \cdot d\mathbf{r}.$$

Naturally, other physical quantities can be expressed in such a way. For example, the electrostatic potential energy gained by moving a charge q along a path C in an electric field **E** is $-q \int_C \mathbf{E} \cdot d\mathbf{r}$. We may also note that Ampère's law concerning the magnetic field **B** associated with a current-carrying wire can be written as

$$\oint_C \mathbf{B} \cdot d\mathbf{r} = \mu_0 I,$$

where I is the current enclosed by a closed path C traversed in a right-handed sense with respect to the current direction.

Magnetostatics also provides a physical example of the third type of line integral in (12.1). If a loop of wire C carrying a current I is placed in a magnetic field **B** then the

---

[3] A result worth remembering, since the squares of sinusoids occur throughout the mathematics of physics and engineering.

force $d\mathbf{F}$ on a small length $d\mathbf{r}$ of the wire is given by $d\mathbf{F} = I\,d\mathbf{r} \times \mathbf{B}$, and so the total (vector) force on the loop is

$$\mathbf{F} = I \oint_C d\mathbf{r} \times \mathbf{B}.$$

12.1.3 Line integrals with respect to a scalar

In addition to those listed in (12.1), we can form other types of line integral, which depend on a particular curve C but for which we integrate with respect to a scalar du, rather than the vector differential $d\mathbf{r}$. This distinction is somewhat arbitrary, however, since we can always rewrite line integrals containing the vector differential $d\mathbf{r}$ as a line integral with respect to some scalar parameter. If the path C along which the integral is taken is described parametrically by $\mathbf{r}(u)$ then

$$d\mathbf{r} = \frac{d\mathbf{r}}{du}\,du,$$

and the second type of line integral in (12.1), for example, can be written as

$$\int_C \mathbf{a} \cdot d\mathbf{r} = \int_C \mathbf{a} \cdot \frac{d\mathbf{r}}{du}\,du.$$

A similar procedure can be followed for the other types of line integral in (12.1).

Commonly occurring special cases of line integrals with respect to a scalar are

$$\int_C \phi\,ds, \qquad \int_C \mathbf{a}\,ds,$$

where s is the arc length along the curve C. We can always represent C parametrically by $\mathbf{r}(u)$, and since

$$(ds)^2 = (dx)^2 + (dy)^2 + (dz)^2 = d\mathbf{r} \cdot d\mathbf{r},$$

we have

$$\left(\frac{ds}{du}\right)^2 = \frac{d\mathbf{r}}{du} \cdot \frac{d\mathbf{r}}{du}.$$

Consequently we may write

$$ds = \sqrt{\frac{d\mathbf{r}}{du} \cdot \frac{d\mathbf{r}}{du}}\,du.$$

The line integrals can therefore be expressed entirely in terms of the parameter u and thence evaluated.

Example Evaluate the line integral $I = \int_C (x - y)^2\,ds$, where C is the semicircle of radius a running from $A = (a, 0)$ to $B = (-a, 0)$ and for which $y \geq 0$.

The semicircular path from A to B can be described in terms of the azimuthal angle ϕ (measured from the x-axis) by

$$\mathbf{r}(\phi) = a\cos\phi\,\mathbf{i} + a\sin\phi\,\mathbf{j},$$

12.2 Connectivity of regions

where ϕ runs from 0 to π. Therefore, the element of arc length is given by

$$ds = \sqrt{\frac{d\mathbf{r}}{d\phi} \cdot \frac{d\mathbf{r}}{d\phi}} \, d\phi = a[(-\sin\phi)^2 + (\cos\phi)^2] \, d\phi = a \, d\phi.$$

Since $(x-y)^2 = a^2(\cos^2\phi - 2\sin\phi\cos\phi + \sin^2\phi) = a^2(1 - \sin 2\phi)$, the line integral becomes

$$I = \int_C (x-y)^2 \, ds = \int_0^\pi a^3(1 - \sin 2\phi) \, d\phi = \pi a^3.$$

As a further illustration of the importance of the integral path chosen, we note that the integral directly along the x-axis, between the same end-points and with the same integrand, has the negative value of $-\frac{2}{3}a^3$ (as the reader may wish to verify). ◀

Finally in this section, we note the form for an element of arc length in three dimensions. As discussed in the previous chapter, the expression (11.53) for its square in general three-dimensional orthogonal curvilinear coordinates u_1, u_2, u_3 is

$$(ds)^2 = h_1^2 \, (du_1)^2 + h_2^2 \, (du_2)^2 + h_3^2 \, (du_3)^2,$$

where h_1, h_2, h_3 are the scale factors of the coordinate system. If a curve C in three dimensions is given parametrically by the equations $u_i = u_i(\lambda)$ for $i = 1, 2, 3$ then the element of arc length along the curve is[4]

$$ds = \sqrt{h_1^2 \left(\frac{du_1}{d\lambda}\right)^2 + h_2^2 \left(\frac{du_2}{d\lambda}\right)^2 + h_3^2 \left(\frac{du_3}{d\lambda}\right)^2} \, d\lambda.$$

EXERCISES 12.1

1. Are the following line integrals scalars or vectors?

$$\int_C (\mathbf{a} \times \mathbf{b}) \, dr, \quad \int_C (\mathbf{a} \times \mathbf{b}) \cdot d\mathbf{r}, \quad \int_C (\mathbf{a} \times \mathbf{b}) \cdot \mathbf{c} \, dr, \quad \int_C (\mathbf{a} \times \mathbf{b}) \cdot \mathbf{c} \, d\mathbf{r}.$$

2. Obtain explicit expressions, in terms of one-dimensional integrals, for $\int_C (\mathbf{a} \times \mathbf{b}) \cdot d\mathbf{r}$ and $\int_C (\mathbf{a} \times \mathbf{b}) \, dr$, where $\mathbf{a} = (1, y, z)$, $\mathbf{b} = (1, x, z)$ and path C is the line defined, between the origin and the point $(1, -1, 1)$, by the intersection of the paraboloid $z = \frac{1}{2}(x^2 + y^2)$ and the half-plane $y = -x$. Evaluate one or both expressions – according to your enthusiasm and stamina!

12.2 Connectivity of regions

In physical systems it is usual to define a scalar or vector field in some region R. In the next and some later sections we will need the concept of the *connectivity* of such a region in both two and three dimensions.

---

4 Express the relationship between ds and $d\phi$ for a spiral given in cylindrical polar coordinates by $\rho = k\phi^2$ and $z = \mu\phi^3$.

Line, surface and volume integrals

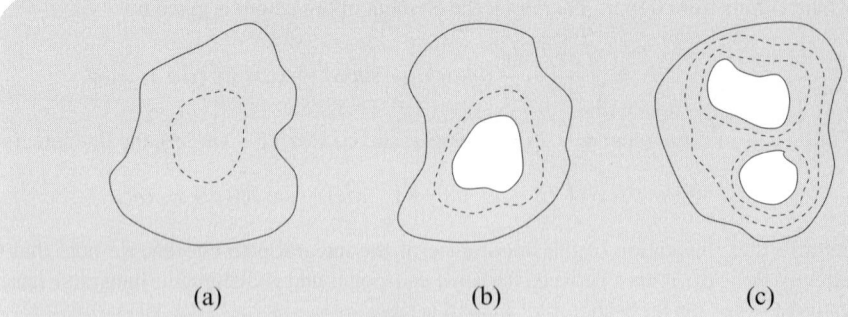

Figure 12.2 (a) A simply connected region; (b) a doubly connected region; (c) a triply connected region.

We begin by discussing planar regions. A plane region R is said to be *simply connected* if every simple closed curve within R can be continuously shrunk to a point without leaving the region [see Figure 12.2(a)]. If, however, the region R contains a hole then there exist simple closed curves that cannot be shrunk to a point without leaving R [see Figure 12.2(b)]. Such a region is said to be doubly connected, since its boundary has two distinct parts.[5] Similarly, a region with $n-1$ holes is said to be *n-fold connected*, or *multiply connected* (the region in Figure 12.2(c) is triply connected).

These ideas can be extended to regions that are not planar, such as general three-dimensional surfaces and volumes. The same criteria concerning the shrinking of closed curves to a point also apply when deciding the connectivity of such regions. In these cases, however, the curves must lie in the surface or volume in question. For example, the interior of a torus is not simply connected, since there exist closed curves in the interior that cannot be shrunk to a point without leaving the torus. The region between two concentric spheres of different radii is simply connected.[6]

12.3 Green's theorem in a plane

In Section 12.1.1 we considered (amongst other things) the evaluation of line integrals for which the path C is closed and lies entirely in the xy-plane. Since the path is closed it will enclose a region R of the plane. We now show how to express the line integral around the loop as a double integral over the enclosed region R.

Suppose the functions $P(x, y)$, $Q(x, y)$ and their partial derivatives are single-valued, finite and continuous inside and on the boundary C of some simply connected region R in the xy-plane. *Green's theorem in a plane* (sometimes called the divergence theorem in

---

[5] Though a more appropriate description might reasonably be thought to be 'doubly disconnected'!
[6] Are the following simply or multiply connected: (a) the glass of a wine glass, (b) the clay of a coffee cup and (c) the clay of a Pythagorean cup with the connecting hole blocked with clay (consult Wikipedia if you are not familiar with one)?

12.3 Green's theorem in a plane

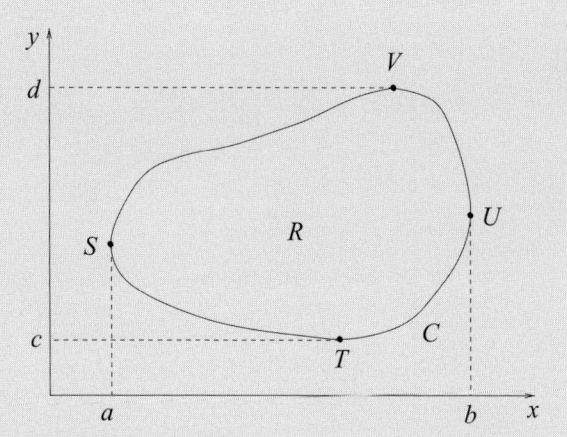

Figure 12.3 A simply connected region R bounded by the curve C.

two dimensions) then states that

$$\oint_C (P\,dx + Q\,dy) = \int\int_R \left(\frac{\partial Q}{\partial x} - \frac{\partial P}{\partial y}\right) dx\,dy, \qquad (12.4)$$

and so relates the line integral around C to a double integral over the enclosed region R. This theorem may be proved straightforwardly in the following way. Consider the simply connected region R in Figure 12.3, and let $y = y_1(x)$ and $y = y_2(x)$ be the equations of the curves STU and SVU respectively. We then write

$$\int\int_R \frac{\partial P}{\partial y} dx\,dy = \int_a^b dx \int_{y_1(x)}^{y_2(x)} dy \frac{\partial P}{\partial y} = \int_a^b dx \Big[P(x,y)\Big]_{y=y_1(x)}^{y=y_2(x)}$$

$$= \int_a^b \Big[P(x, y_2(x)) - P(x, y_1(x))\Big] dx$$

$$= -\int_a^b P(x, y_1(x))\,dx - \int_b^a P(x, y_2(x))\,dx = -\oint_C P\,dx.$$

If we now let $x = x_1(y)$ and $x = x_2(y)$ be the equations of the curves TSV and TUV respectively, we can similarly show that

$$\int\int_R \frac{\partial Q}{\partial x} dx\,dy = \int_c^d dy \int_{x_1(y)}^{x_2(y)} dx \frac{\partial Q}{\partial x} = \int_c^d dy \Big[Q(x,y)\Big]_{x=x_1(y)}^{x=x_2(y)}$$

$$= \int_c^d \Big[Q(x_2(y), y) - Q(x_1(y), y)\Big] dy$$

$$= \int_d^c Q(x_1, y)\,dy + \int_c^d Q(x_2, y)\,dy = \oint_C Q\,dy.$$

Subtracting these two results gives Green's theorem in a plane.[7]

---

[7] Notice that there is no necessary connection between $P(x, y)$ and $Q(x, y)$ and that each result could stand alone. The difference in sign is simply the combined result of the conventional choices for (i) the x- and y-axes and (ii) the positive direction of traversing a closed contour.

Line, surface and volume integrals

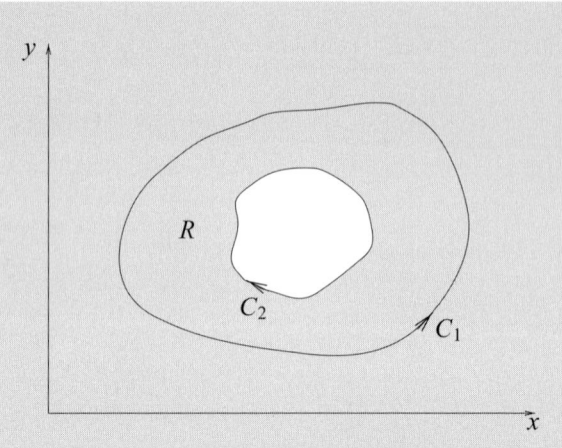

Figure 12.4 A doubly connected region R bounded by the curves C_1 and C_2.

Example Show that the area of a region R enclosed by a simple closed curve C is given by $A = \frac{1}{2}\oint_C (x\,dy - y\,dx) = \oint_C x\,dy = -\oint_C y\,dx$. Hence calculate the area of the ellipse $x = a\cos\phi$, $y = b\sin\phi$.

In Green's theorem (12.4) put $P = -y$ and $Q = x$; then

$$\oint_C (x\,dy - y\,dx) = \int\int_R (1+1)\,dx\,dy = 2\int\int_R dx\,dy = 2A.$$

Therefore the area of the region is $A = \frac{1}{2}\oint_C (x\,dy - y\,dx)$. Alternatively, we could put $P = 0$ and $Q = x$ and obtain $A = \oint_C x\,dy$, or put $P = -y$ and $Q = 0$, which gives $A = -\oint_C y\,dx$.
The area of the ellipse $x = a\cos\phi$, $y = b\sin\phi$ is given by

$$A = \frac{1}{2}\oint_C (x\,dy - y\,dx) = \frac{1}{2}\int_0^{2\pi} ab(\cos^2\phi + \sin^2\phi)\,d\phi$$

$$= \frac{ab}{2}\int_0^{2\pi} d\phi = \pi ab.$$

The parameterisation used here is that for the standard form of an ellipse, $x^2/a^2 + y^2/b^2 = 1$. ◀

It may further be shown that Green's theorem in a plane is also valid for multiply connected regions. In this case, the line integral must be taken over all the distinct boundaries of the region. Furthermore, each boundary must be traversed in the positive direction, so that a person travelling along it in this direction always has the region R on their left. In order to apply Green's theorem to the region R shown in Figure 12.4, the line integrals must be taken over both boundaries, C_1 and C_2, in the directions indicated, and the results added together.[8]

---

[8] A coin has the form of a uniform circular disc of radius a with a central circular hole of radius b removed from it. By setting $Q = xy^2$ and $P = -x^2 y$, show that its moment of inertia about a central axis perpendicular to its plane is $\frac{1}{2}m(a^2 + b^2)$, where m is the mass of the coin.

12.3 Green's theorem in a plane

We may also use Green's theorem in a plane to investigate the path independence (or not) of line integrals when the paths lie in the xy-plane. Let us consider the line integral

$$I = \int_A^B (P\,dx + Q\,dy).$$

For the line integral from A to B to be independent of the path taken, it must have the same value along any two arbitrary paths C_1 and C_2 joining the points. Moreover, if we consider as the path the closed loop C formed by $C_1 - C_2$ then the line integral around this loop must be zero. From Green's theorem in a plane, (12.4), we see that a *sufficient* condition for $I = 0$ is that

$$\frac{\partial P}{\partial y} = \frac{\partial Q}{\partial x}, \tag{12.5}$$

throughout some simply connected region R containing the loop, where we assume that these partial derivatives are continuous in R.

It may be shown that (12.5) is also a *necessary* condition for $I = 0$ and is equivalent to requiring $P\,dx + Q\,dy$ to be an exact differential of some function $\psi(x, y)$ such that $P\,dx + Q\,dy = d\psi$. It follows that $\int_A^B (P\,dx + Q\,dy) = \psi(B) - \psi(A)$ and that $\oint_C (P\,dx + Q\,dy)$ around any closed loop C in the region R is identically zero.[9] These results are special cases of the general results for paths in three dimensions, which are discussed in the next section.

Example Evaluate the line integral

$$I = \oint_C \left[(e^x y + \cos x \sin y)\,dx + (e^x + \sin x \cos y)\,dy\right],$$

around the ellipse $x^2/a^2 + y^2/b^2 = 1$.

Clearly, it is not straightforward to calculate this line integral directly. However, if we let

$$P = e^x y + \cos x \sin y \quad \text{and} \quad Q = e^x + \sin x \cos y,$$

then $\partial P/\partial y = e^x + \cos x \cos y = \partial Q/\partial x$, and so $P\,dx + Q\,dy$ is an exact differential (it is actually the differential of the function $f(x, y) = e^x y + \sin x \sin y$). From the above discussion, we can conclude immediately that $I = 0$. ◂

EXERCISES 12.3

1. Which of the following two-dimensional integrals around a closed loop C, lying in a singly connected region, are necessarily equal to zero?

   (a) $\displaystyle\int_C (4x^3 y^3\,dx + 3x^2 y^4\,dy).$

   (b) $\displaystyle\int_C (3x^2 y^4\,dx + 4x^3 y^3\,dy).$

---

[9] Show that this is the case if $Q = xy^2$ and $P = x^2 y$ (compare with footnote 8) and C is any circular contour of radius r centred on the origin. Identify ψ in this case.

(c) $\int_C \sin x \cos y \, (\cos x \cos y \, dx + \sin x \sin y \, dy)$.

(d) $\int_C \sin x \cos y \, (\cos x \cos y \, dx - \sin x \sin y \, dy)$.

2. Evaluate the line integrals $\oint \mathbf{F} \cdot d\mathbf{r}$ and $\oint \mathbf{G} \cdot d\mathbf{r}$, where C is the unit circle and, in Cartesian coordinates, $\mathbf{F} = (x^3, y^3)$ and $\mathbf{G} = (-y^3, x^3)$.

12.4 Conservative fields and potentials

So far we have made the point that, in general, the value of a line integral between two points A and B depends on the path C taken from A to B. In the previous section, however, we saw that, for paths in the xy-plane, line integrals whose integrands have certain properties are independent of the path taken. We now extend that discussion to the full three-dimensional case.

For line integrals of the form $\int_C \mathbf{a} \cdot d\mathbf{r}$, there exists a class of vector fields for which the line integral between two points is *independent* of the path taken. Such vector fields are called *conservative*. A vector field $\mathbf{a}$ that has continuous partial derivatives in a simply connected region R is conservative if, and only if, any of the following is true.[10]

(i) The integral $\int_A^B \mathbf{a} \cdot d\mathbf{r}$, where A and B lie in the region R, is independent of the path from A to B. Hence the integral $\oint_C \mathbf{a} \cdot d\mathbf{r}$ around any closed loop in R is zero.
(ii) There exists a single-valued function ϕ of position such that $\mathbf{a} = \nabla \phi$.
(iii) $\nabla \times \mathbf{a} = \mathbf{0}$.
(iv) $\mathbf{a} \cdot d\mathbf{r}$ is an exact differential.

The validity or otherwise of any one of these statements implies the same for the other three, as we will now show.

First, let us assume that (i) above is true. If the line integral from A to B is independent of the path taken between the points then its value must be a function only of the positions of A and B. We may therefore write

$$\int_A^B \mathbf{a} \cdot d\mathbf{r} = \phi(B) - \phi(A), \tag{12.6}$$

which defines a single-valued scalar function of position ϕ. If the points A and B are separated by an infinitesimal displacement $d\mathbf{r}$ then (12.6) becomes

$$\mathbf{a} \cdot d\mathbf{r} = d\phi,$$

which shows that we require $\mathbf{a} \cdot d\mathbf{r}$ to be an exact differential: condition (iv). From (11.22) we can write $d\phi = \nabla \phi \cdot d\mathbf{r}$, and so we have

$$(\mathbf{a} - \nabla \phi) \cdot d\mathbf{r} = 0.$$

---

[10] It may be helpful to keep in mind the physical example of a charge q in an electric field $\mathbf{E}$. Then the electrostatic potential is $-\phi$ with $\mathbf{E} = \nabla \phi$, the integral $q \int_A^B \mathbf{E} \cdot d\mathbf{r}$ is the work done on the charge as it moves from A to B (by any route), and $\phi(A) - \phi(B)$ is the increase (or decrease if negative) in the potential energy of the charge.

12.4 Conservative fields and potentials

Since $d\mathbf{r}$ is arbitrary, we find that $\mathbf{a} = \nabla\phi$; this immediately implies $\nabla \times \mathbf{a} = \mathbf{0}$, condition (iii) [see (11.32)].

Alternatively, if we suppose that there exists a single-valued function of position ϕ such that $\mathbf{a} = \nabla\phi$ then $\nabla \times \mathbf{a} = \mathbf{0}$ follows as before. The line integral around a closed loop then becomes

$$\oint_C \mathbf{a} \cdot d\mathbf{r} = \oint_C \nabla\phi \cdot d\mathbf{r} = \oint d\phi.$$

Since we defined ϕ to be single-valued, this integral is zero as required.

Now suppose $\nabla \times \mathbf{a} = \mathbf{0}$. From Stokes' theorem, which is discussed in Section 12.9, we immediately obtain $\oint_C \mathbf{a} \cdot d\mathbf{r} = 0$; then $\mathbf{a} = \nabla\phi$ and $\mathbf{a} \cdot d\mathbf{r} = d\phi$ follow as above.

Finally, let us suppose $\mathbf{a} \cdot d\mathbf{r} = d\phi$. Then immediately we have $\mathbf{a} = \nabla\phi$, and the other results follow as above. This completes the argument that shows that the four conditions stand or fall together.

Example Evaluate the line integral $I = \int_A^B \mathbf{a} \cdot d\mathbf{r}$, where $\mathbf{a} = (xy^2 + z)\mathbf{i} + (x^2 y + 2)\mathbf{j} + x\mathbf{k}$, A is the point (c, c, h) and B is the point $(2c, c/2, h)$, along the different paths

(i) C_1, given by $x = cu$, $y = c/u$, $z = h$,
(ii) C_2, given by $2y = 3c - x$, $z = h$.

Show that the vector field $\mathbf{a}$ is in fact conservative, and find ϕ such that $\mathbf{a} = \nabla\phi$.

Expanding out the integrand, we have

$$I = \int_{(c,c,h)}^{(2c,c/2,h)} \left[(xy^2 + z)\,dx + (x^2 y + 2)\,dy + x\,dz \right], \tag{12.7}$$

which we must evaluate along each of the paths C_1 and C_2.

(i) Along C_1 we have $dx = c\,du$, $dy = -(c/u^2)\,du$, $dz = 0$, and on substituting in (12.7) and finding the limits on u we obtain

$$I = \int_1^2 c\left(h - \frac{2}{u^2}\right) du = c(h-1).$$

(ii) Along C_2 we have $2\,dy = -dx$, $dz = 0$ and, on substituting in (12.7) and using the limits on x, we obtain

$$I = \int_c^{2c} \left(\tfrac{1}{2}x^3 - \tfrac{9}{4}cx^2 + \tfrac{9}{4}c^2 x + h - 1 \right) dx = c(h-1).$$

Hence the line integral has the same value along paths C_1 and C_2. Taking the curl of $\mathbf{a}$, we have

$$\nabla \times \mathbf{a} = (0-0)\mathbf{i} + (1-1)\mathbf{j} + (2xy - 2xy)\mathbf{k} = \mathbf{0},$$

so $\mathbf{a}$ is a conservative vector field, and the line integral between two points *must* be independent of the path taken. Since $\mathbf{a}$ is conservative, we can write $\mathbf{a} = \nabla\phi$. Therefore, ϕ must satisfy

$$\frac{\partial\phi}{\partial x} = xy^2 + z,$$

which implies that $\phi = \tfrac{1}{2}x^2 y^2 + zx + f(y, z)$ for some function f. Secondly, we require

$$\frac{\partial\phi}{\partial y} = x^2 y + \frac{\partial f}{\partial y} = x^2 y + 2,$$

which implies $f = 2y + g(z)$. Finally, since

$$\frac{\partial \phi}{\partial z} = x + \frac{\partial g}{\partial z} = x,$$

we have $g = \text{constant} = k$. It can be seen that we have explicitly constructed the function $\phi = \frac{1}{2}x^2 y^2 + zx + 2y + k$. ◀

The quantity ϕ that figures so prominently in this section is called the *scalar potential function* of the conservative vector field **a** (which satisfies $\nabla \times \mathbf{a} = \mathbf{0}$), and is unique up to an arbitrary additive constant. Scalar potentials that are multivalued functions of position (but in simple ways) are also of value in describing some physical situations, the most obvious example being the scalar magnetic potential associated with a current-carrying wire. When the integral of a field quantity around a closed loop is considered, provided the loop does not enclose a net current, the potential is single-valued and all the above results still hold. If the loop does enclose a net current, however, our analysis is no longer valid and extra care must be taken.

If, instead of being conservative, a vector field **b** satisfies $\nabla \cdot \mathbf{b} = 0$ (i.e. **b** is solenoidal) then it is both possible and useful, for example in the theory of electromagnetism, to define a *vector field* **a** such that $\mathbf{b} = \nabla \times \mathbf{a}$. It may be shown that such a vector field **a** always exists. Further, if **a** is one such vector field then $\mathbf{a}' = \mathbf{a} + \nabla \psi + \mathbf{c}$, where ψ is any scalar function and **c** is any constant vector, also satisfies the above relationship, i.e. $\mathbf{b} = \nabla \times \mathbf{a}'$. This has already been discussed more fully in Section 11.7.2.

EXERCISES 12.4

1. Determine which of the two vector fields

   $\mathbf{a} = (x(3xy - z^2), x^3 + 6y, z(x^2 + 2))$ and $\mathbf{b} = (x(3xy - z^2), x^3 + 6y, -z(x^2 + 2))$

   is conservative. For the one that is, calculate, in a systematic manner rather than by inspection or trial, the potential from which it is derived.

2. For the non-conservative field in the previous exercise, denoted here by **f**, evaluate the line integral $\int \mathbf{f} \cdot d\mathbf{r}$ from the origin to the point (1, 1, 1) along the paths (a) $(0, 0, 0) \to (1, 0, 0) \to (1, 1, 0) \to (1, 1, 1)$, and (b) $(0, 0, 0) \to (0, 1, 0) \to (0, 1, 1) \to (1, 1, 1)$. What would be the values of the corresponding line integrals for the conservative field?

12.5 Surface integrals

As with line integrals, integrals over surfaces can involve vector and scalar fields and, equally, can result in either a vector or a scalar. The simplest case involves entirely scalars and is of the form

$$\int_S \phi \, dS. \tag{12.8}$$

12.5 Surface integrals

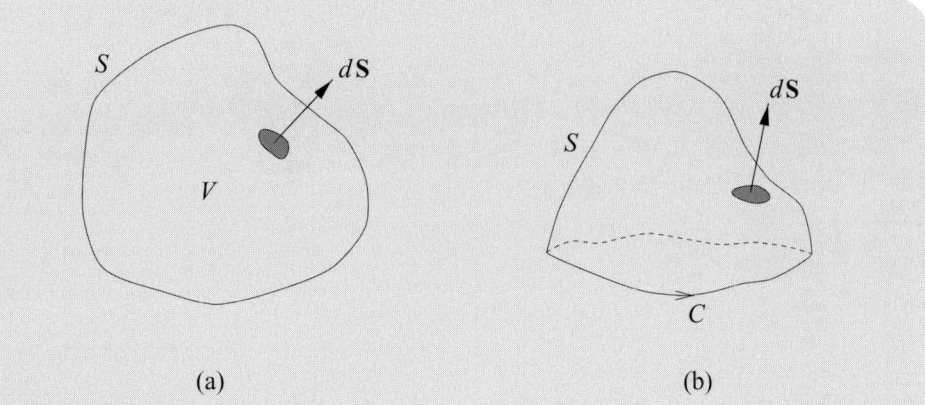

Figure 12.5 (a) A closed surface and (b) an open surface. In each case a normal to the surface is shown: $d\mathbf{S} = \hat{\mathbf{n}}\,dS$.

As analogues of the line integrals listed in (12.1), we may also encounter surface integrals involving vectors, namely

$$\int_S \phi\,d\mathbf{S}, \qquad \int_S \mathbf{a}\cdot d\mathbf{S}, \qquad \int_S \mathbf{a}\times d\mathbf{S}. \tag{12.9}$$

All the above integrals are taken over some surface S, which may be either open or closed, and are therefore, in general, double integrals. Following the notation for line integrals, for surface integrals over a closed surface $\int_S$ is replaced by $\oint_S$.

The vector differential $d\mathbf{S}$ in (12.9) represents a vector area element of the surface S. It may also be written $d\mathbf{S} = \hat{\mathbf{n}}\,dS$, where $\hat{\mathbf{n}}$ is a unit normal to the surface at the position of the element and dS is the scalar area of the element used in (12.8). The convention for the direction of the normal $\hat{\mathbf{n}}$ to a surface depends on whether the surface is open or closed.

A closed surface, see Figure 12.5(a), does not have to be simply connected (for example, the surface of a torus is not), but it does have to enclose a volume V, which may be of infinite extent. The direction of $\hat{\mathbf{n}}$ is taken to point outwards from the enclosed volume as shown.

An open surface, see Figure 12.5(b), spans some perimeter curve C. The direction of $\hat{\mathbf{n}}$ is then given by the right-hand sense with respect to the direction in which the perimeter is traversed, i.e. follows the right-hand screw rule discussed in Section 9.4.2. An open surface does not have to be simply connected, but for our purposes it must be two-sided (a Möbius strip is an example of a one-sided surface).

The formal definition of a surface integral is very similar to that of a line integral. We divide the surface S into N elements of area ΔS_p, $p = 1, 2, \ldots, N$, each with a unit normal $\hat{\mathbf{n}}_p$. If (x_p, y_p, z_p) is any point in ΔS_p then the second type of surface integral in (12.9), for example, is defined as

$$\int_S \mathbf{a}\cdot d\mathbf{S} = \lim_{N\to\infty} \sum_{p=1}^{N} \mathbf{a}(x_p, y_p, z_p)\cdot \hat{\mathbf{n}}_p \Delta S_p,$$

where it is required that all $\Delta S_p \to 0$ as $N \to \infty$.

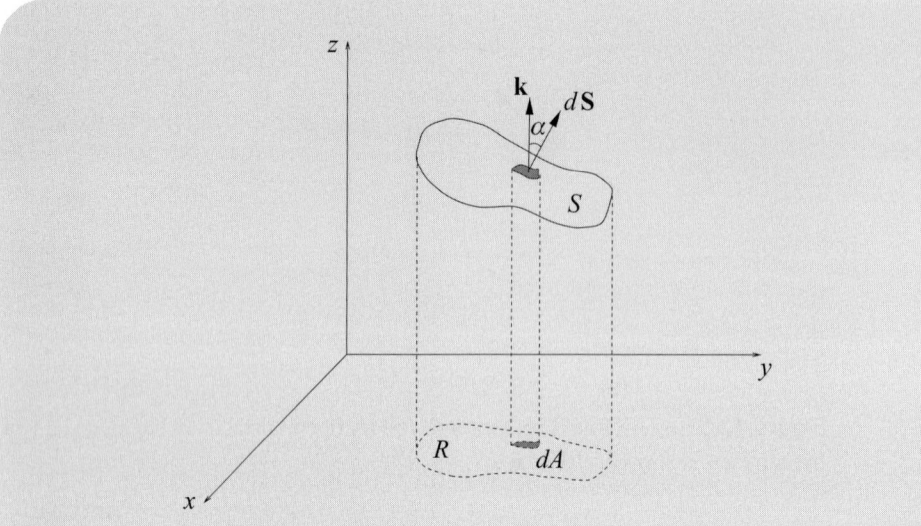

Figure 12.6 A surface S (or part thereof) projected onto a region R in the xy-plane; $d\mathbf{S}$ is a surface element.

12.5.1 Evaluating surface integrals

We now consider how to evaluate surface integrals over some general surface. This involves writing the scalar area element dS in terms of the coordinate differentials of our chosen coordinate system. In some particularly simple cases this is very straightforward. For example, if S is the surface of a sphere of radius a (or some part thereof) then using spherical polar coordinates θ, ϕ on the sphere we have $dS = a^2 \sin\theta \, d\theta \, d\phi$. For a general surface, however, it is not usually possible to represent the surface in a simple way in any particular coordinate system. In such cases, it is usual to work in Cartesian coordinates and consider the projections of the surface onto the coordinate planes.

Consider a surface (or part of a surface) S as in Figure 12.6. The surface S is projected onto a region R of the xy-plane, so that an element of surface area dS projects onto the area element dA. From the figure, we see that $dA = |\cos\alpha| \, dS$, where α is the angle between the unit vector $\mathbf{k}$ in the z-direction and the unit normal $\hat{\mathbf{n}}$ to the surface at P. So, at any given point of S, we have simply

$$dS = \frac{dA}{|\cos\alpha|} = \frac{dA}{|\hat{\mathbf{n}} \cdot \mathbf{k}|}.$$

Now, if the surface S is given by the equation $f(x, y, z) = 0$ then, as shown in Section 11.6.1, the unit normal at any point of the surface is given by $\hat{\mathbf{n}} = \nabla f / |\nabla f|$ evaluated at that point, cf. (11.27). The scalar element of surface area then becomes

$$dS = \frac{dA}{|\hat{\mathbf{n}} \cdot \mathbf{k}|} = \frac{|\nabla f| \, dA}{\nabla f \cdot \mathbf{k}} = \frac{|\nabla f| \, dA}{\partial f / \partial z}, \tag{12.10}$$

where $|\nabla f|$ and $\partial f / \partial z$ are evaluated on the surface S. We can therefore express any surface integral over S as a double integral over the region R in the xy-plane.

12.5 Surface integrals

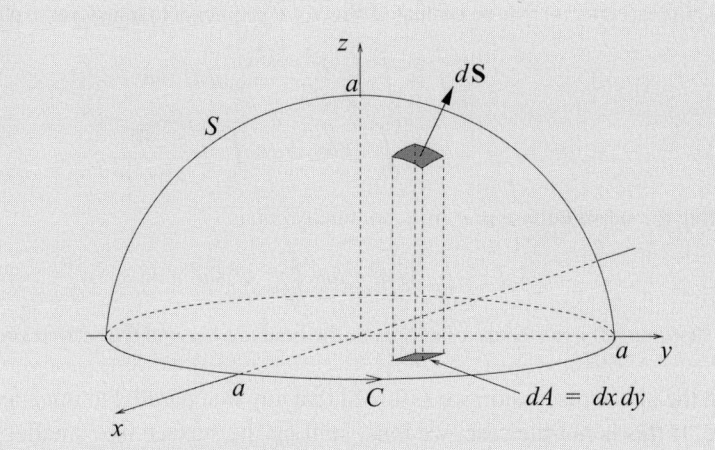

Figure 12.7 The surface of the hemisphere $x^2 + y^2 + z^2 = a^2$, $z \geq 0$.

In the following example both the specific and more general approaches are illustrated; not surprisingly, because of its more universal applicability, the latter is the longer.

Example Evaluate the surface integral $I = \int_S \mathbf{a} \cdot d\mathbf{S}$, where $\mathbf{a} = x\mathbf{i}$ and S is the surface of the hemisphere $x^2 + y^2 + z^2 = a^2$ with $z \geq 0$.

The surface of the hemisphere is shown in Figure 12.7. In this case dS may be easily expressed in spherical polar coordinates as $dS = a^2 \sin\theta \, d\theta \, d\phi$, and the unit normal to the surface at any point is simply $\hat{\mathbf{r}}$. On the surface of the hemisphere we have $x = a\sin\theta\cos\phi$ and so

$$\mathbf{a} \cdot d\mathbf{S} = x(\mathbf{i} \cdot \hat{\mathbf{r}}) \, dS = (a\sin\theta\cos\phi)(\sin\theta\cos\phi)(a^2 \sin\theta \, d\theta \, d\phi).$$

Therefore, inserting the correct limits on θ and ϕ, we have

$$I = \int_S \mathbf{a} \cdot d\mathbf{S} = a^3 \int_0^{\pi/2} d\theta \, \sin^3\theta \int_0^{2\pi} d\phi \, \cos^2\phi = \frac{2\pi a^3}{3}.$$

We could, however, follow the general prescription above and project the hemisphere S onto the region R in the xy-plane that is a circle of radius a centred at the origin. Writing the equation of the surface of the hemisphere as $f(x, y) = x^2 + y^2 + z^2 - a^2 = 0$ and using (12.10), we have

$$I = \int_S \mathbf{a} \cdot d\mathbf{S} = \int_S x(\mathbf{i} \cdot \hat{\mathbf{r}}) \, dS = \int_R x(\mathbf{i} \cdot \hat{\mathbf{r}}) \frac{|\nabla f| \, dA}{\partial f/\partial z}.$$

Now $\nabla f = 2x\mathbf{i} + 2y\mathbf{j} + 2z\mathbf{k} = 2\mathbf{r}$, so on the surface S we have $|\nabla f| = 2|\mathbf{r}| = 2a$. On S we also have $\partial f/\partial z = 2z = 2\sqrt{a^2 - x^2 - y^2}$ and $\mathbf{i} \cdot \hat{\mathbf{r}} = x/a$. Therefore, the integral becomes

$$I = \int\int_R \frac{x^2}{\sqrt{a^2 - x^2 - y^2}} \, dx \, dy.$$

Although this integral may be evaluated directly, it is quicker to transform to plane polar coordinates:

$$I = \int\int_{R'} \frac{\rho^2 \cos^2\phi}{\sqrt{a^2 - \rho^2}} \, \rho \, d\rho \, d\phi$$

$$= \int_0^{2\pi} \cos^2\phi \, d\phi \int_0^a \frac{\rho^3 \, d\rho}{\sqrt{a^2 - \rho^2}}.$$

Making the substitution $\rho = a \sin u$, we finally obtain

$$I = \int_0^{2\pi} \cos^2\phi \, d\phi \int_0^{\pi/2} a^3 \sin^3 u \, du = \frac{2\pi a^3}{3}.$$

The first integral contributes a factor π to the final answer and the second contributes $2a^3/3$. ◀

In the above discussion we assumed that any line parallel to the z-axis intersects S only once. If this is not the case, we must split up the surface into smaller surfaces S_1, S_2 etc. that are of this type. The surface integral over S is then the sum of the surface integrals over S_1, S_2 and so on. This is always necessary for closed surfaces.

Sometimes we may need to project a surface S (or some part of it) onto the zx- or yz-plane, rather than the xy-plane; for such cases, the above analysis is easily modified.

12.5.2 Vector areas of surfaces

The vector area of a surface S is defined as

$$\mathbf{S} = \int_S d\mathbf{S},$$

where the surface integral may be evaluated as above.

Example Find the vector area of the surface of the hemisphere $x^2 + y^2 + z^2 = a^2$ with $z \geq 0$.

As in the previous example, $d\mathbf{S} = a^2 \sin\theta \, d\theta \, d\phi \, \hat{\mathbf{r}}$ in spherical polar coordinates. Therefore, the vector area is given by

$$\mathbf{S} = \int\int_S a^2 \sin\theta \, \hat{\mathbf{r}} \, d\theta \, d\phi.$$

Now, since $\hat{\mathbf{r}}$ varies over the surface S, it also must be integrated. This is most easily achieved by writing $\hat{\mathbf{r}}$ in terms of the constant Cartesian basis vectors. On S we have

$$\hat{\mathbf{r}} = \sin\theta \cos\phi \, \mathbf{i} + \sin\theta \sin\phi \, \mathbf{j} + \cos\theta \, \mathbf{k},$$

so the expression for the vector area becomes

$$\mathbf{S} = \mathbf{i}\left(a^2 \int_0^{2\pi} \cos\phi \, d\phi \int_0^{\pi/2} \sin^2\theta \, d\theta\right) + \mathbf{j}\left(a^2 \int_0^{2\pi} \sin\phi \, d\phi \int_0^{\pi/2} \sin^2\theta \, d\theta\right)$$

$$+ \mathbf{k}\left(a^2 \int_0^{2\pi} d\phi \int_0^{\pi/2} \sin\theta \cos\theta \, d\theta\right)$$

$$= 0 + 0 + \pi a^2 \mathbf{k} = \pi a^2 \mathbf{k}.$$

Note that the magnitude of $\mathbf{S}$ is the projected area of the hemisphere onto the xy-plane, and not the surface area of the hemisphere. ◀

12.5 Surface integrals

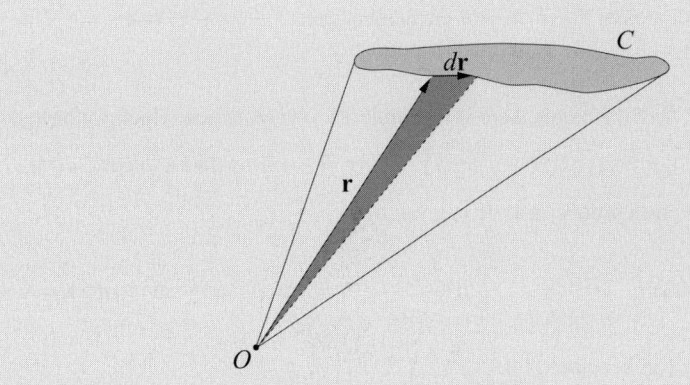

Figure 12.8 The conical surface spanning the perimeter C and having its vertex at the origin.

The hemispherical shell discussed above is an example of an open surface. For a closed surface, however, the vector area is always zero.[11] This may be seen by projecting the surface down onto each Cartesian coordinate plane in turn. For each projection, every positive element of area on the upper surface is cancelled by the corresponding negative element on the lower surface. Therefore, each component of $\mathbf{S} = \oint_S d\mathbf{S}$ vanishes.

An important corollary of this result is that the vector area of an open surface depends only on its perimeter, or boundary curve, C. This may be proved as follows. If surfaces S_1 and S_2 have the same perimeter then $S_1 - S_2$ is a closed surface, for which

$$\oint d\mathbf{S} = \int_{S_1} d\mathbf{S} - \int_{S_2} d\mathbf{S} = \mathbf{0}.$$

Hence $\mathbf{S}_1 = \mathbf{S}_2$. Moreover, we may derive an expression for the vector area of an open surface S solely in terms of a line integral around its perimeter C. Since we may choose any surface with perimeter C, we will consider a cone with its vertex at the origin (see Figure 12.8). The vector area of the elementary triangular region shown in the figure is $d\mathbf{S} = \frac{1}{2} \mathbf{r} \times d\mathbf{r}$.[12] Therefore, the vector area of the cone, and hence of *any* open surface with perimeter C, is given by the line integral[13]

$$\mathbf{S} = \frac{1}{2} \oint_C \mathbf{r} \times d\mathbf{r}.$$

For a surface confined to the xy-plane, $\mathbf{r} = x\mathbf{i} + y\mathbf{j}$ and $d\mathbf{r} = dx\,\mathbf{i} + dy\,\mathbf{j}$, and so, applying the above prescription, we obtain for this special case that the area is $A = \frac{1}{2} \oint_C (x\,dy - y\,dx)$; this is as we found in Section 12.3.

---

[11] Use this result to reduce the solution to the previous worked example to a single sentence.
[12] Note that in this case $\hat{\mathbf{n}}$ points into what, in Figure 12.8, would normally be described as the 'interior' of the hollow cone; however, its direction is in agreement with the convention described on p. 505.
[13] Note that the value obtained for $\mathbf{S}$ does *not* depend upon the position of the surface's perimeter relative to the origin.

Example Find the vector area of the surface of the hemisphere $x^2 + y^2 + z^2 = a^2$, $z \geq 0$, by evaluating the line integral $\mathbf{S} = \frac{1}{2} \oint_C \mathbf{r} \times d\mathbf{r}$ around its perimeter.

The perimeter C of the hemisphere is the circle $x^2 + y^2 = a^2$, on which we have
$$\mathbf{r} = a\cos\phi\,\mathbf{i} + a\sin\phi\,\mathbf{j}, \qquad d\mathbf{r} = -a\sin\phi\,d\phi\,\mathbf{i} + a\cos\phi\,d\phi\,\mathbf{j}.$$
Therefore, the cross product $\mathbf{r} \times d\mathbf{r}$ is given by
$$\mathbf{r} \times d\mathbf{r} = \begin{vmatrix} \mathbf{i} & \mathbf{j} & \mathbf{k} \\ a\cos\phi & a\sin\phi & 0 \\ -a\sin\phi\,d\phi & a\cos\phi\,d\phi & 0 \end{vmatrix} = a^2(\cos^2\phi + \sin^2\phi)\,d\phi\,\mathbf{k} = a^2\,d\phi\,\mathbf{k},$$
and the vector area becomes
$$\mathbf{S} = \tfrac{1}{2}a^2\mathbf{k} \int_0^{2\pi} d\phi = \pi a^2\,\mathbf{k},$$
in agreement with the result of the previous worked example and footnote 11. ◀

12.5.3 Physical examples of surface integrals

There are many examples of surface integrals in the physical sciences. Surface integrals of the form (12.8) occur in computing the total electric charge on a surface or the mass of a shell, $\int_S \rho(\mathbf{r})\,dS$, given the charge or mass density $\rho(\mathbf{r})$. For surface integrals involving vectors, the second form in (12.9) is the most common. For a vector field $\mathbf{a}$, the surface integral $\int_S \mathbf{a} \cdot d\mathbf{S}$ is called the *flux* of $\mathbf{a}$ through S. Examples of physically important flux integrals are numerous.[14] For example, let us consider a surface S in a fluid with density $\rho(\mathbf{r})$ that has a velocity field $\mathbf{v}(\mathbf{r})$. The mass of fluid crossing an element of surface area $d\mathbf{S}$ in time dt is $dM = \rho \mathbf{v} \cdot d\mathbf{S}\,dt$. Therefore, the *net* total mass flux of fluid crossing S is $M = \int_S \rho(\mathbf{r})\mathbf{v}(\mathbf{r}) \cdot d\mathbf{S}$. As another example, the electromagnetic flux of energy out of a given volume V bounded by a surface S is $\oint_S (\mathbf{E} \times \mathbf{H}) \cdot d\mathbf{S}$.

The solid angle, to be defined below, subtended at a point O by a surface (closed or otherwise) can also be represented by an integral of this form, although it is not strictly a flux integral (unless we imagine isotropic rays radiating from O). The integral
$$\Omega = \int_S \frac{\mathbf{r} \cdot d\mathbf{S}}{r^3} = \int_S \frac{\hat{\mathbf{r}} \cdot d\mathbf{S}}{r^2} \qquad (12.11)$$
gives the *solid angle* Ω subtended at O by a surface S if $\mathbf{r}$ is the position vector measured from O of an element of the surface.[15] A little thought will show that (12.11) takes account of all three relevant factors: the size of the element of surface, its inclination to the line joining the element to O and the distance from O. Such a general expression is often useful for computing solid angles when the three-dimensional geometry is complicated. Note that (12.11) remains valid when the surface S is not convex and when a single ray

---

[14] Probably the most familiar is Gauss's theorem, which can be written as $\int_S \mathbf{E} \cdot d\mathbf{S} = \epsilon_0^{-1} \sum_i q_i$ for a system of charges q_i in a vacuum that are contained within a surface S.
[15] Use this result to find an expression for the solid angle enclosed by a cone of half-angle α.

12.6 Volume integrals

from O in certain directions would cut S in more than one place (but we exclude multiply connected regions). In particular, when the surface is closed $\Omega = 0$ if O is outside S and $\Omega = 4\pi$ if O is an interior point.

Surface integrals resulting in vectors occur less frequently. An example is afforded, however, by the total resultant force experienced by a body immersed in a stationary fluid in which the hydrostatic pressure is given by $p(\mathbf{r})$. The pressure is everywhere inwardly directed and the resultant force is $\mathbf{F} = -\oint_S p \, d\mathbf{S}$, taken over the whole surface. The connection with an Archimedean upthrust will be clear!

EXERCISES 12.5

1. Evaluate the surface integrals

$$\text{(a)} \int_S \nabla \cdot \mathbf{f} \, dS, \quad \text{(b)} \int_S \mathbf{f} \cdot d\mathbf{S}, \quad \text{(c)} \int_S \mathbf{f} \times d\mathbf{S},$$

where $\mathbf{f}$ is a vector field with components $(x(y+z), y(z+x), z(x+y))$ and S is the surface of the unit cube that has diametrically opposite corners at $(0, 0, 0)$ and $(1, 1, 1)$. In each case, carry out the calculations for one pair of opposite faces of the cube, and then make use of the cyclic symmetry of the components of $\mathbf{f}$ to find the full surface integral.

2. The surface S is that part of the (inverted) paraboloid of revolution $x^2 + y^2 = 4a(a - z)$ that is above the plane $z = 0$. The vector field $\mathbf{F}$ is given by

$$\mathbf{F} = \frac{F_0}{a}\left(x, y, \frac{x^2 + y^2}{a}\right).$$

(a) Show that the connection between the sizes of an element dS of the surface and its projection dA onto the plane $z = 0$ is $dS = a^{-1}(2a^2 - az)^{1/2} \, dA$.
(b) Find an expression for $\int_S \mathbf{v} \cdot d\mathbf{S}$ for a general vector field $\mathbf{v}$.
(c) Deduce the vector area of S.
(d) Show that $\int_S \mathbf{F} \cdot d\mathbf{S} = 12\pi a^2 F_0$.

12.6 Volume integrals

Volume integrals are defined in an obvious way and are generally simpler than line or surface integrals since the element of volume dV is a scalar quantity. We may encounter volume integrals of the forms

$$\int_V \phi \, dV, \quad \int_V \mathbf{a} \, dV. \quad (12.12)$$

Clearly, the first form results in a scalar, whereas the second form yields a vector. Two closely related physical examples, one of each kind, are provided by the total mass of a fluid contained in a volume V, given by $\int_V \rho(\mathbf{r}) \, dV$, and the total linear momentum of that same fluid, given by $\int_V \rho(\mathbf{r})\mathbf{v}(\mathbf{r}) \, dV$, where $\mathbf{v}(\mathbf{r})$ is the velocity field in the fluid. As a slightly more complicated example of a volume integral we may consider the following.

Example Find an expression for the angular momentum of a solid body rotating with angular velocity $\boldsymbol{\omega}$ about an axis through the origin.

Consider a small volume element dV situated at position $\mathbf{r}$; its linear momentum is $\rho \, dV \dot{\mathbf{r}}$, where $\rho = \rho(\mathbf{r})$ is the density distribution, and its angular momentum about O is $\mathbf{r} \times \rho \dot{\mathbf{r}} \, dV$. Thus for the whole body the angular momentum $\mathbf{L}$ is

$$\mathbf{L} = \int_V (\mathbf{r} \times \dot{\mathbf{r}}) \rho \, dV.$$

Since the body is solid and rotating as a whole, the velocity of an element at position $\mathbf{r}$ is given by $\dot{\mathbf{r}} = \boldsymbol{\omega} \times \mathbf{r}$. Substituting this yields

$$\mathbf{L} = \int_V [\mathbf{r} \times (\boldsymbol{\omega} \times \mathbf{r})] \rho \, dV = \int_V \boldsymbol{\omega} r^2 \rho \, dV - \int_V (\mathbf{r} \cdot \boldsymbol{\omega}) \mathbf{r} \rho \, dV.$$

It should be noted that both integrals produce vectors; the first is necessarily positive and in the direction of $\boldsymbol{\omega}$, but the second could be in any direction. ◂

The evaluation of the first type of volume integral in (12.12) has already been considered in our discussion of multiple integrals in Chapter 8. The evaluation of the second type of volume integral follows directly, since we can write

$$\int_V \mathbf{a} \, dV = \mathbf{i} \int_V a_x \, dV + \mathbf{j} \int_V a_y \, dV + \mathbf{k} \int_V a_z \, dV, \tag{12.13}$$

where a_x, a_y, a_z are the Cartesian components of $\mathbf{a}$. Of course, we could have written $\mathbf{a}$ in terms of the basis vectors of some other coordinate system (e.g. spherical polars) but, since such basis vectors are not, in general, constant, they cannot be taken out of the integral sign as in (12.13) and must be included as part of the integrand.

As discussed in Chapter 8, the volume of a three-dimensional region V is simply $V = \int_V dV$, which may be evaluated directly once the limits of integration have been found. However, the volume of the region obviously depends only on the surface S that bounds it. We should therefore be able to express the volume V in terms of a surface integral over S. This is indeed possible, and the appropriate expression may derived as follows. Referring to Figure 12.9, let us suppose that the origin O is contained within V. The volume of the small shaded cone is $dV = \frac{1}{3} \mathbf{r} \cdot d\mathbf{S}$; the total volume of the region is thus given by

$$V = \frac{1}{3} \oint_S \mathbf{r} \cdot d\mathbf{S}.$$

It may be shown that this expression is valid even when O is not contained in V. Although this surface integral form is available, in practice it is usually simpler to evaluate the volume integral directly.

12.7 Integral forms for grad, div and curl

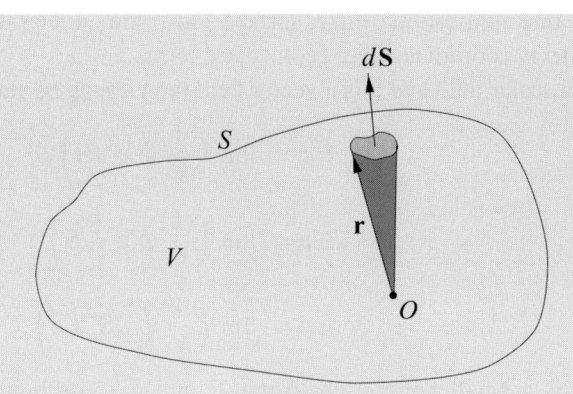

Figure 12.9 A general volume V containing the origin and bounded by the closed surface S.

Example Find the volume enclosed between a sphere of radius a centred on the origin and a circular cone of half-angle α with its vertex at the origin.

The element of vector area $d\mathbf{S}$ on the surface of the sphere is given in spherical polar coordinates by $a^2 \sin\theta \, d\theta \, d\phi \, \hat{\mathbf{r}}$. Now taking the axis of the cone to lie along the z-axis (from which θ is measured) the required volume is given by

$$V = \frac{1}{3} \oint_S \mathbf{r} \cdot d\mathbf{S} = \frac{1}{3} \int_0^{2\pi} d\phi \int_0^{\alpha} a^2 \sin\theta \, \mathbf{r} \cdot \hat{\mathbf{r}} \, d\theta$$

$$= \frac{1}{3} \int_0^{2\pi} d\phi \int_0^{\alpha} a^3 \sin\theta \, d\theta = \frac{2\pi a^3}{3}(1 - \cos\alpha).$$

If the cone is formally 'turned inside out', i.e. α is set equal to π, then the formula for the volume of a complete sphere is recovered. ◀

EXERCISE 12.6

1. For the vector fields $\mathbf{f}$ and $\mathbf{F}$ described in the exercises for Section 12.5 and their associated regions, evaluate the volume integrals

$$\int_V \nabla \cdot \mathbf{f} \, dV, \quad \int_V \nabla \times \mathbf{f} \, dV, \quad \int_V \nabla \cdot \mathbf{F} \, dV, \quad \int_V \nabla \times \mathbf{F} \, dV.$$

12.7 Integral forms for grad, div and curl

In the previous chapter we defined the vector operators grad, div and curl in purely mathematical terms, which depended on the coordinate system in which they were expressed.

Line, surface and volume integrals

An interesting application of line, surface and volume integrals is the expression of grad, div and curl in coordinate-free, geometrical terms.

If ϕ is a scalar field and $\mathbf{a}$ is a vector field then it may be shown that at any point P

$$\nabla \phi = \lim_{V \to 0} \left(\frac{1}{V} \oint_S \phi \, d\mathbf{S} \right), \quad (12.14)$$

$$\nabla \cdot \mathbf{a} = \lim_{V \to 0} \left(\frac{1}{V} \oint_S \mathbf{a} \cdot d\mathbf{S} \right), \quad (12.15)$$

$$\nabla \times \mathbf{a} = \lim_{V \to 0} \left(\frac{1}{V} \oint_S d\mathbf{S} \times \mathbf{a} \right), \quad (12.16)$$

where V is a small volume enclosing P and S is its bounding surface. Indeed, we may consider these equations as the (geometrical) *definitions* of grad, div and curl. An alternative, but equivalent, geometrical definition of $\nabla \times \mathbf{a}$ at a point P, which is often easier to use than (12.16), is given by

$$(\nabla \times \mathbf{a}) \cdot \hat{\mathbf{n}} = \lim_{A \to 0} \left(\frac{1}{A} \oint_C \mathbf{a} \cdot d\mathbf{r} \right), \quad (12.17)$$

where C is a plane contour of area A enclosing the point P and $\hat{\mathbf{n}}$ is the unit normal to the enclosed planar area.

It may be shown, *in any coordinate system*, that all the above equations are consistent with our definitions in Chapter 11, although the difficulty of proof depends on the chosen coordinate system. The most general coordinate system encountered in that chapter was one with orthogonal curvilinear coordinates u_1, u_2, u_3, of which Cartesians, cylindrical polars and spherical polars are all special cases. Although it may be shown that (12.14) leads to the usual expression for grad in curvilinear coordinates, the proof requires complicated manipulations of the derivatives of the basis vectors with respect to the coordinates and is not presented here. In Cartesian coordinates, however, the proof is quite simple.

Example Show that the geometrical definition of grad leads to the usual expression for $\nabla \phi$ in Cartesian coordinates.

Consider the surface S of a small rectangular volume element $\Delta V = \Delta x \, \Delta y \, \Delta z$ that has its faces parallel to the x, y, and z coordinate surfaces; the point P (see above) is at one corner. We must calculate the surface integral (12.14) over each of its six faces. Remembering that the normal to the surface points outwards from the volume on each face, the two faces with $x = $ constant have areas $\Delta \mathbf{S} = -\mathbf{i} \, \Delta y \, \Delta z$ and $\Delta \mathbf{S} = \mathbf{i} \, \Delta y \, \Delta z$ respectively. Furthermore, over each small surface element, we may take ϕ to be constant,[16] so that the net contribution to the surface integral from these two faces is, to first order in Δx,

$$[(\phi + \Delta \phi) - \phi] \Delta y \, \Delta z \, \mathbf{i} = \left(\phi + \frac{\partial \phi}{\partial x} \Delta x - \phi \right) \Delta y \, \Delta z \, \mathbf{i}$$

$$= \frac{\partial \phi}{\partial x} \Delta x \, \Delta y \, \Delta z \, \mathbf{i}.$$

---

16 But, in general, different on the two faces.

12.7 Integral forms for grad, div and curl

The surface integral over the pairs of faces with $y = $ constant and $z = $ constant respectively may be found in a similar way, and we obtain

$$\oint_S \phi \, d\mathbf{S} = \left(\frac{\partial \phi}{\partial x} \mathbf{i} + \frac{\partial \phi}{\partial y} \mathbf{j} + \frac{\partial \phi}{\partial z} \mathbf{k} \right) \Delta x \, \Delta y \, \Delta z.$$

Therefore, $\nabla \phi$ at the point P is given by

$$\nabla \phi = \lim_{\Delta x, \Delta y, \Delta z \to 0} \left[\frac{1}{\Delta x \, \Delta y \, \Delta z} \left(\frac{\partial \phi}{\partial x} \mathbf{i} + \frac{\partial \phi}{\partial y} \mathbf{j} + \frac{\partial \phi}{\partial z} \mathbf{k} \right) \Delta x \, \Delta y \, \Delta z \right]$$

$$= \frac{\partial \phi}{\partial x} \mathbf{i} + \frac{\partial \phi}{\partial y} \mathbf{j} + \frac{\partial \phi}{\partial z} \mathbf{k},$$

which is the same expression as the purely mathematical one for $\nabla \phi$. ◀

We now turn to (12.15) and (12.17). These geometrical definitions may be shown straightforwardly to lead to the usual expressions for div and curl in orthogonal curvilinear coordinates.

Example By considering the infinitesimal volume element $dV = h_1 h_2 h_3 \, \Delta u_1 \, \Delta u_2 \, \Delta u_3$ shown in Figure 12.10, prove that (12.15) leads to the usual expression for $\nabla \cdot \mathbf{a}$ in orthogonal curvilinear coordinates.

Let us write the vector field in terms of its components with respect to the basis vectors of the curvilinear coordinate system as $\mathbf{a} = a_1 \hat{\mathbf{e}}_1 + a_2 \hat{\mathbf{e}}_2 + a_3 \hat{\mathbf{e}}_3$. We consider first the contribution to the RHS of (12.15) from the two faces with $u_1 = $ constant, i.e. $PQRS$ and the face opposite it (see Figure 12.10). Now, the volume element is formed from the orthogonal vectors $h_1 \Delta u_1 \hat{\mathbf{e}}_1$, $h_2 \Delta u_2 \hat{\mathbf{e}}_2$ and $h_3 \Delta u_3 \hat{\mathbf{e}}_3$ at the point P and so for $PQRS$ we have[17]

$$\Delta \mathbf{S} = h_2 h_3 \, \Delta u_2 \, \Delta u_3 \, \hat{\mathbf{e}}_3 \times \hat{\mathbf{e}}_2 = -h_2 h_3 \, \Delta u_2 \, \Delta u_3 \, \hat{\mathbf{e}}_1.$$

Reasoning along the same lines as in the previous example, we conclude that the contribution to the surface integral of $\mathbf{a} \cdot d\mathbf{S}$ over $PQRS$ and its opposite face taken together is given by[18]

$$\frac{\partial}{\partial u_1} (\mathbf{a} \cdot \Delta \mathbf{S}) \, \Delta u_1 = \frac{\partial}{\partial u_1} (a_1 h_2 h_3) \, \Delta u_1 \, \Delta u_2 \, \Delta u_3.$$

The surface integrals over the pairs of faces with $u_2 = $ constant and $u_3 = $ constant respectively may be found in a similar way, and we obtain

$$\oint_S \mathbf{a} \cdot d\mathbf{S} = \left[\frac{\partial}{\partial u_1} (a_1 h_2 h_3) + \frac{\partial}{\partial u_2} (a_2 h_3 h_1) + \frac{\partial}{\partial u_3} (a_3 h_1 h_2) \right] \Delta u_1 \, \Delta u_2 \, \Delta u_3.$$

Therefore, $\nabla \cdot \mathbf{a}$ at the point P is given by

$$\nabla \cdot \mathbf{a} = \lim_{\Delta u_1, \Delta u_2, \Delta u_3 \to 0} \left[\frac{1}{h_1 h_2 h_3 \, \Delta u_1 \, \Delta u_2 \, \Delta u_3} \oint_S \mathbf{a} \cdot d\mathbf{S} \right]$$

$$= \frac{1}{h_1 h_2 h_3} \left[\frac{\partial}{\partial u_1} (a_1 h_2 h_3) + \frac{\partial}{\partial u_2} (a_2 h_3 h_1) + \frac{\partial}{\partial u_3} (a_3 h_1 h_2) \right].$$

This is the same expression for $\nabla \cdot \mathbf{a}$ as that given in Table 11.4. ◀

---

[17] Recall that $\Delta \mathbf{S}$ is in the direction of the outward normal to the volume; hence $\hat{\mathbf{e}}_3 \times \hat{\mathbf{e}}_2$, and not $\hat{\mathbf{e}}_2 \times \hat{\mathbf{e}}_3$.
[18] Note that, since $\Delta \mathbf{S}$ is (anti-)parallel to $\hat{\mathbf{e}}_1$, only the a_1 component of $\mathbf{a}$ contributes to $\mathbf{a} \cdot \Delta \mathbf{S}$.

Line, surface and volume integrals

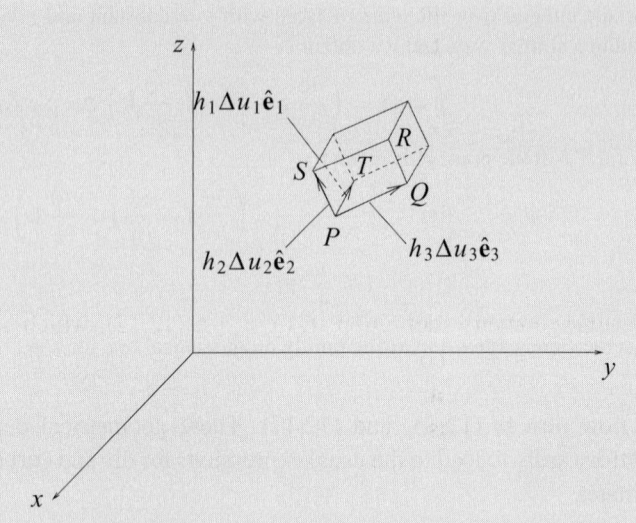

Figure 12.10 A general volume ΔV in orthogonal curvilinear coordinates u_1, u_2, u_3. PT gives the vector $h_1 \Delta u_1 \hat{\mathbf{e}}_1$, PS gives $h_2 \Delta u_2 \hat{\mathbf{e}}_2$ and PQ gives $h_3 \Delta u_3 \hat{\mathbf{e}}_3$.

Example By considering the infinitesimal planar surface element $PQRS$ in Figure 12.10, show that (12.17) leads to the usual expression for $\nabla \times \mathbf{a}$ in orthogonal curvilinear coordinates.

The planar surface $PQRS$ is defined by the orthogonal vectors $h_2 \Delta u_2 \hat{\mathbf{e}}_2$ and $h_3 \Delta u_3 \hat{\mathbf{e}}_3$ at the point P. If we traverse the loop in the direction $PSRQ$ then, by the right-hand convention, the unit normal to the plane is $\hat{\mathbf{e}}_1$. Writing $\mathbf{a} = a_1 \hat{\mathbf{e}}_1 + a_2 \hat{\mathbf{e}}_2 + a_3 \hat{\mathbf{e}}_3$, the line integral around the loop in this direction is given by the sum of four scalar products, each of which has a non-zero contribution from only one of the components of $\mathbf{a}$. The contributions are, in order, $h_2 a_2$, $h_3 a_3$ evaluated at $u_2 + \Delta u_2$, $-h_2 a_2$ evaluated at $u_3 + \Delta u_3$, and $-h_3 a_3$; the negative signs in the final two contributions arise because along RQ and QP the direction of traversal is in the negative $\hat{\mathbf{e}}_2$ and $\hat{\mathbf{e}}_3$ directions, respectively. The line integral is thus

$$\oint_{PSRQ} \mathbf{a} \cdot d\mathbf{r} = a_2 h_2 \Delta u_2 + \left[a_3 h_3 + \frac{\partial}{\partial u_2}(a_3 h_3) \Delta u_2 \right] \Delta u_3$$

$$- \left[a_2 h_2 + \frac{\partial}{\partial u_3}(a_2 h_2) \Delta u_3 \right] \Delta u_2 - a_3 h_3 \Delta u_3$$

$$= \left[\frac{\partial}{\partial u_2}(a_3 h_3) - \frac{\partial}{\partial u_3}(a_2 h_2) \right] \Delta u_2 \Delta u_3.$$

Therefore, from (12.17), the component of $\nabla \times \mathbf{a}$ in the direction $\hat{\mathbf{e}}_1$ at P is given by

$$(\nabla \times \mathbf{a})_1 = \lim_{\Delta u_2, \Delta u_3 \to 0} \left[\frac{1}{h_2 h_3 \Delta u_2 \Delta u_3} \oint_{PSRQ} \mathbf{a} \cdot d\mathbf{r} \right]$$

$$= \frac{1}{h_2 h_3} \left[\frac{\partial}{\partial u_2}(h_3 a_3) - \frac{\partial}{\partial u_3}(h_2 a_2) \right].$$

The other two components are found by cyclically permuting the subscripts 1, 2, 3. Each of the components so found is in accord with the determinantal expression for $\nabla \times \mathbf{a}$ given in Table 11.4.

◀

12.8 Divergence theorem and related theorems

Finally, we note that we can also write the ∇^2 operator as a surface integral by setting $\mathbf{a} = \nabla\phi$ in (12.15), to obtain

$$\nabla^2\phi = \nabla \cdot \nabla\phi = \lim_{V \to 0}\left(\frac{1}{V}\oint_S \nabla\phi \cdot d\mathbf{S}\right).$$

EXERCISE 12.7

1. From the geometric definition in Equation (12.17), which gives the component of curl $\mathbf{a}$ in terms of a contour integral, obtain the usual Cartesian expressions for the components of $\nabla \times \mathbf{a}$. Use a rectangular contour, with its normal in the direction appropriate to the sense in which the contour is traversed.

12.8 Divergence theorem and related theorems

The divergence theorem relates the total flux of a vector field out of a closed surface S to the integral of the divergence of the vector field over the enclosed volume V; it follows almost immediately from our geometrical definition of divergence (12.15).

Imagine a volume V, in which a vector field $\mathbf{a}$ is continuous and differentiable, to be divided up into a large number of small volumes V_i. Using (12.15), we have for each small volume

$$(\nabla \cdot \mathbf{a})V_i \approx \oint_{S_i} \mathbf{a} \cdot d\mathbf{S},$$

where S_i is the surface of the small volume V_i. Summing over i we find that contributions from surface elements interior to S cancel since each surface element appears in two terms with opposite signs, the outward normals in the two terms being equal and opposite. Only contributions from surface elements that are also parts of S survive. If each V_i is allowed to tend to zero then we obtain the *divergence theorem*,

$$\int_V \nabla \cdot \mathbf{a}\, dV = \oint_S \mathbf{a} \cdot d\mathbf{S}. \tag{12.18}$$

We note that the divergence theorem holds for both simply and multiply connected surfaces, provided that they are (i) closed and (ii) enclose some non-zero volume V.

The theorem finds most use as a tool in formal manipulations, but sometimes it is of value in transforming surface integrals of the form $\int_S \mathbf{a} \cdot d\mathbf{S}$ into volume integrals or vice versa. For example, setting $\mathbf{a} = \mathbf{r}$ we immediately obtain

$$\int_V \nabla \cdot \mathbf{r}\, dV = \int_V 3\, dV = 3V = \oint_S \mathbf{r} \cdot d\mathbf{S},$$

which gives the expression for the volume of a region found in Section 12.6. The use of the divergence theorem is further illustrated in the following example.

Example Evaluate the surface integral $I = \int_S \mathbf{a} \cdot d\mathbf{S}$, where $\mathbf{a} = (y - x)\mathbf{i} + x^2 z \mathbf{j} + (z + x^2)\mathbf{k}$ and S is the open surface of the hemisphere $x^2 + y^2 + z^2 = a^2, z \geq 0$.

We could evaluate this surface integral directly, but the algebra is somewhat lengthy. We will therefore evaluate it by use of the divergence theorem. Since the latter only holds for closed surfaces enclosing a non-zero volume V, let us first consider the closed surface $S' = S + S_1$, where S_1 is the circular area in the xy-plane given by $x^2 + y^2 \leq a^2, z = 0$; S' then encloses a hemispherical volume V. By the divergence theorem we have

$$\int_V \nabla \cdot \mathbf{a}\, dV = \oint_{S'} \mathbf{a} \cdot d\mathbf{S} = \int_S \mathbf{a} \cdot d\mathbf{S} + \int_{S_1} \mathbf{a} \cdot d\mathbf{S}.$$

Now $\nabla \cdot \mathbf{a} = -1 + 0 + 1 = 0$, so we can write

$$\int_S \mathbf{a} \cdot d\mathbf{S} = -\int_{S_1} \mathbf{a} \cdot d\mathbf{S}.$$

The surface integral over S_1 is easily evaluated. Remembering that the normal to the surface points outward from the volume, a surface element on S_1 is simply $d\mathbf{S} = -\mathbf{k}\, dx\, dy$. On S_1 we also have $\mathbf{a} = (y - x)\mathbf{i} + x^2 \mathbf{k}$, so that

$$I = -\int_{S_1} \mathbf{a} \cdot d\mathbf{S} = \int\int_R x^2\, dx\, dy,$$

where R is the circular region in the xy-plane given by $x^2 + y^2 \leq a^2$. Transforming to plane polar coordinates we have

$$I = \int\int_{R'} \rho^2 \cos^2 \phi\, \rho\, d\rho\, d\phi = \int_0^{2\pi} \cos^2 \phi\, d\phi \int_0^a \rho^3\, d\rho = \frac{\pi a^4}{4}.$$

Thus the integral $\int \mathbf{a} \cdot d\mathbf{S}$ over a curved surface, for an intricate vector field $\mathbf{a}$, has been evaluated by computing the integral of a much simpler field over an easily specified plane surface. ◀

It is also interesting to consider the two-dimensional version of the divergence theorem. As an example, let us consider a two-dimensional planar region R in the xy-plane bounded by some closed curve C (see Figure 12.11). At any point on the curve the vector $d\mathbf{r} = dx\,\mathbf{i} + dy\,\mathbf{j}$ is a tangent to the curve and the vector $\hat{\mathbf{n}}\, ds = dy\,\mathbf{i} - dx\,\mathbf{j}$ is a normal pointing out of the region R. If the vector field $\mathbf{a}$ is continuous and differentiable in R then the two-dimensional divergence theorem in Cartesian coordinates gives

$$\int\int_R \left(\frac{\partial a_x}{\partial x} + \frac{\partial a_y}{\partial y}\right) dx\, dy = \oint \mathbf{a} \cdot \hat{\mathbf{n}}\, ds = \oint_C (a_x\, dy - a_y\, dx).$$

Letting $P = -a_y$ and $Q = a_x$, we recover Green's theorem in a plane, which was discussed in Section 12.3.

12.8.1 Green's theorems

Consider two scalar functions ϕ and ψ that are continuous and differentiable in some volume V bounded by a surface S. Applying the divergence theorem to the vector field

12.8 Divergence theorem and related theorems

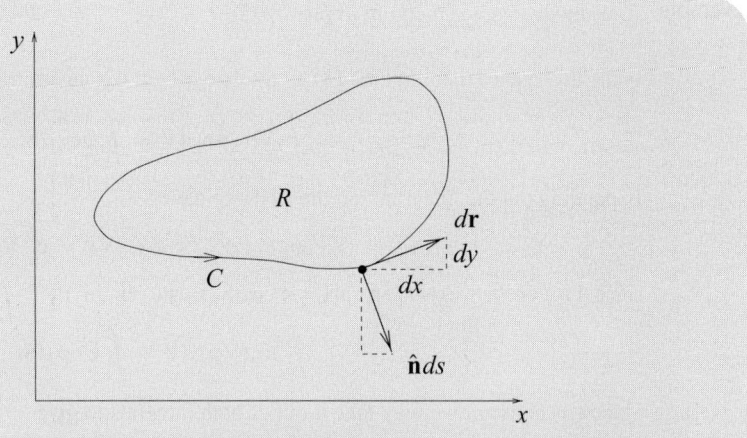

Figure 12.11 A closed curve C in the xy-plane bounding a region R. Vectors tangent and normal to the curve at a given point are also shown.

$\phi\nabla\psi$ we obtain

$$\oint_S \phi\nabla\psi \cdot d\mathbf{S} = \int_V \nabla \cdot (\phi\nabla\psi)\, dV$$
$$= \int_V \left[\phi\nabla^2\psi + (\nabla\phi) \cdot (\nabla\psi)\right] dV. \quad (12.19)$$

Reversing the roles of ϕ and ψ in (12.19) and subtracting the two equations gives

$$\oint_S (\phi\nabla\psi - \psi\nabla\phi) \cdot d\mathbf{S} = \int_V (\phi\nabla^2\psi - \psi\nabla^2\phi)\, dV. \quad (12.20)$$

Equation (12.19) is usually known as Green's first theorem and (12.20) as his second. Green's second theorem is useful in the development of so-called Green's functions that are used in the solution of partial differential equations.

12.8.2 Other related integral theorems

There exist two other integral theorems which are closely related to the divergence theorem and which are of some use in physical applications. If ϕ is a scalar field and $\mathbf{b}$ is a vector field and both ϕ and $\mathbf{b}$ satisfy our usual differentiability conditions in some volume V bounded by a closed surface S, then

$$\int_V \nabla\phi\, dV = \oint_S \phi\, d\mathbf{S}, \quad (12.21)$$

$$\int_V \nabla \times \mathbf{b}\, dV = \oint_S d\mathbf{S} \times \mathbf{b}. \quad (12.22)$$

The first of these is proved in the following example.

Line, surface and volume integrals

Example Use the divergence theorem to prove (12.21).

In the divergence theorem (12.18) let $\mathbf{a} = \phi\mathbf{c}$, where $\mathbf{c}$ is an arbitrary constant vector. We then have

$$\int_V \nabla \cdot (\phi\mathbf{c}) \, dV = \oint_S \phi\mathbf{c} \cdot d\mathbf{S}.$$

Expanding out the integrand on the LHS we have

$$\nabla \cdot (\phi\mathbf{c}) = \phi \nabla \cdot \mathbf{c} + \mathbf{c} \cdot \nabla \phi = \mathbf{c} \cdot \nabla \phi,$$

since $\mathbf{c}$ is constant. Also, $\phi\mathbf{c} \cdot d\mathbf{S} = \mathbf{c} \cdot \phi d\mathbf{S}$, so we obtain

$$\int_V \mathbf{c} \cdot (\nabla \phi) \, dV = \oint_S \mathbf{c} \cdot \phi \, d\mathbf{S}.$$

Since $\mathbf{c}$ is constant we may take it out of both integrals to give

$$\mathbf{c} \cdot \int_V \nabla \phi \, dV = \mathbf{c} \cdot \oint_S \phi \, d\mathbf{S},$$

and since $\mathbf{c}$ is arbitrary we obtain the stated result (12.21).[19] ◀

Equation (12.22) may be proved in a similar way by letting $\mathbf{a} = \mathbf{b} \times \mathbf{c}$ in the divergence theorem, where $\mathbf{c}$ is again a constant vector.

12.8.3 Physical applications of the divergence theorem

The divergence theorem is useful in deriving many of the most important partial differential equations in physics. The basic idea is to use the divergence theorem to convert an integral form, often derived from observation, into an equivalent differential form (used in theoretical statements).

Example For a compressible fluid with time-varying position-dependent density $\rho(\mathbf{r}, t)$ and velocity field $v(\mathbf{r}, t)$, in which fluid is neither being created nor destroyed, show that

$$\frac{\partial \rho}{\partial t} + \nabla \cdot (\rho \mathbf{v}) = 0.$$

For an arbitrary volume V in the fluid, the conservation of mass tells us that the rate of increase or decrease of the mass M of fluid in the volume must equal the net rate at which fluid is entering or leaving the volume, i.e.

$$\frac{dM}{dt} = -\oint_S \rho \mathbf{v} \cdot d\mathbf{S},$$

where S is the surface bounding V. But the mass of fluid in V is simply $M = \int_V \rho \, dV$, so we have

$$\frac{d}{dt} \int_V \rho \, dV + \oint_S \rho \mathbf{v} \cdot d\mathbf{S} = 0.$$

---

[19] Provide a formal proof of '$\mathbf{c} \cdot \mathbf{P} = \mathbf{c} \cdot \mathbf{Q}$ implies that $\mathbf{P} = \mathbf{Q}$ if $\mathbf{c}$ is arbitrary'.

12.8 Divergence theorem and related theorems

Taking the derivative inside the first integral[20] on the LHS and using the divergence theorem to rewrite the second integral, we obtain

$$\int_V \frac{\partial \rho}{\partial t} dV + \int_V \nabla \cdot (\rho \mathbf{v}) dV = \int_V \left[\frac{\partial \rho}{\partial t} + \nabla \cdot (\rho \mathbf{v}) \right] dV = 0.$$

Since the volume V is arbitrary, the integrand (which is assumed continuous) must be identically zero, so we obtain

$$\frac{\partial \rho}{\partial t} + \nabla \cdot (\rho \mathbf{v}) = 0.$$

This is known as the *continuity equation*. It can also be applied to other systems, for example those in which ρ is the density of electric charge or the heat content, etc. For the flow of an incompressible fluid, $\rho = $ constant and the continuity equation becomes simply $\nabla \cdot \mathbf{v} = 0$. ◂

In the previous example we assumed that there were no sources or sinks in the volume V, i.e. that there was no part of V in which fluid was being created or destroyed. We now consider the case where a finite number of *point* sources and/or sinks are present in an incompressible fluid. Let us first consider the simple case where a single source is located at the origin, out of which a quantity of fluid flows radially at a rate Q (m^3 s^{-1}). The velocity field is given by

$$\mathbf{v} = \frac{Q\mathbf{r}}{4\pi r^3} = \frac{Q\hat{\mathbf{r}}}{4\pi r^2}.$$

Now, for a sphere S_1 of radius r centred on the source, the flux across S_1 is

$$\oint_{S_1} \mathbf{v} \cdot d\mathbf{S} = |\mathbf{v}| 4\pi r^2 = Q.$$

Since $\mathbf{v}$ has a singularity at the origin it is not differentiable there, i.e. $\nabla \cdot \mathbf{v}$ is not defined there, but at all other points $\nabla \cdot \mathbf{v} = 0$, as required for an incompressible fluid. Therefore, from the divergence theorem, for any closed surface S_2 that does not enclose the origin we have

$$\oint_{S_2} \mathbf{v} \cdot d\mathbf{S} = \int_V \nabla \cdot \mathbf{v} \, dV = 0.$$

Thus we see that the surface integral $\oint_S \mathbf{v} \cdot d\mathbf{S}$ has value Q or zero depending on whether or not S encloses the source. In order that the divergence theorem is valid for *all* surfaces S, irrespective of whether they enclose the source, we write

$$\nabla \cdot \mathbf{v} = Q\delta(\mathbf{r}),$$

where $\delta(\mathbf{r})$ is the three-dimensional Dirac delta function. An introduction to the properties of this function is given in Section 13.2. The discussion there concentrates on the one-dimensional δ-function, though the underlying notions can be extended to three dimensions

---

[20] The derivative is with respect to time while the integral is with respect to space, and so this interchange is permissible.

without difficulty.[21] For our present purposes we define it to be such that

$$\delta(\mathbf{r} - \mathbf{a}) = 0 \quad \text{for } \mathbf{r} \neq \mathbf{a},$$

$$\int_V f(\mathbf{r})\delta(\mathbf{r} - \mathbf{a})\,dV = \begin{cases} f(\mathbf{a}) & \text{if } \mathbf{a} \text{ lies in } V \\ 0 & \text{otherwise} \end{cases}$$

for any well-behaved function $f(\mathbf{r})$. Therefore, for any volume V containing the source at the origin, we have

$$\int_V \nabla \cdot \mathbf{v}\,dV = Q \int_V \delta(\mathbf{r})\,dV = Q,$$

which is consistent with $\oint_S \mathbf{v} \cdot d\mathbf{S} = Q$ for a closed surface enclosing the source. Hence, by introducing the Dirac delta function the divergence theorem can be made valid even for non-differentiable point sources.

The generalisation to several sources and sinks is straightforward. For example, if a source is located at $\mathbf{r} = \mathbf{a}$ and a sink of equal strength at $\mathbf{r} = \mathbf{b}$, then the velocity field is

$$\mathbf{v} = \frac{(\mathbf{r} - \mathbf{a})Q}{4\pi|\mathbf{r} - \mathbf{a}|^3} - \frac{(\mathbf{r} - \mathbf{b})Q}{4\pi|\mathbf{r} - \mathbf{b}|^3}$$

and its divergence is given by

$$\nabla \cdot \mathbf{v} = Q\delta(\mathbf{r} - \mathbf{a}) - Q\delta(\mathbf{r} - \mathbf{b}).$$

Therefore, the integral $\oint_S \mathbf{v} \cdot d\mathbf{S}$ has the value Q if S encloses the source, $-Q$ if S encloses the sink and zero if S encloses neither the source nor sink or encloses them both. This analysis also applies to other physical systems – for example, in electrostatics we can regard the sources and sinks as positive and negative point charges respectively and replace $\mathbf{v}$ by the electric field $\mathbf{E}$.

EXERCISES 12.8

1. Use the divergence theorem to explain quantitatively the connection between the values of $\int_S \mathbf{F} \cdot d\mathbf{S}$ and $\int_V \nabla \cdot \mathbf{F}\,dV$ obtained in the exercises for Sections 12.5 and 12.6.

2. A vector field $\mathbf{a}$ has components given in spherical polar coordinates by

$$\mathbf{a} = (r^2 \sin^2\theta,\, r^2 \sin^2\theta \cos^2\phi,\, r^2 \sin\theta).$$

   Evaluate the integral $\int_V \nabla \cdot \mathbf{a}\,dV$ over the hemisphere $r \leq R$, $0 \leq \theta \leq \pi/2$, using (a) direct calculation and (b) the divergence theorem.

---

[21] A graph of the one-dimensional Dirac delta function $\delta(x - a)$ can be thought of as an extremely large and very narrow spike centred on $x = a$ with unit area underneath it. The two- and three-dimensional versions, though impossible to draw, follow the same principle.

3. The electrostatic field outside the sphere $r = a$ is given, in spherical polars, by

$$\mathbf{E} = \frac{1}{r^4}(1 - 3\cos^2\theta, -\sin 2\theta, 0).$$

Does the sphere contain any net charge?

12.9 Stokes' theorem and related theorems

Stokes' theorem is the 'curl analogue' of the divergence theorem and relates the integral of the curl of a vector field over an open surface S to the line integral of the vector field around the perimeter C bounding the surface.

Following the same lines as for the derivation of the divergence theorem, we can divide the surface S into many small areas S_i with boundaries C_i and unit normals $\hat{\mathbf{n}}_i$. Using (12.17), we have for each small area

$$(\nabla \times \mathbf{a}) \cdot \hat{\mathbf{n}}_i \, S_i \approx \oint_{C_i} \mathbf{a} \cdot d\mathbf{r}.$$

Summing over i we find that on the RHS all parts of all interior boundaries that are not part of C are included twice, being traversed in opposite directions on each occasion and thus contributing cancelling contributions. Only contributions from line elements that are also parts of C survive. If each S_i is allowed to tend to zero then we obtain Stokes' theorem,

$$\int_S (\nabla \times \mathbf{a}) \cdot d\mathbf{S} = \oint_C \mathbf{a} \cdot d\mathbf{r}. \tag{12.23}$$

We note that Stokes' theorem holds for both simply and multiply connected open surfaces, provided that they are two-sided.

Just as the divergence theorem (12.18) can be used to relate volume and surface integrals for certain types of integrand, Stokes' theorem can be used in evaluating surface integrals of the form $\oint_S (\nabla \times \mathbf{a}) \cdot d\mathbf{S}$ as line integrals or vice versa.

Example Given the vector field $\mathbf{a} = y\mathbf{i} - x\mathbf{j} + z\mathbf{k}$, verify Stokes' theorem for the hemispherical surface $x^2 + y^2 + z^2 = a^2, z \geq 0$.

Let us first evaluate the surface integral

$$\int_S (\nabla \times \mathbf{a}) \cdot d\mathbf{S}$$

over the hemisphere. It is easily shown that $\nabla \times \mathbf{a} = -2\mathbf{k}$, and the surface element is $d\mathbf{S} = a^2 \sin\theta \, d\theta \, d\phi \, \hat{\mathbf{r}}$ in spherical polar coordinates. Therefore

$$\int_S (\nabla \times \mathbf{a}) \cdot d\mathbf{S} = \int_0^{2\pi} d\phi \int_0^{\pi/2} d\theta \, (-2a^2 \sin\theta) \, \hat{\mathbf{r}} \cdot \mathbf{k}$$

$$= -2a^2 \int_0^{2\pi} d\phi \int_0^{\pi/2} \sin\theta \left(\frac{z}{a}\right) d\theta$$

$$= -2a^2 \int_0^{2\pi} d\phi \int_0^{\pi/2} \sin\theta \cos\theta \, d\theta = -2\pi a^2.$$

Line, surface and volume integrals

We now evaluate the line integral around the perimeter curve C of the surface, which is the circle $x^2 + y^2 = a^2$ in the xy-plane. This is given by

$$\oint_C \mathbf{a} \cdot d\mathbf{r} = \oint_C (y\mathbf{i} - x\mathbf{j} + z\mathbf{k}) \cdot (dx\mathbf{i} + dy\mathbf{j} + dz\mathbf{k})$$

$$= \oint_C (y\, dx - x\, dy).$$

Using plane polar coordinates, on C we have $x = a \cos \phi$, $y = a \sin \phi$ so that $dx = -a \sin \phi \, d\phi$, $dy = a \cos \phi \, d\phi$, and the line integral becomes

$$\oint_C (y\, dx - x\, dy) = -a^2 \int_0^{2\pi} (\sin^2 \phi + \cos^2 \phi)\, d\phi = -a^2 \int_0^{2\pi} d\phi = -2\pi a^2.$$

Since the surface and line integrals have the same value,[22] we have verified Stokes' theorem in this case. ◀

The two-dimensional version of Stokes' theorem also yields Green's theorem in a plane. Consider the region R in the xy-plane shown in Figure 12.11, in which a vector field $\mathbf{a}$ is defined. Since $\mathbf{a} = a_x \mathbf{i} + a_y \mathbf{j}$, we have $\nabla \times \mathbf{a} = (\partial a_y/\partial x - \partial a_x/\partial y)\mathbf{k}$, and Stokes' theorem becomes

$$\int\int_R \left(\frac{\partial a_y}{\partial x} - \frac{\partial a_x}{\partial y}\right) dx\, dy = \oint_C (a_x\, dx + a_y\, dy).$$

Letting $P = a_x$ and $Q = a_y$ we recover Green's theorem in a plane, (12.4).

12.9.1 Related integral theorems

As for the divergence theorem, there exist two other integral theorems that are closely related to Stokes' theorem. If ϕ is a scalar field and $\mathbf{b}$ is a vector field, and both ϕ and $\mathbf{b}$ satisfy our usual differentiability conditions on some two-sided open surface S bounded by a closed perimeter curve C, then

$$\int_S d\mathbf{S} \times \nabla \phi = \oint_C \phi \, d\mathbf{r}, \qquad (12.24)$$

$$\int_S (d\mathbf{S} \times \nabla) \times \mathbf{b} = \int_S [\nabla(\mathbf{b} \cdot d\mathbf{S}) - (\nabla \cdot \mathbf{b}) d\mathbf{S}] = \oint_C d\mathbf{r} \times \mathbf{b}. \qquad (12.25)$$

Example Use Stokes' theorem to prove (12.24).

In Stokes' theorem, (12.23), let $\mathbf{a} = \phi \mathbf{c}$, where $\mathbf{c}$ is a constant vector. We then have

$$\int_S [\nabla \times (\phi \mathbf{c})] \cdot d\mathbf{S} = \oint_C \phi \mathbf{c} \cdot d\mathbf{r}. \qquad (12.26)$$

Expanding out the integrand on the LHS we have

$$\nabla \times (\phi \mathbf{c}) = \nabla \phi \times \mathbf{c} + \phi \nabla \times \mathbf{c} = \nabla \phi \times \mathbf{c},$$

---

[22] Note that, since *any* open surface with boundary C will do, the value of the surface integral can be written down immediately if the plane surface $x^2 + y^2 = a^2$, $z = 0$ is used.

12.9 Stokes' theorem and related theorems

since $\mathbf{c}$ is constant, and the scalar triple product on the LHS of (12.26) can therefore be written

$$[\nabla \times (\phi\mathbf{c})] \cdot d\mathbf{S} = (\nabla\phi \times \mathbf{c}) \cdot d\mathbf{S} = \mathbf{c} \cdot (d\mathbf{S} \times \nabla\phi).$$

Substituting this into (12.26) and taking $\mathbf{c}$ out of both integrals because it is constant, we find

$$\mathbf{c} \cdot \int_S d\mathbf{S} \times \nabla\phi = \mathbf{c} \cdot \oint_C \phi \, d\mathbf{r}.$$

Since $\mathbf{c}$ is an *arbitrary* constant vector, result (12.24) follows. ◀

Equation (12.25) may be proved in a similar way, by letting $\mathbf{a} = \mathbf{b} \times \mathbf{c}$ in Stokes' theorem, where $\mathbf{c}$ is again a constant vector. The equality between the two integrands for the surface integral is most easily shown using the summation convention notation (Appendix D).

We also note that by setting $\mathbf{b} = \mathbf{r}$ in (12.25) we find

$$\int_S (d\mathbf{S} \times \nabla) \times \mathbf{r} = \oint_C d\mathbf{r} \times \mathbf{r}.$$

Expanding out the integrand on the LHS gives

$$(d\mathbf{S} \times \nabla) \times \mathbf{r} = \nabla(d\mathbf{S} \cdot \mathbf{r}) - d\mathbf{S}(\nabla \cdot \mathbf{r}) = d\mathbf{S} - 3\,d\mathbf{S} = -2\,d\mathbf{S}.$$

Therefore, as we found in Section 12.5.2, the vector area of an open surface S is given by

$$\mathbf{S} = \int_S d\mathbf{S} = \frac{1}{2} \oint_C \mathbf{r} \times d\mathbf{r}.$$

12.9.2 Physical applications of Stokes' theorem

Like the divergence theorem, Stokes' theorem is useful for converting integral equations into differential ones.

Example From Ampère's law derive Maxwell's equation in the case where the currents are steady, i.e. $\nabla \times \mathbf{B} - \mu_0 \mathbf{J} = \mathbf{0}$.

Ampère's rule for a distributed current with current density $\mathbf{J}$ is

$$\oint_C \mathbf{B} \cdot d\mathbf{r} = \mu_0 \int_S \mathbf{J} \cdot d\mathbf{S},$$

for any circuit C bounding a surface S. Using Stokes' theorem, the LHS can be transformed into $\int_S (\nabla \times \mathbf{B}) \cdot d\mathbf{S}$; hence

$$\int_S (\nabla \times \mathbf{B} - \mu_0 \mathbf{J}) \cdot d\mathbf{S} = 0$$

for *any* surface S. This can only be so if $\nabla \times \mathbf{B} - \mu_0 \mathbf{J} = \mathbf{0}$, which is the required relation. Similarly, from Faraday's law of electromagnetic induction we can derive Maxwell's equation $\nabla \times \mathbf{E} = -\partial \mathbf{B}/\partial t$. ◀

Line, surface and volume integrals

In Section 12.8.3 we discussed the flow of an incompressible fluid in the presence of several sources and sinks. Let us now consider *vortex* flow in an incompressible fluid with a velocity field

$$\mathbf{v} = \frac{1}{\rho}\hat{\mathbf{e}}_\phi,$$

in cylindrical polar coordinates ρ, ϕ, z. For this velocity field $\nabla \times \mathbf{v}$ equals zero everywhere except on the axis $\rho = 0$, where $\mathbf{v}$ has a singularity. Therefore $\oint_C \mathbf{v} \cdot d\mathbf{r}$ equals zero for any path C that does not enclose the vortex line on the axis and 2π if C does enclose the axis. In order for Stokes' theorem to be valid for all paths C, we therefore set

$$\nabla \times \mathbf{v} = 2\pi\,\delta(\rho),$$

where $\delta(\rho)$ is the Dirac delta function.[23] Now, since $\nabla \times \mathbf{v} = \mathbf{0}$, except on the axis $\rho = 0$, there exists a scalar potential ψ such that $\mathbf{v} = \nabla \psi$. It may easily be shown that $\psi = \phi$, the azimuthal angle. Therefore, if C does not enclose the axis then

$$\oint_C \mathbf{v} \cdot d\mathbf{r} = \oint d\phi = 0,$$

and if C does enclose the axis,

$$\oint_C \mathbf{v} \cdot d\mathbf{r} = \Delta\phi = 2\pi n,$$

where n is the number of times we traverse C. Thus ϕ is a multivalued potential.

Similar analyses are valid for other physical systems – for example, in magnetostatics we may replace the vortex lines by current-carrying wires and the velocity field $\mathbf{v}$ by the magnetic field $\mathbf{B}$.

EXERCISES 12.9

1. For the function $\mathbf{f}$ specified in the first exercise of Section 12.5 and the closed plain rectangular contour $C: (0, 0, 0) \to (1, 1, 0) \to (1, 1, 1) \to (0, 0, 1) \to (0, 0, 0)$, show that both $\int_C (\nabla \times \mathbf{f}) \cdot d\mathbf{S}$ and $\int_C \mathbf{f} \cdot d\mathbf{r}$ have zero values (thus satisfying Stokes's theorem), but that the zero values arise from internal cancellations, rather than zero integrands.

2. For a stationary closed circuit C (with or without any physical conducting wire) containing no cells and placed in a magnetic field $\mathbf{B}$, the law of electromagnetic induction states that the (back) e.m.f. induced in the circuit when the magnetic field changes is given by $\mathcal{E} = -\partial\Phi/\partial t$. Here Φ is the total magnetic flux threading the circuit. Given that $\mathcal{E}$ is equal to the line integral $\oint \mathbf{E} \cdot d\mathbf{r}$ around C, where $\mathbf{E}$ is the electric field in the circuit, use Stokes's theorem to deduce Maxwell's equation $\nabla \times \mathbf{E} + \partial \mathbf{B}/\partial t = \mathbf{0}$.

---

[23] See footnote 21.

SUMMARY

1. *Line integral types*

   Scalar type: $\int_C \mathbf{a} \cdot d\mathbf{r}$;   vector type: $\int_C \phi\, d\mathbf{r}$ or $\int_C \mathbf{a} \times d\mathbf{r}$.

2. *Green's theorem in a plane*

$$\oint_C (P\, dx + Q\, dy) = \int\int_R \left(\frac{\partial Q}{\partial x} - \frac{\partial P}{\partial y} \right) dx\, dy.$$

   The theorem is valid for a multiply connected region provided C includes all boundaries and they are traversed in the positive direction.

3. *Conservative fields*

   A vector field $\mathbf{a}$ that has continuous partial derivatives in a simply connected region R is conservative if, and only if, any of the following are true (each implies the other three)

   (i) The integral $\int_A^B \mathbf{a} \cdot d\mathbf{r}$, where A and B lie in the region R, is independent of the path from A to B. Hence the integral $\oint_C \mathbf{a} \cdot d\mathbf{r}$ around any closed loop in R is zero.
   (ii) There exists a single-valued function ϕ of position such that $\mathbf{a} = \nabla \phi$.
   (iii) $\nabla \times \mathbf{a} = \mathbf{0}$.
   (iv) $\mathbf{a} \cdot d\mathbf{r}$ is an exact differential.

4. *Solenoidal fields*

   If $\nabla \cdot \mathbf{b} = 0$, then it always possible to find infinitely many vector fields $\mathbf{a}$ such that $\mathbf{b} = \nabla \times \mathbf{a}$. If $\mathbf{a}$ is one such field, then $\mathbf{a}' = \mathbf{a} + \nabla \psi + \mathbf{c}$ is another, for any scalar ψ and any constant vector $\mathbf{c}$.

5. *Surface integrals*
   - Scalar type: $\int_S \mathbf{a} \cdot d\mathbf{S}$;   vector type: $\int_S \phi\, d\mathbf{S}$ or $\int_S \mathbf{a} \times d\mathbf{S}$.
   - The scalar element of area dS on the surface $f(x, y, z) = 0$ is related to its projection dA on the x–y plane by $dS = \dfrac{|\nabla f|}{\partial f/\partial z}\, dA$.
   - The vector area of a surface, $\mathbf{S} = \int d\mathbf{S}$, is always zero for a closed surface.
   - The vector area of an open surface depends only on its boundary curve C and is given by $\mathbf{S} = \dfrac{1}{2} \oint_C \mathbf{r} \times d\mathbf{r}$.
   - The solid angle Ω subtended at the origin by a surface S is given by $\Omega = \int_S \dfrac{\hat{\mathbf{r}} \cdot d\mathbf{S}}{r^2}$.

Line, surface and volume integrals

6. *Theorems for surface integrals*
   - Stokes' theorem: $\int_S (\nabla \times \mathbf{a}) \cdot d\mathbf{S} = \oint_C \mathbf{a} \cdot d\mathbf{r}$.
   - Other theorems:
   $$\int_S d\mathbf{S} \times \nabla\phi = \oint_C \phi\, d\mathbf{r} \quad \text{and} \quad \int_S [\nabla(\mathbf{b} \cdot d\mathbf{S}) - (\nabla \cdot \mathbf{b})d\mathbf{S}] = \oint_C d\mathbf{r} \times \mathbf{b}.$$

7. *Volume integrals*
   - Scalar type: $\int_V \phi\, dV$; vector type: $\int_V \mathbf{a}\, dV$.
   - The volume of a closed region depends only on its bounding surface S and is given by $V = \dfrac{1}{3} \oint_S \mathbf{r} \cdot d\mathbf{S}$.
   - Grad, div and curl can be represented/defined by integrals over the surface of a (vanishingly) small volume:
   $$\nabla\phi = \lim_{V \to 0}\left(\frac{1}{V} \oint_S \phi\, d\mathbf{S}\right), \quad \nabla \cdot \mathbf{a} = \lim_{V \to 0}\left(\frac{1}{V} \oint_S \mathbf{a} \cdot d\mathbf{S}\right),$$
   $$\nabla \times \mathbf{a} = \lim_{V \to 0}\left(\frac{1}{V} \oint_S d\mathbf{S} \times \mathbf{a}\right).$$

8. *Theorems for volume integrals*
   - Divergence theorem: $\int_V \nabla \cdot \mathbf{a}\, dV = \oint_S \mathbf{a} \cdot d\mathbf{S}$.
   - Green's first theorem: $\oint_S \phi \nabla\psi \cdot d\mathbf{S} = \int_V [\phi \nabla^2 \psi + (\nabla\phi) \cdot (\nabla\psi)]\, dV$.
   - Green's second theorem: $\oint_S (\phi \nabla\psi - \psi \nabla\phi) \cdot d\mathbf{S} = \int_V (\phi \nabla^2 \psi - \psi \nabla^2 \phi)\, dV$.
   - Other theorems: $\int_V \nabla\phi\, dV = \oint_S \phi\, d\mathbf{S}$ and $\int_V \nabla \times \mathbf{b}\, dV = \oint_S d\mathbf{S} \times \mathbf{b}$.

PROBLEMS

12.1. The vector field $\mathbf{F}$ is defined by

$$\mathbf{F} = 2xz\mathbf{i} + 2yz^2\mathbf{j} + (x^2 + 2y^2z - 1)\mathbf{k}.$$

Calculate $\nabla \times \mathbf{F}$ and deduce that $\mathbf{F}$ can be written $F = \nabla\phi$. Determine the form of ϕ.

12.2. A vector field $\mathbf{Q}$ is defined as

$$[3x^2(y+z) + y^3 + z^3]\mathbf{i} + [3y^2(z+x) + z^3 + x^3]\mathbf{j} + [3z^2(x+y) + x^3 + y^3]\mathbf{k}.$$

Show that **Q** is a conservative field, construct its potential function and hence evaluate the integral $J = \int \mathbf{Q} \cdot d\mathbf{r}$ along any line connecting the point A at $(1, -1, 1)$ to B at $(2, 1, 2)$.

12.3. **F** is a vector field $xy^2\mathbf{i} + 2\mathbf{j} + x\mathbf{k}$ and L is a path parameterised by $x = ct$, $y = c/t$, $z = d$ for the range $1 \leq t \leq 2$. Evaluate (a) $\int_L \mathbf{F} \, dt$, (b) $\int_L \mathbf{F} \, dy$ and (c) $\int_L \mathbf{F} \cdot d\mathbf{r}$.

12.4. By making an appropriate choice for the functions $P(x, y)$ and $Q(x, y)$ that appear in Green's theorem in a plane, show that the integral of $x - y$ over the upper half of the unit circle centred on the origin has the value $-\frac{2}{3}$. Show the same result by direct integration in Cartesian coordinates.

12.5. Determine the point of intersection P, in the first quadrant, of the two ellipses

$$\frac{x^2}{a^2} + \frac{y^2}{b^2} = 1 \quad \text{and} \quad \frac{x^2}{b^2} + \frac{y^2}{a^2} = 1.$$

Taking $b < a$, consider the contour L that bounds the area in the first quadrant that is common to the two ellipses. Show that the parts of L that lie along the coordinate axes contribute nothing to the line integral around L of $x \, dy - y \, dx$. Using a parameterisation of each ellipse similar to that employed in the example in Section 12.3, evaluate the two remaining line integrals and hence find the total area common to the two ellipses.

12.6. By using parameterisations of the form $x = a \cos^n \theta$ and $y = a \sin^n \theta$ for suitable values of n, find the area bounded by the curves

$$x^{2/5} + y^{2/5} = a^{2/5} \quad \text{and} \quad x^{2/3} + y^{2/3} = a^{2/3}.$$

12.7. Evaluate the line integral

$$I = \oint_C \left[y(4x^2 + y^2) \, dx + x(2x^2 + 3y^2) \, dy \right]$$

around the ellipse $x^2/a^2 + y^2/b^2 = 1$.

12.8. Criticise the following 'proof' that $\pi = 0$.
(a) Apply Green's theorem in a plane to the functions $P(x, y) = \tan^{-1}(y/x)$ and $Q(x, y) = \tan^{-1}(x/y)$, taking the region R to be the unit circle centred on the origin.
(b) The RHS of the equality so produced is

$$\int\int_R \frac{y - x}{x^2 + y^2} \, dx \, dy,$$

which, either from symmetry considerations or by changing to plane polar coordinates, can be shown to have zero value.

(c) In the LHS of the equality, set $x = \cos\theta$ and $y = \sin\theta$, yielding $P(\theta) = \theta$ and $Q(\theta) = \pi/2 - \theta$. The line integral becomes

$$\int_0^{2\pi} \left[\left(\frac{\pi}{2} - \theta\right) \cos\theta - \theta \sin\theta \right] d\theta,$$

which has the value 2π.

(d) Thus $2\pi = 0$ and the stated result follows.

12.9. A single-turn coil C of arbitrary shape is placed in a magnetic field $\mathbf{B}$ and carries a current I. Show that the couple acting upon the coil can be written as

$$\mathbf{M} = I \int_C (\mathbf{B} \cdot \mathbf{r}) \, d\mathbf{r} - I \int_C \mathbf{B}(\mathbf{r} \cdot d\mathbf{r}).$$

For a planar rectangular coil of sides $2a$ and $2b$ placed with its plane vertical and at an angle ϕ to a uniform horizontal field $\mathbf{B}$, show that $\mathbf{M}$ is, as expected, $4abBI \cos\phi \, \mathbf{k}$.

12.10. Find the vector area $\mathbf{S}$ of the part of the curved surface of the hyperboloid of revolution

$$\frac{x^2}{a^2} - \frac{y^2 + z^2}{b^2} = 1$$

that lies in the region $z \geq 0$ and $a \leq x \leq \lambda a$.

12.11. An axially symmetric solid body with its axis AB vertical is immersed in an incompressible fluid of density ρ_0. Use the following method to show that, whatever the shape of the body, for $\rho = \rho(z)$ in cylindrical polars the Archimedean upthrust is, as expected, $\rho_0 g V$, where V is the volume of the body.

Express the vertical component of the resultant force on the body, $-\int p \, d\mathbf{S}$, where p is the pressure, in terms of an integral; note that $p = -\rho_0 g z$ and that for an annular surface element of width dl, $\mathbf{n} \cdot \mathbf{n}_z \, dl = -d\rho$. Integrate by parts and use the fact that $\rho(z_A) = \rho(z_B) = 0$.

12.12. Show that the expression below is equal to the solid angle subtended by a rectangular aperture, of sides $2a$ and $2b$, at a point on the normal through its centre, and at a distance c from the aperture:

$$\Omega = 4 \int_0^b \frac{ac}{(y^2 + c^2)(y^2 + c^2 + a^2)^{1/2}} \, dy.$$

By setting $y = (a^2 + c^2)^{1/2} \tan\phi$, change this integral into the form

$$\int_0^{\phi_1} \frac{4ac \cos\phi}{c^2 + a^2 \sin^2\phi} \, d\phi,$$

Problems

where $\tan\phi_1 = b/(a^2 + c^2)^{1/2}$, and hence show that

$$\Omega = 4\tan^{-1}\left[\frac{ab}{c(a^2 + b^2 + c^2)^{1/2}}\right].$$

12.13. A vector field $\mathbf{a}$ is given by $-zxr^{-3}\mathbf{i} - zyr^{-3}\mathbf{j} + (x^2 + y^2)r^{-3}\mathbf{k}$, where $r^2 = x^2 + y^2 + z^2$. Establish that the field is conservative (a) by showing that $\nabla \times \mathbf{a} = 0$ and (b) by constructing its potential function ϕ.

12.14. A vector field $\mathbf{a}$ is given by $(z^2 + 2xy)\mathbf{i} + (x^2 + 2yz)\mathbf{j} + (y^2 + 2zx)\mathbf{k}$. Show that $\mathbf{a}$ is conservative and that the line integral $\int \mathbf{a} \cdot d\mathbf{r}$ along any line joining $(1, 1, 1)$ and $(1, 2, 2)$ has the value 11.

12.15. A force $\mathbf{F}(\mathbf{r})$ acts on a particle at $\mathbf{r}$. In which of the following cases can $\mathbf{F}$ be represented in terms of a potential? Where it can, find the potential.

(a) $\mathbf{F} = F_0\left[\mathbf{i} - \mathbf{j} - \frac{2(x-y)}{a^2}\mathbf{r}\right]\exp\left(-\frac{r^2}{a^2}\right)$;

(b) $\mathbf{F} = \frac{F_0}{a}\left[z\mathbf{k} + \frac{(x^2 + y^2 - a^2)}{a^2}\mathbf{r}\right]\exp\left(-\frac{r^2}{a^2}\right)$;

(c) $\mathbf{F} = F_0\left[\mathbf{k} + \frac{a(\mathbf{r} \times \mathbf{k})}{r^2}\right]$.

12.16. One of Maxwell's electromagnetic equations states that all magnetic fields $\mathbf{B}$ are solenoidal (i.e. $\nabla \cdot \mathbf{B} = 0$). Determine whether each of the following vectors could represent a real magnetic field; where it could, try to find a suitable vector potential $\mathbf{A}$, i.e. such that $\mathbf{B} = \nabla \times \mathbf{A}$. (*Hint*: seek a vector potential that is parallel to $\nabla \times \mathbf{B}$.)

(a) $\frac{B_0 b}{r^3}[(x-y)z\mathbf{i} + (x-y)z\mathbf{j} + (x^2 - y^2)\mathbf{k}]$ in Cartesians with $r^2 = x^2 + y^2 + z^2$;

(b) $\frac{B_0 b^3}{r^3}[\cos\theta \cos\phi \, \hat{\mathbf{e}}_r - \sin\theta \cos\phi \, \hat{\mathbf{e}}_\theta + \sin 2\theta \sin\phi \, \hat{\mathbf{e}}_\phi]$ in spherical polars;

(c) $B_0 b^2\left[\frac{z\rho}{(b^2 + z^2)^2}\hat{\mathbf{e}}_\rho + \frac{1}{b^2 + z^2}\hat{\mathbf{e}}_z\right]$ in cylindrical polars.

12.17. The vector field $\mathbf{f}$ has components $y\mathbf{i} - x\mathbf{j} + \mathbf{k}$ and γ is a curve given parametrically by

$$\mathbf{r} = (a - c + c\cos\theta)\mathbf{i} + (b + c\sin\theta)\mathbf{j} + c^2\theta\mathbf{k}, \quad 0 \le \theta \le 2\pi.$$

Describe the shape of the path γ and show that the line integral $\int_\gamma \mathbf{f} \cdot d\mathbf{r}$ vanishes. Does this result imply that $\mathbf{f}$ is a conservative field?

12.18. A vector field $\mathbf{a} = f(r)\mathbf{r}$ is spherically symmetric and everywhere directed away from the origin. Show that $\mathbf{a}$ is irrotational, but that it is also solenoidal only if $f(r)$ is of the form Ar^{-3}.

Line, surface and volume integrals

12.19. Evaluate the surface integral $\int \mathbf{r} \cdot d\mathbf{S}$, where $\mathbf{r}$ is the position vector, over that part of the surface $z = a^2 - x^2 - y^2$ for which $z \geq 0$, by each of the following methods.

(a) Parameterise the surface as $x = a \sin\theta \cos\phi$, $y = a \sin\theta \sin\phi$, $z = a^2 \cos^2\theta$, and show that

$$\mathbf{r} \cdot d\mathbf{S} = a^4 (2 \sin^3\theta \cos\theta + \cos^3\theta \sin\theta) \, d\theta \, d\phi.$$

(b) Apply the divergence theorem to the volume bounded by the surface and the plane $z = 0$.

12.20. Obtain an expression for the value ϕ_P at a point P of a scalar function ϕ that satisfies $\nabla^2 \phi = 0$, in terms of its value and normal derivative on a surface S that encloses it, by proceeding as follows.

(a) In Green's second theorem, take ψ at any particular point Q as $1/r$, where r is the distance of Q from P. Show that $\nabla^2 \psi = 0$, except at $r = 0$.

(b) Apply the result to the doubly connected region bounded by S and a small sphere Σ of radius δ centred on P.

(c) Apply the divergence theorem to show that the surface integral over Σ involving $1/\delta$ vanishes and prove that the term involving $1/\delta^2$ has the value $4\pi \phi_P$.

(d) Conclude that

$$\phi_P = -\frac{1}{4\pi} \int_S \phi \frac{\partial}{\partial n}\left(\frac{1}{r}\right) dS + \frac{1}{4\pi} \int_S \frac{1}{r} \frac{\partial \phi}{\partial n} dS.$$

This important result shows that the value at a point P of a function ϕ that satisfies $\nabla^2 \phi = 0$ everywhere within a closed surface S that encloses P may be expressed *entirely* in terms of its value and normal derivative on S.

12.21. Use result (12.21), together with an appropriately chosen scalar function ϕ, to prove that the position vector $\bar{\mathbf{r}}$ of the centre of mass of an arbitrarily shaped body of volume V and uniform density can be written

$$\bar{\mathbf{r}} = \frac{1}{V} \oint_S \tfrac{1}{2} r^2 \, d\mathbf{S}.$$

12.22. A rigid body of volume V and surface S rotates with angular velocity $\boldsymbol{\omega}$. Show that

$$\boldsymbol{\omega} = -\frac{1}{2V} \oint_S \mathbf{u} \times d\mathbf{S},$$

where $\mathbf{u}(\mathbf{x})$ is the velocity of the point $\mathbf{x}$ on the surface S.

12.23. Demonstrate the validity of the divergence theorem:

(a) by calculating the flux of the vector

$$\mathbf{F} = \frac{\alpha \mathbf{r}}{(r^2 + a^2)^{3/2}}$$

through the spherical surface $|\mathbf{r}| = \sqrt{3}a$;

Problems

(b) by showing that

$$\nabla \cdot \mathbf{F} = \frac{3\alpha a^2}{(r^2 + a^2)^{5/2}}$$

and evaluating the volume integral of $\nabla \cdot \mathbf{F}$ over the interior of the sphere $|\mathbf{r}| = \sqrt{3}a$. The substitution $r = a \tan \theta$ will prove useful in carrying out the integration.

12.24. Prove Equation (12.22) and, by taking $\mathbf{b} = zx^2\mathbf{i} + zy^2\mathbf{j} + (x^2 - y^2)\mathbf{k}$, show that the two integrals

$$I = \int x^2 \, dV \quad \text{and} \quad J = \int \cos^2\theta \sin^3\theta \cos^2\phi \, d\theta \, d\phi,$$

both taken over the unit sphere, must have the same value. Evaluate both directly to show that the common value is $4\pi/15$.

12.25. In a uniform conducting medium with unit relative permittivity, charge density ρ, current density $\mathbf{J}$, electric field $\mathbf{E}$ and magnetic field $\mathbf{B}$, Maxwell's electromagnetic equations take the form (with $\mu_0 \epsilon_0 = c^{-2}$)

(i) $\nabla \cdot \mathbf{B} = 0$,   (ii) $\nabla \cdot \mathbf{E} = \rho/\epsilon_0$,
(iii) $\nabla \times \mathbf{E} + \dot{\mathbf{B}} = 0$,   (iv) $\nabla \times \mathbf{B} - (\dot{\mathbf{E}}/c^2) = \mu_0 \mathbf{J}$.

The density of stored energy in the medium is given by $\frac{1}{2}(\epsilon_0 E^2 + \mu_0^{-1} B^2)$. Show that the rate of change of the total stored energy in a volume V is equal to

$$-\int_V \mathbf{J} \cdot \mathbf{E} \, dV - \frac{1}{\mu_0} \oint_S (\mathbf{E} \times \mathbf{B}) \cdot d\mathbf{S},$$

where S is the surface bounding V.

[The first integral gives the ohmic heating loss, whilst the second gives the electromagnetic energy flux out of the bounding surface. The vector $\mu_0^{-1}(\mathbf{E} \times \mathbf{B})$ is known as the Poynting vector.]

12.26. A vector field $\mathbf{F}$ is defined in cylindrical polar coordinates ρ, θ, z by

$$\mathbf{F} = F_0 \left(\frac{x \cos \lambda z}{a} \mathbf{i} + \frac{y \cos \lambda z}{a} \mathbf{j} + (\sin \lambda z)\mathbf{k} \right) \equiv \frac{F_0 \rho}{a}(\cos \lambda z)\mathbf{e}_\rho + F_0(\sin \lambda z)\mathbf{k},$$

where $\mathbf{i}, \mathbf{j}$ and $\mathbf{k}$ are the unit vectors along the Cartesian axes and $\mathbf{e}_\rho$ is the unit vector $(x/\rho)\mathbf{i} + (y/\rho)\mathbf{j}$.
(a) Calculate, as a surface integral, the flux of $\mathbf{F}$ through the closed surface bounded by the cylinders $\rho = a$ and $\rho = 2a$ and the planes $z = \pm a\pi/2$.
(b) Evaluate the same integral using the divergence theorem.

12.27. The vector field $\mathbf{F}$ is given by

$$\mathbf{F} = (3x^2yz + y^3z + xe^{-x})\mathbf{i} + (3xy^2z + x^3z + ye^x)\mathbf{j} + (x^3y + y^3x + xy^2z^2)\mathbf{k}.$$

Calculate (a) directly and (b) by using Stokes' theorem the value of the line integral $\int_L \mathbf{F} \cdot d\mathbf{r}$, where L is the (three-dimensional) closed contour $OABCDEO$ defined by the successive vertices $(0, 0, 0)$, $(1, 0, 0)$, $(1, 0, 1)$, $(1, 1, 1)$, $(1, 1, 0)$, $(0, 1, 0)$, $(0, 0, 0)$.

12.28. A vector force field $\mathbf{F}$ is defined in Cartesian coordinates by

$$\mathbf{F} = F_0 \left[\left(\frac{y^3}{3a^3} + \frac{y}{a} e^{xy/a^2} + 1 \right) \mathbf{i} + \left(\frac{xy^2}{a^3} + \frac{x+y}{a} e^{xy/a^2} \right) \mathbf{j} + \frac{z}{a} e^{xy/a^2} \mathbf{k} \right].$$

Use Stokes' theorem to calculate

$$\oint_L \mathbf{F} \cdot d\mathbf{r},$$

where L is the perimeter of the rectangle $ABCD$ given by $A = (0, 1, 0)$, $B = (1, 1, 0)$, $C = (1, 3, 0)$ and $D = (0, 3, 0)$.

HINTS AND ANSWERS

12.1. Show that $\nabla \times \mathbf{F} = \mathbf{0}$. The potential $\phi_F(\mathbf{r}) = x^2 z + y^2 z^2 - z$.

12.3. (a) $c^3 \ln 2\, \mathbf{i} + 2\mathbf{j} + (3c/2)\mathbf{k}$; (b) $(-3c^4/8)\mathbf{i} - c\mathbf{j} - (c^2 \ln 2)\mathbf{k}$; (c) $c^4 \ln 2 - c$.

12.5. For P, $x = y = ab/(a^2 + b^2)^{1/2}$. The relevant limits are $0 \le \theta_1 \le \tan^{-1}(b/a)$ and $\tan^{-1}(a/b) \le \theta_2 \le \pi/2$. The total common area is $4ab \tan^{-1}(b/a)$.

12.7. Show that, in the notation of Section 12.3, $\partial Q/\partial x - \partial P/\partial y = 2x^2$; $I = \pi a^3 b/2$.

12.9. $\mathbf{M} = I \int_C \mathbf{r} \times (d\mathbf{r} \times \mathbf{B})$. Show that the horizontal sides in the first term and the whole of the second term contribute nothing to the couple.

12.11. Note that, if $\hat{\mathbf{n}}$ is the outward normal to the surface, $\hat{\mathbf{n}}_z \cdot \hat{\mathbf{n}}\, dl$ is equal to $-d\rho$.

12.13. (b) $\phi = c + z/r$.

12.15. (a) Yes, $F_0(x - y) \exp(-r^2/a^2)$; (b) yes, $-F_0[(x^2 + y^2)/(2a)] \exp(-r^2/a^2)$; (c) no, $\nabla \times \mathbf{F} \ne \mathbf{0}$.

12.17. A spiral of radius c with its axis parallel to the z-direction and passing through (a, b). The pitch of the spiral is $2\pi c^2$. No, because (i) γ is not a closed loop and (ii) the line integral must be zero for *every* closed loop, not just for a particular one. In fact $\nabla \times \mathbf{f} = -2\mathbf{k} \ne \mathbf{0}$ shows that $\mathbf{f}$ is not conservative.

12.19. (a) $d\mathbf{S} = (2a^3 \cos\theta \sin^2\theta \cos\phi\, \mathbf{i} + 2a^3 \cos\theta \sin^2\theta \sin\phi\, \mathbf{j} + a^2 \cos\theta \sin\theta\, \mathbf{k})\, d\theta\, d\phi$.
(b) $\nabla \cdot \mathbf{r} = 3$; over the plane $z = 0$, $\mathbf{r} \cdot d\mathbf{S} = 0$.
The necessarily common value is $3\pi a^4/2$.

12.21. Write $\mathbf{r}$ as $\nabla(\frac{1}{2}r^2)$.

Hints and answers

12.23. The answer is $3\sqrt{3}\pi\alpha/2$ in each case.

12.25. Identify the expression for $\nabla \cdot (\mathbf{E} \times \mathbf{B})$ and use the divergence theorem.

12.27. (a) The successive contributions to the integral are:
$1 - 2e^{-1}, 0, 2 + \frac{1}{2}e, -\frac{7}{3}, -1 + 2e^{-1}, -\frac{1}{2}$.
(b) $\nabla \times \mathbf{F} = 2xyz^2\mathbf{i} - y^2z^2\mathbf{j} + ye^x\mathbf{k}$. Show that the contour is equivalent to the sum of two plane square contours in the planes $z = 0$ and $x = 1$, the latter being traversed in the negative sense. Integral $= \frac{1}{6}(3e - 5)$.

13 Laplace transforms

In Chapter 6 we discussed how complicated functions $f(x)$ may be expressed as power series. Although they were not presented as such, the power series could all be viewed as linear superpositions of the monomial basic set of functions, namely $1, x, x^2, x^3, \ldots, x^n, \ldots$ Natural though this set may seem, they are in many ways far from ideal: for example they possess no mutual orthogonality properties, a characteristic that is generally of great value when it comes to determining, for any particular function, the multiplying constant for each basic function in the sum. Moreover, this particular set of basic functions can only be used to represent continuous functions.

In the case of original functions $f(t)$ that are periodic, some improvement on this situation can be made by using, as the basic set, sine and cosine functions. For a function with period T, say,[1] the set of sine and cosine functions with arguments $2\pi nt/T$, for all $n \geq 0$, form a suitable basic set for expressing f as a series; such a representation is called a *Fourier series*. One great advantage they possess over the monomial functions is that they are mutually orthogonal when integrated over any continuous period of length T, i.e. the integral from t_0 to $t_0 + T$ of the product of any sine and any cosine, or of two sines or cosines with different values of n, is equal to zero.

Unlike Taylor series, a Fourier series can describe functions that are not everywhere continuous and/or differentiable. There are also other advantages in using trigonometric terms. They are easy to differentiate and integrate, their moduli are easily taken and each term contains only one characteristic frequency. This last point is important because Fourier series are often used to represent the response of a system to a periodic input, and this response often depends directly on the frequency content of the input.[2] Fourier series are used in a wide variety of such physical situations, including the vibrations of a finite string, the scattering of light by a diffraction grating and the transmission of an input signal by an electronic circuit.

However, one obvious drawback of Fourier series is their restriction to functions that are periodic. It is clearly desirable, whilst retaining some form of mutual orthogonality property, to recover one of the desirable characteristics of the monomial basic set and be able to obtain a representation for functions that are defined over an infinite interval

---

[1] This is not to imply that the argument of the function is necessarily time; it often is, but periodicity in a spatial dimension is just as important in the physical sciences.

[2] Recall, for example, that the angle through which a glass prism refracts a ray of light depends upon the frequency of that light.

13.1 Laplace transforms

and have no particular periodicity. Such a representation does exist for functions that are defined for all t, provided they tend to zero as $t \to \pm\infty$. It is called a *Fourier transform* and is one of a class of representations called *integral transforms*.

The basic set for Fourier transforms are the exponential functions $e^{i\omega t}$. Here ω is a *continuous* variable, replacing the discrete variable n that appears in both Fourier and power series. A consequent change is that the sum now becomes an integral (over ω which runs from $-\infty$ to $+\infty$); if nothing else, this means that, in general, the whole representation becomes less cumbersome to write out. Additional benefits come from the orthogonality[3] of the functions $e^{i\omega t}$ over $-\infty < t < +\infty$ for *any* two different values of ω; this enables an explicit expression to be given for the 'weight' $\tilde{f}(\omega)$ that $e^{i\omega t}$ has in the integral representing $f(t)$.

Since we are primarily concerned in this book with mathematical tools, rather than the methods that can be built up from them, and both Fourier series and Fourier transforms are essentially methods for tackling particular classes of problems, we will content ourselves with the qualitative outline given above. The reader who wishes to pursue these ideas further should consult a more advanced text.[4]

13.1 Laplace transforms

Notwithstanding what is said in the final paragraph above, we will consider in some detail one form of integral transform that has a number of practical applications, as well as being relevant to the solution of certain types of differential equations considered in Chapter 14. This transform, known as a *Laplace transform*, is essentially a description of the original function for $t > 0$ as the weighted sum of 'decaying exponential functions'. The formal representation of this is beyond the scope of the present book, as it involves the theory of complex variables and line integrals in the complex plane. Despite this, Laplace transforms can be very effective tools in certain circumstances, as we will see later in this chapter.

In physics, and especially in quantum theory, where negative values of the independent variable might represent half of the space occupied by a quantum system, and large negative times are intrinsic to defining its initial state, Fourier methods are often the natural way to proceed. By contrast, we might know the value of a function at $t = 0$ and are interested only in its behaviour for $t > 0$; such initial-value problems are nearly always the case in control engineering and macroscopic physics, where, at all negative times, nothing happens. Other situations in which Fourier transforms are impracticable are those involving functions $f(t)$ for which the Fourier transform does not exist because $f \not\to 0$ as $t \to \infty$, and so the integral defining $\tilde{f}$ does not converge. This would be the case for the function $f(t) = t$, which therefore does not possess a Fourier transform.

---

[3] As the functions are complex, it is the integral $\int f^*(\omega')f(\omega)\,dt = \int_{-\infty}^{\infty} e^{-i\omega't}e^{i\omega t}\,dt$ that is equal to zero if $\omega' \neq \omega$.
[4] Such as that mentioned in the Preface, from which the current text is derived.

Laplace transforms

Initial-value problems of the kind described, or functions with awkward asymptotic properties, lead us to consider the Laplace transform, $\bar{f}(s)$ or $\mathcal{L}[f(t)]$, of $f(t)$, which is defined, for positive s, by

$$\bar{f}(s) \equiv \int_0^\infty f(t) e^{-st}\, dt, \qquad (13.1)$$

provided that the integral exists. Because of the presence of the negative exponential function e^{-st} the integral *will* converge at its upper limit for a very wide range of functions $f(t)$; for it not to converge would require a function $f(t)$ that, by itself, had at least exponential divergence. At the lower limit, the only requirement is that $f(0)$ is finite. We assume here that s is real, but complex values would have to be considered in a more detailed study.[5] In practice, for a given function $f(t)$ there will be some real non-negative number s_0 such that the integral in (13.1) exists for $s > s_0$ but diverges for $s \leq s_0$.

Through (13.1) we define a *linear* transformation $\mathcal{L}$ that converts functions of the variable t to functions of a new variable s. Its linearity is expressed by

$$\mathcal{L}[af_1(t) + bf_2(t)] = a\mathcal{L}[f_1(t)] + b\mathcal{L}[f_2(t)] = a\bar{f}_1(s) + b\bar{f}_2(s). \qquad (13.2)$$

Our first worked example determines the Laplace transforms of some common simple functions.

Example Find the Laplace transforms of the functions (i) $f(t) = 1$, (ii) $f(t) = e^{at}$, (iii) $f(t) = t^n$, for $n = 0, 1, 2, \ldots$

(i) By direct application of the definition of a Laplace transform (13.1), we find

$$\mathcal{L}[1] = \int_0^\infty e^{-st}\, dt = \left[\frac{-1}{s} e^{-st}\right]_0^\infty = \frac{1}{s}, \quad \text{if } s > 0,$$

where the restriction $s > 0$ is required for the integral to exist.

(ii) Again using (13.1) directly, we find

$$\bar{f}(s) = \int_0^\infty e^{at} e^{-st}\, dt = \int_0^\infty e^{(a-s)t}\, dt$$

$$= \left[\frac{e^{(a-s)t}}{a-s}\right]_0^\infty = \frac{1}{s-a}, \quad \text{if } s > a.$$

(iii) Once again using definition (13.1), we have

$$\bar{f}_n(s) = \int_0^\infty t^n e^{-st}\, dt.$$

---

[5] It will be clear that, so far as the convergence of the integral is concerned, it is only the real part of any complex s that matters; any imaginary part could be considered as a part of a redefined $f(t)$, taking the form of a phase factor of unit modulus.

13.1 Laplace transforms

Integrating by parts we find

$$\bar{f}_n(s) = \left[\frac{-t^n e^{-st}}{s}\right]_0^\infty + \frac{n}{s}\int_0^\infty t^{n-1} e^{-st}\, dt$$

$$= 0 + \frac{n}{s}\bar{f}_{n-1}(s), \quad \text{if } s > 0.$$

We now have a recursion relation between successive transforms and by calculating $\bar{f}_0$ we can infer $\bar{f}_1$, $\bar{f}_2$, etc. Since $t^0 = 1$, (i) above gives

$$\bar{f}_0 = \frac{1}{s}, \quad \text{if } s > 0, \tag{13.3}$$

and it then follows from the recurrence relation that

$$\bar{f}_1(s) = \frac{1}{s^2}, \quad \bar{f}_2(s) = \frac{2!}{s^3}, \quad \ldots, \quad \bar{f}_n(s) = \frac{n!}{s^{n+1}} \quad \text{if } s > 0.$$

Thus, in each case (i)–(iii), direct application of the definition of the Laplace transform (13.1) yields the required result.[6] ◀

As noted earlier, unlike that for the Fourier transform, the inversion of the Laplace transform is not an easy operation to perform, since an explicit formula for $f(t)$, given $\bar{f}(s)$, is not straightforwardly obtained from (13.1). Again as noted, the general method for obtaining an inverse Laplace transform makes use of complex variable theory and is outside the scope of this book. However, some progress can be made without having to find an *explicit* inverse, since we can prepare from (13.1) a 'dictionary' of the Laplace transforms of common functions and, when faced with an inversion to carry out, hope to find the given transform (together with its parent function) in the listing. Such a list is given in Table 13.1.

When finding inverse Laplace transforms using Table 13.1, it is useful to note that for all practical purposes the inverse Laplace transform is unique[7] and linear and so

$$\mathcal{L}^{-1}\left[a\bar{f}_1(s) + b\bar{f}_2(s)\right] = af_1(t) + bf_2(t). \tag{13.4}$$

In many practical problems, the function of s for which the inverse Laplace transform is sought is the ratio of two polynomials. In these cases, the method of partial fractions can be used to express the function in terms of entries that appear in the table, as is illustrated below.

---

[6] Verify the linearity of the Laplace transform operation as follows. Write an equation expressing e^{at}, with $a > 0$, as an infinite sum and take the transforms of both sides. Then show, using the binomial theorem, that the resulting equation in s is valid. Verify also that the condition on s for all transforms to be defined is the same as that for the validity of the binomial expansion.

[7] This is not strictly true, since two functions can differ from one another at a finite number of isolated points but have the *same* Laplace transform.

Laplace transforms

Table 13.1 *Standard Laplace transforms. The transforms are valid for $s > s_0$.*

$f(t)$	$\bar{f}(s)$	s_0
c	c/s	0
ct^n	$cn!/s^{n+1}$	0
$\sin bt$	$b/(s^2 + b^2)$	0
$\cos bt$	$s/(s^2 + b^2)$	0
e^{at}	$1/(s - a)$	a
$t^n e^{at}$	$n!/(s - a)^{n+1}$	a
$\sinh at$	$a/(s^2 - a^2)$	$|a|$
$\cosh at$	$s/(s^2 - a^2)$	$|a|$
$e^{at} \sin bt$	$b/[(s - a)^2 + b^2]$	a
$e^{at} \cos bt$	$(s - a)/[(s - a)^2 + b^2]$	a
$t^{1/2}$	$\frac{1}{2}(\pi/s^3)^{1/2}$	0
$t^{-1/2}$	$(\pi/s)^{1/2}$	0
$\delta(t - t_0)$	e^{-st_0}	0
$H(t - t_0) = \begin{cases} 1 & \text{for } t \geq t_0 \\ 0 & \text{for } t < t_0 \end{cases}$	e^{-st_0}/s	0

Example Using Table 13.1 find $f(t)$ if

$$\bar{f}(s) = \frac{s + 3}{s(s + 1)}.$$

Using partial fractions $\bar{f}(s)$ may be written

$$\bar{f}(s) = \frac{3}{s} - \frac{2}{s + 1}.$$

Remembering that $\mathcal{L}[af(x)] = a\mathcal{L}[f(x)]$, and comparing the above RHS with the standard Laplace transforms in Table 13.1, we find that the inverse transform of $3/s$ is 3 for $s > 0$ and the inverse transform of $2/(s + 1)$ is $2e^{-t}$ for $s > -1$, and so

$$f(t) = 3 - 2e^{-t},$$

but only for $s > 0$, so that both conditions on s are satisfied. ◀

There are two functions that appear at the end of Table 13.1 to which no or scant reference has been made previously. They are the Dirac delta function $\delta(t - t_0)$ and the Heaviside step function $H(t - t_0)$. Passing reference was made to the former in Section 12.8, but the latter has been no more than defined in the table. Since both are of considerable value in describing in mathematical terms processes that occur in real systems – the delta function approximates to a finite amount of some physical quantity appearing in an arbitrary small span of space or time, and the step function describes a

13.2 The Dirac δ-function and Heaviside step function

factor that is switched on or switched off at a particular value of its independent variable – we now make a digression to outline their properties and uses.

EXERCISES 13.1

1. Do the following functions have Laplace transforms?

(a) $\ln(at)\,H(t)$, (b) $\dfrac{\sin at}{t}\,H(t)$, (c) $e^{t^2}\,H(t)$, (d) $\dfrac{1}{t}\,H(t-b)$ with $b > 0$.

2. By finding the Laplace transform of $f(t) = e^{ibt}$, prove the transforms quoted in Table 13.1 for $\sin bt$ and $\cos bt$.

13.2 The Dirac δ-function and Heaviside step function

The δ-function is different from most functions encountered in the physical sciences, but we will see that a rigorous mathematical definition exists. It can be visualised as a very sharp narrow pulse (in space, time, density, etc.) which produces an integrated effect having a definite magnitude. The formal properties of the δ-function may be summarised as follows.

The Dirac δ-function has the property that

$$\delta(t) = 0 \quad \text{for } t \neq 0, \tag{13.5}$$

but its fundamental defining property is

$$\int f(t)\delta(t-a)\,dt = f(a), \tag{13.6}$$

provided the range of integration includes the point $t = a$; otherwise the integral equals zero. This leads immediately to two further useful results:

$$\int_{-a}^{b} \delta(t)\,dt = 1 \quad \text{for all } a, b > 0 \tag{13.7}$$

and

$$\int \delta(t-a)\,dt = 1, \tag{13.8}$$

provided the range of integration includes $t = a$.

Equation (13.6) can be used to derive further useful properties of the Dirac δ-function:

$$\delta(t) = \delta(-t), \tag{13.9}$$

$$\delta(at) = \frac{1}{|a|}\delta(t), \tag{13.10}$$

$$t\delta(t) = 0. \tag{13.11}$$

We now prove the second of these.

Example Prove that $\delta(bt) = \delta(t)/|b|$.

Let us first consider the case where $b > 0$. It follows that

$$\int_{-\infty}^{\infty} f(t)\delta(bt)\,dt = \int_{-\infty}^{\infty} f\left(\frac{t'}{b}\right)\delta(t')\frac{dt'}{b} = \frac{1}{b}f(0) = \frac{1}{b}\int_{-\infty}^{\infty} f(t)\delta(t)\,dt,$$

where we have made the substitution $t' = bt$. But $f(t)$ is arbitrary and so we immediately see that $\delta(bt) = \delta(t)/b = \delta(t)/|b|$ for $b > 0$.

Now consider the case where $b = -c < 0$. It follows that

$$\int_{-\infty}^{\infty} f(t)\delta(bt)\,dt = \int_{\infty}^{-\infty} f\left(\frac{t'}{-c}\right)\delta(t')\left(\frac{dt'}{-c}\right) = \int_{-\infty}^{\infty} \frac{1}{c}f\left(\frac{t'}{-c}\right)\delta(t')\,dt'$$

$$= \frac{1}{c}f(0) = \frac{1}{|b|}f(0) = \frac{1}{|b|}\int_{-\infty}^{\infty} f(t)\delta(t)\,dt,$$

where we have made the substitution $t' = bt = -ct$. But $f(t)$ is arbitrary and so

$$\delta(bt) = \frac{1}{|b|}\delta(t),$$

for all b, which establishes the result. ◀

Furthermore, by considering an integral of the form

$$\int f(t)\delta(h(t))\,dt,$$

and making a change of variables to $z = h(t)$, we may show that

$$\delta(h(t)) = \sum_i \frac{\delta(t - t_i)}{|h'(t_i)|}, \tag{13.12}$$

where the t_i are those values of t for which $h(t) = 0$ and $h'(t)$ stands for dh/dt.[8]

The derivative of the delta function, $\delta'(t)$, is defined by

$$\int_{-\infty}^{\infty} f(t)\delta'(t)\,dt = \left[f(t)\delta(t)\right]_{-\infty}^{\infty} - \int_{-\infty}^{\infty} f'(t)\delta(t)\,dt$$

$$= -f'(0), \tag{13.13}$$

and similarly for higher derivatives.[9]

For many practical purposes, effects that are not strictly described by a δ-function may be analysed as such, if they take place in an interval much shorter than the response interval of the system on which they act. For example, the idealised notion of an impulse of magnitude J applied at time t_0 can be represented by

$$j(t) = J\delta(t - t_0). \tag{13.14}$$

---

[8] Results (13.9) and (13.10) are particular examples, in which $h(t) = -t$ and $h(t) = bt$ respectively.
[9] Give an integral expression, involving the function $f(x)$ and the delta function and/or its derivatives, that is equal to the nth derivative of $f(x)$ evaluated at $x = a$.

13.2 The Dirac δ-function and Heaviside step function

Many physical situations are described by a δ-function in space rather than in time. Moreover, we often require the δ-function to be defined in more than one dimension. For example, the charge density of a point charge q at a point $\mathbf{r}_0$ may be expressed as a three-dimensional δ-function

$$\rho(\mathbf{r}) = q\delta(\mathbf{r} - \mathbf{r}_0) = q\delta(x - x_0)\delta(y - y_0)\delta(z - z_0), \quad (13.15)$$

so that a discrete 'quantum' is expressed as if it were a continuous distribution. From (13.15) we see that (as expected) the total charge enclosed in a volume V is given by

$$\int_V \rho(\mathbf{r})\,dV = \int_V q\delta(\mathbf{r} - \mathbf{r}_0)\,dV = \begin{cases} q & \text{if } \mathbf{r}_0 \text{ lies in } V, \\ 0 & \text{otherwise.} \end{cases}$$

Closely related to the Dirac δ-function is the *Heaviside* or *unit step function* $H(t)$, for which

$$H(t) = \begin{cases} 1 & \text{for } t > 0, \\ 0 & \text{for } t < 0. \end{cases} \quad (13.16)$$

This function is clearly discontinuous at $t = 0$ and it is usual to take $H(0) = 1/2$.

One sometimes useful application of the step function is to allow an integral that has finite upper and/or lower limits – a Laplace transform is a prime example – to be represented as an integral from $-\infty$ to $+\infty$, i.e. writing

$$\int_a^b f(t)\,dt \quad \text{as} \quad \int_{-\infty}^{\infty} f(t)H(t - a)H(b - t)\,dt.$$

This then allows multiple integrals to be manipulated, particularly with regard to a change of integration variable, without having to keep detailed track of the consequent changes in the limits; they are 'looked after' by the integrand. This can be of value in relation to multiple convolution-type integrals, where linear changes of integration variable such as 'set $u = t - s$' are commonplace in formal proofs.[10]

A sum, rather than a product, of Heaviside functions can be used as an alternative description of a function that has a constant value over a limited range. For example,

$$f(t) = 3[H(t - a) - H(t - b)]$$

would describe a function that has the value 3 for $a < t < b$ but is zero outside this range.[11]

The Heaviside function and the delta function are related, one being the derivative of the other,

$$H'(t) = \delta(t), \quad (13.17)$$

as we now prove.

---

[10] For example, the proof of the convolution theorem for Laplace transforms given towards the end of Section 13.4 can be rewritten in terms of step functions without the need for a figure such as Figure 13.1.

[11] Using a 'Morse code' in which dots are represented by unit δ-functions and dashes by a value of $+1$ maintained for three time units, construct the function of t that corresponds to the international distress signal 'SOS' ($\cdots - - - \cdots$) starting at $t = 1$. The space between 'sounds' should be one time unit and that between letters should be three units.

Laplace transforms

Example Prove that $H'(t) = \delta(t)$.

Considering, for an *arbitrary* function $f(x)$, the integral

$$\int_{-\infty}^{\infty} f(t)H'(t)\,dt = \left[f(t)H(t)\right]_{-\infty}^{\infty} - \int_{-\infty}^{\infty} f'(t)H(t)\,dt$$

$$= f(\infty) - \int_{0}^{\infty} f'(t)\,dt$$

$$= f(\infty) - \left[f(t)\right]_{0}^{\infty} = f(0),$$

and comparing it with (13.6) when $a = 0$ immediately shows that $H'(t) = \delta(t)$. ◀

EXERCISES 13.2

1. By considering the integral $\int_{-\infty}^{\infty} f(t)t\delta(t)\,dt$ for an arbitrary function $f(t)$ that is everywhere finite, prove the stated result in (13.11) that $t\delta(t) = 0$.

2. Apply result (13.12) to find the value of

$$I = \int_{-\infty}^{\infty} (x^2 + 3x + 6)\delta(x^2 + x - 6)\,dx.$$

3. Express the following functions in terms of Heaviside step functions:
   (a) A single cycle from a square wave function: $f(x) = \begin{cases} \dfrac{x}{|x|} & \text{for } |x| \leq 1, \\ 0 & \text{for } |x| > 1. \end{cases}$
   (b) The function whose graph consists of the x-axis, apart from two straight lines joining $(-a, 0)$ to $(0, 2)$ and $(0, 2)$ to $(a, 0)$.

13.3 Laplace transforms of derivatives and integrals

Following our digression to record the definitions and uses of the Dirac δ-function and Heaviside step function, we return to our study of Laplace transforms. One of the main uses of Laplace transforms is to solve differential equations, particularly linear ones with constant coefficients. Differential equations are the subject of Chapter 14, and we will there employ the transforms as a valuable tool. In the meantime we will derive some of the required basic results, in particular the Laplace transforms of general derivatives and the indefinite integral.

13.3 Laplace transforms of derivatives and integrals

We start with the Laplace transform of the first derivative of $f(t)$, which is given by

$$\mathcal{L}\left[\frac{df}{dt}\right] = \int_0^\infty \frac{df}{dt} e^{-st} \, dt$$

$$= \left[f(t)e^{-st}\right]_0^\infty + s \int_0^\infty f(t) e^{-st} \, dt$$

$$= -f(0) + s\bar{f}(s), \qquad \text{for } s > 0. \tag{13.18}$$

It should be noted how the initial value, $f(0)$, is automatically built into the Laplace transform. Since the evaluation of $\mathcal{L}[df/dt]$ relied only on integration by parts and that can be repeated, the transforms of higher order derivatives may be found in a similar manner. This example finds the transform of the second derivative.

Example Find the Laplace transform of d^2f/dt^2.

Using the definition of the Laplace transform and integrating by parts we obtain

$$\mathcal{L}\left[\frac{d^2 f}{dt^2}\right] = \int_0^\infty \frac{d^2 f}{dt^2} e^{-st} \, dt$$

$$= \left[\frac{df}{dt} e^{-st}\right]_0^\infty + s \int_0^\infty \frac{df}{dt} e^{-st} \, dt$$

$$= -\frac{df}{dt}(0) + s[s\bar{f}(s) - f(0)], \qquad \text{for } s > 0,$$

where (13.18) has been substituted for the integral. This can be written more neatly as

$$\mathcal{L}\left[\frac{d^2 f}{dt^2}\right] = s^2 \bar{f}(s) - sf(0) - \frac{df}{dt}(0), \qquad \text{for } s > 0,$$

showing that the Laplace transform of the second derivative of $f(t)$ automatically has the initial $(t = 0)$ values of the function and its first derivative built into it.

In general the Laplace transform of the nth derivative is given by

$$\mathcal{L}\left[\frac{d^n f}{dt^n}\right] = s^n \bar{f} - s^{n-1} f(0) - s^{n-2} \frac{df}{dt}(0) - \cdots - \frac{d^{n-1} f}{dt^{n-1}}(0), \qquad \text{for } s > 0. \tag{13.19}$$

Again, the initial values of lower derivatives are built into the transform.

We now turn to integration, which is much more straightforward. From the definition (13.1),

$$\mathcal{L}\left[\int_0^t f(u) \, du\right] = \int_0^\infty dt\, e^{-st} \int_0^t f(u) \, du$$

$$= \left[-\frac{1}{s} e^{-st} \int_0^t f(u) \, du\right]_0^\infty + \int_0^\infty \frac{1}{s} e^{-st} f(t) \, dt.$$

Laplace transforms

The first term on the RHS vanishes at both limits,[12] and so

$$\mathcal{L}\left[\int_0^t f(u)\,du\right] = \frac{1}{s}\mathcal{L}[f]. \tag{13.20}$$

EXERCISE 13.3

1. Although differential equations are not formally introduced until Chapter 14, solve the linear differential equation

$$\frac{d^2 f}{dt^2} + 6\frac{df}{dt} + 8f = 2e^t,$$

subject to the initial conditions $f(0) = 1$, $f'(0) = 0$, as follows:
   (a) Take the Laplace transform of each of the four terms (incorporating the initial conditions in the way specified in the text).
   (b) The resulting equation is algebraic. Rearrange it to make $\bar{f}(s)$ its subject.
   (c) Express the polynomial ratio as partial fractions (with three terms).
   (d) Use Table 13.1 to look up the parent function for each of the partial fractions and so construct the solution $f(t)$.
   (e) To close the circle, confirm by substitution that your answer does satisfy both the differential equation and the given initial conditions.

13.4 Other properties of Laplace transforms

From Table 13.1 it will be apparent that multiplying a function $f(t)$ by e^{at} has the effect on its transform that s is replaced by $s - a$. This is easily proved generally:

$$\mathcal{L}\left[e^{at} f(t)\right] = \int_0^\infty f(t) e^{at} e^{-st}\,dt$$

$$= \int_0^\infty f(t) e^{-(s-a)t}\,dt$$

$$= \bar{f}(s - a). \tag{13.21}$$

As it were, multiplying $f(t)$ by e^{at} moves the origin of s by an amount a.

We may now consider the effect of multiplying the Laplace transform $\bar{f}(s)$ by e^{-bs} ($b > 0$). From the definition (13.1),

$$e^{-bs} \bar{f}(s) = \int_0^\infty e^{-s(t+b)} f(t)\,dt$$

$$= \int_b^\infty e^{-sz} f(z - b)\,dz,$$

---

12 Explain why.

13.4 Other properties of Laplace transforms

on putting $t + b = z$. Thus $e^{-bs}\bar{f}(s)$ is the Laplace transform of a function $g(t)$ defined by

$$g(t) = \begin{cases} 0 & \text{for } 0 < t \leq b, \\ f(t-b) & \text{for } t > b. \end{cases}$$

In other words, the function f has been translated to 'later' t (larger values of t) by an amount b.

Further properties of Laplace transforms can be proved in similar ways and are listed below.

(i)
$$\mathcal{L}[f(at)] = \frac{1}{a}\bar{f}\left(\frac{s}{a}\right), \tag{13.22}$$

(ii)
$$\mathcal{L}\left[t^n f(t)\right] = (-1)^n \frac{d^n \bar{f}(s)}{ds^n}, \quad \text{for } n = 1, 2, 3, \ldots, \tag{13.23}$$

(iii)
$$\mathcal{L}\left[\frac{f(t)}{t}\right] = \int_s^\infty \bar{f}(u)\,du, \tag{13.24}$$

provided $\lim_{t \to 0}[f(t)/t]$ exists.

Additional results can be obtained by combining two or more of the properties derived so far, as is now illustrated.

Example Find an expression for the Laplace transform of $t\,d^2f/dt^2$.

From the definition of the Laplace transform we have

$$\mathcal{L}\left[t\frac{d^2f}{dt^2}\right] = \int_0^\infty e^{-st}\, t\, \frac{d^2f}{dt^2}\, dt$$

$$= -\frac{d}{ds}\int_0^\infty e^{-st}\frac{d^2f}{dt^2}\, dt$$

$$= -\frac{d}{ds}[s^2\bar{f}(s) - sf(0) - f'(0)]$$

$$= -s^2\frac{d\bar{f}}{ds} - 2s\bar{f} + f(0).$$

Clearly, any general result, such as this one, is of some value, but here the transform is not very convenient for future manipulation as it contains a derivative with respect to s. ◀

Finally, we mention the convolution theorem for Laplace transforms. If the functions f and g have Laplace transforms $\bar{f}(s)$ and $\bar{g}(s)$ then

$$\mathcal{L}\left[\int_0^t f(u)g(t-u)\,du\right] = \bar{f}(s)\bar{g}(s), \tag{13.25}$$

where the integral in the brackets on the LHS is the *convolution* of f and g, denoted by $f * g$. The convolution defined above[13] is commutative, i.e. $f * g = g * f$, associative,

---

[13] Note that, in general, convolutions involve integration variables that run from $-\infty$ to $+\infty$. The finite upper limit given here reflects the fact that Laplace transforms are concerned with functions that are non-zero only for positive values of their arguments.

Laplace transforms

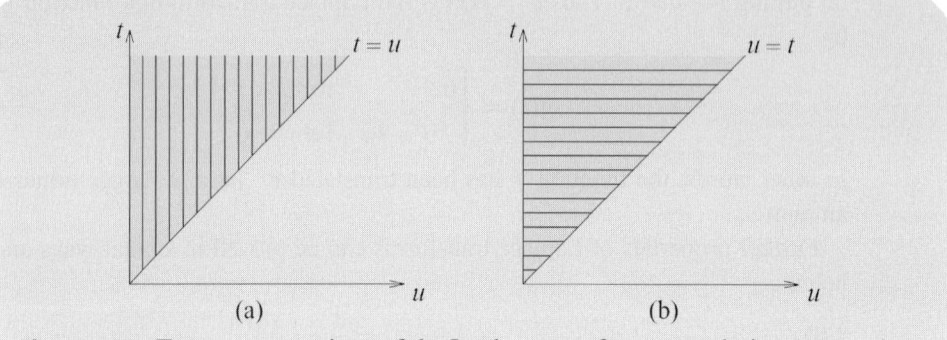

Figure 13.1 Two representations of the Laplace transform convolution (see text).

$(f * g) * h = f * (g * h)$, and distributive, $f * (g + h) = f * g + f * h$. From (13.25) we also see that

$$\mathcal{L}^{-1}\left[\bar{f}(s)\bar{g}(s)\right] = \int_0^t f(u)g(t-u)\,du = f * g.$$

The proof of (13.25) is given in the following worked example.

Example Prove the convolution theorem for Laplace transforms.

From the definition (13.25),

$$\bar{f}(s)\bar{g}(s) = \int_0^\infty e^{-su} f(u)\,du \int_0^\infty e^{-sv} g(v)\,dv$$

$$= \int_0^\infty du \int_0^\infty dv\, e^{-s(u+v)} f(u)g(v).$$

Now letting $u + v = t$ changes the limits on the integrals, with the result that

$$\bar{f}(s)\bar{g}(s) = \int_0^\infty du\, f(u) \int_u^\infty dt\, g(t-u)\,e^{-st}.$$

As shown in Figure 13.1(a) the shaded area of integration may be considered as the sum of vertical strips. However, we may instead integrate over this area by summing over horizontal strips as shown in Figure 13.1(b). Then the integral can be written as

$$\bar{f}(s)\bar{g}(s) = \int_0^t du\, f(u) \int_0^\infty dt\, g(t-u)\,e^{-st}$$

$$= \int_0^\infty dt\, e^{-st} \left\{\int_0^t f(u)g(t-u)\,du\right\}$$

$$= \mathcal{L}\left[\int_0^t f(u)g(t-u)\,du\right],$$

as given in Equation (13.25).

Summary

The properties of the Laplace transform derived in this section can sometimes be useful in finding the Laplace transforms of particular functions.

Example Find the Laplace transform of $f(t) = t \sin bt$.

Although we could calculate the Laplace transform directly, we can use (13.23) to give

$$\bar{f}(s) = (-1)\frac{d}{ds}\mathcal{L}[\sin bt] = -\frac{d}{ds}\left(\frac{b}{s^2+b^2}\right) = \frac{2bs}{(s^2+b^2)^2}, \quad \text{for } s > 0.$$

The direct method of integration by parts yields the same result, as the reader may care to verify. ◀

EXERCISES 13.4

1. Prove the convolution theorem for Laplace transforms [Equation (13.25)] using the Heaviside step functions in the way indicated in footnote 10.

2. Use the properties of Laplace transforms to calculate

   (a) $\mathcal{L}\left[t^{3/2}\right]$, (b) $\mathcal{L}\left[\dfrac{\sin at}{t}\right]$, (c) $\mathcal{L}\left[e^{2t} f(t-4) H(t-4)\right]$,

   without explicitly evaluating any Laplace integrals.

3. Find an expression for the Laplace transform of $t^2 \dfrac{df(at)}{dt}$ in terms of the Laplace transform of $f(t)$. Evaluate it when $f(t) = t^n$ and verify your answer by direct calculation.

SUMMARY

1. *Dirac δ-function*
   - Definition: $\int f(t)\delta(t-a)\,dt = f(a)$ if the integration range includes $t = a$; otherwise the integral is zero.
   - $\int \delta(t-a)\,dt = 1$ if the integration range includes $t = a$.
   - $\delta(-t) = \delta(t)$, $\delta(at) = \dfrac{1}{|a|}\delta(t)$, $t\delta(t) = 0$.
   - $\delta(h(t)) = \sum_i \dfrac{\delta(t-t_i)}{|h'(t_i)|}$, where the t_i are the zeros of $h(t)$.

- The derivatives $\delta^{(n)}(t)$ of the δ-function are defined by

$$\int_{-\infty}^{\infty} f(t)\delta^{(n)}(t)\,dt = (-1)^n f^{(n)}(0).$$

- The Heaviside function $H(t)$, which is defined as $H(t) = 1$ for $t > 0$ and $H(t) = 0$ for $t < 0$, has the property $H'(t) = \delta(t)$.

2. *Laplace transforms*

A Laplace transform $\mathcal{L}[f(t)]$ is a linear transformation $f(t) \to \bar{f}(s)$ given by

$$\bar{f}(s) \equiv \int_0^{\infty} f(t)e^{-st}\,dt \text{ for } s > s_0 \text{ with } s_0 \text{ depending on the form of } f(t).$$

- For standard Laplace transforms, see Table 13.1 on p. 540.
- $\mathcal{L}[t^n f(t)] = (-1)^n \dfrac{d^n \bar{f}(s)}{ds^n}$, for $n = 1, 2, 3 \ldots$

PROBLEMS

13.1. Prove the expressions given in Table 13.1 for the Laplace transforms of $t^{-1/2}$ and $t^{1/2}$, by setting $x^2 = ts$ in the result

$$\int_0^{\infty} \exp(-x^2)\,dx = \tfrac{1}{2}\sqrt{\pi}.$$

13.2. Find the functions $y(t)$ whose Laplace transforms are the following:
  (a) $1/(s^2 - s - 2)$;
  (b) $2s/[(s+1)(s^2+4)]$;
  (c) $e^{-(\gamma+s)t_0}/[(s+\gamma)^2 + b^2]$.

13.3. Use the properties of Laplace transforms to prove the following without evaluating any Laplace integrals explicitly:
  (a) $\mathcal{L}[t^{5/2}] = \tfrac{15}{8}\sqrt{\pi}s^{-7/2}$;
  (b) $\mathcal{L}[(\sinh at)/t] = \tfrac{1}{2}\ln[(s+a)/(s-a)]$, $s > |a|$;
  (c) $\mathcal{L}[\sinh at \cos bt] = a(s^2 - a^2 + b^2)[(s-a)^2 + b^2]^{-1}[(s+a)^2 + b^2]^{-1}$.

13.4. Find the solution (the so-called *impulse response* or *Green's function*) of the equation

$$T\frac{dx}{dt} + x = \delta(t)$$

by proceeding as follows.

Problems

(a) Show by substitution that

$$x(t) = A(1 - e^{-t/T})H(t)$$

is a solution, for which $x(0) = 0$, of

$$T\frac{dx}{dt} + x = AH(t), \qquad (*)$$

where $H(t)$ is the Heaviside step function.

(b) Construct the solution when the RHS of $(*)$ is replaced by $AH(t - \tau)$, with $dx/dt = x = 0$ for $t < \tau$, and hence find the solution when the RHS is a rectangular pulse of duration τ.

(c) By setting $A = 1/\tau$ and taking the limit as $\tau \to 0$, show that the impulse response is $x(t) = T^{-1}e^{-t/T}$.

(d) Obtain the same result much more directly by taking the Laplace transform of each term in the original equation, solving the resulting algebraic equation and then using the entries in Table 13.1.

13.5. This problem is concerned with the limiting behaviour of Laplace transforms.

(a) If $f(t) = A + g(t)$, where A is a constant and the indefinite integral of $g(t)$ is bounded as its upper limit tends to ∞, show that

$$\lim_{s \to 0} s \bar{f}(s) = A.$$

(b) For $t > 0$, the function $y(t)$ obeys the differential equation

$$\frac{d^2y}{dt^2} + a\frac{dy}{dt} + by = c\cos^2 \omega t,$$

where a, b and c are positive constants. Find $\bar{y}(s)$ and show that $s\bar{y}(s) \to c/2b$ as $s \to 0$. Interpret the result in the t-domain.

13.6. By writing $f(x)$ as an integral involving the δ-function $\delta(\xi - x)$ and taking the Laplace transforms of both sides, show that the transform of the solution of the equation

$$\frac{d^4y}{dx^4} - y = f(x)$$

for which y and its first three derivatives vanish at $x = 0$ can be written as

$$\bar{y}(s) = \int_0^\infty f(\xi)\frac{e^{-s\xi}}{s^4 - 1}\,d\xi.$$

Use the properties of Laplace transforms and the entries in Table 13.1 to show that

$$y(x) = \frac{1}{2}\int_0^x f(\xi)[\sinh(x - \xi) - \sin(x - \xi)]\,d\xi.$$

Laplace transforms

13.7. The function $f_a(x)$ is defined as unity for $0 < x < a$ and zero otherwise. Find its Laplace transform $\bar{f}_a(s)$ and deduce that the transform of $xf_a(x)$ is

$$\frac{1}{s^2}\left[1 - (1 + as)e^{-sa}\right].$$

Write $f_a(x)$ in terms of Heaviside functions and hence obtain an explicit expression for

$$g_a(x) = \int_0^x f_a(y) f_a(x - y) \, dy.$$

Use the expression to write $\bar{g}_a(s)$ in terms of the functions $\bar{f}_a(s)$ and $\bar{f}_{2a}(s)$, and their derivatives, and hence show that $\bar{g}_a(s)$ is equal to the square of $\bar{f}_a(s)$, in accordance with the convolution theorem.

13.8. Show that the Laplace transform of $f(t - a)H(t - a)$, where $a \geq 0$, is $e^{-as}\bar{f}(s)$ and that, if $g(t)$ is a periodic function of period T, $\bar{g}(s)$ can be written as

$$\frac{1}{1 - e^{-sT}} \int_0^T e^{-st} g(t) \, dt.$$

(a) Sketch the periodic function defined in $0 \leq t \leq T$ by

$$g(t) = \begin{cases} 2t/T & 0 \leq t < T/2 \\ 2(1 - t/T) & T/2 \leq t \leq T, \end{cases}$$

and, using the previous result, find its Laplace transform.
(b) Show, by sketching it, that

$$\frac{2}{T}\left[tH(t) + 2\sum_{n=1}^{\infty}(-1)^n\left(t - \tfrac{1}{2}nT\right)H\left(t - \tfrac{1}{2}nT\right)\right]$$

is another representation of $g(t)$ and hence derive the relationship

$$\tanh x = 1 + 2\sum_{n=1}^{\infty}(-1)^n e^{-2nx}.$$

HINTS AND ANSWERS

13.1. Prove the result for $t^{1/2}$ by integrating that for $t^{-1/2}$ by parts.

13.3. (a) Use (13.23) with $n = 2$ on $\mathcal{L}[\sqrt{t}]$; (b) use (13.24);
(c) consider $\mathcal{L}[\exp(\pm at)\cos bt]$ and use the translation property, Section 13.4.

13.5. (a) Note that $|\lim \int g(t)e^{-st}\, dt| \leq |\lim \int g(t)\, dt|$.
(b) $(s^2 + as + b)\bar{y}(s) = \{c(s^2 + 2\omega^2)/[s(s^2 + 4\omega^2)]\} + (a + s)y(0) + y'(0)$.

For this damped system, at large t (corresponding to $s \to 0$) rates of change are negligible and the equation reduces to $by = c\cos^2 \omega t$. The average value of $\cos^2 \omega t$ is $\frac{1}{2}$.

13.7. $s^{-1}[1 - \exp(-sa)]$; $g_a(x) = x$ for $0 < x < a$, $g_a(x) = 2a - x$ for $a \leq x \leq 2a$, $g_a(x) = 0$ otherwise.

14
Ordinary differential equations

Differential equations are the group of equations that contain derivatives. There are several different types of differential equations, but here we will be considering only the simplest types. As its name suggests, an ordinary differential equation (ODE) contains only ordinary derivatives (no partial derivatives) and describes the relationship between these derivatives of the *dependent variable*, usually called y, with respect to the *independent variable*, usually called x. The solution to such an ODE is therefore a function of x and is written y(x). For an ODE to have a closed-form solution, it must be possible to express y(x) in terms of the standard elementary functions such as x^2, $\sqrt{x}$, exp x, ln x, sin x, etc. The solutions of some differential equations cannot, however, be written in closed form, but only as an infinite series that carry no special names.

Ordinary differential equations may be separated conveniently into different categories according to their general characteristics. The primary grouping adopted here is by the *order* of the equation. The order of an ODE is simply the order of the highest derivative it contains. Thus, equations containing dy/dx, but no higher derivatives, are called first order, those containing d^2y/dx^2 are called second order and so on. In this chapter we consider first-order equations and some of the more straightforward equations of second order.

Ordinary differential equations may be classified further according to *degree*. The degree of an ODE is the power to which the highest order derivative is raised, after the equation has been rationalised to contain only integer powers of derivatives. Hence the ODE

$$\frac{d^3y}{dx^3} + x\left(\frac{dy}{dx}\right)^{3/2} + x^2y = 0$$

is of third order and second degree, since after rationalisation it contains the term $(d^3y/dx^3)^2$.

The *general solution* to an ODE is the most general function y(x) that satisfies the equation; it will contain *constants of integration* which may be determined by the application of some suitable *boundary conditions*. For example, we may be told that for a certain first-order differential equation, the solution y(x) is equal to zero when the independent variable x is equal to unity; this allows us to determine the value of the constant of integration.

The *general solutions* to nth-order ODEs will contain n (essential) arbitrary constants of integration and therefore we will need n (independent and self-consistent) boundary conditions if these constants are to be determined (see Section 14.1).[1] When the boundary conditions have been applied, and the constants found, we are left with a *particular solution* to the ODE, which obeys the given boundary conditions.

Some ODEs of degree greater than unity also possess *singular solutions*, which are solutions that contain no arbitrary constants and cannot be found from the general solution; singular solutions are discussed in more detail in Section 14.3. When any solution to an ODE has been found, it is always possible to check its validity by substitution into the original equation and verification that any given boundary conditions are met.

In this chapter we are initially concerned with various types of first-order ODE, including those of higher degree that can be solved in closed form. We then move on to second-order equations, but restrict our attention to those in which the coefficients multiplying the various derivatives are *constants*; second-order ODEs with variable coefficients lie beyond the scope of this introductory book. The chapter concludes with a brief study of linear recurrence relations, which are the discrete analogue of linear ODEs. However, in the next section, we begin with a discussion of the general form of the solutions of ODEs; this discussion is relevant to both first- and higher-order ODEs.

14.1 General form of solution

It is helpful when considering the general form of the solution to an ODE to consider the inverse process, namely that of obtaining an ODE from a given group of functions, each one of which is a solution of the ODE. Suppose the members of the group can be written as

$$y = f(x, a_1, a_2, \ldots, a_n), \tag{14.1}$$

each member being specified by a different set of values[2] of the parameters a_i. For example, consider the group of functions

$$y = a_1 \sin x + a_2 \cos x; \tag{14.2}$$

here $n = 2$.

Since an ODE is required for which *any* of the group is a solution, it clearly must not contain any of the a_i. As there are n of the a_i in expression (14.1), we must obtain $n + 1$ equations involving them in order that, by elimination, we can obtain one final equation without them.

Initially we have only (14.1), but if this is differentiated n times, a total of $n + 1$ equations is obtained from which (in principle) all the a_i can be eliminated, to give one ODE satisfied by all the group. As a result of the n differentiations, $d^n y/dx^n$ will be present in one of the $n + 1$ equations and hence in the final equation, which will therefore be of nth order.

---

[1] For the first-order ODE considered in the final sentence of the previous paragraph there is only one constant of integration and so only one boundary condition is needed.

[2] This does not preclude some values being the same in two different sets, but does require that at least one of the a_i is different for any pair of members.

Ordinary differential equations

In the case of (14.2), we have

$$\frac{dy}{dx} = a_1 \cos x - a_2 \sin x,$$

$$\frac{d^2y}{dx^2} = -a_1 \sin x - a_2 \cos x.$$

Here the elimination of a_1 and a_2 is trivial (because of the similarity of the forms of y and d^2y/dx^2), resulting in

$$\frac{d^2y}{dx^2} + y = 0,$$

a second-order equation.[3]

Thus, to summarise, a group of functions (14.1) with n parameters satisfies an nth-order ODE in general (although in some degenerate cases an ODE of less than nth order is obtained). The intuitive converse of this is that the general solution of an nth-order ODE contains n arbitrary parameters (constants); for our purposes, this will be assumed to be valid although a totally general proof is difficult.

As mentioned earlier, external factors affect a system described by an ODE, by fixing the values of the dependent variables for particular values of the independent ones. These externally imposed (or *boundary*) conditions on the solution are thus the means of determining the parameters and so of specifying precisely which function is the required solution. It is apparent that the number of boundary conditions should match the number of parameters and hence the order of the equation, if a unique solution is to be obtained. Fewer independent boundary conditions than this will lead to a number of undetermined parameters in the solution, whilst an excess will usually mean that no acceptable solution is possible.

For an nth-order equation the required n boundary conditions can take many forms, for example the value of y at n different values of x, or the value of any $n - 1$ of the n derivatives $dy/dx, d^2y/dx^2, \ldots, d^ny/dx^n$ together with that of y, all for the same value of x, or many intermediate combinations.

EXERCISES 14.1

1. What are the orders and degrees of the following differential equations?

   (a) $\left(\dfrac{d^2y}{dx^2}\right)^2 + \left(\dfrac{dy}{dx}\right)^3 = 0.$

   (b) $\dfrac{d^3y}{dx^3} + 2\left(\dfrac{d^2y}{dx^2}\right)^{1/2} + y = 0.$

   (c) $\left(\dfrac{d^2y}{dx^2}\right)^{1/2} + \omega^2 y = 0.$

3. Find the differential equation satisfied by all functions of the form $y(x) = ax^3 + be^{-x}$, where a and b are arbitrary constants. Verify your answer by re-substituting $y = x^3$ and $y = e^{-x}$ separately, formally corresponding to the cases $a = 1, b = 0$ and $a = 0, b = 1$, respectively.

14.2 First-degree first-order equations

2. Show that a differential equation satisfied, for all a and b, by functions of the form $y(x) = a \sin bx$ is

$$\sin^2\left(\sqrt{\frac{-y''}{y}} \, x\right) = \frac{yy''}{yy'' - (y')^2}.$$

Try to derive the equation, rather than merely substitute in it.

14.2 First-degree first-order equations

First-degree first-order ODEs contain only dy/dx equated to some function of x and y, and can be written in either of two equivalent standard forms,

$$\frac{dy}{dx} = F(x, y), \qquad A(x, y)\,dx + B(x, y)\,dy = 0,$$

where $F(x, y) = -A(x, y)/B(x, y)$, and $F(x, y)$, $A(x, y)$ and $B(x, y)$ are in general functions of both x and y. Which of the two above forms is the more useful for finding a solution depends on the type of equation being considered. There are several different types of first-degree first-order ODEs that are of interest in the physical sciences. These equations and their respective solutions are discussed below.

14.2.1 Separable-variable equations

A separable-variable equation is one which may be written in the conventional form

$$\frac{dy}{dx} = f(x)g(y), \tag{14.3}$$

where $f(x)$ and $g(y)$ are functions of x and y respectively, including cases in which $f(x)$ or $g(y)$ is simply a constant. Rearranging this equation so that the terms depending on x and on y appear on opposite sides (i.e. are separated), and integrating, we obtain

$$\int \frac{dy}{g(y)} = \int f(x)\,dx.$$

Finding the solution $y(x)$ that satisfies (14.3) then depends only on the ease with which the integrals in the above equation can be evaluated. It is also worth noting that ODEs that at first sight do not appear to be of the form (14.3) can sometimes be made separable by an appropriate factorisation. Here is a simple example.

Example Solve

$$\frac{dy}{dx} = x + xy.$$

Since the RHS of this equation can be factorised to give $x(1 + y)$, it can be put into a separable form yielding

$$\int \frac{dy}{1 + y} = \int x\,dx.$$

Now integrating both sides separately, we find

$$\ln(1+y) = \frac{x^2}{2} + c,$$

and so

$$1 + y = \exp\left(\frac{x^2}{2} + c\right) = A \exp\left(\frac{x^2}{2}\right),$$

where c, and hence $A = e^c$, is an arbitrary constant.[4]

◀

Solution method. *Factorise the equation so that it becomes separable. Rearrange it so that the terms depending on x and those depending on y appear on opposite sides and then integrate directly. Remember the constant of integration, which can be evaluated if further information is given.*

14.2.2 Exact equations

An *exact* first-degree first-order ODE is one of the form

$$A(x, y)\, dx + B(x, y)\, dy = 0, \quad \text{and for which} \quad \frac{\partial A}{\partial y} = \frac{\partial B}{\partial x}. \tag{14.4}$$

In this case $A(x, y)\, dx + B(x, y)\, dy$ is an exact differential, $dU(x, y)$ say (see Section 7.3). For this conclusion to follow requires that

$$A\, dx + B\, dy = dU = \frac{\partial U}{\partial x}\, dx + \frac{\partial U}{\partial y}\, dy,$$

from which we must have

$$A(x, y) = \frac{\partial U}{\partial x}, \tag{14.5}$$

$$B(x, y) = \frac{\partial U}{\partial y}. \tag{14.6}$$

Since $\partial^2 U/\partial x \partial y = \partial^2 U/\partial y \partial x$, the requirement becomes

$$\frac{\partial A}{\partial y} = \frac{\partial B}{\partial x}, \tag{14.7}$$

i.e. as stated in (14.4). If (14.7) holds then (14.4) can be written $dU(x, y) = 0$, which has the solution $U(x, y) = c$, where c is a constant, and from (14.5) $U(x, y)$ is given by

$$U(x, y) = \int A(x, y)\, dx + F(y). \tag{14.8}$$

The function $F(y)$ can be found from (14.6) by differentiating (14.8) with respect to y and equating the result to $B(x, y)$, as in the next worked example.

---

[4] Determine explicit expressions for $y(x)$, (a) if $y(0) = 1$ and (b) if $y(0) = -1$.

14.2 First-degree first-order equations

Example Solve
$$x\frac{dy}{dx} + 3x + y = 0.$$

Rearranging into the form (14.4) we have
$$(3x + y)\,dx + x\,dy = 0,$$

i.e. $A(x, y) = 3x + y$ and $B(x, y) = x$. Since $\partial A/\partial y = 1 = \partial B/\partial x$, the equation is exact, and by (14.8) the solution is given by

$$U(x, y) = \int (3x + y)\,dx + F(y) = c_1 \quad \Rightarrow \quad U(x, y) = \frac{3x^2}{2} + yx + F(y) = c_1.$$

Differentiating $U(x, y)$ with respect to y gives $0 + x + dF/dy$ and then equating this to $B(x, y) = x$ we obtain $dF/dy = 0$, which integrates immediately to give $F(y) = c_2$. Therefore, letting $c = c_1 - c_2$, the solution to the original ODE is

$$\frac{3x^2}{2} + xy = c.$$

As expected, and required, of a first-order equation, there is only one constant of integration in the solution. ◀

Solution method. *Check whether the equation is an exact differential using (14.7); if it is, then solve using (14.8). Find the function $F(y)$ by differentiating (14.8) with respect to y and using (14.6).*

14.2.3 Inexact equations: integrating factors

Equations that may be written in the form

$$A(x, y)\,dx + B(x, y)\,dy = 0, \quad \text{but for which} \quad \frac{\partial A}{\partial y} \neq \frac{\partial B}{\partial x}, \qquad (14.9)$$

are known as inexact equations. However, the differential $A\,dx + B\,dy$ can always be made exact by multiplying by an *integrating factor* (IF) $\mu(x, y)$, which is such that

$$\frac{\partial(\mu A)}{\partial y} = \frac{\partial(\mu B)}{\partial x}. \qquad (14.10)$$

Unfortunately, if the appropriate integrating factor is a function of both x and y, i.e. $\mu = \mu(x, y)$, there exists no general method for finding it; in such cases it may sometimes be found by inspection or an 'educated guess'.[5] If, however, an integrating factor exists that is a function of either x or y alone then (14.10) can be solved to find it. For example, if we assume that the integrating factor is a function of x alone, i.e. $\mu = \mu(x)$, then (14.10) reads

$$\mu\frac{\partial A}{\partial y} = \mu\frac{\partial B}{\partial x} + B\frac{d\mu}{dx}.$$

---

[5] For example, if $A(x, y) = 3y^3 - 4x^{-1}y^5$ and $B(x, y) = xy^2 - 6y^4 + 2x^{-2}y^2$ then the equation is not exact. However, an integrating factor of the form $\mu(x, y) = x^n y^m$ converts it to an exact equation with solution $x^3 y - 2x^2 y^3 + 2y = c$. The reader should verify each assertion made here.

Ordinary differential equations

Rearranging this expression we find

$$\frac{d\mu}{\mu} = \frac{1}{B}\left(\frac{\partial A}{\partial y} - \frac{\partial B}{\partial x}\right) dx = f(x) dx,$$

where we require $f(x)$ also to be a function of x only; indeed this provides a general method of determining whether the integrating factor μ is a function of x alone. This integrating factor is then given by

$$\mu(x) = \exp\left\{\int f(x) dx\right\} \quad \text{where} \quad f(x) = \frac{1}{B}\left(\frac{\partial A}{\partial y} - \frac{\partial B}{\partial x}\right). \quad (14.11)$$

Similarly, if $\mu = \mu(y)$ then

$$\mu(y) = \exp\left\{\int g(y) dy\right\} \quad \text{where} \quad g(y) = \frac{1}{A}\left(\frac{\partial B}{\partial x} - \frac{\partial A}{\partial y}\right). \quad (14.12)$$

We next illustrate this general approach.

Example Solve

$$\frac{dy}{dx} = -\frac{2}{y} - \frac{3y}{2x}.$$

Rearranging this into the form (14.9), we have

$$(4x + 3y^2) dx + 2xy\, dy = 0, \quad (14.13)$$

i.e. $A(x, y) = 4x + 3y^2$ and $B(x, y) = 2xy$. Now

$$\frac{\partial A}{\partial y} = 6y, \qquad \frac{\partial B}{\partial x} = 2y,$$

so the ODE is not exact in its present form. However, we see that

$$\frac{1}{B}\left(\frac{\partial A}{\partial y} - \frac{\partial B}{\partial x}\right) = \frac{2}{x},$$

a function of x alone. Therefore, an integrating factor exists that is also a function of x alone and, ignoring the arbitrary constant of integration, is given by

$$\mu(x) = \exp\left\{2\int \frac{dx}{x}\right\} = \exp(2\ln x) = x^2.$$

Multiplying (14.13) through by $\mu(x) = x^2$ we obtain

$$(4x^3 + 3x^2 y^2) dx + 2x^3 y\, dy = 4x^3\, dx + (3x^2 y^2\, dx + 2x^3 y\, dy) = 0.$$

By inspection this integrates immediately to give the solution $x^4 + y^2 x^3 = c$, where c is a constant. ◂

Solution method. *Examine whether the quantities denoted by $f(x)$ and $g(y)$ in Equations (14.11) and (14.12), respectively, are, in fact, functions of x only or y only. If so, then the required integrating factor is a function of either x or y only, and is given by the integral in the relevant equation. If the integrating factor is a function of both x and y, then sometimes it may be found by inspection or by trial and error. In any case, the*

14.2 First-degree first-order equations

integrating factor μ must satisfy (14.10). Once the equation has been made exact, solve by the method of Section 14.2.2.

14.2.4 Linear equations

Linear first-order ODEs are a special case of inexact ODEs (discussed in the previous subsection) and can be written in the conventional form

$$\frac{dy}{dx} + P(x)y = Q(x). \tag{14.14}$$

Such equations can be made exact by multiplying through by an appropriate integrating factor in a similar manner to that discussed above. In this case, however, the integrating factor is always a function of x alone and may be expressed in a particularly simple form. An integrating factor $\mu(x)$ must be such that

$$\mu(x)\frac{dy}{dx} + \mu(x)P(x)y = \frac{d}{dx}[\mu(x)y] = \mu(x)Q(x); \tag{14.15}$$

the second equality may then be integrated directly to give[6]

$$\mu(x)y = \int \mu(x)Q(x)\,dx. \tag{14.16}$$

The required integrating factor $\mu(x)$ is determined by the first equality in (14.15), i.e.

$$\frac{d}{dx}(\mu y) = \mu\frac{dy}{dx} + \frac{d\mu}{dx}y = \mu\frac{dy}{dx} + \mu Py,$$

which immediately gives the simple relation[7]

$$\frac{d\mu}{dx} = \mu(x)P(x) \quad \Rightarrow \quad \mu(x) = \exp\left\{\int P(x)\,dx\right\}. \tag{14.17}$$

The following illustrates the procedure.

Example Solve

$$\frac{dy}{dx} + 2xy = 4x.$$

The integrating factor is given immediately by

$$\mu(x) = \exp\left\{\int 2x\,dx\right\} = \exp x^2.$$

Multiplying through the ODE by $\mu(x) = \exp x^2$ gives

$$\frac{dy}{dx}\exp x^2 + y(2x \exp x^2) = \frac{d}{dx}(y \exp x^2) = 4x \exp x^2.$$

---

[6] Recall that, despite it being the standard notation, the x that appears under the integral sign on the RHS is not the same as the x on the LHS. The x on the RHS is a dummy variable of integration, whilst that on the left is the (implied) variable upper limit of the integral on the right.

[7] Note that no constant of integration is included in the integral giving the integrating factor; to include one is equivalent to multiplying (14.15) all through by a non-zero constant, which would not change the essential content of the equation.

Ordinary differential equations

Integrating the second equality, we have

$$y \exp x^2 = 4 \int x \exp x^2 \, dx = 2 \exp x^2 + c.$$

The solution to the ODE is therefore given by $y = 2 + c \exp(-x^2)$. ◀

Solution method. *Rearrange the equation into the form (14.14) and multiply by the integrating factor $\mu(x)$ given by (14.17). The left- and right-hand sides can then be integrated directly, giving y from (14.16).*

14.2.5 Homogeneous equations

Homogeneous equation are ODEs that may be written in the form

$$\frac{dy}{dx} = \frac{A(x, y)}{B(x, y)} = F\left(\frac{y}{x}\right), \tag{14.18}$$

where $A(x, y)$ and $B(x, y)$ are homogeneous functions of the same degree. A function $f(x, y)$ is homogeneous of degree n if, for any λ, it obeys

$$f(\lambda x, \lambda y) = \lambda^n f(x, y).$$

For example, if $A = x^2 y - xy^2$ and $B = x^3 + y^3$ then A and B are both homogeneous functions of degree 3. For a general pair of functions, consisting of the sums of products of powers of x and y, the requirement that each is homogeneous and both are of the same degree means that the sum of the powers in x and y in *each* term of *both* functions has to be the same (in this example equal to 3). The RHS of a homogeneous ODE can be written as a function of y/x. The equation may then be solved by making the substitution $y = vx$, with the consequence that

$$\frac{dy}{dx} = v + x\frac{dv}{dx} = F(v).$$

This is now a separable equation and can be integrated directly to give

$$\int \frac{dv}{F(v) - v} = \int \frac{dx}{x}. \tag{14.19}$$

The following worked example includes on the RHS of the equation a term that is not the ratio of two homogeneous functions – but it is a function of y/x, which is the essential criterion for applying the method.

14.2 First-degree first-order equations

Example Solve
$$\frac{dy}{dx} = \frac{y}{x} + \tan\left(\frac{y}{x}\right).$$

Substituting $y = vx$ we obtain
$$v + x\frac{dv}{dx} = v + \tan v.$$

Cancelling v on both sides, rearranging and integrating gives
$$\int \cot v \, dv = \int \frac{dx}{x} = \ln x + c_1.$$

But
$$\int \cot v \, dv = \int \frac{\cos v}{\sin v} \, dv = \ln(\sin v) + c_2,$$

so the solution to the ODE is $y = x \sin^{-1} Ax$, where A is a constant.[8]

Solution method. *Check to see whether the equation is homogeneous. If so, make the substitution $y = vx$, separate variables as in (14.19) and then integrate directly. Finally, replace v by y/x to obtain the solution.*

14.2.6 Bernoulli's equation

Bernoulli's equation has the form
$$\frac{dy}{dx} + P(x)y = Q(x)y^n \quad \text{where } n \neq 0 \text{ or } 1. \tag{14.20}$$

This equation is very similar in form to the linear equation (14.14), but is in fact non-linear due to the extra y^n factor on the RHS. However, the equation can be made linear by substituting $v = y^{1-n}$ and correspondingly
$$\frac{dy}{dx} = \left(\frac{y^n}{1-n}\right)\frac{dv}{dx}.$$

Substituting this into (14.20) and dividing through by $y^n/(1-n)$, we find
$$\frac{dv}{dx} + (1-n)P(x)v = (1-n)Q(x),$$

which is a linear equation and may be solved by the method described in Section 14.2.4.

---

8 Carry through the final steps of this calculation and express A in terms of c_1 and c_2.

Ordinary differential equations

Example Solve
$$\frac{dy}{dx} + \frac{y}{x} = 2x^3 y^4.$$

Here $n = 4$, and if we let $v = y^{1-4} = y^{-3}$ then
$$\frac{dy}{dx} = -\frac{y^4}{3}\frac{dv}{dx}.$$

Substituting this into the ODE and rearranging, we obtain
$$\frac{dv}{dx} - \frac{3v}{x} = -6x^3,$$

which is linear and may be solved by multiplying through by the integrating factor (see Section 14.2.4)
$$\exp\left\{-3\int\frac{dx}{x}\right\} = \exp(-3\ln x) = \frac{1}{x^3}.$$

The resulting equation,
$$\frac{1}{x^3}\frac{dv}{dx} - \frac{3v}{x^4} = \frac{d}{dx}\left(\frac{v}{x^3}\right) = -6,$$

can now be integrated and yields the solution
$$\frac{v}{x^3} = -6x + c.$$

Remembering that $v = y^{-3}$, we obtain $y^{-3} = -6x^4 + cx^3$, from which y can be obtained.[9] ◀

Solution method. *Rearrange the equation into the form (14.20) and make the substitution $v = y^{1-n}$. This leads to a linear equation in v, which can be solved by the method of Section 14.2.4. Then replace v by y^{1-n} to obtain the solution.*

EXERCISES 14.2

1. Solve the following equations:

   (a) $\dfrac{dy}{dx} = xy + x + y + 1$, (b) $\dfrac{dy}{dx} = \dfrac{1}{xy + x + y + 1}$, (c) $\dfrac{dy}{dx} + 3x^2 y^3 = 2axy^3$.

2. Show that the equations below are exact or can be made so and solve them.

   (a) $\dfrac{dy}{dx} = -\dfrac{1 + 6xy}{3x^2 + 4y}$.

   (b) $\cos x \cos y\, dx - \sin x \sin y\, dy = 0$.

   (c) $\left(3y + \dfrac{2y^2}{x}\right)dx + (x + 2y)\,dy = 0$.

   (d) $\dfrac{x^2 + y^2 - 1}{x^2}\,dx + \dfrac{x^2 y - y + y^3}{x(1 - y^2)}\,dy = 0$.

---

[9] (a) Check this solution by direct re-substitution and (b), if $y(1) = \frac{1}{2}$, find the range of x over which a positive solution for y is defined.

14.3 Higher degree first-order equations

3. Solve the following linear equations:

$$\text{(a)} \quad \frac{dy}{dx} - y \tan x = x^2, \qquad \text{(b)} \quad \frac{dy}{dx} + y \ln x = x^{-x}.$$

4. Solve the following homogeneous equations:

$$\text{(a)} \quad x^2 \frac{dy}{dx} = x^2 + xy + y^2, \qquad \text{(b)} \quad \frac{dy}{dx} = \frac{xy}{x^2 - y^2}.$$

5. Find a solution $y(x)$ of the equation

$$\frac{dy}{dx} + 2x^2 y = x^5 \sqrt{y}$$

that has $y(0) = 1$. Why are two values possible for the constant of integration?

14.3 Higher degree first-order equations

First-order equations of degree higher than the first do not occur often in the description of physical systems, since squared and higher powers of first-order derivatives usually arise from resistive or driving mechanisms, when an acceleration or other higher order derivative is also present. They do sometimes appear in connection with geometrical problems, however.

Higher degree first-order equations can be written as $F(x, y, dy/dx) = 0$. The most general standard form is

$$p^n + a_{n-1}(x, y)p^{n-1} + \cdots + a_1(x, y)p + a_0(x, y) = 0, \qquad (14.21)$$

where for ease of notation we write $p = dy/dx$. If the equation can be solved for x or y or p, then either an explicit or a parametric solution can sometimes be obtained. We discuss the main types of such equations below.

14.3.1 Equations soluble for p

Sometimes the LHS of (14.21) can be factorised into the form

$$(p - F_1)(p - F_2) \cdots (p - F_n) = 0, \qquad (14.22)$$

where $F_i = F_i(x, y)$. We are then left with solving the n first-degree equations $p = F_i(x, y)$. Writing the solutions to these first-degree equations as $G_i(x, y) = 0$, the general solution to (14.22) is given by the product

$$G_1(x, y)G_2(x, y) \cdots G_n(x, y) = 0. \qquad (14.23)$$

A somewhat contrived example is as follows.

Example Solve
$$(x^3 + x^2 + x + 1)p^2 - (3x^2 + 2x + 1)yp + 2xy^2 = 0. \qquad (14.24)$$

This equation may be factorised to give
$$[(x+1)p - y][(x^2+1)p - 2xy] = 0.$$

Taking each bracket in turn we have
$$(x+1)\frac{dy}{dx} - y = 0,$$
$$(x^2+1)\frac{dy}{dx} - 2xy = 0,$$

which have the solutions $y - c(x+1) = 0$ and $y - c(x^2+1) = 0$ respectively (see Section 14.2 on first-degree first-order equations). The general solution to (14.24) is therefore given by
$$[y - c_1(x+1)]\left[y - c_2(x^2+1)\right] = 0.$$

Note that the arbitrary constants in these two solutions could be written differently (as above) or be taken to be the same, because the original equation is satisfied if *either* first-order equation is satisfied, and this requires only one constant of integration. ◀

Solution method. *If the equation can be factorised into the form (14.22) then solve the first-order ODE $p - F_i = 0$ for each factor and write the solution in the form $G_i(x, y) = 0$. The solution to the original equation is then given by the product (14.23).*

14.3.2 Equations soluble for x

Equations that can be solved for x, i.e. such that they may be written in the form
$$x = F(y, p), \qquad (14.25)$$

can be reduced to first-degree first-order equations in p by differentiating both sides with respect to y, so that
$$\frac{dx}{dy} = \frac{1}{p} = \frac{\partial F}{\partial y} + \frac{\partial F}{\partial p}\frac{dp}{dy}.$$

This results in an equation of the form $G(y, p) = 0$, which can be used together with (14.25) to eliminate p and give the general solution. Note that often a singular solution (see the introduction to this chapter) to the equation will be found at the same time, as is the case in the following worked example.

14.3 Higher degree first-order equations

Example Solve
$$6y^2 p^2 + 3xp - y = 0. \tag{14.26}$$

This equation can be solved for x explicitly to give $3x = (y/p) - 6y^2 p$. Differentiating both sides with respect to y, we find

$$3\frac{dx}{dy} = \frac{3}{p} = \frac{1}{p} - \frac{y}{p^2}\frac{dp}{dy} - 6y^2 \frac{dp}{dy} - 12yp,$$

which factorises to give

$$(1 + 6yp^2)\left(2p + y\frac{dp}{dy}\right) = 0. \tag{14.27}$$

Setting the factor containing dp/dy equal to zero gives a first-degree first-order equation in p, which may be solved to give $py^2 = c$. Substituting for p in (14.26) then yields the general solution of (14.26):

$$y^3 = 3cx + 6c^2. \tag{14.28}$$

If we now consider the first factor in (14.27), we find $6p^2 y = -1$ as a possible solution. Substituting for p in (14.26) we find the singular solution

$$8y^3 + 3x^2 = 0.$$

Note that the singular solution contains no arbitrary constants and cannot be found from the general solution (14.28) by any choice of the constant c.[10] ◀

Solution method. *Write the equation in the form (14.25) and differentiate both sides with respect to y. Rearrange the resulting equation into the form $G(y, p) = 0$, which can be used together with the original ODE to eliminate p and so give the general solution. If $G(y, p)$ can be factorised then the factor containing dp/dy should be used to eliminate p and give the general solution. Using the other factors in this fashion will instead lead to singular solutions.*

14.3.3 Equations soluble for y

Equations that can be solved for y, i.e. are such that they may be written in the form

$$y = F(x, p), \tag{14.29}$$

can be reduced to first-degree first-order equations in p by differentiating both sides with respect to x, so that

$$\frac{dy}{dx} = p = \frac{\partial F}{\partial x} + \frac{\partial F}{\partial p}\frac{dp}{dx}.$$

---

[10] Show that, in fact, the singular solution is the envelope of the family of general solutions and find the coordinates of the general point of contact.

568   Ordinary differential equations

This results in an equation of the form $G(x, p) = 0$, which can be used together with (14.29) to eliminate p and give the general solution. As for equations soluble for x, an additional (singular) solution to the equation is sometimes found.

Example   Solve

$$xp^2 + 2xp - y = 0. \tag{14.30}$$

This equation can be solved trivially for y to give $y = xp^2 + 2xp$. Differentiating both sides with respect to x, we find

$$\frac{dy}{dx} = p = 2xp\frac{dp}{dx} + p^2 + 2x\frac{dp}{dx} + 2p,$$

which after factorising gives

$$(p+1)\left(p + 2x\frac{dp}{dx}\right) = 0. \tag{14.31}$$

To obtain the general solution of (14.30), we consider the factor containing dp/dx. This first-degree first-order equation in p has the solution $xp^2 = c$ (see Section 14.3.1), which we then use to eliminate p from (14.30). Thus we find that the general solution to (14.30) is[11,12]

$$(y - c)^2 = 4cx. \tag{14.32}$$

If instead, we set the other factor in (14.31) equal to zero, we obtain the very simple solution $p = -1$. Substituting this into (14.30) then gives

$$x + y = 0,$$

which is a singular solution to (14.30). ◀

Solution method. *Write the equation in the form (14.29) and differentiate both sides with respect to x. Rearrange the resulting equation into the form $G(x, p) = 0$, which can be used together with the original ODE to eliminate p and so give the general solution. If $G(x, p)$ can be factorised then the factor containing dp/dx should be used to eliminate p and give the general solution. Using the other factors in this fashion will instead lead to singular solutions.*

EXERCISES 14.3

1. Solve the non-linear equation

$$x^3\left(\frac{dy}{dx}\right)^2 + x^2(1 - y - x^2)\frac{dy}{dx} - x(y + x^2) = 0.$$

---

11 Whilst it has nothing to do with the particular method of solution, it provides good practice at curve sketching to make a rough plot of the solutions as c is varied, including both positive and negative values. Find algebraically the point(s) of intersection of the curves with parameters c_1 and c_2.
12 Prove that, as in the previous example, the singular solution is (part of) the envelope of the general family of solutions. On your plot from the previous footnote, note how each curve touches the line $x + y = 0$ at $(c, -c)$ and the line $x = 0$ at $(0, c)$.

2. The equation

$$x = 2\frac{dy}{dx}y^3$$

can be easily solved directly to give $x^2 = y^4 + k$. However, it is instructive to solve it using the two methods given in Sections 14.3.2 and 14.3.3, so that the workings of those methods can be seen without being obscured by a mass of complicated calculation. Solve the equation using each of the methods; in both cases you should obtain a Bernoulli equation which can be solved using an integrating factor and the original equation.

14.4 Higher order linear ODEs

Following on from the discussion of first-order ODEs given in previous sections, we now examine equations of second and higher order, and, in particular, linear equations with constant coefficients.

Linear equations are of paramount importance in the description of physical processes. Moreover, it is an empirical fact that, when put into mathematical form, many natural processes appear as higher order linear ODEs, most often as second-order equations. Although we could restrict our attention to those of second order, nth-order equations require no additional fundamental considerations, and so for our general discussion we will consider this case.

A linear ODE of general order n has the form

$$a_n(x)\frac{d^n y}{dx^n} + a_{n-1}(x)\frac{d^{n-1} y}{dx^{n-1}} + \cdots + a_1(x)\frac{dy}{dx} + a_0(x)y = f(x). \quad (14.33)$$

If $f(x) = 0$ then the equation is called *homogeneous*; otherwise it is *inhomogeneous*. The first-order linear equation studied in Section 14.2.4 is a special case of (14.33). As discussed in Section 14.1, the general solution to (14.33) will contain n arbitrary constants, which may be determined if n boundary conditions are also provided.[13]

In order to solve any equation of the form (14.33), we need first to find the general solution of the *complementary equation*, i.e. the equation formed by setting $f(x) = 0$:

$$a_n(x)\frac{d^n y}{dx^n} + a_{n-1}(x)\frac{d^{n-1} y}{dx^{n-1}} + \cdots + a_1(x)\frac{dy}{dx} + a_0(x)y = 0. \quad (14.34)$$

To determine the general solution of (14.34), we must find n linearly independent functions that satisfy it. Once we have found these solutions, the general solution is given by a linear superposition of these n functions. In other words, if the n solutions of (14.34) are $y_1(x), y_2(x), \ldots, y_n(x)$, then the general solution is given by the linear superposition

$$y_c(x) = c_1 y_1(x) + c_2 y_2(x) + \cdots + c_n y_n(x), \quad (14.35)$$

---

[13] They must be both independent and self-consistent.

Ordinary differential equations

where the c_m are arbitrary constants that may be determined if n boundary conditions are provided.[14] The linear combination $y_c(x)$ is called the *complementary function* of (14.33).

The question naturally arises how we establish that any n individual solutions to (14.34) are indeed linearly independent.[15] For n functions to be linearly independent over an interval, there must not exist *any* set of constants $c_1, c_2, \ldots, c_n$ such that

$$c_1 y_1(x) + c_2 y_2(x) + \cdots + c_n y_n(x) = 0 \tag{14.36}$$

over the interval in question, except for the trivial case $c_1 = c_2 = \cdots = c_n = 0$.

A statement equivalent to (14.36), which is perhaps more useful for the practical determination of linear independence, can be found by repeatedly differentiating (14.36), $n-1$ times in all, to obtain n simultaneous equations for $c_1, c_2, \ldots, c_n$:

$$\begin{aligned}
c_1 y_1(x) + c_2 y_2(x) + \cdots + c_n y_n(x) &= 0, \\
c_1 y_1'(x) + c_2 y_2'(x) + \cdots + c_n y_n'(x) &= 0, \\
&\vdots \\
c_1 y_1^{(n-1)}(x) + c_2 y_2^{(n-1)}(x) + \cdots + c_n y_n^{(n-1)}(x) &= 0,
\end{aligned} \tag{14.37}$$

where the primes denote differentiation with respect to x. Referring to the discussion of simultaneous linear equations given in Chapter 10, if the determinant of the coefficients of $c_1, c_2, \ldots, c_n$ is non-zero then the only solution to equations (14.37) is the trivial solution $c_1 = c_2 = \cdots = c_n = 0$. In other words, the n functions $y_1(x), y_2(x), \ldots, y_n(x)$ are linearly independent over an interval if

$$W(y_1, y_2, \ldots, y_n) = \begin{vmatrix} y_1 & y_2 & \cdots & y_n \\ y_1' & y_2' & & \vdots \\ \vdots & & \ddots & \vdots \\ y_1^{(n-1)} & & \cdots & y_n^{(n-1)} \end{vmatrix} \neq 0 \tag{14.38}$$

over that interval; $W(y_1, y_2, \ldots, y_n)$ is called the *Wronskian* of the set of functions. It should be noted, however, that the converse, the vanishing of the Wronskian, does *not* guarantee that the functions are linearly dependent.[16]

If the original Equation (14.33) has $f(x) = 0$ (i.e. it is homogeneous) then of course the complementary function $y_c(x)$ in (14.35) is already the general solution. If, however, the equation has $f(x) \neq 0$ (i.e. it is inhomogeneous) then $y_c(x)$ is only one part of the

---

[14] See footnote 13.

[15] For $n = 2$ this is trivial, as the requirement is that one is not a simple multiple of the other. However, for higher values of n, determination by inspection becomes increasingly more difficult and a mechanistic procedure is needed.

[16] Consider the functions $f(x) = x^5$ and $g(x) = |x^5|$, defined as x^5 for $x \geq 0$ and $-x^5$ for $x < 0$. Show by considering the solutions of $af(x) + bg(x) = 0$ at $x = \pm 1$ that they are linearly independent, but, by evaluating it, that their Wronskian is everywhere zero.

14.4 Higher order linear ODEs

solution. The general solution of (14.33) is then given by

$$y(x) = y_c(x) + y_p(x), \tag{14.39}$$

where $y_p(x)$ is the *particular integral*, which can be *any* function that satisfies (14.33) directly, provided it is linearly independent of $y_c(x)$. It should be emphasised for practical purposes that *any* such function, no matter how simple (or complicated), is equally valid in forming the general solution (14.39).

It is important to realise that the above method for finding the general solution to an ODE by superposing particular solutions assumes crucially that the ODE is linear. For non-linear equations this method cannot be used, and indeed it is often impossible to find closed-form solutions to such equations.

Before we leave the general properties of linear equations, there is an essential point to be made in connection with fitting boundary conditions for inhomogeneous equations. Making the general solution fit the given boundary conditions determines the unknown constants that appear as part of the complementary function. However, it is crucial that the conditions are incorporated *after* the particular integral has been included in the solution. As an illustration of this, consider the following example (in which the statements made about the forms of solutions may be checked by re-substitution).

The complementary function solution of the equation

$$\frac{d^2y}{dx^2} - \frac{dy}{dx} - 2y = x$$

is $y_c(x) = Ae^{2x} + Be^{-x}$ and a particular integral is $y_p(x) = \frac{1}{4} - \frac{1}{2}x$. Suppose that the given boundary conditions are $y(0) = 1$ and $y'(0) = 0$.

If these conditions are (mistakenly) fitted to the complementary function alone, we obtain

$$A + B = 1 \text{ and } 2A - B = 0 \quad \Rightarrow \quad A = \tfrac{1}{3} \text{ and } B = \tfrac{2}{3}.$$

The subsequent addition of the particular integral then yields as the (incorrect) full solution

$$y(x) = \tfrac{1}{3}e^{2x} + \tfrac{2}{3}e^{-x} + \tfrac{1}{4} - \tfrac{1}{2}x.$$

Re-substitution will confirm that, although this $y(x)$ is a solution of the differential equation, it does *not* satisfy the boundary conditions.

The correct procedure is to take the general solution, including the particular integral,

$$y(x) = Ae^{2x} + Be^{-x} + \tfrac{1}{4} - \tfrac{1}{2}x,$$

and make *this* fit the boundary conditions. They then require

$$A + B + \tfrac{1}{4} = 1 \text{ and } 2A - B - \tfrac{1}{2} = 0 \quad \Rightarrow \quad A = \tfrac{5}{12} \text{ and } B = \tfrac{1}{3}.$$

The correct full solution is therefore

$$y(x) = \tfrac{5}{12}e^{2x} + \tfrac{1}{3}e^{-x} + \tfrac{1}{4} - \tfrac{1}{2}x,$$

as can be confirmed, if necessary, by calculating $y(0)$ and $y'(0)$.

Ordinary differential equations

EXERCISES 14.4

1. Calculate the Wronskian of the functions e^x, e^{-x} and xe^x and hence show that they are linearly independent.

2. Find the complementary functions for the following equations:
   (a) $\dfrac{d^2y}{dx^2} + 8\dfrac{dy}{dx} + 7y = 0$,   (b) $\dfrac{d^2y}{dx^2} + 6\dfrac{dy}{dx} - 9y = 0$,
   (c) $\dfrac{d^2y}{dx^2} + 2\dfrac{dy}{dx} + 6y = 0$,   (d) $\dfrac{d^2y}{dx^2} + 6\dfrac{dy}{dx} + 9y = 0$,

3. Find the explicit solutions of
$$\frac{d^2y}{dx^2} + 8\frac{dy}{dx} + 7y = f(x)$$
that satisfy $y(0) = 1$ and $y'(0) = 0$ for the three cases (a) $f(x) = e^x$, (b) $f(x) = \sin x$ and (c) $f(x) = e^{-x}$.

14.5 Linear equations with constant coefficients

If the a_m in (14.33) are constants rather than functions of x then we have

$$a_n \frac{d^n y}{dx^n} + a_{n-1} \frac{d^{n-1} y}{dx^{n-1}} + \cdots + a_1 \frac{dy}{dx} + a_0 y = f(x). \tag{14.40}$$

Equations of this sort are very common throughout the physical sciences and engineering, and the method for solving them falls into two parts as discussed in the previous section, i.e. finding the complementary function $y_c(x)$ and finding the particular integral $y_p(x)$. If $f(x) = 0$ in (14.40) then we do not have to find a particular integral, and the complementary function is by itself the general solution.[17]

14.5.1 Finding the complementary function $y_c(x)$

The complementary function must satisfy

$$a_n \frac{d^n y}{dx^n} + a_{n-1} \frac{d^{n-1} y}{dx^{n-1}} + \cdots + a_1 \frac{dy}{dx} + a_0 y = 0 \tag{14.41}$$

and contain n arbitrary constants [see Equation (14.35)]. The standard method for finding $y_c(x)$ is to try a solution of the form $y = Ae^{\lambda x}$, substituting this into (14.41). After dividing the resulting equation through by $Ae^{\lambda x}$, we are left with a polynomial equation in λ of order n; this is the *auxiliary equation* and reads

$$a_n \lambda^n + a_{n-1} \lambda^{n-1} + \cdots + a_1 \lambda + a_0 = 0. \tag{14.42}$$

In general the auxiliary equation has n roots, say $\lambda_1, \lambda_2, \ldots, \lambda_n$. In certain cases, some of these roots may be repeated and some may be complex. The three main cases are as follows.

---

[17] Formally, we can think of the solution $y(x) = 0$ for all x as a (particularly simple, but perfectly acceptable) particular integral $y_p(x)$. This is then added to $y_c(x)$ to give the general solution.

14.5 Linear equations with constant coefficients

(i) *All roots real and distinct.* In this case the n solutions to (14.41) are $y(x) = \exp(\lambda_m x)$ for $m = 1$ to n. It is easily shown by calculating the Wronskian (14.38) of these functions that if all the λ_m are distinct then these solutions are linearly independent. We can therefore linearly superpose them, as in (14.35), to form the complementary function

$$y_c(x) = c_1 e^{\lambda_1 x} + c_2 e^{\lambda_2 x} + \cdots + c_n e^{\lambda_n x}. \tag{14.43}$$

(ii) *Some roots complex.* For the special (but usual) case that all the coefficients a_m in (14.41) are real, if one of the roots of the auxiliary equation (14.42) is complex, say $\alpha + i\beta$, then its complex conjugate $\alpha - i\beta$ is also a root (see the end of Section 5.4.3). In this case we can write

$$c_1 e^{(\alpha + i\beta)x} + c_2 e^{(\alpha - i\beta)x} = e^{\alpha x}(d_1 \cos \beta x + d_2 \sin \beta x)$$

$$= A e^{\alpha x} \begin{Bmatrix} \sin \\ \cos \end{Bmatrix} (\beta x + \phi), \tag{14.44}$$

where A and ϕ are arbitrary constants.

(iii) *Some roots repeated.* If, for example, λ_1 occurs k times ($k > 1$) as a root of the auxiliary equation, then we have not found n linearly independent solutions of (14.41); formally the Wronskian (14.38) of these solutions, having two or more identical columns, is equal to zero. We must therefore find $k - 1$ further solutions that are linearly independent of those already found and also of each other. By direct substitution into (14.41) it is found that[18]

$$x e^{\lambda_1 x}, \quad x^2 e^{\lambda_1 x}, \quad \ldots, \quad x^{k-1} e^{\lambda_1 x}$$

are also solutions, and by calculating the Wronskian it can be shown that they, together with the solutions already found, form a linearly independent set of n functions. Therefore, the complementary function is given by

$$y_c(x) = (c_1 + c_2 x + \cdots + c_k x^{k-1}) e^{\lambda_1 x} + c_{k+1} e^{\lambda_{k+1} x} + c_{k+2} e^{\lambda_{k+2} x} + \cdots + c_n e^{\lambda_n x}. \tag{14.45}$$

If more than one root is repeated then the above argument is easily extended. For example, suppose as before that λ_1 is a k-fold root of the auxiliary equation and, further, that λ_2 is an l-fold root (of course, $k > 1$ and $l > 1$). Then, from the above argument, the complementary function reads

$$y_c(x) = (c_1 + c_2 x + \cdots + c_k x^{k-1}) e^{\lambda_1 x} + (c_{k+1} + c_{k+2} x + \cdots + c_{k+l} x^{l-1}) e^{\lambda_2 x}$$
$$+ c_{k+l+1} e^{\lambda_{k+l+1} x} + c_{k+l+2} e^{\lambda_{k+l+2} x} + \cdots + c_n e^{\lambda_n x}. \tag{14.46}$$

The following is a simple example.

---

[18] A general algebraic proof of this is rather messy, as it involves Leibnitz's theorem and multiple summations.

Ordinary differential equations

Example Find the complementary function of the equation

$$\frac{d^2y}{dx^2} - 2\frac{dy}{dx} + y = e^x. \tag{14.47}$$

Setting the RHS to zero, substituting $y = Ae^{\lambda x}$ and dividing through by $Ae^{\lambda x}$ we obtain the auxiliary equation

$$\lambda^2 - 2\lambda + 1 = 0.$$

The root $\lambda = 1$ occurs twice and so, although e^x is a solution to (14.47), we must find a further solution to the equation that is linearly independent of e^x. From the above discussion, we deduce that xe^x is such a solution, and so the full complementary function is given by the linear superposition

$$y_c(x) = (c_1 + c_2 x)e^x.$$

This can be checked by re-substitution; only the $c_2 x e^x$ actually needs to be checked, as the $c_1 e^x$ term is a proved solution, rather than merely a stated one. ◀

Solution method. *Set the RHS of the ODE to zero (if it is not already so), and substitute $y = Ae^{\lambda x}$. After dividing through the resulting equation by $Ae^{\lambda x}$, obtain an nth-order polynomial equation in λ [the auxiliary equation, see (14.42)]. Solve the auxiliary equation to find the n roots, $\lambda_1, \lambda_2, \ldots, \lambda_n$, say. If all these roots are real and distinct then $y_c(x)$ is given by (14.43). If, however, some of the roots are complex or repeated then $y_c(x)$ is given by (14.44) or (14.45), or the extension (14.46) of the latter, respectively.*

14.5.2 Finding the particular integral $y_p(x)$

There is no generally applicable method for finding the particular integral $y_p(x)$ but, for linear ODEs with constant coefficients and a simple RHS, $y_p(x)$ can often be found by inspection or by assuming a parameterised form similar to $f(x)$. The latter method is sometimes called the *method of undetermined coefficients*. If $f(x)$ contains only polynomial, exponential, or sine and cosine terms then, by assuming a trial function for $y_p(x)$ of similar form but one which contains a number of undetermined parameters and substituting this trial function into (14.41), the parameters can be found and an acceptable $y_p(x)$ deduced.[19] Standard trial functions are as follows.

(i) If $f(x) = ae^{rx}$ then try

$$y_p(x) = be^{rx}.$$

(ii) If $f(x) = a_1 \sin rx + a_2 \cos rx$ (a_1 or a_2 may be zero) then try

$$y_p(x) = b_1 \sin rx + b_2 \cos rx.$$

---

[19] It should always be borne in mind that *any* valid particular integral will do; the difference between that and any other particular integral can always be made up for by a different choice of constants in the complementary function.

14.5 Linear equations with constant coefficients

(iii) If $f(x) = a_0 + a_1 x + \cdots + a_N x^N$ (some a_m may be zero) then try
$$y_p(x) = b_0 + b_1 x + \cdots + b_N x^N.$$

(iv) If $f(x)$ is the sum or product of any of the above then try $y_p(x)$ as the sum or product of the corresponding individual trial functions.

It should be noted that this method fails if any term in the assumed trial function is also contained within the complementary function $y_c(x)$. In such a case the trial function should be multiplied by the smallest integer power of x such that it will then contain no term that already appears in the complementary function. The undetermined coefficients in the trial function can now be found by substitution into (14.40).[20]

The next worked example illustrates this point – in duplicate, it may be said!

Example Find a particular integral of the equation
$$\frac{d^2 y}{dx^2} - 2\frac{dy}{dx} + y = e^x.$$

From the above discussion our first guess at a trial particular integral would be $y_p(x) = be^x$. However, since the complementary function of this equation is $y_c(x) = (c_1 + c_2 x)e^x$ (as in the previous subsection), we see that e^x is already contained in it, as indeed is xe^x. Multiplying our first guess by the lowest integer power of x such that the result does not appear in $y_c(x)$, we therefore try $y_p(x) = bx^2 e^x$. Substituting this into the ODE, we find that $b = 1/2$, so the particular integral is given by $y_p(x) = x^2 e^x / 2$. ◀

For equations with RHSs whose forms are not amongst those listed earlier, more sophisticated methods of finding a particular integral $y_p(x)$ are available. For the record, we might mention those based on Green's functions, the variation of parameters, and a change in the dependent variable using knowledge of the complementary function. However, the development of these methods, which are also applicable to linear equations with variable coefficients, are beyond the scope of this book.

Solution method. *If the RHS of an ODE contains only functions mentioned at the start of this subsection then the appropriate trial function should be substituted into it, thereby fixing the undetermined parameters. If, however, the RHS of the equation is not of this form then a more advanced text should be consulted and one of the more general methods mentioned above employed.*

14.5.3 Constructing the general solution $y_c(x) + y_p(x)$

As stated earlier, the full solution to the ODE (14.40) is found by adding together the complementary function and any particular integral. In order to illustrate further the

---

[20] It is important to recognise that the coefficient in the particular integral is *not* arbitrary, unlike those in the complementary function. Thus the number of boundary conditions needed to determine the unknown coefficients in a general solution is not altered by the inclusion of a particular integral.

Ordinary differential equations

material discussed in the last two subsections, let us find the general solution to a new example, starting from the beginning.

Example Solve

$$\frac{d^2y}{dx^2} + 4y = x^2 \sin 2x. \tag{14.48}$$

First we set the RHS to zero and assume the trial solution $y = Ae^{\lambda x}$. Substituting this into (14.48) leads to the auxiliary equation

$$\lambda^2 + 4 = 0 \quad \Rightarrow \quad \lambda = \pm 2i. \tag{14.49}$$

Therefore the complementary function is given by

$$y_c(x) = c_1 e^{2ix} + c_2 e^{-2ix} = d_1 \cos 2x + d_2 \sin 2x. \tag{14.50}$$

We must now turn our attention to the particular integral $y_p(x)$. Consulting the list of standard trial functions in the previous subsection, we find that a first guess at a suitable trial function for this case should be

$$(ax^2 + bx + c)\sin 2x + (dx^2 + ex + f)\cos 2x. \tag{14.51}$$

However, we see that this trial function contains terms in $\sin 2x$ and $\cos 2x$, both of which already appear in the complementary function (14.50). We must therefore multiply (14.51) by the smallest integer power of x which ensures that none of the resulting terms appears in $y_c(x)$. Since multiplying by x will suffice, we finally assume the trial function

$$(ax^3 + bx^2 + cx)\sin 2x + (dx^3 + ex^2 + fx)\cos 2x. \tag{14.52}$$

Substituting this into (14.48) to fix the constants appearing in (14.52), we find the particular integral to be[21]

$$y_p(x) = -\frac{x^3}{12} \cos 2x + \frac{x^2}{16} \sin 2x + \frac{x}{32} \cos 2x. \tag{14.53}$$

The general solution to (14.48) then reads

$$y(x) = y_c(x) + y_p(x)$$

$$= d_1 \cos 2x + d_2 \sin 2x - \frac{x^3}{12} \cos 2x + \frac{x^2}{16} \sin 2x + \frac{x}{32} \cos 2x,$$

with d_1 and d_2 undetermined until two boundary conditions are imposed. ◂

14.5.4 Laplace transform method

A further method available for solving ODEs with constant coefficients is that of Laplace transforms. Taking the Laplace transform of such an equation turns it into a purely *algebraic* equation in which the unknown is the Laplace transform of the required solution. Once the algebraic equation has been solved for this Laplace transform, the general solution to the original ODE can be obtained by performing an inverse Laplace transform.

---

[21] Carry out this substitution, using Leibnitz's theorem to obtain the second derivatives, and show that the equations to be satisfied are $12a = 0$, $8b + 6d = 0$, $4c = 0$, $-12d = 1$, $6a - 8e = 0$ and $2b - 4f = 0$.

14.5 Linear equations with constant coefficients

One advantage of this method is that, for given boundary conditions, it provides the solution in just one step, instead of having to find the complementary function and particular integral separately.

In order to apply the method we need only two results from the Laplace transform theory of the previous chapter. First, the Laplace transform of a function $f(x)$ is defined by

$$\bar{f}(s) \equiv \int_0^\infty e^{-sx} f(x)\, dx. \tag{14.54}$$

The second result, given by Equation (13.19), concerns the Laplace transform of the nth derivative of $f(x)$:

$$\overline{f^{(n)}}(s) = s^n \bar{f}(s) - s^{n-1} f(0) - s^{n-2} f'(0) - \cdots - s f^{(n-2)}(0) - f^{(n-1)}(0), \tag{14.55}$$

where the primes and superscripts in parentheses denote differentiation with respect to x. Using these relations, along with Table 13.1, on p. 540, which gives Laplace transforms of standard functions, we are in a position to solve a linear ODE with constant coefficients. The following worked example illustrates the method.

Example Solve

$$\frac{d^2y}{dx^2} - 3\frac{dy}{dx} + 2y = 2e^{-x}, \tag{14.56}$$

subject to the boundary conditions $y(0) = 2$, $y'(0) = 1$.

Taking the Laplace transform of (14.56), incorporating the boundary conditions as in (14.55), and using the table of standard results, we obtain

$$s^2 \bar{y}(s) - sy(0) - y'(0) - 3[s\bar{y}(s) - y(0)] + 2\bar{y}(s) = \frac{2}{s+1},$$

which reduces to

$$(s^2 - 3s + 2)\bar{y}(s) - 2s + 5 = \frac{2}{s+1}. \tag{14.57}$$

Solving this algebraic equation for $\bar{y}(s)$, the Laplace transform of the required solution to (14.56), we obtain

$$\bar{y}(s) = \frac{2s^2 - 3s - 3}{(s+1)(s-1)(s-2)} = \frac{1}{3(s+1)} + \frac{2}{s-1} - \frac{1}{3(s-2)}, \tag{14.58}$$

where in the final step we have used partial fractions. Taking the inverse Laplace transform of (14.58), again using Table 13.1, we find the specific solution to (14.56) to be

$$y(x) = \tfrac{1}{3}e^{-x} + 2e^x - \tfrac{1}{3}e^{2x}.$$

That the given initial conditions are satisfied by this solution can be directly verified. ◀

It should be noted that, for an equation with constant coefficients, the factor multiplying $\bar{y}(s)$ in the transformed equation (in Equation (14.57) it is $s^2 - 3s + 2$) is the same as the polynomial that is equated to zero in the auxiliary equation for the ODE. This ensures, as

Ordinary differential equations

it must, that, when the expression for $\bar{y}(s)$ is written in partial fractions, it consists of a sum of terms of the general form $(s - \lambda_i)^{-1}$, where the λ_i are the roots of the auxiliary equation; each of these terms gives rise to a corresponding term of the form $e^{\lambda_i x}$ in the resulting solution for $y(x)$. It should also be noted that if the boundary conditions in a problem are given as symbols, rather than just numbers, then the step involving partial fractions can often involve a considerable amount of algebra.

The Laplace transform method is at its most useful when sets of *simultaneous* linear ODEs with constant coefficients are to be solved. It saves having to explicitly differentiate the equations further in order to eliminate, by substitution, all but one of the dependent variables. We now illustrate this with an example.

Example Two electrical circuits, both of negligible resistance, each consist of a coil having self-inductance L and a capacitor having capacitance C. The mutual inductance of the two circuits is M. There is no source of e.m.f. in either circuit. Initially the second capacitor is given a charge CV_0, the first capacitor being uncharged, and at time $t = 0$ a switch in the second circuit is closed to complete the circuit. Find the subsequent current in the first circuit.

Subject to the initial conditions $q_1(0) = \dot{q}_1(0) = \dot{q}_2(0) = 0$ and $q_2(0) = CV_0 = V_0/G$, say, we have to solve

$$L\ddot{q}_1 + M\ddot{q}_2 + Gq_1 = 0,$$
$$M\ddot{q}_1 + L\ddot{q}_2 + Gq_2 = 0.$$

On taking the Laplace transform of the above equations, we obtain

$$(Ls^2 + G)\bar{q}_1 + Ms^2\bar{q}_2 = sMV_0C,$$
$$Ms^2\bar{q}_1 + (Ls^2 + G)\bar{q}_2 = sLV_0C.$$

The terms on the RHSs arise from incorporation of the initial value of q_2 into the transform of $\ddot{q}_2$. Eliminating $\bar{q}_2$ and rewriting as an equation for $\bar{q}_1$, we find

$$\bar{q}_1(s) = \frac{MV_0 s}{[(L+M)s^2 + G][(L-M)s^2 + G]}$$

$$= \frac{V_0}{2G}\left[\frac{(L+M)s}{(L+M)s^2 + G} - \frac{(L-M)s}{(L-M)s^2 + G}\right].$$

Using Table 13.1,

$$q_1(t) = \tfrac{1}{2}V_0 C(\cos\omega_1 t - \cos\omega_2 t),$$

where $\omega_1^2(L+M) = G$ and $\omega_2^2(L-M) = G$. Differentiating this result gives

$$i_1(t) = \tfrac{1}{2}V_0 C(\omega_2 \sin\omega_2 t - \omega_1 \sin\omega_1 t)$$

as the current in the first circuit. As expected, the initial value $i_1(0) = 0$ is automatically built into the solution. ◀

Solution method. *Perform a Laplace transform, as defined in (14.54), on the entire equation, using (14.55) to calculate the transform of the derivatives. Then solve the*

14.6 Linear recurrence relations

resulting algebraic equation for $\bar{y}(s)$, the Laplace transform of the required solution to the ODE. By using the method of partial fractions and consulting a table of Laplace transforms of standard functions, calculate the inverse Laplace transform. The resulting function $y(x)$ is the solution of the ODE that obeys the given boundary conditions.

EXERCISES 14.5

1. Find the general solutions of
   (a)
   $$\frac{d^2y}{dx^2} + 5\frac{dy}{dx} + 4y = 8x^2,$$
   (b)
   $$\frac{d^2y}{dx^2} + 4\frac{dy}{dx} + 5y = 8x^2.$$

2. Solve $\dfrac{d^2y}{dx^2} + 6\dfrac{dy}{dx} + 9y = 3e^{3x}$ subject to $y(0) = y'(0) = 0$.

3. The equations governing a coupled system are
   $$\ddot{x} + x + 2y = \cos t,$$
   $$\ddot{y} + 2x + y = \sin t,$$
   and it starts from its equilibrium position, i.e. $x(0) = y(0) = \dot{x}(0) = \dot{y}(0) = 0$. Find an expression for the Laplace transform of $x(t)$, reduced to its simplest form with no common factor in the numerator and denominator. Do not attempt to invert the transform, but, by inspecting its denominator, state the sinusoidal and exponential terms you expect to be present in $x(t)$.

14.6 Linear recurrence relations

Before leaving our discussion of higher order ODEs, we take this opportunity to discuss the discrete analogues of differential equations, which are called *recurrence relations* (or sometimes *difference equations*). Whereas a differential equation gives a prescription, in terms of current values, for the new value of a dependent variable at a point only infinitesimally far away, a recurrence relation describes how the next in a sequence of values u_n, defined only at (non-negative) integer values of the 'independent variable' n, is to be calculated.

In its most general form a recurrence relation expresses the way in which u_{n+1} is to be calculated from all the preceding values $u_0, u_1, \ldots, u_n$. Just as the most general differential equations are intractable, so are the most general recurrence relations, and we will limit ourselves to analogues of the types of differential equations studied earlier in this chapter, namely those that are linear, have constant coefficients and possess simple functions on the RHS. Such equations occur over a broad range of engineering and statistical physics as well as in the realms of finance, business planning and gambling! They form the basis of many numerical methods, particularly those concerned with the numerical solution of ordinary and partial differential equations.

A general recurrence relation is exemplified by the formula

$$u_{n+1} = \sum_{r=0}^{N-1} a_r u_{n-r} + k, \qquad (14.59)$$

where N and the a_r are fixed and k is a constant or a simple function of n. Such an equation, involving terms of the series whose indices differ by up to N (ranging from $n-N+1$ to n), is called an Nth-order recurrence relation. It is clear that, given values for $u_0, u_1, \ldots, u_{N-1}$, this is a definitive scheme for generating the series and therefore has a unique solution.

Parallelling the nomenclature of differential equations, if the term not involving any u_n is absent, i.e. $k = 0$, then the recurrence relation is called *homogeneous*. The parallel continues with the form of the general solution of (14.59). If v_n is the general solution of the homogeneous relation, and w_n is *any* solution of the full relation, then

$$u_n = v_n + w_n$$

is the most general solution of the complete recurrence relation. This is straightforwardly verified as follows:

$$\begin{aligned} u_{n+1} &= v_{n+1} + w_{n+1} \\ &= \sum_{r=0}^{N-1} a_r v_{n-r} + \sum_{r=0}^{N-1} a_r w_{n-r} + k \\ &= \sum_{r=0}^{N-1} a_r(v_{n-r} + w_{n-r}) + k \\ &= \sum_{r=0}^{N-1} a_r u_{n-r} + k. \end{aligned}$$

Of course, if $k = 0$ then $w_n = 0$ for all n is a trivial particular solution and the complementary solution, v_n, is itself the most general solution.

14.6.1 First-order recurrence relations

First-order relations, for which $N = 1$, are exemplified by

$$u_{n+1} = a u_n + k, \qquad (14.60)$$

14.6 Linear recurrence relations

with u_0 specified. The solution to the homogeneous relation is immediate,

$$u_n = Ca^n,$$

and, if k is a constant, the particular solution is equally straightforward: $w_n = K$ for all n, provided K is chosen to satisfy

$$K = aK + k,$$

i.e. $K = k(1 - a)^{-1}$. This will be sufficient unless $a = 1$, in which case $u_n = u_0 + nk$ is obvious by inspection.

Thus the general solution of (14.60) is

$$u_n = \begin{cases} Ca^n + k/(1-a) & a \neq 1, \\ u_0 + nk & a = 1. \end{cases} \tag{14.61}$$

If u_0 is specified for the case of $a \neq 1$ then C must be chosen as $C = u_0 - k/(1-a)$, resulting in the equivalent form

$$u_n = u_0 a^n + k \frac{1-a^n}{1-a}. \tag{14.62}$$

We now illustrate this method with a worked example.

Example A house-buyer borrows capital B from a bank that charges a fixed annual rate of interest $R\%$. If the loan is to be repaid over Y years, at what value should the fixed annual payments P, made at the end of each year, be set? For a loan over 25 years at 6%, what percentage of the first year's payment goes towards paying off the capital?

Let u_n denote the outstanding debt at the end of year n, and write $R/100 = r$. Then the relevant recurrence relation is

$$u_{n+1} = u_n(1+r) - P$$

with $u_0 = B$. From (14.62) we have

$$u_n = B(1+r)^n - P \frac{1-(1+r)^n}{1-(1+r)}.$$

As the loan is to be repaid over Y years, $u_Y = 0$ and thus

$$P = \frac{Br(1+r)^Y}{(1+r)^Y - 1}.$$

The first year's interest is rB and so the fraction of the first year's payment going towards capital repayment is $(P - rB)/P$, which, using the above expression for P, is equal to $(1+r)^{-Y}$. With the given figures, this is (only) 23%. ◂

With only small modifications, the method just described can be adapted to handle recurrence relations in which the constant k in (14.60) is replaced by $k\alpha^n$, i.e. the relation is

$$u_{n+1} = au_n + k\alpha^n. \tag{14.63}$$

Ordinary differential equations

As for an inhomogeneous linear differential equation (see Section 14.5.2), we may try as a potential particular solution a form which resembles the term that makes the equation inhomogeneous. Here, the presence of the term $k\alpha^n$ indicates that a particular solution of the form $u_n = A\alpha^n$ should be tried. Substituting this into (14.63) gives

$$A\alpha^{n+1} = aA\alpha^n + k\alpha^n,$$

from which it follows that $A = k/(\alpha - a)$ and that there is a particular solution having the form $u_n = k\alpha^n/(\alpha - a)$, provided $\alpha \neq a$. For the special case $\alpha = a$, the reader can readily verify that a particular solution of the form $u_n = An\alpha^n$ is appropriate. This mirrors the corresponding situation for linear differential equations when the RHS of the differential equation is contained in the complementary function of its LHS.

In summary, the general solution to (14.63) is

$$u_n = \begin{cases} C_1 a^n + k\alpha^n/(\alpha - a) & \alpha \neq a, \\ C_2 a^n + kn\alpha^{n-1} & \alpha = a, \end{cases} \qquad (14.64)$$

with $C_1 = u_0 - k/(\alpha - a)$ and $C_2 = u_0$.

14.6.2 Second-order recurrence relations

We consider next recurrence relations that involve u_{n-1} in the prescription for u_{n+1} and treat the general case in which the intervening term, u_n, is also present. A typical equation is thus

$$u_{n+1} = au_n + bu_{n-1} + k. \qquad (14.65)$$

As previously, the general solution of this is $u_n = v_n + w_n$, where v_n satisfies

$$v_{n+1} = av_n + bv_{n-1} \qquad (14.66)$$

and w_n is *any* particular solution of (14.65); the proof follows the same lines as that given earlier.

We have already seen for a first-order recurrence relation that the solution to the homogeneous equation is given by terms forming a geometric series, and we consider a corresponding series of powers in the present case. Setting $v_n = A\lambda^n$ in (14.66) for some λ, as yet undetermined, gives the requirement that λ should satisfy

$$A\lambda^{n+1} = aA\lambda^n + bA\lambda^{n-1}.$$

Dividing through by $A\lambda^{n-1}$ (assumed non-zero) shows that λ could be either of the roots, λ_1 and λ_2, of

$$\lambda^2 - a\lambda - b = 0, \qquad (14.67)$$

which is known as the *characteristic equation* of the recurrence relation.

That there are two possible series of terms of the form $A\lambda^n$ is consistent with the fact that two initial values (boundary conditions) have to be provided before the series

14.6 Linear recurrence relations

can be calculated by repeated use of (14.65). These two values are sufficient to determine the appropriate coefficient A for each of the series. Since (14.66) is both linear and homogeneous, and is satisfied by both $v_n = A\lambda_1^n$ and $v_n = B\lambda_2^n$, its general solution is

$$v_n = A\lambda_1^n + B\lambda_2^n,$$

for arbitrary values of A and B.[22]

If the coefficients a and b are such that (14.67) has two equal roots, i.e. $a^2 = -4b$, then, as in the analogous case of repeated roots for differential equations [see Section 14.5.1(iii)], the second term of the general solution is replaced by $Bn\lambda_1^n$ to give

$$v_n = (A + Bn)\lambda_1^n.$$

A further possibility is that the roots of the characteristic equation are complex, in which case the general solution of the homogeneous equation takes the form

$$v_n = A\mu^n e^{in\theta} + B\mu^n e^{-in\theta} = \mu^n (C \cos n\theta + D \sin n\theta).$$

Finding a particular solution is straightforward if k is a constant: a trivial but adequate solution is $w_n = k(1 - a - b)^{-1}$ for all n. As with first-order equations, particular solutions can be found for other simple forms of k by trying functions similar to k itself. Thus, particular solutions for the cases $k = Cn$ and $k = D\alpha^n$ can be found by trying $w_n = E + Fn$ and $w_n = G\alpha^n$ respectively.

Example Find the value of u_{16} if the series u_n satisfies

$$u_{n+1} + 4u_n + 3u_{n-1} = n$$

for $n \geq 1$, with $u_0 = 1$ and $u_1 = -1$.

We first solve the characteristic equation,

$$\lambda^2 + 4\lambda + 3 = 0,$$

to obtain the roots $\lambda = -1$ and $\lambda = -3$. Thus the complementary function is

$$v_n = A(-1)^n + B(-3)^n.$$

In view of the form of the RHS of the original relation, we try

$$w_n = E + Fn$$

as a particular solution and obtain

$$E + F(n+1) + 4(E + Fn) + 3[E + F(n-1)] = n,$$

yielding $F = 1/8$ and $E = 1/32$.

---

[22] Of which second-order recurrence relation and initial values would $u_n = 3(-2)^n - 2(-3)^n$ be the unique solution? Evaluate u_4, (a) directly using your recurrence relation and (b) by using the given solution.

Ordinary differential equations

Thus the complete general solution is

$$u_n = A(-1)^n + B(-3)^n + \frac{n}{8} + \frac{1}{32},$$

and now using the given values for u_0 and u_1 determines A as $7/8$ and B as $3/32$. Thus

$$u_n = \frac{1}{32}\left[28(-1)^n + 3(-3)^n + 4n + 1\right].$$

Finally, substituting $n = 16$ gives $u_{16} = 4\,035\,633$, a value the reader may (or may not) wish to verify by repeated application of the initial recurrence relation. ◀

14.6.3 Higher order recurrence relations

It will be apparent that linear recurrence relations of order $N > 2$ do not present any additional difficulty in principle, though two obvious practical difficulties are (i) that the characteristic equation is of order N and in general will not have roots that can be written in closed form and (ii) that a correspondingly large number of given values is required to determine the N otherwise arbitrary constants in the solution. The algebraic labour needed to solve the set of simultaneous linear equations that determines them increases rapidly with N. We do not give specific examples here, but some are included in the later problems at the end of the chapter.

EXERCISES 14.6

1. Find the general term of the series with first term u_0 that satisfies the recurrence relation

$$u_{n+1} - u_n = \alpha^n + 1.$$

2. Find an expression for the general term u_n of the series generated by the first-order recurrence relation

$$u_{n+1} = 3u_n + n$$

with $u_0 = 1$. Check the first four directly computed terms against your expression.

3. Evaluate u_{10} for the series that has $u_0 = 0$ and $u_1 = 1$ and satisfies

$$u_{n+1} - 3u_n + 2u_{n-1} = 0.$$

4. A series generated by a recurrence relation of the form

$$u_{n+1} + au_n + bu_{n-1} = cn, \qquad n \geq 1,$$

has as its first five terms

$$u_0 = 2, \quad u_1 = -7, \quad u_2 = 16, \quad u_3 = -17, \quad u_4 = 30.$$

Find an expression for a general term of the series.

SUMMARY

1. *First-order equation types and their solution methods*

$$\text{General form} \quad p = \frac{dy}{dx} = F(x, y) \quad \text{or} \quad A(x, y)\, dx + B(x, y)\, dy = 0.$$

Name	Typical form	Solution method
Separable	$F = f(x)g(y)$	$\int \frac{dy}{g(y)} = \int f(x)\,dx.$
Exact	$h(x, y) = \frac{\partial A}{\partial y} - \frac{\partial B}{\partial x} = 0$	$U(x, y) = \int A(x, y)\,dx + V(y);$ $V(y)$ is such that $\frac{\partial U}{\partial y} = B(x, y).$
Inexact	$h(x, y) = \frac{\partial A}{\partial y} - \frac{\partial B}{\partial x} \neq 0$	If $f = \frac{h}{B} \neq f(y)$, then $\mu(x) = \int f(x)\,dx$ is an IF. If $g = \frac{h}{A} \neq g(x)$, then $\mu(y) = -\int g(y)\,dy$ is an IF.
Linear	$\frac{dy}{dx} + P(x)y = Q(x)$	$\mu(x) = \exp\{\int P(x)\,dx\}$ is an IF.
Homogeneous	$\frac{dy}{dx} = H\left(\frac{y}{x}\right)$	Put $y = vx$ to obtain separated $\int \frac{dv}{H(v) - v} = \ln x + c.$
Bernoulli	$\frac{dy}{dx} + P(x)y = Q(x)y^n$ $(n \neq 0,\ n \neq 1)$	Put $v = y^{1-n}$ and obtain linear $\frac{dv}{dx} + (1-n)P(x)v = (1-n)Q(x).$
Higher degree, soluble for p	$\prod_{i=1}^{n}(p - F_i) = 0$	Solve $p - F_i = 0$ for $G_i(x, y) = 0$. Then $\prod_{i=1}^{n} G_i(x, y) = 0$.
*Higher degree, soluble for x	$x = H(y, p)$	Solve $\frac{1}{p} = \frac{\partial H}{\partial y} + \frac{\partial H}{\partial p}\frac{dp}{dy}$ for $G(y, p) = 0$. Eliminate p between this and $x = H(y, p)$.
*Higher degree, soluble for y	$y = H(x, p)$	Solve $p = \frac{\partial H}{\partial x} + \frac{\partial H}{\partial p}\frac{dp}{dx}$ for $G(x, p) = 0$. Eliminate p between this and $y = H(x, p)$.

- The higher degree equations marked * may give rise to singular solutions.
- One boundary condition is needed for each equation.

2. *General considerations for higher order ODEs*
   - A set of n functions is linearly independent over an interval if the Wronskian

   $$W(y_1, y_2, \ldots, y_n) = \begin{vmatrix} y_1 & y_2 & \cdots & y_n \\ y_1' & y_2' & & \vdots \\ \vdots & & \ddots & \vdots \\ y_1^{(n-1)} & \cdots & \cdots & y_n^{(n-1)} \end{vmatrix}$$

   is not identically zero over that interval; the vanishing of the Wronskian does *not* guarantee linear dependence.
   - An nth-order homogeneous linear ODE has n linearly independent solutions, $y_i(x)$ for $i = 1, 2, \ldots, n$ and the complementary function (CF) is $y_c(x) = \sum_i c_i y_i(x)$.
   - The complete solution to an inhomogeneous linear equation is $y(x) = y_c(x) + y_p(x)$, where $y_p(x)$ is *any* particular solution (however simple) of the ODE.
   - An nth-order equation requires n independent and self-consistent boundary conditions (BC) for a unique solution.
   - *Warning*: the BC must be applied to $y_c(x) + y_p(x)$ as a whole (and *not* to y_c alone, with y_p added later).

3. *Linear equations with constant coefficients*

   $$a_n \frac{d^n y}{dx^n} + a_{n-1} \frac{d^{n-1} y}{dx^{n-1}} + \cdots + a_1 \frac{dy}{dx} + a_0 y = f(x). \quad (*)$$

   - With $f(x)$ set equal to zero, a trial solution of the form $y = e^{\lambda x}$ gives an nth-degree polynomial in λ with
     (i) each real distinct root λ_i giving a solution $e^{\lambda_i x}$,
     (ii) pairs of complex roots $\alpha \pm i\beta$ giving solutions $e^{\alpha x}(d_1 \cos \beta x + d_2 \sin \beta x)$,
     (iii) a k-repeated root λ_i giving k (of the n) solutions as $e^{\lambda_i x}, xe^{\lambda_i x}, \ldots, x^{k-1} e^{\lambda_i x}$.
     The CF is a linear combination of the solutions so found.
   - A particular integral (PI) can be found by trying a multiple of $f(x)$. If $f(x)$ is proportional to a term in the CF, then the PI is $y_m(x) = Ax^m f(x)$, where m is the lowest positive integer such that y_m does not appear in the CF.
   - Taking the Laplace transform of $(*)$ converts it to an algebraic equation for $\bar{y}(s)$, which can often be inverse transformed to $y(x)$, using partial fractions and a table of Laplace transforms.

4. *Linear recurrence relations* with constant coefficients
   The general Nth-order recurrence relation is

   $$u_{n+1} = \sum_{r=0}^{N-1} a_r u_{n-r} + k(n). \quad (**)$$

- If v_n is the general solution of (∗∗) when $k = 0$, and w_n is *any* solution of (∗∗), then the general solution is $u_n = v_n + w_n$.
- Setting $v_n = A\lambda^n$ in (∗∗) with $k = 0$ gives the characteristic equation $\lambda^N = \sum_{r=0}^{N-1} a_r \lambda^{N-1-r}$, an Nth-degree polynomial equation.
- If the N roots of the characteristic equation are λ_i, then $v_n = \sum_{i=1}^{N} A_i \lambda_i^n$.
- If two of the roots are complex conjugates $\alpha \pm i\beta$, then two of the terms in v_n are replaced by $r^n(A \cos n\phi + B \sin n\phi)$, where $\tan \phi = \beta/\alpha$ and $r = \sqrt{\alpha^2 + \beta^2}$.
- If λ_i is a k-fold root, k of the terms in v_n are replaced by $(A_1 + A_2 n + \cdots + A_{k-1} n^{k-1})\lambda_i^n$.
- A particular solution w_n is sought by trying forms similar to $k(n)$.
- The coefficients in $u_n = v_n + w_n$ are determined by the given initial values $u_0, u_1, \ldots, u_N$.

PROBLEMS

14.1. A radioactive isotope decays in such a way that the number of atoms present at a given time, $N(t)$, obeys the equation

$$\frac{dN}{dt} = -\lambda N.$$

If there are initially N_0 atoms present, find $N(t)$ at later times.

14.2. Solve the following equations by separation of the variables:
(a) $y' - xy^3 = 0$;
(b) $y' \tan^{-1} x - y(1 + x^2)^{-1} = 0$;
(c) $x^2 y' + xy^2 = 4y^2$.

14.3. Show that the following equations either are exact or can be made exact, and solve them:
(a) $y(2x^2 y^2 + 1)y' + x(y^4 + 1) = 0$;
(b) $2xy' + 3x + y = 0$;
(c) $(\cos^2 x + y \sin 2x)y' + y^2 = 0$.

14.4. Find the values of α and β that make

$$dF(x, y) = \left(\frac{1}{x^2 + 2} + \frac{\alpha}{y}\right) dx + (xy^\beta + 1) dy$$

an exact differential. For these values solve $F(x, y) = 0$.

14.5. By finding suitable IFs, solve the following equations:
(a) $(1 - x^2)y' + 2xy = (1 - x^2)^{3/2}$;
(b) $y' - y \cot x + \operatorname{cosec} x = 0$;
(c) $(x + y^3)y' = y$ (treat y as the independent variable).

Ordinary differential equations

14.6. By finding an appropriate IF, solve
$$\frac{dy}{dx} = -\frac{2x^2 + y^2 + x}{xy}.$$

14.7. Find, in the form of an integral, the solution of the equation
$$\alpha \frac{dy}{dt} + y = f(t)$$
for a general function $f(t)$. Find the specific solutions for
(a) $f(t) = H(t)$,
(b) $f(t) = \delta(t)$,
(c) $f(t) = \beta^{-1} e^{-t/\beta} H(t)$ with $\beta < \alpha$.
For case (c), what happens if $\beta \to 0$?

14.8. A series electric circuit contains a resistance R, a capacitance C and a battery supplying a time-varying electromotive force $V(t)$. The charge q on the capacitor therefore obeys the equation
$$R \frac{dq}{dt} + \frac{q}{C} = V(t).$$
Assuming that initially there is no charge on the capacitor, and given that $V(t) = V_0 \sin \omega t$, find the charge on the capacitor as a function of time.

14.9. Using tangential–polar coordinates (see Problem 3.22), consider a particle of mass m moving under the influence of a force f directed towards the origin O. By resolving forces along the instantaneous tangent and normal and making use of the result of Problem 3.22 for the instantaneous radius of curvature, prove that
$$f = -mv \frac{dv}{dr} \quad \text{and} \quad mv^2 = fp \frac{dr}{dp}.$$
Show further that $h = mpv$ is a constant of the motion and that the law of force can be deduced from
$$f = \frac{h^2}{mp^3} \frac{dp}{dr}.$$

14.10. Use the result of Problem 14.9 to find the law of force, acting towards the origin, under which a particle must move so as to describe the following trajectories:
(a) A circle of radius a that passes through the origin;
(b) An equiangular spiral, which is defined by the property that the angle α between the tangent and the radius vector is constant along the curve.

14.11. Solve
$$(y - x) \frac{dy}{dx} + 2x + 3y = 0.$$

Problems

14.12. By changing the dependent variable to $v = x + 2y$, solve
$$\frac{dy}{dx} = \frac{1}{x + 2y + 1}.$$

14.13. Using a technique similar to that employed in the previous problem, solve
$$\frac{dy}{dx} = -\frac{x + y}{3x + 3y - 4}.$$

14.14. If $u = 1 + \tan y$, calculate $d(\ln u)/dy$; hence find the general solution of
$$\frac{dy}{dx} = \tan x \cos y \, (\cos y + \sin y).$$

14.15. Find the curve with the property that at each point on it the sum of the intercepts on the x- and y-axes of the tangent to the curve (taking account of sign) is equal to unity.

14.16. The action of the control mechanism on a particular system for an input $f(t)$ is described, for $t \geq 0$, by the coupled first-order equations:
$$\dot{y} + 4z = f(t),$$
$$\dot{z} - 2z = \dot{y} + \tfrac{1}{2}y.$$

Use Laplace transforms to find the response $y(t)$ of the system to a unit step input, $f(t) = H(t)$, given that $y(0) = 1$ and $z(0) = 0$.

Questions 14.17 to 14.25 are intended to give the reader practice in choosing an appropriate method. The level of difficulty varies within the set; if necessary, the hints may be consulted for an indication of the most appropriate approach.

14.17. Find the general solutions of the following:
(a) $\dfrac{dy}{dx} + \dfrac{xy}{a^2 + x^2} = x$;   (b) $\dfrac{dy}{dx} = \dfrac{4y^2}{x^2} - y^2$.

14.18. Solve the following first-order equations for the boundary conditions given:
(a) $y' - (y/x) = 1$, $y(1) = -1$;
(b) $y' - y \tan x = 1$, $y(\pi/4) = 3$;
(c) $y' - y^2/x^2 = 1/4$, $y(1) = 1$;
(d) $y' - y^2/x^2 = 1/4$, $y(1) = 1/2$.

14.19. An electronic system has two inputs, to each of which a constant unit signal is applied, but starting at different times. The equations governing the system thus take the form
$$\dot{x} + 2y = H(t),$$
$$\dot{y} - 2x = H(t - 3).$$

Initially (at $t = 0$), $x = 1$ and $y = 0$; find $x(t)$ at later times.

Ordinary differential equations

14.20. Solve the differential equation

$$\sin x \frac{dy}{dx} + 2y \cos x = 1,$$

subject to the boundary condition $y(\pi/2) = 1$.

14.21. A reflecting mirror is made in the shape of the surface of revolution generated by revolving the curve $y(x)$ about the x-axis. In order that light rays emitted from a point source at the origin are reflected back parallel to the x-axis, the curve $y(x)$ must obey

$$\frac{y}{x} = \frac{2p}{1 - p^2},$$

where $p = dy/dx$. By solving this equation for x, find the curve $y(x)$.

14.22. Find a parametric solution of

$$x \left(\frac{dy}{dx}\right)^2 + \frac{dy}{dx} - y = 0$$

as follows.
(a) Write an equation for y in terms of $p = dy/dx$ and show that

$$p = p^2 + (2px + 1)\frac{dp}{dx}.$$

(b) Using p as the independent variable, arrange this as a linear first-order equation for x.
(c) Find an appropriate integrating factor to obtain

$$x = \frac{\ln p - p + c}{(1 - p)^2},$$

which, together with the expression for y obtained in (a), gives a parameterisation of the solution.
(d) Reverse the roles of x and y in steps (a) to (c), putting $dx/dy = p^{-1}$, and show that essentially the same parameterisation is obtained.

14.23. Find the solution $y = y(x)$ of

$$x\frac{dy}{dx} + y - \frac{y^2}{x^{3/2}} = 0,$$

subject to $y(1) = 1$.

14.24. Find the solution of

$$(2 \sin y - x)\frac{dy}{dx} = \tan y,$$

if (a) $y(0) = 0$, and (b) $y(0) = \pi/2$.

Problems

14.25. Find the family of solutions of

$$\frac{d^2y}{dx^2} + \left(\frac{dy}{dx}\right)^2 + \frac{dy}{dx} = 0$$

that satisfy $y(0) = 0$.

14.26. Solve the equation

$$\frac{d^2y}{dx^2} + 7\frac{dy}{dx} + 12y = 6,$$

subject to the boundary conditions $y(0) = 0$ and $y(1/3) = \frac{1}{2}(1 - e^{-1})$. What is the value of $y(1)$?

14.27. A simple harmonic oscillator, of mass m and natural frequency ω_0, experiences an oscillating driving force $f(t) = ma \cos \omega t$. Therefore, its equation of motion is

$$\frac{d^2x}{dt^2} + \omega_0^2 x = a \cos \omega t,$$

where x is its position. Given that at $t = 0$ we have $x = dx/dt = 0$, find the function $x(t)$. Describe the solution if ω is approximately, but not exactly, equal to ω_0.

14.28. Find the roots of the auxiliary equation for the following. Hence solve them for the boundary conditions stated.

(a) $\dfrac{d^2f}{dt^2} + 2\dfrac{df}{dt} + 5f = 0,$ with $f(0) = 1$, $f'(0) = 0$.

(b) $\dfrac{d^2f}{dt^2} + 2\dfrac{df}{dt} + 5f = e^{-t} \cos 3t,$ with $f(0) = 0$, $f'(0) = 0$.

14.29. The theory of bent beams shows that at any point in the beam the 'bending moment' is given by K/ρ, where K is a constant (that depends upon the beam material and cross-sectional shape) and ρ is the radius of curvature at that point. Consider a light beam of length L whose ends, $x = 0$ and $x = L$, are supported at the same vertical height and which has a weight W suspended from its centre. Verify that at any point x ($0 \le x \le L/2$ for definiteness) the net magnitude of the bending moment (bending moment = force × perpendicular distance) due to the weight and support reactions, evaluated on either side of x, is $Wx/2$.

If the beam is only slightly bent, so that $(dy/dx)^2 \ll 1$, where $y = y(x)$ is the downward displacement of the beam at x, show that the beam profile satisfies the approximate equation

$$\frac{d^2y}{dx^2} = -\frac{Wx}{2K}.$$

By integrating this equation twice and using physically imposed conditions on your solution at $x = 0$ and $x = L/2$, show that the downward displacement at the centre of the beam is $WL^3/(48K)$.

14.30. Solve the differential equation

$$\frac{d^2 f}{dt^2} + 6\frac{df}{dt} + 9f = e^{-t},$$

subject to the conditions $f = 0$ and $df/dt = \lambda$ at $t = 0$.

Find the equation satisfied by the positions of the turning points of $f(t)$ and hence, by drawing suitable sketch graphs, determine the number of turning points the solution has in the range $t > 0$ if (a) $\lambda = 1/4$, and (b) $\lambda = -1/4$.

14.31. The function $f(t)$ satisfies the differential equation

$$\frac{d^2 f}{dt^2} + 8\frac{df}{dt} + 12f = 12e^{-4t}.$$

For the following sets of boundary conditions determine whether it has solutions, and, if so, find them:
(a) $f(0) = 0$, $f'(0) = 0$, $f(\ln \sqrt{2}) = 0$;
(b) $f(0) = 0$, $f'(0) = -2$, $f(\ln \sqrt{2}) = 0$.

14.32. Determine the values of α and β for which the following four functions are linearly dependent:

$$y_1(x) = x \cosh x + \sinh x,$$
$$y_2(x) = x \sinh x + \cosh x,$$
$$y_3(x) = (x + \alpha)e^x,$$
$$y_4(x) = (x + \beta)e^{-x}.$$

You will find it convenient to work with those linear combinations of the $y_i(x)$ that can be written the most compactly.

14.33. A solution of the differential equation

$$\frac{d^2 y}{dx^2} + 2\frac{dy}{dx} + y = 4e^{-x}$$

takes the value 1 when $x = 0$ and the value e^{-1} when $x = 1$. What is its value when $x = 2$?

14.34. The two functions $x(t)$ and $y(t)$ satisfy the simultaneous equations

$$\frac{dx}{dt} - 2y = -\sin t,$$
$$\frac{dy}{dt} + 2x = 5\cos t.$$

Problems

Find explicit expressions for $x(t)$ and $y(t)$, given that $x(0) = 3$ and $y(0) = 2$. Sketch the solution trajectory in the xy-plane for $0 \leq t < 2\pi$, showing that the trajectory crosses itself at $(0, 1/2)$ and passes through the points $(0, -3)$ and $(0, -1)$ in the negative x-direction.

14.35. Find the general solutions of

(a) $\dfrac{d^3 y}{dx^3} - 12\dfrac{dy}{dx} + 16y = 32x - 8,$

(b) $\dfrac{d}{dx}\left(\dfrac{1}{y}\dfrac{dy}{dx}\right) + (2a \coth 2ax)\left(\dfrac{1}{y}\dfrac{dy}{dx}\right) = 2a^2,$ where a is a constant.

14.36. Use the method of Laplace transforms to solve

(a) $\dfrac{d^2 f}{dt^2} + 5\dfrac{df}{dt} + 6f = 0, \qquad f(0) = 1, \; f'(0) = -4,$

(b) $\dfrac{d^2 f}{dt^2} + 2\dfrac{df}{dt} + 5f = 0, \qquad f(0) = 1, \; f'(0) = 0.$

14.37. The quantities $x(t)$, $y(t)$ satisfy the simultaneous equations

$$\ddot{x} + 2n\dot{x} + n^2 x = 0,$$
$$\ddot{y} + 2n\dot{y} + n^2 y = \mu\dot{x},$$

where $x(0) = y(0) = \dot{y}(0) = 0$ and $\dot{x}(0) = \lambda$. Show that

$$y(t) = \tfrac{1}{2}\mu\lambda t^2 \left(1 - \tfrac{1}{3}nt\right)\exp(-nt).$$

14.38. Use Laplace transforms to solve, for $t \geq 0$, the differential equations

$$\ddot{x} + 2x + y = \cos t,$$
$$\ddot{y} + 2x + 3y = 2\cos t,$$

which describe a coupled system that starts from rest at the equilibrium position. Show that the subsequent motion takes place along a straight line in the xy-plane and determine the time dependences of x and y.

14.39. Two unstable isotopes A and B and a stable isotope C have the following decay rates per atom present: $A \to B$, 3 s^{-1}; $A \to C$, 1 s^{-1}; $B \to C$, 2 s^{-1}. Initially a quantity x_0 of A is present, but there are no atoms of the other two types. Using Laplace transforms, find the amount of C present at a later time t.

14.40. For a lightly damped ($\gamma < \omega_0$) harmonic oscillator driven at its undamped resonance frequency ω_0, the displacement $x(t)$ at time t satisfies the equation

$$\dfrac{d^2 x}{dt^2} + 2\gamma\dfrac{dx}{dt} + \omega_0^2 x = F \sin \omega_0 t.$$

Use Laplace transforms to find the displacement at a general time if the oscillator starts from rest at its equilibrium position.

(a) Show that ultimately the oscillation has amplitude $F/(2\omega_0\gamma)$, with a phase lag of $\pi/2$ relative to the driving force per unit mass F.

(b) By differentiating the original equation, conclude that if $x(t)$ is expanded as a power series in t for small t, then the first non-vanishing term is $F\omega_0 t^3/6$. Confirm this conclusion by expanding your explicit solution.

14.41. The 'golden mean', which is said to describe the most aesthetically pleasing proportions for the sides of a rectangle (e.g. the ideal picture frame), is given by the limiting value of the ratio of successive terms of the Fibonacci series u_n, which is generated by

$$u_{n+2} = u_{n+1} + u_n,$$

with $u_0 = 0$ and $u_1 = 1$. Find an expression for the general term of the series and verify that the golden mean is equal to the larger root of the recurrence relation's characteristic equation.

14.42. In a particular scheme for numerically modelling one-dimensional fluid flow, the successive values, u_n, of the solution are connected for $n \geq 1$ by the difference equation

$$c(u_{n+1} - u_{n-1}) = d(u_{n+1} - 2u_n + u_{n-1}),$$

where c and d are positive constants. The boundary conditions are $u_0 = 0$ and $u_M = 1$. Find the solution to the equation, and show that successive values of u_n will have alternating signs if $c > d$.

14.43. The first few terms of a series u_n, starting with u_0, are 1, 2, 2, 1, 6, -3. The series is generated by a recurrence relation of the form

$$u_n = Pu_{n-2} + Qu_{n-4},$$

where P and Q are constants. Find an expression for the general term of the series and show that, in fact, the series consists of two interleaved series given by

$$u_{2m} = \tfrac{2}{3} + \tfrac{1}{3}4^m,$$
$$u_{2m+1} = \tfrac{7}{3} - \tfrac{1}{3}4^m,$$

for $m = 0, 1, 2, \ldots$

14.44. Find an explicit expression for the u_n satisfying

$$u_{n+1} + 5u_n + 6u_{n-1} = 2^n,$$

given that $u_0 = u_1 = 1$. Deduce that $2^n - 26(-3)^n$ is divisible by 5 for all non-negative integers n.

14.45. Find the general expression for the u_n satisfying

$$u_{n+1} = 2u_{n-2} - u_n$$

with $u_0 = u_1 = 0$ and $u_2 = 1$, and show that they can be written in the form

$$u_n = \frac{1}{5} - \frac{2^{n/2}}{\sqrt{5}} \cos\left(\frac{3\pi n}{4} - \phi\right),$$

where $\tan \phi = 2$.

14.46. Consider the seventh-order recurrence relation

$$u_{n+7} - u_{n+6} - u_{n+5} + u_{n+4} - u_{n+3} + u_{n+2} + u_{n+1} - u_n = 0.$$

Find the most general form of its solution, and show that:
(a) if only the four initial values $u_0 = 0$, $u_1 = 2$, $u_2 = 6$ and $u_3 = 12$, are specified, then the relation has one solution that cycles repeatedly through this set of four numbers;
(b) but if, in addition, it is required that $u_4 = 20$, $u_5 = 30$ and $u_6 = 42$ then the solution is unique, with $u_n = n(n+1)$.

HINTS AND ANSWERS

14.1. $N(t) = N_0 \exp(-\lambda t)$.

14.3. (a) exact, $x^2 y^4 + x^2 + y^2 = c$; (b) IF $= x^{-1/2}$, $x^{1/2}(x+y) = c$; (c) IF $= \sec^2 x$, $y^2 \tan x + y = c$.

14.5. (a) IF $= (1-x^2)^{-2}$, $y = (1-x^2)(k + \sin^{-1} x)$; (b) IF $= \operatorname{cosec} x$, leading to $y = k \sin x + \cos x$; (c) exact equation is $y^{-1}(dx/dy) - xy^{-2} = y$, leading to $x = y(k + y^2/2)$.

14.7. $y(t) = e^{-t/\alpha} \int^t \alpha^{-1} e^{t'/\alpha} f(t') dt'$; (a) $y(t) = 1 - e^{-t/\alpha}$; (b) $y(t) = \alpha^{-1} e^{-t/\alpha}$; (c) $y(t) = (e^{-t/\alpha} - e^{-t/\beta})/(\alpha - \beta)$. It becomes case (b).

14.9. Note that if the angle between the tangent and the radius vector is α, then $\cos \alpha = dr/ds$ and $\sin \alpha = p/r$.

14.11. Homogeneous equation, put $y = vx$ to obtain $(1-v)(v^2 + 2v + 2)^{-1} dv = x^{-1} dx$; write $1 - v$ as $2 - (1+v)$, and $v^2 + 2v + 2$ as $1 + (1+v)^2$; $A[x^2 + (x+y)^2] = \exp\{4 \tan^{-1}[(x+y)/x]\}$.

14.13. Set $v = x+y$; $x + 3y + 2 \ln(x+y-2) = A$.

14.15. The curve must satisfy $y = (1 - p^{-1})^{-1}(1 - x + px)$, which has solution $x = (p-1)^{-2}$, leading to $y = (1 \pm \sqrt{x})^2$ or $x = (1 \pm \sqrt{y})^2$; the singular solution $p' = 0$ gives straight lines joining $(\theta, 0)$ and $(0, 1-\theta)$ for any θ.

Ordinary differential equations

14.17. (a) Integrating factor is $(a^2 + x^2)^{1/2}$, $y = (a^2 + x^2)/3 + A(a^2 + x^2)^{-1/2}$;
(b) separable, $y = x(x^2 + Ax + 4)^{-1}$.

14.19. Use Laplace transforms; $\bar{x}s(s^2 + 4) = s + s^2 - 2e^{-3s}$; $x(t) = \frac{1}{2}\sin 2t + \cos 2t - \frac{1}{2}H(t-3) + \frac{1}{2}\cos(2t - 6)H(t - 3)$.

14.21. Eliminate y to obtain $p(p^2 - 1) = 2xp'$, then separate variables. Obtain $p = \pm(1 - Bx)^{-1/2}$. Substitute for p, giving $y^2 = 4A^2 - 4Ax$, a parabola with apex at $x = A$.

14.23. Bernoulli's equation with $n = 2$; $y(x) = 5x^{3/2}/(2 + 3x^{5/2})$.

14.25. Show that $p = (Ce^x - 1)^{-1}$, where $p = dy/dx$; $y = \ln[C - e^{-x})/(C - 1)]$ or $\ln[D - (D - 1)e^{-x}]$ or $\ln(e^{-K} + 1 - e^{-x}) + K$.

14.27. The function is $a(\omega_0^2 - \omega^2)^{-1}(\cos \omega t - \cos \omega_0 t)$; for moderate t, $x(t)$ is a sine wave of linearly increasing amplitude $(t \sin \omega_0 t)/(2\omega_0)$; for large t it shows beats of maximum amplitude $2(\omega_0^2 - \omega^2)^{-1}$.

14.29. Ignore the term y'^2, compared with 1, in the expression for ρ. $y = 0$ at $x = 0$. From symmetry, $dy/dx = 0$ at $x = L/2$.

14.31. General solution $f(t) = Ae^{-6t} + Be^{-2t} - 3e^{-4t}$. (a) No solution, inconsistent boundary conditions; (b) $f(t) = 2e^{-6t} + e^{-2t} - 3e^{-4t}$.

14.33. The auxiliary equation has repeated roots and the RHS is contained in the complementary function. The solution is $y(x) = (A + Bx)e^{-x} + 2x^2 e^{-x}$. $y(2) = 5e^{-2}$.

14.35. (a) The auxiliary equation has roots 2, 2, -4; $(A + Bx)\exp 2x + C\exp(-4x) + 2x + 1$; (b) multiply through by $\sinh 2ax$ and note that $\int \text{cosech } 2ax \, dx = (2a)^{-1}\ln(|\tanh ax|)$; $y = B(\sinh 2ax)^{1/2}(|\tanh ax|)^A$.

14.37. Use Laplace transforms; write $s(s+n)^{-4}$ as $(s+n)^{-3} - n(s+n)^{-4}$.

14.39. $\mathcal{L}[C(t)] = x_0(s+8)/[s(s+2)(s+4)]$, yielding $C(t) = x_0[1 + \frac{1}{2}\exp(-4t) - \frac{3}{2}\exp(-2t)]$.

14.41. The characteristic equation is $\lambda^2 - \lambda - 1 = 0$. $u_n = [(1 + \sqrt{5})^n - (1 - \sqrt{5})^n]/(2^n \sqrt{5})$.

14.43. From u_4 and u_5, $P = 5$, $Q = -4$. $u_n = 3/2 - 5(-1)^n/6 + (-2)^n/4 + 2^n/12$.

14.45. The general solution is $A + B2^{n/2}\exp(i3\pi n/4) + C2^{n/2}\exp(i5\pi n/4)$. The initial values imply that $A = 1/5$, $B = (\sqrt{5}/10)\exp[i(\pi - \phi)]$ and $C = (\sqrt{5}/10)\exp[i(\pi + \phi)]$.

15

Elementary probability

All scientists will know the importance of experiment and observation and, equally, be aware that the results of some experiments depend to a degree on chance. For example, in an experiment to measure the heights of a random sample of people, we would not be in the least surprised if all the heights were found to be different; but, if the experiment were repeated often enough, we would expect to find some sort of regularity in the results. Statistical methods are concerned with the analysis of real experimental data of this sort.

In this final chapter we discuss the subject of probability, which is the theoretical basis for most statistical methods. Our development of probability will be with an eye to its eventual applications in statistics, with little emphasis on the axioms and theorems approach favoured by most pure mathematicians.

We first discuss the terminology required, with particular reference to the convenient graphical representation of experimental results as Venn diagrams. The concepts of random variables and distributions of random variables are then introduced. It is here that the connection with statistics is made; we assert that the results of many experiments are random variables and that those results have some sort of regularity, represented by a distribution. Finally, the defining equations for some important distributions, together with some useful quantities that characterise them, are introduced and discussed.

15.1 Venn diagrams

We call a single performance of an experiment a *trial* and each possible result an *outcome*. The *sample space* S of the experiment is then the set of all possible outcomes of an individual trial. For example, if we throw a six-sided die then there are six possible outcomes that together form the sample space of the experiment. At this stage we are not concerned with how likely a particular outcome might be (we will return to the probability of an outcome in due course) but rather will concentrate on the classification of possible outcomes. It is clear that some sample spaces are finite (e.g. the outcomes of throwing a die) whilst others are infinite (e.g. the outcomes of measuring people's heights). Most often, one is not interested in individual outcomes but in whether an outcome belongs to a given subset A (say) of the sample space S; these subsets are called *events*. For example, we might be interested in whether a person is taller or shorter than 180 cm, in which

Elementary probability

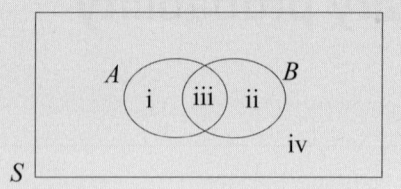

Figure 15.1 A Venn diagram.

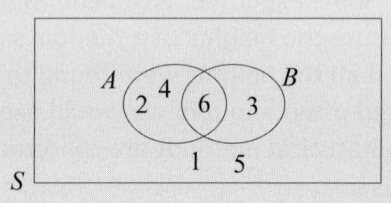

Figure 15.2 The Venn diagram for the outcomes of the die-throwing trials described in the worked example.

case we divide the sample space into just two events: namely, that the outcome (height measured) is (i) greater than 180 cm or (ii) less than 180 cm.

A common graphical representation of the outcomes of an experiment is the *Venn diagram*. A Venn diagram usually consists of a rectangle, the interior of which represents the sample space, together with one or more closed curves inside it. The interior of each closed curve then represents an event. Figure 15.1 shows a typical Venn diagram representing a sample space S and two events A and B. Every possible outcome is assigned to an appropriate region; in this example there are four regions to consider (marked i to iv in Figure 15.1):

(i) outcomes that belong to event A but not to event B;
(ii) outcomes that belong to event B but not to event A;
(iii) outcomes that belong to both event A and event B;
(iv) outcomes that belong to neither event A nor event B.

As a concrete example, consider the following.

Example A six-sided die is thrown. Let event A be 'the number obtained is divisible by 2' and event B be 'the number obtained is divisible by 3'. Draw a Venn diagram to represent these events.

It is clear that the outcomes 2, 4, 6 belong to event A and that the outcomes 3, 6 belong to event B. Of these, 6 belongs to both A and B. The remaining outcomes, 1, 5, belong to neither A nor B. These observations are recorded schematically in the Venn diagram shown in Figure 15.2. ◂

In the above example, one outcome, 6, is divisible by both 2 and 3 and so belongs to both A and B. This outcome is placed in region iii of Figure 15.1, which is called the *intersection* of A and B and is denoted by $A \cap B$ [see Figure 15.3(a)]. If no events lie in the region of intersection then A and B are said to be *mutually exclusive* or *disjoint*.

15.1 Venn diagrams

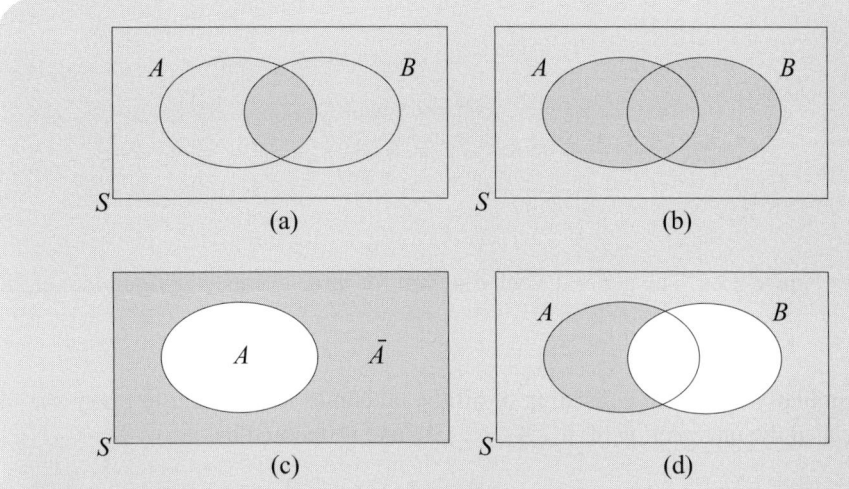

Figure 15.3 Venn diagrams: the shaded regions show (a) $A \cap B$, the intersection of two events A and B, (b) $A \cup B$, the union of events A and B, (c) the complement $\bar{A}$ of an event A, (d) $A - B$, those outcomes in A that do not belong to B.

In this case, the Venn diagram is often (re-)drawn so that the closed curves representing the events A and B do not overlap, so as to make graphically explicit the fact that A and B are disjoint. It is not necessary, however, to draw the diagram in this way, since we may simply assign zero outcomes to the shaded region in Figure 15.3(a). An event that contains no outcomes is called the *empty event* and denoted by $\emptyset$.

The event comprising all the elements that belong to either A or B, or to both, is called the *union* of A and B and is denoted by $A \cup B$ [see Figure 15.3(b)]. In the previous example, $A \cup B = \{2, 3, 4, 6\}$. It is sometimes convenient to talk about those outcomes that do *not* belong to a particular event. The set of outcomes that do not belong to A is called the *complement* of A and is denoted by $\bar{A}$ [see Figure 15.3(c)]; this can also be written as $\bar{A} = S - A$. It is clear that $A \cup \bar{A} = S$ and $A \cap \bar{A} = \emptyset$.

The above notation can be extended in an obvious way, so that $A - B$ denotes the outcomes in A that do not belong to B. It is clear from Figure 15.3(d) that $A - B$ can also be written as $A \cap \bar{B}$. Finally, when *all* the outcomes in event B (say) also belong to event A, but A may contain, in addition, outcomes that do not belong to B, then B is called a *subset* of A, a situation that is denoted by $B \subset A$; alternatively, one may write $A \supset B$, which states that A contains B. In this case, the closed curve representing the event B is often drawn lying completely within the closed curve representing the event A.[1]

The operations $\cup$ and $\cap$ are extended straightforwardly to more than two events. If there exist n events $A_1, A_2, \ldots, A_n$, in some sample space S, then the event consisting of all those outcomes that belong to *one or more* of the A_i is the *union* of $A_1, A_2, \ldots, A_n$ and is denoted by

$$A_1 \cup A_2 \cup \cdots \cup A_n. \tag{15.1}$$

---

[1] What would be the meaning of $A - B$ if $B \supset A$?

Elementary probability

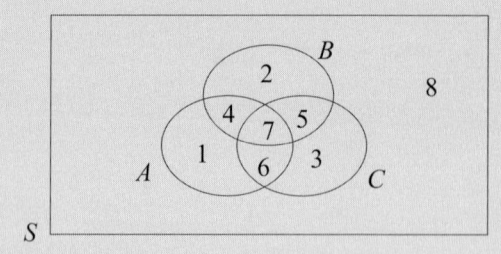

Figure 15.4 The general Venn diagram for three events is divided into eight regions.

Similarly, the event consisting of all the outcomes that belong to *every one* of the A_i is called the *intersection* of $A_1, A_2, \ldots, A_n$ and is denoted by

$$A_1 \cap A_2 \cap \cdots \cap A_n. \tag{15.2}$$

If, for *any* pair of values i, j with $i \neq j$,

$$A_i \cap A_j = \emptyset \tag{15.3}$$

then the events A_i and A_j are said to be *mutually exclusive* or *disjoint*.

Consider three events A, B and C with a Venn diagram such as is shown in Figure 15.4. It will be clear that, in general, the diagram will be divided into eight regions and they will be of four different types. Three regions correspond to a single event; three regions are each the intersection of exactly two events; one region is the threefold intersection of all three events; and finally one region corresponds to none of the events. Let us now consider the numbers of different regions in a general n-event Venn diagram.

For one-event Venn diagrams there are two regions, for the two-event case there are four regions and, as we have just seen, for the three-event case there are eight. In the general n-event case there are 2^n regions, as is clear from the fact that any particular region R lies either inside or outside the closed curve of any particular event. With two choices (inside or outside) for each of n closed curves, there are 2^n different possible combinations with which to characterise R. Once n gets beyond three it becomes impossible to draw a simple two-dimensional Venn diagram, but this does not change the results.

The 2^n regions will break down into $n + 1$ types, with the numbers of each type as follows:[2]

$$\begin{aligned}
\text{no events,} \quad & {}^nC_0 = 1; \\
\text{one event but no intersections,} \quad & {}^nC_1 = n; \\
\text{twofold intersections,} \quad & {}^nC_2 = \tfrac{1}{2}n(n-1); \\
\text{threefold intersections,} \quad & {}^nC_3 = \tfrac{1}{3!}n(n-1)(n-2); \\
& \vdots \\
\text{an } n\text{-fold intersection,} \quad & {}^nC_n = 1.
\end{aligned}$$

---

[2] The symbols ${}^nC_i$, for $i = 0, 1, 2, \ldots, n$, are a convenient notation for combinations; their general definitions are given in the Summary at the end of the chapter.

15.1 Venn diagrams

That this makes a total of 2^n can be checked by considering the binomial expansion
$$2^n = (1+1)^n = 1 + n + \tfrac{1}{2}n(n-1) + \cdots + 1.$$

Using Venn diagrams, it is straightforward to show that the operations $\cap$ and $\cup$ obey the following algebraic laws:

commutativity, $\quad A \cap B = B \cap A, \quad A \cup B = B \cup A;$
associativity, $\quad (A \cap B) \cap C = A \cap (B \cap C), \quad (A \cup B) \cup C = A \cup (B \cup C);$
distributivity, $\quad A \cap (B \cup C) = (A \cap B) \cup (A \cap C),$
$\quad\quad\quad\quad\quad A \cup (B \cap C) = (A \cup B) \cap (A \cup C);$
idempotency, $\quad A \cap A = A, \quad A \cup A = A.$

The following illustrates the operation of this algebra.

Example Show that (i) $A \cup (A \cap B) = A \cap (A \cup B) = A$, (ii) $(A - B) \cup (A \cap B) = A$.

(i) Using the distributivity and idempotency laws above, we see that
$$A \cup (A \cap B) = (A \cup A) \cap (A \cup B) = A \cap (A \cup B).$$
By sketching a Venn diagram[3] it is immediately clear that both expressions are equal to A. Nevertheless, we here proceed in a more formal manner in order to deduce this result algebraically. Let us begin by writing
$$X = A \cup (A \cap B) = A \cap (A \cup B), \quad\quad (15.4)$$
from which we want to deduce a simpler expression for the event X. Using the first equality in (15.4) and the algebraic laws for $\cap$ and $\cup$, we may write
$$A \cap X = A \cap [A \cup (A \cap B)]$$
$$= (A \cap A) \cup [A \cap (A \cap B)]$$
$$= A \cup (A \cap B) = X.$$
Since $A \cap X = X$ we must have $X \subset A$. Now, using the second equality in (15.4) in a similar way, we find
$$A \cup X = A \cup [A \cap (A \cup B)]$$
$$= (A \cup A) \cap [A \cup (A \cup B)]$$
$$= A \cap (A \cup B) = X,$$
from which we deduce that $A \subset X$. Thus, since $X \subset A$ and $A \subset X$, we must conclude that $X = A$.
(ii) Since we do not know how to deal with compound expressions containing a minus sign, we begin by writing $A - B = A \cap \bar{B}$ as mentioned above. Then, using the distributivity law, we obtain
$$(A - B) \cup (A \cap B) = (A \cap \bar{B}) \cup (A \cap B)$$
$$= A \cap (\bar{B} \cup B)$$
$$= A \cap S = A.$$
As noted earlier, both of these results can be proved trivially by drawing appropriate Venn diagrams. ◂

---

[3] Make this sketch and confirm both of the stated results.

Elementary probability

Further useful results may be derived from Venn diagrams. In particular, it is simple to show that the following rules hold:

(i) if $A \subset B$ then $\bar{A} \supset \bar{B}$;
(ii) $\overline{A \cup B} = \bar{A} \cap \bar{B}$;
(iii) $\overline{A \cap B} = \bar{A} \cup \bar{B}$.

Statements (ii) and (iii) are known jointly as *de Morgan's laws* and are sometimes useful in simplifying logical expressions, as in the following worked example.

Example There exist two events A and B such that

$$\overline{(X \cup A)} \cup \overline{(X \cup \bar{A})} = B.$$

Find an expression for the event X in terms of A and B.

We begin by taking the complement of both sides of the above expression: applying de Morgan's laws we obtain

$$\bar{B} = (X \cup A) \cap (X \cup \bar{A}).$$

We may then use the algebraic laws obeyed by $\cap$ and $\cup$ to yield

$$\bar{B} = X \cup (A \cap \bar{A}) = X \cup \emptyset = X.$$

Thus, we find that $X = \bar{B}$. ◀

EXERCISES 15.1

1. Assign the letters of the alphabet to a Venn diagram in which event A is that the letter is a vowel and event B is that the upper case form of the letter contains at least one curved line.

2. Simplify the following expressions:
    (a) $[B \cap \overline{(B \cup C)}] \cup [(B \cup C) \cup \bar{C}]$;
    (b) $[(A \cup B) \cap \bar{B}] \cup \overline{[A - B]}$;
    (c) $[(A \cap C) \cup (A \cap \bar{B})] \cap B$.

15.2 Probability

In the previous section we discussed Venn diagrams, which are graphical representations of the possible outcomes of experiments. We did not, however, give any indication of how likely each outcome or event might be when any particular experiment is performed. Most experiments show some regularity. By this we mean that the relative frequency of an event is approximately the same on each occasion that a set of trials is performed. For example, if we throw a die N times then we expect that a six will occur approximately $N/6$ times

15.2 Probability

(assuming, of course, that the die is not biased). The regularity of outcomes allows us to define the *probability*, Pr(A), as the expected relative frequency of event A in a large number of trials. More quantitatively, if an experiment has a total of n_S outcomes in the sample space S and n_A of these outcomes correspond to the event A, then the probability that event A will occur is

$$\Pr(A) = \frac{n_A}{n_S}. \tag{15.5}$$

15.2.1 Axioms and theorems

From (15.5) we may deduce the following properties of the probability Pr(A).

(i) For any event A in a sample space S,

$$0 \leq \Pr(A) \leq 1. \tag{15.6}$$

If $\Pr(A) = 1$ then A is a certainty; if $\Pr(A) = 0$ then A is an impossibility.

(ii) For the entire sample space S we have

$$\Pr(S) = \frac{n_S}{n_S} = 1, \tag{15.7}$$

which simply states that we are certain to obtain one of the possible outcomes.

(iii) If A and B are two events in S then, from the Venn diagrams in Figure 15.3, we see that

$$n_{A \cup B} = n_A + n_B - n_{A \cap B}, \tag{15.8}$$

the final subtraction arising because the outcomes in the intersection of A and B are counted twice when the outcomes of A are added to those of B. Dividing both sides of (15.8) by n_S, we obtain the *addition rule* for probabilities

$$\Pr(A \cup B) = \Pr(A) + \Pr(B) - \Pr(A \cap B). \tag{15.9}$$

However, if A and B are *mutually exclusive* events ($A \cap B = \emptyset$) then $\Pr(A \cap B) = 0$ and we obtain the special case

$$\Pr(A \cup B) = \Pr(A) + \Pr(B). \tag{15.10}$$

(iv) If $\bar{A}$ is the complement of A then $\bar{A}$ and A are mutually exclusive events. Thus, from (15.7) and (15.10) we have

$$1 = \Pr(S) = \Pr(A \cup \bar{A}) = \Pr(A) + \Pr(\bar{A}),$$

from which we obtain the *complement law*

$$\Pr(\bar{A}) = 1 - \Pr(A). \tag{15.11}$$

This is particularly useful for problems in which evaluating the probability of the complement is easier than evaluating the probability of the event itself.

Our next worked example is a simple application of the addition rule.

Elementary probability

Example Calculate the probability of drawing an ace or a spade from a pack of cards.

Let A be the event that an ace is drawn and B the event that a spade is drawn. It immediately follows that $\Pr(A) = \frac{4}{52} = \frac{1}{13}$ and $\Pr(B) = \frac{13}{52} = \frac{1}{4}$. The intersection of A and B consists of only the ace of spades and so $\Pr(A \cap B) = \frac{1}{52}$. Thus, from (15.9)

$$\Pr(A \cup B) = \tfrac{1}{13} + \tfrac{1}{4} - \tfrac{1}{52} = \tfrac{4}{13}.$$

In this case it is just as simple to recognise that there are 16 cards in the pack that satisfy the required condition (13 spades plus 3 other aces) and so the probability is $\frac{16}{52}$.[4] ◀

The above theorems can be extended to a greater number of events. For example, if $A_1, A_2, \ldots, A_n$ are mutually exclusive events then (15.10) becomes

$$\Pr(A_1 \cup A_2 \cup \cdots \cup A_n) = \Pr(A_1) + \Pr(A_2) + \cdots + \Pr(A_n). \quad (15.12)$$

Furthermore, if $A_1, A_2, \ldots, A_n$ (whether mutually exclusive or not) *exhaust* S, i.e. are such that $A_1 \cup A_2 \cup \cdots \cup A_n = S$, then

$$\Pr(A_1 \cup A_2 \cup \cdots \cup A_n) = \Pr(S) = 1. \quad (15.13)$$

The die described in the next and several later worked examples is so blatantly biased that it could never be used in any notionally fair game of chance, but it *is* convenient for mathematical illustration!

Example A biased six-sided die has probabilities $\frac{1}{2}p, p, p, p, p, 2p$ of showing 1, 2, 3, 4, 5, 6 respectively. Calculate p.

Given that the individual events are mutually exclusive, (15.12) can be applied to give

$$\Pr(1 \cup 2 \cup 3 \cup 4 \cup 5 \cup 6) = \tfrac{1}{2}p + p + p + p + p + 2p = \tfrac{13}{2}p.$$

The union of all possible outcomes on the LHS of this equation is clearly the sample space, S, and so

$$\Pr(S) = \tfrac{13}{2}p.$$

Now (15.7) requires that $\Pr(S) = 1$, and so $p = \frac{2}{13}$. ◀

When the possible outcomes of a trial correspond to more than two events, and those events are *not* mutually exclusive, the calculation of the probability of the union of a number of events is more complicated, and the generalisation of the addition law (15.9) requires further work. Let us begin by considering the union of three events A_1, A_2 and A_3, which need not be mutually exclusive. We first define the event $B = A_2 \cup A_3$ and,

---

[4] Suppose that the card drawing were scored and the drawn card replaced in the pack, at random, before a subsequent draw. What would be the expected average score for N such drawings (a) with one point scored if the card is an ace or a spade, zero otherwise, and (b) with one point scored if the card is an ace and one point scored if the card is a spade, zero otherwise?

15.2 Probability

using the addition law (15.9), we obtain

$$\Pr(A_1 \cup A_2 \cup A_3) = \Pr(A_1 \cup B) = \Pr(A_1) + \Pr(B) - \Pr(A_1 \cap B). \quad (15.14)$$

However, we may write $\Pr(A_1 \cap B)$ as

$$\begin{aligned}\Pr(A_1 \cap B) &= \Pr[A_1 \cap (A_2 \cup A_3)] \\ &= \Pr[(A_1 \cap A_2) \cup (A_1 \cap A_3)] \\ &= \Pr(A_1 \cap A_2) + \Pr(A_1 \cap A_3) - \Pr(A_1 \cap A_2 \cap A_3).\end{aligned}$$

Substituting this expression, and that for $\Pr(B)$ obtained from (15.9), into (15.14) we obtain the probability addition law for three general events,

$$\begin{aligned}\Pr(A_1 \cup A_2 \cup A_3) = {} & \Pr(A_1) + \Pr(A_2) + \Pr(A_3) - \Pr(A_2 \cap A_3) - \Pr(A_1 \cap A_3) \\ & - \Pr(A_1 \cap A_2) + \Pr(A_1 \cap A_2 \cap A_3), \quad (15.15)\end{aligned}$$

which we now apply to a card-drawing problem.

Example Calculate the probability of drawing from a pack of cards one that is an ace or is a spade or shows an even number (2, 4, 6, 8, 10).

If, as previously, A is the event that an ace is drawn, $\Pr(A) = \frac{4}{52}$. Similarly, the event B, that a spade is drawn, has $\Pr(B) = \frac{13}{52}$. The further possibility C, that the card is even (but not a picture card), has $\Pr(C) = \frac{20}{52}$. The two-fold intersections have probabilities

$$\Pr(A \cap B) = \frac{1}{52}, \quad \Pr(A \cap C) = 0, \quad \Pr(B \cap C) = \frac{5}{52}.$$

There is no threefold intersection as events A and C are mutually exclusive. Hence

$$\Pr(A \cup B \cup C) = \frac{1}{52}[(4 + 13 + 20) - (1 + 0 + 5) + (0)] = \frac{31}{52}.$$

The reader should identify the 31 cards involved. ◀

When the probabilities are combined to calculate the probability for the union of the n general events, the result, which may be proved by induction upon n, is

$$\Pr(A_1 \cup A_2 \cup \cdots \cup A_n) = \sum_i \Pr(A_i) - \sum_{i,j} \Pr(A_i \cap A_j) + \sum_{i,j,k} \Pr(A_i \cap A_j \cap A_k)$$
$$- \cdots + (-1)^{n+1} \Pr(A_1 \cap A_2 \cap \cdots \cap A_n). \quad (15.16)$$

Each summation runs over all possible sets of subscripts, except those in which any two subscripts in a set are the same; each pair of unequal subscripts must be counted only once. The number of terms in the summation of probabilities of m-fold intersections of the n events is given by ^nC_m (as discussed in Section 15.1). Equation (15.9) is a special case of (15.16) in which $n = 2$ and only the first two terms on the RHS survive. We now illustrate this result with a worked example that has $n = 4$ and includes a fourfold intersection.

Example Find the probability of drawing from a pack a card that has at least one of the following properties:
A, it is an ace;
B, it is a spade;
C, it is a black honour card (ace, king, queen, jack or 10);
D, it is a black ace.

Measuring all probabilities in units of $\frac{1}{52}$, the single-event probabilities are

$$\Pr(A) = 4, \quad \Pr(B) = 13, \quad \Pr(C) = 10, \quad \Pr(D) = 2.$$

The twofold intersection probabilities, measured in the same units, are

$$\Pr(A \cap B) = 1, \quad \Pr(A \cap C) = 2, \quad \Pr(A \cap D) = 2,$$
$$\Pr(B \cap C) = 5, \quad \Pr(B \cap D) = 1, \quad \Pr(C \cap D) = 2.$$

The threefold intersections have probabilities

$$\Pr(A \cap B \cap C) = 1, \quad \Pr(A \cap B \cap D) = 1, \quad \Pr(A \cap C \cap D) = 2, \quad \Pr(B \cap C \cap D) = 1.$$

Finally, the fourfold intersection, requiring all four conditions to hold, is satisfied only by the ace of spades, and hence (again in units of $\frac{1}{52}$)

$$\Pr(A \cap B \cap C \cap D) = 1.$$

Substituting in (15.16) gives

$$P = \frac{1}{52}[(4 + 13 + 10 + 2) - (1 + 2 + 2 + 5 + 1 + 2) + (1 + 1 + 2 + 1) - (1)] = \frac{20}{52}$$

for the probability that the drawn card has at least one of the listed properties. ◀

We conclude this section on basic theorems by deriving a useful general expression for the probability $\Pr(A \cap B)$ that two events A and B both occur in the case where A (say) is the union of a set of n *mutually exclusive* events A_i. In this case

$$A \cap B = (A_1 \cap B) \cup \cdots \cup (A_n \cap B),$$

where the events $A_i \cap B$ are also mutually exclusive. Thus, from the addition law (15.12) for mutually exclusive events, we find

$$\Pr(A \cap B) = \sum_i \Pr(A_i \cap B). \tag{15.17}$$

Moreover, in the special case where the events A_i *exhaust* the sample space S, we have $A \cap B = S \cap B = B$, and we obtain the *total probability law*

$$\Pr(B) = \sum_i \Pr(A_i \cap B). \tag{15.18}$$

15.2.2 Conditional probability

So far we have defined only probabilities of the form 'what is the probability that event A happens?'. In this section we turn to *conditional probability*, the probability that a particular event occurs *given* the occurrence of another, possibly related, event. For example, we may wish to know the probability of event B, drawing an ace from a pack of cards from

15.2 Probability

which one card has already been removed, given that event A, the card already removed was itself an ace, has occurred.

We denote this probability by $\Pr(B|A)$ and may obtain a formula for it by considering the total probability $\Pr(A \cap B) = \Pr(B \cap A)$ that both A and B will occur. This may be written in two ways, i.e.

$$\Pr(A \cap B) = \Pr(A)\Pr(B|A)$$
$$= \Pr(B)\Pr(A|B).$$

From this we obtain

$$\Pr(A|B) = \frac{\Pr(A \cap B)}{\Pr(B)} \tag{15.19}$$

and

$$\Pr(B|A) = \frac{\Pr(B \cap A)}{\Pr(A)}. \tag{15.20}$$

In terms of Venn diagrams, we may think of $\Pr(B|A)$ as the probability of B in the reduced sample space defined by A. Thus, if two events A and B are mutually exclusive then

$$\Pr(A|B) = 0 = \Pr(B|A). \tag{15.21}$$

When an experiment consists of drawing objects at random from a given set of objects, it is termed *sampling a population*. We need to distinguish between two different ways in which such a *sampling experiment* may be performed. After an object has been drawn at random from the set it may either be put aside or returned to the set before the next object is randomly drawn. The former is termed 'sampling without replacement', the latter 'sampling with replacement'. The following demonstrates the difference.

> **Example** Find the probability of drawing two aces at random from a pack of cards (i) when the first card drawn is replaced at random into the pack before the second card is drawn and (ii) when the first card is put aside after being drawn.

Let A be the event that the first card is an ace and B the event that the second card is an ace. Now

$$\Pr(A \cap B) = \Pr(A)\Pr(B|A),$$

and for both (i) and (ii) we know that $\Pr(A) = \frac{4}{52} = \frac{1}{13}$.

(i) If the first card is replaced in the pack before the next is drawn then $\Pr(B|A) = \Pr(B) = \frac{4}{52} = \frac{1}{13}$, since A and B are independent events. We then have

$$\Pr(A \cap B) = \Pr(A)\Pr(B) = \frac{1}{13} \times \frac{1}{13} = \frac{1}{169}.$$

(ii) If the first card is put aside and the second then drawn, A and B are not independent and $\Pr(B|A) = \frac{3}{51}$. We then have

$$\Pr(A \cap B) = \Pr(A)\Pr(B|A) = \frac{1}{13} \times \frac{3}{51} = \frac{1}{221}$$

for the combined probability of the two events.[5]

◀

---

[5] Would the answer be different if the first card had been replaced but the requirement had been for two different aces?

Elementary probability

Two events A and B are *statistically independent* if $\Pr(A|B) = \Pr(A)$ [or equivalently if $\Pr(B|A) = \Pr(B)$]. In words, the probability of A given B is then the same as the probability of A regardless of whether B occurs. For example, if we throw a coin and a die at the same time, we would normally expect that the probability of throwing a six was independent of whether a head was thrown. If A and B are statistically independent then it follows that

$$\Pr(A \cap B) = \Pr(A)\Pr(B). \tag{15.22}$$

In fact, on the basis of intuition and experience, (15.22) may be regarded as the *definition* of the statistical independence of two events.

The idea of statistical independence is easily extended to an arbitrary number of events $A_1, A_2, \ldots, A_n$. The events are said to be (mutually) independent if

$$\Pr(A_i \cap A_j) = \Pr(A_i)\Pr(A_j),$$
$$\Pr(A_i \cap A_j \cap A_k) = \Pr(A_i)\Pr(A_j)\Pr(A_k),$$
$$\vdots$$
$$\Pr(A_1 \cap A_2 \cap \cdots \cap A_n) = \Pr(A_1)\Pr(A_2)\cdots\Pr(A_n),$$

for all combinations of indices i, j and k for which no two indices are the same. Even if all n events are not mutually independent, any two events for which $\Pr(A_i \cap A_j) = \Pr(A_i)\Pr(A_j)$ are said to be *pairwise independent*.

We now derive two results that often prove useful when working with conditional probabilities. Let us suppose that an event A is the union of n *mutually exclusive* events A_i. If B is some other event then from (15.17) we have

$$\Pr(A \cap B) = \sum_i \Pr(A_i \cap B).$$

Dividing both sides of this equation by $\Pr(B)$, and using (15.19), we obtain

$$\Pr(A|B) = \sum_i \Pr(A_i|B), \tag{15.23}$$

which is the *addition law for conditional probabilities*.

Furthermore, if the set of mutually exclusive events A_i exhausts the sample space S then, from the *total probability law* (15.18), the probability $\Pr(B)$ of some event B in S can be written as

$$\Pr(B) = \sum_i \Pr(A_i)\Pr(B|A_i). \tag{15.24}$$

The following very artificial car-routing problem illustrates this point.

15.2 Probability

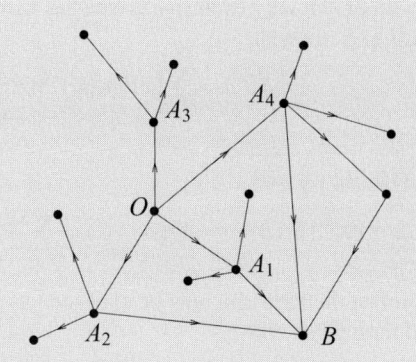

Figure 15.5 A collection of traffic islands connected by one-way roads.

Example A collection of traffic islands connected by a system of one-way roads is shown in Figure 15.5. At any given island a car driver chooses a direction at random from those available. What is the probability that a driver starting at O will arrive at B?

In order to leave O the driver must pass through one of A_1, A_2, A_3 or A_4, which thus form a complete set of mutually exclusive events. Since at each island (including O) the driver chooses a direction at random from those available, we have that $\Pr(A_i) = \frac{1}{4}$ for $i = 1, 2, 3, 4$. From Figure 15.5, we see also that

$$\Pr(B|A_1) = \tfrac{1}{3}, \quad \Pr(B|A_2) = \tfrac{1}{3}, \quad \Pr(B|A_3) = 0, \quad \Pr(B|A_4) = \tfrac{2}{4} = \tfrac{1}{2}.$$

Now the total probability law, (15.24), gives

$$\Pr(B) = \sum_i \Pr(A_i) \Pr(B|A_i) = \tfrac{1}{4}\left(\tfrac{1}{3} + \tfrac{1}{3} + 0 + \tfrac{1}{2}\right) = \tfrac{7}{24}.$$

as the probability of arriving at B. ◀

Finally, we note that the concept of conditional probability may be straightforwardly extended to several compound events. For example, in the case of three events A, B, C, we may write $\Pr(A \cap B \cap C)$ in several ways. Some of the possibilities are

$$\Pr(A \cap B \cap C) = \begin{cases} \Pr(C)\Pr(A \cap B|C) \\ \Pr(B \cap C)\Pr(A|B \cap C) \\ \Pr(C)\Pr(B|C)\Pr(A|B \cap C) \end{cases}.$$

The result of the following example is sometimes useful when a set of mutually exclusive events that exhaust the sample space can be identified.

Example Suppose $\{A_i\}$ is a set of mutually exclusive events that exhausts the sample space S. If B and C are two other events in S, show that

$$\Pr(B|C) = \sum_i \Pr(A_i|C) \Pr(B|A_i \cap C).$$

Using (15.19) and (15.17), we may write

$$\Pr(C)\Pr(B|C) = \Pr(B \cap C) = \sum_i \Pr(A_i \cap B \cap C). \qquad (15.25)$$

Each term in the sum on the RHS can now be expanded as a product of conditional probabilities. Following the third form given above:

$$\Pr(A_i \cap B \cap C) = \Pr(B \cap A_i \cap C) = \Pr(C)\Pr(A_i|C)\Pr(B|A_i \cap C).$$

Substituting this form into (15.25) and dividing through by $\Pr(C)$ gives the required result. ◄

15.2.3 Bayes' theorem

In the previous section we saw that the probability that both an event A and a related event B will occur can be written either as $\Pr(A)\Pr(B|A)$ or $\Pr(B)\Pr(A|B)$. Hence

$$\Pr(A)\Pr(B|A) = \Pr(B)\Pr(A|B),$$

from which we obtain *Bayes' theorem*,

$$\Pr(A|B) = \frac{\Pr(A)}{\Pr(B)} \Pr(B|A). \qquad (15.26)$$

This theorem clearly shows that $\Pr(B|A) \neq \Pr(A|B)$, unless $\Pr(A) = \Pr(B)$. It is sometimes useful to rewrite $\Pr(B)$, if it is not known directly, as

$$\Pr(B) = \Pr(A)\Pr(B|A) + \Pr(\bar{A})\Pr(B|\bar{A})$$

so that Bayes' theorem becomes

$$\Pr(A|B) = \frac{\Pr(A)\Pr(B|A)}{\Pr(A)\Pr(B|A) + \Pr(\bar{A})\Pr(B|\bar{A})}. \qquad (15.27)$$

The next example illustrates its use.

Example Suppose that the blood test for some disease is reliable in the following sense: for people who are infected with the disease the test produces a positive result in 99.99% of cases; for people not infected a positive test result is obtained in only 0.02% of cases. Furthermore, assume that in the general population one person in 10 000 people is infected. A person is selected at random and found to test positive for the disease. What is the probability that the individual is actually infected?

Let A be the event that the individual is infected and B be the event that the individual tests positive for the disease. Using Bayes' theorem the probability that a person who tests positive is actually infected is

$$\Pr(A|B) = \frac{\Pr(A)\Pr(B|A)}{\Pr(A)\Pr(B|A) + \Pr(\bar{A})\Pr(B|\bar{A})}.$$

Now $\Pr(A) = 1/10\,000 = 1 - \Pr(\bar{A})$, and we are told that $\Pr(B|A) = 9999/10\,000$ and $\Pr(B|\bar{A}) = 2/10\,000$. Thus we obtain

$$\Pr(A|B) = \frac{1/10\,000 \times 9999/10\,000}{(1/10\,000 \times 9999/10\,000) + (9999/10\,000 \times 2/10\,000)} = \frac{1}{3}.$$

Thus, there is only a one in three chance that a person chosen at random, who tests positive for the disease, is actually infected.

At first glance, this answer may seem a little surprising, but the reason for the counter-intuitive result is that the 0.02% chance of an erroneous positive test for an uninfected person is of the same order of magnitude as the 0.01% chance of selecting somebody who is actually infected. ◀

We note that (15.27) may be written in a more general form if S is not simply divided into A and $\bar{A}$ but, rather, into *any* set of mutually exclusive events A_i that exhaust S. Using the total probability law (15.24), we may then write

$$\Pr(B) = \sum_i \Pr(A_i) \Pr(B|A_i),$$

so that Bayes' theorem takes the form

$$\Pr(A|B) = \frac{\Pr(A)\Pr(B|A)}{\sum_i \Pr(A_i)\Pr(B|A_i)}, \qquad (15.28)$$

where the event A need not coincide with any of the A_i.

As a final point, we comment that sometimes we are concerned only with the *relative* probabilities of two events A and C (say), given the occurrence of some other event B. From (15.26) we then obtain a different form of Bayes' theorem,

$$\frac{\Pr(A|B)}{\Pr(C|B)} = \frac{\Pr(A)\Pr(B|A)}{\Pr(C)\Pr(B|C)}, \qquad (15.29)$$

which does not contain $\Pr(B)$ at all. This is the case in the following example

Example The routine in a TV game-show is that contestants are presented with three boxes, one of which contains a large cash prize; the other two are empty. The contestant chooses one of the boxes but, before he or she is allowed to open it, the game-show host, who knows the location of each particular prize, opens one of the other boxes – naturally, it is empty. He then offers the contestants the chance to change their choice to the third box. Should they accept the offer?

At first sight it might seem that the originally chosen and the third boxes are equally likely to contain the prize and that there is nothing to be gained by making the swap. However, this is not so, as we now show.

Denote the selected box by 'red', the one opened by the host by 'blue', and the third box by 'green'. Then the three mutually exclusive but exhaustive events, in which the prize is in the red (A_1), blue (A_2) and green (A_3) box, each have a probability $\Pr(A_i) = \frac{1}{3}$.

Elementary probability

The probabilities that the box opened by the host is blue (event B) or is green (event G) for each A_i are

$$\Pr(B|A_1) = \tfrac{1}{2}, \qquad \Pr(G|A_1) = \tfrac{1}{2},$$
$$\Pr(B|A_2) = 0, \qquad \Pr(G|A_2) = 1,$$
$$\Pr(B|A_3) = 1, \qquad \Pr(G|A_3) = 0.$$

We can now apply Bayes' theorem in the form (15.29) to calculate the relative values of $\Pr(A_1|B)$ and $\Pr(A_3|B)$.

$$\frac{\Pr(A_1|B)}{\Pr(A_3|B)} = \frac{\Pr(A_1)\Pr(B|A_1)}{\Pr(A_3)\Pr(B|A_3)} = \frac{\tfrac{1}{3} \times \tfrac{1}{2}}{\tfrac{1}{3} \times 1} = \frac{1}{2},$$

Since $\Pr(A_3|B) = 2 \times \Pr(A_1|B)$, the contestants should clearly accept the offer to change to the third box. ◀

EXERCISES 15.2

1. A single card is drawn from a normal pack of cards. Construct the simplest possible Venn diagram for the situation in which event A is that the card is red (a heart or a diamond), event B is that it is black (a spade or a club) and event C is that it is an honour card (ace, king, queen, jack or ten).
   (a) Label each distinct region of the diagram with its probability.
   (b) Identify and shade the region R given by $(A \cap \bar{B}) \cup C$. Describe in words the outcomes it contains.
   (c) Calculate the probability that the outcome of a single trial belongs to R.

2. Three dice, red, blue and green, are thrown. Event A_1 is that the product of the numbers shown on the green and blue dice is odd; events A_2 and A_3 are similarly defined for the red and green, and the red and blue dice, respectively. Calculate the probability that at least one of these three events occurred.

3. Tennis player W has a 70% chance of beating player X, a 50% chance of beating Y, but only a 20% chance of beating Z. He was eliminated from a knock-out competition by one of these three players. Because of the seeding arrangements his chances of meeting each of these opponents were not all equal, but in the ratio $X : Y : Z = 3 : 2 : 1$. What is the probability that his conqueror was player Y?

15.3 Permutations and combinations

In Equation (15.5) we defined the probability of an event A in a sample space S as

$$\Pr(A) = \frac{n_A}{n_S},$$

where n_A is the number of outcomes belonging to event A and n_S is the total number of possible outcomes. It is therefore necessary to be able to count the number of possible outcomes in various common situations.

15.3 Permutations and combinations

15.3.1 Permutations

Let us first consider a set of n objects that are all different. We may ask in how many ways these n objects may be arranged, i.e. how many *permutations* of these objects exist. This is straightforward to deduce, as follows: the object in the first position may be chosen in n different ways, that in the second position in $n - 1$ ways, and so on until the final object is positioned. The number of possible arrangements is therefore

$$n(n-1)(n-2)\cdots(1) = n! \tag{15.30}$$

Generalising (15.30) slightly, let us suppose we choose only k ($< n$) objects from n. The number of possible permutations of these k objects selected from n is given by

$$\underbrace{n(n-1)(n-2)\cdots(n-k+1)}_{k \text{ factors}} = \frac{n!}{(n-k)!} \equiv {^nP_k}. \tag{15.31}$$

In calculating the number of permutations of the various objects we have so far assumed that the objects are sampled *without replacement* – i.e. once an object has been drawn from the set it is put aside. As mentioned previously, however, we may instead replace each object before the next is chosen. The number of permutations of k objects from n *with replacement* may be calculated very easily since the first object can be chosen in n different ways, as can the second, the third, etc. Therefore, the number of permutations is simply n^k. This may also be viewed as the number of permutations of k objects from n where repetitions are allowed, i.e. each object may be used as often as one likes.

The following worked example requires both sorts of calculation and produces a mildly surprising result.

Example Find the probability that in a group of k people at least two have the same birthday (ignoring 29 February).

This is a situation in which it is easier first to calculate the probability p that no two people share a birthday and then to calculate the required probability that at least two do, as $q = 1 - p$.

Firstly, we imagine each of the k people in turn pointing to their birthday on a year planner. Thus, we are sampling the 365 days of the year 'with replacement' and so the total number of possible outcomes is $(365)^k$.

Now (for the moment) we assume that no two people share a birthday and imagine the process being repeated, except that as each person points out their birthday it is crossed off the planner. In this case, we are sampling the days of the year 'without replacement', and so the possible number of outcomes for which all the birthdays are different is, as in (15.31),

$$^{365}P_k = \frac{365!}{(365-k)!}.$$

Hence the probability that all the birthdays are different is

$$p = \frac{365!}{(365-k)!\,365^k}.$$

Elementary probability

Now using the complement rule (15.11), the probability q that two or more people have the same birthday is simply

$$q = 1 - p = 1 - \frac{365!}{(365-k)!\, 365^k}.$$

This expression may be conveniently evaluated using Stirling's approximation for $n!$ when n is large, namely

$$n! \sim \sqrt{2\pi n}\left(\frac{n}{e}\right)^n,$$

to give[6]

$$q \approx 1 - e^{-k}\left(\frac{365}{365-k}\right)^{365-k+0.5}.$$

It is interesting to note that if $k = 23$ the probability is a little greater than a half that at least two people have the same birthday, and if $k = 50$ the probability rises to 0.970. This can prove a good bet at a party of non-mathematicians! ◂

So far we have assumed that all n objects are different (or *distinguishable*). Let us now consider n objects of which n_1 are identical and of type 1, n_2 are identical and of type 2, ..., n_m are identical and of type m (clearly $n = n_1 + n_2 + \cdots + n_m$). From (15.30) the number of permutations of these n objects is again $n!$. However, the number of *distinguishable* permutations is only

$$\frac{n!}{n_1!\, n_2! \cdots n_m!}, \tag{15.32}$$

since the ith group of identical objects can be rearranged in $n_i!$ ways without changing the distinguishable permutation.

Example A set of snooker balls consists of a white, a yellow, a green, a brown, a blue, a pink, a black and 15 reds. How many distinguishable permutations of the balls are there?

In total there are 22 balls, the 15 reds being indistinguishable. Thus from (15.32) the number of distinguishable permutations is

$$\frac{22!}{(1!)(1!)(1!)(1!)(1!)(1!)(1!)(15!)} = \frac{22!}{15!} = 859\,541\,760.$$

Although a factor of $(1!)$ does not change a computed value, it is advisable to write and account for each one in a calculation such as this. That the sum of the 'arguments' of the factorial functions appearing in the denominator equals the 'argument' of the factorial function in the numerator gives a useful check that nothing has been overlooked. ◂

15.3.2 Combinations

We now consider the number of *combinations* of various objects when their order is immaterial. Assuming all the objects to be distinguishable, from (15.31) we see that the

---

[6] Carry through this calculation.

15.3 Permutations and combinations

number of permutations of k objects chosen from n is $^nP_k = n!/(n-k)!$. Now, since we are no longer concerned with the order of the chosen objects, which can be internally arranged in $k!$ different ways, the number of combinations of k objects from n is

$$\frac{n!}{(n-k)!k!} \equiv {}^nC_k \equiv \binom{n}{k} \quad \text{for } 0 \leq k \leq n, \tag{15.33}$$

where ^nC_k is called the *binomial coefficient* since it also appears [see Equation (1.41)] in the binomial expansion for positive integer n, namely

$$(a+b)^n = \sum_{k=0}^{n} {}^nC_k a^k b^{n-k}. \tag{15.34}$$

Result (15.33) is used several times in the following card-dealing calculation.

Example A hand of 13 playing cards is dealt from a well-shuffled pack of 52. What is the probability that the hand contains two aces?

Since the order of the cards in the hand is immaterial, the total number of distinct hands is simply equal to the number of combinations of 13 objects drawn from 52, i.e. $^{52}C_{13}$. However, the number of hands containing two aces is equal to the number of ways, $^4C_2$, in which the two aces can be drawn from the four available, multiplied by the number of ways, $^{48}C_{11}$, in which the remaining 11 cards in the hand can be drawn from the 48 cards that are not aces. Thus the required probability is given by

$$\frac{{}^4C_2 \, {}^{48}C_{11}}{{}^{52}C_{13}} = \frac{4!}{2!2!} \frac{48!}{11!37!} \frac{13!39!}{52!}$$
$$= \frac{(3)(4)}{2} \frac{(12)(13)(38)(39)}{(49)(50)(51)(52)} = 0.213.$$

This calculation is easily adapted for any number of aces.[7]

Another useful result that may be derived using the binomial coefficients is the number of ways in which n distinguishable objects can be divided into m piles, with n_i objects in the ith pile, $i = 1, 2, \ldots, m$ (the ordering of objects within each pile being unimportant). This may be straightforwardly calculated as follows. We may choose the n_1 objects in the first pile from the original n objects in $^nC_{n_1}$ ways. The n_2 objects in the second pile can then be chosen from the $n - n_1$ remaining objects in $^{n-n_1}C_{n_2}$ ways, etc. We may continue in this fashion until we reach the $(m-1)$th pile, which may be formed in $^{n-n_1-\cdots-n_{m-2}}C_{n_{m-1}}$ ways. The remaining objects then form the mth pile and so can only be 'chosen' in one way. Thus the total number of ways of dividing the original n objects into m piles is given

---

[7] Is it more likely that a particular hand contains exactly one king or that it contains no kings?

Elementary probability

by the product

$$N = {}^nC_{n_1} {}^{n-n_1}C_{n_2} \cdots {}^{n-n_1-\cdots-n_{m-2}}C_{n_{m-1}}$$

$$= \frac{n!}{n_1!(n-n_1)!} \frac{(n-n_1)!}{n_2!(n-n_1-n_2)!} \cdots \frac{(n-n_1-n_2-\cdots-n_{m-2})!}{n_{m-1}!(n-n_1-n_2-\cdots-n_{m-2}-n_{m-1})!}$$

$$= \frac{n!}{n_1!(n-n_1)!} \frac{(n-n_1)!}{n_2!(n-n_1-n_2)!} \cdots \frac{(n-n_1-n_2-\cdots-n_{m-2})!}{n_{m-1}!n_m!}$$

$$= \frac{n!}{n_1!n_2!\cdots n_m!}. \qquad (15.35)$$

These numbers are called *multinomial coefficients* since (15.35) is the coefficient of $x_1^{n_1} x_2^{n_2} \cdots x_m^{n_m}$ in the multinomial expansion of $(x_1 + x_2 + \cdots + x_m)^n$, i.e. for positive integer n

$$(x_1 + x_2 + \cdots + x_m)^n = \sum_{\substack{n_1,n_2,\ldots,n_m \\ n_1+n_2+\cdots+n_m=n}} \frac{n!}{n_1!n_2!\cdots n_m!} x_1^{n_1} x_2^{n_2} \cdots x_m^{n_m}.$$

For the case $m = 2$, $n_1 = k$, $n_2 = n - k$, (15.35) reduces to the binomial coefficient ${}^nC_k$. Furthermore, we note that the multinomial coefficient (15.35) is identical to the expression (15.32) for the number of distinguishable permutations of n objects, n_i of which are identical and of type i (for $i = 1, 2, \ldots, m$ and $n_1 + n_2 + \cdots + n_m = n$). A few moments' thought should convince the reader that the two expressions (15.35) and (15.32) must be identical.

The following example applies these ideas to a further card-dealing problem.

> **Example** In the card game of bridge, each of four players is dealt 13 cards from a full pack of 52. What is the probability that each player is dealt an ace?
>
> From (15.35), the total number of distinct bridge deals is $52!/(13!13!13!13!)$. However, the number of ways in which the four aces can be distributed with one in each hand is $4!/(1!1!1!1!) = 4!$; the remaining 48 cards can then be dealt out in $48!/(12!12!12!12!)$ ways. Thus the probability that each player receives an ace is
>
> $$4! \frac{48!}{(12!)^4} \frac{(13!)^4}{52!} = \frac{24(13)^4}{(49)(50)(51)(52)} = 0.105.$$
>
> Note that result (15.35) has been applied three times in this calculation. ◂

As in the case of permutations, we might ask how many combinations of k objects can be chosen from n *with replacement* (repetition). To calculate this, we may imagine the n (distinguishable) objects set out on a table. Each combination of k objects can then be made by pointing to k of the n objects in turn (with repetitions allowed). These k equivalent selections distributed amongst n different but re-choosable objects are strictly analogous to the placing of k indistinguishable 'balls' in n different boxes with no restriction on the number of balls in each box. A particular selection in the case $k = 7$, $n = 5$ may be symbolised as

$$xxx| \quad |x|xx|x.$$

15.3 Permutations and combinations

This denotes three balls in the first box, none in the second, one in the third, two in the fourth and one in the fifth. We therefore need only consider the number of (distinguishable) ways in which k crosses and $n - 1$ vertical lines can be arranged, i.e. the number of permutations of $k + n - 1$ objects of which k are identical crosses and $n - 1$ are identical lines. This is given by (15.33) as

$$\frac{(k+n-1)!}{k!(n-1)!} = {}^{n+k-1}C_k. \tag{15.36}$$

We note that this expression also occurs in the binomial expansion for negative integer powers [see Equations (1.45) and (1.46)]. If n is a positive integer, then

$$(a+b)^{-n} = \sum_{k=0}^{\infty} (-1)^k \, {}^{n+k-1}C_k \, a^{-n-k} b^k,$$

where a is taken to be larger than b in magnitude.

We illustrate these results with a further example taken from the snooker table.

Example A set of snooker balls consists of 15 identical red balls and 7 balls which are individually coloured (including black and white). A snooker table has six pockets, labelled by their positions around the table. Find (i) the number of distinct arrangements with n_i balls in the ith pocket and (ii) the *total* number of such distinct arrangements, for (a) the coloured balls and, separately, (b) the red balls. In each case, every ball has to be assigned to a pocket.

(a) The coloured balls. (i) The balls are distinguishable but since the order in which the n_i balls in the ith pocket were placed there does not matter, the number of distinct arrangements is given by (15.35) as

$$\frac{7!}{n_1! n_2! \cdots n_6!}.$$

(ii) The total number of distinct arrangements is formally obtained by summing this result over all possible sets of positive values of $n_1, n_2, \ldots, n_6$, where $\sum_i n_i = 7$, but is much more easily obtained by noting that there are six possible pockets for each of the seven balls; the total is therefore $6^7 = 279\,936$.

(b) The red balls. (i) As the balls are indistinguishable, for each set of n_i there is, by definition, only one distinct arrangement (whatever the actual values of the individual n_i). (ii) The total number of distinct arrangements (equal to the number of distinct sets $\{n_i\}$) is given by (15.36) as ${}^{6+15-1}C_{15} = 15\,504$. ◂

EXERCISES 15.3

1. From a catalogue of 12 different video games, presents are to be selected for four boys and two girls. In how many distinct ways can the presents be chosen and distributed under each of the following conditions?
   (a) There are no restrictions – each should receive what seems most appropriate.
   (b) All children get the same game – then there can be no squabbling.
   (c) No two children should get the same game – so that they can share and enjoy them all.

(d) The boys should all get the same; the two girls should get the same as each other, but different from the boys – so that neither sex can acquire bragging rights.
(e) No two boys should get the same, and the girls' presents should differ from each other – as for (c), in principle.

2. A hand of 13 cards is dealt from a well-shuffled pack. Calculate the probability that it contains
   (a) no card above a nine,[8]
   (b) a void, i.e. has no cards belonging to at least one suit,
   (c) exactly four spades and three of each of the other three suits.

3. The coins in a cash-box are one £1 coin, two 50-pence coins, five 20-pence coins and ten 10-pence coins; all coins of any particular denomination are identical. In how many distinguishable ways can the coins be laid out in a single line that contains all of them?

4. Denoting $n = 3$ different objects by a, b and c, write down a typical member of each of the various possible patterns for selecting, with replacement, $k = 5$ objects. Determine how many distinct examples of each pattern are possible, given that three different objects are available. Hence find the total number of distinct selections possible and confirm that it agrees with value given by Equation (15.36), namely $^7C_5$. Repeat the exercise with $n = 4$ and $k = 4$.

15.4 Random variables and distributions

Suppose an experiment has an outcome sample space S. A real variable X that is defined for all possible outcomes in S (so that a real number – not necessarily unique – is assigned to each possible outcome) is called a *random variable* (RV). The outcome of the experiment may already be a real number and hence a random variable, e.g. the number of heads obtained in 10 throws of a coin, or the sum of the values if two dice are thrown. However, more arbitrary assignments are possible, e.g. the assignment of a 'quality' rating to each successive item produced by a manufacturing process. Furthermore, assuming that a probability can be assigned to all possible outcomes in a sample space S, it is possible to assign a *probability distribution* to any random variable. Random variables may be divided into two classes, discrete and continuous, and we now examine each of these in turn.

15.4.1 Discrete random variables

A random variable X that takes only discrete values $x_1, x_2, \ldots, x_n$, with probabilities $p_1, p_2, \ldots, p_n$, is called a *discrete* random variable. The number of values n for which X has a non-zero probability is finite or at most countably infinite. As mentioned above, an example of a discrete random variable is the number of heads obtained in 10 throws of a coin. If X is a discrete random variable, we can define a *probability function* (PF) $f(x)$

---

[8] Such a hand is known as a yarborough, after an Earl of Yarborough who is said to have offered odds of 1000 to 1 against it. There should not have been many takers!

15.4 Random variables and distributions

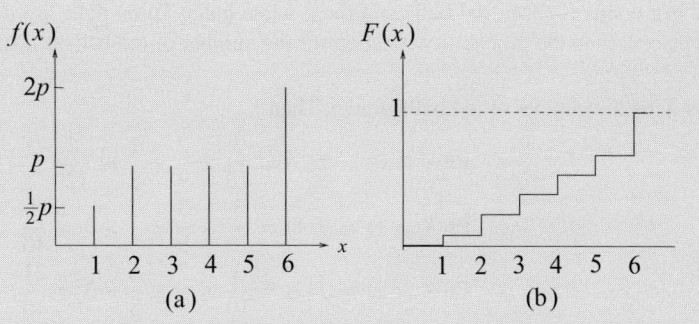

Figure 15.6 (a) A typical probability function for a discrete distribution, that for the biased die discussed earlier. Since the probabilities must sum to unity we require $p = 2/13$. (b) The cumulative probability function for the same discrete distribution. [Note that a different scale has been used for (b).]

that assigns probabilities to all the distinct values that X can take, such that

$$f(x) = \Pr(X = x) = \begin{cases} p_i & \text{if } x = x_i, \\ 0 & \text{otherwise.} \end{cases} \quad (15.37)$$

A typical PF (see Figure 15.6) thus consists of spikes, at *valid values* of X, whose height at x corresponds to the probability that $X = x$. Since the probabilities must sum to unity, we require

$$\sum_{i=1}^{n} f(x_i) = 1. \quad (15.38)$$

We may also define the *cumulative probability function* of X, $F(x)$, whose value gives the probability that $X \leq x$, so that

$$F(x) = \Pr(X \leq x) = \sum_{x_i \leq x} f(x_i). \quad (15.39)$$

Hence $F(x)$ is a step function that has upward jumps of p_i at $x = x_i$, $i = 1, 2, \ldots, n$, and is constant between possible values of X. We may also calculate the probability that X lies between two limits, l_1 and l_2 ($l_1 < l_2$); this is given by

$$\Pr(l_1 < X \leq l_2) = \sum_{l_1 < x_i \leq l_2} f(x_i) = F(l_2) - F(l_1), \quad (15.40)$$

i.e. it is the sum of all the probabilities for which x_i lies within the relevant interval.

We now give an example of how a set of $f(x_1)$, denoted in the calculation by $f(i)$, might be calculated for a process whose outcome is a random variable.

Elementary probability

Example A bag contains seven red balls and three white balls. Three balls are drawn at random and not replaced. Find the probability function for the number of red balls drawn.

Let X be the number of red balls drawn. Then

$$\Pr(X = 0) = f(0) = \frac{3}{10} \times \frac{2}{9} \times \frac{1}{8} = \frac{1}{120},$$

$$\Pr(X = 1) = f(1) = \frac{3}{10} \times \frac{2}{9} \times \frac{7}{8} \times 3 = \frac{7}{40},$$

$$\Pr(X = 2) = f(2) = \frac{3}{10} \times \frac{7}{9} \times \frac{6}{8} \times 3 = \frac{21}{40},$$

$$\Pr(X = 3) = f(3) = \frac{7}{10} \times \frac{6}{9} \times \frac{5}{8} = \frac{7}{24}.$$

The factors of 3 that appear for $X = 1$ and $X = 2$ should be noted and explained.[9] It should also be noted that, as expected, $\sum_{i=0}^{3} f(i) = 1$.

◀

15.4.2 Continuous random variables

A random variable X is said to have a *continuous* distribution if X is defined for a continuous range of values between given limits (often $-\infty$ to ∞). An example of a continuous random variable is the height of a person drawn from a population, which can take *any* value (within limits!). We can define the *probability density function* (PDF) $f(x)$ of a continuous random variable X such that

$$\Pr(x < X \leq x + dx) = f(x)\,dx,$$

i.e. $f(x)\,dx$ is the probability that X lies in the interval $x < X \leq x + dx$. Clearly $f(x)$ must be a real function that is everywhere ≥ 0. If X can take only values between the limits l_1 and l_2 then, in order for the sum of the probabilities of all possible outcomes to be equal to unity, we require

$$\int_{l_1}^{l_2} f(x)\,dx = 1.$$

Often X can take any value between $-\infty$ and ∞ and so

$$\int_{-\infty}^{\infty} f(x)\,dx = 1.$$

The probability that X lies in the interval $a < X \leq b$ is then given by

$$\Pr(a < X \leq b) = \int_{a}^{b} f(x)\,dx, \qquad (15.41)$$

i.e. $\Pr(a < X \leq b)$ is equal to the area under the curve of $f(x)$ between these limits (see Figure 15.7).

---

[9] For $X = 1$, say, consider each relevant scenario, WWR, WRW and RWW, and show that although the contributing probability ratios are not the same in each case, their products are.

15.4 Random variables and distributions

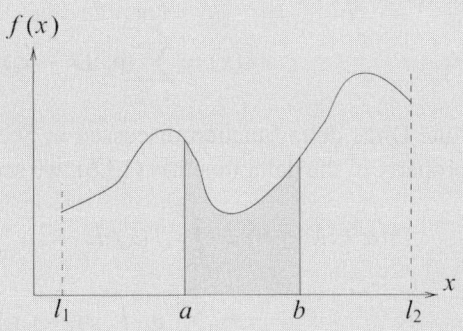

Figure 15.7 The probability density function for a continuous random variable X that can take values only between the limits l_1 and l_2. The shaded area under the curve gives $\Pr(a < X \leq b)$, whereas the total area under the curve, between the limits l_1 and l_2, is equal to unity.

We may also define the cumulative probability function $F(x)$ for a continuous random variable by

$$F(x) = \Pr(X \leq x) = \int_{l_1}^{x} f(u)\,du, \tag{15.42}$$

where u is a (dummy) integration variable. We can then write

$$\Pr(a < X \leq b) = F(b) - F(a).$$

From (15.42) it is clear that $f(x) = dF(x)/dx$. These ideas are illustrated by the following example.

Example A random variable X has a PDF $f(x)$ given by Ae^{-x} in the interval $0 < x < \infty$ and zero elsewhere. Find the value of the constant A and hence calculate the probability that X lies in the interval $1 < X \leq 2$.

We require the integral of $f(x)$ between 0 and ∞ to equal unity. Evaluating this integral, we find

$$\int_0^\infty Ae^{-x}\,dx = \left[-Ae^{-x}\right]_0^\infty = A,$$

and hence $A = 1$. From (15.41), we then obtain

$$\Pr(1 < X \leq 2) = \int_1^2 f(x)\,dx = \int_1^2 e^{-x}\,dx = -e^{-2} - (-e^{-1}) = 0.23$$

as the probability that X lies in the interval $1 < X \leq 2$. ◀

It is worth mentioning here that a *discrete* RV can in fact be treated as continuous and assigned a corresponding probability density function. If X is a discrete RV that takes only the values $x_1, x_2, \ldots, x_n$ with probabilities $p_1, p_2, \ldots, p_n$ then we may describe X

Elementary probability

as a continuous RV with PDF

$$f(x) = \sum_{i=1}^{n} p_i \delta(x - x_i), \qquad (15.43)$$

where $\delta(x)$ is the Dirac delta function discussed in Section 13.2. From (15.41) and the fundamental property of the delta function (13.6), we see that

$$\Pr(a < X \leq b) = \int_a^b f(x)\,dx,$$

$$= \sum_{i=1}^{n} p_i \int_a^b \delta(x - x_i)\,dx = \sum_i p_i,$$

where the final sum extends over those values of i for which $a < x_i \leq b$.

15.4.3 Sets of random variables

It is common in practice to consider two or more random variables simultaneously. For example, one might be interested in both the height and weight of a person drawn at random from a population. In the general case, these variables may depend on one another and are described by *joint probability density functions*. Although there is no formal difficulty in dealing with three or more random variables, we will limit our discussion to two such variables, and furthermore consider only cases in which they are both discrete or both continuous.

If we have two random variables X and Y, then, by analogy with the single-variable case, we define their joint probability density function $f(x, y)$ in such a way that, if X and Y are discrete RVs,

$$\Pr(X = x_i,\ Y = y_j) = f(x_i, y_j),$$

or, if X and Y are continuous RVs,

$$\Pr(x < X \leq x + dx,\ y < Y \leq y + dy) = f(x, y)\,dx\,dy.$$

In many circumstances, however, random variables do not depend on one another, i.e. they are *independent*. As an example, for a person drawn at random from a population, we might expect height and IQ to be independent random variables. Let us suppose that X and Y are two random variables with probability density functions $g(x)$ and $h(y)$ respectively. In mathematical terms, X and Y are independent RVs if their joint probability density function is given by $f(x, y) = g(x)h(y)$. Thus, for independent RVs, if X and Y are both discrete then

$$\Pr(X = x_i,\ Y = y_j) = g(x_i)h(y_j)$$

or, if X and Y are both continuous, then

$$\Pr(x < X \leq x + dx,\ y < Y \leq y + dy) = g(x)h(y)\,dx\,dy.$$

The important point in each case is that the RHS is simply the product of the individual probability density functions (compare with the expression for $\Pr(A \cap B)$ in (15.22) for statistically independent events A and B). In the following worked example both RVs are continuous.

15.5 Properties of distributions

Example The independent random variables X and Y have the PDFs $g(x) = e^{-x}$ and $h(y) = 2e^{-2y}$ respectively. Calculate the probability that X lies in the interval $1 < X \leq 2$ and Y lies in the interval $0 < Y \leq 1$.

Since X and Y are independent RVs, the joint PDF is the product of the two separate PDFs and

$$\Pr(1 < X \leq 2, \ 0 < Y \leq 1) = \int_1^2 g(x)\,dx \int_0^1 h(y)\,dy$$

$$= \int_1^2 e^{-x}\,dx \int_0^1 2e^{-2y}\,dy$$

$$= [-e^{-x}]_1^2 \times [-e^{-2y}]_0^1 = 0.23 \times 0.86 = 0.20$$

gives the required probability that both variables lie in the specified ranges. ◀

EXERCISES 15.4

1. Two unbiased six-sided dice are thrown and the two numbers shown are noted.
   (a) Calculate and plot the cumulative probability function for their sum.
   (b) Calculate and plot the cumulative probability function for the magnitude of their difference.
   (c) Calculate and plot the probability function for their product.

2. The Cauchy distribution, discussed in Section 15.8.2, has a PDF given by

$$f(x) = \frac{a}{\pi} \frac{1}{a^2 + x^2}, \qquad -\infty < x < \infty.$$

   Find the probability that x lies in the region $|x| \leq a$.

3. Two continuous RVs, X and Y, have a joint PDF given by

$$f(x, y) = A \sin \tfrac{1}{2}(x + y) \quad \text{with} \quad 0 \leq x \leq \pi, \quad 0 \leq y \leq \pi,$$

   where A is a constant. Show that the probability that X and Y are both ≤ 1 is 0.059. Are X and Y independent random variables?

15.5 Properties of distributions

For a single random variable X, the probability density function $f(x)$ contains all possible information about how the variable is distributed. However, for the purposes of comparison, it is conventional and useful to characterise $f(x)$ by certain of its properties. Most of these standard properties are defined in terms of *averages* or *expectation values* of various functions of X. In the most general case, the expectation value $E[g(X)]$ of any function $g(X)$ of the random variable X is defined as

$$E[g(X)] = \begin{cases} \sum_i g(x_i) f(x_i) & \text{for a discrete distribution,} \\ \int g(x) f(x)\,dx & \text{for a continuous distribution,} \end{cases} \tag{15.44}$$

Elementary probability

where the sum or integral is over all allowed values of X. It is assumed that the series is absolutely convergent or that the integral exists, as the case may be. From its definition it is straightforward to show that the expectation value has the following properties:

(i) if a is a constant then $E[a] = a$;
(ii) if a is a constant then $E[ag(X)] = aE[g(X)]$;
(iii) if $g(X) = s(X) + t(X)$ then $E[g(X)] = E[s(X)] + E[t(X)]$.

It should be noted that the expectation value is not a function of any particular value of X, but is instead a number that depends on the form of the probability density function $f(x)$ and the function $g(x)$; for such expectation values, 'X' acts merely as a label showing to which RV the functions f and g relate. We now consider some of these standard quantities.

15.5.1 Mean

The property most commonly used to characterise a probability distribution is its *mean*, which is defined simply as the expectation value $E[X]$ of the variable X itself. Thus, the mean is given by

$$E[X] = \begin{cases} \sum_i x_i f(x_i) & \text{for a discrete distribution,} \\ \int x f(x)\, dx & \text{for a continuous distribution.} \end{cases} \quad (15.45)$$

The alternative notations μ and $\langle x \rangle$ are also commonly used to denote the mean. If in (15.45) the series is not absolutely convergent, or the integral does not exist, we say that the distribution does not have a mean, but this is very rare in physical applications. Our worked example is taken from quantum physics.

Example The probability of finding a 1s electron in a hydrogen atom in a given infinitesimal volume dV is $\psi^*\psi\, dV$, where the quantum mechanical wavefunction ψ is given by

$$\psi = Ae^{-r/a_0}.$$

Find the value of the real constant A and thereby deduce the mean distance of the electron from the origin.

Let us consider the random variable $R = $ 'distance of the electron from the origin'. Since the 1s orbital has no θ- or ϕ-dependence (it is spherically symmetric), we may consider the infinitesimal volume element dV as the spherical shell with inner radius r and outer radius $r + dr$. Thus, $dV = 4\pi r^2\, dr$ and the PDF of R is simply

$$\Pr(r < R \leq r + dr) \equiv f(r)\, dr = 4\pi r^2 A^2 e^{-2r/a_0}\, dr.$$

The value of A is found by requiring the total probability (i.e. the probability that the electron is *somewhere*) to be unity. Since R must lie between zero and infinity, we require that

$$A^2 \int_0^\infty e^{-2r/a_0} 4\pi r^2\, dr = 1.$$

15.5 Properties of distributions

Integrating by parts we find that we must have $A = 1/(\pi a_0^3)^{1/2}$. Now, using the definition of the mean (15.45), we find

$$E[R] = \int_0^\infty r f(r)\, dr = \frac{4}{a_0^3} \int_0^\infty r^3 e^{-2r/a_0}\, dr.$$

The integral on the RHS may be also integrated by parts and takes the value $3a_0^4/8$; consequently we find that $E[R] = 3a_0/2$.[10] ◀

15.5.2 Mode and median

Although the mean discussed in the last section is the most common measure of the 'average' of a distribution, two other measures, which do not rely on the concept of expectation values, are frequently encountered.

The *mode* of a distribution is the value of the random variable X at which the probability (density) function $f(x)$ has its greatest value. If there is more than one value of X for which this is true then each value may equally be called the mode of the distribution.

The *median* M of a distribution is the value of the random variable X at which the cumulative probability function $F(x)$ takes the value $\frac{1}{2}$, i.e. $F(M) = \frac{1}{2}$. Related to the median are the lower and upper quartiles Q_l and Q_u of the PDF, which are defined to be such that

$$F(Q_l) = \tfrac{1}{4}, \qquad F(Q_u) = \tfrac{3}{4}.$$

Thus the median and lower and upper quartiles divide the PDF into four regions each containing one quarter of the probability. Smaller subdivisions are also possible, e.g. the nth percentile, P_n, of a PDF is defined by $F(P_n) = n/100$.

Example Find the mode of the PDF for the distance from the origin of the electron whose wavefunction was given in the previous example.

We found in the previous example that the PDF for the electron's distance from the origin was given by

$$f(r) = \frac{4r^2}{a_0^3} e^{-2r/a_0}. \tag{15.46}$$

As we need the value of r that makes $f(r)$ maximal, we differentiate $f(r)$ with respect to r and obtain

$$\frac{df}{dr} = \frac{8r}{a_0^3}\left(1 - \frac{r}{a_0}\right) e^{-2r/a_0}.$$

The derivative has zero value at $r = 0$, $r = a_0$ and $r = \infty$. Since $f(0) = f(\infty) = 0$ and $f(a_0)$ is clearly positive, the maximum must occur at $r = a_0$. Moreover, it is a global maximum (as opposed to just a local one). Thus the mode of $f(r)$ occurs at $r = a_0$. ◀

---

[10] a_0 is known as the Bohr radius; it can be expressed in terms of fundamental constants as $(\epsilon_0 h^2)/(\pi m_e e^2)$ and has a value of 52.9×10^{-12} m. The reader should confirm that the given expression does indeed have the dimensions of a length.

15.5.3 Variance and standard deviation

The *variance* of a distribution, $V[X]$, also written σ^2, is defined by

$$V[X] = E\left[(X - \mu)^2\right] = \begin{cases} \sum_j (x_j - \mu)^2 f(x_j) & \text{for a discrete distribution,} \\ \int (x - \mu)^2 f(x) \, dx & \text{for a continuous distribution.} \end{cases}$$

(15.47)

Here μ has been written for the expectation value $E[X]$ of X. As in the case of the mean, unless the series and the integral in (15.47) converge the distribution does not have a variance. From the definition (15.47) we may easily derive the following useful properties of $V[X]$. If a and b are constants then

(i) $V[a] = 0$,
(ii) $V[aX + b] = a^2 V[X]$.

The variance of a distribution is always positive; its positive square root is known as the *standard deviation* of the distribution and is often denoted by σ. Roughly speaking, σ measures the spread (about $x = \mu$) of the values that X can assume. For a distribution with a finite variance it can be shown that, for any positive constant c,

$$\Pr(|X - \mu| \geq c) \leq \frac{\sigma^2}{c^2}.$$

(15.48)

Thus, for *any* such distribution $f(x)$ we have, for example,

$$\Pr(|X - \mu| \geq 2\sigma) \leq \frac{1}{4} \quad \text{and} \quad \Pr(|X - \mu| \geq 3\sigma) \leq \frac{1}{9}.$$

We finish this section with a variance calculation for a continuous distribution.

► **Example** Find the standard deviation of the PDF for the distance from the origin of the electron whose wavefunction was discussed in the previous two examples.

Inserting the expression (15.46) for the PDF $f(r)$ into (15.47), the variance of the random variable R is given by

$$V[R] = \int_0^\infty (r - \mu)^2 \frac{4r^2}{a_0^3} e^{-2r/a_0} \, dr = \frac{4}{a_0^3} \int_0^\infty (r^4 - 2r^3 \mu + r^2 \mu^2) e^{-2r/a_0} \, dr,$$

where the mean $\mu = E[R] = 3a_0/2$. Integrating each term in the integrand by parts we obtain

$$V[R] = 3a_0^2 - 3\mu a_0 + \mu^2 = \frac{3a_0^2}{4}.$$

Thus the standard deviation of the distribution is $\sigma = \sqrt{3}a_0/2$. ◄

15.5.4 Moments

The mean (or expectation) of X is sometimes called the *first moment* of X, since it is defined as the sum or integral of the probability density function multiplied by the first

15.5 Properties of distributions

power of x. By a simple extension the kth moment of a distribution is defined by

$$\mu_k \equiv E[X^k] = \begin{cases} \sum_j x_j^k f(x_j) & \text{for a discrete distribution,} \\ \int x^k f(x)\, dx & \text{for a continuous distribution.} \end{cases} \quad (15.49)$$

For notational convenience, we have introduced the symbol μ_k to denote $E[X^k]$, the kth moment of the distribution. Clearly, the mean of the distribution is then denoted by μ_1, often abbreviated simply to μ, as in the previous subsection, as this rarely causes confusion.

A useful result that relates the second moment, the mean and the variance of a distribution[11] is proved using the properties of the expectation operator:

$$\begin{aligned} V[X] &= E\left[(X-\mu)^2\right] \\ &= E\left[X^2 - 2\mu X + \mu^2\right] \\ &= E\left[X^2\right] - 2\mu E[X] + \mu^2 \\ &= E\left[X^2\right] - 2\mu^2 + \mu^2 \\ &= E\left[X^2\right] - \mu^2. \end{aligned} \quad (15.50)$$

In alternative notations, this result can be written

$$\langle (x-\mu)^2 \rangle = \langle x^2 \rangle - \langle x \rangle^2 \qquad \text{or} \qquad \sigma^2 = \mu_2 - \mu_1^2.$$

We now use the biased die introduced in a previous example to illustrate the numerical workings of the definitions we have made so far.

Example A biased die has probabilities $p/2, p, p, p, p, 2p$ of showing 1, 2, 3, 4, 5, 6 respectively. Find (i) the mean, (ii) the second moment and (iii) the variance of this probability distribution.

As shown previously, by demanding that the sum of the probabilities equals unity we require $p = 2/13$. Now, using the definition of the mean (15.45) for a discrete distribution,

$$E[X] = \sum_j x_j f(x_j) = 1 \times \tfrac{1}{2}p + 2 \times p + 3 \times p + 4 \times p + 5 \times p + 6 \times 2p$$

$$= \frac{53}{2}p = \frac{53}{2} \times \frac{2}{13} = \frac{53}{13}.$$

Similarly, using the definition of the second moment (15.49),

$$E[X^2] = \sum_j x_j^2 f(x_j) = 1^2 \times \tfrac{1}{2}p + 2^2 p + 3^2 p + 4^2 p + 5^2 p + 6^2 \times 2p$$

$$= \frac{253}{2}p = \frac{253}{13}.$$

---

[11] This relationship is the basis of variance and standard deviation calculations when calculators are used to process a set of input values.

Elementary probability

Finally, using the definition of the variance (15.47), with $\mu = 53/13$, we obtain

$$V[X] = \sum_j (x_j - \mu)^2 f(x_j)$$
$$= (1-\mu)^2 \tfrac{1}{2}p + (2-\mu)^2 p + (3-\mu)^2 p + (4-\mu)^2 p + (5-\mu)^2 p + (6-\mu)^2 2p$$
$$= \left(\frac{3120}{169}\right) p = \frac{480}{169}.$$

It is easy to verify that $V[X] = E[X^2] - (E[X])^2$. ◀

The calculation of higher order moments ($k > 2$) rapidly becomes laborious and it is often simpler to use an elegant device known as a *moment generating function* to evaluate them, and sometimes even use it to find means and variances. However, limited space and topic balance mean that, within the present volume, we are unable to develop the necessary basic properties of moment generating functions, or those of the closely related *probability generating functions* which are appropriate in the case of discrete RVs. Some of the results that are quoted later in connection with particular distributions can be obtained concisely only by using the moment generating function approach, and will therefore have to be taken on trust.[12]

EXERCISES 15.5

1. The continuous RV X has a PDF proportional to $x^3 e^{-x/a}$ for $x \geq 0$. Find the mean and mode of the distribution and an algebraic equation satisfied by m, the median value of X.

2. Find the variance and standard deviation of the RV defined in the previous exercise and hence determine the minimum value of c for which (15.48) leads to a non-trivial conclusion.

3. Calculate directly the first three moments μ_1, μ_2 and μ_3 of the PF $f(x_i)$ for a single unbiased six-sided die.
   (a) Determine the mean and variance of the distribution.
   (b) Verify numerically that $\mu_3 - 3\mu_1\mu_2 + 2\mu_1^3 = 0$.
   (c) By expanding the expression $\sum_i (x_i - \mu_1)^3 p_i$, and by also considering its physical meaning, explain why result (b) is to be expected for this PF.

15.6 Functions of random variables

Suppose X is some random variable for which the probability density function $f(x)$ is known. In many cases, we are more interested in a related random variable $Y = Y(X)$,

---

[12] A full development of the generating function approach can be found in Section 30.7 of the third edition of MMPE.

15.6 Functions of random variables

where $Y(X)$ is some function of X. What is the probability density function $g(y)$ for the new random variable Y? We now discuss how to obtain this function, but restrict our discussion to continuous random variables and to cases in which $Y(X)$ has a single-valued inverse, i.e. only one x-value corresponds to any particular y-value. Discrete RVs and functions with multi-valued inverses can also be treated, but usually involve articulating all possible cases individually.

15.6.1 Continuous random variables

If X is a continuous RV, then so too is the new random variable $Y = Y(X)$. The probability that Y lies in the range y to $y + dy$ is given by

$$g(y)\,dy = \int_{dS} f(x)\,dx, \qquad (15.51)$$

where dS corresponds to all values of x for which Y lies in the range y to $y + dy$. We may write, to first order in dy, that

$$g(y)\,dy = \left|\int_{x(y)}^{x(y+dy)} f(x')\,dx'\right| = \int_{x(y)}^{x(y)+\left|\frac{dx}{dy}\right|dy} f(x')\,dx',$$

from which we obtain[13]

$$g(y) = f(x(y))\left|\frac{dx}{dy}\right|. \qquad (15.52)$$

As an example, consider the following.

Example *A lighthouse is situated at a distance L from a straight coastline, opposite a point O, and sends out a narrow continuous beam of light simultaneously in opposite directions. The beam rotates with constant angular velocity. If the random variable Y is the distance along the coastline, measured from O, of the spot that the light beam illuminates, find its probability density function.*

The situation is illustrated in Figure 15.8. Since the light beam rotates at a constant angular velocity, θ is distributed uniformly between $-\pi/2$ and $\pi/2$, and so $f(\theta) = 1/\pi$. Now $y = L\tan\theta$, which possesses the single-valued inverse $\theta = \tan^{-1}(y/L)$, provided that θ lies between $-\pi/2$ and $\pi/2$. Since $dy/d\theta = L\sec^2\theta = L(1+\tan^2\theta) = L[1+(y/L)^2]$, from (15.52) we find

$$g(y) = \frac{1}{\pi}\left|\frac{d\theta}{dy}\right| = \frac{1}{\pi L[1+(y/L)^2]} \quad \text{for } -\infty < y < \infty.$$

A distribution of this form is called a *Cauchy distribution* and is discussed in Section 15.8.2. ◄

Our discussion may be extended to cases in which the new random variable is a function of *several* other random variables. However, the only generally important result is that the PDF $p(z)$ of the sum Z of two independent continuous random variables, X and Y, in the range $-\infty$ to ∞, with PDFs $g(x)$ and $h(y)$ respectively, is given by

$$p(z)\,dz = \left(\int_{-\infty}^{\infty} g(x)h(z-x)\,dx\right)dz.$$

---

[13] Show that, if a continuous random variable X has a probability density function $f(x)$ and the corresponding cumulative probability function is $F(x)$, then the random variable $Y = F(X)$ is uniformly distributed between 0 and 1.

Elementary probability

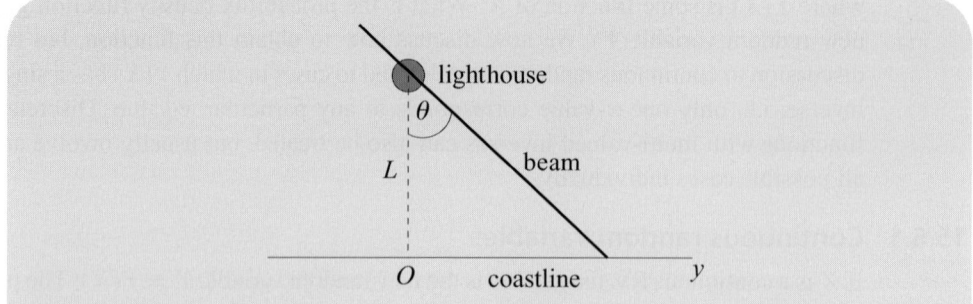

Figure 15.8 The illumination of a coastline by the beam from a lighthouse.

This states that the PDF of $Z = X + Y$ is given by the *convolution* of the PDFs of X and Y (symbolically, $p = g * h$; see the end of Section 13.4).

15.6.2 Expectation values and variances

In some cases, one is interested only in the expectation value or the variance of the new variable Z rather than in its full probability density function. For definiteness, let us again consider the random variable $Z = Z(X, Y)$, which is a function of two RVs X and Y with a known joint distribution $f(x, y)$; the results we will obtain are readily generalised to more (or fewer) variables.

It is clear that $E[Z]$ and $V[Z]$ can be obtained, in principle, by first using the methods discussed above to obtain $p(z)$ and then evaluating the appropriate sums or integrals. The intermediate step of calculating $p(z)$ is not necessary, however, since it is straightforward to obtain expressions for $E[Z]$ and $V[Z]$ in terms of the variables X and Y. For example, if X and Y are continuous RVs then the expectation value of Z is given by

$$E[Z] = \int z p(z) \, dz = \iint Z(x, y) f(x, y) \, dx \, dy. \tag{15.53}$$

An analogous result exists for discrete random variables.

Integrals of the form (15.53) are often difficult to evaluate. Nevertheless, we may use (15.53) to derive an important general result concerning expectation values. If X and Y are *any* two random variables and a and b are arbitrary constants, then by letting $Z = aX + bY$ we find

$$E[aX + bY] = aE[X] + bE[Y].$$

Furthermore, we may use this result to obtain an *approximate* expression for the expectation value $E[Z(X, Y)]$ of any arbitrary function of X and Y. Letting $\mu_X = E[X]$ and $\mu_Y = E[Y]$, and provided $Z(X, Y)$ can be reasonably approximated by the linear terms of its Taylor expansion about the point (μ_X, μ_Y), we have

$$Z(X, Y) \approx Z(\mu_X, \mu_Y) + \left(\frac{\partial Z}{\partial X}\right)(X - \mu_X) + \left(\frac{\partial Z}{\partial Y}\right)(Y - \mu_Y), \tag{15.54}$$

15.6 Functions of random variables

where the partial derivatives are evaluated at $X = \mu_X$ and $Y = \mu_Y$. Taking the expectation values of both sides, we find

$$E[Z(X,Y)] \approx Z(\mu_X, \mu_Y) + \left(\frac{\partial Z}{\partial X}\right)(E[X] - \mu_X) + \left(\frac{\partial Z}{\partial Y}\right)(E[Y] - \mu_Y).$$

Now, since $E[X] = \mu_X$ and $E[Y] = \mu_Y$, this gives the approximate result

$$E[Z(X,Y)] \approx Z(\mu_X, \mu_Y). \tag{15.55}$$

By analogy with (15.53), the variance of $Z = Z(X, Y)$ is given by

$$V[Z] = \int (z - \mu_Z)^2 p(z)\, dz = \iint [Z(x, y) - \mu_Z]^2 f(x, y)\, dx\, dy, \tag{15.56}$$

where $\mu_Z = E[Z]$. Application of this expression yields a second useful result. If X and Y are two *independent* random variables,[14] implying that $f(x, y) = g(x)h(y)$, and a, b and c are constants, then by setting $Z = aX + bY + c$ in (15.56) we obtain

$$V[aX + bY + c] = a^2 V[X] + b^2 V[Y]. \tag{15.57}$$

From (15.57) we also obtain the important special cases[15]

$$V[X + Y] = V[X - Y] = V[X] + V[Y].$$

Provided X and Y are indeed independent random variables, we may obtain an approximate expression for $V[Z(X, Y)]$, for any arbitrary function $Z(X, Y)$, in a similar manner to that used in approximating $E[Z(X, Y)]$ above. Taking the variance of both sides of (15.54), and using (15.57), we find

$$V[Z(X,Y)] \approx \left(\frac{\partial Z}{\partial X}\right)^2 V[X] + \left(\frac{\partial Z}{\partial Y}\right)^2 V[Y], \tag{15.58}$$

the partial derivatives being evaluated at $X = \mu_X$ and $Y = \mu_Y$.

EXERCISES 15.6

1. The RV X is uniformly distributed on $[0, 1]$. The related RV Y is given by $Y(X) = \sin^{-1} X$, defined to make the inverse $X = X(Y)$ single-valued with $Y(0) = 0$. Determine the PDF for Y and find the means of X and Y. Compare your set of answers with the single-variable analogue of approximation (15.55).

2. The RV Z is the sum of two samples of the RV X of the previous exercise. By writing the PDF of X as $f(x) = H(x)H(1 - x)$, find the PDF for Z. You will need to consider four separate intervals for z.

---

[14] If the RVs are not independent, then an additional term involving their covariance is needed in (15.57), as in Equation (15.89).
[15] That is, it does not matter whether X and Y are added or subtracted to form a new RV, the variance of the new RV is always formed by *adding* those of X and Y.

Elementary probability

Table 15.1 *Some important discrete probability distributions.*

Distribution	Probability law $f(x)$	$E[X]$	$V[X]$
binomial	${}^nC_x p^x q^{n-x}$	np	npq
negative binomial	${}^{r+x-1}C_x p^r q^x$	$\dfrac{rq}{p}$	$\dfrac{rq}{p^2}$
geometric	$q^{x-1} p$	$\dfrac{1}{p}$	$\dfrac{q}{p^2}$
hypergeometric	$\dfrac{(Np)!(Nq)!n!(N-n)!}{x!(Np-x)!(n-x)!(Nq-n+x)!N!}$	np	$\dfrac{N-n}{N-1}npq$
Poisson	$\dfrac{\lambda^x}{x!} e^{-\lambda}$	λ	λ

3. (a) Show by direct calculation that the variance of an RV uniformly distributed on the interval $[a, b]$ is $\frac{1}{12}(b-a)^2$.
   (b) Independent random variables X and Y are uniformly distributed on $[0, 3]$ and $[2, 4]$ respectively. The derived RV Z is given by $Z = X^2(Y+1)$. Find approximate values for the mean and variance of Z.

15.7 Important discrete distributions

Having discussed some general properties of distributions, we now consider the more important discrete distributions encountered in physical applications. These are discussed in detail below, and summarised for convenience in Table 15.1; we refer the reader to the relevant section below for an explanation of the symbols used.

15.7.1 The binomial distribution

Perhaps the most important discrete probability distribution is the *binomial distribution*. This distribution describes processes that consist of a number of independent identical *trials* with two possible outcomes, A and $B = \bar{A}$. We may call these outcomes 'success' and 'failure' respectively. If the probability of a success is $\Pr(A) = p$ then the probability of a failure is $\Pr(B) = q = 1 - p$. If we perform n trials then the discrete random variable

$$X = \text{number of times } A \text{ occurs}$$

can take the values $0, 1, 2, \ldots, n$; its distribution amongst these values is described by the *binomial distribution*.

We now calculate the probability that in n trials we obtain x successes (and so $n - x$ failures). One way of obtaining such a result is to have x successes followed by $n - x$ failures. Since the trials are assumed independent, the probability of this is

$$\underbrace{pp \cdots p}_{x \text{ times}} \times \underbrace{qq \cdots q}_{n - x \text{ times}} = p^x q^{n-x}.$$

15.7 Important discrete distributions

This is, however, just one permutation of x successes and $n - x$ failures. The total number of permutations of n objects, of which x are identical and of type 1 and $n - x$ are identical and of type 2, is given by (15.33) as

$$\frac{n!}{x!(n-x)!} \equiv {}^nC_x.$$

Therefore, the total probability of obtaining x successes from n trials is

$$f(x) = \Pr(X = x) = {}^nC_x \, p^x q^{n-x} = {}^nC_x \, p^x (1-p)^{n-x}, \quad (15.59)$$

which is the *binomial probability distribution formula*.

When a random variable X follows the binomial distribution for n trials, with a probability of success p, we write $X \sim \text{Bin}(n, p)$; such a random variable X is often referred to as a binomial *variate*.

As a simple example of a binomial calculation, consider the following.

Example If an unbiased single six-sided die is rolled five times, what is the probability that a six is thrown exactly three times?

Here the number of 'trials' $n = 5$, and we are interested in the random variable

$$X = \text{number of sixes thrown}.$$

Since the probability of a 'success' is $p = \frac{1}{6}$, Equation (15.59) gives

$$\Pr(X = 3) = \frac{5!}{3!(5-3)!} \left(\frac{1}{6}\right)^3 \left(\frac{5}{6}\right)^{(5-3)} = 0.032$$

as the probability of obtaining exactly three sixes in five throws. ◂

For evaluating binomial probabilities a useful result is the binomial recurrence formula

$$\Pr(X = x + 1) = \frac{p}{q}\left(\frac{n-x}{x+1}\right)\Pr(X = x), \quad (15.60)$$

which enables successive probabilities $\Pr(X = x + k)$, $k = 1, 2, \ldots$, to be calculated once $\Pr(X = x)$ is known; it is often quicker to use than (15.59) if several values are needed.

We note that, as required, the binomial distribution satisfies

$$\sum_{x=0}^{n} f(x) = \sum_{x=0}^{n} {}^nC_x \, p^x q^{n-x} = (p+q)^n = 1. \quad (15.61)$$

Furthermore, from the definitions of $E[X]$ and $V[X]$ for a discrete distribution, we may show that for the binomial distribution $E[X] = np$ and $V[X] = npq$. The direct summations involved are rather cumbersome and the results follow more simply from the probability or moment generating functions of the distribution. However, we now give proofs of these two results based directly on their definitions.[16]

---

[16] Though it must be admitted that the algebra has been somewhat shortened as a result of knowing the expected answer.

Elementary probability

Considering first the mean of the binomial distribution, $\mu = E[X]$. From its definition:

$$\mu = \sum_{x=0}^{n} x(^nC_x \, p^x q^{n-x}) = \sum_{x=0}^{n} x \, \frac{n!}{x!\,(n-x)!} \, p^x q^{n-x}$$

$$= np \sum_{x=1}^{n} \frac{(n-1)!}{(x-1)!\,(n-x)!} \, p^{x-1} q^{n-x}.$$

In the second line we have omitted the $x = 0$ term, as, containing a factor x, it is always zero. Now we change the summation index by setting $x = y + 1$, where the new index y runs from 0 to $n - 1$:

$$\mu = np \sum_{y=0}^{n-1} \frac{(n-1)!}{y!\,(n-1-y)!} \, p^y q^{n-1-y}$$

$$= np \sum_{y=0}^{n-1} {}^{n-1}C_y \, p^y q^{n-1-y} = np(p+q)^{n-1} = np.$$

In the final line we have used the identity (15.61), but with n replaced by $n - 1$. Thus the mean of the binomial distribution has the (not unexpected) value of np.

The most convenient way to evaluate the variance $V[X]$ of the binomial distribution is to first find the expectation value of $X(X - 1)$. Following a similar line to that used above, of omitting zero terms ($x = 0$ and $x = 1$ in this case), changing the summation index, and applying identity (15.61), we have

$$E[X(X-1)] = \sum_{x=0}^{n} x(x-1) \, \frac{n!}{x!\,(n-x)!} \, p^x q^{n-x}$$

$$= n(n-1)p^2 \sum_{x=2}^{n} \frac{(n-2)!}{(x-2)!\,(n-x)!} \, p^{x-2} q^{n-x}$$

$$= n(n-1)p^2 \sum_{y=0}^{n-2} \frac{(n-2)!}{y!\,(n-2-y)!} \, p^y q^{n-2-y}$$

$$= n(n-1)p^2 \sum_{y=0}^{n-2} {}^{n-2}C_y \, p^y q^{n-2-y}$$

$$= n(n-1)p^2(p+q)^{n-2} = n(n-1)p^2.$$

Now, from (15.50),

$$V[X] = E[X^2] - (E[X])^2 = E[X^2 - X] + E[X] - (E[X])^2$$
$$= n(n-1)p^2 + np - n^2 p^2$$
$$= np - np^2 = npq.$$

It is worth noting that the variance is zero if $p = 0$ or $p = 1$ and that, for a given number of trials, the variance is largest when $p = q = 0.5$.

15.7 Important discrete distributions

Multiple binomial distributions

Suppose that X_i are two or more *independent* random variables, all of which are described by binomial distributions with a common probability of success p, but with (in general) different numbers of trials. With the X_i, $i = 1, 2, \ldots, N$, distributed as $X_i \sim \text{Bin}(n_i, p)$, it can be shown that their sum $Z = X_1 + X_2 + \cdots + X_N$ is distributed as $Z \sim \text{Bin}(n_1 + n_2 + \cdots + n_N, p)$. This result is as expected, since the result of $\sum_i n_i$ trials cannot depend on how they are split up.

The corresponding result for two independent RVs with a common value of n, but different values of p, is that the mean and variance of their sum are $n(p_1 + p_2)$ and $n(p_1 q_1 + p_2 q_2)$, respectively.

Unfortunately, no equivalent simple results exist for the probability distributions of the *difference* between binomially distributed variables; clearly, this negative aspect also applies to linear sums of RVs containing one or more minus signs.

15.7.2 The multinomial distribution

The binomial distribution describes the probability of obtaining x 'successes' from n independent trials, where each trial has only two possible outcomes. This may be generalised to the case where each trial has k possible outcomes with respective probabilities p_1, $p_2, \ldots, p_k$. If we consider the random variables X_i, $i = 1, 2, \ldots, n$, to be the number of outcomes of type i in n trials then we may calculate their joint probability function

$$f(x_1, x_2, \ldots, x_k) = \Pr(X_1 = x_1, \ X_2 = x_2, \ \ldots, \ X_k = x_k),$$

where we must have $\sum_{i=1}^{k} x_i = n$. In n trials the probability of obtaining x_1 outcomes of type 1, followed by x_2 outcomes of type 2, etc. is given by

$$p_1^{x_1} p_2^{x_2} \cdots p_k^{x_k}.$$

However, the number of distinguishable permutations of this result is

$$\frac{n!}{x_1! x_2! \cdots x_k!},$$

and thus

$$f(x_1, x_2, \ldots, x_k) = \frac{n!}{x_1! x_2! \cdots x_k!} p_1^{x_1} p_2^{x_2} \cdots p_k^{x_k}. \tag{15.62}$$

This is the *multinomial probability distribution*.

If $k = 2$ then the multinomial distribution reduces to the familiar binomial distribution. Although in this form the binomial distribution appears to be a function of two random variables, it must be remembered that, in fact, since $p_2 = 1 - p_1$ and $x_2 = n - x_1$, the distribution of X_1 is entirely determined by the parameters p_1 (or p_2) and n. That X_1 has a *binomial* distribution follows from recalling that it represents the number of objects of a particular type obtained from sampling with replacement, the very process that led to the original definition of the binomial distribution. In fact, any of the random variables X_i has a binomial distribution, with parameters n and p_i. It immediately follows that

$$E[X_i] = np_i \quad \text{and} \quad V[X_i]^2 = np_i(1 - p_i). \tag{15.63}$$

The following worked example applies the multinomial distribution several times.

Example At a village fête patrons were invited, for a 10p entry fee, to pick without looking six tickets from a drum containing equal large numbers of red, blue and green tickets. If five or more of the tickets were of the same colour a prize of 100p was awarded. A consolation award of 40p was made if two tickets of each colour were picked. Was a good time had by all?

In this case, all types of outcome (red, blue and green) have the same probabilities. The probability of obtaining any given combination of tickets is given by the multinomial distribution with $n = 6$, $k = 3$ and $p_i = \frac{1}{3}$, $i = 1, 2, 3$.

(i) The probability of picking six tickets of the same colour is given by

$$\text{Pr (six of the same colour)} = 3 \times \frac{6!}{6!0!0!} \left(\frac{1}{3}\right)^6 \left(\frac{1}{3}\right)^0 \left(\frac{1}{3}\right)^0 = \frac{1}{243}.$$

The factor of 3 is present because there are three different colours.

(ii) The probability of picking five tickets of one colour and one ticket of another colour is

$$\text{Pr(five of one colour; one of another)} = 3 \times 2 \times \frac{6!}{5!1!0!} \left(\frac{1}{3}\right)^5 \left(\frac{1}{3}\right)^1 \left(\frac{1}{3}\right)^0 = \frac{4}{81}.$$

The factors of 3 and 2 are included because there are three ways to choose the colour of the five matching tickets, and then two ways to choose the colour of the remaining ticket.

(iii) Finally, the probability of picking two tickets of each colour is

$$\text{Pr (two of each colour)} = \frac{6!}{2!2!2!} \left(\frac{1}{3}\right)^2 \left(\frac{1}{3}\right)^2 \left(\frac{1}{3}\right)^2 = \frac{10}{81}.$$

Thus the expected return to any patron was, in pence,

$$100 \left(\frac{1}{243} + \frac{4}{81}\right) + \left(40 \times \frac{10}{81}\right) = 10.29.$$

A good time was had by all but the stall holder! ◀

15.7.3 The geometric and negative binomial distributions

A special case of the binomial distribution occurs when instead of the number of successes we consider the discrete random variable

$$X = \text{number of trials required to obtain the first success.}$$

The probability that x trials are required in order to obtain the first success, is simply the probability of obtaining $x - 1$ failures followed by one success. If the probability of a success on each trial is p, then for $x > 0$

$$f(x) = \Pr(X = x) = (1 - p)^{x-1} p = q^{x-1} p,$$

where $q = 1 - p$. This distribution is sometimes called the *geometric distribution*. Its mean and variance can be shown to be

$$E[X] = \frac{1}{p}, \qquad V[X] = \frac{q}{p^2}.$$

The former is as would probably be expected; the latter is more difficult to anticipate.

15.7 Important discrete distributions

Another distribution closely related to the binomial is the *negative binomial distribution*. This describes the probability distribution of the random variable

$$X = \text{number of failures before the } r\text{th success.}$$

One way of obtaining x failures before the rth success is to have $r - 1$ successes followed by x failures followed by the rth success, for which the probability is

$$\underbrace{pp \cdots p}_{r-1 \text{ times}} \times \underbrace{qq \cdots q}_{x \text{ times}} \times p = p^r q^x.$$

However, the first $r + x - 1$ factors constitute just one permutation of $r - 1$ successes and x failures. The total number of permutations of these $r + x - 1$ objects, of which $r - 1$ are identical and of type 1 and x are identical and of type 2, is $^{r+x-1}C_x$. Therefore, the total probability of obtaining x failures before the rth success is

$$f(x) = \Pr(X = x) = {}^{r+x-1}C_x p^r q^x,$$

which is called the *negative binomial distribution* (see the related discussion on p. 617). Using moment generating functions, it is relatively straightforward to show that its mean and variance are given by

$$E[X] = \frac{rq}{p} \quad \text{and} \quad V[X] = \frac{rq}{p^2}.$$

15.7.4 The hypergeometric distribution

In Section 15.7.1 we saw that the probability of obtaining x successes in n *independent* trials was given by the binomial distribution. Suppose that these n 'trials' actually consist of drawing at random n balls, from a set of N such balls of which M are red and the rest white. Let us consider the random variable $X =$ number of red balls drawn.

On the one hand, if the balls are drawn *with replacement* then the trials are independent and the probability of drawing a red ball is $p = M/N$ each time. Therefore, the probability of drawing x red balls in n trials is given by the binomial distribution as

$$\Pr(X = x) = \frac{n!}{x!(n-x)!} p^x (1-p)^{n-x}.$$

On the other hand, if the balls are drawn *without replacement* the trials are not independent and the probability of drawing a red ball depends on how many red balls have already been drawn. We can, however, still derive a general formula for the probability of drawing x red balls in n trials, as follows.

The number of ways of drawing x red balls from M is ${}^M C_x$, and the number of ways of drawing $n - x$ white balls from $N - M$ is ${}^{N-M} C_{n-x}$. Therefore, the total number of ways to obtain x red balls in n trials is ${}^M C_x \, {}^{N-M} C_{n-x}$. However, the total number of ways of drawing n objects from N is simply ${}^N C_n$. Hence the probability of obtaining x red balls

Elementary probability

in n trials is

$$\Pr(X = x) = \frac{{}^M C_x \, {}^{N-M}C_{n-x}}{{}^N C_n}$$

$$= \frac{M!}{x!(M-x)!} \frac{(N-M)!}{(n-x)!(N-M-n+x)!} \frac{n!(N-n)!}{N!} \quad (15.64)$$

$$= \frac{(Np)!(Nq)!\,n!(N-n)!}{x!(Np-x)!(n-x)!(Nq-n+x)!\,N!}, \quad (15.65)$$

where in the last line $p = M/N$ and $q = 1 - p$. This is called the *hypergeometric distribution*.

By performing the relevant summations directly, it may be shown that the hypergeometric distribution has mean

$$E[X] = n\frac{M}{N} = np$$

and variance[17]

$$V[X] = \frac{nM(N-M)(N-n)}{N^2(N-1)} = \frac{N-n}{N-1}npq.$$

Note that if n, the number of trials (balls drawn), is small compared with each of N, M and $N - M$ then not replacing the balls is of little consequence, and we may approximate the hypergeometric distribution by the binomial distribution (with $p = M/N$); this is much easier to evaluate.

An application of the hypergeometric distribution that is of interest (at least in the UK) far beyond the realms of physics and engineering, is the calculation of the odds against winning the National Lottery.

Example In the UK National Lottery each participant chooses six different numbers between 1 and 49. In each weekly draw six numbered winning balls are subsequently drawn. Find the probabilities that a participant correctly predicts 0, 1, 2, 3, 4, 5, 6 winning numbers.

The probabilities are given by a hypergeometric distribution with N (the total number of balls) = 49, M (the number of winning balls drawn) = 6, and n (the number of numbers chosen by each participant) = 6. Thus, substituting in (15.64), we find

$$\Pr(0) = \frac{{}^6C_0 \, {}^{43}C_6}{{}^{49}C_6} = \frac{1}{2.29}, \quad \Pr(1) = \frac{{}^6C_1 \, {}^{43}C_5}{{}^{49}C_6} = \frac{1}{2.42},$$

$$\Pr(2) = \frac{{}^6C_2 \, {}^{43}C_4}{{}^{49}C_6} = \frac{1}{7.55}, \quad \Pr(3) = \frac{{}^6C_3 \, {}^{43}C_3}{{}^{49}C_6} = \frac{1}{56.6},$$

$$\Pr(4) = \frac{{}^6C_4 \, {}^{43}C_2}{{}^{49}C_6} = \frac{1}{1032}, \quad \Pr(5) = \frac{{}^6C_5 \, {}^{43}C_1}{{}^{49}C_6} = \frac{1}{54\,200},$$

$$\Pr(6) = \frac{{}^6C_6 \, {}^{43}C_0}{{}^{49}C_6} = \frac{1}{13.98 \times 10^6}.$$

---

[17] Note that, for choosing just one ball, the variance is the same as that for a binomial trial, namely $npq = pq$, but if all N are chosen there is no variance in the results – the number of red balls included will always be M.

15.7 Important discrete distributions

It can easily be seen that:

(i) $\sum_{i=0}^{6} \Pr(i) = 0.44 + 0.41 + 0.13 + 0.02 + O(10^{-3}) = 1$, as expected;
(ii) as the stake money is £1, the prize of £10 for three correct predictions, and a typical jackpot share of about £2M for six correct predictions are both less than 20% of a 'fair return'.

Perhaps a little surprising is the fact that the chance of no correct predictions is only a little larger than the chance of one correct – but neither gets you a prize! ◀

15.7.5 The Poisson distribution

We have seen that the binomial distribution describes the number of successful outcomes in a certain number of trials n. The Poisson distribution also describes the probability of obtaining a given number of successes but for situations in which the number of 'trials' cannot be enumerated; rather it describes the situation in which discrete events occur in a continuum. Typical examples of discrete random variables X described by a Poisson distribution are the number of telephone calls received by a switchboard in a given interval, or the number of stars above a certain brightness in a particular area of the sky. Given a mean rate of occurrence λ of these events in the relevant interval or area, the Poisson distribution gives the probability $\Pr(X = x)$ that exactly x events will occur.

The form of the Poisson distribution can be derived as the limit of the binomial distribution when the number of trials $n \to \infty$ and the probability of 'success' $p \to 0$, in such a way that $np = \lambda$ remains finite. However, we now derive it directly by establishing the differential equation it satisfies.

Let us consider the example of a telephone switchboard. Suppose that the probability that x calls have been received in a time interval t is $P_x(t)$. If the average number of calls received in a unit time is λ then in a further small time interval Δt the probability of receiving a call is $\lambda \Delta t$, provided Δt is short enough that the probability of receiving two or more calls in this small interval is negligible. Similarly, the probability of receiving no call during the same small interval is simply $1 - \lambda \Delta t$.

Thus, for $x > 0$, the probability of receiving exactly x calls in the total interval $t + \Delta t$ is given by

$$P_x(t + \Delta t) = P_x(t)(1 - \lambda \Delta t) + P_{x-1}(t)\lambda \Delta t.$$

Rearranging the equation, dividing through by Δt and letting $\Delta t \to 0$, we obtain the differential recurrence equation

$$\frac{dP_x(t)}{dt} = \lambda P_{x-1}(t) - \lambda P_x(t). \tag{15.66}$$

For $x = 0$ (i.e. no calls received), however, (15.66) simplifies to

$$\frac{dP_0(t)}{dt} = -\lambda P_0(t),$$

which may be integrated to give $P_0(t) = P_0(0)e^{-\lambda t}$. But since the probability of receiving no calls in a zero time interval must equal unity, we have $P_0(0) = 1$ and $P_0(t) = e^{-\lambda t}$. This expression for $P_0(t)$ may then be substituted back into (15.66) with $x = 1$ to obtain

Elementary probability

a differential equation for $P_1(t)$ that has the solution $P_1(t) = \lambda t e^{-\lambda t}$. We may repeat this process to obtain expressions for $P_2(t), P_3(t), \ldots, P_x(t)$, and we find[18]

$$P_x(t) = \frac{(\lambda t)^x}{x!} e^{-\lambda t}. \tag{15.67}$$

By setting $t = 1$ in (15.67), we obtain the Poisson distribution,

$$f(x) = \Pr(X = x) = \frac{e^{-\lambda} \lambda^x}{x!}, \tag{15.68}$$

for obtaining exactly x calls in a unit time interval.

If a discrete random variable is described by a Poisson distribution of mean λ then we write $X \sim \text{Po}(\lambda)$. As it must be, the sum of the probabilities is unity:

$$\sum_{x=0}^{\infty} \Pr(X = x) = e^{-\lambda} \sum_{x=0}^{\infty} \frac{\lambda^x}{x!} = e^{-\lambda} e^{\lambda} = 1.$$

From (15.68) we may also derive the *Poisson recurrence formula*,

$$\Pr(X = x + 1) = \frac{\lambda}{x + 1} \Pr(X = x) \quad \text{for } x = 0, 1, 2, \ldots, \tag{15.69}$$

which enables successive probabilities to be calculated easily once one is known.

Example A person receives on average one email message per half-hour interval. Assuming that the emails are received randomly in time, find the probabilities that in any particular hour 0, 1, 2, 3, 4, 5 messages are received.

Let X = number of emails received per hour. Clearly the mean number of emails per hour is two, and so X follows a Poisson distribution with $\lambda = 2$, i.e.

$$\Pr(X = x) = \frac{2^x}{x!} e^{-2}.$$

Thus $\Pr(X = 0) = e^{-2} = 0.135$, $\Pr(X = 1) = 2e^{-2} = 0.271$, $\Pr(X = 2) = 2^2 e^{-2}/2! = 0.271$, $\Pr(X = 3) = 2^3 e^{-2}/3! = 0.180$, $\Pr(X = 4) = 2^4 e^{-2}/4! = 0.090$, $\Pr(X = 5) = 2^5 e^{-2}/5! = 0.036$. These values are most conveniently found using the recurrence formula (15.69). ◀

The above example illustrates the point that a Poisson distribution typically rises and then falls. It either has a maximum when x is equal to the integer part of λ or, if λ happens to be an integer, has equal maximal values at $x = \lambda - 1$ and $x = \lambda$. The Poisson distribution always has a long 'tail' towards higher values of X, but the higher the value of the mean the more symmetric the distribution becomes. Some typical Poisson distributions are shown in Figure 15.9.

Using the definitions of mean and variance, we may show that, for the Poisson distribution, $E[X] = \lambda$ and $V[X] = \lambda$. As in the case of the binomial distribution, these results are most elegantly proved using the moment generating function, though direct proof of

---

[18] Verify, by substitution, that these solutions do satisfy (15.66).

15.7 Important discrete distributions

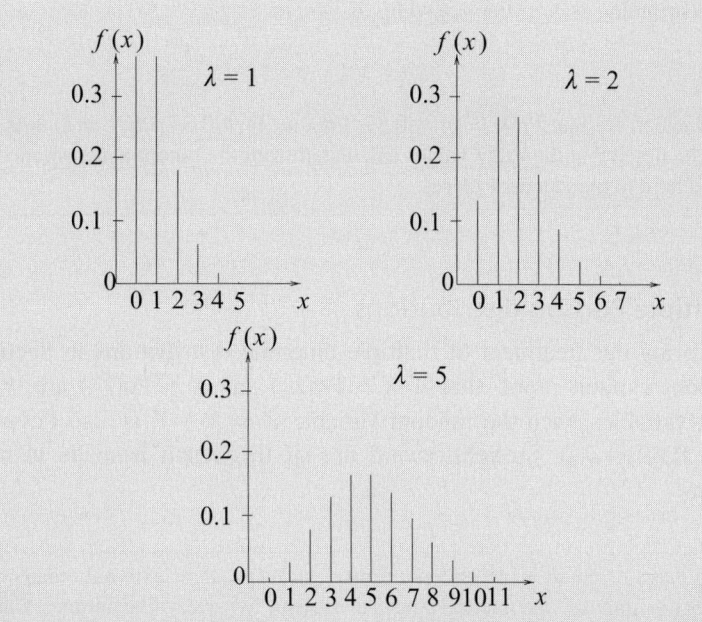

Figure 15.9 Three Poisson distributions for different values of the parameter λ.

the mean is straightforward:

$$\mu = \sum_{x=0}^{\infty} x \frac{e^{-\lambda}\lambda^x}{x!} = \lambda e^{-\lambda} \sum_{x=1}^{\infty} \frac{\lambda^{x-1}}{(x-1)!} = \lambda e^{-\lambda} \sum_{y=0}^{\infty} \frac{\lambda^y}{y!} = \lambda e^{-\lambda} e^{\lambda} = \lambda.$$

The Poisson approximation to the binomial distribution

As stated earlier, the Poisson distribution can be derived as the limit of the binomial distribution when $n \to \infty$ and $p \to 0$ in such a way that $np = \lambda$ remains finite, where λ is the mean of the Poisson distribution. It is not surprising, therefore, that the Poisson distribution is a very good approximation to the binomial distribution for large n (≥ 50, say) and small p (≤ 0.1, say). Moreover, it is easier to calculate, as it involves fewer factorials.

Example In a large batch of light bulbs, the probability that a bulb is defective is 0.5%. For a sample of 200 bulbs taken at random, find the approximate probabilities that none, one and two of the bulbs respectively are defective.

Let the random variable X = number of defective bulbs in a sample. This is distributed as $X \sim$ Bin(200, 0.005), implying that $\lambda = np = 1.0$. Since n is large and p small, we may approximate the distribution as $X \sim \text{Po}(1)$, giving

$$\Pr(X = x) \approx e^{-1}\frac{1^x}{x!},$$

from which we find $\Pr(X = 0) \approx 0.37$, $\Pr(X = 1) \approx 0.37$, $\Pr(X = 2) \approx 0.18$. For comparison, it may be noted that the exact values calculated from the binomial distribution are identical to those found here to two decimal places. ◂

Multiple Poisson distributions

Mirroring our treatment of multiple binomial distributions in Section 15.7.1, we note, without explicit proof, that if $X \sim \text{Po}(\lambda_1)$ and $Y \sim \text{Po}(\lambda_2)$ are two *independent* random variables, then the random variable $Z = X + Y$ is also Poisson distributed, with $Z \sim \text{Po}(\lambda_1 + \lambda_2)$. Straightforward use of this result is made in the worked example below.

Example Two types of email arrive independently and at random: external emails at a mean rate of one every five minutes and internal emails at a rate of two every five minutes. Calculate the probability of receiving two or more emails in any two-minute interval.

Let

$$X = \text{number of external emails per two-minute interval},$$
$$Y = \text{number of internal emails per two-minute interval}.$$

Since we expect on average one external email and two internal emails every five minutes we have, when considering a two-minute interval, that $X \sim \text{Po}(0.4)$ and $Y \sim \text{Po}(0.8)$. Letting $Z = X + Y$ we have $Z \sim \text{Po}(0.4 + 0.8) = \text{Po}(1.2)$. Now

$$\Pr(Z \geq 2) = 1 - \Pr(Z < 2) = 1 - \Pr(Z = 0) - \Pr(Z = 1)$$

and

$$\Pr(Z = 0) = e^{-1.2} = 0.301,$$
$$\Pr(Z = 1) = e^{-1.2}\frac{1.2}{1} = 0.361.$$

Hence $\Pr(Z \geq 2) = 1 - 0.301 - 0.361 = 0.338$ gives the probability of receiving at least two emails within the two-minute interval. ◂

The above result can be extended, of course, to any number of Poisson processes, so that if $X_i = \text{Po}(\lambda_i)$, $i = 1, 2, \ldots, n$ then the random variable $Z = X_1 + X_2 + \cdots + X_n$ is distributed as $Z \sim \text{Po}(\lambda_1 + \lambda_2 + \cdots + \lambda_n)$.

EXERCISES 15.7

1. A bag contains large numbers of 20p, 10p, 5p and 1p coins in equal monetary amounts. A sequence of seven coins is drawn at random from the bag, each coin being replaced

15.8 Important continuous distributions

Table 15.2 *Some important continuous probability distributions.*

Distribution	Probability law $f(x)$	$E[X]$	$V[X]$
Gaussian	$\dfrac{1}{\sigma\sqrt{2\pi}} \exp\left[-\dfrac{(x-\mu)^2}{2\sigma^2}\right]$	μ	σ^2
exponential	$\lambda e^{-\lambda x}$	$\dfrac{1}{\lambda}$	$\dfrac{1}{\lambda^2}$
chi-squared	$\dfrac{1}{2^{n/2}\Gamma(n/2)} x^{(n/2)-1} e^{-x/2}$	n	$2n$
uniform	$\dfrac{1}{b-a}$	$\dfrac{a+b}{2}$	$\dfrac{(b-a)^2}{12}$

before the next is drawn. Show that the chance that the total value of the coins selected is exactly 58p is a little greater than 1 in 400.

2. For a series of binomial trials, show that the expected numbers of trials needed to obtain the *r*th success deduced from (a) the geometric distribution and (b) the negative binomial distribution are consistent.

3. In an alternative form of the UK National Lottery, known as *Thunderball*, the contestant has to attempt to predict the five winning numbered balls out of a pool of 34, as well as the one winning numbered ball out of a separate pool of 14 (the so-called thunderball). Show that the chances of selecting four correct numbers and the thunderball are about 26 866 to 1.

4. Premium bonds are a form of savings certificate for which the aggregated interest from all issued certificates is distributed as a number of prizes. The draw to determine the winning bonds takes place monthly. The holder of a (fixed) number of bonds finds that over a long period they can expect to win a prize twice a year. What are the chances of winning two or more prizes in a single draw?

5. A chess board has 300 small beads scattered at random on its surface. *Estimate* the chances of finding a white square that has no bead on it, if the square (a) was nominated before the beads were scattered and (b) selected after the scattering. Explain why the scattering procedure could be described in terms of trials governed by a binomial distribution.

15.8 Important continuous distributions

Having discussed the most commonly encountered discrete probability distributions, we now consider some of the more important continuous probability distributions. These are summarised for convenience in Table 15.2; we refer the reader to the relevant subsection below for an explanation of the symbols used.

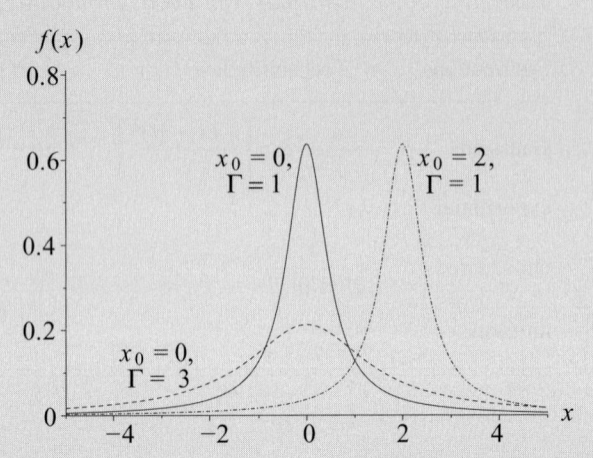

Figure 15.10 The PDF $f(x)$ for the Breit–Wigner distribution for different values of the parameters x_0 and Γ.

15.8.1 The uniform distribution

Firstly we mention the very simple, but common, *uniform distribution*, which describes a continuous random variable that has a constant PDF over its allowed range of values. If the limits on X are a and b then

$$f(x) = \begin{cases} 1/(b-a) & \text{for } a \leq x \leq b, \\ 0 & \text{otherwise.} \end{cases}$$

Its mean and variance are found by direct calculation to be

$$E[X] = \frac{a+b}{2}, \qquad V[X] = \frac{(b-a)^2}{12}.$$

15.8.2 The Cauchy and Breit–Wigner distributions

A random variable X (in the range $-\infty$ to ∞) that obeys the *Cauchy distribution* is described by the PDF

$$f(x) = \frac{1}{\pi} \frac{1}{1+x^2}.$$

This is a special case of the *Breit–Wigner distribution*

$$f(x) = \frac{1}{\pi} \frac{\frac{1}{2}\Gamma}{\frac{1}{4}\Gamma^2 + (x-x_0)^2},$$

which is encountered in the study of nuclear and particle physics. In Figure 15.10, we plot some examples of the Breit–Wigner distribution for several values of the parameters x_0 and Γ.

We see from the figure that the peak (or mode) of the distribution occurs at $x = x_0$. It is also straightforward to show that the parameter Γ is equal to the width of the peak at half the maximum height.

Although the Breit–Wigner distribution is not generally as important as the other continuous distributions discussed here, it is mentioned because it posts a warning. Despite being normalisable, it does not formally possess a mean. This is because the integrals $\int_{-\infty}^{0} x f(x)\,dx$ and $\int_{0}^{\infty} x f(x)\,dx$ both diverge. Similar divergences occur for all higher moments of the distribution.[19]

15.8.3 The Gaussian distribution

By far the most important continuous probability distribution is the *Gaussian* or *normal* distribution. The reason for its importance is that a great many random variables of interest, in all areas of the physical sciences and beyond, are described either exactly or approximately by a Gaussian distribution. Moreover, the Gaussian distribution can be used to approximate other, more complicated, probability distributions.

The probability density function for a Gaussian distribution of a random variable X, with mean $E[X] = \mu$ and variance $V[X] = \sigma^2$, takes the form

$$f(x) = \frac{1}{\sigma\sqrt{2\pi}} \exp\left[-\frac{1}{2}\left(\frac{x-\mu}{\sigma}\right)^2\right]. \tag{15.70}$$

The factor $1/\sqrt{2\pi}$ arises from the normalisation requirement of the distribution, i.e. that

$$\int_{-\infty}^{\infty} f(x)\,dx = 1.$$

The Gaussian distribution is symmetric about the point $x = \mu$ and has the characteristic 'bell' shape shown in Figure 15.11. The width of the curve is described by the standard deviation σ; if σ is large then the curve is broad, and if σ is small then the curve is narrow (see the figure). At $x = \mu \pm \sigma$, $f(x)$ falls to $e^{-1/2} \approx 0.61$ of its peak value; these points are points of inflection, where $d^2 f/dx^2 = 0$. When a random variable X follows a Gaussian distribution with mean μ and variance σ^2, we write $X \sim N(\mu, \sigma^2)$.

The effects of changing μ and σ are only to shift the curve along the x-axis or to broaden or narrow it, respectively. Thus all Gaussians are equivalent in that a change of origin and scale can reduce them to a standard form. We therefore consider the random variable $Z = (X - \mu)/\sigma$, for which the PDF takes the form

$$\phi(z) = \frac{1}{\sqrt{2\pi}} \exp\left(-\frac{z^2}{2}\right), \tag{15.71}$$

which is called the *standard Gaussian distribution* and has mean $\mu = 0$ and variance $\sigma^2 = 1$. The random variable Z is called the *standard variable*.

---

[19] Using a change of variable and the normal definition of an infinite integral over $-\infty$ to $+\infty$, the distribution could be said to have a mean of x_0, but even this device fails when applied to a calculation of the variance.

Elementary probability

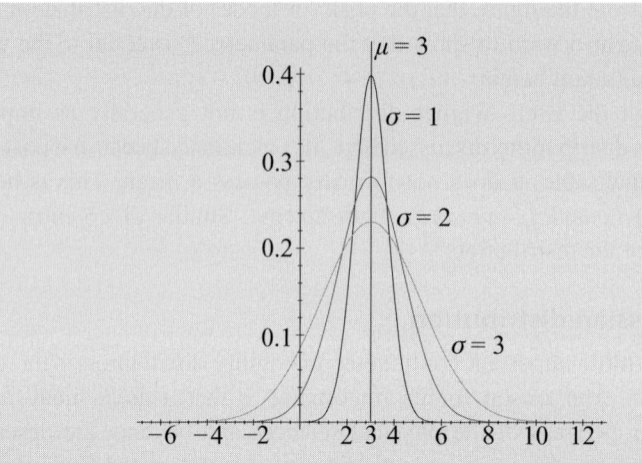

Figure 15.11 The Gaussian or normal distribution for mean $\mu = 3$ and various values of the standard deviation σ.

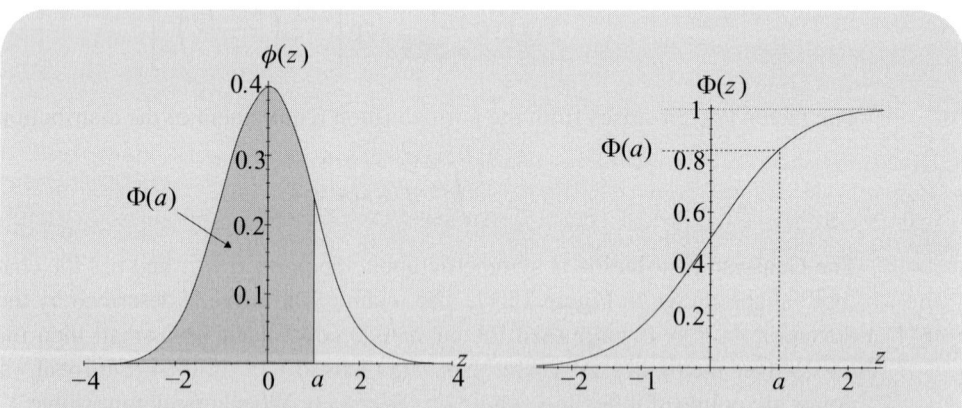

Figure 15.12 On the left, the standard Gaussian distribution $\phi(z)$; the shaded area gives $\Pr(Z < a) = \Phi(a)$. On the right, the cumulative probability function $\Phi(z)$ for a standard Gaussian distribution $\phi(z)$.

From (15.70) we can define the cumulative probability function for a Gaussian distribution as

$$F(x) = \Pr(X < x) = \frac{1}{\sigma\sqrt{2\pi}} \int_{-\infty}^{x} \exp\left[-\frac{1}{2}\left(\frac{u-\mu}{\sigma}\right)^2\right] du, \tag{15.72}$$

where u is a (dummy) integration variable. Unfortunately, this (indefinite) integral cannot be evaluated analytically. It is therefore standard practice to tabulate values of the cumulative probability function for the standard Gaussian distribution (see Figure 15.12),

15.8 Important continuous distributions

i.e.
$$\Phi(z) = \Pr(Z < z) = \frac{1}{\sqrt{2\pi}} \int_{-\infty}^{z} \exp\left(-\frac{u^2}{2}\right) du. \tag{15.73}$$

It is usual only to tabulate $\Phi(z)$ for $z > 0$, since it can be seen easily, from Figure 15.12 and the symmetry of the Gaussian distribution, that $\Phi(-z) = 1 - \Phi(z)$; see Table 15.3. Using such a table it is then straightforward to evaluate the probability that Z lies in a given range of z-values. For example, for a and b constant,

$$\Pr(Z < a) = \Phi(a),$$
$$\Pr(Z > a) = 1 - \Phi(a),$$
$$\Pr(a < Z \leq b) = \Phi(b) - \Phi(a).$$

Remembering that $Z = (X - \mu)/\sigma$ and comparing (15.72) and (15.73), we see that

$$F(x) = \Phi\left(\frac{x - \mu}{\sigma}\right),$$

and so we may also calculate the probability that the original random variable X lies in a given x-range. For example,

$$\Pr(a < X \leq b) = \frac{1}{\sigma\sqrt{2\pi}} \int_{a}^{b} \exp\left[-\frac{1}{2}\left(\frac{u - \mu}{\sigma}\right)^2\right] du \tag{15.74}$$
$$= F(b) - F(a) \tag{15.75}$$
$$= \Phi\left(\frac{b - \mu}{\sigma}\right) - \Phi\left(\frac{a - \mu}{\sigma}\right). \tag{15.76}$$

We now derive some generally useful standard indicators of how close to its mean any individual RV value is likely to be.

Example If X is described by a Gaussian distribution of mean μ and variance σ^2, calculate the probabilities that X lies within 1σ, 2σ and 3σ of the mean.

From (15.76)
$$\Pr(\mu - n\sigma < X \leq \mu + n\sigma) = \Phi(n) - \Phi(-n) = \Phi(n) - [1 - \Phi(n)],$$
and so from Table 15.3
$$\Pr(\mu - \sigma < X \leq \mu + \sigma) = 2\Phi(1) - 1 = 0.6826 \approx 68.3\%,$$
$$\Pr(\mu - 2\sigma < X \leq \mu + 2\sigma) = 2\Phi(2) - 1 = 0.9544 \approx 95.4\%,$$
$$\Pr(\mu - 3\sigma < X \leq \mu + 3\sigma) = 2\Phi(3) - 1 = 0.9974 \approx 99.7\%.$$

Thus we expect X to be distributed in such a way that about two-thirds of the values will lie between $\mu - \sigma$ and $\mu + \sigma$, 95% will lie within 2σ of the mean and 99.7% will lie within 3σ of the mean. These limits are called the one-, two- and three-sigma limits respectively; it is particularly important to note that they are independent of the actual values of the mean and variance.[20] ◂

---

[20] And are worth committing to memory for the purpose of challenging rash statistical claims!

Elementary probability

Table 15.3 The cumulative probability function $\Phi(z)$ for the standard Gaussian distribution, as given by (15.73). The units and the first decimal place of z are specified in the column under $\Phi(z)$ and the second decimal place is specified by the column headings. Thus, for example, $\Phi(1.23) = 0.8907$.

$\Phi(z)$	.00	.01	.02	.03	.04	.05	.06	.07	.08	.09
0.0	.5000	.5040	.5080	.5120	.5160	.5199	.5239	.5279	.5319	.5359
0.1	.5398	.5438	.5478	.5517	.5557	.5596	.5636	.5675	.5714	.5753
0.2	.5793	.5832	.5871	.5910	.5948	.5987	.6026	.6064	.6103	.6141
0.3	.6179	.6217	.6255	.6293	.6331	.6368	.6406	.6443	.6480	.6517
0.4	.6554	.6591	.6628	.6664	.6700	.6736	.6772	.6808	.6844	.6879
0.5	.6915	.6950	.6985	.7019	.7054	.7088	.7123	.7157	.7190	.7224
0.6	.7257	.7291	.7324	.7357	.7389	.7422	.7454	.7486	.7517	.7549
0.7	.7580	.7611	.7642	.7673	.7704	.7734	.7764	.7794	.7823	.7852
0.8	.7881	.7910	.7939	.7967	.7995	.8023	.8051	.8078	.8106	.8133
0.9	.8159	.8186	.8212	.8238	.8264	.8289	.8315	.8340	.8365	.8389
1.0	.8413	.8438	.8461	.8485	.8508	.8531	.8554	.8577	.8599	.8621
1.1	.8643	.8665	.8686	.8708	.8729	.8749	.8770	.8790	.8810	.8830
1.2	.8849	.8869	.8888	.8907	.8925	.8944	.8962	.8980	.8997	.9015
1.3	.9032	.9049	.9066	.9082	.9099	.9115	.9131	.9147	.9162	.9177
1.4	.9192	.9207	.9222	.9236	.9251	.9265	.9279	.9292	.9306	.9319
1.5	.9332	.9345	.9357	.9370	.9382	.9394	.9406	.9418	.9429	.9441
1.6	.9452	.9463	.9474	.9484	.9495	.9505	.9515	.9525	.9535	.9545
1.7	.9554	.9564	.9573	.9582	.9591	.9599	.9608	.9616	.9625	.9633
1.8	.9641	.9649	.9656	.9664	.9671	.9678	.9686	.9693	.9699	.9706
1.9	.9713	.9719	.9726	.9732	.9738	.9744	.9750	.9756	.9761	.9767
2.0	.9772	.9778	.9783	.9788	.9793	.9798	.9803	.9808	.9812	.9817
2.1	.9821	.9826	.9830	.9834	.9838	.9842	.9846	.9850	.9854	.9857
2.2	.9861	.9864	.9868	.9871	.9875	.9878	.9881	.9884	.9887	.9890
2.3	.9893	.9896	.9898	.9901	.9904	.9906	.9909	.9911	.9913	.9916
2.4	.9918	.9920	.9922	.9925	.9927	.9929	.9931	.9932	.9934	.9936
2.5	.9938	.9940	.9941	.9943	.9945	.9946	.9948	.9949	.9951	.9952
2.6	.9953	.9955	.9956	.9957	.9959	.9960	.9961	.9962	.9963	.9964
2.7	.9965	.9966	.9967	.9968	.9969	.9970	.9971	.9972	.9973	.9974
2.8	.9974	.9975	.9976	.9977	.9977	.9978	.9979	.9979	.9980	.9981
2.9	.9981	.9982	.9982	.9983	.9984	.9984	.9985	.9985	.9986	.9986
3.0	.9987	.9987	.9987	.9988	.9988	.9989	.9989	.9989	.9990	.9990
3.1	.9990	.9991	.9991	.9991	.9992	.9992	.9992	.9992	.9993	.9993
3.2	.9993	.9993	.9994	.9994	.9994	.9994	.9994	.9995	.9995	.9995
3.3	.9995	.9995	.9995	.9996	.9996	.9996	.9996	.9996	.9996	.9997
3.4	.9997	.9997	.9997	.9997	.9997	.9997	.9997	.9997	.9997	.9998

15.8 Important continuous distributions

There are many other ways in which the Gaussian distribution may be used. We now illustrate some of the uses in more complicated examples.

> **Example** Sawmill A produces boards whose distribution of lengths is well approximated by a Gaussian with mean 209.4 cm and standard deviation 5.0 cm. A board is accepted if it is longer than 200 cm but is rejected otherwise. Show that 3% of boards are rejected.
>
> Sawmill B produces boards of the same standard deviation but of mean length 210.1 cm. Find the proportion of boards rejected if they are drawn at random from the outputs of A and B in the ratio $3 : 1$.

Let X = length of boards from A, so that $X \sim N(209.4, (5.0)^2)$ and

$$\Pr(X < 200) = \Phi\left(\frac{200 - \mu}{\sigma}\right) = \Phi\left(\frac{200 - 209.4}{5.0}\right) = \Phi(-1.88).$$

But, since $\Phi(-z) = 1 - \Phi(z)$ we have, using Table 15.3,

$$\Pr(X < 200) = 1 - \Phi(1.88) = 1 - 0.9699 = 0.0301,$$

i.e. 3.0% of boards are rejected.

Now let Y = length of boards from B, so that $Y \sim N(210.1, (5.0)^2)$ and

$$\Pr(Y < 200) = \Phi\left(\frac{200 - 210.1}{5.0}\right) = \Phi(-2.02)$$
$$= 1 - \Phi(2.02)$$
$$= 1 - 0.9783 = 0.0217.$$

Therefore, when taken alone, only 2.2% of boards from B are rejected. If, however, boards are drawn at random from A and B in the ratio $3 : 1$ then

$$\tfrac{1}{4}(3 \times 0.030 + 1 \times 0.022) = 0.028 = 2.8\%$$

gives the percentage of boards rejected. ◀

Gaussian approximation to other distributions

An alternative view of the Gaussian distribution is as the limit of the binomial distribution when the number of trials $n \to \infty$ but the probability of a success p remains finite. The mean of the Gaussian, equal to np, the expected total number of successes, also tends to infinity.[21] The standard deviation of the distribution, σ in the Gaussian prescription, retains its binomial value of $\sqrt{np(1-p)}$. In other words, a Gaussian distribution results when an experiment with a finite probability of success is repeated a large number of times, with both its mean and variance increasing linearly with the number of trials increases.

Thus we see that the *value* of the Gaussian *probability density function* $f(x)$ is a good approximation to the *probability* of obtaining x successes in n trials. This approximation

---

[21] This contrasts with the Poisson distribution, which corresponds to the limit $n \to \infty$ and $p \to 0$ with $np = \lambda$ remaining finite.

Table 15.4 *Comparison of the binomial distribution for $n = 10$ and $p = 0.6$ with its Gaussian approximation.*

x	$f(x)$ (binomial)	$f(x)$ (Gaussian)
0	0.0001	0.0001
1	0.0016	0.0014
2	0.0106	0.0092
3	0.0425	0.0395
4	0.1115	0.1119
5	0.2007	0.2091
6	0.2508	0.2575
7	0.2150	0.2091
8	0.1209	0.1119
9	0.0403	0.0395
10	0.0060	0.0092

is actually very good even for relatively small n. For example, if $n = 10$ and $p = 0.6$ then the Gaussian approximation to the binomial distribution is (15.70) with $\mu = 10 \times 0.6 = 6$ and $\sigma = \sqrt{10 \times 0.6(1 - 0.6)} = 1.549$. The probability functions $f(x)$ for the binomial and associated Gaussian distributions for these parameters are given in Table 15.4, and it can be seen that the Gaussian approximation is a good one.[22]

Strictly speaking, however, since the Gaussian distribution is continuous and the binomial distribution is discrete, we should use the integral of $f(x)$ for the Gaussian distribution in the calculation of approximate binomial probabilities. More specifically, we should apply a *continuity correction* so that the discrete integer x in the binomial distribution becomes the interval $[x - 0.5, x + 0.5]$ in the Gaussian distribution. Explicitly,

$$\Pr(X = x) \approx \frac{1}{\sigma\sqrt{2\pi}} \int_{x-0.5}^{x+0.5} \exp\left[-\frac{1}{2}\left(\frac{u - \mu}{\sigma}\right)^2\right] du.$$

The Gaussian approximation is particularly useful for estimating the binomial probability that X lies between the (integer) values x_1 and x_2,

$$\Pr(x_1 < X \leq x_2) \approx \frac{1}{\sigma\sqrt{2\pi}} \int_{x_1-0.5}^{x_2+0.5} \exp\left[-\frac{1}{2}\left(\frac{u - \mu}{\sigma}\right)^2\right] du,$$

as we now illustrate with a worked example.

---

[22] What are the chances of throwing 60 or more heads in 100 tosses of an unbiased coin?

15.8 Important continuous distributions

Example A manufacturer makes computer chips of which 10% are defective. For a random sample of 200 chips, find the approximate probability that more than 15 are defective.

We first define the random variable

$$X = \text{number of defective chips in the sample},$$

which has a binomial distribution $X \sim \text{Bin}(200, 0.1)$. Therefore, the mean and variance of this distribution are

$$E[X] = 200 \times 0.1 = 20 \quad \text{and} \quad V[X] = 200 \times 0.1 \times (1 - 0.1) = 18,$$

and we may approximate the binomial distribution with a Gaussian distribution such that $X \sim N(20, 18)$. The standard variable is

$$Z = \frac{X - 20}{\sqrt{18}},$$

and so, using $X = 15.5$ to allow for the continuity correction,

$$\Pr(X > 15.5) = \Pr\left(Z > \frac{15.5 - 20}{\sqrt{18}}\right) = \Pr(Z > -1.06)$$
$$= \Pr(Z < 1.06) = 0.86.$$

The final probability in this calculation was obtained from Table 15.3. ◀

Since the Poisson distribution can be considered as the limit of the binomial distribution for $n \to \infty$ and $p \to 0$, taken in such a way that $np = \lambda$ remains finite, it should come as no surprise that the Gaussian distribution can also be used to approximate the Poisson distribution when the mean λ becomes large. In this case the Gaussian form becomes

$$f(x) \approx \frac{1}{\sqrt{2\pi\lambda}} \exp\left[-\frac{(x-\lambda)^2}{2\lambda}\right],$$

in which both the mean and standard deviation are determined by a single parameter λ.

The larger the value of λ, the better is the Gaussian approximation to the Poisson distribution; the approximation is reasonable even for $\lambda = 5$, but $\lambda \geq 10$ is safer. As in the case of the Gaussian approximation to the binomial distribution, a continuity correction is necessary since the Poisson distribution is discrete.

It is actually the case that almost all probability distributions tend towards a Gaussian when the numbers involved become large – that this should happen is required by the central limit theorem, which we now state without formal proof.

Central limit theorem. *Suppose that X_i, $i = 1, 2, \ldots, n$, are independent random variables, each of which is described by a probability density function $f_i(x)$ (these may all be different) with a mean μ_i and a variance σ_i^2. The random variable $Z = \left(\sum_i X_i\right)/n$, i.e. the 'mean' of the X_i, has the following properties:*

(i) *its expectation value is given by $E[Z] = \left(\sum_i \mu_i\right)/n$;*
(ii) *its variance is given by $V[Z] = \left(\sum_i \sigma_i^2\right)/n^2$;*
(iii) *as $n \to \infty$ the probability function of Z tends to a Gaussian with corresponding mean and variance.*

Elementary probability

We emphasise that, for the theorem to hold, the probability density functions $f_i(x)$ need not be Gaussian, but they must possess formal means and variances. Thus, for example, if any of the X_i were described by a Cauchy distribution (see Section 15.8.2) then the theorem would not apply.

For the particular case in which we consider Z to be the mean of n independent measurements of the *same* random variable X (so that $X_i = X$ for $i = 1, 2, \ldots, n$) we have that, as $n \to \infty$, Z has a Gaussian distribution with mean μ and variance σ^2/n.

Multiple Gaussian distributions

As we did previously for binomial- and Poisson-distributed RVs, consider two *independent* Gaussian-distributed random variables, X and Y, with distributions $X \sim N(\mu_1, \sigma_1^2)$ and $Y \sim N(\mu_2, \sigma_2^2)$. Their sum, $Z = X + Y$, can be proved to have a Gaussian distribution $Z \sim N(\mu_1 + \mu_2, \sigma_1^2 + \sigma_2^2)$.[23]

One obvious extension of this is to the sum, and hence the arithmetic average, of n RVs all drawn from the same population; if the population is $\sim N(\mu, \sigma^2)$, then the average is $\sim N(\mu, \sigma^2/n)$.

A similar calculation may be performed to calculate the PDF of the random variable $W = X - Y$. If we introduce the variable $\tilde{Y} = -Y$ then $W = X + \tilde{Y}$, where $\tilde{Y} \sim N(-\mu_1, \sigma_1^2)$. Thus, using the result above, we find $W \sim N(\mu_1 - \mu_2, \sigma_1^2 + \sigma_2^2)$.

The following example makes no allowance for '(the wrong sort of) leaves on the line' or for flat tyres!

Example An executive travels home from her office every evening. Her journey consists of a train ride, followed by a bicycle ride. The time spent on the train is Gaussian distributed with mean 52 minutes and standard deviation 1.8 minutes, while the time for the bicycle journey is Gaussian distributed with mean 8 minutes and standard deviation 2.6 minutes. Assuming these two factors are independent, estimate the percentage of occasions on which the *whole* journey takes more than 65 minutes.

We first define the random variables

$$X = \text{time spent on train}, \quad Y = \text{time spent on bicycle},$$

so that $X \sim N(52, (1.8)^2)$ and $Y \sim N(8, (2.6)^2)$. Since X and Y are independent, the total journey time $T = X + Y$ is distributed as

$$T \sim N(52 + 8, \ (1.8)^2 + (2.6)^2) = N(60, (3.16)^2).$$

The standard variable is thus

$$Z = \frac{T - 60}{3.16},$$

and the required probability is given by

$$\Pr(T > 65) = \Pr\left(Z > \frac{65 - 60}{3.16}\right) = \Pr(Z > 1.58) = 1 - 0.943 = 0.057.$$

Thus the total journey time exceeds 65 minutes on 5.7% of occasions. ◀

---

[23] Note that this is an exact result, and its proof is not linked with any aspect of the central limit theorem.

15.8 Important continuous distributions

The above results may be extended. For example, if the random variables X_i, $i = 1, 2, \ldots, n$, are distributed as $X_i \sim N(\mu_i, \sigma_i^2)$ then the random variable $Z = \sum_i c_i X_i$ (where the c_i are constants) is distributed as $Z \sim N(\sum_i c_i \mu_i, \sum_i c_i^2 \sigma_i^2)$.

15.8.4 The exponential distribution

The exponential distribution with positive parameter λ is given by

$$f(x) = \begin{cases} \lambda e^{-\lambda x} & \text{for } x > 0, \\ 0 & \text{for } x \leq 0 \end{cases} \quad (15.77)$$

and satisfies $\int_{-\infty}^{\infty} f(x)\,dx = 1$ as required. The exponential distribution occurs naturally if we consider the distribution of the length of intervals between successive events in a Poisson process or, equivalently, the distribution of the interval (i.e. the waiting time) before the first event. If the average number of events per unit interval is λ then on average there are λx events in interval x, so that from the Poisson distribution the probability that there will be no events in this interval is given by

$$\text{Pr(no events in interval } x) = e^{-\lambda x}.$$

The probability that an event occurs in the next infinitesimal interval $[x, x + dx]$ is given by $\lambda\,dx$, so that

$$\text{Pr(the first event occurs in interval } [x, x + dx]) = e^{-\lambda x} \lambda\,dx.$$

Hence the required probability density function is given by

$$f(x) = \lambda e^{-\lambda x}.$$

The expectation and variance of the exponential distribution can be evaluated from their definitions by straightforward integration as $1/\lambda$ and $(1/\lambda)^2$ respectively.[24]

15.8.5 The chi-squared distribution

Because of the extreme importance of its applications in *statistics*, one final continuous distribution that needs to be mentioned here is the so-called chi-squared (χ^2) distribution. Its derivation and the proofs of its properties lie beyond the scope of the present volume and so we merely record them.

Let us consider n independent Gaussian random variables $X_i \sim N(\mu_i, \sigma_i^2)$, $i = 1, 2, \ldots, n$, and define the new variable

$$\chi_n^2 = \sum_{i=1}^{n} \frac{(X_i - \mu_i)^2}{\sigma_i^2}. \quad (15.78)$$

This variable (to be treated as a single symbol, despite the appearance of the squared superscript) has the distribution

$$f(\chi_n^2) = \frac{1}{2^{n/2}\Gamma(\tfrac{1}{2}n)} (\chi_n^2)^{(n/2)-1} \exp(-\tfrac{1}{2}\chi_n^2), \quad (15.79)$$

---

[24] The above discussion can be generalised to obtain the PDF for the interval between every rth event in a Poisson process or, equivalently, the interval (waiting time) before the rth event. The resulting distribution is known as the *gamma* distribution of order r with parameter λ.

Elementary probability

which is known as the *chi-squared distribution* of (integer) order n. The factor $\Gamma(\frac{1}{2}n)$ that appears in the denominator is the generalised gamma function (see footnote 8 of Section 1.2); for even n it is given by $(\frac{1}{2}n - 1)!$ and for odd n it is a well-defined multiple of $\sqrt{\pi}$. The expectation value of χ_n^2 can be shown to be n, whilst its variance has the value $2n$. It is these two quantities that in statistical analyses are made the basis of tests to determine whether or not sets of experimental data fit stated hypotheses.

An important generalisation occurs when the n Gaussian variables X_i are *not* linearly independent but are instead required to satisfy a linear constraint of the form

$$c_1 X_1 + c_2 X_2 + \cdots + c_n X_n = 0, \tag{15.80}$$

in which the constants c_i are not all zero. In this case, it may be shown that the variable χ_n^2 defined in (15.78) is still described by a chi-squared distribution, but one of order $n - 1$. Indeed, this result may be extended to show that if the n Gaussian variables X_i satisfy m linear constraints of the form (15.80) then the variable χ_n^2 defined in (15.78) is described by a chi-squared distribution of order $n - m$.

EXERCISES 15.8

1. Of the Cauchy and Gaussian distributions for a standard variable, which has the larger value for its PDF when the relevant variable is equal to $\sqrt{2}$?

2. For a RV that $\sim N(100, (10)^2)$ what, to 2 d.p., are its quartiles?

3. At a certain collegiate university, the 25 colleges are ranked annually according to the examination results of their students. The scoring system is such that the college scores are Gaussian distributed with mean 300 and variance 400. A particular college has a disappointing score of 270; how many other colleges are likely to be below it in the table?

4. The initial claims section of an insurance company receives, on average, 30 collision damage, 40 accidental damage, and 20 burglary claims each day. Its regular staff can handle up to 100 claims in a normal day, but have to work overtime when this number is exceeded. On what fraction of their working days can they expect to be working overtime?

5. The marks obtained by the students in a particular class for English and mathematics are independent RVs distributed as $\sim N(60, (12)^2)$ and $\sim N(50, (15)^2)$, respectively. To obtain an overall ranking, all English marks are scaled to make their mean the same as that for mathematics and then the two marks for each student are added together. A particular student had a raw English mark of 72; estimate the minimum mathematics mark she needs in order to make it probable that she will be placed in the top one-sixth of the class.

6. The average lifetime of a continuously lit light bulb is 500 hours. A set of seven such bulbs is switched on. Estimate how long it will be before the number of functioning bulbs has been reduced to three.

7. It is claimed that the following readings are all taken, at random, from a normal distribution that has unit mean and variance:

$$1.2, \quad 0.6, \quad 1.3, \quad -0.2, \quad 2.8, \quad 0.9, \quad -0.6.$$

What is the corresponding value of χ^2? Without making any attempt to calculate or look up the relevant distribution values, comment on whether the claim seems plausible.

15.9 Joint distributions

As mentioned briefly in Section 15.4.3, it is common in the physical sciences to consider simultaneously two or more random variables that are not independent, in general, and are thus described by *joint probability density functions*. We will restrict ourselves to *bivariate* distributions, i.e. distributions of only two random variables. Furthermore, we will consider only the cases where X and Y are either both discrete or both continuous random variables.

15.9.1 Bivariate distributions

In direct analogy with the one-variable (univariate) case, if X and Y are the RVs, then the probability function of the joint distribution is defined by

$$f(x, y) = \begin{cases} \Pr(X = x_i, Y = y_j) & \text{for } x = x_i, y = y_j, \\ 0 & \text{otherwise,} \end{cases}$$

in the discrete case, or by

$$f(x, y) \, dx \, dy = \Pr(x < X \le x + dx, \ y < Y \le y + dy), \qquad (15.81)$$

in the continuous one.

The normalisation of $f(x, y)$ implies that either

$$\sum_i \sum_j f(x_i, y_j) = 1, \qquad (15.82)$$

where the sums over i and j take all valid pairs of values, or that

$$\int_{-\infty}^{\infty} \int_{-\infty}^{\infty} f(x, y) \, dx \, dy = 1,$$

as the case may be.

The cumulative probability function $F(x, y)$ can be defined in an obvious way, with the probability that the RVs take values in certain ranges being given by combinations of its values at different values of it arguments, e.g. the probability that X lies in the range $[a_1, a_2]$ and Y lies in the range $[b_1, b_2]$ is given by

$$\Pr(a_1 < X \le a_2, \ b_1 < Y \le b_2) = F(a_2, b_2) - F(a_1, b_2) - F(a_2, b_1) + F(a_1, b_1).$$

Finally, though we will be more concerned with dependent RVs, as before, we define X and Y to be *independent* if we can write their joint distribution in the form

$$f(x, y) = f_X(x) f_Y(y), \qquad (15.83)$$

i.e. as the product of two univariate distributions.

Elementary probability

Two independent random variables are needed to describe the outcome of any one trial in the simple experiment analysed in the following worked example.

Example A flat table is ruled with parallel straight lines a distance D apart, and a thin needle of length $l < D$ is tossed onto the table at random. What is the probability that the needle will cross a line?

Let θ be the angle that the needle makes with the lines and let x be the distance from the centre of the needle to the nearest line. Since the needle is tossed 'at random' onto the table, the angle θ is uniformly distributed in the interval $[0, \pi]$, and the distance x is uniformly distributed in the interval $[0, D/2]$. Assuming that θ and x are independent, their joint distribution is just the product of their individual distributions, and is given by

$$f(\theta, x) = \frac{1}{\pi} \frac{1}{D/2} = \frac{2}{\pi D}.$$

The needle will cross a line if the distance x of its centre from that line is less than $\tfrac{1}{2} l \sin \theta$. Thus the required probability is

$$\frac{2}{\pi D} \int_0^\pi \int_0^{\frac{1}{2} l \sin \theta} dx\, d\theta = \frac{2}{\pi D} \frac{l}{2} \int_0^\pi \sin \theta\, d\theta = \frac{2l}{\pi D}.$$

This gives an experimental (but cumbersome) method of determining π. ◀

15.9.2 Properties of joint distributions

The probability density function $f(x, y)$ contains all the information on the joint probability distribution of two random variables X and Y. In a similar manner to that presented for univariate distributions, however, it is conventional to characterise $f(x, y)$ by certain of its properties, which we now discuss. Once again, most of these properties are based on the concept of expectation values, which are defined for joint distributions in an analogous way to those for single-variable distributions (15.45). Thus, the expectation value of any function $g(X, Y)$ of the random variables X and Y is given by

$$E[g(X, Y)] = \begin{cases} \sum_i \sum_j g(x_i, y_j) f(x_i, y_j) & \text{for the discrete case,} \\ \int_{-\infty}^\infty \int_{-\infty}^\infty g(x, y) f(x, y)\, dx\, dy & \text{for the continuous case.} \end{cases}$$

15.9.3 Means

The means of X and Y are defined respectively as the expectation values of the variables X and Y. Thus, the mean of X is given by

$$E[X] = \mu_X = \begin{cases} \sum_i \sum_j x_i f(x_i, y_j) & \text{for the discrete case,} \\ \int_{-\infty}^\infty \int_{-\infty}^\infty x f(x, y)\, dx\, dy & \text{for the continuous case.} \end{cases} \quad (15.84)$$

$E[Y]$ is obtained in a similar manner and a simple, but not unexpected, result involving both is now proved.

15.9 Joint distributions

Example Show that if X and Y are independent random variables then $E[XY] = E[X]E[Y]$.

Let us consider the case where X and Y are continuous random variables. Since X and Y are independent $f(x, y) = f_X(x) f_Y(y)$, and so

$$E[XY] = \int_{-\infty}^{\infty} \int_{-\infty}^{\infty} xy f_X(x) f_Y(y) \, dx \, dy = \int_{-\infty}^{\infty} x f_X(x) \, dx \int_{-\infty}^{\infty} y f_Y(y) \, dy = E[X]E[Y].$$

An analogous proof exists for the discrete case. ◀

15.9.4 Variances

The definitions of the variances of X and Y are analogous to those for the single-variable case (15.47), i.e. the variance of X is given by

$$V[X] = \sigma_X^2 = \begin{cases} \sum_i \sum_j (x_i - \mu_X)^2 f(x_i, y_j) & \text{for the discrete case,} \\ \int_{-\infty}^{\infty} \int_{-\infty}^{\infty} (x - \mu_X)^2 f(x, y) \, dx \, dy & \text{for the continuous case.} \end{cases} \quad (15.85)$$

Equivalent definitions exist for the variance of Y.

15.9.5 Covariance and correlation

Means and variances of joint distributions provide useful information about the distributions of the two individual RVs, but do not give any indication of the relationship between them. If we were to measure the heights and weights of a sample of people, we would not be surprised to find a tendency for tall people to be heavier than short people and vice versa; what is needed is a quantitative measure of this tendency. We will show in this section that two measures, the *covariance* and the *correlation*, that can be defined for a bivariate distribution may be used to characterise the relationship between the two random variables.

The *covariance* of two random variables X and Y is defined by

$$\text{Cov}[X, Y] = E[(X - \mu_X)(Y - \mu_Y)], \quad (15.86)$$

where μ_X and μ_Y are the expectation values of X and Y respectively. It will be clear that contributions to the covariance will be positive if corresponding values of X and Y are either both greater, or both smaller, than their respective means. If one is greater but the other is smaller, the covariance receives a negative contribution.

Clearly related to the covariance is the *correlation* of the two random variables, defined by

$$\text{Corr}[X, Y] = \frac{\text{Cov}[X, Y]}{\sigma_X \sigma_Y}, \quad (15.87)$$

where σ_X and σ_Y are the standard deviations of X and Y respectively. It can be shown that the correlation function lies between -1 and $+1$. If the value assumed is negative, X and Y are said to be *negatively correlated*, if it is positive they are said to be *positively correlated* and if it is zero they are said to be *uncorrelated*. We will now justify the use of these terms.

Elementary probability

One particularly useful consequence of its definition is that the covariance of two *independent* variables, X and Y, is zero. It immediately follows from (15.87) that their correlation is also zero, and this justifies the use of the term 'uncorrelated' for two such variables. To show this extremely important property we first note that

$$\begin{aligned} \text{Cov}\,[X, Y] &= E[(X - \mu_X)(Y - \mu_Y)] \\ &= E[XY - \mu_X Y - \mu_Y X + \mu_X \mu_Y] \\ &= E[XY] - \mu_X E[Y] - \mu_Y E[X] + \mu_X \mu_Y \\ &= E[XY] - \mu_X \mu_Y. \end{aligned} \quad (15.88)$$

Now, if X and Y are independent then $E[XY] = E[X]E[Y] = \mu_X \mu_Y$ and so $\text{Cov}\,[X, Y] = 0$.

It is important to note that the converse of this result is not necessarily true; two variables dependent on each other can still be uncorrelated. In other words, it is possible (and not uncommon) for two variables X and Y to be described by a joint distribution $f(x, y)$ that *cannot* be factorised into a product of the form $g(x)h(y)$, but for which $\text{Corr}\,[X, Y] = 0$. Indeed, from the definition (15.86), we see that for any joint distribution $f(x, y)$ that is symmetric in x about μ_X (or similarly in y) we have $\text{Corr}\,[X, Y] = 0$.

We have already asserted that if the correlation of two random variables is positive (negative) they are said to be positively (negatively) correlated. We have also stated that the correlation lies between -1 and $+1$. The terminology suggests that if the two RVs are identical (i.e. $X = Y$) then they are completely correlated and that their correlation should be $+1$. Likewise, if $X = -Y$ then the functions are completely anticorrelated and their correlation should be -1. Values of the correlation function between these extremes show the existence of some degree of correlation. In fact it is not necessary that $X = Y$ for $\text{Corr}\,[X, Y] = 1$; it is sufficient that Y is a linear function of X, i.e. $Y = aX + b$ (with a positive). If a is negative then $\text{Corr}\,[X, Y] = -1$. To show this we first note that $\mu_Y = a\mu_X + b$. Now

$$Y = aX + b = aX + \mu_Y - a\mu_X \quad \Rightarrow \quad Y - \mu_Y = a(X - \mu_X),$$

and so, using the definition of the covariance (15.86),

$$\text{Cov}\,[X, Y] = aE[(X - \mu_X)^2] = a\sigma_X^2.$$

It follows from the properties of the variance (Section 15.5.3) that $\sigma_Y = |a|\sigma_X$ and so, using the definition (15.87) of the correlation,

$$\text{Corr}\,[X, Y] = \frac{a\sigma_X^2}{|a|\sigma_X^2} = \frac{a}{|a|},$$

which is the stated result.

It should be noted that, even if the possibilities of X and Y being non-zero are mutually exclusive, $\text{Corr}\,[X, Y]$ need not have value ± 1.

The biased die of several previous worked examples is now used to demonstrate these definitions in action.

15.9 Joint distributions

Example A biased die gives probabilities $\frac{1}{2}p, p, p, p, p, 2p$ of throwing 1, 2, 3, 4, 5, 6 respectively. If the random variable X is the number shown on the die and the random variable Y is defined as X^2, calculate the covariance and correlation of X and Y.

We have already calculated in Sections 15.2.1 and 15.5.4 that

$$p = \frac{2}{13}, \quad E[X] = \frac{53}{13}, \quad E[X^2] = \frac{253}{13}, \quad V[X] = \frac{480}{169}.$$

Using (15.88), we obtain

$$\text{Cov}[X, Y] = \text{Cov}[X, X^2] = E[X^3] - E[X]E[X^2].$$

Now $E[X^3]$ is given by

$$E[X^3] = 1^3 \times \tfrac{1}{2}p + (2^3 + 3^3 + 4^3 + 5^3)p + 6^3 \times 2p$$
$$= \frac{1313}{2}p = 101,$$

and so the covariance of X and Y is given by

$$\text{Cov}[X, Y] = 101 - \frac{53}{13} \times \frac{253}{13} = \frac{3660}{169}.$$

The correlation is defined by $\text{Corr}[X, Y] = \text{Cov}[X, Y]/\sigma_X \sigma_Y$. The standard deviation of Y may be calculated from the definition of the variance. Letting $\mu_Y = E[X^2] = \frac{253}{13}$ gives

$$\sigma_Y^2 = \frac{p}{2}\left(1^2 - \mu_Y\right)^2 + p\left(2^2 - \mu_Y\right)^2 + p\left(3^2 - \mu_Y\right)^2 + p\left(4^2 - \mu_Y\right)^2$$
$$+ p\left(5^2 - \mu_Y\right)^2 + 2p\left(6^2 - \mu_Y\right)^2$$
$$= \frac{187\,356}{169}p = \frac{28\,824}{169}.$$

We deduce that

$$\text{Corr}[X, Y] = \frac{3660}{169}\sqrt{\frac{169}{28\,824}}\sqrt{\frac{169}{480}} \approx 0.984.$$

Thus the random variables X and Y display a strong degree of positive correlation, as we would expect.[25] ◀

The covariance of two random variables plays a part in a variety of circumstances. For example, if X and Y are *not* independent then the variance of $X + Y$ is given by

$$V[X + Y] = E\left[(X + Y)^2\right] - (E[X + Y])^2$$
$$= E\left[X^2\right] + 2E[XY] + E\left[Y^2\right] - \{(E[X])^2 + 2E[X]E[Y] + (E[Y])^2\}$$
$$= V[X] + V[Y] + 2(E[XY] - E[X]E[Y])$$
$$= V[X] + V[Y] + 2\,\text{Cov}[X, Y].$$

More generally, we find (for a, b and c constant)

$$V[aX + bY + c] = a^2 V[X] + b^2 V[Y] + 2ab\,\text{Cov}[X, Y]. \quad (15.89)$$

---

[25] But note that they are not 100% correlated.

Elementary probability

Note that if X and Y are in fact independent then Cov $[X, Y] = 0$ and we recover the expression (15.57) in Section 15.6.2.

We may use (15.89) to obtain an approximate expression for $V[f(X, Y)]$ for any arbitrary function f, even when the random variables X and Y are correlated. Approximating $f(X, Y)$ by the linear terms of its Taylor expansion about the point (μ_X, μ_Y), we have

$$f(X, Y) \approx f(\mu_X, \mu_Y) + \left(\frac{\partial f}{\partial X}\right)(X - \mu_X) + \left(\frac{\partial f}{\partial Y}\right)(Y - \mu_Y), \quad (15.90)$$

where the partial derivatives are evaluated at $X = \mu_X$ and $Y = \mu_Y$. Taking the variance of both sides, and using (15.89),[26] we find

$$V[f(X, Y)] \approx \left(\frac{\partial f}{\partial X}\right)^2 V[X] + \left(\frac{\partial f}{\partial Y}\right)^2 V[Y] + 2\left(\frac{\partial f}{\partial X}\right)\left(\frac{\partial f}{\partial Y}\right) \text{Cov}[X, Y]. \quad (15.91)$$

Clearly, if Cov $[X, Y] = 0$, we recover the result (15.58) derived in Section 15.6.2. We note that (15.90) and, hence, also (15.91) are exact if $f(X, Y)$ is linear in X and Y.

EXERCISES 15.9

All six exercises refer to the random variables defined in Exercise 1.

1. Two unbiased six-sided dice, one red and one blue, are thrown and the numbers they show, R and B, are recorded. The RVs X and Y are defined by

   $$X = R + B \text{ if } B \div 2, \text{ but } X = R, \text{ otherwise,}$$
   $$Y = B + R \text{ if } R \div 3, \text{ but } Y = B, \text{ otherwise.}$$

   Draw up, as a table, the joint PDF for X and Y and find the probability that

   $$\text{both} \quad 4 \leq X \leq 7 \quad \text{and} \quad 4 \leq Y \leq 7.$$

2. Show that the means, μ_X and μ_Y, are 5.5 and 5, respectively. The values of the variances are $\sigma_X^2 = 8.25$ and $\sigma_Y^2 = 8.17$; verify one of these by direct calculation.

3. Represent the PDF as a scatter-plot using ordinary graph paper. Sketch an envelope for the plot and, by examining it, say whether you think that X and Y are positively correlated, negatively correlated or uncorrelated.

4. Use the form (15.88) to calculate the correlation coefficient for X and Y and compare it with your previous answer. Give a qualitative explanation of why the definitions of the RVs should be expected to lead to this outcome.

5. Defining a new random variable by $Z = X - Y$, deduce its mean and variance.

6. The function $f(X, Y)$ is defined by $f(X, Y) = X(X - Y)^2$. Calculate the approximate mean and variance of f.

---

[26] With $a = \partial f / \partial X$, $b = \partial f / \partial Y$ and constant $c = f(\mu_X, \mu_Y) - \mu_X(\partial f / \partial X) - \mu_Y(\partial f / \partial Y)$.

SUMMARY

1. *Venn diagrams, unions* ($\cup$), *and intersections* ($\cap$)
   S is the sample space, $\emptyset$ is the empty event and the *complement* of event A is denoted by $\bar{A}$.
   - $A \cup \bar{A} = S$ and $A \cap \bar{A} = \emptyset$.
   - The operations $\cup$ and $\cap$ are commutative, associative, idempotent, and have distribution laws
   $$A \cap (B \cup C) = (A \cap B) \cup (A \cap C),$$
   $$A \cup (B \cap C) = (A \cup B) \cap (A \cup C).$$
   - De Morgan's laws: $\overline{A \cup B} = \bar{A} \cap \bar{B}$ and $\overline{A \cap B} = \bar{A} \cup \bar{B}$.
   - The Venn diagram for n events has 2^n regions, with ^nC_r r-fold intersection regions.

2. *Probability*, $\Pr(A)$
   (i) For a single event, $\Pr(\bar{A}) = 1 - \Pr(A)$.
   (ii) For two events, the *conditional* probability for event A, given that event B has already occurred, is $\Pr(A|B) = \Pr(A \cap B)/\Pr(B)$.

Relationship	$\Pr(A \cup B)$	$\Pr(A \cap B)$	$\Pr(A\|B)$
Statistically independent	$\Pr(A) + \Pr(B) - \Pr(A \cap B)$	$\Pr(A)\Pr(B)$	$\Pr(A)$
Mutually exclusive	$\Pr(A) + \Pr(B)$	0	0

   (iii) For n events A_i
   - The probability of their union, $\Pr(A_1 \cup A_2 \cup \cdots \cup A_n)$, is given by (sums run over all sets of unrepeated subscripts)
   $$\sum_i \Pr(A_i) - \sum_{i,j} \Pr(A_i \cap A_j) + \sum_{i,j,k} \Pr(A_i \cap A_j \cap A_k)$$
   $$- \cdots + (-1)^{n+1} \Pr(A_1 \cap A_2 \cap \cdots \cap A_n).$$
   - If a set of mutually exclusive events A_i exhaust S, then
   $$\Pr(B) = \sum_i \Pr(A_i) \Pr(B|A_i).$$
   - *Bayes' theorem*
   $$\Pr(A|B) = \frac{\Pr(A) \Pr(B|A)}{\sum_i \Pr(A_i) \Pr(B|A_i)}.$$
   Here (a) A may coincide with one of the A_i, but does not have to, and (b) in many applications $n = 2$ with $A_1 = A$ and $A_2 = \bar{A}$.

3. *Permutations, combinations and statistical counts*
   - The number of ways of selecting k objects (order immaterial) from n without replacement is $^nC_k = \dfrac{n!}{(n-k)!\,k!}$; with replacement it is $^{n+k-1}C_k = \dfrac{(n+k-1)!}{(n-1)!\,k!}$.
   - Multinomial distribution: with $\sum_{i=1}^{m} n_i = n$, the multinomial coefficient is
   $$M(n;n_i) = \dfrac{n!}{n_1!\,n_2!\cdots n_m!}.$$
     (i) It gives the number of distinguishable permutations of the n objects when there are n_i objects of type i, for $i = 1, 2, \ldots, m$.
     (ii) It gives the number of ways n distinguishable objects can be divided into m piles, with n_i objects in the ith pile.
     (iii) It is the coefficient of $x_1^{n_1} x_2^{n_2} \cdots x_m^{n_m}$ in the (multinomial) expansion of $(x_1 + x_2 + \cdots + x_m)^n$.
     (iv) When x_i is the probability that a single Bernoulli trial will result in an outcome of type i, with $\sum_i x_i = 1$, $M(n;n_i) x_1^{n_1} x_2^{n_2} \cdots x_M^{n_m}$ is the probability that n trials will contain exactly n_i outcomes of type i for each i.

4. *Central limit theorem*[27]
   If X_i, $i = 1, 2, \ldots, n$, are IRVs, each described by its own probability density function $f_i(x)$, with a mean μ_i and a variance σ_i^2, then the random variable $Z = \left(\sum_i X_i\right)/n$, has the following properties:
   (i) its expectation value is given by $E[Z] = \left(\sum_i \mu_i\right)/n$;
   (ii) its variance is given by $V[Z] = \left(\sum_i \sigma_i^2\right)/n^2$;
   (iii) as $n \to \infty$ the probability function of Z tends to a Gaussian with corresponding mean and variance.

5. *Probability distributions*
   In the table below, X is a random variable and $\Pr(X = x_i) = p_i$, $\Pr(x < X < x + dx) = f(x)\,dx$ and $E[U]$ is the expectation value of RV U. Only the continuous case is given; discrete PDFs can be replaced by $f(x) = \sum_{i=1}^{n} p_i \delta(x - x_i)$.

Property or definition	Formula	Conditions or notes
Normalisation	$\int_{-\infty}^{\infty} f(x)\,dx = 1$	
Cumulative PF	$F(x) = \int_{-\infty}^{x} f(x)\,dx$	$F(\infty) = 1$
Independent RVs	$h(x, y) = f(x) g(y)$	

---

[27] See the table in 5 below for the definitions of terms.

Summary

Property or definition	Formula	Conditions or notes
Mean, $E[X]$ or μ	$\mu = \int_{-\infty}^{\infty} x f(x)\,dx$	
Median	$M = F^{-1}(\tfrac{1}{2})$	M is such that $F(M) = \tfrac{1}{2}$
nth percentile	$P_n = F^{-1}(n/100)$	
Variance, $V[X]$	$\int_{-\infty}^{\infty}(x-\mu)^2 f(x)\,dx$	$V[aX+b] = a^2 V[X]$
Standard deviation	$\sigma = +\sqrt{\text{variance}}$	$\sigma^2 = E[X^2] - \mu^2$
Moments, $E[x^k]$	$\mu_k = \int_{-\infty}^{\infty} x^k f(x)\,dx$	
Function, $Y(X)$	$g(y) = f(x(y))\left|\dfrac{dx}{dy}\right|$	Unique inverse $X = X(Y)$
PDF for $Z = X + Y$	$p(Z) = f * g$	Convolution of $f(X)$ and $g(Y)$
$E[aX + bY]$	$aE[X] + bE[Y]$	
$V[aX + bY]$	$a^2 V[X] + b^2 V[Y]$	X and Y independent RVs
$E[W(X, Y)]$	$\approx W(\mu_X, \mu_Y)$	X and Y independent RVs
$^\dagger V[W(X, Y)]$	$\approx \left(\dfrac{\partial W}{\partial X}\right)^2 V[X]$ $+ \left(\dfrac{\partial W}{\partial Y}\right)^2 V[Y]$	X and Y independent RVs

† In the expression for $V[W(X, Y)]$ the two partial derivatives are both to be evaluated at $x = \mu_X$ and $y = \mu_Y$.

- For the parameters of important discrete distributions, see Table 15.1 on p. 632.
- For the parameters of important continuous distributions, see Table 15.2 on p. 643.

6. Bivariate distributions $f(X, Y)$

   For a general function $g(X, Y)$, its expectation

$$E[g(X, Y)] = \int_{-\infty}^{\infty}\int_{-\infty}^{\infty} g(x, y) f(x, y)\,dx\,dy.$$

Elementary probability

Parameter	Calculation
Normalisation	$\int_{-\infty}^{\infty} \int_{-\infty}^{\infty} f(x, y)\, dx\, dy = 1$
Mean, μ_X	$E[X]$
Variance, $V[X]$ or σ_X^2	$E[(X - \mu_X)^2]$
$E[XY]$ for IRVs	$E[X]E[Y]$
Covariance, Cov $[X, Y]$	$E[(X - \mu_X)(Y - \mu_Y)] = E[XY] - \mu_X \mu_Y$
Covariance of IRVs	0
Correlation, Corr $[X, Y]$	$r = \dfrac{\text{Cov}\,[X, Y]}{\sigma_X \sigma_Y}, \quad -1 \le r \le +1$

PROBLEMS

15.1. By shading or numbering Venn diagrams, determine which of the following are valid relationships between events. For those that are, prove the relationship using de Morgan's laws.
(a) $\overline{(\bar{X} \cup Y)} = X \cap \bar{Y}$.
(b) $\bar{X} \cup \bar{Y} = \overline{(X \cup Y)}$.
(c) $(X \cup Y) \cap Z = (X \cup Z) \cap Y$.
(d) $X \cup \overline{(Y \cap Z)} = (X \cup \bar{Y}) \cap \bar{Z}$.
(e) $X \cup \overline{(Y \cap Z)} = (X \cup \bar{Y}) \cup \bar{Z}$.

15.2. Given that events X, Y and Z satisfy

$$(X \cap Y) \cup (Z \cap X) \cup \overline{(\bar{X} \cup \bar{Y})} = \overline{(Z \cup \bar{Y})} \cup \{[\overline{(\bar{Z} \cup \bar{X})} \cup (\bar{X} \cap Z)] \cap Y\},$$

prove that $X \supset Y$, and that either $X \cap Z = \emptyset$ or $Y \supset Z$.

15.3. A and B each have two unbiased four-faced dice, the four faces being numbered 1, 2, 3, 4. Without looking, B tries to guess the sum x of the numbers on the bottom faces of A's two dice after they have been thrown onto a table. If the guess is correct B receives x^2 euros, but if not he loses x euros.
   Determine B's expected gain per throw of A's dice when he adopts each of the following strategies:
(a) He selects x at random in the range $2 \le x \le 8$.
(b) He throws his own two dice and guesses x to be whatever they indicate.

Problems

(c) He takes your advice and always chooses the same value for x. Which number would you advise?

15.4. $X_1, X_2, \ldots, X_n$ are independent, identically distributed, random variables drawn from a uniform distribution on [0, 1]. The random variables A and B are defined by

$$A = \min(X_1, X_2, \ldots, X_n), \qquad B = \max(X_1, X_2, \ldots, X_n).$$

For any fixed k such that $0 \leq k \leq \frac{1}{2}$, find the probability, p_n, that both

$$A \leq k \quad \text{and} \quad B \geq 1 - k.$$

Check your general formula by considering directly the cases (a) $k = 0$, (b) $k = \frac{1}{2}$, (c) $n = 1$ and (d) $n = 2$.

15.5. Two duellists, A and B, take alternate shots at each other, and the duel is over when a shot (fatal or otherwise!) hits its target. Each shot fired by A has a probability α of hitting B, and each shot fired by B has a probability β of hitting A. Calculate the probabilities P_1 and P_2, defined as follows, that A will win such a duel: P_1, A fires the first shot; P_2, B fires the first shot.

If they agree to fire simultaneously, rather than alternately, what is the probability P_3 that A will win, i.e. hit B without being hit himself?

15.6. This problem shows that the odds are hardly ever 'evens' when it comes to dice rolling.
   (a) Gamblers A and B each roll a fair six-faced die, and B wins if his score is strictly greater than A's. Show that the odds are 7 to 5 in A's favour.
   (b) Calculate the probabilities of scoring a total T from two rolls of a fair die for $T = 2, 3, \ldots, 12$. Gamblers C and D each roll a fair die twice and score respective totals T_C and T_D, D winning if $T_D > T_C$. Realising that the odds are not equal, D insists that C should increase her stake for each game. C agrees to stake £1.10 per game, as compared to D's £1.00 stake. Who will show a profit?

15.7. An electronics assembly firm buys its microchips from three different suppliers; half of them are bought from firm X, whilst firms Y and Z supply 30% and 20%, respectively. The suppliers use different quality-control procedures and the percentages of defective chips are 2%, 4% and 4% for X, Y and Z, respectively. The probabilities that a defective chip will fail two or more assembly-line tests are 40%, 60% and 80%, respectively, whilst all defective chips have a 10% chance of escaping detection. An assembler finds a chip that fails only one test. What is the probability that it came from supplier X?

15.8. As every student of probability theory will know, Bayesylvania is awash with natives, not all of whom can be trusted to tell the truth, and lost, and apparently somewhat deaf, travellers who ask the same question several times in an attempt to get directions to the nearest village.

One such traveller finds himself at a T-junction in an area populated by the Asciis and Bisciis in the ratio 11 to 5. As is well known, the Biscii always lie, but the Ascii tell the truth three-quarters of the time, giving independent answers to all questions, even to immediately repeated ones.

(a) The traveller asks one particular native twice whether he should go to the left or to the right to reach the local village. Each time he is told 'left'. Should he take this advice, and, if he does, what are his chances of reaching the village?

(b) The traveller then asks the same native the same question a third time, and for a third time receives the answer 'left'. What should the traveller do now? Have his chances of finding the village been altered by asking the third question?

15.9. A boy is selected at random from amongst the children belonging to families with n children. It is known that he has at least two sisters. Show that the probability that he has $k - 1$ brothers is

$$\frac{(n-1)!}{(2^{n-1} - n)(k-1)!(n-k)!},$$

for $1 \leq k \leq n - 2$ and zero for other values of k. Assume that boys and girls are equally likely.

15.10. Villages A, B, C and D are connected by overhead telephone lines joining AB, AC, BC, BD and CD. As a result of severe gales, there is a probability p (the same for each link) that any particular link is broken.

(a) Show that the probability that a call can be made from A to B is

$$1 - p^2 - 2p^3 + 3p^4 - p^5.$$

(b) Show that the probability that a call can be made from D to A is

$$1 - 2p^2 - 2p^3 + 5p^4 - 2p^5.$$

15.11. A set of $2N + 1$ rods consists of one of each integer length $1, 2, \ldots, 2N, 2N + 1$. Three, of lengths a, b and c, are selected, of which a is the longest. By considering the possible values of b and c, determine the number of ways in which a non-degenerate triangle (i.e. one of non-zero area) can be formed (i) if a is even and (ii) if a is odd. Combine these results appropriately to determine the total number of non-degenerate triangles that can be formed using three of the $2N + 1$ rods, and hence show that the probability that such a triangle can be formed from a random selection (without replacement) of three rods is

$$\frac{(N-1)(4N+1)}{2(4N^2 - 1)}.$$

Problems

15.12. A certain marksman never misses his target, which consists of a disc of unit radius with centre O. The probability that any given shot will hit the target within a distance t of O is t^2, for $0 \leq t \leq 1$. The marksman fires n independent shots at the target, and the random variable Y is the radius of the smallest circle with centre O that encloses all the shots. Determine the PDF for Y and hence find the expected area of the circle.

The shot that is furthest from O is now rejected and the corresponding circle determined for the remaining $n - 1$ shots. Show that its expected area is

$$\frac{n-1}{n+1}\pi.$$

15.13. The duration (in minutes) of a telephone call made from a public call-box is a random variable T. The probability density function of T is

$$f(t) = \begin{cases} 0 & t < 0, \\ \frac{1}{2} & 0 \leq t < 1, \\ ke^{-2t} & t \geq 1, \end{cases}$$

where k is a constant. To pay for the call, 20 pence has to be inserted at the beginning, and a further 20 pence after each subsequent half-minute. Determine by how much the average cost of a call exceeds the cost of a call of average length charged at 40 pence per minute.

15.14. Kittens from different litters do not get on with each other, and fighting breaks out whenever two kittens from different litters are present together. A cage initially contains x kittens from one litter and y from another. To quell the fighting, kittens are removed at random, one at a time, until peace is restored. Show, by induction, that the expected number of kittens finally remaining is

$$N(x, y) = \frac{x}{y+1} + \frac{y}{x+1}.$$

15.15. A tennis tournament is arranged on a straight knockout basis for 2^n players, and for each round, except the final, opponents for those still in the competition are drawn at random. The quality of the field is so even that in any match it is equally likely that either player will win. Two of the players have surnames that begin with 'Q'. Find the probabilities that they play each other
(a) in the final,
(b) at some stage in the tournament.

15.16. A particle is confined to the one-dimensional space $0 \leq x \leq a$, and classically it can be in any small interval dx with equal probability. However, quantum mechanics gives the result that the probability distribution is proportional to $\sin^2(n\pi x/a)$, where n is an integer. Find the variance in the particle's position in both the classical and quantum mechanical pictures, and show that, although they differ, the latter tends to the former in the limit of large n, in agreement with the correspondence principle of physics.

15.17. A point P is chosen at random on the circle $x^2 + y^2 = 1$. The random variable X denotes the distance of P from $(1, 0)$. Find the mean and variance of X and the probability that X is greater than its mean.

15.18. As assistant to a celebrated and imperious newspaper proprietor, you are given the job of running a lottery, in which each of his five million readers will have an equal independent chance, p, of winning a million pounds; you have the job of choosing p. However, if nobody wins it will be bad for publicity, whilst if more than two readers do so, the prize cost will more than offset the profit from extra circulation – in either case you will be sacked! Show that, however you choose p, there is more than a 40% chance you will soon be clearing your desk.

15.19. The number of errors needing correction on each page of a set of proofs follows a Poisson distribution of mean μ. The cost of the first correction on any page is α and that of each subsequent correction on the same page is β. Prove that the average cost of correcting a page is

$$\alpha + \beta(\mu - 1) - (\alpha - \beta)e^{-\mu}.$$

15.20. In the game of Blackball, at each turn Muggins draws a ball at random from a bag containing five white balls, three red balls and two black balls; after being recorded, the ball is replaced in the bag. A white ball earns him \$1, whilst a red ball gets him \$2; in either case, he also has the option of leaving with his current winnings or of taking a further turn on the same basis. If he draws a black ball the game ends and he loses all he may have gained previously. Find an expression for Muggins' expected return if he adopts the strategy of drawing up to n balls, provided he has not been eliminated by then.

Show that, as the entry fee to play is \$3, Muggins should be dissuaded from playing Blackball, but, if that cannot be done, what value of n would you advise him to adopt?

15.21. The probability distribution for the number of eggs in a clutch is Po(λ), and the probability that each egg will hatch is p (independently of the size of the clutch). Show by direct calculation that the probability distribution for the number of chicks that hatch is Po(λp).

15.22. A shopper buys 36 items at random in a supermarket, where, because of the sales tax imposed, the final digit (the number of pence) in the price is uniformly and randomly distributed from 0 to 9. Instead of adding up the bill exactly, she rounds each item to the nearest 10 pence, rounding up or down with equal probability if the price ends in a '5'. Should she suspect a mistake if the cashier asks her for 23 pence more than she estimated?

Problems

15.23. Under EU legislation on harmonisation, all kippers are to weigh 0.2000 kg, and vendors who sell underweight kippers must be fined by their government. The weight of a kipper is normally distributed, with a mean of 0.2000 kg and a standard deviation of 0.0100 kg. They are packed in cartons of 100 and large quantities of them are sold.

Every day, a carton is to be selected at random from each vendor and tested according to one of the following schemes, which have been approved for the purpose.

(a) The entire carton is weighed, and the vendor is fined 2500 euros if the average weight of a kipper is less than 0.1975 kg.
(b) Twenty-five kippers are selected at random from the carton; the vendor is fined 100 euros if the average weight of a kipper is less than 0.1980 kg.
(c) Kippers are removed one at a time, at random, until one has been found that weighs *more* than 0.2000 kg; the vendor is fined $4n(n-1)$ euros, where n is the number of kippers removed.

Which scheme should the Chancellor of the Exchequer be urging his government to adopt?

15.24. In a certain parliament, the government consists of 75 New Socialites and the opposition consists of 25 Preservatives. Preservatives never change their mind, always voting against government policy without a second thought; New Socialites vote randomly, but with probability p that they will vote for their party leader's policies.

Following a decision by the New Socialites' leader to drop certain manifesto commitments, N of his party decide to vote consistently with the opposition. The leader's advisors reluctantly admit that an election must be called if N is such that, at any vote on government policy, the chance of a simple majority in favour would be less than 80%. Given that $p = 0.8$, estimate the lowest value of N that would precipitate an election.

15.25. A practical-class demonstrator sends his 12 students to the storeroom to collect apparatus for an experiment, but forgets to tell each which type of component to bring. There are three types, A, B and C, held in the stores (in large numbers) in the proportions 20%, 30% and 50%, respectively, and each student picks a component at random. In order to set up one experiment, one unit each of A and B and two units of C are needed. Let $\Pr(N)$ be the probability that at least N experiments can be set up.
(a) Evaluate $\Pr(3)$.
(b) Find an expression for $\Pr(N)$ in terms of k_1 and k_2, the numbers of components of types A and B respectively selected by the students. Show that $\Pr(2)$ can be written in the form

$$\Pr(2) = (0.5)^{12} \sum_{i=2}^{6} {}^{12}C_i \, (0.4)^i \sum_{j=2}^{8-i} {}^{12-i}C_j \, (0.6)^j.$$

(c) By considering the conditions under which no experiments can be set up, show that $\Pr(1) = 0.9145$.

15.26. A husband and wife decide that their family will be complete when it includes two boys and two girls – but that this would then be enough! The probability that a new baby will be a girl is p. Ignoring the possibility of identical twins, show that the expected size of their family is

$$2\left(\frac{1}{pq} - 1 - pq\right),$$

where $q = 1 - p$.

15.27. The continuous random variables X and Y have a joint PDF proportional to $xy(x - y)^2$ with $0 \leq x \leq 1$ and $0 \leq y \leq 1$. Show that X and Y are negatively correlated with correlation coefficient $-\frac{2}{3}$.

15.28. A discrete random variable X takes integer values $n = 0, 1, \ldots, N$ with probabilities p_n. A second random variable Y is defined as $Y = (X - \mu)^2$, where μ is the expectation value of X. Prove that the covariance of X and Y is given by

$$\text{Cov}[X, Y] = \sum_{n=0}^{N} n^3 p_n - 3\mu \sum_{n=0}^{N} n^2 p_n + 2\mu^3.$$

Now suppose that X takes all of its possible values with equal probability, and hence demonstrate that two random variables can be uncorrelated, even though one is defined in terms of the other.

15.29. Two continuous random variables X and Y have a joint probability distribution

$$f(x, y) = A(x^2 + y^2),$$

where A is a constant and $0 \leq x \leq a$, $0 \leq y \leq a$. Show that X and Y are negatively correlated with correlation coefficient $-15/73$. By sketching a rough contour map of $f(x, y)$ and marking off the regions of positive and negative correlation, convince yourself that this (perhaps counter-intuitive) result is plausible.

HINTS AND ANSWERS

15.1. (a) Yes, (b) no, (c) no, (d) no, (e) yes.

15.3. Show that, if $p_x/16$ is the probability that the total will be x, then the corresponding gain is $[p_x(x^2 + x) - 16x]/16$. (a) A loss of 0.36 euros; (b) a gain of 27/64 euros; (c) a gain of 46/16 euros, provided he takes your advice and guesses '6' each time.

Hints and answers

15.5. $P_1 = \alpha(\alpha + \beta - \alpha\beta)^{-1}$; $P_2 = \alpha(1 - \beta)(\alpha + \beta - \alpha\beta)^{-1}$; $P_3 = P_2$.

15.7. The relative probabilities are $X : Y : Z = 50 : 36 : 8$ (in units of 10^{-4}); $25/47$.

15.9. Take A_j as the event that a family consists of j boys and $n - j$ girls, and B as the event that the boy has at least two sisters. Apply Bayes' theorem.

15.11. (i) For a even, the number of ways is $1 + 3 + 5 + \cdots + (a - 3)$; (ii) for a odd it is $2 + 4 + 6 + \cdots + (a - 3)$. Combine the results for $a = 2m$ and $a = 2m + 1$, with m running from 2 to N, to show that the total number of non-degenerate triangles is given by $N(4N + 1)(N - 1)/6$. The number of possible selections of a set of three rods is $(2N + 1)(2N)(2N - 1)/6$.

15.13. Show that $k = e^2$ and that the average duration of a call is 1 minute. Let p_n be the probability that the call ends during the interval $0.5(n - 1) \le t < 0.5n$ and $c_n = 20n$ be the corresponding cost. Prove that $p_1 = p_2 = \frac{1}{4}$ and that $p_n = \frac{1}{2}e^2(e - 1)e^{-n}$, for $n \ge 3$. It follows that the average cost is

$$E[C] = \frac{30}{2} + 20 \frac{e^2(e-1)}{2} \sum_{n=3}^{\infty} n e^{-n}.$$

The arithmetico-geometric series has sum $(3e^{-1} - 2e^{-2})/(e - 1)^2$ and the total charge is $5(e + 1)/(e - 1) = 10.82$ pence more than the 40 pence a uniform rate would cost.

15.15. If p_r is the probability that before the rth round both players are still in the tournament (and therefore have not met each other), show that

$$p_{r+1} = \frac{1}{4} \frac{2^{n+1-r} - 2}{2^{n+1-r} - 1} p_r \quad \text{and hence that} \quad p_r = \left(\frac{1}{2}\right)^{r-1} \frac{2^{n+1-r} - 1}{2^n - 1}.$$

(a) The probability that they meet in the final is $p_n = 2^{-(n-1)}(2^n - 1)^{-1}$.
(b) The probability that they meet at some stage in the tournament is given by the sum $\sum_{r=1}^{n} p_r (2^{n+1-r} - 1)^{-1} = 2^{-(n-1)}$.

15.17. Mean $= 4/\pi$. Variance $= 2 - (16/\pi^2)$. Probability that X exceeds its mean $= 1 - (2/\pi)\sin^{-1}(2/\pi) = 0.561$.

15.19. Consider separately, 0, 1 and ≥ 2 errors on a page.

15.21. $\Pr(k \text{ chicks hatching}) = \sum_{n=k}^{\infty} \text{Po}(n, \lambda) \text{Bin}(n, p)$.

15.23. There is not much to choose between the schemes. In (a) the critical value of the standard variable is -2.5 and the average fine would be 15.5 euros. For (b) the corresponding figures are -1.0 and 15.9 euros. Scheme (c) is governed by a geometric distribution with $p = q = \frac{1}{2}$, and leads to an expected fine of $\sum_{n=1}^{\infty} 4n(n-1)(\frac{1}{2})^n$. The sum can be evaluated by differentiating the result $\sum_{n=1}^{\infty} p^n = p/(1-p)$ with respect to p, and gives the expected fine as 16 euros.

15.25. (a) $[12!(0.5)^6(0.3)^3(0.2)^3]/(6! \, 3! \, 3!) = 0.0624$.

Elementary probability

15.27. You will need to establish the normalisation constant for the distribution (36), the common mean value (3/5) and the common standard deviation (3/10). The separate probability distributions are $f(x) = 3x(6x^2 - 8x + 3)$, and the same function of y. The covariance has the value $-3/50$, yielding a correlation of $-2/3$.

15.29. $A = 3/(24a^4)$; $\mu_X = \mu_Y = 5a/8$; $\sigma_X^2 = \sigma_Y^2 = 73a^2/960$; $E[XY] = 3a^2/8$; $\text{Cov}[X, Y] = -a^2/64$.

A The base for natural logarithms

In this appendix, elementary calculus is used to prove that

$$\exp(x) = e^x \tag{A.1}$$

and to establish the particular properties of the exponential function that make e the natural choice for a logarithmic base.

We start by determining the first derivative of $\exp(x)$ with respect to x. To do this we differentiate the definition (1.22) term by term and, remembering that the derivative of the constant term (the $n = 0$ term) is zero, obtain

$$\frac{d\,\exp(x)}{dx} = \frac{d}{dx}\left(\sum_{n=0}^{\infty} \frac{x^n}{n!}\right) = \sum_{n=1}^{\infty} \frac{nx^{n-1}}{n!} = \sum_{n=1}^{\infty} \frac{x^{n-1}}{(n-1)!}.$$

In the final summation, in which index n now runs from 1 to ∞, we set $n = m + 1$ with m running from 0 to ∞. This yields

$$\frac{d\,\exp(x)}{dx} = \sum_{m=0}^{\infty} \frac{x^m}{m!}.$$

But this final expression is exactly the definition of $\exp(x)$. Thus $\exp(x)$ has the special property that it is equal to its own derivative. As a formula,

$$\frac{d\,\exp(x)}{dx} = \exp(x). \tag{A.2}$$

It follows that the second derivative of $\exp(x)$ is also equal to the original function – and, in fact, that all derivatives of $\exp(x)$ are equal to $\exp(x)$.

We continue by determining the derivative of $\ln x$, with $\ln x$ defined by the second equality in (1.27); clearly it is given by dy/dx. To obtain an expression for the latter, we consider the derivative with respect to x of $\exp(y)$, using the first equality in definition (1.27) and the chain rule:

$$1 = \frac{dx}{dx} = \frac{d[\exp(y)]}{dx}$$

$$= \frac{d[\exp(y)]}{dy}\frac{dy}{dx}$$

$$= \exp(y)\frac{dy}{dx}.$$

To obtain the final line we have used result (A.2).

Appendix A

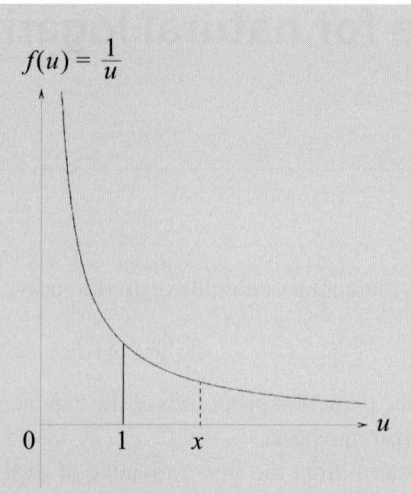

Figure A.1 The definition of $\ln x$ as an area under a curve.

But, as we have previously noted, $dy/dx = d(\ln x)/dx$, and so we can write

$$\frac{d(\ln x)}{dx} = \frac{dy}{dx} = \frac{1}{\exp(y)} = \frac{1}{x}.$$

Thus, we have the important intermediate result that

$$\frac{d(\ln x)}{dx} = \frac{1}{x}. \tag{A.3}$$

We now use the connection between a derivative and the function from which it is derived to invert relationship (A.3) so as to read

$$\ln x = \int_c^x \frac{1}{u}\, du,$$

where c is some constant. Because, from (1.17), we must have $\ln 1 = 0$, c must take the value 1, giving as one representation of $\ln x$,

$$\ln x = \int_1^x \frac{1}{u}\, du. \tag{A.4}$$

Thus, $\ln x$ can be interpreted graphically as the area, between $u = 1$ and $u = x$, under the (hyperbolic) curve $f(u) = 1/u$ for positive u. This is illustrated in Figure A.1. If $0 < x < 1$, then both the area and $\ln x$ are negative.

The next step in the proof that

$$e^z = \exp(z)$$

is to show that, for a general real variable z,

$$\ln x^z = z \ln x.$$

Appendix A

Using representation (A.4), we have that

$$\ln x^z = \int_1^{x^z} \frac{1}{u}\, du.$$

Making the change of variable $u = v^z$, with $du = zv^{z-1}\, dv$, the expression for $\ln x^z$ becomes

$$\ln x^z = \int_1^x \frac{1}{v^z} zv^{z-1}\, dv = z\int_1^x \frac{1}{v}\, dv = z\ln x.$$

This establishes the stated result.

Finally, we set $x = \exp(1) = e$ in this result and use (1.18), obtaining

$$\ln[\exp(1)^z] = z\ln[\exp(1)] = z\ln e = z,$$
$\Rightarrow \qquad \ln(e^z) = z,$
$\Rightarrow \qquad e^z = \exp(z).$

To obtain the last line, we again used (1.27), with the second equality implying the first; in (1.27) z replaced y and e^z replaced x. Since z is a general variable, result (A.1) is established, and, consequently, so is statement (1.26).

As e^x, rather than $\exp(x)$, is the form most commonly used, we restate result (A.2) explicitly as

$$\frac{de^x}{dx} = e^x \quad \text{and} \quad \frac{d^n(e^x)}{dx^n} = e^x \text{ for all } n. \tag{A.5}$$

B Sinusoidal definitions

In this appendix the connection is made between the geometrical definitions of the sine and cosine functions, as given by Figure 1.2 and Equation (1.50), and the algebraic power series definitions of the same two functions given in (1.53) and (1.54), namely

$$\sin\theta = \sum_{n=0}^{\infty} \frac{(-1)^n \theta^{2n+1}}{(2n+1)!} = \theta - \frac{\theta^3}{3!} + \frac{\theta^5}{5!} - \cdots, \tag{B.1}$$

and

$$\cos\theta = \sum_{n=0}^{\infty} \frac{(-1)^n \theta^{2n}}{(2n)!} = 1 - \frac{\theta^2}{2!} + \frac{\theta^4}{4!} - \cdots \tag{B.2}$$

In the series definitions the angle θ *must* be expressed in radians.

The proof, which has similarities to a method used in differential calculus, depends on the fact that if two functions, $f(\theta, \phi)$ and $g(\theta, \phi)$, each consisting of a sum of terms of the form $\theta^p \phi^q$ have equal values over finite ranges of values of θ and ϕ, then the coefficients of corresponding product powers of θ and ϕ are equal in f and g, on a term-by-term basis.

We start from results (1.59) and (1.60) for the sine and cosine of the sum of two angles:

$$\sin(A + B) = \sin A \cos B + \cos A \sin B, \tag{B.3}$$
$$\cos(A + B) = \cos A \cos B - \sin A \sin B, \tag{B.4}$$

and take A as θ and B as ϕ. Both angles are measured in radians and ϕ will later be taken to be $\ll 1$.

We now suppose that $\sin x$ and $\cos x$ can each be represented for all x by a power series in x, i.e.

$$\sin x = \sum_{n=0}^{\infty} s_n x^n \quad \text{and} \quad \cos x = \sum_{n=0}^{\infty} c_n x^n,$$

for some sets of coefficients, s_n and c_n, that are independent of x.

Now, when ϕ is small and measured in radians, the quantity y in Figure 1.2 is very nearly equal to the arc length corresponding to ϕ, namely, $1 \times \phi = \phi$. Consequently, for $\phi \ll 1$, $\sin \phi \approx \phi$. The corresponding value of x, and hence of $\cos \phi$, can be obtained from Pythagoras' theorem, Equation (1.56), and the binomial expansion with $n = 1/2$:

$$x = \sqrt{1 - y^2} \approx (1 - \phi^2)^{1/2} = 1 - \frac{\frac{1}{2}\phi^2}{1!} + \cdots$$

Appendix B

There is no linear term in ϕ in this expansion and so $\cos\phi \approx 1$. These two observations allow us to determine the first two coefficients in each of the power series. In the sine series, $s_0 = 0$ and $s_1 = 1$, whilst in the cosine series, $c_0 = 1$ and $c_1 = 0$.

Before we substitute the series forms of sine and cosine into (B.3) and (B.4), we note that $\sin(\theta + \phi)$ can be expanded, using the binomial expansion, in the form

$$\sin(\theta + \phi) = \sum_{n=0}^{\infty} s_n(\theta + \phi)^n = \sum_{n=0}^{\infty} s_n \theta^n \left(1 + \frac{\phi}{\theta}\right)^n = \sum_{n=0}^{\infty} s_n \theta^n \left(1 + \frac{n\phi}{\theta} + \cdots\right),$$

and that, in a similar way,

$$\cos(\theta + \phi) = \sum_{n=0}^{\infty} c_n \theta^n \left(1 + \frac{n\phi}{\theta} + \cdots\right).$$

On substituting into (B.3), we obtain

$$\sum_{n=0}^{\infty} s_n \theta^n \left(1 + \frac{n\phi}{\theta} + \cdots\right) = \sum_{n,m=0}^{\infty} s_n c_m \theta^n \phi^m + \sum_{n,m=0}^{\infty} c_n s_m \theta^n \phi^m.$$

Equating the coefficients of θ^n on the two sides gives the correct, but unhelpful, identity $s_n = s_n c_0 + c_n s_0 = s_n \cdot 1 + c_n \cdot 0 = s_n$, but equating those of $\theta^{n-1}\phi$ yields the significant relationship

$$n s_n = s_{n-1} c_1 + c_{n-1} s_1 = s_{n-1} \cdot 0 + c_{n-1} \cdot 1 = c_{n-1}. \quad (*)$$

Similarly, substitution in (B.4) yields an identity for the coefficients of θ^n, but the relationship

$$n c_n = c_{n-1} c_1 - s_{n-1} s_1 = c_{n-1} \cdot 0 - s_{n-1} \cdot 1 = -s_{n-1} \quad (**)$$

for the coefficients of $\theta^{n-1}\phi$.

Results (*) and (**) can now be combined to give recurrence relations for s_n and c_n, which, taken with the known values of s_0, s_1, c_0 and c_1, give complete prescriptions for the coefficients in the two series. Using first (**), and then (*) with n replaced by $n - 1$, gives

$$c_n = -\frac{s_{n-1}}{n} = -\frac{1}{n}\frac{c_{n-2}}{(n-1)} = \frac{-c_{n-2}}{n(n-1)}.$$

Since $c_1 = 0$, it follows that c_n is zero for all odd n. For n even and equal to $2m$, we have

$$c_{2m} = \frac{-c_{2m-2}}{2m(2m-1)} = \frac{c_{2m-4}}{2m(2m-1)(2m-2)(2m-3)} = \cdots$$
$$= \frac{(-1)^m c_0}{(2m)!} = \frac{(-1)^m}{(2m)!}.$$

A similar recurrence relation can be obtained for the s_n, but it is quicker to recall that $s_{n-1} = -n c_n$, and so (i) the fact that $c_{2m+1} = 0$ for all m means that $s_{2m} = 0$ for all m and (ii)

$$s_{2m+1} = -(2m+2)c_{2m+2} = -(2m+2)\frac{(-1)^{2m+2}}{(2m+2)!} = \frac{(-1)^m}{(2m+1)!}.$$

Appendix B

This completes the specification of the coefficients in both power series, and they are seen to be in accord with those given in (B.1) and (B.2). What we have not done is to guarantee that the coefficients of all terms of the form $\theta^p \phi^q$ balance on the two sides of the equations; this can be shown using complex numbers and Euler's theorem (see Chapter 5), but here must be taken on trust.

In summary, the requirement that the *algebraic* power series coefficients should be such that the *geometric* addition formulae are satisfied provides the link between the two approaches, and uniquely determines what the coefficients in the power series must be if the two definitions of a sine or cosine are to be compatible.

C Leibnitz's theorem

In this appendix we give the mathematical proof of Leibnitz's theorem concerning the nth derivative, $f^{(n)}$, of a function f that can be written as the product of two functions u and v, i.e. $f = uv$. The independent variable x, of which u, v and f are all functions, and with respect to which all derivatives are taken, has been suppressed in order to improve the clarity of the layout. The theorem states that

$$f^{(n)} = \sum_{r=0}^{n} \frac{n!}{r!(n-r)!} u^{(r)} v^{(n-r)} = \sum_{r=0}^{n} {}^nC_r u^{(r)} v^{(n-r)}, \quad (C.1)$$

where the fraction $n!/[r!(n-r)!]$ is identified with the binomial coefficient ${}^nC_r$ (see Section 1.4.1). To prove the theorem, we use the method of induction as follows.

Assume that (C.1) is valid for n equal to some integer N. Then, using rule (3.8) for differentiating a product, we have

$$f^{(N+1)} = \sum_{r=0}^{N} {}^NC_r \frac{d}{dx}\left(u^{(r)} v^{(N-r)}\right)$$

$$= \sum_{r=0}^{N} {}^NC_r [u^{(r)} v^{(N-r+1)} + u^{(r+1)} v^{(N-r)}]$$

$$= \sum_{s=0}^{N} {}^NC_s u^{(s)} v^{(N+1-s)} + \sum_{s=1}^{N+1} {}^NC_{s-1} u^{(s)} v^{(N+1-s)},$$

where we have substituted summation index s for r in the first summation and for $r+1$ in the second. Now, from our earlier discussion of binomial coefficients, Equation (1.43), we have

$$ {}^NC_s + {}^NC_{s-1} = {}^{N+1}C_s $$

and so, after separating out the first term of the first summation and the last term of the second, obtain

$$f^{(N+1)} = {}^NC_0 u^{(0)} v^{(N+1)} + \sum_{s=1}^{N} {}^{N+1}C_s u^{(s)} v^{(N+1-s)} + {}^NC_N u^{(N+1)} v^{(0)}.$$

Appendix C

But $^N C_0 = 1 = {}^{N+1}C_0$ and $^N C_N = 1 = {}^{N+1}C_{N+1}$, and so we may write

$$f^{(N+1)} = {}^{N+1}C_0 u^{(0)} v^{(N+1)} + \sum_{s=1}^{N} {}^{N+1}C_s u^{(s)} v^{(N+1-s)} + {}^{N+1}C_{N+1} u^{(N+1)} v^{(0)}$$

$$= \sum_{s=0}^{N+1} {}^{N+1}C_s u^{(s)} v^{(N+1-s)}.$$

But this is just (C.1) with n set equal to $N + 1$. Thus, assuming the validity of (C.1) for $n = N$ implies its validity for $n = N + 1$. However, when $n = 1$, Equation (C.1) is simply the product rule, and this we have already proved directly. These results taken together establish the validity of (C.1) for all n and prove Leibnitz's theorem.

D Summation convention

In the main text, particularly when dealing with matrices and vector calculus in three dimensions, we often need to take a sum over a number of terms which are all of the same general form, and differ only in the value of an indexing subscript or the symbol associated with each of the three different Cartesian coordinates. Thus the prescription for the elements of the product **P** of two matrices, **A** and **B**, might be written as

$$P_{ij} = \sum_{k=1}^{N} A_{ik} B_{kj} \tag{D.1}$$

or the expression for the divergence of a vector **a** (see Chapter 11) given in the form

$$\nabla \cdot \mathbf{a} = \frac{\partial a_x}{\partial x} + \frac{\partial a_y}{\partial y} + \frac{\partial a_z}{\partial z}. \tag{D.2}$$

Sometimes, for example in a multiple matrix product, several sums appear in an expression, each with its own explicit summation sign.

Such calculations can be significantly compacted, and in some cases simplified, if the Cartesian coordinates x, y and z are replaced symbolically by the indexed coordinates x_i, where i takes the values 1, 2 and 3, and the so-called *summation convention* is adopted. We will also find that when the convention is used, together with two particular indexed quantities, the Kronecker delta δ_{ij} and the Levi-Civita symbol ϵ_{ijk},[1] many of the equalities appearing in vector algebra and calculus can be established very straightforwardly.

The convention is that any *lower-case* alphabetic subscript that appears *exactly* twice in any term of an expression is understood to be summed over all the values that a subscript in that position can take (unless the contrary is specifically stated). The subscripted quantities may appear in the numerator and/or the denominator of a term in an expression. This naturally implies that any such pair of repeated subscripts must occur only in subscript positions that have the same range of values. Sometimes the ranges of values have to be specified but usually they are apparent from the context. As specific examples, for vector calculus in ordinary three-dimensional space the range is 1 to 3, and for the multiplication of $n \times n$ matrices it is 1 to n.

In this notation (D.1) becomes

$$P_{ij} = A_{ik} B_{kj} \qquad \text{i.e. without the explicit summation sign,}$$

and (D.2) is even more compacted to $\nabla \cdot \mathbf{a} = \dfrac{\partial a_i}{\partial x_i}$.

---

[1] These two quantities are technically isotropic Cartesian tensors, though this aspect will not be part of our considerations here.

Appendix D

The following simple examples illustrate further what is meant (in the three-dimensional case):

(i) $a_i x_i$ stands for $a_1 x_1 + a_2 x_2 + a_3 x_3$, the scalar product $\mathbf{a} \cdot \mathbf{x}$;

(ii) a_{ii} stands for $a_{11} + a_{22} + a_{33}$, the trace of $\mathbf{A}$;

(iii) $a_{ij} b_{jk} c_k$ stands for $\sum_{j=1}^{3} \sum_{k=1}^{3} a_{ij} b_{jk} c_k$, the ith component of the vector $\mathbf{ABc}$;

(iv) $\dfrac{\partial^2 \phi}{\partial x_i \partial x_i}$ stands for $\dfrac{\partial^2 \phi}{\partial x_1^2} + \dfrac{\partial^2 \phi}{\partial x_2^2} + \dfrac{\partial^2 \phi}{\partial x_3^2}$, ∇^2 of scalar ϕ;

(v) $\dfrac{\partial^2 v_j}{\partial x_i \partial x_i}$ stands for $\dfrac{\partial^2 v_j}{\partial x_1^2} + \dfrac{\partial^2 v_j}{\partial x_2^2} + \dfrac{\partial^2 v_j}{\partial x_3^2}$, ∇^2 of the jth component of vector $\mathbf{v}$.

Subscripts that are summed over are called *dummy subscripts* and the others *free subscripts*. For example, in (v) above, i is a dummy subscript, but j is a free subscript. It is worth remarking that, when introducing a dummy subscript into an expression, care should be taken not to use one that is already present, either as a free or as a dummy subscript. For example, $a_{ij} b_{jk} c_{kl}$ cannot, and must not, be replaced by $a_{ij} b_{jj} c_{jl}$ or by $a_{il} b_{lk} c_{kl}$, but could be replaced by $a_{im} b_{mk} c_{kl}$ or by $a_{im} b_{mn} c_{nl}$. Naturally, free subscripts must not be changed at all unless the working calls for it.

Intrinsic to the application of the summation convention is the subscripted quantity the Kronecker delta, δ_{ij}, which is defined by

$$\delta_{ij} = \begin{cases} 1 & \text{if } i = j, \\ 0 & \text{otherwise.} \end{cases}$$

When the summation convention has been adopted, the main effect of δ_{ij} is to replace one subscript by another in certain expressions. Examples might include

$$b_j \delta_{ij} = b_i,$$

and

$$a_{ij} \delta_{jk} = a_{ij} \delta_{kj} = a_{ik}. \tag{D.3}$$

In the second of these the dummy index shared by both terms on the left-hand side (namely j) has been replaced by the free index carried by the Kronecker delta (namely k), and the delta symbol has disappeared. In matrix language, (D.3) can be written as $\mathbf{AI} = \mathbf{A}$, where $\mathbf{A}$ is the matrix with elements a_{ij} and $\mathbf{I}$ is the unit matrix having the same dimensions as $\mathbf{A}$ – its elements are simply δ_{ij}, i.e. ones on the leading diagonal where $i = j$ and zero elsewhere where $i \neq j$.

In some expressions we may use the Kronecker delta to replace indices in a number of different ways, e.g.

$$a_{ij} b_{jk} \delta_{ki} = a_{ij} b_{ji} \quad \text{or} \quad a_{kj} b_{jk},$$

where the two expressions on the RHS are totally equivalent to one another, both being equal to the trace of $\mathbf{AB}$.

Appendix D

Use of the summation convention and the Levi-Civita symbol, defined by

$$\epsilon_{ijk} = \begin{cases} +1 & \text{if } i, j, k \text{ is an even permutation of 1, 2, 3,} \\ -1 & \text{if } i, j, k \text{ is an odd permutation of 1, 2, 3,} \\ 0 & \text{otherwise,} \end{cases}$$

allow very compact representations of some expressions that arise frequently in matrix and vector algebras. For example, the ith component of a vector product given in (9.33) as

$$\mathbf{a} \times \mathbf{b} = (a_y b_z - a_z b_y)\mathbf{i} + (a_z b_x - a_x b_z)\mathbf{j} + (a_x b_y - a_y b_x)\mathbf{k},$$

can be written using the summation convention as $\epsilon_{ijk} a_j b_k$, whilst the whole of the RHS can be written as $\epsilon_{ijk} a_j b_k \hat{\mathbf{e}}_i$ in a notation in which $\hat{\mathbf{e}}_i$ represents $\mathbf{i}$, $\mathbf{j}$ and $\mathbf{k}$ for i equal to 1, 2 and 3, respectively. In a similar way, the scalar triple product $[\mathbf{a} \cdot (\mathbf{b} \times \mathbf{c})]$ in (9.36) can be expressed as $\epsilon_{ijk} a_i b_j c_k$ and the determinant given in (10.46) as $\epsilon_{ijk} a_{1i} a_{2j} a_{3k}$.

It should be noted that

$$\epsilon_{ijk} = \epsilon_{jki} = \epsilon_{kij} = -\epsilon_{ikj} = -\epsilon_{kji} = -\epsilon_{jik}.$$

One useful identity connecting the Kronecker delta and the Levi-Civita symbol, which we quote here without proof, is

$$\epsilon_{ijk} \epsilon_{k\ell m} = \delta_{i\ell} \delta_{jm} - \delta_{im} \delta_{j\ell}. \tag{D.4}$$

An expression involving both the Levi-Civita symbol and partial differentiation is that for the ith component of the curl of a vector:

$$(\nabla \times \mathbf{a})_i = \frac{\partial a_k}{\partial x_j} - \frac{\partial a_j}{\partial x_k},$$

where i, j and k are a cyclic permutation of 1, 2 and 3 (in that order). Since $\epsilon_{ijk} = 0$ if any two of i, j and k are equal, this same expression can be extended to read

$$(\nabla \times \mathbf{a})_i = \frac{\partial a_k}{\partial x_j} - \frac{\partial a_j}{\partial x_k} = \epsilon_{ijk} \frac{\partial a_k}{\partial x_j},$$

with the sum taken over all j and k, as implied by the summation convention. There are formally nine terms in the sum, but, for a given i, only two of them are non-zero.

It should be noted that whilst the expression for a vector component contains one free subscript, namely i, the other three examples given here contain only dummy subscripts; the 1, 2 and 3 appearing in the expression for a determinant are not lower-case alphabetic subscripts and are not classed as either.

E Physical constants

Speed of light in a vacuum, $c = 3.0 \times 10^8$ m s^{-1}.
Elementary charge, $e = 1.60 \times 10^{-19}$ C.
Mass of electron, $m_e = 9.1 \times 10^{-31}$ kg.
Mass of proton, $m_p = 1.67 \times 10^{-27}$ kg.
Avogadro constant, $N_A = 6.0 \times 10^{23}$ mol^{-1}.
Planck constant, $h = 6.6 \times 10^{-34}$ J s.
Boltzmann constant, $k = 1.38 \times 10^{-23}$ J K^{-1}.
Stefan–Boltzmann constant, $\sigma = 5.7 \times 10^{-8}$ W m^{-2} K^{-4}.
Gravitational constant, $G = 6.7 \times 10^{-11}$ N kg^{-2} m^2.
Gravitational acceleration $g \approx 9.8$ m s^{-2}.
Permeability of a vacuum, $\mu_0 = 4\pi \times 10^{-7}$ H m^{-1}.
Permittivity of a vacuum, $\epsilon_0 = 8.8 \times 10^{-12}$ F m^{-1}.

F Footnote answers

This appendix contains short answers to those footnotes that are in the form of a question. They have been deliberately placed away from the questions so as to encourage the reader to formulate their own response, as they would be expected to do in a supervision or tutorial, before seeking confirmation. It should be remembered that the questions, typically requiring only brief answers, are normally designed to test whether a particular point in the main text, which may in itself be a relatively small one, has been correctly grasped. Thus some answers may seem trivial to the reader – if they do, so much the better!

1. Arithmetic and geometry

1 (i) $a \odot b = a^2 + b^2 = b \odot a$. $(a \odot b) \odot c = (a^2 + b^2)^2 + c^2 \neq a^2 + (b^2 + c^2)^2 = a \odot (b \odot c)$. Commutative but not associative.
(ii) $a \odot b = \sqrt{a^2 + b^2} = b \odot a$. $(a \odot b) \odot c = \sqrt{a^2 + b^2} \odot c = \sqrt{a^2 + b^2 + c^2} = a \odot (b \odot c)$. Commutative and associative.
(iii) $a \odot b = a^b \neq b^a = b \odot a$. $(a \odot b) \odot c = (a^b) \odot c = (a^b)^c \neq a^{(b^c)} = a \odot (b \odot c)$. For example, $(2 \odot 3) \odot 4 = 8^4 = 4096$ but $2 \odot (3 \odot 4) = 2^{81} \approx 2.4 \times 10^{24}$. Neither commutative nor associative.

5 For e or f to be infinite would require $c^2 - d^2 p = 0$. This would imply that $\sqrt{p} = \pm c/d$, i.e. that $\sqrt{p}$ is rational, which would be an internal contradiction.

6 Binary: $32 + 0 + 8 + 0 + 2 + 0 = 42$. Octal: $(5 \times 8) + (2 \times 1) = 42$. Hexadecimal: $(2 \times 16) + (1 \times A) = 42$, where $A = 10, B = 11, \ldots, F = 15$.

9 Let $z = x \ln a$. Then $y = a^x \Rightarrow \ln y = x \ln a = z \Rightarrow y = e^z$.

12 (a) 10^{-28} m^2, (b) 1.11×10^{-7} kg m^{-1}, (c) 1.016×10^{-1} m, (d) 0.902 cd, (e) 1.11×10^{-9} F, (f) 1.421×10^{-4} m^3, (g) 5.03 m, (h) 10^{-8} s, (i) 10^{-52} m^2, (j) 14.59 kg, (k) 10^{-1} W^{-1} m^2 K.

13 The spring constant k has dimensions $[\text{N m}^{-1}] = MLT^{-2}L^{-1}$. Therefore $[\frac{1}{2}kx^2] = [\frac{1}{2}][k][x^2] = MT^{-2}L^2 = [\text{Energy}]$.

14 We have $u = \frac{1}{2}\epsilon_0 E^2$, where u is an energy density with dimensions $[J]L^{-3}$, and $[E] = [V]L^{-1} = [J](IT)^{-1}L^{-1}$. Hence, $[\epsilon_0] = [J]^{-1}(IT)^2 L^{-1} = M^{-1}L^{-3}T^4I^2$, since $[J] = ML^2T^{-2}$.

16 $x^5 + 5x^4y + 10x^3y^2 + 10x^2y^3 + 5xy^4 + y^5$.

19 Continuous interest: $x = (0.05)^{-1} \ln 2 = 13.863$ years. Annual interest: $x = \ln 2 / \ln 1.05 = 14.207$ years. But the annual interest is not paid until 15 years have elapsed, which is $1.137 \times 365 = 415$ days later.

20 If M is the mid-point of secant RP, then MR has length $OR \sin \theta/2$, and the whole secant is twice this length.

Appendix F

21 (i) A little, for at least checking the 'calculated' value. (ii) Quite a lot – somebody who can construct this answer probably did so with (a) tongue in check and (b) quite a good understanding!

22 $u = [\cos x]^{-1}$ and $x = \sin v$. Therefore, $u = [\cos x]^{-1} = [\cos(\sin v)]^{-1}$. Further, $\cos x = u^{-1} \Rightarrow x = \cos^{-1} u^{-1}$, and so $v = \sin^{-1} x = \sin^{-1}(\cos^{-1} u^{-1})$.

23 Divide $\sin^2 \theta + \cos^2 \theta = 1$ through by $\cos \theta \sin \theta$ giving $\dfrac{\sin \theta}{\cos \theta} + \dfrac{\cos \theta}{\sin \theta} = \dfrac{1}{\cos \theta} \dfrac{1}{\sin \theta}$, i.e. $\tan \theta + \cot \theta = \sec \theta \csc \theta$.

24 LHS of (1.56) $= \cos^2(A \pm B) + \sin^2(A \pm B)$

$= \left[\cos^2 A \cos^2 B \mp 2 \cos A \cos B \sin A \sin B + \sin^2 A \sin^2 B\right]$
$+ \left[\sin^2 A \cos^2 B \pm 2 \sin A \cos B \cos A \sin B + \cos^2 A \sin^2 B\right]$
$= \cos^2 A(\cos^2 B + \sin^2 B) + \sin^2 A(\sin^2 B + \cos^2 B)$
$= \cos^2 A + \sin^2 A = 1.$

25 The given line has $m_1 = 2$; the value 5 affects where the lines meet, but not the angle at which they meet. For a 45° intersection, $\tan \theta_{12} = \pm 1$. If the line through the origin is $y = mx$, then

$$\pm 1 = \dfrac{2 - m}{1 + 2m} \Rightarrow 3m = 1 \text{ or } m = -3.$$

The two lines are $y = \tfrac{1}{3}x$ and $y = -3x$.

26 If $t = \tan \pi/12$, then $(1 - t^2)/(1 + t^2) = \cos \pi/6 = \sqrt{3}/2$. This rearranges to $t^2 = (2 - \sqrt{3})/(2 + \sqrt{3}) = (2 - \sqrt{3})^2/(4 - 3)$. Hence $t = +(2 - \sqrt{3})$.

28 The result in the text implies that $ax^n + bx^{-n} \geq 2\sqrt{ax^n bx^{-n}} = 2\sqrt{ab}$, with this minimum value being reached when $ax^n = bx^{-n}$, i.e. $x = (b/a)^{1/2n}$.

2. Preliminary algebra

3 The value of a_0 determines the vertical position of a graph of $f(x)$ but does not affect its shape; in particular, it does not affect its slope at any point. The derivative measures the slope and is therefore independent of a_0.

5 Since $f(-\infty) = -\infty$ and $f(x)$ is increasing as x decreases through zero from positive to negative values, there must be (at least) one negative value of x at which a maximum occurs. As x becomes increasingly more negative it may have further maxima, but between each pair of maxima there must be a minimum. So, if there are n real maxima in all for negative x, there are $2n - 1$ turning points in the same range. But the equation determining these is a quintic which has, at most, five real roots. Thus $n \leq 3$.

7 Considered as a quadratic equation in x, $0 = x^2 - y^2 = (x + y)(x - y)$ has $a = 1$, $b = 0$ and $c = -y^2$. It has roots $x = -y$ and $x = y$. The sum of these is $-y + y = 0$, equal to the (absent) coefficient of the term linear in x. Their product is $-y^2$, equal to c/a.

10 See Figure F.1.

11 Parameterisation check: $a^4 b^2 \sin^2 2t = 4b^2 a^2 \sin^2 t \, a^2 \cos^2 t$, i.e. $(\sin 2t)^2 = (2 \sin t \cos t)^2$. For the sketch, see Figure F.2.

12 See Figure F.3. Without the convention the lower circle is generated; with it the upper circle is traversed a second time for $\pi < \phi < 2\pi$.

Appendix F

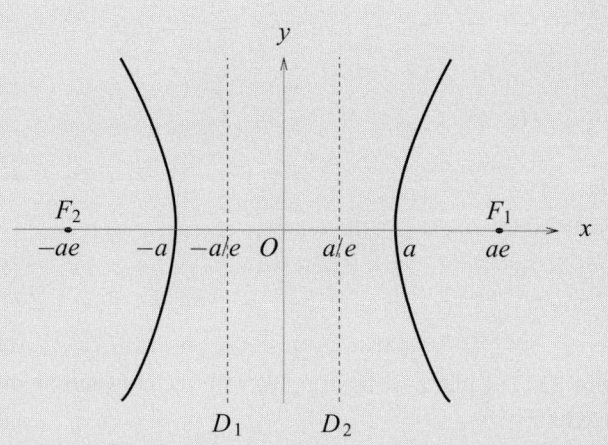

Figure F.1 A standard hyperbola $x^2/a^2 - y^2/b^2 = 1$ with eccentricity e given by $e^2 = 1 + b^2/a^2$. The foci, F_1 and F_2, and their corresponding directrices, D_1 and D_2, are shown.

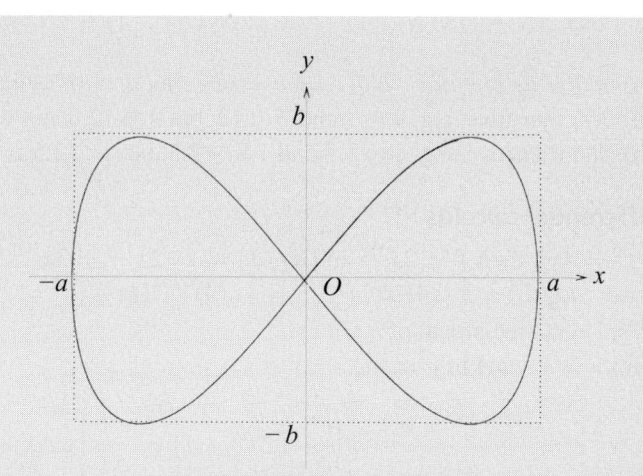

Figure F.2 The curve $a^4 y^2 = 4b^2 x^2 (a^2 - x^2)$, parameterised by $x = a \sin t$ and $y = b \sin 2t$.

13 Factorising the four given quadratic functions: $x^2 + 2x - 3 = (x+3)(x-1)$; $x^2 + 4x + 3 = (x+3)(x+1)$; $x^2 + 6x + 9 = (x+3)^2$; $x^2 - 1 = (x-1)(x+1)$. Thus the l.c.m. is $(x+3)^2(x-1)(x+1)$ or $(x^2 + 6x + 9)(x^2 - 1) = x^4 + 6x^3 + 8x^2 - 6x - 9$ in unfactored form.

15 The denominator factorises as $(x+2)(x+1)$. So, setting

$$\frac{x^2 + 2x + 1}{x^2 + 3x + 2} = \frac{A}{x+2} + \frac{B}{x+1} = \frac{A(x+1) + B(x+2)}{x^2 + 3x + 2}$$

gives $A + B = 2$ and $A + 2B = 1$, but the numerator of the RHS contains no term in x^2 whereas the LHS does.

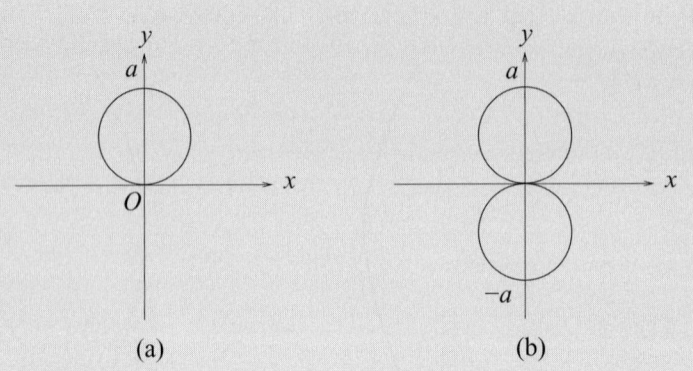

Figure F.3 The plot $\rho = a\sin\phi$, (a) with the convention and (b) without the convention.

17 Note that $(N-1)$, N and $(N+1)$ are three *consecutive* integers and so one of them must divide by 3; consequently so does their product. $9N$ clearly divides by 3. Thus it follows that $2(N-1)N(N+1)+9N = 2N(N^2-1)+9N = 2N^3+7N$ divides by 3.

18 (i) $a\odot b \Rightarrow b = ma$; $b\odot c \Rightarrow c = nb$. So, $c = nb = nma \Rightarrow a\odot c$. Transitive.
 (ii) *Not* transitive, e.g. $9\odot 7$ and $7\odot 12$, but $9\odot 12$ is not valid.
 (iii) *Not* transitive, e.g. $1\odot 1.8$ and $1.8\odot 2.5$, but $1\odot 2.5$ is not valid.

3. Differential calculus

2 If $f^2 = 4x^3$, then $f = 2x^{3/2}$ and $df/dx = \frac{3}{2}\cdot 2x^{1/2} = 3x^{1/2}$. Further, $x = (4)^{-1/3}f^{2/3}$ and so $dx/df = \frac{2}{3}\cdot(4)^{-1/3}f^{-1/3} = \frac{2}{3}\cdot(4)^{-1/3}(4x^3)^{-1/6} = \frac{1}{3}\cdot(4)^{[1/2-1/3-1/6]}x^{-3/6} = \frac{1}{3}x^{-1/2}$, i.e. the reciprocal of df/dx.

3 With $y = e^{ax}$ and $\ln y = ax$,

$$\frac{\Delta y}{\Delta x} = \frac{e^{a(x+\Delta x)} - e^{ax}}{\Delta x} = \frac{e^{ax}}{\Delta x}\left(e^{a\Delta x} - 1\right) = \frac{e^{ax}}{\Delta x}\left[a\Delta x + \frac{(a\Delta x)^2}{2!} + \cdots\right].$$

Taking the limit $\Delta x \to 0$, $dy/dx = ae^{ax}$. Further,

$$\frac{\Delta \ln y}{\Delta y} = \frac{a(x+\Delta x) - ax}{e^{ax}[a\Delta x + (a\Delta x)^2/2! + \cdots]} = \frac{1}{e^{ax}(1 + a\Delta x/2! + \cdots)}.$$

Taking the limit $\Delta x \to 0$, $d(\ln y)/dy = (e^{ax})^{-1} = 1/y$.

4 Denote the derivative of $g(x)$ by g'. Then $[(x^2)(x\sin x)]' = x^2(x\sin x)' + (x^2)'(x\sin x) = x^2[x(\sin x)' + (x)'\sin x] + (2x)(x\sin x) = x^2(x\cos x + 1\cdot\sin x) + 2x^2\sin x = x^3\cos x + 3x^2\sin x$, i.e. as in the text.

5 $f(x) = (3+x^2)^3 = 3^3 + 3\cdot 3^2 x^2 + 3\cdot 3x^4 + x^6 = x^6 + 9x^4 + 27x^2 + 27$, with derivative $f' = 6x^5 + 36x^3 + 54x$. Now, $6x(3+x^2)^2 = 6x(9 + 6x^2 + x^4) = f'$.

6 $dy/dx = dy/d\phi \div dx/d\phi = b\cos\phi \div [-a\sin\phi] = b(x/a) \div [-ay/b] = -b^2 x/a^2 y$.

Appendix F

7 With $x = a\cos\phi$ and $y = b\sin\phi$, the equation of the ellipse is $x^2/a^2 + y^2/b^2 = 1$. Implicit differentiation gives $2x/a^2 + 2y(dy/dx)/b^2 = 0$, from which $dy/dx = -b^2x/a^2y$.

9 For $(uv)^{(n)}$ there are $n+1$ terms; so, eight terms for a seventh derivative. However, if $v = 3x^5$, its sixth and seventh derivatives will vanish and $(uv)^{(7)}$ will have only six terms.

10 For $f(x) = \sin x$: $f'(x) = \cos x$; $f''(x) = -\sin x$. For $f(x) = \cos x$: $f'(x) = -\sin x$; $f''(x) = -\cos x$. In either case, $f''(x) = -f(x)$. For a maximum at a turning point x_0, we require $f''(x_0) < 0$, i.e. $f(x_0) > 0$. Similarly, for a minimum at x_0 we require $f''(x_0) > 0$, i.e. $f(x_0) < 0$.

11 Since $f(3) = -79$ and $f(-2) = 46$ have opposite signs, there are three real roots.

12 $f(-2) = (-8+7)/(4-8+3) = 1$. $f(2) = (8+7)/(4+8+3) = 1$. Differentiation of the quotient gives $f'(x) = h(x)/(x^2 + 4x + 3)^2$, where $h(x) = [(x^2+4x+3)4 - (4x+7)(2x+4)] = -4x^2 - 14x - 16 = -g(x)$ and, as stated, $g(x)$ has no real zeros. However, as a factorisation of its denominator and a rough sketch will show, $f(x)$ has infinite discontinuities at $x = -1$ and $x = -3$. The first of these falls in the given range and, as $f(x)$ is not differentiable there, Rolle's theorem cannot be applied.

13 The slope at a general point x is $3x^2$. This has the same slope as the chord when $(c^3 - a^3)/(c - a) = 3x^2$. But $c^3 - a^3 = (c-a)(c^2 + ac + a^2)$ and so the stated result follows. Taking $c > a$: $a = (3a^2)^{1/2}/\sqrt{3} < (c^2 + ac + a^2)^{1/2}/\sqrt{3} < (3c^2)^{1/2}/\sqrt{3} = c$, i.e. $a < x < c$, as expected.

14 The mean value theorem applied to $\cos x$, with $0 < x < \pi/2$, gives $(\cos x - \cos 0)/(x - 0) = -\sin b$, with $0 < b < x$. Hence $1 - \cos x = x\sin b < x\sin x$ for $0 < x < \pi/2$, or $2\sin^2 x/2 < 2x \sin x/2 \cos x/2$. Thus $\tan x/2 < x$ for $0 < x < \pi/2$, or $\tan c < 2c$ for $0 < c < \pi/4$. That $\tan c > c$ has already been shown.

15 Somebody has a sense of humour – or is going to need one!

17 In each case, B is the value of the original numerator calculated at $x = 3$.

4. Integral calculus

3 By sketching both inverse functions in the range $-a \leq x \leq a$, and remembering that both are multi-valued (or by applying the general formulae for $\cos(A+B)$ and $\sin(A+B)$ to its LHS), convince yourself that $\cos^{-1}(x/a) = -\sin^{-1}(x/a) + \pi/2$. Thus, if the constant in the first integral is written as c', then $c' = -c - \pi/2$.

5 Following the model in the text,

$$\int \tan ax \, dx = \int \frac{\sin ax}{\cos ax} dx = -\frac{1}{a}\int \frac{(\cos ax)'}{\cos ax} dx = -\frac{1}{a}\ln(\cos ax) = \frac{1}{a}\ln(\sec ax).$$

6 $f(x) = x^4 + 2x^3 + 5x^2 + 8x + 4$ cannot have any positive zeros since all of its coefficients are positive. The coefficient pattern (the sum of those of even powers = the sum of those of odd powers) indicates that $x = -1$ is a zero, and hence that $x + 1$ is a factor. So, $f(x) = (x+1)(x^3 + x^2 + 4x + 4) = (x+1)(x+1)(x^2+4)$. Hence

$$\int \frac{1}{f(x)} dx = \int \left[\frac{A}{(x+1)^2} + \frac{B}{x+1} + \frac{Cx+D}{x^2+4}\right] dx$$

$$= \frac{E}{x+1} + F\ln(x+1) + G\ln(x^2+4) + H\tan^{-1}\frac{x}{2} + I.$$

Appendix F

9 Using the t-substitution:
$$\int \sin 2x \, dx = \int 2\sin x \cos x \, dx$$
$$= 2\int \frac{t}{\sqrt{1+t^2}} \frac{1}{\sqrt{1+t^2}} \frac{dt}{1+t^2} = \int \frac{2t}{(1+t^2)^2} dt$$
$$= -\frac{1}{1+t^2} + A = -\frac{1}{2}(2\cos^2 x - 1) + B = -\frac{1}{2}\cos 2x + B.$$

10 $\int_0^1 \sin^{-1} x \, dx = \left[x \sin^{-1} x\right]_0^1 - \int_0^1 \frac{1}{\sqrt{1-x^2}} x \, dx = \frac{\pi}{2} + \left[\sqrt{1-x^2}\right]_0^1 = \frac{\pi}{2} - 1.$

13 $\int \frac{2a^2b^2}{b^2 \cos^2 \phi + a^2 \sin^2 \phi} d\phi = \int \frac{2a^2b^2}{[b^2(1+t^2)^{-1} + a^2t^2(1+t^2)^{-1}]} \frac{dt}{1+t^2}$
$$= \int \frac{2a^2b^2}{b^2 + a^2t^2} dt = 2b^2 \int \frac{1}{(b/a)^2 + t^2} dt.$$

15 The equation of the thread is $z = h\phi/2\pi$, $\rho = a$. Now, $(ds)^2 = (d\rho)^2 + (\rho \, d\phi)^2 + (dz)^2 = 0 + (a \, d\phi)^2 + (h/2\pi \, d\phi)^2$. The length of thread per turn is
$$\int_0^{2\pi} \sqrt{a^2 + \frac{h^2}{4\pi^2}} d\phi = 2\pi \sqrt{a^2 + \frac{h^2}{4\pi^2}} = \sqrt{4\pi^2 a^2 + h^2}.$$

17 The bowl is $x^2 = 4ay$ in shape and has a base of radius $2a$ and a height of h. Its radius is $2a$ when $(2a)^2 = 4ay$, i.e. $y = a$. Therefore its volume is
$$\int_a^{a+h} \pi x^2 \, dy = \int_a^{a+h} \pi 4ay \, dy = 2\pi a \left[y^2\right]_a^{a+h}$$
$$= 2\pi a(a^2 + 2ah + h^2 - a^2) = 2\pi ah(2a + h).$$

5. Complex numbers and hyperbolic functions

2 $x(x - 1) + ix(x + 4) = 6 - 3i$. If x is real we can equate the real and imaginary parts on both sides: $x^2 - x - 6 = 0$ and $x^2 + 4x + 3 = 0$. Factorising gives $(x - 3)(x - 2) = 0$ and $(x + 1)(x + 3) = 0$. The first equation requires $x = 3$ or $x = 2$, whilst the second needs either $x = -1$ or $x = -3$. Thus, there is no real x that satisfies both and so no real solution to the original equation.

4 From (5.6),
$$|z_1 z_2|^2 = (x_1 x_2 - y_1 y_2)^2 + (x_1 y_2 + y_1 x_2)^2$$
$$= x_1^2 x_2^2 - 2x_1 x_2 y_1 y_2 + y_1^2 y_2^2 + x_1^2 y_2^2 + 2x_1 y_2 y_1 x_2 + y_1^2 x_2^2$$
$$= x_1^2(x_2^2 + y_2^2) + y_1^2(y_2^2 + x_2^2)$$
$$= (x_1^2 + y_1^2)(x_2^2 + y_2^2) = |z_1|^2 |z_2|^2. \quad \text{Hence the result.}$$

For the arguments:
$$\tan[\arg z_1 z_2] = \frac{x_1 y_2 + y_1 x_2}{x_1 x_2 - y_1 y_2} = \frac{(y_2/x_2) + (y_1/x_1)}{1 - (y_1/x_1)(y_2/x_2)}$$
$$= \frac{\tan[\arg z_2] + \tan[\arg z_1]}{1 - (\tan[\arg z_1])(\tan[\arg z_2])} = \tan[\arg z_1 + \arg z_2].$$

Thus $\arg(z_1 z_2) = \arg z_1 + \arg z_2.$

Appendix F

5 If $z^* = x_2 + iy_2$, then $|z^*| = |z_1|$ implies that $x_2^2 + y_2^2 = x_1^2 + y_1^2$, and $\text{Im}(z_1 z^*) = 0$ implies that $x_1 y_2 + y_1 x_2 = 0$. Consequently,

$$x_1^2 + y_1^2 = x_2^2 + \frac{y_1^2 x_2^2}{x_1^2} = \frac{x_2^2}{x_1^2}(x_1^2 + y_1^2) \quad \Rightarrow \quad x_1^2 = x_2^2 \text{ and hence } y_1^2 = y_2^2.$$

If $x_1 = -x_2$, then $y_2 = -y_1 x_2/x_1 = y_1$ and $\text{Re}(z_1 z^*) = x_1 x_2 - y_1 y_2$ has both terms negative and so *is not* > 0. However, if $x_1 = +x_2$, then $y_2 = -y_1 x_2/x_1 = -y_1$ and $\text{Re}(z_1 z^*) = x_1 x_2 - y_1 y_2$ has both terms positive and so *is necessarily* > 0. Hence, from the alternative definition of a complex conjugate, $x_2 = x_1$ and $y_2 = -y_1$ with $z^* = x_1 - iy_1$.

6 Both $z = x + iy$ and $z^* = x - iy$ have modulus $(x^2 + y^2)^{1/2}$ and so their quotient must have modulus 1. Explicitly, from (5.16),

$$\left|\frac{z}{z^*}\right|^2 = \frac{(x^2 - y^2)^2}{(x^2 + y^2)^2} + \frac{(2xy)^2}{(x^2 + y^2)^2} = \frac{x^4 - 2x^2 y^2 + y^4 + 4x^2 y^2}{(x^2 + y^2)^2} = \frac{(x^2 + y^2)^2}{(x^2 + y^2)^2} = 1.$$

7 Given

$$z = \frac{z_1}{z_2} = \frac{z_3}{|z_2|^2} \quad \Rightarrow \quad z_3 = z_1 z_2^* \quad \Rightarrow \quad z_3^* = z_1^* z_2.$$

Hence, $z^{-1} = \frac{z_2}{z_1} = \frac{z_2 z_1^*}{|z_1|^2} = \frac{z_3^*}{|z_1|^2}$.

8 Recalling that if $z = x + iy$ then $z^* = x - iy$: $e^z \times e^{z^*} = e^{[(x+iy)+(x-iy)]} = e^{2x}$, which is real; $e^z/e^{z^*} = e^{[(x+iy)-(x-iy)]} = e^{2iy}$, which has unit modulus.

9 To the nearest 0.5 mm, $r_1 = 14.5$ mm, $r_2 = 9.5$ mm and $r_1 r_2 = 28.5$ mm. If unity corresponds to u mm, then $(14.5/u) \times (9.5/u) = 28.5/u$, leading to $u = 4.83$ mm, the radius of the unit circle missing from the figure.

10 $(\cos\theta + i\sin\theta)^i = (e^{i\theta})^i = e^{-\theta}$, which is real if θ is, and is < 1 if θ is positive.

11 $\sin 4\theta$ is equal to the imaginary part of $(\cos\theta + i\sin\theta)^4$. On expansion, this will contain two terms involving odd powers of i; consequently, two imaginary terms are expected in the expansion and the same number therefore in the expression for $\sin 4\theta$.

$(\cos\theta + i\sin\theta)^4 = \cos^4\theta + 4i\cos^3\theta\sin\theta - 6\cos^2\theta\sin^2\theta - 4i\cos\theta\sin^3\theta + \sin^4\theta,$

$$\sin 4\theta = 4\cos^3\theta\sin\theta - 4\cos\theta\sin^3\theta$$
$$= 4\cos\theta\sin\theta(\cos^2\theta - \sin^2\theta)$$
$$= \cos\theta(4\sin\theta - 8\sin^3\theta).$$

A square root is needed to convert the $\cos\theta$ factor into a function of $\sin\theta$, and so $\sin 4\theta$ *cannot* be expressed as a *polynomial* in $\sin\theta$.

12 If n is odd, the expansion of $\frac{1}{2^n}\left(z - \frac{1}{z}\right)^n$, whose terms are of the form $z^{n-r} \times \left(\frac{1}{z}\right)^r = (z^{n-r}) \times (z^{-r})$, contains none that is a constant, since this would require $n - r - r = 0$, i.e. $n = 2r$ making n even.

13 If n is odd, the expansion of $\frac{1}{2^n}\left(z + \frac{1}{z}\right)^n$ has no term in z^0 (see the previous footnote). When n is even, the coefficient of z^0 is $\frac{1}{2^n}\,{}^nC_{n/2} = \frac{n!}{2^n(n/2)!\,(n/2)!}$.

15 Denote the original equation by $\sum_{r=0}^{6} a_r z^r = 0$, with the six roots $2^{1/3}$, $2^{1/3}\frac{1}{2}(-1 \pm \sqrt{3}i)$, $\pm 2i$ and 1. The sum of these is $2^{1/3} - \frac{2}{2}2^{1/3} + 2i - 2i + 1 = 1$; this is equal to $-a_5/a_6$, as it should be. The product of the roots is $2^{1/3} \cdot (2^{1/3})^2 \frac{1}{4}(1+3) \cdot 4 \cdot 1 = 8$; this is $(-1)^6 a_0/a_6$, again as it should be.

16 Using the definition of a complex power, $1^z = e^{z \operatorname{Ln} 1} = e^{z[\ln 1 + i(0 + 2\pi n)]} = e^{2\pi n i z}$. This is equal to 1 if $n = 0$, i.e. $\operatorname{Ln} z$ has its principal value. However, if $n \neq 0$, $1^z = e^{2\pi nai} e^{-2\pi nb}$, where $z = a + ib$. Thus, if $\operatorname{Ln} z$ does not have its principal value and z has a non-zero imaginary part, 1^z does not have unit modulus.

17 $20^{i/2\pi} = e^{(i/2\pi)\operatorname{Ln} 20} = e^{(i/2\pi)[\ln 20 + i(0 + 2\pi n)]} = Ae^{-n}$, where $A = e^{i \ln 20/2\pi}$. These values of z lie on a radial line that makes an angle of $\ln 20/2\pi = 27.3°$ with the x-axis. Their radial distances are $\ldots, e^{-3}, e^{-2}, e^{-1}, 1, e, e^2, e^3, \ldots$

19 $(e^{3x} \cos 4x)' = (e^{3x})' \cos 4x + e^{3x}(\cos 4x)' = 3e^{3x} \cos 4x - 4e^{3x} \sin 4x$.

20 Remembering that $\tanh z$ contains a 'hidden' $\sinh z$ and so a product of two tanh functions has an additional minus sign, compared with the product of two tan functions in a sinusoidal formula: $\tanh(u - v) = (\tanh u - \tanh v)/(1 - \tanh u \tanh v)$.

21 The equation reads $\frac{1}{2}a(e^x + e^{-x}) + \frac{1}{2}b(e^x - e^{-x}) = c$, which rearranges to (i) $(a+b)e^{2x} - 2ce^x + (a-b) = 0$ or to (ii) $(a-b)e^{-2x} - 2ce^{-x} + (a+b) = 0$.

(i) gives $e^x = \dfrac{c \pm \sqrt{c^2 - (a+b)(a-b)}}{a+b}$, (ii) gives $e^{-x} = \dfrac{c \pm \sqrt{c^2 - (a-b)(a+b)}}{a-b}$.

The *same* solution for x corresponds to *opposite* choices of the sign of the square root. So, for example,

$$e^x e^{-x} = \frac{c + \sqrt{c^2 - (a+b)(a-b)}}{a+b} \times \frac{c - \sqrt{c^2 - (a+b)(a-b)}}{a-b}$$

$$= \frac{c^2 - (c^2 - a^2 + b^2)}{a^2 - b^2} = \frac{a^2 - b^2}{a^2 - b^2} = 1.$$

Similarly, $e^x e^{-x} = 1$ if the choice of signs is reversed.

22 This is almost an 'inverse' rationalisation and brings the square root into the denominator.

$$\ln(x - \sqrt{x^2 - 1}) = \ln\left[\frac{(x - \sqrt{x^2 - 1})(x + \sqrt{x^2 - 1})}{(x + \sqrt{x^2 - 1})}\right]$$

$$= \ln\left[\frac{x^2 - (x^2 - 1)}{x + \sqrt{x^2 - 1}}\right]$$

$$= \ln\left(\frac{1}{x + \sqrt{x^2 - 1}}\right) = -\ln(x + \sqrt{x^2 - 1}).$$

23 $\sinh 2x = 2 \sinh x \cosh x \Rightarrow 2 \cosh 2x = 2 \sinh^2 x + 2 \cosh^2 x$. Hence $\cosh 2x = \cosh^2 x + \sinh^2 x$. Now, using basic definitions instead, and starting from the RHS:

$$\tfrac{1}{4}(e^x + e^{-x})^2 + \tfrac{1}{4}(e^x - e^{-x})^2 = \tfrac{1}{4}(e^{2x} + 2 + e^{-2x} + e^{2x} - 2 + e^{-2x}) = \tfrac{1}{2}(e^{2x} + e^{-2x}),$$

which is the definition of $\cosh 2x$.

Appendix F

6. Series and limits

1 $u_n = x^n/n!$ with $u_{n+1} = [x/(n+1)]u_n$; $u_n = (-1)^n x^n/n!$ with $u_{n+1} = -[x/(n+1)]u_n$.

3 The given series has first term 4 and last term $4 + (N-1)$ and so $2695 = (N/2)(4 + 4 + N - 1)$, or $N^2 + 7N - 5390 = 0$, from which $N = 70$.

5 If a fraction r of its kinetic energy is retained at each bounce, the successive distances travelled after the first bounce are $2hr$, $2hr^2$, $2hr^3$, ... The total distance is therefore $h + 2hr/(1-r)$. Thus $\alpha = 1 + 2r/(1-r)$ $\Rightarrow$ $r = (\alpha - 1)/(\alpha + 1)$.

6 Since the sum is zero, $a + rd/(1-r) = 0$ $\Rightarrow$ $r = a/(a-d)$. Also, since the series converges, $|r| < 1$ and so $|a| < |a-d|$. Thus, either $d < 0$ or $d > 2a$.

7 The four terms $f(N)$, $f(N-1)$, $-f(0)$ and $-f(-1)$ come from u_N, u_{N-1}, u_2 and u_1, respectively.

8 Consider the total coefficient T_p of $f(p)$ in the sum. When u_n is added to the sum, it contributes a_{n-p} to T_p, so long as $0 \leq n - p \leq m$. Values of n greater than $p + m$ contribute nothing to T_p, which therefore is equal to $a_0 + a_1 + a_2 + \cdots + a_m$. For all terms [except for the finite number involving $f(-m+1)$, $f(-m+2)$, ..., $f(-1)$ and $f(N-m+1)$, $f(N-m+2)$, ..., $f(N)$] to vanish requires that $T_p = \sum_{n=p}^{p+m} a_{n-p} = 0$, i.e. $a_0 + a_1 + \cdots + a_m = 0$.

10 $f(N) = N(2N^2 + 15N + 31) = 2N(N+1)(N+2) + aN(N+1) + bN$ implies that $15 = 6 + a$ and that $31 = 4 + a + b$. Thus $a = 9$ and $b = 18$, with $f(N) = 2N(N+1)(N+2) + 9N(N+1) + 18N$. The first term on the RHS contains three consecutive integers and must therefore divide by both 2 and 3; the second term clearly divides by 3 and, as N and $N+1$ are consecutive integers, one of them must divide by 2; in the final term 18 divides by 6. Thus all three terms divide by 6 and so therefore does $f(N)$.

11 Set $x = e^{-y}$ with $y > 0$, i.e. $x < 1$. Then $S = \sum_{s=1}^{\infty} sx^s = x/(1-x)^2$, using the result from the worked example that $\sum_{n=0}^{\infty}(n+1)x^n = (1-x)^{-2}$. Thus $S = e^{-y}/(1-e^{-y})^2 = e^{-y}e^y/(e^{y/2} - e^{-y/2})^2 = 1/[4\sinh^2(y/2)]$.

12 $S(0) = [\exp(\cos 0)][\cos(\sin 0)] = e^1 \cos 0 = e \cdot 1 = e$. As a series

$$S(0) = 1 + 1 + \frac{1}{2!} + \frac{1}{3!} + \cdots = \sum_{n=0}^{\infty} \frac{1}{n!} = e.$$

13 The sum of the moduli is $\sum_{n=0}^{\infty} |x|^n/n!$, which is equal to $e^{|x|}$ by definition. $|x|$ is positive, and so the sum of the original series is equal to this if $-x$ is positive, i.e. $x \leq 0$.

17 (a) $u_{n+1}/u_n = x^2/[(2n+3)(2n+1)] \to 0$ as $n \to \infty$ for any x. So, convergent for all x.

(b) $u_{n+1}/u_n = xn/(n+1) \to x$ as $n \to \infty$. So, convergent if $0 < x < 1$; divergent if $x > 1$; no conclusion can be drawn from the ratio test for $x = 1$, though $\sum 1/n$ is actually divergent (see the main text).

(c)
$$\frac{u_{n+1}}{u_n} = \frac{n+1}{n} x^{[(n+1)^{-1} - n^{-1}]} = \left(1 + \frac{1}{n}\right) x^{-1/n(n+1)}.$$

This $\to 1$ as $n \to \infty$ for all x and so no conclusion is possible from the ratio test. However, $nx^{1/n}$ does not $\to 0$ for any x, and so the series must diverge.

Appendix F

18 We can always chose r_1 such that $0 < r < r_1 < 1$. Take $v_n = r_1^n$, then $\sum v_n$ is convergent [to $1/(1-r_1)$] and $v_{n+1}/v_n = r_1$. Now,

$$\frac{u_{n+1}}{u_n} = \frac{(n+1)^k r^{n+1}}{n^k r^n} = \left(1 + \frac{1}{n}\right)^k r.$$

This is $< r_1$ for all $n > [(r_1/r)^{1/k} - 1]^{-1}$, and so, by the ratio comparison test, $\sum u_n$ is convergent.

19 Using the Maclaurin series for $\ln(1+x)$:

$$u_n = n \ln\left(\frac{n+1}{n-1}\right) - 2 = n \ln\left(\frac{1+n^{-1}}{1-n^{-1}}\right) - 2$$

$$= n\left(\frac{1}{n} - \frac{1}{2n^2} + \frac{1}{3n^3} - \cdots\right) - n\left(-\frac{1}{n} - \frac{1}{2n^2} - \frac{1}{3n^3} - \cdots\right) - 2$$

$$= \frac{2}{3n^2} + O\left(\frac{1}{n^4}\right).$$

Now, taking $v_n = n^{-2}$, the quotient $u_n/v_n = 2/3 + O(n^{-2})$. This tends to the finite but non-zero limit of $\frac{2}{3}$ as $n \to \infty$ and so, since $\sum v_n$ converges, so does $\sum u_n$. For the definition of $O(x)$, see Section 6.5.

21 With $u_n = (n+a)a^n e^{-n}$, the Cauchy root test requires $\rho = (n+a)^{1/n} ae^{-1}$ to be < 1 for convergence. Now, $(n+a)^{1/n} \to 1$ as $n \to \infty$, and so the requirement is that $ae^{-1} < 1$, i.e. $a < e$.

24 Evaluating each series at $x = \pi/2$:
(a) $S(x) = \sum_{n=1}^{\infty} n^{-1} \sin nx = \sin x + \frac{1}{2} \sin 2x + \frac{1}{3} \sin 3x + \cdots = 1 - \frac{1}{3} + \frac{1}{5} - \cdots$
Convergent (alternating signs).
(b) $dS/dx = \sum_{n=1}^{\infty} \cos nx = \cos x + \cos 2x + \cos 3x + \cdots = 0 - 1 + 0 + 1 + 0 - \cdots$ Oscillates.
(c) $\int S(x)\,dx = -\sum_{n=1}^{\infty} n^{-2} \cos nx = \frac{1}{2^2} - \frac{1}{4^2} + \frac{1}{6^2} - \cdots$ Absolutely convergent.

25 Expanding each function using its Maclaurin series:

$$f(x) = \sinh x - \sin x - \frac{x^3}{3}$$

$$= \left(x + \frac{x^3}{3!} + \frac{x^5}{5!} + \frac{x^7}{7!} + \cdots\right) - \left(x - \frac{x^3}{3!} + \frac{x^5}{5!} - \frac{x^7}{7!} + \cdots\right) - \frac{x^3}{3}$$

$$= \frac{2x^3}{3!} + \frac{2x^7}{7!} + \cdots - \frac{x^3}{3} = \frac{2x^7}{7!} + \cdots = O(x^7) \text{ as } x \to 0.$$

26 If $g(x) = 2x^7$, then $f \approx lg$ with $l = 1/7!$.

29 For $f(z) = 1 - z^2/2 + z^4/4 - z^6/8 + \cdots$, the limit, as $n \to \infty$, of $|a_{n+2}/a_n|$ is $1/2$. Thus, the radius of convergence is given by $z^2 = (1/2)^{-1}$, i.e. $z = \sqrt{2}$. The series is a geometric one with common ratio $-z^2/2$ and so $f(z) = 1/[1-(-z^2/2)] = 1/(1 + \frac{1}{2}z^2)$. This function has singularities at $z = \pm i\sqrt{2}$; these are (both) a distance $\sqrt{2}$ from the origin and, as expected, lie on the circle of convergence.

Appendix F

30 The calculation is straightforward, though a bit lengthy. The reader should find that the successive coefficients, up to that of x^5, are calculated as

$$1, \; 1, \; \frac{1}{2!}, \; \left(-\frac{1}{3!} + \frac{1}{3!}\right), \; \left(-\frac{2}{2!\,3!} + \frac{1}{4!}\right), \; \left(\frac{1}{5!} - \frac{3}{3!\,3!} + \frac{1}{5!}\right).$$

31 A constant of integration must be added; it has value -1 if $\cos x$ is defined as having $\cos 0 = 1$.

33 For the function

$$f^{(n)}(x) = \sin\left(x + \frac{n\pi}{2}\right) = \sin x \cos \frac{n\pi}{2} + \cos x \sin \frac{n\pi}{2},$$

with n odd ($n = 2m + 1$), $\cos n\pi/2 = 0$, $\sin n\pi/2 = (-1)^m$ and $f^{(n)}(x) = (-1)^m \cos x$. For n even ($n = 2m$), $\sin n\pi/2 = 0$, $\cos n\pi/2 = (-1)^m$ and $f^{(n)}(x) = (-1)^m \sin x$. These agree with the usual (separate) results in all cases.

34 (a) $\cos^2 x - \sin^2 x$ about $x = 0$: an even function of x, hence only even powers of x present.
(b) $\cos x$ about $x = \pi/2$: $\cos(\pi/2 + y) = \cos(\pi/2)\cos y - \sin(\pi/2)\sin y = -\sin y$. Only odd powers of $(x - \pi/2)$ present.
(c) $\sin 2x$ about $\pi/4$: $\sin(2 \cdot \pi/4 + 2y) = \sin(\pi/2 + 2y) = \cos 2y$. Only even powers of $(2x - \pi/2)$ present. This could be expressed as even powers of $(x - \pi/4)$.

35 Including the quartic term, $+x^4/4!$, adds 0.002 604, making the partial sum 0.877 604, which is now 0.000 02(16) larger than the accurate value; this is an error of about 0.0025%. The omitted sixth-power term has a value of $-0.000\,021\,7$.

36 The infinity symbol ∞ must not be treated as an object that has a particular value, but as a description of the outcome of a particular limiting operation. The given example should be treated as

$$\lim_{x \to 0}\left(\frac{1}{x} - \frac{1}{x^2}\right) = \lim_{x \to 0}\frac{x-1}{x^2} = \frac{\lim_{x \to 0}(x-1)}{\lim_{x \to 0} x^2} = \frac{-1}{0} = -\infty.$$

38

$$\lim_{x \to \pi/2}\frac{\tan^n x}{\sec^m x} = \lim_{x \to \pi/2}\frac{\sin^n x}{\cos^n x}\cos^m x = \lim_{x \to \pi/2}\sin^n x \cos^{m-n} x$$

$$= \begin{cases} 0 & \text{for } m > n, \\ 1 & \text{for } m = n, \\ \infty & \text{for } m < n. \end{cases}$$

7. Partial differentiation

3 There are $n(n-1)$ ways of choosing two *different* variables, but, since $\partial^2 f/\partial x_i \partial x_j = \partial^2 f/\partial x_j \partial x_i$, only $\frac{1}{2}n(n-1)$ are independent. There are also n ways of choosing a repeated variable. Thus there is a total of $\frac{1}{2}n(n-1) + n = \frac{1}{2}n(n+1)$ *independent* second partial derivatives.

5 Multiplying $x\,dy + 3y\,dx$ by x^2 gives $x^3\,dy + 3x^2 y\,dx$. By inspection $x^3 = \partial(x^3 y)/\partial y$ and $3x^2 y = \partial(x^3 y)/\partial x$. Thus, x^2 is a suitable integrating factor and the corresponding $f(x, y) = x^3 y$.

Appendix F

6 With $f(x, y) = \int^x A(u, y)\, du$, we have

$$df = \left(\frac{\partial f}{\partial x}\right) dx + \left(\frac{\partial f}{\partial y}\right) dy = A(x, y)\, dx + \left(\int^x \frac{\partial A(u, y)}{\partial y}\, du\right) dy$$

$$= A(x, y)\, dx + \left(\int^x \frac{\partial B(u, y)}{\partial u}\, du\right) dy$$

$$= A(x, y)\, dx + B(x, y)\, dy.$$

Thus (7.9), which was used to replace $\partial A(u, y)/\partial y$ by $\partial B(u, y)/\partial u$, is a sufficient condition to make df an exact differential.

7 Starting from $\tan z = y/x$:

(a) Keeping x fixed; $x \sec^2 z = \left(\dfrac{\partial y}{\partial z}\right)_x \Rightarrow \left(\dfrac{\partial y}{\partial z}\right)_x = x(1 + \tan^2 z) = \dfrac{x^2 + y^2}{x}$.

(b) Keeping y fixed; $\sec^2 z \left(\dfrac{\partial z}{\partial x}\right)_y = -\dfrac{y}{x^2} \Rightarrow \left(\dfrac{\partial z}{\partial x}\right)_y = -\dfrac{y}{x^2} \dfrac{x^2}{x^2 + y^2} = \dfrac{-y}{x^2 + y^2}$.

(c) Keeping z fixed; $0 = -\dfrac{y}{x^2} \left(\dfrac{\partial x}{\partial y}\right)_z + \dfrac{1}{x} \Rightarrow \left(\dfrac{\partial x}{\partial y}\right)_z = \dfrac{x}{y}$.

$$\left(\frac{\partial x}{\partial y}\right)_z \left(\frac{\partial y}{\partial z}\right)_x \left(\frac{\partial z}{\partial x}\right)_y = \frac{x}{y} \cdot \frac{x^2 + y^2}{x} \cdot \frac{-y}{x^2 + y^2} = -1.$$

9 To maximise $T(x, y) = 1 + xy$ with $x^2 + y^2 = 1$, set $S(x) = 1 \pm x\sqrt{1 - x^2}$ and maximise with respect to x:

$$\frac{dS}{dx} = \pm\sqrt{1 - x^2} \pm \frac{\frac{1}{2}x(-2x)}{\sqrt{1 - x^2}} = \pm\frac{(1 - x^2) - x^2}{\sqrt{1 - x^2}}.$$

For $dS/dx = 0$ we must have $1 - 2x^2 = 0$, i.e. $x = \pm 1/\sqrt{2}$, which also implies that $y = \pm 1/\sqrt{2}$ (since $x^2 + y^2 = 1$). The rest of the example is as in the main text.

11 With $E_k = (k - 1)E_0$, the two equations give

$$N = \sum_{k=1}^{\infty} C e^{\mu(k-1)E_0} = \sum_{n=0}^{\infty} C e^{\mu n E_0} = \frac{C}{1 - e^{\mu E_0}},$$

and, substituting for C as needed,

$$E = \sum_{k=1}^{\infty} C(k - 1)E_0 e^{\mu(k-1)E_0} = C E_0 \sum_{n=0}^{\infty} n e^{\mu n E_0} = N(1 - e^{\mu E_0})E_0 \sum_{n=0}^{\infty} n e^{\mu n E_0}.$$

By evaluating $\sum_{n=0}^{\infty} n e^{\mu n E_0}$, the enthusiastic reader may show that $\mu = \dfrac{1}{E_0} \ln\left(\dfrac{E}{E + NE_0}\right)$.

12 A typical circle is $f(x, y, a) = (x - 2a)^2 + y^2 - a^2 = 0$. Thus the envelope is given by

$$0 = \frac{\partial f}{\partial a} = -4(x - 2a) - 2a \quad \Rightarrow \quad a = \frac{2x}{3}.$$

Appendix F

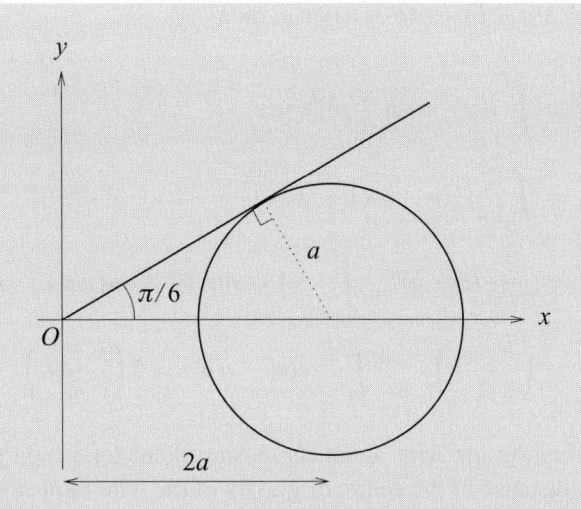

Figure F.4 A typical circle and tangent. Note that $\sin^{-1}(a/2a) = \tan^{-1}(1/\sqrt{3}) = \pi/6$.

Substituting in the equation for the corresponding circle:

$$\left(x - \frac{4x}{3}\right)^2 + y^2 - \frac{4x^2}{9} = 0 \;\Rightarrow\; y^2 = \frac{x^2}{3} \;\Rightarrow\; y = \pm\frac{x}{\sqrt{3}}.$$

These are lines through the origin making angles of $\pm\pi/6$ with the x-axis. For a typical circle to which these lines are tangents, see Figure F.4.

14 $\displaystyle\int_0^T e^{-\alpha t}\,dt = \left[\frac{e^{-\alpha t}}{-\alpha}\right]_0^T = \frac{1}{\alpha}(1 - e^{-\alpha T}).$

Now differentiate both sides of this result with respect to α:

$$-\int_0^T t e^{-\alpha t}\,dt = -\frac{1}{\alpha^2}(1 - e^{-\alpha T}) + \frac{T e^{-\alpha T}}{\alpha} = -\frac{1}{\alpha^2}\left[1 - (1 + \alpha T)e^{-\alpha T}\right].$$

Hence the stated result.

8. Multiple integrals

1 $\displaystyle\int \frac{x^2(1-x)^2}{2}\,dx = \frac{1}{2}\int_0^1 (x^2 - 2x^3 + x^4)\,dx = \frac{1}{2}\left(\frac{1}{3} - \frac{2}{4} + \frac{1}{5}\right) = \frac{1}{60}.$

$\displaystyle\int \frac{(1-y)^3 y}{3}\,dy = \frac{1}{3}\int_0^1 (y - 3y^2 + 3y^3 - y^4)\,dy = \frac{1}{3}\left(\frac{1}{2} - \frac{3}{3} + \frac{3}{4} - \frac{1}{5}\right) = \frac{1}{60}.$

2 A regular octahedron consists of eight tetrahedra of the type considered, but with $a = b = c$. Their total volume is therefore $8 \times a^3/6 = 4a^3/3$. The enclosing sphere has radius a and a volume of $4\pi a^3/3$. Thus, the octahedron volume : sphere volume $= 1 : \pi$, a factor of more than three.

3 Set $y = \frac{1}{2}(a+b) + \frac{1}{2}(b-a)\sin u$, then

$$I = \int_a^b [(y-a)(b-y)]^{n/2}\,dy$$

$$= \int_{-\pi/2}^{\pi/2} [\tfrac{1}{2}(b-a)(1+\sin u)\tfrac{1}{2}(b-a)(1-\sin u)]^{n/2} \tfrac{1}{2}(b-a)\cos u\,du$$

$$= \frac{1}{2^{n+1}}(b-a)^{n+1} \int_{-\pi/2}^{\pi/2} (1-\sin^2 u)^{n/2} \cos u\,du$$

$$= \left(\frac{b-a}{2}\right)^{n+1} 2 \int_0^{\pi/2} \cos^{n+1} u\,du = 2\left(\frac{b-a}{2}\right)^{n+1} J(n+1,0).$$

7 Since rotating the wire about its closing diameter would generate a sphere of area $4\pi a^2$, the distance of the centre of gravity of the wire (which is not actually on the wire) from that diameter must be $(4\pi a^2 \div \pi a) \div 2\pi$. Therefore, on suspension, the closing diameter of the wire would make an angle with the vertical of $\tan^{-1} 2a/\pi \div a = 0.5669$ rad $= 32.5°$.

8 In this case, the integral's limits are $\pm\frac{1}{2}a$ and its value is $[\sigma bx^3/3]_{-a/2}^{a/2} = \sigma ba^3/12$.

9. Vector algebra

1 If the body is in equilibrium, the resultant force acting upon it must be zero, represented in the vector diagram by the zero vector $\mathbf{0}$. But the resultant of all those forces that do act is represented by their vector sum. If this is $\mathbf{0}$ then the n-sided polygon (different-shaped n-sided polygons are possible, depending on the order in which the n forces are added) must be a closed polygon.

2 For $\mu > \lambda$, by writing $\mathbf{p}$ as $(\mu - \lambda)^{-1}(\mu \mathbf{a} - \lambda \mathbf{a} + \lambda \mathbf{a} - \lambda \mathbf{b}) = \mathbf{a} + [\lambda/(\mu - \lambda)](\mathbf{a} - \mathbf{b})$, P is seen to be a point on the extension (beyond A) of the line BA. The extension is of length λ if the length of BA is $\lambda - \mu$, i.e. the length BP is μ on the same scale. If $\lambda > \mu$, the extension is beyond B and, again, $AP : BP = \lambda : \mu$.

3 F is the mid-point of BC and so $\mathbf{f} = \frac{1}{2}(\mathbf{b} + \mathbf{c})$. If G is to lie on AF it must be possible to write $\mathbf{g} = \frac{1}{3}(\mathbf{a} + \mathbf{b} + \mathbf{c})$ as $v\mathbf{a} + (1-v)\frac{1}{2}(\mathbf{b}+\mathbf{c})$ for some v. Clearly, this occurs if $v = \frac{1}{3}$ and $\mathbf{g}$ takes the form $\mathbf{g} = \frac{1}{3}\mathbf{a} + \frac{2}{3}\mathbf{f}$, showing that $AG : FG = \frac{2}{3} : \frac{1}{3}$.

6 With $\mathbf{a} = (1+i, 2i, 3)$ and $\mathbf{b} = (2-i, 3+i, 1-i)$,

$$\mathbf{a} \cdot \mathbf{b} = (1-i)(2-i) + (-2i)(3+i) + (3)(1-i)$$
$$= (1-3i) + (2-6i) + (3-3i) = 6 - 12i,$$
$$\mathbf{b} \cdot \mathbf{a} = (2+i)(1+i) + (3-i)(2i) + (1+i)(3)$$
$$= (1+3i) + (2+6i) + (3+3i) = 6 + 12i.$$

Confirmation: $\mathbf{a} \cdot \mathbf{b} = (6 - 12i) = (6 + 12i)^* = (\mathbf{b} \cdot \mathbf{a})^*$.
$|\mathbf{a}|^2 = (1-i)(1+i) + (-2i)(2i) + (3)(3) = 15 \Rightarrow |\mathbf{a}| = \sqrt{15}$,
$|\mathbf{b}|^2 = (2+i)(2-i) + (3-i)(3+i) + (1+i)(1-i) = 17 \Rightarrow |\mathbf{b}| = \sqrt{17}$.

Appendix F

7 For the special case

$$(\mathbf{a} \times \mathbf{b}) \times \mathbf{c} = \mathbf{b}(\mathbf{a} \cdot \mathbf{c}) - \mathbf{a}(\mathbf{b} \cdot \mathbf{c}) = \mathbf{b}(\mathbf{a} \cdot \mathbf{c}),$$
$$\mathbf{a} \times (\mathbf{b} \times \mathbf{c}) = \mathbf{b}(\mathbf{a} \cdot \mathbf{c}) - \mathbf{c}(\mathbf{a} \cdot \mathbf{b}) = \mathbf{b}(\mathbf{a} \cdot \mathbf{c}),$$

since $\mathbf{b} \cdot \mathbf{c}$ and $\mathbf{a} \cdot \mathbf{b}$ are both zero.

8 $\mathbf{a} \times \mathbf{c} = \mathbf{b} \times \mathbf{c}$ $\Rightarrow$ $(\mathbf{a} - \mathbf{b}) \times \mathbf{c} = 0$, and hence the truth of at least one of the following: (i) $\mathbf{c} = 0$, (ii) $\mathbf{a} = \mathbf{b}$ and (iii) $\mathbf{a} - \mathbf{b}$ and $\mathbf{c}$ are parallel vectors.

9 As the Arctic Circle is at 66.5° N, the Earth's axis is tilted at $\theta = 23.5°$ to the plane of the ecliptic. The angular speed of the Earth's rotation is $2\pi/(24 \times 3600) = 7.27 \times 10^{-5}$ rad s^{-1}. With the x-axis in the direction of the Earth at the winter solstice, at which time the North Pole is leaning directly away from the Sun, the components of the angular velocity vector are $(\omega \sin\theta, 0, \omega \cos\theta)$; they remain the same throughout the year. So, for (a), (b) and (c), the vector has components $(2.90, 0, 6.67)$ measured in 10^{-5} rad s^{-1}.

11 $A = ab \sin\theta_{ab} = ab\sqrt{1 - \cos^2\theta_{ab}} = \sqrt{a^2b^2 - a^2b^2 \cos^2\theta_{ab}} = \sqrt{a^2b^2 - (\mathbf{a} \cdot \mathbf{b})^2}$. For the current case $a = |\mathbf{a}| = \sqrt{14}$, $b = |\mathbf{b}| = \sqrt{77}$ and $\mathbf{a} \cdot \mathbf{b} = 4 + 10 + 18 = 32$. Hence $A = \sqrt{14 \times 77 - (32)^2} = \sqrt{54}$, in agreement with the calculation in the main text.

12 With coplanar vectors, we can write

$$\mathbf{a} \cdot (\mathbf{b} \times \mathbf{c}) = (\lambda\mathbf{b} + \mu\mathbf{c}) \cdot (\mathbf{b} \times \mathbf{c}) = \lambda\mathbf{b} \cdot (\mathbf{b} \times \mathbf{c}) + \mu\mathbf{c} \cdot (\mathbf{b} \times \mathbf{c}) = 0\lambda + 0\mu = 0.$$

13
$$\mathbf{s} = \mathbf{a} \times (\mathbf{b} \times \mathbf{c}) + \mathbf{b} \times (\mathbf{c} \times \mathbf{a}) + \mathbf{c} \times (\mathbf{a} \times \mathbf{b})$$
$$= \mathbf{b}(\mathbf{a} \cdot \mathbf{c}) - \mathbf{c}(\mathbf{a} \cdot \mathbf{b}) + \mathbf{c}(\mathbf{b} \cdot \mathbf{a}) - \mathbf{a}(\mathbf{b} \cdot \mathbf{c}) + \mathbf{a}(\mathbf{c} \cdot \mathbf{b}) - \mathbf{b}(\mathbf{c} \cdot \mathbf{a})$$
$$= \mathbf{a}[(\mathbf{c} \cdot \mathbf{b} - \mathbf{b} \cdot \mathbf{c})] + \mathbf{b}[(\mathbf{a} \cdot \mathbf{c} - \mathbf{c} \cdot \mathbf{a})] + \mathbf{c}[(\mathbf{b} \cdot \mathbf{a} - \mathbf{a} \cdot \mathbf{b})]$$
$$= 0, \text{ since } \mathbf{c} \cdot \mathbf{b} = \mathbf{b} \cdot \mathbf{c}, \text{ etc. for real vectors.}$$

16 Any point on the line will suffice, so let $z = 0$. Then $x + 3y = 5$ and $2x - 2y = 3$ give $x = 19/8$, $y = 7/8$. Thus a general point on the line is $\left(\frac{19}{8} + 10\mu, \frac{7}{8} - 6\mu, -8\mu\right)$. To get the stated form, we set the x-coordinates equal, $\frac{19}{8} + 10\mu = 3 + 5\lambda$, i.e. $\mu = \frac{1}{10}\left(\frac{5}{8} + 5\lambda\right)$. In terms of the new parameter λ, the coordinates become

$$\left(3 + 5\lambda, \frac{7}{8} - \frac{6}{10}\left[\frac{5}{8} + 5\lambda\right], -\frac{8}{10}\left[\frac{5}{8} + 5\lambda\right]\right) = \left(3 + 5\lambda, \frac{1}{2} - 3\lambda, -\frac{1}{2} - 4\lambda\right).$$

17 Taking $\mathbf{c}$ as point P, and $\mathbf{r} = \mathbf{a} + \lambda(\mathbf{b} - \mathbf{a})$ as the line, the given expression yields

$$d_c = \left|(\mathbf{c} - \mathbf{a}) \times \frac{(\mathbf{b} - \mathbf{a})}{|\mathbf{b} - \mathbf{a}|}\right|.$$

From this

$$d_c|\mathbf{b} - \mathbf{a}| = |\mathbf{c} \times \mathbf{b} - \mathbf{a} \times \mathbf{b} - \mathbf{c} \times \mathbf{a} + 0| = |\mathbf{c} \times \mathbf{b} + \mathbf{a} \times \mathbf{c} + \mathbf{b} \times \mathbf{a}| = s, \text{ say}.$$

From the cyclic symmetry of the expression for s, it is clear that $d_a|\mathbf{c} - \mathbf{b}|$ and $d_b|\mathbf{a} - \mathbf{c}|$ also have the value s. The three points $\mathbf{a}$, $\mathbf{b}$ and $\mathbf{c}$ define a triangle lying in a plane and each of the three equal expressions can be interpreted as the product of the length of one side of the triangle and the perpendicular distance to the opposite vertex. Thus s is equal to twice the area of the triangle.

Appendix F

18 For *general* points $(\lambda, \lambda, \lambda)$ on the first line and $(\mu, 2\mu, 1+3\mu)$ on the second, the minimum distance between the lines is given by

$$d = \frac{1}{\sqrt{6}}[(\mu - \lambda)\mathbf{i} + (2\mu - \lambda)\mathbf{j} + (1 + 3\mu - \lambda)\mathbf{k}] \cdot (\mathbf{i} - 2\mathbf{j} + \mathbf{k})]$$

$$= \frac{1}{\sqrt{6}}(\mu - \lambda - 4\mu + 2\lambda + 1 + 3\mu - \lambda) = \frac{1}{\sqrt{6}} \quad \text{(for any } \lambda \text{ and } \mu\text{)}.$$

10. Matrices and vector spaces

3 Denoting Equation (10.12) by (i), Equation (10.13) by (ii), and the general properties of complex conjugation by (iii):

$$\langle \lambda \mathbf{a} + \mu \mathbf{b} | \mathbf{c} \rangle \stackrel{(i)}{=} \langle \mathbf{c} | \lambda \mathbf{a} + \mu \mathbf{b} \rangle^* \stackrel{(ii)}{=} (\langle \mathbf{c} | \lambda \mathbf{a} \rangle + \langle \mathbf{c} | \mu \mathbf{b} \rangle)^*$$

$$\stackrel{(iii)}{=} \langle \mathbf{c} | \lambda \mathbf{a} \rangle^* + \langle \mathbf{c} | \mu \mathbf{b} \rangle^* \stackrel{(ii)}{=} (\lambda \langle \mathbf{c} | \mathbf{a} \rangle)^* + (\mu \langle \mathbf{c} | \mathbf{b} \rangle)^*$$

$$\stackrel{(iii)}{=} \lambda^* \langle \mathbf{c} | \mathbf{a} \rangle^* + \mu^* \langle \mathbf{c} | \mathbf{b} \rangle^* \stackrel{(i)}{=} \lambda^* \langle \mathbf{a} | \mathbf{c} \rangle + \mu^* \langle \mathbf{b} | \mathbf{c} \rangle. \quad \text{(iv)}$$

$$\langle \lambda \mathbf{a} | \mu \mathbf{b} \rangle \stackrel{(iv)}{=} \lambda^* \langle \mathbf{a} | \mu \mathbf{b} \rangle \stackrel{(ii)}{=} \lambda^* \mu \langle \mathbf{a} | \mathbf{b} \rangle.$$

4 With $\mathcal{A}\mathbf{x} = (2x_1 + x_2, x_2)$, $\mathcal{B}\mathbf{x} = (x_1, x_1 + 2x_2)$ and $\mathcal{C}\mathbf{x} = (x_1 - x_2, 2x_2)$, we have, taking a general vector as $\mathbf{x} = (x, y)$,

$$\mathcal{AC}\mathbf{x} = \mathcal{A}(x - y, 2y) = [2(x - y) + 2y, 2y] = (2x, 2y),$$
$$\mathcal{CA}\mathbf{x} = \mathcal{C}(2x + y, y) = [(2x + y) - y, 2y] = (2x, 2y),$$

i.e. $\mathcal{A}$ and $\mathcal{C}$ commute. However,

$$\mathcal{AB}\mathbf{x} = \mathcal{A}(x, x + 2y) = [2x + (x + 2y), x + 2y] = (3x + 2y, x + 2y),$$
$$\mathcal{BA}\mathbf{x} = \mathcal{B}(2x + y, y) = [2x + y, (2x + y) + 2y] = (2x + y, 2x + 3y),$$

showing that $\mathcal{A}$ and $\mathcal{B}$ do not commute.

6 The relevant matrices are

$$\mathsf{A} = \begin{pmatrix} 2 & 1 \\ 0 & 1 \end{pmatrix}, \quad \mathsf{B} = \begin{pmatrix} 1 & 0 \\ 1 & 2 \end{pmatrix}, \quad \mathsf{C} = \begin{pmatrix} 1 & -1 \\ 0 & 2 \end{pmatrix}.$$

$$\mathsf{AC} = \begin{pmatrix} 2 & 1 \\ 0 & 1 \end{pmatrix}\begin{pmatrix} 1 & -1 \\ 0 & 2 \end{pmatrix} = \begin{pmatrix} 2 & 0 \\ 0 & 2 \end{pmatrix} = \begin{pmatrix} 1 & -1 \\ 0 & 2 \end{pmatrix}\begin{pmatrix} 2 & 1 \\ 0 & 1 \end{pmatrix} = \mathsf{CA}.$$

$$\mathsf{AB} = \begin{pmatrix} 2 & 1 \\ 0 & 1 \end{pmatrix}\begin{pmatrix} 1 & 0 \\ 1 & 2 \end{pmatrix} = \begin{pmatrix} 3 & 2 \\ 1 & 2 \end{pmatrix}, \quad \text{whilst}$$

$$\mathsf{BA} = \begin{pmatrix} 1 & 0 \\ 1 & 2 \end{pmatrix}\begin{pmatrix} 2 & 1 \\ 0 & 1 \end{pmatrix} = \begin{pmatrix} 2 & 1 \\ 2 & 3 \end{pmatrix}, \quad \text{which is not the same.}$$

$$\mathsf{BC} = \begin{pmatrix} 1 & 0 \\ 1 & 2 \end{pmatrix}\begin{pmatrix} 1 & -1 \\ 0 & 2 \end{pmatrix} = \begin{pmatrix} 1 & -1 \\ 1 & 3 \end{pmatrix}, \quad \text{but}$$

$$\mathsf{CB} = \begin{pmatrix} 1 & -1 \\ 0 & 2 \end{pmatrix}\begin{pmatrix} 1 & 0 \\ 1 & 2 \end{pmatrix} = \begin{pmatrix} 0 & -2 \\ 2 & 4 \end{pmatrix}, \quad \text{which is not the same.}$$

So, $\mathcal{BC} \neq \mathcal{CB}$.

Appendix F

7 (i) If A and B commute, then

$$(A + B)(A - B) = A^2 + BA - AB - B^2 = A^2 + BA - BA - B^2 = A^2 - B^2.$$

(ii) If $(A + B)(A - B) = A^2 - B^2$, then

$$A^2 - B^2 = (A + B)(A - B) = A^2 + BA - AB - B^2 \quad \Rightarrow \quad BA - AB = 0,$$

i.e. A and B commute.

8 If $A = \begin{pmatrix} 1 & 0 & 0 \\ 0 & -1 & 0 \\ 0 & 0 & 1 \end{pmatrix}$, then $A^{2m} = \begin{pmatrix} 1 & 0 & 0 \\ 0 & 1 & 0 \\ 0 & 0 & 1 \end{pmatrix}$ and $A^{2m+1} = \begin{pmatrix} 1 & 0 & 0 \\ 0 & -1 & 0 \\ 0 & 0 & 1 \end{pmatrix}$.

$$\exp(iA) = \sum_{n=0}^{\infty} \frac{1}{n!}(i^n A^n) = \sum_{m=0}^{\infty} \frac{(-1)^m}{(2m)!} A^{2m} + i \sum_{m=0}^{\infty} \frac{(-1)^m}{(2m+1)!} A^{2m+1}$$

$$= \cos 1 \begin{pmatrix} 1 & 0 & 0 \\ 0 & 1 & 0 \\ 0 & 0 & 1 \end{pmatrix} + i \sin 1 \begin{pmatrix} 1 & 0 & 0 \\ 0 & -1 & 0 \\ 0 & 0 & 1 \end{pmatrix},$$

$\text{Tr } \exp(iA) = \cos 1 \times 3 + i \sin 1 \times 1 = 3 \cos 1 + i \sin 1$.

13 A^{-1} is defined by $A^{-1}A = I$. Define A_R^{-1} by $AA_R^{-1} = I$ and consider

$$A^{-1} = A^{-1}I = A^{-1}(AA_R^{-1}) = (A^{-1}A)A_R^{-1} = IA_R^{-1} = A_R^{-1},$$

thus establishing that the left inverse is also a right inverse.

14 For example, the cofactor of the '3' in the a_{23} position is given by $(-1)^{3+2}[(2)(-2) - (1)(3)] = +7$.

16 A matrix all of whose entries are equal has rank 1 (whatever the value of N).
(a) As all columns are the same, any two are linearly dependent and so the matrix has only *one* linearly independent column. Thus, the matrix has rank 1.
(b) Any 2×2 submatrix has a determinant of the form $(\lambda \times \lambda) - (\lambda \times \lambda) = 0$. The determinant of any larger submatrix can be expressed as a sum of the determinants of 2×2 submatrices; thus all such larger submatrices have zero determinant. The largest submatrix with a non-zero determinant is a 1×1 submatrix with determinant λ, and so the matrix has rank 1.

20 The determinant $(=0)$ is equal to the product of the diagonal elements; the trace $(=0)$ is equal to their sum. Thus, one element must be zero and the two remaining (non-zero) elements must have opposite signs, i.e. it has the form $A = \text{diag}(0, \lambda, -\lambda)$ but, of course, the zero could appear as any one of the three entries.

21 A has determinant 1. As this $\neq 0$, A has an inverse:

$$A = \begin{pmatrix} 1 & 0 & a \\ 0 & 1 & b \\ 0 & 0 & 1 \end{pmatrix} \quad A^{-1} = \begin{pmatrix} 1 & 0 & -a \\ 0 & 1 & -b \\ 0 & 0 & 1 \end{pmatrix},$$

and $A + A^{-1} = 2I$. Multiplying on the right by A gives $A^2 + A^{-1}A = 2IA$; hence, $A^2 - 2A + I = 0$. This can be written $(A - I)^2 = 0$. However, this does *not* imply that $A - I = 0$; two $N \times N$ matrices can multiply to give the zero matrix without either being the zero

Appendix F

matrix. Explicitly, in this case,

$$(A - I)^2 = \begin{pmatrix} 0 & 0 & a \\ 0 & 0 & b \\ 0 & 0 & 0 \end{pmatrix} \begin{pmatrix} 0 & 0 & a \\ 0 & 0 & b \\ 0 & 0 & 0 \end{pmatrix} = \begin{pmatrix} 0 & 0 & 0 \\ 0 & 0 & 0 \\ 0 & 0 & 0 \end{pmatrix},$$

yet neither matrix in the central expression is the zero matrix.

23 The matrices are

$$S_x = \begin{pmatrix} 0 & 1 \\ 1 & 0 \end{pmatrix}, \quad S_y = \begin{pmatrix} 0 & -i \\ i & 0 \end{pmatrix}, \quad S_z = \begin{pmatrix} 1 & 0 \\ 0 & -1 \end{pmatrix}.$$

S_x and S_z: real, symmetric, Hermitian, orthogonal, unitary.
S_y: antisymmetric, Hermitian, unitary.

27 If one of the eigenvalues of A is zero, then the determinant of A (equal to the product of the eigenvalues) is also zero and A is singular. Therefore, the inverse of A does not exist.

28 If A is real and orthogonal, i.e. $A^T A = I$, and has eigenvalue λ, then $Ax = \lambda x$ and $x^T A^T = \lambda x^T$ for some non-zero x. Thus,

$$(x^T A^T)(Ax) = (\lambda x^T)(\lambda x) \quad \Rightarrow \quad x^T I x = \lambda^2 x^T x \quad \Rightarrow \quad x^T x (1 - \lambda^2) = 0.$$

But $x^T x \neq 0$ and so $\lambda^2 = 1$ and $\lambda = \pm 1$.

29 Following the suggested procedure:

$$|A - \lambda I| = \begin{vmatrix} 1-\lambda & 1 & 3 \\ 1 & 1-\lambda & -3 \\ 3 & -3 & -3-\lambda \end{vmatrix} = \begin{vmatrix} 2-\lambda & 1 & 3 \\ 2-\lambda & 1-\lambda & -3 \\ 0 & -3 & -3-\lambda \end{vmatrix}$$

$$= (2-\lambda) \begin{vmatrix} 1 & 1 & 3 \\ 1 & 1-\lambda & -3 \\ 0 & -3 & -3-\lambda \end{vmatrix} = (2-\lambda) \begin{vmatrix} 1 & 1 & 3 \\ 0 & -\lambda & -6 \\ 0 & -3 & -3-\lambda \end{vmatrix}$$

$$= (2-\lambda)(1)[(-\lambda)(-3-\lambda) - 18]$$

$$= (2-\lambda)(\lambda^2 + 3\lambda - 18) = (2-\lambda)(\lambda+6)(\lambda-3).$$

30 If all three eigenvectors are equal, then *any* vector is an eigenvector. Therefore, choose the simplest orthonormal set, $(1, 0, 0)$, $(0, 1, 0)$ and $(0, 0, 1)$.

33 If two of the eigenvalues are zero, the equation of the quadratic surface will contain only one of its coordinates, e.g. $\lambda_3 x_3^2 = 1$. This means $x_3 = \pm \lambda^{-1/2}$ and the surface consists of two parallel planes with their common normal in the direction of the eigenvector corresponding to the non-zero eigenvalue. The same situation can also be thought of formally as an ellipsoid with two infinitely long perpendicular semi-axes.

11. Vector calculus

1 With $\mathbf{r}(t) = R \hat{\mathbf{e}}_\rho$, we have $\mathbf{v}(t) = R\dot{\phi} \hat{\mathbf{e}}_\phi$ and $\mathbf{a}(t) = R\ddot{\phi} \hat{\mathbf{e}}_\phi - R\dot{\phi}^2 \hat{\mathbf{e}}_\rho$. With $\phi = \omega t + \phi_0$, $\ddot{\phi} = 0$ and the (inward) acceleration has magnitude $R\omega^2$.

3 $\mathbf{r}(\rho, \phi) = \rho \cos\phi \, \mathbf{i} + \rho \sin\phi \, \mathbf{j} + (\rho^2/4a) \, \mathbf{k}$, leading to $\mathbf{n} = (\partial \mathbf{r}/\partial \rho) \times (\partial \mathbf{r}/\partial \phi) = -(\rho/2a)[\cos\phi \, \mathbf{i} + \sin\phi \, \mathbf{j}] + \mathbf{k}$, which has magnitude $[(\rho/2a)^2 + 1]^{1/2}$. Hence $dS = (2a)^{-1}[\rho^2 + 4a^2]^{1/2} \, d\rho \, d\phi$.

Appendix F

4 (i) $2x^3y^2z^2$, (ii) $2xy^2z^2(x^2+y^2+z^2)$, (iii) $2xy^2z^2(2x^2+y^2+z^2)$.

5 $\nabla\phi = -\dfrac{nA}{r^{n+1}}\left(\dfrac{\partial r}{\partial x}, \dfrac{\partial r}{\partial y}, \dfrac{\partial r}{\partial z}\right) = -\dfrac{nA}{r^{n+1}}\left(\dfrac{x}{r}, \dfrac{y}{r}, \dfrac{z}{r}\right)$, i.e. directly towards the origin.

$$|\nabla\phi\cdot\hat{\mathbf{r}}| = \left|-\dfrac{nA}{r^{n+1}}\dfrac{x^2+y^2+z^2}{r^2}\right| = \dfrac{nA}{r^{n+1}}.$$

6 $2z(x^2+y^2)[2xz, 2yz, x^2+y^2]$.

7 If $z=0$ or $x=y=0$, i.e. on the x–y plane and on the z-axis.

8 $\displaystyle\sum_{x,y,z}\dfrac{\partial(\phi x)}{\partial x} = \sum_{x,y,z}\left(\phi + x\dfrac{\partial\phi}{\partial r}\dfrac{\partial r}{\partial x}\right) = \sum_{x,y,z}\left(\phi + x\dfrac{\partial\phi}{\partial r}\dfrac{x}{r}\right) = 3\phi + \dfrac{r^2}{r}\dfrac{\partial\phi}{\partial r}$.

13
$$\dfrac{1}{r^2}\dfrac{\partial}{\partial r}\left(r^2\dfrac{\partial\Phi}{\partial r}\right) = \dfrac{1}{r^2}\left[2r\dfrac{\partial\Phi}{\partial r} + r^2\dfrac{\partial^2\Phi}{\partial r^2}\right] = \dfrac{\partial^2\Phi}{\partial r^2} + \dfrac{2}{r}\dfrac{\partial\Phi}{\partial r},$$

$$\dfrac{1}{r}\dfrac{\partial^2}{\partial r^2}(r\phi) = \dfrac{1}{r}\dfrac{\partial}{\partial r}\left[\Phi + r\dfrac{\partial\Phi}{\partial r}\right] = \dfrac{1}{r}\dfrac{\partial\Phi}{\partial r} + \dfrac{1}{r}\dfrac{\partial\Phi}{\partial r} + \dfrac{\partial^2\Phi}{\partial r^2}.$$

14 Circular cylinders centred on the z-axis; half-planes containing the z-axis; planes with normals along the z-axis.

12. Line, surface and volume integrals

1 $\mathbf{i}\left(\displaystyle\int_C a_y\,dz - \int_C a_z\,dy\right) + \mathbf{j}\left(\int_C a_z\,dx - \int_C a_x\,dz\right) + \mathbf{k}\left(\int_C a_x\,dy - \int_C a_y\,dx\right)$.

2 The integral is $\displaystyle\int_1^4 (x + \tfrac{1}{2}\sqrt{x} + \tfrac{1}{2})\,dx$.

4 $ds = \phi\sqrt{4k^2 + \phi^2(k^2+9\mu^2)}\,d\phi$.

6 (a) Simply; (b) multiply – consider a closed curve in the form of a 'D' with the curved part lying within the handle and the straight part within the cylindrical part of the cup; (c) simply – though the physics of the cup is more interesting than its connectivity!

8
$$I = \int_R \sigma(x^2+y^2)\,dx\,dy = \int_{r=a} \sigma(xy^2\,dy - x^2y\,dx) - \int_{r=b} \sigma(xy^2\,dy - x^2y\,dx).$$

Now set $x = r\cos\phi$ and $y = r\sin\phi$ leading to $I = \sigma(a^4-b^4)\int_0^{2\pi}\tfrac{1}{2}\sin^2(2\phi)\,d\phi$. The mass $m = \sigma\pi(a^2-b^2)$.

9 The integrand is $r^4(\sin^2\phi\cos^2\phi - \cos^2\phi\sin^2\phi) = 0$. The potential function $\psi(x,y) = \tfrac{1}{2}x^2y^2$.

11 Since the vector area of the closed hemisphere is $\mathbf{0}$ and that of its base is clearly $-\pi a^2\mathbf{k}$, the vector area of its curved surface must be $+\pi a^2\mathbf{k}$.

15
$$\Omega = \int_S \dfrac{r^2\sin\theta\,d\theta\,d\phi}{r^2}\hat{\mathbf{r}}\cdot\hat{\mathbf{r}} = \int_0^{2\pi} d\phi \int_0^\alpha \sin\theta\,d\theta = 2\pi(1-\cos\alpha).$$

19 $\mathbf{c}\cdot\mathbf{P} = \mathbf{c}\cdot\mathbf{Q} \Rightarrow \mathbf{c}\cdot(\mathbf{P}-\mathbf{Q}) = 0$. Since $\mathbf{c}$ is arbitrary, take it as $\mathbf{P}-\mathbf{Q}$. Then $0 = (\mathbf{P}-\mathbf{Q})\cdot(\mathbf{P}-\mathbf{Q}) = |\mathbf{P}-\mathbf{Q}|^2$. The only vector with zero modulus is the zero vector $\mathbf{0}$. Thus $\mathbf{P}-\mathbf{Q} = \mathbf{0}$ and hence $\mathbf{P} = \mathbf{Q}$.

13. Laplace transforms

6 The transform of e^{at} is $(s-a)^{-1}$. But we also have

$$\mathcal{L}\left[\sum_{n=0}^{\infty}\frac{a^n t^n}{n!}\right]=\sum_{n=0}^{\infty}\frac{a^n n!}{n! s^{n+1}}=\frac{1}{s}\sum_{n=0}^{\infty}\frac{a^n}{s^n}=\frac{1}{s}\left(1-\frac{a}{s}\right)^{-1}=\frac{1}{s-a}.$$

The Laplace transformations require $s > a$ and $s > 0$; the binomial expansion requires $s > a$.

9 $f^{(n)}(a) = (-1)^n \int_{-\infty}^{\infty} f(t)\delta^{(n)}(t-a)\,dt$.

11 $\sum_i \delta(t-a_i) + \sum_j H(t-b_j) - \sum_k H(t-c_k)$, where $a_i = 1, 2, 3, 20, 21, 22$, $b_j = 6, 10, 14$ and $c_k = 9, 13, 17$. Technically, there should be no spaces between the letters, as the signal is meant to be sent as a continuous sequence of nine sounds.

12 At $t = 0$ the integral has zero range and so has zero value. At $t = \infty$ the factor $e^{-st} \to 0$ for $s > 0$.

14. Ordinary differential equations

3 With $y = ax^3 + be^{-x}$, $y' = 3ax^2 - be^{-x}$, $y'' = 6ax + be^{-x}$; eliminating a and b yields $x(x+3)y'' + (x^2 - 6)y' - 3(x+2)y = 0$.

4 $1 + y = A\exp(x^2/2)$. (a) $y(0) = 1 \Rightarrow A = 2$ and $y = 2\exp(x^2/2) - 1$. (b) $y(0) = -1 \Rightarrow A = 0$ and $y(x) = -1$ for all x.

5 $\partial A/\partial y - \partial B/\partial x = 8y^2 - 20x^{-1}y^4 + 4x^{-3}y^2$, i.e. not a function of x or y alone. For $\mu(x,y) = x^n y^m$ to be a valid IF requires a self-consistent solution of $2(n-2) = 0$, $3(m+3) = n+1$ and $4(m+5) = 6n$. It is $n = 2$, $m = -2$ and $\mu(x,y) = x^2 y^{-2}$. $\partial(\mu A)/\partial y = 3x^2 - 12xy^2 = \partial(\mu B)/\partial x$.

8 $A = e^{(c_1-c_2)}$.

9 (a) Differentiating the solution and eliminating c gives $-\dfrac{3}{y^4}\dfrac{dy}{dx} = -6x^3 + \dfrac{3}{xy^3}$.
(b) $c = 14$ and $0 < x < 7/3$ is the only valid range.

10 With $f(x, y, c) = y^3 - 3cx - 6c^2$, eliminate c between $f = 0$ and $\partial f/\partial c = 0$. The contact point is $(-4c, -(6c^2)^{1/3})$.

11 A set of parabolas, all with horizontal axes and touching the y-axis; those touching above the x-axis lie to the right of the y-axis, and those touching below it lie to the left. The curves with parameters c_1 and c_2 intersect at $x = \frac{1}{4}(c_1 + c_2 \mp 2\sqrt{c_1 c_2})$ and $y = \pm\sqrt{c_1 c_2}$, provided c_1 and c_2 have the same sign.

12 Eliminating c between $f(x, y) = (y-c)^2 - 4cx = 0$ and $\partial f/\partial c = -2(y-c) - 4x = 0$ gives $4x(x+y) = 0$. Hence $x = 0$ and $x+y = 0$ form the envelope to the set of curves.

16 At $x = 1$, $a + b = 0$, whilst at $x = -1$, $-a + b = 0$; hence $a = b = 0$. $W = \pm(5x^9 - 5x^9) = 0$.

21 First determine the combination of functions that multiply each constant, e.g. for a it is $-4x^3 \sin 2x + 12x^2 \cos 2x + 6x \sin 2x + 4x^3 \sin 2x$. Then regroup all such terms, collecting together the combination of constants that multiply any particular function; the first two equations given are those corresponding to $x^2 \cos 2x$ and $x \cos 2x$.

22 The sum of -2 and -3 is -5 and their product is 6; the characteristic equation is therefore $\lambda^2 + 5\lambda + 6 = 0$ and the recurrence relation is $u_{n+1} + 5u_n + 6u_{n-1} = 0$ for

Appendix F

$n \geq 1$, with $u_0 = 1$ and $u_1 = 0$.
  (a) $u_2 = -6$, $u_3 = 30$ and $u_4 = -114$. (b) $u_4 = 48 - 162 = -114$.

15. Elementary probability

1 $A - B = A \cap \bar{B} = \emptyset$, because $B \supset A$ and so A and $\bar{B}$ do not intersect.
4 (a) $16N/52$. (b) $17N/52$.
5 Yes, the replaced card would increase the number that could be drawn, but would not count as a success. The probability would now be $(4/52) \times (3/52) = 3/676$.
7 $(^4C_0\ ^{48}C_{13})/^{52}C_{13} = 0.304$, but $(^4C_1\ ^{48}C_{12})/^{52}C_{13} = 0.439$ and so one king is more likely.
9 In the order of the sequences given, the factors are $(3/10)(2/9)(7/8)$, $(3/10)(7/9)(2/8)$ and $(7/10)(3/9)(2/8)$.
13 With $0 \leq Y \leq 1$, $dY/dX = f(X)$ and hence, by (15.52), $g(Y) = f(X) \times |dX/dY| = f(X) \times [1/f(X)] = 1$.
22 The mean is 50 and, using the Gaussian approximation, $\sigma = \sqrt{100 \times 0.5 \times 0.5} = 5$. So 60 is two standard deviations from the mean and the chances of equalling or exceeding it is $1 - \Phi(2) = 0.0228$, i.e. about 2.3%.

Index

absolute convergence of series, 225
acceleration vector, 449
addition rule for probabilities, 603, 608
adjoint, *see* Hermitian conjugate
algebra of
   complex numbers, 177–8
   matrices, 378–9
   power series, 236
   series, 233
   unions and intersections, 601
   vectors, 332–3
      in a vector space, 369–70
      in component form, 337–8
alternating series test, 231
Ampère's rule (law), 495, 525
angle between two vectors, 341
angular momentum
   of particles, 451
   of solid body, 512
   vector representation, 365
angular velocity, vector representation, 344, 365, 464
anti-Hermitian matrices, 410
   eigenvalues, 414
   eigenvectors, 414
anticommutativity of vector or cross product, 342
antisymmetric functions, 126
antisymmetric matrices, 409
   general properties, *see* anti-Hermitian matrices
approximately equal ≈, definition, 234
arbitrary parameters for ODE, 555
arc length of plane curve, 162–4
arccosech, arccosh, arccoth, arcsech, arcsinh,
      arctanh, *see* hyperbolic functions, inverses
Archimedean upthrust, 511, 530
area element in
   Cartesian coordinates, 301
   plane polars, 318
area of
   circle, 159
   ellipse, 159–60, 325
   parallelogram, 343
   region, as a line integral, 500

region, using multiple integrals, 306–8
surfaces, 457
   as vector, 508–10, 525
arg, argument of a complex number, 178–9
Argand diagram, 175
arithmetic mean, 35
arithmetic series, 215
arithmetico-geometric series, 217
arrays, *see* matrices
associative law for
   addition
      in a vector space of finite dimensionality, 369–70
      of complex numbers, 177
      of matrices, 378
      of vectors, 332
   convolution, 548
   linear operators, 375
   multiplication
      of a matrix by a scalar, 378
      of a vector by a scalar, 333–4
      of complex numbers, 180
      of matrices, 380
   multiplication by a scalar
      in a vector space of finite dimensionality, 370
associativity, 2
asymptotes, 126
atomic orbitals, s-states, 624
auxiliary equation, 572
   repeated roots, 573
average value, *see* mean value

base
   for logarithms, 9
   for natural logarithms, 10, 673
basis vectors, 336–8, 370–1
   linear independence, 337
   non-orthogonal, 372
   orthonormal, 372
   required properties, 337
Bayes' theorem, 610–12
Bernoulli equation, 563
Bessel inequality, 373

Index

bilinear transformation, general, 208
binomial coefficient ^nC_k, 22–3, 615–17
  elementary properties, 21
  in Leibnitz's theorem, 113, 679
  negative n, 22
  non-integral n, 22
binomial distribution Bin(n, p), 632–5
  and Gaussian distribution, 649
  and Poisson distribution, 639, 641
  mean and variance, 634
  recurrence formula, 633
binomial expansion, 20–4, 244
  and the exponential function, 23
  proof, 22
birthdays, different, 613
bivariate distributions, 655–60
  correlation, 657–60
    and independence, 658
    positive and negative, 657
    uncorrelated, 657
  covariance, 657–60
  expectation (mean), 656
  independent, 655, 657
  variance, 657
Boltzmann distribution, 280
Born approximation, 257
boundary conditions for ODE, 554, 556, 577
Bragg formula, 364
Breit–Wigner distribution, 644

calculus, elementary, 102–67
card drawing, 603, 605, 607, 615, 616, *see* probability
Cartesian coordinates, 337–8
Cauchy
  distribution, 629
  product, 233
  root test, 230
central limit theorem, 651
central moments, *see* moments, central
centre of mass, 309–10
  of hemisphere, 309
  of semicircular lamina, 312
centroid, 309
  of plane area, 310
  of plane curve, 311
  of triangle, 335–6
CF, *see* complementary function
chain rule for functions of
  one real variable, 108–9
  several real variables, 267–8
change of basis, *see* similarity transformations
change of variables

  and coordinate systems, 268–70
  in multiple integrals, 315–23
    evaluation of Gaussian integral, 318–20
    general properties, 322–3
change of variables in RVD, 628–31
characteristic equation, 418
  of recurrence relation, 582
charge (point), Dirac δ-function representation, 543
charged particle in electromagnetic fields, 485
chi-squared (χ^2) distribution, 653
Cholesky decomposition, 404, 441
circle
  area of, 159
  equation for, 66
cofactor of a matrix element, 387
column matrix, 376
column vector, 376
combinations (probability), 612–17
common ratio in geometric series, 216
commutative law for
  addition
    in a vector space of finite dimensionality, 369–70
    of complex numbers, 177
    of matrices, 378
    of vectors, 332
  complex scalar or dot product, 342
  convolution, 548
  inner product, 371
  multiplication
    of a vector by a scalar, 333–4
    of complex numbers, 180
  scalar or dot product, 340
commutativity, 2
commutator of two matrices, 439
comparison test, 226
complement, 599
  probability for, 603
complementary equation, 569
complementary function (CF), 570
  for ODE, 572
  repeated roots of auxiliary equation, 573
completeness of basis vectors, 371
completing the square
  as a means of integration, 151–2
  for quadratic equations, 54
complex conjugate
  z^*, of complex number, 181–4
  of a matrix, 384–5
  of scalar or dot product, 342
  properties of, 182–3
complex exponential function, 185

Index

complex logarithms, 194–6
  principal value of, 194
complex numbers, 174–212
  addition and subtraction of, 177–8
  applications to differentiation and integration, 196–7
  argument of, 178–9
  associativity of
    addition and subtraction, 177
    multiplication, 180
  commutativity of
    addition, 177
    multiplication, 180
  complex conjugate of, *see* complex conjugate
  components of, 175
  de Moivre's theorem, *see* de Moivre's theorem
  division of, 184, 187
  from roots of polynomial equations, 174
  imaginary part of, 174–5
  modulus of, 178–9
  multiplication of, 179–81, 187–8
    as rotation in the Argand diagram, 180–1
  notation, 174–5
  phase, 178n
  polar representation of, 185–8
  real part of, 174–5
  trigonometric representation of, 186
complex power series, 235
complex powers, 194–6
components
  of a complex number, 175
  of a vector, 336–8
    in a non-orthogonal basis, 358–9
    uniqueness, 371
compound-angle identities, 28–32
conditional convergence, 225
conditional probability, *see* probability, conditional
cone
  surface area of, 165–6
  volume of, 167
conic sections, 65
  eccentricity, 67
  parametric forms, 70
  standard forms, 66
conjugate roots of polynomial equations, 193
connectivity of regions, 497
conservative fields, 502–4
  necessary and sufficient conditions, 502–4
  potential (function), 504
constant coefficients in ODE, 572–9
  auxiliary equation, 572
constants of integration, 145, 554

constraints, stationary values under, *see* Lagrange undetermined multipliers
continuity correction for discrete RV, 650
continuity equation, 521
contradiction, proof by, 87–8
convergence of infinite series
  absolute, 225
  complex power series, 235
  conditional, 225
  necessary condition, 226
  power series, 234
    under various manipulations, *see* power series, manipulation
  rearrangement of terms, 225
  tests for convergence, 226–33
    alternating series test, 231
    comparison test, 226
    grouping terms, 230
    integral test, 229
    quotient test, 228
    ratio comparison test, 228
    ratio test (D'Alembert), 227, 234
    root test (Cauchy), 230
convolution
  Laplace transforms, *see* Laplace transforms, convolution
convolution theorem for Laplace transforms, 547
coordinate geometry, 64–73
  conic sections, 65
  straight line, 64
coordinate systems, *see* Cartesian, curvilinear, cylindrical polar, plane polar *and* spherical polar coordinates
coordinate transformations
  and integrals, *see* change of variables
  and matrices, *see* similarity transformations
coplanar vectors, 347
correlation of bivariate distributions, 657–60
correspondence principle in quantum mechanics, 668
cosh, cosh(x), *see also* hyperbolic functions
  hyperbolic cosine, 198
  Maclaurin series for, 244
cosine, cos(x)
  geometrical and algebraic definitions, 676
  geometrical definition, 25
  in terms of exponential functions, 199
  Maclaurin series for, 243
  reciprocal and inverse, 27
covariance of bivariate distributions, 657–60
Cramer determinant, 405
Cramer's rule, 404–6

Index

cross product, *see* vector product
crystal lattice, 256
cube roots of unity, 191
curl of a vector field, 463
    as a determinant, 464
    as integral, 514, 516
    curl curl, 468
    in curvilinear coordinates, 480
    in cylindrical polars, 472
    in spherical polars, 475
    Stokes' theorem, 523–6
curvature, 116–19
    circle of, 117
    of a function, 116
    radius of, 117
curves, *see* plane curves
curvilinear coordinates, 476–81
    basis vectors, 477
    length and volume elements, 478
    scale factors, 477
    surfaces and curves, 477
    vector operators, 479–81
cyclic relation for partial derivatives, 266
cycloid, 485
cylindrical polar coordinates, 469–73
    area element, 471
    basis vectors, 470
    length element, 471
    vector operators, 470–3
    volume element, 471

δ-function (Dirac), *see* Dirac δ-function
δ_{ij}, *see* Kronecker delta, δ_{ij}
D'Alembert's ratio test, 227
    in convergence of power series, 234
damped harmonic oscillators, 366
de Moivre's theorem, 189
    applications, 189–93
        finding the nth roots of unity, 191–2
        solving polynomial equations, 192–3
        trigonometric identities, 189–91
de Morgan's laws, 602
defective matrices, 416, 442
degenerate eigenvalues, 415, 420–1
degree
    of ODE, 554
    of polynomial equation, 53
del ∇, *see* gradient operator (grad)
del squared ∇^2 (Laplacian), 463
    as integral, 517
    in curvilinear coordinates, 480
    in cylindrical polar coordinates, 472
    in spherical polar coordinates, 475

del squared ∇^2 (Laplacian), 269
delta function (Dirac), *see* Dirac δ-function
dependent random variables, 655–60
derivative, *see also* differentiation
    Laplace transform of, 544
    normal, 461
    of basis vectors, 450–1
    of composite vector expressions, 451–2
    of function of a function, 108–9
    of hyperbolic functions, 202–5
    of products, 106–8, 112–13
    of quotients, 109–10
    of simple functions, 106
    of vectors, 448
    ordinary, first, second and nth, 103–5
    partial, *see* partial differentiation
    total, 263
determinant form for curl, 464
determinants, 386–91
    adding rows or columns, 390
    and singular matrices, 392
    as product of eigenvalues, 426
    evaluation using Laplace expansion, 387
    identical rows or columns, 390
    in terms of cofactors, 387
    interchanging two rows or columns, 390
    Jacobian representation, 317, 321, 323
    notation, 386
    of Hermitian conjugate matrices, 389
    of order three, in components, 388
    of transpose matrices, 389
    product rule, 390
    properties, 389–91, 444
    relationship with rank, 396–7
    removing factors, 390
    secular, 418
diagonal matrices, 408
diagonalisation of matrices, 424–6
    normal matrices, 425–6
    properties of eigenvalues, 426
diamond, unit cell, 361
dice throwing, 604, 627, 633, 658
die throwing, *see* probability
difference method for summation of series, 217–19
differentiable function of a real variable, 103
differential
    definition, 104
    exact and inexact, 264–5
    of vector, 452, 455
    total, 262
differential equations, *see* ordinary differential equations

Index

differential equations, particular
   Bernoulli, 563
differentiation, *see also* derivative
   as gradient, 103
   as rate of change, 102
   chain rule, 108–9
   from first principles, 102–6
   implicit, 110–11
   logarithmic, 111
   notation, 105
   of integrals, 288–90
   of power series, 237
   partial, *see* partial differentiation
   product rule, 106–8, 112–13
   quotient rule, 109–10
   theorems, 120–1
   using complex numbers, 196
dimensionality of vector space, 370
dimensions, physical, 15–19
   dimensional analysis, 16
dipole matrix elements, 326
Dirac δ-function, 521, 541–4
   definition, 541
   impulses, 542
   point charges, 543
   properties, 541
   relation to Heaviside (unit step) function, 543
   three-dimensional, 543
direction cosines, 342
directrix, of a conic section, 66
disc, moment of inertia, 327
disjoint events, *see* mutually exclusive events
distance from a
   line to a line, 355–6
   line to a plane, 356–7
   point to a line, 353–4
   point to a plane, 354–5
distributive law for
   addition of matrix products, 381
   convolution, 548
   inner product, 371
   linear operators, 375
   multiplication
      of a matrix by a scalar, 378
      of a vector by a complex scalar, 342
      of a vector by a scalar, 333–4
   multiplication by a scalar
      in a vector space of finite dimensionality, 370
   scalar or dot product, 340
   vector or cross product, 342
divergence of vector fields, 462
   as integral, 514, 515
   in curvilinear coordinates, 479
   in cylindrical polars, 472
   in spherical polars, 475
divergence theorem
   for vectors, 517–18
   in two dimensions, 499
   physical applications, 520–2
   related theorems, 519
division of complex numbers, 184
dot product, *see* scalar product
double integrals, *see* multiple integrals
double-angle identities, 30
dummy variable, 142

ϵ_{ijk}, *see* Levi-Civita symbol, ϵ_{ijk}
e^x, *see* exponential function
eccentricity, of conic sections, 67
eigenvalues, 412–21
   characteristic equation, 418
   definition, 412
   degenerate, 420–1
   determination, 418–21
   notation, 413
   of anti-Hermitian matrices, *see* anti-Hermitian matrices
   of general square matrices, 415–16
   of Hermitian matrices, *see* Hermitian matrices
   of linear operators, 412
   of unitary matrices, 415
   under similarity transformation, 426
eigenvectors, 412–21
   characteristic equation, 418
   definition, 412
   determination, 418–21
   normalisation condition, 413
   notation, 413
   of anti-Hermitian matrices, *see* anti-Hermitian matrices
   of commuting matrices, 416
   of general square matrices, 415–16
   of Hermitian matrices, *see* Hermitian matrices
   of linear operators, 412
   of unitary matrices, 415
   stationary properties for quadratic and Hermitian forms, 429–30
electromagnetic fields
   flux, 510
   Maxwell's equations, 488, 525
ellipse
   area of, 159–60, 325, 500
   as section of quadratic surface, 431
   equation for, 66
ellipsoid, volume of, 326
empty event ∅, 599

Index

end-points, of a range, 33
envelopes, 282–4
   equations of, 282
   to a family of curves, 282
equating real and imaginary parts, 176
equivalence transformations, *see* similarity transformations
error terms in Taylor series, 242–3
Euler equation, trigonometric, 186
even functions, *see* symmetric functions
events, 597
   complement of, 599
   empty ∅, 599
   intersection of ∩, 598
   mutually exclusive, 607
   statistically independent, 608
   union of ∪, 599
exact differentials, 264–5
exact equations, 558
   condition for, 558
expectation values, *see* probability distributions, mean
exponent, of a power, 1
exponential distribution, 653
   from Poisson, 653
exponential function
   and a general power, 4
   and logarithms, 7
   and the natural logarithmic base, 9, 673
   definition, 10
   equivalence of $\exp(x)$ and e^x, 673
   from the binomial expansion, 23
   Maclaurin series for, 243
   of a complex variable, 185
   properties, 11
   relation with hyperbolic functions, 198

Fabry–Pérot interferometer, 254
factorial function for a positive integer, 10
factorisation, of a polynomial equation, 60
Fibonacci series, 594
fields
   conservative, 502–4
   scalar, 458
   vector, 458
first law of thermodynamics, 285
first-order differential equations, *see* ordinary differential equations
fluids
   Archimedean upthrust, 511, 530
   continuity equation, 521

flux, 510
   irrotational flow, 464
   sources and sinks, 521–2
   velocity potential, 526
   vortex flow, 526
focus, of a conic section, 66
function of a matrix, 382
functions of one real variable
   differentiation of, 102–13
   integration of, 141–61
   limits, *see* limits
   maxima and minima of, 114–15
   stationary values of, 114–15
   Taylor series, *see* Taylor series
functions of several real variables
   chain rule, 267–8
   differentiation of, 259–90
   integration of, *see* multiple integrals, evaluation
   rates of change, 261–3
   Taylor series, 270–1
functions of two real variables
   maxima and minima, 272–5
   points of inflection, 272–5
   saddle points, 272–5
   stationary values, 272–5
fundamental theorem of
   algebra, 174, 175
   calculus, 143–5
   complex numbers, *see* de Moivre's theorem

Gaussian (normal) distribution $N(\mu, \sigma^2)$, 645–53
   and binomial distribution, 649
   and central limit theorem, 651
   and Poisson distribution, 651
   continuity correction, 650
   cumulative probability function, 646
      tabulation, 648
   integration with infinite limits, 318–20
   mean and variance, 645–9
   multiple, 652–3
   sigma limits, 647
   standard variable, 645
Gaussian elimination with interchange, 399–401
geometric distribution, 636
geometric mean, 35
geometric series, 215
Gibbs' free energy, 287
golden mean, 594
gradient of a function of
   one variable, 103
   several real variables, 261–3
gradient of scalar, 459–62

Index

gradient operator (grad), 458
  as integral, 514
  in curvilinear coordinates, 479
  in cylindrical polars, 472
  in spherical polars, 475
graph papers, logarithmic, 65
graphs, 124–32
  and approximate solutions, 52
  general considerations, 125
  horizontal asymptote, 127
  vertical asymptote, 126
  worked examples, 127–32
gravitation, Newton's law, 453
Green's function for ODE, 298
Green's theorems
  in a plane, 498–501, 524
  in three dimensions, 518
grouping terms as a test for convergence, 230

half-angle identities, 31, 150
harmonic oscillators, damped, 366
Heaviside function, 543
  relation to Dirac δ-function, 543
Helmholtz potential, 287
hemisphere, centre of mass and centroid, 309
Hermitian conjugate, 384–5
  and inner product, 385
  product rule, 384
Hermitian forms, 427–31
  positive definite and semi-definite, 428–9
  stationary properties of eigenvectors, 429–30
Hermitian matrices, 410
  eigenvalues, 414–15
    reality, 414
  eigenvectors, 414–15
    orthogonality, 414–15
higher order differential equations, *see* ordinary differential equations
homogeneous
  differential equations, 569
  dimensionally consistent, 562–3
  simultaneous linear equations, 397
hydrogen atom
  s-states, 624
hydrogen atom, electron wavefunction, 326
hyperbola
  as section of quadratic surface, 431
  equation for, 66
hyperbolic functions, 197–205
  calculus of, 202–5
  definitions, 198
  graphs, 198
  identities, 200
  in equations, 200–1
  inverses, 201–2
    graphs, 202
  trigonometric analogies, 199–200
hypergeometric distribution, 637–9
  mean and variance, 638

i, j, k (unit vectors), 338
i, square root of -1, 175
identity matrices, 381, 382
identity operator, 375
imaginary part or term of a complex number, 174–5
improper integrals, 156
impulses, δ-function representation, 542
independent random variables, 631, 658
index, of a power, 1
induction, proof by, 85–6
inequalities
  algebraic, 32–9
  amongst integrals, 160
  Bessel, 373
  Schwarz, 373
  triangle, 373
inexact differentials, 264–5
inexact equation, 559
infinite integrals, 156
infinite series, *see* series
inflection
  general points of, 115
  stationary points of, 114–15
inhomogeneous
  differential equations, 569
  simultaneous linear equations, 397
inner product in a vector space, *see also* scalar product
  of finite dimensionality, 371–2
  and Hermitian conjugate, 385
  commutativity, 371
  distributivity over addition, 371
integral test for convergence of series, 229
integrals, *see also* integration
  definite, 141
  double, *see* multiple integrals
  improper, 156
  indefinite, 145
  inequalities, 160
  infinite, 156
  Laplace transform of, 545
  limits
    containing variables, 302
    fixed, 141
    variable, 143

Index

line, *see* line integrals
multiple, *see* multiple integrals
 of vectors, 453
 properties, 143
 triple, *see* multiple integrals
 undefined, 142
integrand, 141
integrating factor (IF)
 first-order ODE, 559–62
integration, *see also* integrals
 applications, 161–7
  finding the length of a curve, 162–4
  mean value of a function, 161–2
  surfaces of revolution, 164–6
  volumes of revolution, 166–7
 as area under a curve, 141–2
 as the inverse of differentiation, 143–5
 formal definition, 142
 from first principles, 142–3
 in plane polar coordinates, 159–60
 logarithmic, 148
 multiple, *see* multiple integrals
 of functions of several real variables, *see*
  multiple integrals
 of hyperbolic functions, 202–5
 of power series, 237
 of simple functions, 146
 of singular functions, 156
 of sinusoidal functions, 146–8
integration constant, 145
integration, methods for
 by inspection, 146
 by parts, 152–4
 by substitution, 149–52
  t substitution, 150–1
 change of variables, *see* change of variables
 completing the square, 151–2
 partial fractions, 148–9
 reduction formulae, 155–6
 trigonometric expansions, 146–8
 using complex numbers, 196–7
intersection ∩
 algebra of, 601
intervals, open and closed, 33
inverse hyperbolic functions, 201–2
inverse Laplace transforms, 539
 uniqueness, 539
inverse matrices, 392–4
 elements, 392
 in solution of simultaneous linear equations, 401
 left and right, 392n
 product rule, 394
 properties, 394
inverse of a linear operator, 375
irrotational vectors, 464
isotope decay, 587, 593

j, square root of -1, 175
Jacobians
 analogy with derivatives, 323
 and change of variables, 322–3
 definition in
  two dimensions, 316
  three dimensions, 321
 general properties, 322–3
 in terms of a determinant, 317, 321, 323
joint distributions, *see* bivariate distributions *and*
  multivariate distributions

Kronecker delta, δ_{ij}, 682
 and orthogonality, 372

L'Hôpital's rule, 246–8
Lagrange undetermined multipliers, 276–81
 and stationary properties of the eigenvectors of
  quadratic forms, 429
 for functions of more than two variables, 277–81
 in deriving the Boltzmann distribution, 280–1
 with several constraints, 277–81
Lagrange's identity, 345
lamina: mass, centre of mass and centroid, 308–9
Laplace expansion, 387
Laplace transforms, 536–49
 convolution
  associativity, commutativity, distributivity, 548
  definition, 547
 convolution theorem, 547
 definition, 538
 for ODE with constant coefficients, 576–9
 inverse, 539
  uniqueness, 539
 properties: translation, exponential
  multiplication, etc., 546
 table for common functions, 540
Laplace transforms, examples
 constant, 538
 derivatives, 544
 exponential function, 538
 integrals, 545
 polynomial, 538
Laplacian, *see* del squared ∇^2 (Laplacian)
Leibnitz's rule for differentiation of integrals, 289
Leibnitz's theorem, 112–13, 679

Index

length of
  a vector, 338
  plane curves, 162–4
Levi-Civita symbol, ϵ_{ijk}, 681
limits, 244–8
  definition, 245
  L'Hôpital's rule, 246–8
  of functions containing exponents, 246
  of integrals, 141
    containing variables, 302
  of products, 245
  of quotients, 245, 246–8
  of sums, 245
line integrals
  and Stokes' theorem, 523–6
  of scalars, 491–501
  of vectors, 491–504
  physical examples, 495
  round closed loop, 501
line, vector equation of, 349
linear dependence and independence
  definition in a vector space, 370
  of basis vectors, 337
  relationship with rank, 395–6
linear equations, differential
  first-order ODE, 561
  general ODE, 569–79
  ODE with constant coefficients, 572–9
linear equations, simultaneous, *see* simultaneous linear equations
linear independence of functions, 570
  Wronskian test, 570
linear operators, 374–5
  associativity, 375
  distributivity over addition, 375
  eigenvalues and eigenvectors, 412
  in a particular basis, 374–5
  inverse, 375
  non-commutativity, 375
  particular: identity, null or zero, singular and non-singular, 375
  properties, 375
linear vector spaces, *see* vector spaces
Ln of a complex number, 194–6
ln (natural logarithm)
  choice of base, 673
  Maclaurin series for, 244
  of a complex number, 194–6
logarithmic graph papers, 65
logarithms, 7–14
  and data analysis, 12
  and practical calculations, 8, 12
  and the value of 0^0, 14
  choice of base, 9
  definition, 8
  nomenclature, 9
  properties, 9
lottery (UK), and hypergeometric distribution, 638
lower triangular matrices, 408
LU decomposition, 401–4

Maclaurin series, 240
  standard expressions, 243
Madelung constant, 256
magnetic dipole, 340
magnitude of a vector, 338
  in terms of scalar or dot product, 342
mass of non-uniform bodies, 308
matrices, 369–431
  as a vector space, 379
  as arrays of numbers, 376
  as representation of a linear operator, 376
  column, 376
  elements, 376
    minors and cofactors, 387
  identity or unit, 381
  row, 376
  zero or null, 381
matrices, algebra of
  addition, 378–9
  change of basis, 421–4
  Cholesky decomposition, 404, 441
  diagonalisation, *see* diagonalisation of matrices
  LU decomposition, 401–4
  multiplication, 379–81
    and common eigenvalues, 416
    commutator, 439
    non-commutativity, 381
  multiplication by a scalar, 378–9
  similarity transformations, *see* similarity transformations
  simultaneous linear equations, *see* simultaneous linear equations
  subtraction, 378
matrices, derived
  adjoint, 384–5
  complex conjugate, 384–5
  Hermitian conjugate, 384–5
  inverse, *see* inverse matrices
  table of, 433
  transpose, 376
matrices, properties of
  anti- or skew-symmetric, 409
  anti-Hermitian, *see* anti-Hermitian matrices
  determinant, *see* determinants

Index

diagonal, 408
eigenvalues, *see* eigenvalues
eigenvectors, *see* eigenvectors
Hermitian, *see* Hermitian matrices
normal, *see* normal matrices
order, 376
orthogonal, 410
rank, 395
square, 376
symmetric, 409
trace or spur, 385
triangular, 408
unitary, *see* unitary matrices
maxima and minima (local) of a function of
   constrained variables, *see* Lagrange
      undetermined multipliers
   one real variable, 114–15
     sufficient conditions, 115
   two real variables, 272–5
     sufficient conditions, 274
Maxwell's
   electromagnetic equations, 488, 525
   thermodynamic relations, 285–7
mean μ
   of RVD, 624–5
mean value of a function
   of one variable, 161–2
   of several variables, 313
mean value theorem, 120–1
median of RVD, 625
minor of a matrix element, 387
mode of RVD, 625
modulus
   of a complex number, 178–9
   of a vector, *see* magnitude of a vector
moments (of distributions)
   of RVD, 626
moments (of forces), vector representation of, 344
moments of inertia
   definition, 312
   of disc, 327
   of rectangular lamina, 313
   of right circular cylinder, 327
   of sphere, 321
   perpendicular axes theorem, 327
multinomial distribution, 635–6
multiple angles, trigonometric formulae, 24
multiple integrals
   application in finding
     area and volume, 306–8
     mass, centre of mass and centroid, 308–10
     mean value of a function of several variables, 313
     moments of inertia, 312–13
   change of variables
     double integrals, 315–20
     general properties, 322–3
     triple integrals, 320–1
   definitions of
     double integrals, 301
     triple integrals, 304
   evaluation, 302–5
   notation, 301, 303, 304
   order of integration, 302–3, 305
multivariate distributions
   multinomial, 635–6
mutually exclusive events, 598, 607

nabla ∇, *see* gradient operator (grad)
natural logarithm, *see* ln *and* Ln
natural numbers, in series, 85, 220–1
necessary and sufficient conditions, 88–90
negative (semi-)definite function, 36
negative binomial distribution, 636
negative powers, 3
negative vector, 370
Newton's law of gravitation, *see* gravitation, Newton's law
non-Cartesian coordinates, *see* curvilinear, cylindrical polar, plane polar *and* spherical polar coordinates
norm of vector, 372
normal
   to coordinate surface, 478
   to plane, 350
   to surface, 456, 461, 505
normal derivative, 461
normal distribution, *see* Gaussian (normal) distribution
normal matrices, 411
normalisation of
   eigenvectors, 413
   vectors, 338
null (zero)
   matrix, 381, 382
   operator, 375
   vector, 333, 370

$O(x)$, order of, 234
observables in quantum mechanics, 414
odd functions, *see* antisymmetric functions
ODE, *see* ordinary differential equations (ODE)
operators linear, *see* linear operators
order of
   approximation in Taylor series, 240n
   ODE, 554

Index

ordinary differential equations (ODE), *see also* differential equations, particular
  boundary conditions, 554, 556, 577
  complementary function, 570
  degree, 554
  dimensionally homogeneous, 562
  exact, 558
  first-order, 554–68
  first-order higher degree, 565–8
    soluble for p, 565
    soluble for x, 566
    soluble for y, 567
  general form of solution, 554–6
  higher order, 569–79
  homogeneous, 569
  inexact, 559
  linear, 561, 569–79
  order, 554
  particular integral (solution), 555, 571, 574–5
  singular solution, 555, 566, 568
ordinary differential equations, methods for
  equations with constant coefficients, 572–9
  integrating factors, 559–62
  Laplace transforms, 576–9
  separable variables, 557
  undetermined coefficients, 574
orthogonal lines, condition for, 30
orthogonal matrices, 410
  general properties, *see* unitary matrices
orthogonal systems of coordinates, 477
orthogonality of
  eigenvectors of an Hermitian matrix, 414–15
  vectors, 339, 371
orthonormal basis vectors, 372
  under unitary transformation, 423
outcome, of trial, 597

Pappus's theorems, 310–12
parabola, equation for, 66
parallel axis theorem, 366
parallel vectors, 343
parallelepiped, volume of, 346, 347
parallelogram equality, 373
parallelogram, area of, 343, 345
parametric equations
  of conic sections, 69
  of cycloid, 485
  of surfaces, 455
partial derivative, *see* partial differentiation
partial differentiation, 259–90
  as gradient of a function of several real variables, 259
  chain rule, 267–8
  change of variables, 268–70
  definitions, 259–61
  properties, 266–7
    cyclic relation, 266
    reciprocity relation, 266
partial fractions, 74–83
  and degree of numerator, 79
  as a means of integration, 148–9
  complex roots, 81
  in inverse Laplace transforms, 540, 577
  repeated roots, 81
partial sum, 213
particular integrals (PI), 555, *see also* ordinary differential equation, methods for
parts, integration by, 152–4
path integrals, *see* line integrals
PDFs, 620
permutations, 612–17
  distinguishable, 614
  symbol ^nP_k, 612
perpendicular axes theorem, 327
perpendicular vectors, 339, 371
PF, *see* probability functions
phase, of a complex number, 178n
physical constants, values, 684
physical dimensions, 15–19
  derived quantities, 15n, 16
  dimensional analysis, 16
PI, *see* particular integrals
plane curves, length of, 162–4
  in Cartesian coordinates, 162
  in plane polar coordinates, 164
plane polar coordinates, 72, 159, 450
  arc length, 164, 473
  area element, 318, 473
  basis vectors, 450
  velocity and acceleration, 450
planes
  and simultaneous linear equations, 406–7
  vector equation of, 349–51
point charges, δ-function representation, 543
points of inflection of a function of
  one real variable, 114–16
  two real variables, 272–5
Poisson distribution Po(λ), 639–42
  and Gaussian distribution, 651
  as limit of binomial distribution, 639, 641
  mean and variance, 641
  multiple, 642
  recurrence formula, 640
polar coordinates, *see* plane polar *and* cylindrical polar *and* spherical polar coordinates
polar representation of complex numbers, 185–8

Index

polynomial equations, 53–63
  conjugate roots, 193
  factorisation, 60
  multiplicities of roots, 56
  number of roots, 174, 175
  properties of roots, 62
  real roots, 53
  solution of, using de Moivre's theorem, 192–3
positive (semi-)definite function, 36
positive (semi-)definite quadratic and Hermitian forms, 428
positive semi-definite norm, 372
potential energy of
  ion in a crystal lattice, 256
  magnetic dipole in a field, 340
potential function
  and conservative fields, 504
  vector, 504
potential, thermodynamic, 286
power series
  interval of convergence, 234
  Maclaurin, *see* Maclaurin series
  manipulation: difference, differentiation, integration, product, substitution, sum, 236–8
  Taylor, *see* Taylor series
power series in a complex variable, 235
  circle and radius of convergence, 235
powers
  combining, 1
  complex, 194–6
  general, 4
  index or exponent, 1
  negative, 3
  rational, 3
  real, 1
prime, non-existence of largest, 88
principal axes of quadratic surfaces, 430
principal value of
  complex logarithms, 194
  improper integral, 158n12
probability, 602–60
  axioms, 603
  conditional, 606–12
    Bayes' theorem, 610–12
    combining, 608
  definition, 603
  for intersection ∩, 598
  for union ∪, 599, 603–6
probability distributions, 618, *see also individual distributions*
  bivariate, *see* bivariate distributions
  change of variables, 628–31
  mean μ, 624–5
  mean of functions, 625
  mode, median and quartiles, 625
  moments, 626–8
  multivariate, *see* multivariate distributions
  standard deviation σ, 626
  table of
    continuous distributions, 643
    discrete distributions, 632
  variance σ^2, 626
probability distributions, individual
  binomial Bin(n, p), 632–5
  Cauchy (Breit–Wigner), 644
  chi-squared (χ^2), 653
  exponential, 653
  Gaussian (normal) $N(\mu, \sigma^2)$, 645–53
  geometric, 636
  hypergeometric, 637
  multinomial, 635–6
  negative binomial, 636
  Poisson Po(λ), 639–42
  uniform (rectangular), 644
probability functions (PFs), 619
  cumulative, 619, 620
  density functions (PDFs), 620
product notation, 62n6
product rule for differentiation, 106–8, 112–13

quadratic equations
  complex roots of, 174
  properties of roots, 62
  roots of, 53
quadratic forms, 427–31
  positive definite and semi-definite, 428–9
  quadratic surfaces, 430–1
  removing cross terms, 427–8
  stationary properties of eigenvectors, 429–30
quartiles, of RVD, 625
quotient rule for differentiation, 109–10
quotient test for series, 228

radian, 25
radius of convergence, 235
radius of curvature, of plane curves, 117
random variable distributions, *see* probability distributions
random variables (RV), 618–23
  continuous, 620–3
  dependent, 655–60
  discrete, 618–20
  independent, 631, 658
  uncorrelated, 658

Index

rank of matrices, 395
  and determinants, 396–7
  and linear dependence, 395–6
rate of change of a function of
  one real variable, 102
  several real variables, 261–3
ratio comparison test, 228
ratio test (D'Alembert), 227
  in convergence of power series, 234
ratio theorem, 334
  and centroid of a triangle, 335–6
rational functions, 125
rational powers, 3
rationalisation, involving surds, 5
real part or term of a complex number, 174–5
real roots, of a polynomial equation, 53
reciprocal vectors, 357–9
reciprocity relation for partial derivatives, 266
rectangular distribution, 644
recurrence relations (series), 579–84
  characteristic equation, 582
  first-order, 580
  second-order, 582
  higher order, 584
reduction formulae for integrals, 155–6
relative velocities, 338
remainder term in Taylor series, 240
repeated roots of auxiliary equation, 573
rhomboid, volume of, 364
Riemann theorem for conditional convergence, 225
Riemann zeta series, 229, 230
right-hand screw rule, 342
Rolle's theorem, 57, 120
root test (Cauchy), 230
roots
  of a polynomial equation, 53
    properties, 62
  of a real variable, 3
  of unity, 191–2
rotation of a vector, *see* curl of a vector field
row matrix, 376
RV, *see* random variables
RVD (random variable distributions), *see* probability distributions

saddle points, 273
  sufficient conditions, 274
sampling
  space, 597
  with or without replacement, 607
scalar fields, 458
  derivative along a space curve, 459
  gradient, 459–62
  line integrals, 491–501
  rate of change, 459
scalar product, 339–42
  and inner product, 371
  and perpendicular vectors, 339, 371
  for vectors with complex components, 342
  in Cartesian coordinates, 341
scalar triple product, 346–7
  cyclic permutation of, 347
  in Cartesian coordinates, 347
    determinant form, 347
  interchange of dot and cross, 347
scalars, 331
scale factors, 471, 474, 477
Schwarz inequality, 373
second-order differential equations, *see* ordinary differential equations
secular determinant, 418
semicircle, angle in, 69
semicircular lamina, centre of mass, 312
separable variables in ODE, 557
series, 213–44
  convergence of, *see* convergence of infinite series
  differentiation of, 233
  finite and infinite, 214
  integration of, 233
  multiplication by a scalar, 233
  multiplication of (Cauchy product), 233
  notation, 214
  operations, 232
  summation, *see* summation of series
series, particular
  arithmetic, 215
  arithmetico-geometric, 217
  geometric, 215
  Maclaurin, 240, 243
  power, *see* power series
  powers of natural numbers, 220–1
  Riemann zeta, 229, 230
  Taylor, *see* Taylor series
similarity transformations, 421–4
  properties of matrix under, 423
  unitary transformations, 423–4
simultaneous linear equations, 397–407
  and intersection of planes, 406–7
  homogeneous and inhomogeneous, 397
  number of solutions, 398–9
  solution using
    Cramer's rule, 404–6
    Gaussian elimination, 399–401

Index

inverse matrix, 401
    LU decomposition, 401–4
sine, $\sin(x)$
    geometrical and algebraic definitions, 676
    geometrical definition, 25
    in terms of exponential functions, 199
    Maclaurin series for, 243
    reciprocal and inverse, 27
singular and non-singular
    linear operators, 375
    matrices, 392
singular integrals, *see* improper integrals
singular solution of ODE, 555, 566, 568
sinh, $\sinh(x)$, *see also* hyperbolic functions
    hyperbolic sine, 198
    Maclaurin series for, 243
sinusoidal functions
    common values, 26
    identities, 28
skew-symmetric matrices, 409
solenoidal vectors, 463, 504
solid angle
    as surface integral, 510
    subtended by rectangle, 530
solid: mass, centre of mass and centroid, 308–9
spaces, *see* vector spaces
span of a set of vectors, 370
sphere, vector equation of, 351
spherical polar coordinates, 473–6
    area element, 474
    basis vectors, 474
    length element, 474
    vector operators, 473–6
    volume element, 321, 474
spur of a matrix, *see* trace of a matrix
square matrices, 376
standard deviation σ, 626
stationary values
    of functions of
        one real variable, 114–15
        two real variables, 272–5
    under constraints, *see* Lagrange undetermined multipliers
Stokes' theorem, 503, 523–6
    physical applications, 525
    related theorems, 524
submatrices, 396–7
subscripts
    dummy, 682
    free, 682
    summation convention, 681
substitution, integration by, 149–52

summation convention, 432, 681–3
summation of series, 215–23
    arithmetic, 215
    arithmetico-geometric, 217
    difference method, 217–19
    geometric, 215
    powers of natural numbers, 220–1
    transformation methods, 221–3
        differentiation, 221
        integration, 221
        substitution, 223
surd, 5
surface area, as a vector, 508–10
    as a line integral, 509
surface integrals
    and divergence theorem, 517
    Archimedean upthrust, 511, 530
    of scalars, vectors, 504–11
    physical examples, 510
surfaces, 455–7
    area of, 457
        cone, 165–6
        solid, and Pappus's theorem, 310–12
        sphere, 457
    coordinate curves, 456
    normal to, 456, 461
    of revolution, 164–6
    parametric equations, 455
    quadratic, 430–1
    tangent plane, 456
symmetric functions, 126
symmetric matrices, 409
    general properties, *see* Hermitian matrices

t substitution, 150–1
$\tan^{-1} x$, Maclaurin series for, 243
tangent planes to surfaces, 456
tangent, $\tan(x)$
    geometrical definition, 25
    Maclaurin series for, 243
tanh, $\tanh(x)$, *see* hyperbolic functions
Taylor series, 238–44
    and Taylor's theorem, 239–42
    approximation errors, 242–3
    for functions of several real variables, 270–1
    remainder term, 240
    required properties, 239
    standard forms, 240
tetrahedron
    mass of, 308
    volume of, 306
thermodynamic potential, 286

Index

thermodynamics
  first law of, 285
  Maxwell's relations, 285–7
torque, vector representation of, 344
total derivative, 263
total differential, 262
trace of a matrix, 385–6
  as sum of eigenvalues, 418, 426
    invariance under similarity transformations, 423
transformation matrix, 422, 427–8
transformations
  similarity, *see* similarity transformations
transforms, Laplace, *see* Laplace transforms
transpose of a matrix, 376, 383
  product rule, 383
trial functions, for PI of ODE, 574
trials, 597
triangle inequality, 373
triangle, centroid of, 335–6
triangular matrices, 401, 408
trigonometric identities, 24–32
triple integrals, *see* multiple integrals
triple scalar product, *see* scalar triple product
triple vector product, *see* vector triple product
turning point, 114

undetermined coefficients, method of, 574
undetermined multipliers, *see* Lagrange undetermined multipliers
uniform distribution, 644
union ∪
  algebra of, 601
unit step function, *see* Heaviside function
unit vectors, 338
unitary matrices, 410
  eigenvalues and eigenvectors, 415
unitary transformations, 423–4
upper triangular matrices, 408

variable, dummy, 142
variance σ^2, 626
  of dependent RV, 659
vector operators, 458–81
  acting on sums and products, 465–7
  combinations of, 467–9
  curl, 463, 480
  del ∇, 458
  del squared ∇^2, 463
  divergence (div), 462
  geometrical definitions, 513–17
  gradient operator (grad), 459–62, 479
  identities, 467
  Laplacian, 463, 480
  non-Cartesian, 469–81
vector product, 342–5
  anticommutativity, 342
  definition, 342
  determinant form, 345
  in Cartesian coordinates, 345
  non-associativity, 342
vector spaces, 370–3
  associativity of addition, 369–70
  basis vectors, 370–1
  commutativity of addition, 369–70
  complex, 370
  defining properties, 370
  dimensionality, 370
  inequalities: Bessel, Schwarz, triangle, 373
  matrices as an example, 379
  parallelogram equality, 373
  real, 370
  span of a set of vectors in, 370
vector triple product, 348
  identities, 348
  non-associativity, 348
vectors
  as geometrical objects, 369
  base, 450
  column, 376
  compared with scalars, 331
  component form, 336–8
  examples of, 331
  graphical representation of, 331
  irrotational, 464
  magnitude of, 338
  non-Cartesian, 450, 470, 474
  notation, 331
  solenoidal, 463, 504
  span of, 370
vectors, algebra of, 331–59
  addition and subtraction, 332–3
    in component form, 337–8
  angle between, 341
  associativity of addition and subtraction, 332
  commutativity of addition and subtraction, 332
  multiplication by a complex scalar, 342
  multiplication by a scalar, 333–4
  multiplication of, *see* scalar product *and* vector product
vectors, applications
  centroid of a triangle, 335–6
  equation of a line, 349
  equation of a plane, 349–51
  equation of a sphere, 351
  finding distance from a

Index

    line to a line, 355–6
    line to a plane, 356–7
    point to a line, 353–4
    point to a plane, 354–5
  intersection of two planes, 350, 351
vectors, calculus of, 448–81
  differentiation, 448–52, 454
  integration, 453–4
  line integrals, 491–504
  surface integrals, 504–11
  volume integrals, 511–13
vectors, derived quantities
  curl, 463
  derivative, 448
  differential, 452, 455
  divergence (div), 462
  reciprocal, 357–9
  vector fields, 458
    curl, 523
    divergence, 462
    flux, 510
    rate of change, 461
vectors, physical
  acceleration, 449
  angular momentum, 365
  angular velocity, 344, 365, 464
  area, 508–10, 525
  area of parallelogram, 343, 345
  force, 331, 332, 340
  moment or torque of a force, 344
  velocity, 449
velocity vectors, 449
Venn diagrams, 597–602
volume elements
  curvilinear coordinates, 478
  cylindrical polars, 471
  spherical polars, 321, 474
volume integrals, 511–13
  and divergence theorem, 517
volume of
  cone, 167
  ellipsoid, 326
  parallelepiped, 346
  rhomboid, 364
  tetrahedron, 306
volumes
  as surface integrals, 512, 517
  of regions, using multiple integrals, 306–8
volumes of revolution, 166–7
  and surface area and centroid, 310–12

wave equation, from Maxwell's equations, 488
wavefunction of electron in hydrogen atom, 326
wedge product, *see* vector product
work done
  by force, 495
  vector representation, 340
Wronskian test for linear independence, 570

X-ray scattering, 364

z, as a complex number, 174
z^*, as complex conjugate, 181–4
zero (null)
  matrix, 381, 382
  operator, 375
  vector, 333, 370
zeros, of a polynomial, 53
zeta series (Riemann), 229, 230
z-plane, *see* Argand diagram